Artificial Intelligence Technologies for Education: Advancements, Challenges, and Impacts

Artificial Intelligence Technologies for Education: Advancements, Challenges, and Impacts

Guest Editors

Yu Liang
Wenjun Wu
Ying Li

MDPI

Basel • Beijing • Wuhan • Barcelona • Belgrade • Novi Sad • Cluj • Manchester

Guest Editors
Yu Liang
College of Computer Science
Beijing University of Technology
Beijing
China

Wenjun Wu
Institute of Artificial Intelligence
Beihang University
Beijing
China

Ying Li
School of Computer Science and Engineering
Beihang University
Beijing
China

Editorial Office
MDPI AG
Grosspeteranlage 5
4052 Basel, Switzerland

This is a reprint of the Special Issue, published open access by the journal *Applied Sciences* (ISSN 2076-3417), freely accessible at: https://www.mdpi.com/journal/applsci/special_issues/A10EVOIBT7.

For citation purposes, cite each article independently as indicated on the article page online and as indicated below:

Lastname, A.A.; Lastname, B.B. Article Title. *Journal Name* **Year**, *Volume Number*, Page Range.

ISBN 978-3-7258-5037-2 (Hbk)
ISBN 978-3-7258-5038-9 (PDF)
https://doi.org/10.3390/books978-3-7258-5038-9

Contents

About the Editors

Yu Liang

Yu Liang is an Assistant Professor and Master's Supervisor at the College of Computer Science, Beijing University of Technology. He received his B.Eng. and Ph.D. from Beihang University in 2017 and 2022, and conducted visiting research at Arizona State University and the University of Cambridge. His research focuses on artificial intelligence, deep learning, human–computer interaction, multimodal learning, and educational information technology, with work on intelligent image and video detection, multi-source signal analysis, cognitive modeling, and intelligent tutoring. He has led projects funded by the National Natural Science Foundation of China, the Beijing Natural Science Foundation, and industry partners, and published nearly 30 papers. He serves as an Executive Member of the CCF Technical Committee on Intelligent Robot and as a Guest Editor of *Applied Sciences*, holds patents, has contributed to standards, and has received multiple awards.

Wenjun Wu

Wenjun Wu is a Professor and Doctoral Supervisor at the Institute of Artificial Intelligence, Beihang University. He earned his master's and doctoral degrees under Academician Wei Li, and from 2002 to 2010 he studied and worked at Indiana University and Argonne National Laboratory in the United States. His research interests include advanced scientific computing, collaborative network platforms, swarm intelligence, multi-agent systems, cognitive modeling, intelligent tutoring, and intelligent software engineering. He has published over 120 papers in international journals and conferences and authored a monograph on swarm software engineering. His collaborative software systems have been deployed in over 100 research institutions worldwide. He has led and participated in more than 20 projects, including the National Key R&D Program, National Natural Science Foundation key projects, and national 863 and 973 programs.

Ying Li

Ying Li is an Associate Professor at the School of Computer Science, Beihang University. Her research focuses on smart education, innovative educational resources, and the integration of science, technology, and education. She was Guest Editor of this *Applied Sciences* Special Issue and a key contributor to two national-level and three university-level first-class undergraduate courses. She has published over 30 teaching-related papers in journals and conferences, authored one Chinese and one English textbook, and contributed to six others. She has led national-level projects funded by the National Natural Science Foundation of China, the Ministry of Education, and the Ministry of Science and Technology, as well as industry–academia collaboration projects. Her work has earned more than 25 honors, including the Special Prize in the National University Teaching Innovation Competition and the First Prize in the Beijing Computer Teaching Video Competition.

Preface

This Reprint, "Artificial Intelligence Technologies for Education: Advancements, Challenges and Impacts", brings together a diverse collection of research and review articles that examine the transformative role of artificial intelligence (AI) in modern education. The scope of this Reprint spans theoretical foundations, novel algorithms, and practical applications, paying particular attention to personalized learning, adaptive assessment, intelligent tutoring, student and cognitive modeling, educational data mining, and human–computer interactions in teaching and learning contexts.

Our aim is to provide a comprehensive overview of the latest developments in this rapidly evolving field, highlighting both the opportunities and challenges of integrating AI into educational systems and inspiring new lines of inquiry. The motivation for compiling this Reprint lies in the growing demand for effective, scalable, and equitable AI-driven solutions that can support learners and educators alike while respecting the human-centered values that underpin quality education.

This work is addressed to researchers, practitioners, policymakers, and advanced students interested in the intersection of AI and education. We believe the contributions gathered here will serve as a valuable resource for understanding current trends, guiding future research, and informing the design of innovative educational technologies that can shape the future of learning.

Yu Liang, Wenjun Wu, and Ying Li
Guest Editors

applied sciences

MDPI

Article

Augmented Reality for Learning Algorithms: Evaluation of Its Impact on Students' Emotions Using Artificial Intelligence

Mónica Gómez-Ríos [1,2,*], Maximiliano Paredes-Velasco [2], J. Ángel Velázquez-Iturbide [2] and Miguel Ángel Quiroz Martínez [1]

1 Computer Science Department, Universidad Politécnica Salesiana, Guayaquil 090101, Ecuador; mquiroz@ups.edu.ec
2 Department of Computer Science and Statistics, Universidad Rey Juan Carlos, 28933 Móstoles, Madrid, Spain; maximiliano.paredes@urjc.es (M.P.-V.); angel.velazquez@urjc.es (J.Á.V.-I.)
* Correspondence: mgomezr@ups.edu.ec or md.gomez.2018@alumnos.urjc.es

Abstract: Augmented reality is an educational technology mainly used in disciplines with a strong physical component, such as architecture or engineering. However, its application is much less common in more abstract fields, such as programming and algorithms. Some augmented reality apps for algorithm education exist, but their effect remains insufficiently assessed. In particular, emotions are an important factor for learning, and the emotional impact of augmented reality should be determined. This article inquires about the impact on students' emotions of an augmented reality tool for learning Dijkstra's algorithm. This investigation uses an artificial intelligence tool that detects emotions in real time through facial recognition. The data captured with this tool show that students' positive emotions increased significantly, statistically surpassing negative emotions, and that some negative emotions, such as fear, were considerably reduced. The results show the same trend as those obtained with psychometric questionnaires, but both positive and negative emotions registered with questionnaires were significantly greater than those registered with the artificial intelligence tool. The contribution of this article is twofold. Firstly, it reinforces previous findings on the positive emotional impact of augmented reality on students. Secondly, it shows an alignment of different instruments to measure emotions, but to varying degrees.

Keywords: augmented reality; artificial intelligence; emotions; algorithms

1. Introduction

Augmented reality (AR) has shown great potential for learning by combining virtual objects with the real environment [1]. AR is an innovative tool that fosters creativity [2], facilitates the development of key digital skills [3], and increases student motivation and engagement by allowing them to explore educational content interactively, maintaining their long-term interest [2]. This impact is visible at all educational levels [4]: from early childhood education, where concepts are often introduced playfully, to higher education, where it provides practical experiences in different disciplines [5].

Likewise, emotions play an important role in students' academic performance, as educational experiences are deeply tied to students' emotional states. In addition to improving the understanding of abstract concepts, AR can influence students' emotional states during the learning process. Evaluating this impact is key to optimizing the educational experience [6] and thus contributing to a more appropriate learning experience design.

AR has been less widely used in virtual disciplines such as programming or algorithms than in other disciplines, such as history or engineering. However, AR also has educational potential for these fields. Thus, it can facilitate the understanding of algorithms by allowing the visualization of data structures and computational processes. Experiments with university students have shown that AR-based learning systems make it possible to reduce the cognitive load and improve the efficiency of algorithm learning, allowing for better visualization and understanding of structures such as trees, graphs, and optimization strategies [2]. However, challenges have been identified in the accessibility of the technology, the availability of suitable devices, and the need for teacher training for its effective implementation. As AR continues to evolve, it is crucial to further investigate its impact at different educational levels and to design pedagogical strategies that maximize its benefits in teaching algorithms [2]. Although AR has proven useful in teaching abstract virtual concepts, its impact on students has been insufficiently explored. This article presents a study of students' emotions while using the AR tool DARA, aimed at learning Dijkstra's algorithm.

This study addresses two main objectives. The first is to analyze the impact of using an augmented-reality-based tool (DARA) on students' emotional responses during the learning of Dijkstra's algorithm. The second objective is to compare the effectiveness of two emotional assessment techniques—facial recognition based on artificial intelligence and the AEQ (achievement emotions questionnaire) psychometric questionnaire—for measuring students' emotions before, during, and after the learning experience.

Based on these objectives, the following are the main contributions of this study. Firstly, the positive impact of an AR app on students' emotions in algorithm education was replicated. The replication of results is relevant, as a different measurement instrument was used. Secondly, the degree of increase or decrease in emotions was statistically different between the two instruments. This divergence was not expected and can be relevant for researchers focusing on the emotional impact of technology in education; thus, it deserves further research.

This article is structured as follows: Section 2 presents related work, Sections 3 and 4 describe the technological resources and the methodology used, respectively, Section 5 analyzes the results obtained, Section 6 presents the discussion, and, finally, Section 7 reports the conclusions and future lines of research.

2. Related Work

This section presents the role of emotions in learning and reviews the instruments for their assessment.

2.1. Emotions in Learning

Learning is influenced by multiple factors, including context, teaching methodology, and students' emotional state [7]. Various studies have shown that emotions play a crucial role in motivation and attention, directly impacting students' ability to process and retain information [7–9]. In this sense, the relationship between emotions and learning has been studied, where positive emotions have been shown to enhance comprehension and academic performance, while negative emotions can interfere with concentration and generate barriers to learning [10]. An emotionally favorable environment can improve students' commitment and willingness to study [11]. Therefore, the use of technologies, such as AR, has been explored as a way to make learning more motivating and emotionally enriching [10].

If we distinguish between types of emotions, positive emotions can mediate the relationship between teacher support and student engagement, while negative emotions can interfere with commitment and concentration [10]. Likewise, intense emotional experiences, whether positive or negative, facilitate long-term information retention [7].

Emotions such as curiosity, interest, and confidence enhance academic performance, while frustration and anxiety can motivate or hinder performance, depending on the context [12]. Given their direct impact on learning, measuring student emotions has become a key objective for educators and researchers.

2.2. Assessment of Emotions

The current literature includes several methods and instruments for assessing the impact of technology on emotions in the educational setting. Psychometric questionnaires are probably the most common tool, where students typically rate different questions or statements on how they feel about the learning experience. For example, the AEQ [12] and PANAS (Positive and Negative Affect Schedule) [13] questionnaires assess the positive and negative emotions experienced during learning in this way. Other questionnaires used are DASS (Depression Anxiety Stress Scale) and POMS (Profile of Mood States) [14], which measure student tension, fatigue, anxiety, and stress. Complementary instruments include the CD-RISC (Connor–Davidson Resilience Scale) [15], which evaluates emotional resilience, analyzing the student's ability to face challenging situations in the educational environment.

Psychometric questionnaires have limitations due to participant subjectivity and the difficulty in capturing emotions in real time. For example, students may not fully understand the questions posed and, therefore, may answer inaccurately. Or, if questionnaires are completed after the learning experience, students respond based on what they remember, which can lead to a discrepancy between the feelings they experienced and what they expressed [16,17]. These limitations motivate the need to explore complementary assessment methods that offer a more complete and accurate view of students' emotions. Computer technology makes it possible to measure the emotions experienced while learning more objectively and continuously [18]. Biometric methods, such as physiological sensors and electroencephalography (EEG) devices, make it possible to capture emotions in real time in an objective manner, providing more detailed and accurate data than that obtained through traditional questionnaires [19]. Physiological sensors make it possible to monitor emotions in real time to analyze emotional responses in educational settings [17]. Electroencephalography has been used to study brain activity linked to emotions. It has proven effective in research that analyzes how different technological resources influence the emotional state of students [20]. It is also applied in studies that relate physical activity with emotions through Frontal Alpha Asymmetry (FAA) [21]. Recent advances have used neural networks to improve the accuracy of emotion recognition through EEG signals [22]. However, biometric instruments have the problem of their invasive nature, as they force the student to wear sensors and devices with limited usability that prevent free interaction [19]. AI combined with non-invasive devices is a recent line of research that attempts to solve this problem, for example, through facial emotion recognition [18]. Examples of these systems are found in work environments and e-learning environments [23]. Some systems use traditional algorithms for facial recognition. Algorithms such as Eigenfaces [24] and Viola–Jones [25] have been successfully applied in emotion detection, achieving high accuracy in educational environments. Their use has shown that their combination with deep learning models allows for accurate facial recognition in real time [8]. These technologies make it possible to generate detailed reports on students' emotional behavior, enabling adaptive education [26]. Emotions have also been identified in online learning using models based on convolutional neural networks (CNNs), providing valuable feedback regarding the emotional impact on academic performance [27]. However, these systems have limitations in terms of scalability, accuracy, or applicability in face-to-face educational environments. There are some other techniques, although they are less commonly used. Techniques such as text analysis using natural language processing (NLP) have made it possible to infer emotions from student writing, such as essays or forum discussions [27]. Eye tracking has made it possible to monitor students' attention through visual patterns, correlating this data with emotions such as curiosity and engagement [28]. AI-based voice analysis techniques have also proven effective in identifying emotions such as anxiety and excitement in verbal interactions, providing a deeper interpretation of the emotions felt during educational activities. Finally, hybrid approaches that combine several of these techniques are gaining popularity, allowing for comprehensive emotion monitoring [29].

3. Technological Resources: DARA and Emolive

This section describes the two technological resources used in this research: the AR educational tool used to explain Dijkstra's algorithm and the facial recognition tool used to assess students' emotions.

3.1. DARA in Learning Greedy Algorithms with Augmented Reality

The greedy design technique solves optimization problems in stages. At each stage, a candidate (an element available or calculable from the problem's input data) is chosen and incorporated into the solution, with no subsequent reversal of the choice. This strategy can be expressed in a high-level schema with several functions: selection of the next candidate, solution feasibility, objective function, etc. Although the greedy schema is simple, its application to specific problems is not obvious [30]. In some cases, the functions that define the greedy schema may be implemented differently, may not be necessary, or may even be merged. This high variability usually is not explained [31] and makes it difficult for students to understand and apply the greedy technique.

DARA is an AR educational application designed to facilitate learning the well-known Dijkstra's greedy algorithm [32]. In addition to its emotional impact, the effectiveness of DARA in supporting knowledge acquisition has been evaluated in previous studies, showing improvements in student understanding and learning outcomes [33], particularly in relation to Dijkstra's algorithm, which solves the problem of calculating the shortest paths from one node to the other nodes in a weighted graph. It uses a smartphone camera to capture code fragments, onto which it overlays visualizations and textual explanations. Screenshots of these augmented visualizations and DARA's interface are provided in Supplementary Material S2. Some interface elements are in Spanish, as the app was originally developed in that language.

A student uses DARA as follows. Initially, they read the problem statement and an explanation of the greedy schema. In the second phase, the smartphone camera is activated, and the student focuses on the Java source code for Dijkstra's algorithm from the book [30] (either on paper or digitally captured). This phase of the application uses AR. The contextual explanations appear over the detected code (Figure 1b). On the other hand, if the camera captures the code fragment corresponding to the declaration of the graph as an adjacency matrix, a 3D visualization of the graph is generated (Figure 1a). These explanations help the student to interpret the different parts of the source code and relate them to the elements that make up the greedy schema. The student can move through the code and receive augmented feedback in real time.

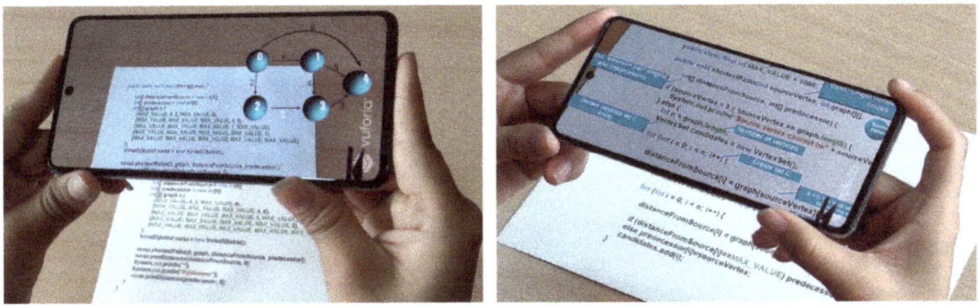

(**a**) (**b**)

Figure 1. When the smartphone camera detects the Java code corresponding to the adjacency matrix declaration, DARA displays a 3D visualization of the graph to help students understand its structure, as shown in (**a**). When detecting other code segments, DARA overlays textual explanations directly onto the source code, allowing the student to relate programming instructions to the components of the greedy algorithm, as illustrated in (**b**). The student moves the camera across the code to explore different fragments and their contextual annotations.

Finally, the student can activate the third phase of the application, which displays a 2D interactive animation of the algorithm. This animation shows a table with the main variables of the algorithm implemented in Java, where the edges and nodes selected by the algorithm can be consulted. It is possible to move forward or backward through the execution to observe the evolution of the variables.

In summary, augmented reality is implemented in DARA through virtual overlays that display contextual explanations and 3D visualizations of the algorithm via a smartphone camera. Its non-immersive nature and compatibility with mobile devices facilitate its integration into educational environments without requiring additional equipment.

3.2. Emotion Detection with Emolive

Emolive is a tool that allows users to track their emotions in real time using AI facial recognition. Emolive can detect six emotions: anger, disgust, fear, happiness, surprise, and sadness. If the detected facial expression is not recognized as any of these emotions, Emolive categorizes it as "neutral", just like other facial recognition systems [34].

Emolive uses deep learning techniques to accurately detect emotions. In particular, it uses a convolutional neural network trained with the FER2013 dataset, which is widely used in emotion recognition tasks.

The tool can detect the emotions of up to 12 students simultaneously with a single camera. It also allows teachers to create student groups and generate follow-up reports.

Using Emolive is straightforward. The teacher registers the group of students and can either start monitoring immediately or schedule specific activation periods (start time and duration). During monitoring, the teacher can pause and resume the system as needed. The dashboard displays each student's emotions through average bar charts and time series plots, which are automatically updated every two minutes.

This tool has been used in classroom settings and has demonstrated an average accuracy of 85% in detecting students' emotions (ranging from 81% to 93%, depending on the emotion) [33]. As it is open-source and compatible with commonly available devices, it is suitable for use in resource-constrained environments.

A representative screenshot of the Emolive interface is shown in Figure 2, and additional examples are provided in Supplementary Material S3.

Figure 2. Real-time emotion detection in the classroom using the Emolive system. The application identifies the emotions of multiple students simultaneously through webcam input. In this screenshot, different emotional states are detected and labeled above the corresponding students.

4. Methodology

This section describes the methodology for classroom intervention with DARA.

4.1. Hypotheses

The two objectives of the study were outlined in the introduction. We coherently propose the following two hypotheses:

H1. *Students who use AR for learning greedy algorithms experience significantly more positive emotions than negative ones.*

H2. *The assessment of positive and negative emotions using AR does not vary when using psychometric questionnaire techniques or facial recognition techniques.*

4.2. Participants

The total sample size for this study was 48 students, divided into a group of 30 students and a group of 18. All of them consented to participating in this study.

Let us call the first group GEmolive, since their emotions were identified using the Emolive application. This group is made up of 30 s-year students of the Computer Science Engineering degree program Universidad Politécnica Salesiana, University, in Guayaquil, Ecuador, during the 2023–2024 academic year. The students were taking the Data Structures course, which studies data structures (both linear and nonlinear) and the design and analysis of basic manipulation algorithms of these data structures.

We call the second group GEAQ since their emotions were identified using the validated AEQ, a psychometric instrument designed to measure a wide range of academic emotions (see Supplementary Material S1 for the full list of AEQ items used). This group consists of 18 students from a previous study conducted in the 2019–2020 academic year [35], which serves as a control group to validate hypothesis H2. The students in this second group had the same profile (subject, course, degree, and university) as the members of the GEmolive group, thus eliminating possible cultural or educational biases.

Although these data were collected in a previous academic year, they were ethically obtained, anonymized, and stored for academic purposes. Their use in the present research was limited to statistical comparison, with no additional contact or intervention involving the original participants. Furthermore, the current study was reviewed and authorized by the Research Vice-Rectorate of the Universidad Politécnica Salesiana in 2024, confirming compliance with ethical standards.

4.3. Procedure

Participants in both groups worked on the same learning task (learning Dijkstra's algorithm), used the same version of DARA, and followed a very similar working methodology. The only difference was the instrument each group used to measure emotions.

The classroom intervention with the students of the GEmolive group was organized in three phases:

1. Getting Started. Fifteen minutes were spent providing students with instructions on how to use the DARA app and installing it on their mobile devices.
2. Development. Thirty minutes were allocated to complete the work assignment. Students had the Java source code for Dijkstra's algorithm on paper and used DARA to read the problem statement and an explanation of the greedy scheme. DARA then activated the smartphone camera, and students focused on the parts of the source code they found the most complex or did not understand, to obtain additional explanations on the smartphone screen. The students then manipulated an animation of the algorithm for a given graph.
3. Finalization. This final phase, which lasted 30 min, was used to address students' questions and concerns.

In Phases 1 and 3, students were organized into 3 shifts of 10 students each, ensuring that each shift did not exceed the maximum number of 12 users supported by Emolive. All three shifts received the same treatment and were taught by the same teacher. Each shift convened at a different time. While one shift of students was in Phases 2 and 3, the remaining students were given free time outside the classroom.

The students in the GAEQ group followed the same methodology, with only two variations. Firstly, students completed the task of Phase 2 in a single turn (there was no need to limit the number to 12 students because Emolive was not used). Second, an additional 20 min was allocated between Phases 1 and 3 to allow the students to complete the questionnaires (in Phase 2, questionnaires were not provided so as not to interrupt the task). A visual summary of the classroom intervention methodology, including group assignments, intervention phases, and the emotion assessment timeline, is shown in Figure 3.

GEmolive group participants were assessed using the Emolive tool. Emolive allows for continuous, real-time measurement of emotions, but it was decided to record emotions at only three moments: at the beginning, during, and at the end of the learning task. This way, the analysis is focused on some of the key moments of the session and avoids an overload of irrelevant data. Therefore, the six non-neutral emotions detected by Emolive (anger, disgust, fear, happiness, surprise, and sadness) were measured at all three moments. For each of the three measurement moments, two average variables were computed, one for positive emotions (happy and surprised) and another for negative emotions (anger, fear, disgust, and sadness). An anonymized example of the Emolive emotion detection log is available in Supplementary Material S4.

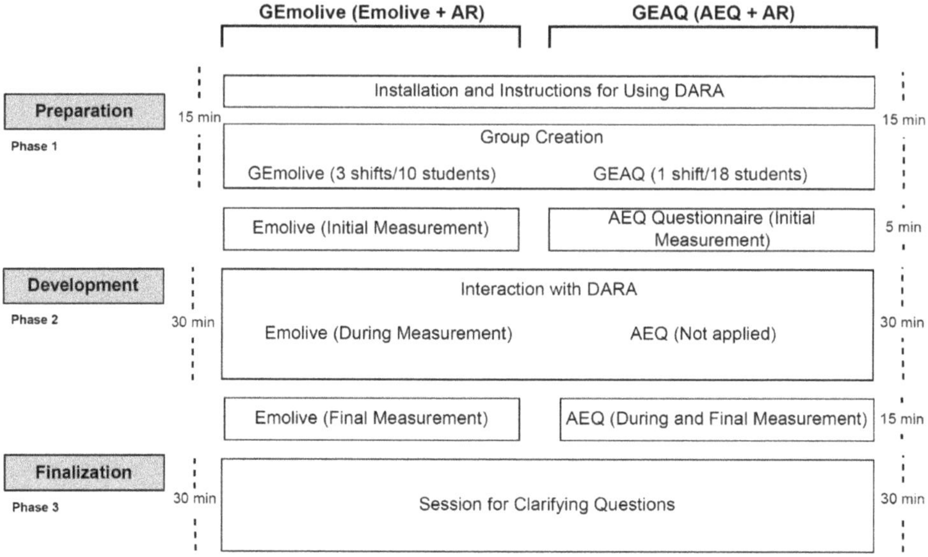

Figure 3. Methodological design of the classroom intervention comparing the GEmolive and GEAQ groups. The process is organized into three main phases: Preparation, Development, and Finalization. The duration of each stage is indicated, along with the specific instruments and procedures applied in each group.

The emotions of the GEAQ student group were measured with the psychometric AEQ questionnaire. The questionnaire presents statements about different emotional states, and students must indicate the degree to which they identify with those emotions using a 5-point Likert scale, from "very little" (scored with 1) to "very much" (scored 5). This questionnaire measures eight possible emotions: enjoyment, hope, pride, anger, anxiety, shame, hopelessness, and boredom. It consists of 75 items that are organized into three sections that allow for the emotional state to be assessed at three different times: before, during, and after completing the learning task.

Note that some specific emotions are not captured by both instruments; for example, one measures happiness while the other one measures enjoyment. To compare the measures from both instruments, emotions were grouped into positive and negative, as has been done in other studies [36].

5. Results

Statistical analyses, whether descriptive or inferential, were performed with SPSS Statistics (version 29.0.2.0), with a 95% confidence level for inferential procedures.

5.1. Analysis of H1: Different Between Positive and Negative Emotions

Table 1 shows the descriptive statistics for the emotions measured in the GEmolive group. It can be observed that the emotion "disgusted" was not experienced at any point during the experience (Mean = 0.00), while the emotion "happy" increased continuously from the beginning of the session to the end. Likewise, the emotions "anger" and "fear" decreased throughout the experience. Finally, the emotions "sad" and "surprised" fluctuated throughout the experience.

Figure 4 shows the means of these variables together with the neutral emotion, where it can be seen that the positive and negative emotions started with a similar mean value, but throughout the experience, the former increased, while the latter decreased.

Table 1. Descriptive statistics of emotions at three moments of the experience.

Emotion (Type)	Before (N = 30)				During (N = 30)				After (N = 30)			
	Min.	Max.	Mean	SD	Min.	Max.	Mean	SD	Min.	Max.	Mean	SD
Anger (negative)	0.00	5.00	2.03	2.01	0.00	4.00	1.20	1.69	0.00	5.00	1.10	1.67
Upset (negative)	0.00	0.00	0.00	0.00	0.00	0.00	0.00	0.00	0.00	0.00	0.00	0.00
Fearful (negative)	3.00	5.00	4.50	0.68	2.00	5.00	4.03	1.03	0.00	5.00	2.77	1.63
Happy (positive)	0.00	5.00	2.40	1.83	0.00	5.00	3.00	1.98	1.00	5.00	3.57	1.10
Neutral (neutral)	0.00	5.00	4.00	1.51	0.00	5.00	2.77	1.70	2.00	5.00	3.83	1.15
Sad (negative)	1.00	5.00	3.83	1.29	0.00	4.00	1.53	1.20	0.00	3.00	2.20	1.10
Surprised (positive)	0.00	4.00	2.70	1.64	0.00	5.00	2.97	1.59	0.00	5.00	2.90	1.79

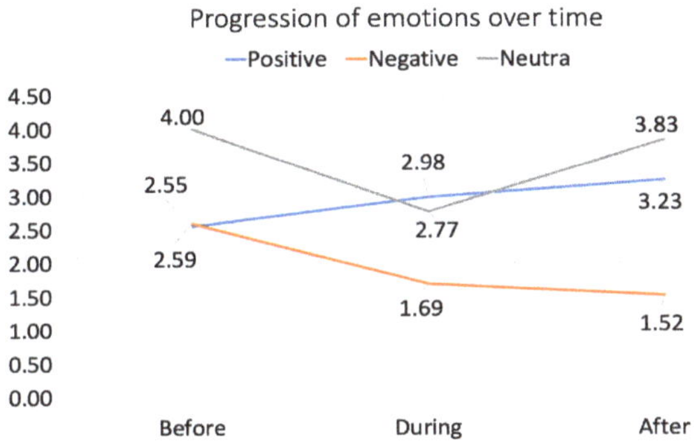

Figure 4. Evolution of emotions during the experience detected with Emolive.

To determine whether the difference between the means of positive and negative emotions at the beginning and end was significant, an inferential statistical analysis was performed. First, due to the sample size being less than 50, the Shapiro–Wilk test with Lilliefors significance correction was applied to determine whether the sample followed a normal distribution. The result was that none of the variables followed a normal distribution. Consequently, a parametric test for comparing means, the Wilcoxon signed-rank test for related samples, was applied. The results are shown in Table 2. Significant differences are bolded across the three instants (before, during, and after).

Table 2. Wilcoxon Signed-Rank Test.

Variable	Sig.
Positive-negative emotions (before)	0.925
Positive-negative emotions (during)	<0.001
Positive-negative emotions (after)	<0.001

5.2. Analysis of H2: Emotional Variation Through Psycometric Questioinnaire and Facial Recognition

The emotion scores measured with Emolive and those measured with AEQ were compared. Table 3 shows the descriptive statistics for the positive and negative emotions measured at the three time points for both groups.

The sample distribution was checked. It was observed that all variables in the GAEQ group followed a normal distribution. However, this was not the case for the GEmolive group (Shapiro–Wilk with Lilliefors significance correction, $p < 0.05$). Consequently, the nonparametric Mann–Whitney U test for unrelated samples was applied. Table 4 shows the hypotheses tested by this test and its results (significant differences found are marked in bold). Significant differences were found in the positive emotions of both groups at the beginning and end of the experience, as well as in the negative emotions of both groups at the end.

Table 3. Descriptive statistics of the emotions detected by AEQ and Emolive.

Variable (Emotions)	Group (N)	Min.	Max.	Mean	SD
Positive before	GAEQ (18)	2.50	5.00	3.93	0.88
	GEmolive (30)	0.00	3.50	2.55	0.98
Negative before	GAEQ (18)	1.00	4.13	2.28	0.78
	GEmolive (30)	1.25	3.75	2.59	0.78
Positive before	GAEQ (18) GEmolive(30)	2.59 0.00	4.95 4.00	3.85 2.98	0.65 1.63
Negative during	GAEQ (18) GEmolive (30)	1.00 0.50	4.23 2.75	2.03 1.69	0.79 0.69
Positive after	GAEQ (18) GEmolive (30)	2.25 2.50	5.00 4.50	3.69 3.23	0.85 0.91
Negative after	GAEQ (18) GEmolive (30)	1.44 0.00	4.19 3.00	2.44 1.52	0.85 0.94

Table 4. Hypothesis contrast of equality of means between groups.

H.	Null Hypothesis	Sig. (U of Mann–Whitney)	Decision
1	The distribution of positive emotions before is the same across group categories (AEQ/Emolive)	**<0.001**	Reject the null hypothesis
2	The distribution of negative emotions before is the same across group categories (AEQ/Emolive)	0.140	Fail to reject the null hypothesis
3	The distribution of positive emotions is the same across group categories (AEQ/Emolive)	0.344	Fail to reject the null hypothesis
4	The distribution of negative emotions is the same across group categories (AEQ/Emolive)	0.151	Fail to reject the null hypothesis
5	The distribution of positive emotions after is the same across group categories (AEQ/Emolive)	**0.042**	Reject the null hypothesis
6	The distribution of negative emotions after is the same across group categories (AEQ/Emolive)	**0.006**	Reject the null hypothesis

6. Discussion

6.1. Hypothesis 1

The first hypothesis was stated as follows:

H1. *Students who use AR for learning greedy algorithms experience significantly more positive emotions than negative ones.*

The results from Emolive indicate that, at the beginning of the experience, the positive and negative emotions experienced by the students were at a similar level (varying within

0.4 points on a scale of 1 to 5). However, as they participated in the experience, positive emotions increased while negative emotions decreased, with this difference being statistically significant at the end of the experience. Therefore, hypothesis H1 of the intervention can be affirmed, confirming that students who used AR to learn greedy algorithms experienced significantly more positive emotions than negative ones.

Previous research has found evidence along these lines. For example, in a study on the relationship between activity and emotions in algorithm learning with AR [35], it was found that participants reported higher levels of positive emotions than negative emotions during the AR experience. The decrease in negative emotions (anger, disgust, fear, and sadness) found in this study is in line with the findings of other studies analyzing emotions using facial recognition. Specifically, it had been found [35] that 86% and 87% of students felt that using the AR tool was neither unpleasant nor depressing, respectively.

Among negative emotions, the following finding is striking. One of the negative emotions that commonly manifests with the use of AR is anxiety [37]. Emolive does not allow for anxiety to be measured directly, but it does measure fear, which is strongly related to anxiety [10,35]. Therefore, one would expect this emotion to evolve similarly to anxiety due to the use of AR. However, the results found in the present study indicate otherwise. Initially, the fear experienced by the students was high (4.5 points out of 5), and by the end, it had decreased by 38% (practically reaching 2.7 out of 5). It is worth noting that the previous study used as a control group in the present study [35] found an increase in anxiety, using the same DARA tool and the same task as those used in the present study. Therefore, it would be advisable to conduct additional studies to shed light on the divergence between the increase in anxiety and the reduction in fear.

6.2. Hypothesis 2

The second hypothesis to be validated was stated as follows:

H2. *The assessment of positive and negative emotions using AR does not vary when using psychometric questionnaire techniques and facial recognition techniques.*

The results showed significant differences between the measurements collected using the EAQ questionnaire and the Emolive system. Thus, at the end of the experience, the positive and negative emotions detected by the psychometric questionnaire were significantly greater than those automatically detected by facial recognition with Emolive. The same difference was found in the positive emotions at the beginning of the experience. Consequently, hypothesis H2 is rejected, and it cannot be stated that the assessment of positive and negative emotions using AR does not vary with different measurement techniques. It is unclear what may explain this divergence between the two types of emotion recognition. Some studies have identified limitations in the use of psychometric questionnaire techniques. For example, a student may express a false memory of an emotion, and emotional distortion may occur when responding based on memories of past emotions [12,38]. These phenomena could explain the divergence found in this study between the two emotion assessment techniques. However, the evidence found in this study is not conclusive enough to make this conclusion, and a more in-depth analysis of the specific emotions assessed by each technique is necessary.

The differences observed between these techniques may be attributed to both psychological and technical factors. From the psychological perspective, questionnaires such as the AEQ can be subject to memory or social desirability biases [39]. In contrast, AI-based facial recognition tools like Emolive assess emotions in real time, but may be limited by technical conditions such as lighting or facial visibility [40]. Therefore, some researchers argue that both methods can be complementary and provide a more holistic understanding of learners' emotional states [41].

6.3. Limitations

Studying has several limitations for the results to be generalizable. The first limitation stems from the simultaneous use of the two software applications. Sometimes, difficulties were encountered in recognizing emotions due to the students' body posture during their interaction with DARA. In these cases, Emolive did not correctly capture emotions due to the students' forward tilt of the face when interacting with the mobile AR device, which affected the quality of the facial expression analysis. In these cases, Emolive informed the teacher of this situation, and the teacher reminded the students to adopt a more appropriate posture.

A second limitation relates to the constraints of both emotional measurement instruments. Although the AEQ is a validated psychometric tool, and the Emolive system has demonstrated an accuracy of 85%, both approaches present inherent limitations that may affect the results. Additionally, the assessment was conducted with a limited number of students (only 48 participants) and within a specific academic context—second-year undergraduate students in a computer science program—which limits the generalizability of the findings to other populations or educational settings.

A third limitation lies in the use of historical data for the GEAQ group, which corresponds to a previous academic year. This introduces potential confounding variables, such as differences in group characteristics or educational conditions. Furthermore, the absence of a control group within the same academic cycle reduces the level of experimental control.

7. Conclusions

This article explores the emotional impact of using an augmented reality tool for learning Dijkstra's algorithm, called DARA. The evolution of the emotional state of 48 s-year students enrolled in the Computer Science degree during the use of DARA was analyzed. The emotional state was monitored using an emotion detection tool through facial recognition with AI, called Emolive. The main result is that students experienced significantly more positive emotions than negative ones after using augmented reality. On the other hand, it was found that some unpleasant emotions such as fear were reduced by 38% in intensity at the end of the experience. The results of this study were compared with those of a previous study based on the AEQ psychometric questionnaire. The trend recorded with both approaches is the same. However, statistically significant differences were found in the measures of both methods, both for positive and negative emotions.

For the future, we propose two directions for further research. First, to analyze the causes of the divergence found between the two techniques in emotion detection, which may be due to psychological biases in self-report tools and technical limitations in AI-based recognition. Second, to explore the integration of both techniques into a joint emotion assessment model. The differences identified in this study, though only briefly addressed, suggest the potential value of combining both approaches for a more comprehensive emotional assessment. A more detailed analysis of these psychological and technical factors is suggested as a future research direction.

Supplementary Materials: The following supporting information can be downloaded at: https://www.mdpi.com/article/10.3390/app15147745/s1, Material S1: Supplementary_AEQ_items.pdf; Material S2: Supplementary_DARA_screenshots.pdf; Material S3: Supplementary_Emolive_interface.pdf; Material S4: Supplementary_Session_Log_Example_Anonymized.xlsx.

Author Contributions: Conceptualization, M.G.-R.; Methodology, M.G.-R.; Software, M.G.-R., M.Á.Q.M., and M.P.-V.; Formal analysis, M.P.-V.; Investigation, M.G.-R.; Resources, M.G.-R.; Data curation, M.G.-R.; Writing—original draft preparation, M.G.-R.; Writing—review and editing, M.G.-R., M.P.-V., and J.Á.V.-I.; Visualization, M.G.-R.; Supervision, M.P.-V. and J.Á.V.-I.; Project administration, M.G.-R. All authors have read and agreed to the published version of the manuscript.

Funding: This research was supported by Universidad Politécnica Salesiana, Ecuador, which covered the article processing charges (APCs). No external funding was received.

Institutional Review Board Statement: This study was reviewed and approved by the Vicerrectorate for Research of the Universidad Politécnica Salesiana (UPS), and a signed statement was issued by the Vicerrector for Research (Ref. UPS/VINV/2024/05/10), confirming that the study complies with the applicable ethical principles for academic research activities.

Informed Consent Statement: Informed consent was obtained from all participants involved in this study. Participation was voluntary, and students were informed of the study's purpose, procedures, and their right to withdraw at any time without consequences.

Data Availability Statement: The data presented in this study are openly available at https://doi.org/10.21950/FZF6YA.

Acknowledgments: The authors would like to thank the Universidad Politécnica Salesiana of Ecuador for supporting the publication of this article. No external funding was received.

Conflicts of Interest: The authors declare no conflicts of interest.

References

1. Altinpulluk, H. Determining the Trends of Using Augmented Reality in Education between 2006–2016. *Educ. Inf. Technol.* **2019**, *24*, 1089–1114. [CrossRef]
2. Avila-Garzon, C.; Bacca-Acosta, J.; Duarte, J.; Betancourt, J. Augmented Reality in Education: An Overview of Twenty-Five Years of Research. *Contemp. Educ. Technol.* **2021**, *13*, ep312. [CrossRef]
3. Moreno-Fernández, O.; Solís-Espallargas, C.; Moreno-Crespo, P.; Ferreras-Listán, M. Augmented Reality and Primary Education: Linkage, Potentiality and Applicability from the Perspective of Teachers in Initial Training. *Hacet. Univ. J. Educ.* **2023**, *38*, 388–389. [CrossRef]
4. Huang, T.-C.; Tseng, H.-P. Extended Reality in Applied Sciences Education: A Systematic Review. *Appl. Sci.* **2025**, *15*, 4038. [CrossRef]
5. Cabero-Almenara, J.; Llorente-Cejudo, C.; Martinez-Roig, R. The Use of Mixed, Augmented and Virtual Reality in History of Art Teaching: A Case Study. *Appl. Syst. Innov.* **2022**, *5*, 44. [CrossRef]
6. Ibáñez, M.B.; Uriarte Portillo, A.; Zatarain Cabada, R.; Barrón, M.L. Impact of Augmented Reality Technology on Academic Achievement and Motivation of Students from Public and Private Mexican Schools. A Case Study in a Middle-School Geometry Course. *Comput. Educ.* **2020**, *145*, 103734. [CrossRef]
7. Vistorte, A.O.R.; Deroncele-Acosta, A.; Ayala, J.L.M.; Barrasa, A.; López-Granero, C.; Martí-González, M. Integrating Artificial Intelligence to Assess Emotions in Learning Environments: A Systematic Literature Review. *Front. Psychol.* **2024**, *15*, 1387089. [CrossRef]
8. Kaur, M.; Kumar, M. Facial Emotion Recognition: A Comprehensive Review. *Expert Syst.* **2024**, *41*, e13670. [CrossRef]
9. Kalateh, S.; Estrada-Jimenez, L.A.; Nikghadam-Hojjati, S.; Barata, J. A Systematic Review on Multimodal Emotion Recognition: Building Blocks, Current State, Applications, and Challenges. *IEEE Access* **2024**, *12*, 103976–104019. [CrossRef]
10. Gómez-Rios, M.D.; Paredes-Velasco, M.; Hernández-Beleño, R.D.; Fuentes-Pinargote, J.A. Analysis of Emotions in the Use of Augmented Reality Technologies in Education: A Systematic Review. *Comput. Appl. Eng. Educ.* **2023**, *31*, 216–234. [CrossRef]
11. Ibáñez, M.-B.; Delgado-Kloos, C. Augmented Reality for STEM Learning: A Systematic Review. *Comput. Educ.* **2018**, *123*, 109–123. [CrossRef]
12. Pekrun, R.; Goetz, T.; Frenzel, A.C.; Barchfeld, P.; Perry, R.P. Measuring Emotions in Students' Learning and Performance: The Achievement Emotions Questionnaire (AEQ). *Contemp. Educ. Psychol.* **2011**, *36*, 36–48. [CrossRef]
13. Watson, D.; Clark, L.A.; Tellegen, A. Development and Validation of Brief Measures of Positive and Negative Affect: The PANAS Scales. *J. Pers. Soc. Psychol.* **1988**, *54*, 1063. [CrossRef]
14. Petrowski, K.; Albani, C.; Zenger, M.; Brähler, E.; Schmalbach, B. Revised Short Screening Version of the Profile of Mood States (POMS) From the German General Population. *Front. Psychol.* **2021**, *12*, 631668. [CrossRef] [PubMed]

15. Connor, K.M.; Davidson, J.R.T. Development of a New Resilience Scale: The Connor-Davidson Resilience Scale (CD-RISC). *Depress. Anxiety* **2003**, *18*, 76–82. [CrossRef]
16. Bosch, N.; D'Mello, S.; Baker, R.; Ocumpaugh, J.; Shute, V.; Ventura, M.; Wang, L.; Zhao, W. Automatic Detection of Learning-Centered Affective States in the Wild. In Proceedings of the 20th International Conference on Intelligent User Interfaces, Atlanta, GA, USA, 29 March–1 April 2015; ACM: New York, NY, USA, 2015; pp. 379–388.
17. Xie, J.; Luo, Y.; Wang, S.; Liu, G. Electroencephalography-Based Recognition of Six Basic Emotions in Virtual Reality Environments. *Biomed. Signal Process. Control* **2024**, *93*, 106189. [CrossRef]
18. Flores, E.; Solis-Fonseca, J.-P.; Cuba-Aguilar, C.-R.; Rosales-Fernandez, J.-H.; Barahona-Altao, Y.-G. Emotional State of the Classroom through Artificial Intelligence to Improve the Teaching-Learning Process in a University. In Proceedings of the 21th LACCEI International Multi-Conference for Engineering, Education and Technology, Buenos Aires, Argentina, 17–21 July 2023; Latin American and Caribbean Consortium of Engineering Institutions: Boca Raton, FL, USA, 2023. [CrossRef]
19. Bogicevic Sretenovic, M.; Milenkovic, I.; Jovanovic, B.; Simic, D.; Minovic, M.; Milovanovic, M. Bringing Biometric Sensors to the Classroom: A Fingerprint Acquisition Laboratory for Improving Student Motivation and Commitment. *Appl. Sci.* **2020**, *10*, 880. [CrossRef]
20. Wang, X.; Ren, Y.; Luo, Z.; He, W.; Hong, J.; Huang, Y. Deep Learning-Based EEG Emotion Recognition: Current Trends and Future Perspectives. *Front. Psychol.* **2023**, *14*, 1126994. [CrossRef]
21. Wilhelm, R.A.; Lacey, M.F.; Masters, S.L.; Breeden, C.J.; Mann, E.; MacDonald, H.V.; Gable, P.A.; White, E.J.; Stewart, J.L. Greater Weekly Physical Activity Linked to Left Resting Frontal Alpha Asymmetry in Women: A Study on Gender Differences in Highly Active Young Adults. *Psychol. Sport. Exerc.* **2024**, *74*, 102679. [CrossRef]
22. Wang, Z.; Wang, Y.; Tang, Y.; Pan, Z.; Zhang, J. Knowledge Distillation Based Lightweight Domain Adversarial Neural Network for Electroencephalogram-Based Emotion Recognition. *Biomed. Signal Process. Control* **2024**, *95*, 106465. [CrossRef]
23. Canazas, A.P.; Blaz, J.J.R.; Martínez, P.D.T.; Mamani, X.J. Sistema de Identificación de Emociones a Través de Reconocimiento Facial Utilizando Inteligencia Artificial. *Innovación Softw.* **2022**, *3*, 140–150. [CrossRef]
24. Mukhopadhyay, S.; Sharma, S. Real Time Facial Expression and Emotion Recognition Using Eigen Faces, LBPH and Fisher Algorithms. In Proceedings of the 2020 10th International Conference on Cloud Computing, Data Science & Engineering (Confluence), Noida, India, 29–31 January 2020; IEEE: Piscataway, NJ, USA, 2020; pp. 212–220. [CrossRef]
25. Aouani, H.; Ben Ayed, Y. Deep Facial Expression Detection Using Viola-Jones Algorithm, CNN-MLP and CNN-SVM. *Soc. Netw. Anal. Min.* **2024**, *14*, 65. [CrossRef]
26. Grassi, D. Supporting Developers' Emotional Awareness: From Self-Reported Emotions to Biometrics. In Proceedings of the 28th International Conference on Evaluation and Assessment in Software Engineering, Salerno, Italy, 18–21 June 2024; ACM: New York, NY, USA, 2024; pp. 500–504.
27. Liu, J.; Yan, C.; Liu, W.; Liu, X.; Ding, Y.; Zhou, Y. Research on Learner "Emotion-Behavior-Ability" Characteristics Based on MOOC Online Education User Profiles. *Inf. Process Manag.* **2025**, *62*, 104026. [CrossRef]
28. Huangfu, Q.; Li, H.; Ban, Y.; He, J. An Eye-Tracking Study on the Effects of Displayed Teacher Enthusiasm on Students' Learning Procedural Knowledge of Chemistry in Video Lectures. *J. Chem. Educ.* **2024**, *101*, 259–269. [CrossRef]
29. Liu, T.; Wang, M.; Yang, B.; Liu, H.; Yi, S. ESERNet: Learning Spectrogram Structure Relationship for Effective Speech Emotion Recognition with Swin Transformer in Classroom Discourse Analysis. *Neurocomputing* **2025**, *612*, 128711. [CrossRef]
30. Velázquez-Iturbide, J.Á. The Design and Coding of Greedy Algorithms Revisited. In Proceedings of the 16th Annual Joint Conference on Innovation and Technology in Computer Science Education, Darmstadt, Germany, 27–29 June 2011; ACM: New York, NY, USA, 2011; pp. 8–12.
31. Velázquez-Iturbide, J. ángel Reflections on Teaching Algorithm Courses. In Proceedings of the 56th ACM Technical Symposium on Computer Science Education V. 1, Pittsburgh, PA, USA, 26 February–1 March 2025; pp. 1148–1154.
32. Sahni, S. *Data Structures, Algorithms, and Applications in Java*; Universities Press: Ibadan, Nigeria, 2000; ISBN 8173715238.
33. Quiroz-Martinez, M.A.; Díaz-Fernández, S.; Aguirre-Sánchez, K.; Gomez-Rios, M.D. Analysis of Students' Emotions in Real-Time during Class Sessions through an Emotion Recognition System. In *X Congreso Internacional de Ciencia, Tecnología e Innovación para la Sociedad (CITIS 2024)*; Springer: Cham, Switzerland, 2024. [CrossRef]
34. Salloum, S.A.; Alomari, K.M.; Alfaisal, A.M.; Aljanada, R.A.; Basiouni, A. Emotion Recognition for Enhanced Learning: Using AI to Detect Students' Emotions and Adjust Teaching Methods. *Smart Learn. Environ.* **2025**, *12*, 21. [CrossRef]
35. Paredes-Velasco, M.; Velázquez-Iturbide, J.Á.; Gómez-Ríos, M. Augmented Reality with Algorithm Animation and Their Effect on Students' Emotions. *Multimed. Tools Appl.* **2023**, *82*, 11819–11845. [CrossRef]
36. Fernández Herrero, J.; Gómez Donoso, F.; Roig Vila, R. The First Steps for Adapting an Artificial Intelligence Emotion Expression Recognition Software for Emotional Management in the Educational Context. *Br. J. Educ. Technol.* **2023**, *54*, 1939–1963. [CrossRef]
37. Poitras, E.G.; Harley, J.M.; Liu, Y.S. Achievement Emotions with Location-based Mobile Augmented Reality: An Examination of Discourse Processes in Simulated Guided Walking Tours. *Br. J. Educ. Technol.* **2019**, *50*, 3345–3360. [CrossRef]
38. D'Mello, S.; Graesser, A. Dynamics of Affective States during Complex Learning. *Learn. Instr.* **2012**, *22*, 145–157. [CrossRef]

39. Robinson, M.D.; Clore, G.L. Episodic and Semantic Knowledge in Emotional Self-Report: Evidence for Two Judgment Processes. *J. Pers. Soc. Psychol.* **2002**, *83*, 198–215. [CrossRef] [PubMed]
40. Barrett, L.F.; Adolphs, R.; Marsella, S.; Martinez, A.M.; Pollak, S.D. Emotional Expressions Reconsidered: Challenges to Inferring Emotion From Human Facial Movements. *Psychol. Sci. Public Interest* **2019**, *20*, 1–68. [CrossRef] [PubMed]
41. Calvo, R.A.; D'Mello, S. Affect Detection: An Interdisciplinary Review of Models, Methods, and Their Applications. *IEEE Trans. Affect. Comput.* **2010**, *1*, 18–37. [CrossRef]

applied sciences

MDPI

Article

Using Generative AI to Support UX Design Students in Web Development Courses

Félix Buendía-García [1,*] and Javier Piris-Ruano [2]

[1] Department of Computer Engineering, Universitat Politècnica de Valencia, 46022 Valencia, Spain
[2] Department of Computer Systems and Computation, Universitat Politècnica de Valencia, 46022 Valencia, Spain; jpiris@dsic.upv.es
* Correspondence: fbuendia@disca.upv.es; Tel.: +34-682347392

Abstract: This work explores the integration of Generative AI (GenAI) tools into web development educational settings, with a focus on enhancing the user experience (UX) design process and supporting students with limited technical backgrounds. The democratization of GenAI has allowed non-technical users to engage in the creation of computing applications. However, its adoption among UX-focused learners remains limited. In this study, we propose an innovative instructional approach designed to facilitate the effective use of GenAI tools in web development courses for UX designers. This approach is based on incorporating interaction mechanisms that allow students to prompt GenAI tools using an incremental strategy. Moreover, this scaffolding process enables the definition of benchmarks that can be used as a reference for the development tasks proposed and their assessment by comparing the students' outcomes with the benchmarks provided. The results obtained from applying the proposed approach in a web development course using GitHub Copilot environments show its potential to address the challenges UX design students face in these educational settings.

Keywords: Generative Artificial Intelligence; web development; user experience design; prompt engineering

1. Introduction

Large Language Models (LLMs) are emerging as influential platforms in the development of computing applications, and the use of Generative Artificial Intelligence (GenAI) tools is becoming increasingly common among software practitioners [1]. These tools have enabled more effective and faster software development by accelerating most of the phases of this process [2–5]. In this context, the incorporation of these platforms and tools plays a crucial role in software-development-related educational settings [6], highlighting their ability to generate code and assess students' perceptions about how they can be helped with their programming tasks. Nevertheless, examinations have revealed some pitfalls regarding the use of LLMs as coding assistants for computer science students [7].

There are multiple studies related to the educational application of GenAI in CS (or computing education) disciplines [8–10], and Raihan et al. [11] recently performed a systematic review of the use of LLMs in computer science education. Moreover, GenAI tools have been deployed in several computing areas, ranging from introductory programming [12] to more specific fields such as software engineering education [13] and database administration courses [14]. In particular, attention has been paid to novice programmers who use LLM-based code generators [15], including the benefits and harms they can bring about, and GenAI's impact on novice programmer metacognition aspects [16]. However, the democratization of Generative AI tools has not yet translated into widespread adoption by non-technical users for developing computing applications. This is the case for UX (user experience) design students taking web development (WD) courses, the main target in our research. These students often face issues mainly precipitated by technical barriers, limited integration of design tools, and concerns around creative control [17].

In some cases, UX designers turn to Low-Code Development Platforms (LCDPs) [18], which offer easy-to-use visual environments, allowing these students to develop software applications without a solid programming background. Moreover, cooperation between UX designers and developers has long been promoted [19], and their collaboration has been boosted to improve efficiency in transformation processes from design to development through LCDPs [20]. The intention behind the different perspective considered in this work is to provide UX design students with tools enabling them to actively participate in the development process. This aspect is particularly important for students enrolled in creative technology courses whose instructional settings combine design skills with a basic knowledge of end-user development.

End-user development has long been linked with web applications [21], and Barricelli et al. [22] carried out a systematic mapping study concerning its relationships with end-user programming and software engineering. There are multiple aspects to be considered in this context. Recently, Paternó [23] addressed the contribution of AI technologies to the development of end-user applications. Assessing the impact of these technologies, particularly concerning the integration of GenAI when teaching WD courses, is one of the main goals of this work. The aim of such integration is genuine synergy between web design knowledge and development programming practices, thereby empowering students to leverage GenAI tools and facilitate this convergence. In line with this research idea, three main questions are proposed:

- RQ1: How can GenAI tools be effectively used in or integrated into web development courses tailored to UX designers?
- RQ2: What new interaction mechanisms can be explored to foster the practical application of these tools in web development educational settings?
- RQ3: How can these GenAI-based interactive learning settings be assessed and evaluated?

The remainder of the work is structured as follows. The first section reviews related work in consideration of the research questions listed above. Section 3 describes the materials and methods deployed in the current research, including a description of the target courses and the methods used to apply GenAI tools in this context. Section 4 reveals the results obtained after applying the proposed methods to the target courses. Section 5 provides a discussion on these results in consideration of the three research questions addressed in this work. Finally, some conclusions and ideas for further work are provided.

2. Related Works

As mentioned earlier, web applications have a strong relationship with end-user development [21], especially when geared towards users with low levels of programming skills. In this context, LLM platforms and GenAI tools can provide invaluable support for UX designers interested in developing these types of applications. The use of LLMs for end-user website generation has been endorsed [24], with emphasis placed on their ability to generate websites by abstracting away concerns about the underlying code. Furthermore, the impact of GenAI on front-end development processes has been thoroughly examined [25], revealing that it promotes automated code generation and dynamic content creation [26]. In this vein, the WebDiffusion tool [27] allows web developers to harness

Generative AI to develop text and images. Web Sculptor [28] is another example of a GenAI-based development framework that improves code generation in web applications.

Despite the general success of these AI-enabled design support tools, several obstacles have been identified regarding UX practitioners. The first gap was observed by Lu et al. [29], who stated that many AI-based design tools are based on graphical interface elements, neglecting a design-thinking-oriented view. Additionally, some issues regarding overreliance and perceived lack of control over the automated code generated have been reported [30]. Hsiao & Tank [17] sought to apply GenAI tools in assisting the user experience design process, combining image prototype generators with prompts that allowed interaction with ChatGpt's DALL·E, moving from abstract to more specific requirements. The idea of using incremental prompts was also exploited by Calo & de Russis [24], who employed LLMs to allow users to refine their prompt outcomes based on predefined templates when generating websites. A similar strategy was incorporated in Kodless [31], an LLM-based tool designed to automatically build web applications through an iterative prompt refinement process. These methods of organizing iterative prompts can be extremely useful in WD educational settings, and they are in line with the mechanisms proposed for interacting with GenAI tools in this study.

There are several initiatives relevant to our research focus, i.e., assessing the impact of GenAI tools in WD educational settings. York [32] evaluated ChatGPT's use as a GenAI tool capable of assisting UX and WD students in three main areas: ideation, design, and coding. This evaluation focused on examining prompts, which could be used in these areas in a UX course context. The results of this evaluation were based on judging the quality of prompt outcomes, lacking a quantitative assessment. Lively et al. [33] were also interested in integrating GenAI tools into web design education by enhancing students' aesthetic and creative abilities. In this case, the authors assessed the artifacts generated by students along with a collection of their impressions derived from the application of GenAI tools. In a similar way, students were asked to assess improvements in their critical thinking skills in a web programming module using this kind of tool [34]. A self-evaluation questionnaire was employed to collect students' opinions after each practice activity, with the results showing limited critical-thinking improvement. The use of Generative AI by students as a support tool in an advanced web development course was studied in [35]. The authors analyzed the use of GenAI-based prompts in different educational tasks, ranging from ideation processes and explaining concepts to helping students generate code. They also accounted for the results after the formulation of prompts, differentiating their types together with the outcomes obtained.

3. Materials and Methods

3.1. Case Study

We conducted a case study on a *Web Applications* (WA) course during the 2nd semester of the academic year 2024–2025. This course is part of a bachelor's degree titled "Design and Creative Technologies" in the Arts School at UPV (Universitat Politècnica de València). The Degree in Design and Creative Technologies focuses on the professional application of artistic processes in cultural industries, emphasizing the creation, production, and commercialization of creative content. It prepares students to develop innovative goods and services in the creative sector by endowing them with a deep understanding of visual communication and functional and aesthetic design solutions, as well as through the integration of various techniques, including emerging technologies. In this context, students assume the role of UX designers by applying their artistic and technological skills to create user-centered digital experiences. It is important to note the combination of these artistic and technological skills, including those linked with coding and development capabilities, which enable designers to complement their creative profiles.

That is the case for the WA course, which was taught by lecturers in two different computing departments: one teaching programming language topics and the other addressing technological aspects in web infrastructure issues. Furthermore, this WA course was scheduled in the third year of the four-year degree mentioned earlier, as shown in Figure 1, which displays a timeline of the course and its relationships with other courses. Figure 1 shows the temporal evolution of courses relevant to the development of web applications. During the first year, there was a compulsory course called *Programming Basics* that introduced some basic notions of programming using *Processing* [36] as the language for creative coding. *Processing*, which is based on the Java language/framework, provides a free graphics library and integrated development environment (IDE) tailored to electronic arts and visual design. It also allows non-expert programmers to work with the fundamentals of computer programming in a visual context.

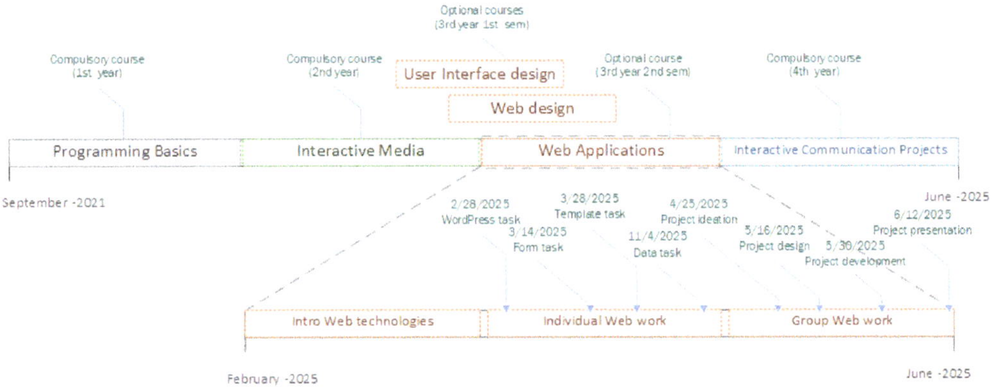

Figure 1. Timeline of the creative design courses.

In the second year, the *Interactive Media* course helped students acquire knowledge and skills required for interacting with narrative artistic elements using web technologies such as HTML5, CSS, and JavaScript and physical art installations based on Arduino components. In the 3rd year, the students were taught several elective courses dealing with the creation of web artifacts from different perspectives. During the first semester, the *User Interface Design* and *Web Design* courses provided a front-end perspective focused on visual aspects and introducing UI (User Interface) notions. The *Web Applications* (WA) course took place during the second semester and included topics related to back-end aspects that complemented the previously presented front-end perspective in the *Web Design* course (considered a recommended course). In the 4th year, there was a final compulsory course that promoted the integration of web and XR (extended reality) technologies into the production of interactive UX (user experience) projects. Naturally, other courses were taken over the course of this degree. The courses mentioned in Figure 1 were strongly connected with the WA course, which is the main research target.

Figure 1 also provides a graphical representation of the WA course curriculum, which was divided into three main parts. The first part introduced some basic topics on technologies used to maintain the server infrastructure required to host a web application and some examples of these applications, highlighting the role of Content Management Systems (CMSs) such as *WordPress* or *Joomla* [37]. The second part consisted of a practical teaching strategy involving students in individual activities, whose corresponding assignments needed to be submitted by the dates shown in Figure 1. These proposed activities covered different aspects of the front- or back-end views in a web application. For example, a global perspective of both views could be achieved by building a sample WordPress website. Other activities required the students to write PHP scripts that input data via visual interfaces and connected to a database in order to store and manage these data items (back-end). The third part of the WA course was oriented towards compelling the students to work collaboratively via a final web project. The intervention in this case study involved a cohort of 21 students who participated in the application of GenAI tools during a 2nd-semester course. The following subsections outline the ways these tools were integrated as part of the teaching methods used in the course, the interaction framework for applying GenAI tools in specific course tasks, and, finally, the data analysis instruments used to verify the outcomes yielded.

3.2. Teaching Methods

The WA course, part of the *Design* and *Creative Technologies* Degree since 2021, was taught via a traditional teaching approach combining theoretical topics related to web applications with development activities in lab sessions. Last year, degree students started working with ChatGPT-4, which has become a popular AI tool for solving technical problems related to implementing complex PHP scripts or managing specific database operations. The lecturers, who are also the authors of this study, decided to adapt their teaching methods in order to integrate Generative AI techniques while maintaining the original structure of the course. Individual activities, scheduled as shown in Figure 1, had precise instructions, including a rubric document that contained a set of associated assessment criteria. These types of activities represented a suitable scenario for incorporating GenAI tools and guiding students in their application. For example, the *Template-Session* task (the 3rd individual activity) required the students to write a PHP script allowing a user to log in to a web page. Once the user signed in, he or she could initialize a PHP session to acquire a private view of the page. This strict sequence of steps could be structured as a prompt template, helping the students to interact with GenAI tools in an incremental way and leading them through a controlled learning path.

Moreover, the rubric associated with each task represented a way of informing the students about the steps to be followed during a task's completion and how these steps would be assessed and scored. Table 1 shows a sample rubric attached to the *Template-Session* task and displaying six assessment criteria, which were graded from A (the highest score) to D (the lowest score). The first three criteria were linked with the "public view," and they were used to grade questions related to the completion of template page structures, the number of relevant multimedia resources, or the visual display of the provided graphical interface. The second set of assessment criteria referred to the sign-in/login identification process, the management of private session information once the user has been identified, or the user's ability to interact with some content items on a web page. Following an established sequence of steps, the students not only had an accurate idea of their assessments but also an appropriate guide with which to formulate AI-based iterative prompts, serving as one of the innovative teaching methods proposed in this work.

The second part of the teaching methodology employed in the current work was based on group activities. This type of activity invited the students to collaborate on the development of website projects, which could include some of the functionalities previously implemented through individual activities. Project tasks could encompass typical UX stages, ranging from initial ideation and the analysis of its potential target users to final testing and project evaluation. The intermediate stages dealt with tasks such as designing the project information architecture, prototyping user web interfaces in the form of wireframes or high-fidelity mockups, or coding and developing different aspects such as User Interface (UI) components related to the front-end side or access to information stored in a database or multimedia resources in the back-end server. Some of these UX stages have already been scrutinized with the assistance of the Generative AI tools [38]; this study focuses on the potential of these tools to support the implementation of complex programming tasks.

Table 1. Template task rubric.

Assessment Criteria/Scores	A 100–80%	B 80–60%	C 60–40%	D 40–0%
Page structure	The page is complete and well structured based on the right template.	The page is not complete, but it is appropriately structured.	The page is not complete, and its code is not sufficiently structured.	The page is not complete, and its code lacks a template-based structure.
Multimedia content	Suitable multimedia resources have been used.	The number of multimedia resources is sufficient but could be improved.	Relevant multimedia resources are lacking.	No multimedia content has been included.
Graphical interface	The graphical interface is clear and offers unified content display in public mode.	The graphical interface is correct, but the public content display has some flaws.	The graphical interface is correct, but the public content display is not adequate.	The graphical interface is poorly designed.
Private access	The code allows user identification and access to the main website page.	The code allows user identification, but the user name is not shown on the page.	The code allows user identification, but the main page is not connected.	The code does not allow user identification or access to other pages.
Session management	The page uses sessions to show user ID during private access and includes a logout function.	The page uses sessions to show user ID during private access but lacks a logout feature.	The page uses sessions for private access, but user ID is not shown.	Sessions are not used.
Content interaction	The identified user can add and view comments on content items after logging in.	The identified user can add comments after logging in, but these comments cannot be viewed.	The identified user can add limited comments after logging in.	No special interaction occurs after logging in.

At this juncture, it is important to highlight that innovative teaching methods may encounter challenges when addressing the development of web projects with moderate to high complexity, even when supported by Generative AI technologies. The next subsections describe some of the tools and instruments employed to support the use of these technologies and evaluate their application in the context of web development courses. This context entailed confronting the challenge of using these technologies in complex software projects that usually involve the cooperation of multiple actors. However, a collaborative use of GenAI tools is beyond the scope of the current research, but it should be considered for future work.

3.3. GenAI Interaction Framework

In line with the introduction of Generative AI tools and the support they could offer in addressing our research questions, a framework named AI4WD (Artificial Intelligence for Web Development) was established. In a previous work [38], a prompt-engineering strategy based on the exchange of Voila AI-assistant messages [39] was formulated to facilitate UX ideation processes and prototype design tasks. In this research, a similar strategy was employed by providing task instructions, which should help students generate iterative prompts in order to develop certain web functionalities. The purpose of AI4WD was not to deliver a specific sequence of prompts so as to automatically produce a website project, but to help students generate their own prompts that could interact with tools such as ChatGPT-4 or GitHub Copilot. Moreover, the instruments included in AI4WD enabled us to assess the suitability of the generated prompts and monitor their outcomes.

The idea of using GenAI techniques and tools for effective SW development has received widespread attention (see the Section 1). Li et al. [4] advocated for the use of ChatGPT along with prompt-engineering techniques for rapid code development. They noted that current LLM applications "lack guidance for practical software development process" and proposed an environment that was structured as three main services: (i) inputting the original prompt containing the main functional requirements and including references to some technological components; (ii) configuring this original prompt; and (iii) employing a self-debugging optimization process to improve the results. This proposal is somewhat similar to the one offered in the AI4WD framework, despite some differences in how the sequences of required prompts are built. Originally, a skeleton of the prompt template was provided to the students to assess their knowledge of web UI concepts. They were invited to use ChatGPT to complete this skeleton with their own prompt contributions, test the outcomes obtained, and return them to the lecturers, as described in the Section 4. This preliminary step was considered very useful for gaining an initial insight into the students' perspectives, although ChatGPT demonstrated that, at least in the free version, it lacked mechanisms with which to easily export and manage the outcomes produced. Instead, an alternative GenAI tool (GitHub Copilot) was evaluated for incorporation in the AI4WD context. GitHub Copilot (GC) [40] is a code completion and automatic programming tool that assists developers via Visual Code editors. Within these editors acting as IDEs (integrated development environments), GC can be integrated with several LLMs through plugin and extension mechanisms. Figure 2 shows a screenshot that displays the use of GC with the chat extension [41], represented on the left side of the image, to ask prompts in a programming context (right side).

The students, who were initially unaware of GC, rapidly realized how easy it was to use and quickly discovered its capacity to generate code automatically. The first observation consisted of confirming the need to assist *Creative Design* students in the formulation of prompts to guide web development tasks. The students usually employed the "trial and error" technique, constituting a major reason for facilitating the preparation of prompt templates, serving as one of the main AI4WD pillars. After a short tutorial on the GC features (particularly those related to chat prompt mechanisms), the students were invited to carefully read a set of task instructions (e.g., those related to the *Template-Session* task) and organize the sequence of prompts they should formulate to accomplish the task. These task instructions were important not only for guiding the students in their academic work but also for providing a sort of benchmark that could be used for comparison with the prompt sequences submitted by the students. This benchmark summarized the instructions provided for each task and served as a useful analysis instrument. In this sense, GC's ability to export/import chat logs in JSON format [42] was invaluable for tracking and processing student interactions, another fundamental pillar within the AI4WD framework.

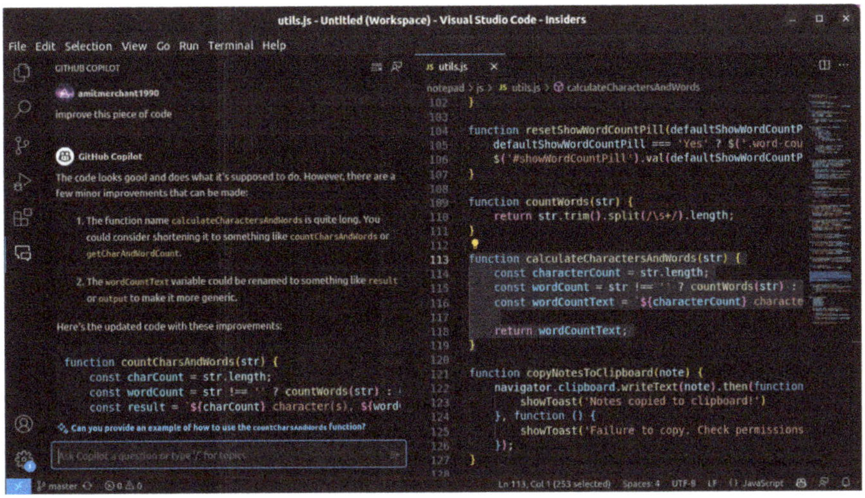

Figure 2. GitHub Copilot with the chat extension. This image was reproduced with permission from Amit Merchant.

3.4. Data Analysis

We obtained chat logs from GC–student interactions, composed of a variety of data corresponding to the user prompts formulated and their resulting outcomes in the form of generated pieces of code or technical reports explaining their generation. The analysis of these logs was extremely useful in uncovering the details of human–AI interactions, but the logs had to be cleaned and processed beforehand. Figure 3 shows the different stages of the entire analysis process, starting with the filtering of chat information by using Python 3.9 scripts to convert the chat logs into MD (MarkDown) notation to allow improved display and extracting student requests via regex expressions [43]. After the filtering stage, the logs were translated to English and summarized for subsequent processing. R and Python scripts were then used to load a udpipe (trainable pipe) model [44] that allowed for the annotation of the incoming text and the extraction of terms such as nouns and verbs. This stage played a crucial role in analyzing the actions discussed by the students (through verbs) and the objects (nouns) of such actions.

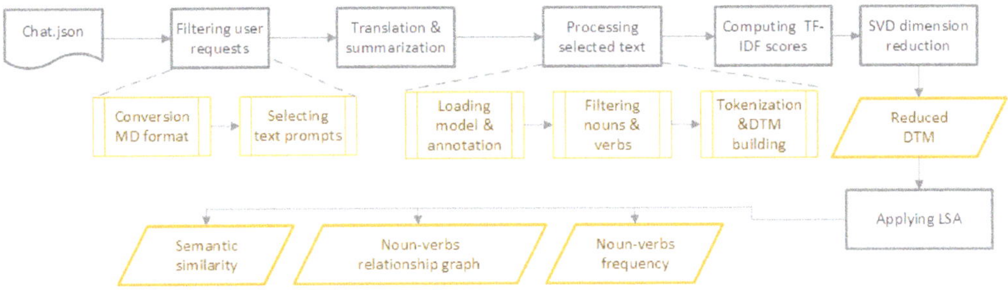

Figure 3. AI4WD processing stages.

Once nouns and verbs were extracted as lemmatized terms, we retrieved a corpus of tokenized documents to build DTM (Document–Term Matrix) structures [45] in order to create a vector-space representation of these documents. These structures were scored by using TF-IDF (Term Frequency-Inverse Document Frequency) measures to retrieve the most meaningful terms and obtain a compact DTM. After these processing steps, LSA (Latent Semantic Analysis) procedures [46] were applied to achieve several relevant outcomes, such as a graphical display of term relationships and their semantic similarity. We used LSA, a statistical technique that models word usage patterns to measure semantic relationships and compare the degree of similarity between different pieces of text through these relationships [47]. A Python script [48] was adapted to compute this semantic similarity value by comparing logs produced by the students against the proposed benchmark in each experiment case.

In addition to these text analytics techniques, other methods were employed in assessing the outcomes obtained, such as usability tests and user questionnaires. The usability tests were oriented towards evaluating certain aspects of the web pages designed by the students [49], while the user questionnaires were prepared to collect the students' opinions regarding the use of GenAI tools in their courses.

4. Results

Several experiments were performed in order to apply the methods and tools introduced in the previous section. First, the integration of GenAI techniques into the proposed WA course was examined. A second batch of experiments processed student interactions gathered as chat.json logs in different stages of the WA course. Finally, a comprehensive evaluation of student scores, usability tests, and student opinions was implemented.

4.1. GenAI Integration

The first set of experiments was used to assess the integration of GenAI mechanisms into key aspects of instructional tasks within the WA course. One of the most interesting aspects for creative design students was the elaboration of visual interfaces based on templates and their implementation as website pages. This question was also challenging since further stages would require the students to develop more examples of webpages and connect their content items to database records.

Figure 4 shows an example of a generic mockup based on *templatetemo* designs (https://templatemo.com/, accessed on 28 May 2025) that displays a sample web homepage composed of two main parts. In the upper area, the header contains a "Logo" title on the left side and a horizontal navigation menu with options such as *Home*, *Browse*, *How it Works*, *Contact*, and *Pages*, along with a user icon (enabling sign-in actions). The central part of the *Homepage* features a *Hero* section with a main title and a subtitle complemented by a search component and some card samples. The students could download this mockup image in a graphic format. The first experiment conducted to test the use of GenAI tools involved providing the students with a prompt template skeleton with some relevant web design aspects they were tasked with completing. Table 2 shows a list of aspects the students could use to prompt ChatGPT, a well-known GenAI tool.

Figure 5 shows a plot of four bar charts that display the frequency of terms categorized by the items defined in Table 2. These charts were obtained from the responses of nine students who experimented using the proposed prompt skeleton and completed the main web components described in the template. Regarding the *Structure* category, the most mentioned term by the students was *Card*, which represented a suitable unit for organizing content items. In the case of the *Style* category, the students were mostly concerned with the visual appearance of buttons, and *Focus* was selected as the favorite concept for capturing user attention. The last category was *Content*, in which *Text* was the recurrent term.

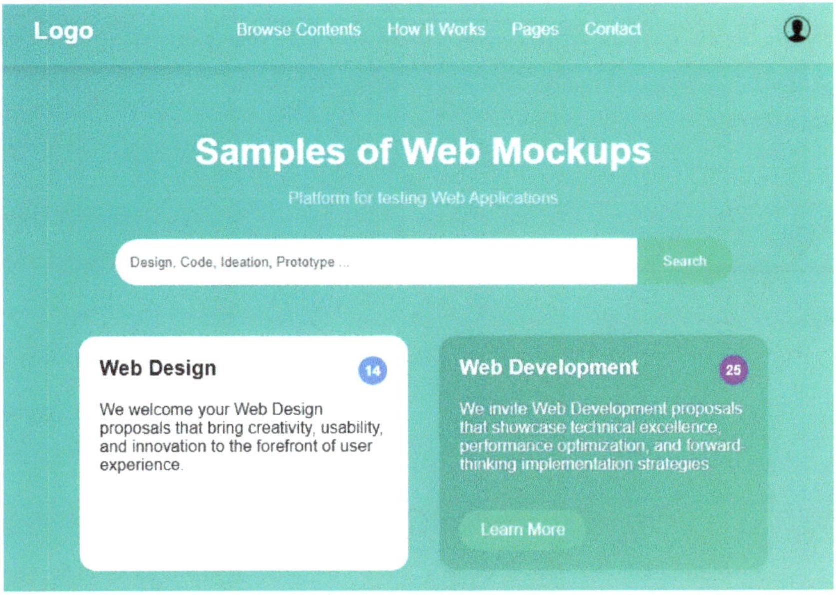

Figure 4. Example of a web mockup.

Table 2. Design prompt templates.

Category	Description
Preamble	I have a screenshot of a UI design, and I want you to analyze it and create a detailed html document based on the following criteria.
Structure	Identify the specific components within the provided design, such as [to be completed], and break down the structure of each component using aspects such as [to be completed]. Focus on individual component layouts rather than large sections like headers or footers.
Style	Describe the visual styles, including [to be completed]. Mention any noticeable visual elements that could be included in the current design.
Interaction	Identify any potential interactions or animations, such as [to be completed]. Specify what should happen when the identified elements are hovered over or clicked or focused on.
Content	Provide details about the content within the design, including [to be completed with any visible or potential data items]. Describe attributes, such as [to be completed], of these data items.

In further experiments, we used this starting point to provide a sample UI specification from the mockup displayed in Figure 4 that the students could use to complete their course tasks. In these new experiments, GenAI techniques were applied based on using the GC tool to develop several tasks (either individual or group-led) and gather information from student interactions. The first set of tasks was proposed to help the students develop a front-end UI view based on PHP templates, and it was complemented by a "sign-in" login procedure. Furthermore, the students were invited to work with back-end mechanisms concerning database connections or access operations to retrieve data using SQL expressions. A second set of tasks dealt with group activities that involved linking UI components with data items. These kinds of activities were more sophisticated since they required a higher degree of coordination between the students in charge of developing front-end modules and those involved in the database (back-end) access task. They also represented a more challenging task from the perspectives of GenAI integration and the preparation of suitable benchmarks for analysis. The next subsection describes the experiments conducted to collect data logs from the students' interactions with the GC Chatchat in these task examples.

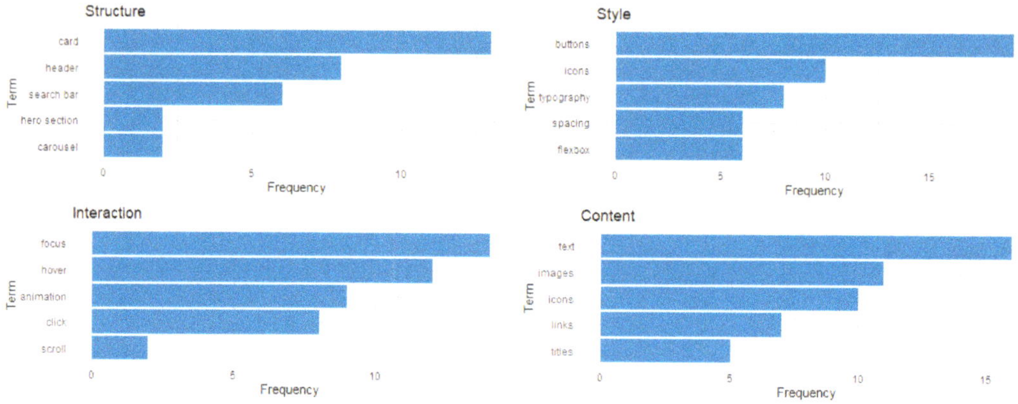

Figure 5. Graphical display of web component term frequency.

4.2. GenAI Interaction

A second set of results was obtained by gathering the outcomes derived from interactions between the students and the GenAI mechanisms (GitHub Copilot-assisted). These outcomes were based on two data sources: (i) the set of development artifacts produced by the students during their learning activities and (ii) the collection of chat logs derived from the GenAI interactions and exported as JSON files. These AI-based learning activities were a form of support for task completion (scheduled in Figure 1), and the rate of completion of these tasks was also computed. Before reviewing the associated academic results, a careful analysis of chat logs was performed, beginning with the visualization of extracted terms (e.g., nouns and verbs) and their relationships. Figure 6 shows an example of a graph displaying the relationships between the verbs represented in the prompt benchmark (green points) and the entity nouns involved (blue points) for the *Template-Session* task benchmark (see Appendix A). The resulting graphs displayed two different areas. The first, located in the upper -left corner, showed a dense network of concepts with generic entities such as "html", "php", or "function (located on the right-hand side of this area) and references to web components such as "header", "section", "navigation", and "menu". The second main area was located in the lower part of Figure 6, displaying more sparsely distributed relationships among entities that are more specific in the context of the target task, such as "session", "username", and "login". Importantly, this type of graph could be useful as a support for comparing the relationship patterns exhibited by student–AI interactions (through their associated chats) and the task benchmark, a visual reference of which is displayed in Figure 6.

However, this graphical representation was difficult to use as an instrument for measuring the quantitative distance between the provided benchmark (which, in this case, is associated with the *Template-Session* task) and the outcomes obtained by the students in their chat logs. Instead, alternative instruments were used, such as those reported in the *Data Analysis* subsection (see Figure 3). For example, the computed term frequencies gathered from student chats were compared with the terms found in the task benchmark. This process is depicted in Figure 7, which shows the distribution of differences in the frequency of terms identified as lemmas by comparing the frequency of lemmas in student chats against the selected terms in the *Template-Session* task benchmark. This comparison is divided by lemma POS (Part of Speech) entities, such as nouns (displayed in blue) and verbs (in green). The accumulated frequency values for verbs did not show remarkable differences in student-based outcomes compared to the provided benchmark. However, notable discrepancies emerged for terms such as "session" or, especially, "login" with high negative accumulation (fewer than −20 occurrences), indicating that this particular concept was underestimated or neglected by the students in their chats.

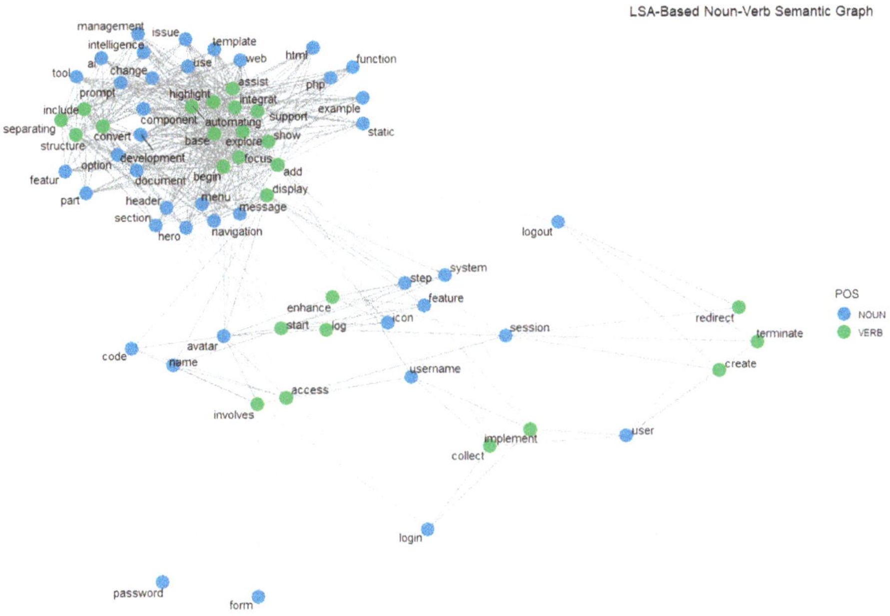

Figure 6. Graphical display of terms' semantic relationships.

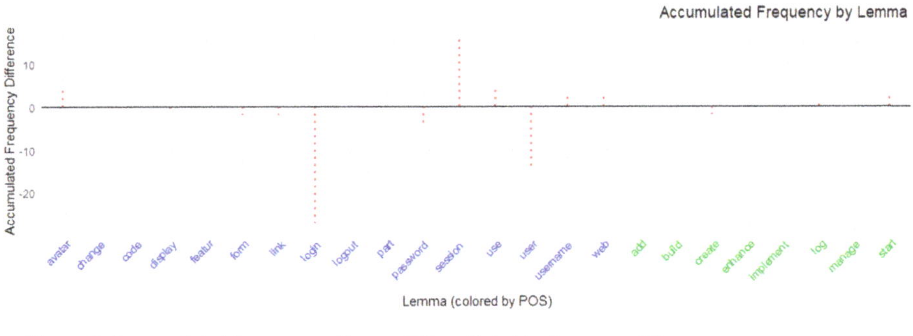

Figure 7. Accumulated frequency of terms differentiated by lemma entities.

4.3. Final Assessment of GenAI Outcomes

In previous experiments, an initial attempt to analyze students' interactions with GenAI tools like GC was made. The analysis of term frequency in a specific task provided a first impression of how close the prompts submitted by students were in comparison with those provided as benchmark references. A second instrument, based on evaluating the semantic similarity of student chat logs against provided benchmarks, could contribute to a more accurate measurement. In this case, two examples of tasks were assessed in order to analyze the percentage of semantic similarity and the task completion rate. The first task consisted of the individual activity comprising assessing *Template-Session* concepts (already mentioned in previous sections), while the second one referred to a group activity based on defining an initial UI to access a prototype of project data items without a database connection. Figure 8 shows two charts comparing semantic similarity with task completion percentages in both activities. In the individual activity case (a), there were 11 student answers showing values of semantic similarity ranging from 62% to 86%. However, the level of correlation between the similarity and completion percentages was rather low (approximately 21%). The second analyzed case, in Figure 8b (a group-led activity with seven answers), displayed closer semantic similarity values (exceeding 80%), with a correlation rate of 60% between the similarity and completion rates.

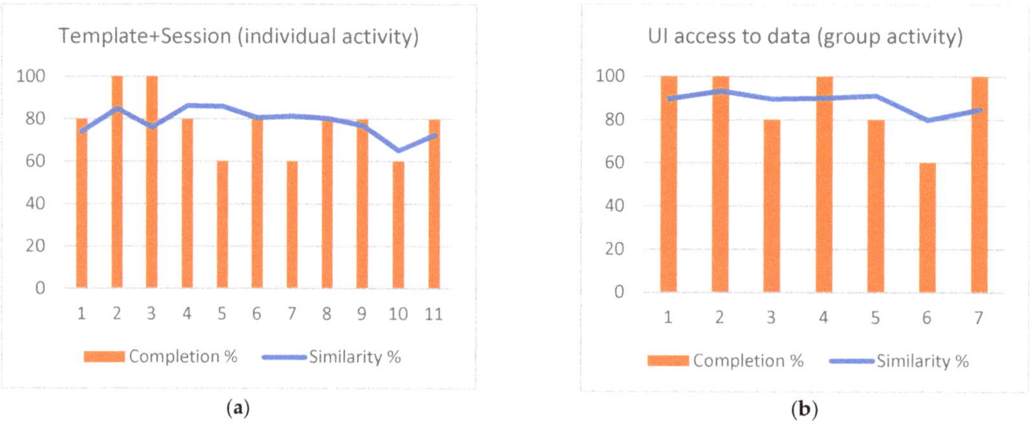

Figure 8. Comparison of semantic similarity rates and task completion degrees. (**a**) Template + Session (individual activity); (**b**) UI access to data (group activity).

Two additional group activities were analyzed concerning the work involving database items. The first activity was completed by seven groups that exhibited good performance in basic database operations, but the corresponding computed semantic similarity was quite irregular, with low values ranging from 30 to 60%. The second proposed activity was based on employing several PHP scripts to develop a sort of database administrator based on multiple CRUD (Create/Read/Update/Delete) operations. This was a new activity for the WA course and difficult to complete without the help of GenAI tools since it required a high level of expertise in database management. The students involved in this activity spent several hours completing it as a kind of final project/test, and the analysis of their chat logs revealed the difficulties these students encountered. These chat logs contained multiple messages indicating several error issues, and student interactions involving the management of these issues were difficult to track. This observation confirmed the general impression of the use of GenAI tools, which require careful prompt organization in order to yield effective results, especially as task complexity grows.

As the final part of this assessment stage, two questionnaires were used to collect feedback from users. The first one was administered to students in the *Interactive Communication Projects* (ICP) course (a fourth-year course) who examined web prototypes developed by WA groups. There were more students (about forty) in this course, and their projects were unrelated to those worked on by the WA students. Seven web projects were tested by 27 students (with a range of three to five evaluations per project). Figure 9a displays a bar chart that shows the results for four of the assessment criteria used for testing the web prototypes developed. These results were obtained by averaging the ICP students' answers, revealing assessment values close to 80% in three categories, although the *Interaction* category received a lower score. The full list of questions is available in Appendix B. A second questionnaire (see Appendix C) was used to gather the WA students' perceptions regarding the application of GenAI tools in their course. The bar chart in Figure 9b displays the average answers from fourteen students in four assessment categories that measured the suitability of applying GenAI, the ease of producing prompts in the course context, and the usefulness of and trust levels regarding the GenAI outcomes obtained. None of the assessed categories reached 80% agreement, and only 62% of the students, on average, provided positive opinions regarding the suitability of GenAI use during the course.

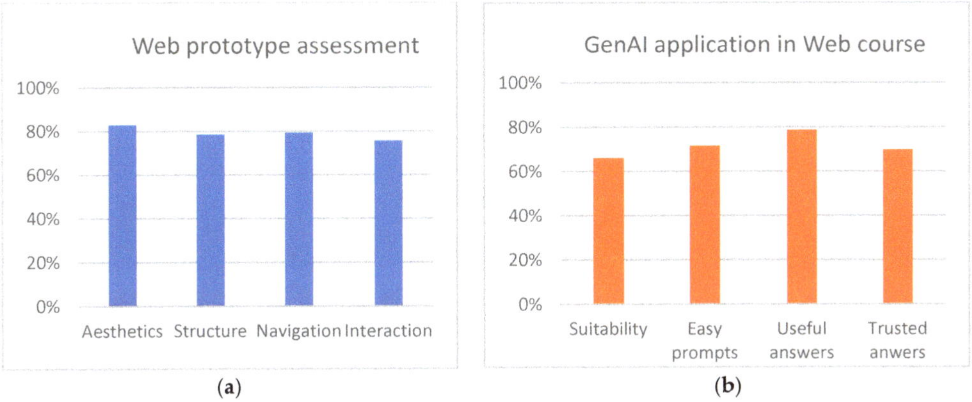

Figure 9. Summary of user questionnaire results. (**a**) Web prototype assessment; (**b**) GenAI application in Web course.

5. Discussion

This section reports on the various ways Generative AI tools could impact a set of learning activities within web development courses. In this case, the courses were tailored to students whose affinity for design was stronger than that for computing, so they required special consideration for their creative profiles and limited programming skills. The aspects discussed are organized according to the previously introduced research questions.

The first question (RQ1) deals with the way GenAI-based technology could be effectively used in or integrated into WD courses for designers. As mentioned in the *Introduction*, multiple studies have addressed the use of these technologies in several web development areas, although these studies mainly focused on ideation and prototyping stages [32]. As shown in the questionnaire results in Figure 9, the Creative Design students analyzed in this study often expressed their concerns about the impact of AI applications on these areas, but they were also fully aware of the benefits of using this kind of technology in terms of the dissemination of ideas or their rapid prototyping, particularly for web development, which involves close collaboration between designers and programmers, requiring their combined efforts. The students in the *Creative Design* program realized that their designs needed "to be moved forward", and WD courses represented a good chance to assume a new hybrid role. This hybridization phenomenon regarding web designers and developers has revealed multiple challenges and opportunities [50]. In this case, the degree curriculum taught to students during the first years included *Programming Basics* and *Interactive Media* courses, providing them with a sound background of basic programming skills and HTML/CSS knowledge. However, they usually lack deeper notions related to database management, asynchronous access to web content (AJAX), and the usage of widespread API services. These educational programs currently lack information on these advanced programming technologies, but the deployment of GenAI tools could help provide students with a suitable introduction to innovative web mechanisms. In this study, we tested the use of the GitHub Copilot (GC) environment with respect to students who had limited practice with ChatGPT or Copilot tools. Non-expert students explored the possibilities of this kind of environment using their chat capabilities in technical tasks related to the creation of "workplaces" integrating multiple CSS files and HTML documents, pieces of PHP code to dynamically generate and organize web pages, or working with SQL statements to retrieve database items. GC also provided invaluable help in explaining pieces of generated code [51] or fixing programming issues that appeared during the development process in small-scale tasks. Moreover, the instructors reported that the GC tool led to a reduction in the frequency of minor technical inquiries among the students, thereby freeing them to help the students in other areas more aligned with a creative profile. As the students were improving their programming abilities, they requested involvement in projects with a higher level of complexity. Therefore, there was a continuous challenge regarding the integration of GenAI tools in this development context and the need to promote the application of these tools in completing more complex tasks. The projects also required close collaboration between front-end designers and back-end programmers, so new tools may be required to support such collaboration.

Before dealing with this collaboration challenge, we considered a second research question (RQ2) concerning the determination of what new interaction mechanisms could be considered in a GenAI context, with the students requesting prompts to develop increasingly more complex tasks. As mentioned before, the *Creative Design* students in this study were well aware of the potential of tools such as ChatGPT or Copilot, but they occasionally interacted with them in a disordered or unsystematic manner. They usually submitted prompts without a predefined plan, and one of the main goals of this study was to guide the students toward a more systematic and rigorous approach. Accordingly, an investigation into the creative skills required for prompt engineering should be considered to deal with this situation [52]. Experiments with GC Chats in individual WD tasks were conducted by providing the students with a set of instructions rather than a detailed list of "mechanical" prompts to be submitted. This idea was complemented by an incremental method to develop pieces of code by adding functionalities in a given order while avoiding a rigidly constrained plan. Importantly, the combination of the development instructions provided and the use of incremental methods helped to build benchmarks that could be used as a reference for comparison with further interactions. The analysis of the results gathered from the chat logs revealed that the students started to provide their prompts using a more organized strategy that was closer to the proposed benchmarks. The studied outcomes, such as the graphical display of web topic maps derived from these student interactions or the frequency counts of the used terms, indicated at least the willingness of students in this direction. Interestingly, the application of semantic similarity techniques highlighted that the students followed the instructions provided to formulate prompts, helping them achieve their academic endeavors. Alternative ways of analyzing the closeness between student prompts and benchmark proposals were explored, such as knowledge-based text analysis methods [53], but their values were less representative than those returned by the corpus LSA-based techniques. Nevertheless, the problem of comparing the semantic similarity of texts persists, and new opportunities could emerge with the use of LLMs to address this issue [54]. The way GenAI–user interactions could improve students' performance in web or other programming areas is another challenging field, and more research should be carried out to address it, particularly when collaborative tasks are involved.

The final research question (RQ3) focused on how the use of these GenAI-based interactive learning technologies could be evaluated. As mentioned before, there were many

AI–student interactions, which were measured in several ways, aimed at either evaluating individual or group-led learning activities. These interactions were gathered from GC Chat export capabilities, though employing manual procedures entailed a significant time investment. Automatic methods of collecting such chat logs should be implemented in order to obtain a more effective way of facilitating their evaluation. Despite the paucity of chat logs collected in each activity, they allowed us to observe how the students interacted with GenAI tools such as GC and helped them to solve the tasks they were assigned. In the case of individual tasks, semantic similarity measures offered a method of comparing GC Chat outcomes with a given benchmark, highlighting the moderately high values of the matching percentages obtained (about 78%, on average, for the *Template-Session* task). A group-based activity was also evaluated using this semantic matching method, yielding even higher percentages (close to 90%, on average), although it was difficult to differentiate each group member's interactions. Thus, researchers should pay attention to collaboration mechanisms, which enabled a more meaningful evaluation of these GenAI–student-group interaction patterns. Furthermore, correlation indices between semantic similarity and task completion percentages were measured. In the case of the *Template-Session* task, the correlation index was rather low (about 21%), highlighting the difficulty of establishing a direct link between GenAI–student interactions and students' academic performance. This research question thus inspires reflection on the potential split between the activities assisted by GenAI that could be evaluated using a formative assessment perspective and other complementary activities based on summative methods. Ethical issues regarding the impact of GenAI tools could be incorporated to determine the extent to which academic grades were based on AI-assisted tasks. In addition to discussion about formative vs. summative assessments, other evaluation criteria could be considered to analyze the intervention of GenAI tools. The first impression inspired by this course experience is that the students needed less time to finish their tasks, indicating, at least partially, the GenAI tool's effectiveness. Furthermore, the decrease in the number of technical questions the students asked during these tasks also indicated greater autonomy in their learning processes. A more systematic evaluation of these aspects could be performed in order to measure student productivity. Finally, qualitative measurement methods, such as assessing the quality of web projects developed by students, should also be considered. In this sense, the usability tests used in collaboration with students from other courses were an interesting instrument for acquiring an initial assessment of this project's quality.

6. Conclusions

This work presents several important aspects regarding the use of GenAI tools to support UX design students in web development courses. We focused on a case study in a Web Applications (WA) course for students pursuing a *Design* and *Creative Technologies* degree, who were primarily UX designers with a combination of artistic and technological skills. In this context, the adoption of tools such as GitHub Copilot and ChatGPT represents a remarkable step toward democratizing access to programming resources and fostering their inclusion in different instructional settings. However, the full pedagogical potential of these tools has yet to be realized, especially in areas in which creativity, user experience aspects, and basic programming skills are central to the learning process. This latency was evident in the WA course, which, during the first semester of 2025, registered twenty-one students who participated in a set of individual and group-based activities dealing with web development topics.

One of the primary focuses of this work was exploring how these students were able to incorporate this type of GenAI tool through their instructional pathways. GenAI integration played a crucial role in supporting students in coding and developing web artifacts related to front- and back-end topics. Most of the students were familiar with ChatGPT technologies, facilitating the integration of new tools such as GitHub Copilot and their chat plugins integrated with the Visual Code Editor. They realized GC's potential to generate code in response to simple prompts, even for organizing a set of files in which the generated code could be easily structured. The outcomes obtained by the students in the form of chat.json logs demonstrate seamless GenAI integration during learning activities. At least for individual, small-scale tasks, the observations in these logs show how GC provided significant help with code generation, explaining code, and fixing programming issues. The instructors also noticed a significant decrease in the number of minor technical questions asked by the students, suggesting increased autonomy in their learning processes. Nevertheless, challenges remain in applying GenAI for more complex tasks, such as in group activities, and the need to articulate AI-supported collaboration mechanisms between front-end designers and back-end programmers has become evident.

Regarding the introduction of innovative interaction mechanisms, an ordered application of these mechanisms in the form of prompt skeletons and instruction benchmarks has revealed multiple benefits that can guide some of the proposed instructional activities. GenAI played a pivotal role in scaffolding the learning experience through several stages in the web development process, helping the students deal with increasingly complex tasks. For example, the students experienced a reduction in cognitive overload when managing multiple design ideas and exhibited the ability to build web prototypes and test database operations in a faster, yet controlled, manner. An analysis of student–AI chat logs, including graphical displays of term relationships and frequency accounting of the terms, indicated that the students have begun to adopt a more organized prompting strategy closer to the proposed benchmarks. Semantic similarity techniques further highlighted that the students followed the instructions provided, aiding them in completing their academic tasks. Moreover, the use of such benchmarks provided a suitable instrument for assessing GenAI's impact on students' activities. Several measurement techniques were used to assess and evaluate this impact, such as counting the frequency of selected terms and computing the semantic similarity between student prompts and the benchmarks provided. These measurements were combined with a review of task completion rates as well as questionnaires submitted to students in a fourth-year course, who tested the usability of the prototypes developed by the WA groups. This combination of results confirmed the suitability of the current instructional approach and the benefits of using GenAI in this educational setting. Nevertheless, open issues regarding the study of users' interactions with GenAI tools should be taken into account, including issues of overreliance on these tools or a lack of reflection, constituting pitfalls into which students could fall when overusing these tools.

In future work, we plan to extend the use of GenAI tools to new courses dealing with web topics ranging from basic design concepts to more advanced development platforms. These experiences could include new programming methods, including those based on agile methodologies, and incorporate developer platforms like GitHub, potentially helping to foster better collaborative skills. Consequently, the application of GenAI tools that can be shared should be examined, as they are required to fulfill such teamwork requirements. Future research should focus on collaboration mechanisms to allow a more meaningful evaluation of GenAI–student group interaction patterns. In this sense, automatic methods of collecting chat logs should be employed to make these logs' evaluation easier and more effective. Additional research is required in order to reveal how GenAI–student interactions can improve students' performance in web development or other programming areas. Therefore, a more systematic evaluation of these interactions should include student productivity metrics or incorporate measurements assessing the quality of web projects developed by students.

In conclusion, integrating GenAI into design-oriented web development courses is both feasible and potentially transformative. With careful pedagogical framing, collaborative task design, and critical engagement, GenAI can serve not just as a coding assistant but also as a catalyst for broader student participation in developing web artifacts, especially among learners traditionally marginalized by technical barriers.

Author Contributions: Conceptualization, F.B.-G. and J.P.-R.; methodology, F.B.-G.; software, J.P.-R.; validation, F.B.-G. and J.P.-R.; formal analysis, F.B.-G.; investigation, F.B.-G.; resources, J.P.-R.; writing—original draft preparation, F.B.-G. and J.P.-R.; writing—review and editing, F.B.-G. and J.P.-R.; All authors have read and agreed to the published version of the manuscript.

Funding: This research received no external funding.

Institutional Review Board Statement: Not applicable.

Informed Consent Statement: Informed consent was obtained from the subjects involved in this study.

Data Availability Statement: The datasets presented in this article are not readily available due to time limitations. Requests to access the datasets should be directed to corresponding author.

Conflicts of Interest: The authors declare no conflicts of interest.

Abbreviations

The following abbreviations are used in this manuscript:

GenAI	Generative AI
UX	User Experience
XR	Extended Reality
LCDPs	Low-Code Development Platforms
WD	Web Development
WAs	Web Applications
ICPs	Interactive Communication Projects
IDE	Integrated Development Environment
LLM	Large Language Model
CMSs	Content Management Systems
GC	GitHub Copilot
POS	Part of Speech
DTM	Document–Term Matrix
LSA	Latent Semantic Analysis
TF-IDF	Term Frequency-Inverse Document Frequency
MD	MarkDown
CRUD	Create/Read/Update/Delete
AI4WD	Artificial Intelligence for Web Development

Appendix A

A sample summarized instruction benchmark for the *Template-Session* task.

> The document explores the use of Artificial Intelligence tools, particularly GitHub Copilot, to support the development of PHP-based websites using templates and session management.
> It begins with a simplified web template focusing on two components: a header with a navigation menu and user avatar, and a hero section displaying a main message. AI is first used to convert a static HTML example into a structured PHP template by separating components using the include function.

The next step involves adding user session features, starting with a login system accessed via the avatar icon. A modal login form is implemented to collect the username and password from users. After logging in, the code is enhanced to start a session and display the logged-in user's name next to the avatar. A display issue prompts further changes to correctly show the username and add a logout option. A logout link is created to terminate the session and redirect the user. The document highlights how AI can assist in automating parts of web development and integrating interactive features. Overall, it demonstrates the practical application of AI to build and manage basic dynamic websites.

Appendix B

A user questionnaire for webpage testing.

Purpose. The Web page example is easy to understand, and its purpose is clearly conveyed. *Recognition.* The product or brand presented in the context of the Web page is easily identifiable. *Interest.* The team has created an attractive presentation on their Web page that captured my interest. *Style.* The visual resources used on the Web page contribute positively to conveying a specific goal or purpose. *Structure.* The structure used to organize the content on the Web page can be considered appropriate. *Navigation.* The navigation schemes or mechanisms used are suitable. *Interaction.* The interactive elements provided on the Web page encourage user engagement. Any suggestions for the team that developed the Web page?

Appendix C

A user questionnaire for GenAI student evaluation.

Support. As a student interested in Creative Design topics, to what extent do you think Web Application Development can help you enhance or support those topics? *Improvement.* Do you think the use of Generative Artificial Intelligence (GenAI) tools can contribute to better training in the field of Creative Design? *Applicability.* To what extent have you used or are using these types of tools throughout your degree program? *Suitability.* When you have used GenAI tools, how suitable do you find them to apply within your degree field? *Proficiency.* Do you feel that your knowledge of such tools allows you to make good use of them in tasks related to Web Application Development? *Easy prompts.* Do you think it is easy to formulate prompts for GenAI that can help you carry out those tasks? *Useful answers.* Have the answers obtained from the prompts you formulated for GenAI been useful for the assigned tasks? *Trusted answers.* Based on the answers provided by GenAI to solve the proposed tasks, how much confidence do those answers give you, including their quality? *Continuity.* Finally, to what extent do you believe you will continue using GenAI tools in the future for tasks related to Design or the Development of various types of applications?

References

1. Ebert, C.; Louridas, P. Generative AI for software practitioners. *IEEE Softw.* **2023**, *40*, 30–38. [CrossRef]
2. Nguyen-Duc, A.; Abrahamsson, P.; Khomh, F. (Eds.) *Generative AI for Effective Software Development*; Springer Nature: Cham, Switzerland, 2024.
3. Coutinho, M.; Marques, L.; Santos, A.; Dahia, M.; França, C.; de Souza Santos, R. The role of generative ai in software development productivity: A pilot case study. In *AIware 2024, Proceedings of the 1st ACM International Conference on AI-Powered Software, Porto de Galinhas, Brazil, 15–16 July 2024*; Association for Computing Machinery: New York, NY, USA, 2024; pp. 131–138.
4. Li, Y.; Shi, J.; Zhang, Z. An approach for rapid source code development based on ChatGPT and prompt engineering. *IEEE Access* **2024**, *12*, 53074–53087. [CrossRef]
5. Rajbhoj, A.; Somase, A.; Kulkarni, P.; Kulkarni, V. Accelerating software development using generative ai: Chatgpt case study. In *ISEC '24, Proceedings of the 17th Innovations in Software Engineering Conference, Bangalore, India, 22–24 February 2024*; Association for Computing Machinery: New York, NY, USA, 2024; pp. 1–11.
6. Petrovska, O.; Clift, L.; Moller, F.; Pearsall, R. Incorporating generative ai into software development education. In *CEP '24, Proceedings of the 8th Conference on Computing Education Practice, Durham, UK, 5 January 2024*; Association for Computing Machinery: New York, NY, USA, 2024; pp. 37–40.
7. Pirzado, F.A.; Ahmed, A.; Mendoza-Urdiales, R.A.; Terashima-Marin, H. Navigating the pitfalls: Analyzing the behavior of LLMs as a coding assistant for computer science students-a systematic review of the literature. *IEEE Access* **2024**, *12*, 112605–112625. [CrossRef]
8. Prather, J.; Denny, P.; Leinonen, J.; Becker, B.A.; Albluwi, I.; Craig, M.; Keuning, H.; Kiesler, N.; Kohn, T.; Luxton-Reilly, A.; et al. The robots are here: Navigating the generative ai revolution in computing education. In *ITiCSE-WGR '23, Proceedings of the 2023 Working Group Reports on Innovation and Technology in Computer Science Education, Turku, Finland, 7–12 July 2023*; Association for Computing Machinery: New York, NY, USA, 2023; pp. 108–159.
9. Denny, P.; Prather, J.; Becker, B.A.; Finnie-Ansley, J.; Hellas, A.; Leinonen, J.; Luxton-Reilly, A.; Reeves, B.N.; Santos, E.A.; Sarsa, S. Computing education in the era of generative AI. *Commun. ACM* **2024**, *67*, 56–67. [CrossRef]
10. Sheard, J.; Denny, P.; Hellas, A.; Leinonen, J.; Malmi, L.; Simon. Instructor Perceptions of AI Code Generation Tools—A Multi-Institutional Interview Study. In *SIGCSE 2024, Proceedings of the 55th ACM Technical Symposium on Computer Science Education, Portland, OR, USA, 20–23 March 2024*; Association for Computing Machinery: New York, NY, USA, 2024; Volume 1, pp. 1223–1229.
11. Raihan, N.; Siddiq, M.L.; Santos, J.C.; Zampieri, M. Large language models in computer science education: A systematic literature review. In *SIGCSETS 2025, Proceedings of the 56th ACM Technical Symposium on Computer Science Education, Pittsburgh, PA, USA, 26 February–1 March 2025*; Association for Computing Machinery: New York, NY, USA, 2025; Volume 1, pp. 938–944.
12. Becker, B.A.; Craig, M.; Denny, P.; Keuning, H.; Kiesler, N.; Leinonen, J.; Luxton-Reilly, A.; Prather, J.; Quille, K. Generative AI in introductory programming. In *Computer Science Curricula 2023*; Association for Computing Machinery: New York, NY, USA, 2024; pp. 438–439.
13. Kirova, V.D.; Ku, C.S.; Laracy, J.R.; Marlowe, T.J. Software engineering education must adapt and evolve for an llm environment. In *SIGCSE 2024, Proceedings of the 55th ACM Technical Symposium on Computer Science Education, Portland, OR, USA, 20–23 March 2024*; Association for Computing Machinery: New York, NY, USA, 2024; Volume 1, pp. 666–672.
14. López-Fernández, D.; Vergaz, R. ChatGPT in Computer Science Education: A Case Study on a Database Administration Course. *Appl. Sci.* **2025**, *15*, 985. [CrossRef]
15. Kazemitabaar, M.; Hou, X.; Henley, A.; Ericson, B.J.; Weintrop, D.; Grossman, T. How novices use LLM-based code generators to solve CS1 coding tasks in a self-paced learning environment. In *Koli Calling '23, Proceedings of the 23rd Koli Calling International Conference on Computing Education Research, Koli, Finland, 13–18 November 2023*; Association for Computing Machinery: New York, NY, USA, 2024; pp. 1–12.
16. Prather, J.; Reeves, B.N.; Leinonen, J.; MacNeil, S.; Randrianasolo, A.S.; Becker, B.A.; Kimmel, B.; Wright, J.; Briggs, B. The widening gap: The benefits and harms of generative ai for novice programmers. In *ICER '24, Proceedings of the 2024 ACM Conference on International Computing Education Research, Melbourne, Australia, 13–15 August 2024*; Association for Computing Machinery: New York, NY, USA, 2024; Volume 1, pp. 469–486.
17. Hsiao, H.L.; Tang, H.H. A Study on the Application of Generative AI Tools in Assisting the User Experience Design Process. In *Artificial Intelligence in HCI, Proceedings of the International Conference on Human-Computer Interaction, Washington, DC, USA, 29 June–4 July 2024*; Springer Nature: Cham, Switzerland, 2024; pp. 175–189.
18. Sahay, A.; Indamutsa, A.; Di Ruscio, D.; Pierantonio, A. Supporting the understanding and comparison of low-code development platforms. In Proceedings of the 2020 46th Euromicro Conference on Software Engineering and Advanced Applications (SEAA), Portoroz, Slovenia, 26–28 August 2020; IEEE: Piscataway, NJ, USA, 2020; pp. 171–178.
19. Ferreira, J.; Sharp, H.; Robinson, H. User experience design and agile development: Managing cooperation through articulation work. *Softw. Pract. Exp.* **2011**, *41*, 963–974. [CrossRef]

20. Pacheco, J.; Garbatov, S.; Goulão, M. Improving collaboration efficiency between UX/UI designers and developers in a low-code platform. In Proceedings of the 2021 ACM/IEEE International Conference on Model Driven Engineering Languages and Systems Companion (MODELS-C), Fukuoka, Japan, 10–15 October 2021; IEEE: Piscataway, NJ, USA, 2021; pp. 138–147.
21. Rode, J.; Rosson, M.B.; Quinones, M.A.P. End user development of web applications. In *End User Development*; Springer: Dordrecht, The Netherlands, 2006; pp. 161–182.
22. Barricelli, B.R.; Cassano, F.; Fogli, D.; Piccinno, A. End-user development, end-user programming and end-user software engineering: A systematic mapping study. *J. Syst. Softw.* **2019**, *149*, 101–137. [CrossRef]
23. Paternò, F. End-user development. In *Handbook of Human Computer Interaction*; Springer International Publishing: Cham, Switzerland, 2023; pp. 1–27.
24. Calò, T.; De Russis, L. Leveraging large language models for end-user website generation. In *End-User Development, Proceedings of the International Symposium on End User Development, Cagliari, Italy, 6–8 June 2023*; Springer Nature: Cham, Switzerland, 2023; pp. 52–61.
25. Dobbala, M.K. Rise of Generative AI: Impacts on Frontend Development. *J. Technol. Innov.* **2023**, *4*, 3.
26. Avdić, A. Generative AI Tools in Web Design. In *Proceedings of the Sinteza 2024-International Scientific Conference on Information Technology, Computer Science, and Data Science, Belgrade, Serbia, 16 May 2024*; Singidunum University: Belgrade, Serbia, 2024; pp. 392–397.
27. AlDahoul, N.; Hong, J.; Varvello, M.; Zaki, Y. Towards a World Wide Web powered by generative AI. *Sci. Rep.* **2025**, *15*, 7251. [CrossRef] [PubMed]
28. Ratani, S.; Phatta, D.; Phalke, D.; Varun, N.; Aher, A. Web Sculptor-Generative AI Based Comprehensive Web Development Framework. In Proceedings of the 2024 IEEE International Conference on Computing, Power and Communication Technologies (IC2PCT), Greater Noida, India, 9–10 February 2024; IEEE: Piscataway, NJ, USA, 2024; Volume 5, pp. 1729–1732.
29. Lu, Y.; Zhang, C.; Zhang, I.; Li, T.J.J. Bridging the Gap between UX Practitioners' work practices and AI-enabled design support tools. In *CHI EA '22, Proceedings of the CHI Conference on Human Factors in Computing Systems Extended Abstracts, New Orleans, LA, USA, 29 April–5 May 2022*; Association for Computing Machinery: New York, NY, USA, 2022; pp. 1–7.
30. Luo, Y. Designing with AI: A Systematic Literature Review on the Use, Development, and Perception of AI-Enabled UX Design Tools. *Adv. Hum.-Comput. Interact.* **2025**, *2025*, 3869207. [CrossRef]
31. Voronin, D.N. Development and Evaluation of an LLM-Based Tool for Automatically Building Web Applications. Doctoral Dissertation, Massachusetts Institute of Technology, Cambridge, MA, USA, 2024.
32. York, E. Evaluating chatgpt: Generative ai in ux design and web development pedagogy. In *SIGDOC '23, Proceedings of the 41st ACM International Conference on Design of Communication, Orlando, FL, USA, 26–28 October 2023*; Association for Computing Machinery: New York, NY, USA, 2023; pp. 197–201.
33. Lively, J.; Hutson, J.; Melick, E. Integrating AI-generative tools in web design education: Enhancing student aesthetic and creative copy capabilities using image and text-based AI generators. *J. Artif. Intell. Robot.* **2023**, *1*, 23–33.
34. Naeem, U.; Styve, A.; Virkki, O.T. Stimulating Critical Thinking in a Web Programming Module with Generative AI Tools. In Proceedings of the 2025 IEEE Global Engineering Education Conference (EDUCON), London, UK, 22–25 April 2025; IEEE: Piscataway, NJ, USA, 2025; pp. 1–7.
35. Alpizar-Chacon, I.; Keuning, H. Student's Use of Generative AI as a Support Tool in an Advanced Web Development Course. *arXiv* **2025**, arXiv:2503.15684.
36. Greenberg, I. *Processing: Creative Coding and Computational Art*; Apress: Berkeley, CA, USA, 2007.
37. Patel, S.K.; Rathod, V.R.; Parikh, S. Joomla, Drupal and WordPress—A statistical comparison of open source CMS. In Proceedings of the 3rd International Conference on Trendz in Information Sciences & Computing (TISC2011), Chennai, India, 8–9 December 2011; IEEE: Piscataway, NJ, USA, 2012; pp. 182–187.
38. Buendía-García, F. Using UX Methods and AI Tools to Enhance Collaborative Scenarios in Creative Technology Courses. In Proceedings of the 2024 International Symposium on Computers in Education (SIIE), A Coruna, Spain, 19–21 June 2024; IEEE: Piscataway, NJ, USA, 2024; pp. 1–6.
39. Voila. AI Assistant for Your Team. 2024. Available online: https://www.getvoila.ai/chatgpt-for-teams (accessed on 28 May 2025).
40. Wermelinger, M. Using GitHub Copilot to solve simple programming problems. In *SIGCSE 2023, Proceedings of the 54th ACM Technical Symposium on Computer Science Education, Toronto, ON, Canada, 15–18 March 2023*; Association for Computing Machinery: New York, NY, USA, 2023; Volume 1, pp. 172–178.
41. Merchant, A. I Got Access to GitHub Copilot Chat. 2024. Available online: https://www.amitmerchant.com/github-copilot-chat/ (accessed on 28 May 2025).
42. Pezoa, F.; Reutter, J.L.; Suarez, F.; Ugarte, M.; Vrgoč, D. Foundations of JSON schema. In *WWW'16, Proceedings of the 25th International Conference on World Wide Web, Montreal, QC, Canada, 11–15 April 2016*; International World Wide Web Conferences Steering Committee, Republic and Canton of Geneva: Geneva, Switzerland, 2016; pp. 263–273.

43. Chapman, C.; Wang, P.; Stolee, K.T. Exploring regular expression comprehension. In Proceedings of the 2017 32nd IEEE/ACM International Conference on Automated Software Engineering (ASE), Urbana, IL, USA, 30 October–3 November 2017; IEEE: Piscataway, NJ, USA, 2017; pp. 405–416.
44. Straka, M.; Hajic, J.; Straková, J. UDPipe: Trainable pipeline for processing CoNLL-U files performing tokenization, morphological analysis, pos tagging and parsing. In Proceedings of the Tenth International Conference on Language Resources and Evaluation (LREC'16), Portorož, Slovenia, 23–28 May 2016; pp. 4290–4297.
45. Anandarajan, M.; Hill, C.; Nolan, T.; Anandarajan, M.; Hill, C.; Nolan, T. Term-document representation. In *Practical Text Analytics: Maximizing the Value of Text Data*; Springer: Cham, Switzerland, 2019; pp. 61–73.
46. Landauer, T.K.; Foltz, P.W.; Laham, D. An introduction to latent semantic analysis. *Discourse Process.* **1998**, *25*, 259–284. [CrossRef]
47. Foltz, P.W. Latent semantic analysis for text-based research. *Behav. Res. Methods Instrum. Comput.* **1996**, *28*, 197–202. [CrossRef]
48. Gururangan, S. Latent Semantic Analysis Project. 2015. Available online: https://github.com/kernelmachine/pyLSA (accessed on 28 May 2025).
49. Brinck, T.; Gergle, D.; Wood, S.D. *Usability for the Web: Designing Web Sites That Work*; Elsevier: Amsterdam, The Netherlands, 2001.
50. Zhang, S.; Zhang, T.; Cheng, J.; Zhou, S. Who is to Blame: A Comprehensive Review of Challenges and Opportunities in Designer-Developer Collaboration. *arXiv* **2025**, arXiv:2501.11748. [CrossRef]
51. Nam, D.; Macvean, A.; Hellendoorn, V.; Vasilescu, B.; Myers, B. Using an llm to help with code understanding. In *ICSE '24, Proceedings of the IEEE/ACM 46th International Conference on Software Engineering, Lisbon, Portugal, 14–20 April 2024*; Association for Computing Machinery: New York, NY, USA, 2024; pp. 1–13.
52. Oppenlaender, J.; Linder, R.; Silvennoinen, J. Prompting AI art: An investigation into the creative skill of prompt engineering. *Int. J. Hum.–Comput. Interact.* **2024**, 1–23. [CrossRef]
53. Mihalcea, R.; Corley, C.; Strapparava, C. Corpus-based and knowledge-based measures of text semantic similarity. In Proceedings of the Twenty-First National Conference on Artificial Intelligence and the Eighteenth Innovative Applications of Artificial Intelligence Conference, Boston, MA, USA, 16–20 July 2006; Volume 6, pp. 775–780.
54. Gatto, J.; Sharif, O.; Seegmiller, P.; Bohlman, P.; Preum, S.M. Text encoders lack knowledge: Leveraging generative llms for domain-specific semantic textual similarity. *arXiv* **2023**, arXiv:2309.06541.

applied sciences

MDPI

Article

Enhancing Online Learning Through Multi-Agent Debates for CS University Students

Jing Du [1], Guangtao Xu [2], Wenhao Liu [3], Dibin Zhou [3,4] and Fuchang Liu [3,*]

1 Department of Media and Communication, Kangwon National University, Chuncheon 24341, Republic of Korea; dj@kangwon.ac.kr
2 Jing Hengyi School of Education, Hangzhou Normal University, Hangzhou 311121, China; xuguangtao@hznu.edu.cn
3 School of Information Science and Technology, Hangzhou Normal University, Hangzhou 311121, China; 20110039@hznu.edu.cn (W.L.); dibinz@zju.edu.cn (D.Z.)
4 Center for Engineering and Scientific Computation, Zhejiang University, Hangzhou 310058, China
* Correspondence: liufc@hznu.edu.cn

Abstract: As recent advancements in large language models enhance reasoning across various domains, educators are increasingly exploring their use in conversation-based tutoring systems. However, since LLMs are black-box models to users and lack human-like problem-solving strategies, users are hardly convinced by the answers provided by LLMs. This lack of trust can potentially undermine the effectiveness of learning in educational scenarios. To address these issues, we introduce a novel approach that integrates multi-agent debates into a lecture video Q&A system, aiming to assist computer science (CS) university students in self-learning by using LLMs to simulate debates between affirmative and negative debaters and a judge to reach a final answer and presenting the entire process to users for review. This approach is expected to lead to better learning outcomes and the improvement of students' critical thinking. To validate the effectiveness of this approach, we carried out a user study through a prototype system and conducted preliminary experiments based on video lecture learning involving 90 CS students from three universities. The study compared different conditions and demonstrated that students who had access to a combination of video-based Q&A and multi-agent debates performed significantly better on quizzes compared to those who only had access to the video or video-based Q&A. These findings indicate that integrating multi-agent debates with lecture videos can substantially enhance the learning experience, which is also beneficial for the development of students' high-order thinking abilities in the future.

Keywords: online learning; multi-agent; large language models

1. Introduction

With the increasing availability of online courses, a growing number of undergraduate and graduate students are turning to these resources for learning. Millions now access free educational videos on platforms like Khan Academy, Coursera, edX, Udacity, MIT OpenCourseWare, YouTube, and Chinese MOOC platforms. This trend underscores how Internet technology is enhancing educational access and broadening the range of available courses, particularly serving as a valuable alternative for students unable to attend in-person classes during the COVID-19 pandemic [1]. However, most current online courses still rely primarily on one-way lectures from an instructor, supplemented by a few post-lecture quizzes or self-regulated learning prompts [2,3]. There is a lack of necessary interaction and discussion, making them not significantly different from traditional, teacher-centered instruction, except that they are delivered through the Internet. Previous research on human–computer interaction has focused on enhancing educational videos through various techniques, such as interactive transcripts, word clouds, keyword search, and storyboard highlights [4]. Other studies have segmented videos to display key moments alongside snapshots and transcripts [5]. Additionally, some researchers have developed video digests structured in a textbook-like format, organized into chapters, titles, and sections with accompanying text summaries [6]. Moreover, Shimada et al. [7] found that students who used summarized slide previews achieved higher pre-quiz scores and required less time compared to those who previewed the original learning materials. Therefore, using large language models (LLMs) to summarize video content and subsequently perform retrieval-augmented generation (RAG) may also provide an effective human–computer interaction (HCI) approach to enhance students' learning efficiency.

The advent of LLMs, such as Generative Pre-trained Transformer 4 (GPT-4) [8], has provided significant technological advancements in various educational activities, including engineering education [9], language teaching and learning [10], and personalized learning [11]. Despite their potential, LLMs still face challenges [12,13] such as hallucination, reliance on static training data, lack of real-time knowledge, limited problem-solving capabilities, memory continuity issues, ethical and safety concerns, and insufficient personalization. Careful use of LLMs poses challenges for both teachers and learners [14]. However, these limitations can potentially be mitigated by advanced prompt engineering techniques, such as prompt engineering [15], chain of thought (CoT) [16], self-consistency (SC) [17], and retrieval-augmented generation (RAG) [18]. Ongoing advancements in generative AI, including RAG, are progressively enhancing the factual reliability and responsiveness of large language models (LLMs). Autonomous agents have long been a central research area in both academia and industry. Large language models have been applied as autonomous agents in many fields, especially in educational assistance [19,20]. This advancement suggests a promising pathway for implementing multi-agent systems based on large language models, enabling novel online teaching and learning experiences.

Accordingly, we present an online learning-assisted scheme that offers an intuitive and user-friendly solution. This scheme supports both question-and-answer sessions and topic discussions through video-based multi-agent debates. The purpose of this paper is to explore the role of open-source LLM-based agents in facilitating self-learning among undergraduate students, particularly in the field of computer science.

The research question for this study is the following:

- RQ. How can video-based multi-agent debates enhance online learning outcomes for CS students?

Regarding the above-mentioned research question, we further elaborate and propose the following hypotheses to explore the effects of multi-agents on students.

- H1. Video-based multi-agent debates are more effective than video-only and video-based Q&A approaches for online learning among CS students in terms of objective questions.
- H2. Video-based multi-agent debates are more effective than video-only and video-based Q&A approaches for online learning among CS students in terms of enriching the answer dimensions of open-ended questions.

2. Literature Review

In this section, we conducted a review of the relevant literature. Prior to conducting this review, we first carried out a literature survey on Clarivate Analytics' Web of Science using the keywords "online learning and video", resulting in 3057 documents published since 2012. Keyword analysis reveals that it is relatively rare to use multi-agent debates to enhance online learning in a video-based environment, as shown in Table 1. We performed a topic analysis on the abstracts of relevant literature, as illustrated in Figure 1. The findings reveal that cutting-edge technologies such as "virtual reality" and "artificial intelligence" have garnered significant attention in recent years, while research on online courses and online learning environments has consistently remained a central focus in the field. The contribution of this paper lies in presenting a novel approach to online learning. It leverages LLMs as multiple debaters to generate answers and, through this process, aims to inspire students' high-order thinking abilities. The aforementioned literature highlights the application and transformative impact of "online learning and video" in education. Therefore, we summarize the prior literature about online learning from three perspectives: massive open online courses (MOOCs), online learning environments, and LLM-based multi-agents.

Table 1. The top ten most relevant keywords were extracted from 3057 documents retrieved from Clarivate Analytics' Web of Science using the keywords "online learning and video", constrained to the educational research field.

	Author Keywords	Articles	Keywords Plus	Articles
1	ONLINE LEARNING	270	EDUCATION	264
2	LEARNING	227	STUDENTS	231
3	E-LEARNING	189	ONLINE	186
4	EDUCATION	172	PERFORMANCE	145
5	FLIPPED CLASSROOM	137	IMPACT	141
6	HIGHER EDUCATION	135	TECHNOLOGY	131
7	BLENDED LEARNING	121	VIDEO	114
8	ONLINE	121	ENGAGEMENT	113
9	VIDEO	112	MOTIVATION	95
10	TECHNOLOGY	103	PERCEPTIONS	95

Figure 1. A topic analysis on the abstracts of relevant literature from 3057 documents retrieved from Clarivate Analytics' Web of Science. The figure was generated using Biblioshiny, a tool within the Bibliometrix package, by importing bibliographic data from Clarivate Analytics' Web of Science.

2.1. MOOCs

MOOCs represent a widely studied innovation in distance education, first introduced in 2008 [21]. They gained prominence as a popular learning format in 2012, a year often referred to as the "Year of the MOOC" [22]. MOOCs are distinguished by their openness, scalability, and ability to cater to diverse audiences. Openness facilitates free enrollment and unrestricted access to course materials, thereby advancing the democratization of education [23]. Scalability allows instructors to interact with thousands of learners simultaneously, surpassing the constraints of traditional classroom settings [24]. The global pandemic of COVID-19 underscored the role of MOOCs in delivering educational and learning opportunities to a wide audience [25–27]. A particularly appealing feature of MOOCs for adult learners is the high degree of control and autonomy they offer over the learning process [28,29]. However, despite their significant potential, MOOCs continue to face challenges related to high dropout rates and low completion rates, which hinder their broader adoption [30].

Despite offering convenient online learning, MOOCs face notable limitations. Completion rates typically range between 5% and 15%, with longer course durations correlating with lower completion rates [31]. Margaryan et al. [32] highlighted shortcomings in traditional MOOCs' learning outcomes, attributing them to a lack of personalized support, limited feedback mechanisms, and insufficient opportunities for deep learning. The one-size-fits-all approach of MOOCs often fails to address the diverse learning styles and needs of participants [33]. Additionally, the absence of personalized feedback and direct interaction with instructors remains a critical challenge [34].

To address the aforementioned issues, educators and technologists have explored various interactive teaching strategies. The integration of interactive elements into MOOCs has been shown to significantly enhance learning outcomes. Interactive approaches in MOOCs include discussion forums, gamification, and interactive videos. MOOC discussion forums serve as a vital channel for communication between instructors and learners, with their effectiveness shaped by various factors. Wei et al. [35] demonstrated that small group discussions, particularly those with clear objectives and assessment criteria, enhance learner engagement. Similarly, Neha et al. [36] found that structured topics encourage participation, while immediate feedback helps sustain engagement. Galikyan et al. [37] highlighted the dual role of forums in facilitating knowledge exchange and fostering social connections, contributing to deeper learning when instructors actively guide discussions and incorporate peer review. Gamification in MOOCs has demonstrated significant positive impacts. Freitas et al [38] found that gamification enhances motivation and engagement, with achievement badges serving as powerful incentives. Similarly, Meng et al. [39] showed that a well-designed scoring system promotes skill development and multidimensional engagement but cautioned against the risk of goal misalignment if gamification is overused. Maintaining clear rules and mechanisms is essential to ensure that gamification remains aligned with educational objectives. MOOCs incorporate interactive videos with various elements to enrich the learning experience. Embedded quizzes, implemented by platforms such as Coursera and Udacity [40], serve as checkpoints for self-assessment of knowledge. Furthermore, Chen et al. [41] introduced time-anchored discussions, enabling learners to engage in conversations tied to specific video timestamps, thereby fostering interaction and enhancing content understanding. Some platforms have further innovated by combining linear discussions with live comments, providing more flexible and dynamic interaction opportunities.

Integrating large language models (LLMs) into MOOCs represents a transformative development. These models enable personalized learning experiences, as exemplified by Khan Academy's Khanmigo, which supports students through adaptive, progressive questioning. Similarly, the Massive AI-Empowered Course (MAIC) [42] establishes an intelligent classroom environment, leveraging AI-driven roles to create personalized learning paths that were previously unattainable in traditional MOOCs.

2.2. Online Learning Environment

The online learning environment is a diverse and digital educational ecosystem that provides learners with flexible educational platforms that transcend geographical, temporal, and cultural boundaries through the Internet, cloud computing, and advanced digital interaction technologies [43]. This complex learning ecosystem is not merely a technological medium but an innovative paradigm that integrates education, cognitive science, and information technology, with the core value of offering highly personalized, adaptive, and inclusive learning experiences [44]. Educational theories provide the essential theoretical foundation and normative guidance for the cognitive framework and practical paradigms of online learning. Self-Determination Theory (SDT) extensively analyzes the inherent mechanisms of learners' basic psychological needs—autonomy, competence, and relatedness—regarding learning motivation and performance, offering significant theoretical insights into learner agency in online learning contexts [45]. The Cognitive–Affective Immersive Learning (CAMIL) model deconstructs the immersive experience of learners in digital learning environments in various dimensions, elucidating the complex dialectical relationships between cognitive engagement, emotional involvement, and interaction experiences throughout the learning process [46]. The Inquiry-Based Learning (IBL) framework provides a structured methodology for online learning, emphasizing the cultivation of learners' autonomy in exploration, critical thinking, and problem-solving skills within the knowledge construction process [47]. Self-Regulated Learning (SRL) strategies reveal the complex pathways through which learners regulate their behaviors and achieve goals during autonomous learning [48,49]. Deeper investigations indicate that learners' emotional trajectories and levels of interaction critically influence learning outcomes, not only impacting cognitive knowledge acquisition but also reflecting the development of learners' metacognitive skills and learning adaptability [50]. This study demonstrates that cognitive presence, social presence, and teaching presence are significantly correlated with autonomy, competence, relevance, and perceived learning in the context of online educational settings in China [51].

The COVID-19 pandemic, in particular, has accelerated the application of digital technologies in education, leading to a growing number of technological innovations in online learning platforms. For example, virtual reality (VR) technology provides a revolutionary learning paradigm for disciplines that require high levels of practicality and interactivity by creating highly immersive, interactive learning environments [46,52]. At the same time, artificial intelligence (AI) optimizes learning processes and dynamically adapts learning paths through intelligent algorithms tailored to the individual needs of learners [53].

By breaking the structural limitations of traditional education, online learning environments achieve global sharing of educational resources, promote learner autonomy, and implement the concept of lifelong learning, thereby advancing educational equity and knowledge democratization in a globalized context. Despite advancements in technology, online education environments still face significant structural challenges. First, traditional online courses primarily rely on one-way lecturing models, which closely mirror conventional classrooms by simply transferring content to the Internet, thus lacking substantial pedagogical innovation. As a result, students remain in a highly passive state during the learning process, hindering their ability to think independently and explore knowledge actively. Second, the lack of interactivity leads to substantial declines in student satisfaction and learning outcomes due to insufficient opportunities for immediate questioning and deep communication [54]. Existing online learning platforms often fail to adapt precisely to learners' cognitive characteristics and learning paces, resulting in a general lack of personalized learning paths and, consequently, low efficiency and engagement and poor learning outcomes [55]. More critically, these platforms tend to overlook the multidimensional nature of learning, failing to adequately stimulate learners' cognitive autonomy and emotional engagement, which causes online education to become a mechanical and soulless tool for knowledge transmission. These inherent limitations severely restrict the developmental potential of online education and hinder the genuine transformation of the digital learning ecosystem.

2.3. LLM-Based Learning Systems

LLMs have recently demonstrated remarkable potential, achieving reasoning and planning capabilities comparable to those of humans. This aligns precisely with human expectations for autonomous agents capable of perceiving their surroundings, making decisions, and taking appropriate actions in response [56,57]. Building on the impressive capabilities of single LLM-based agents, LLM-based multi-agent systems have been proposed to harness collective intelligence and specialized skills across multiple agents. Compared to systems employing a single LLM-powered agent, multi-agent systems offer enhanced functionalities by (1) specializing LLMs into distinct agents, each with unique capabilities, and (2) facilitating interactions among these agents to effectively simulate complex real-world environments. Recent studies [58] have shown promising results in leveraging LLM-based multi-agents for diverse tasks, such as software development, multi-robot systems, societal simulations, policy modeling, and game simulations, with particular success in simulating human behavior. Notable examples include Generative Agents [59], Ghost in the Minecraft [60], and GPT-Bargaining [61].

As LLMs continue to advance, intense discussions have arisen regarding methodologies in this field [62,63]. A key focus lies in fully harnessing the capabilities of large models to simulate real classrooms with multiple agents for automated teaching. LLMs have already been applied to various intelligent educational tasks to support teaching. Ref. [64] introduced a multi-agent classroom simulation framework involving user participation, demonstrating that LLMs can effectively replicate traditional classroom interaction patterns while enhancing the user experience. This work pioneered the use of LLM-powered multi-agent systems in virtual classroom teaching.

In summary, how to effectively utilize AI technology to advance education has long been a key research focus, with scholars exploring this topic from various perspectives [9]. For example, Huber et al. [65] explored the use of LLMs through playful and game-based learning methods, while [66,67] implemented LLM prompt-based personalized tutoring. This study approaches the topic from the perspective of video-assisted online learning. AI may negatively impact independent thinking by fostering student dependency, leading to a lack of initiative in problem-solving. This can hinder deeper memory retention of knowledge and the development of critical thinking skills. Mohamed et al. [68] examined the benefits and drawbacks of integrating AI tools into language instruction, while [69] explored various issues caused by ChatGPT (gpt-3) in a broader range of educational scenarios. Zhang et al. [70] discussed students' dependency on AI and identified the top five negative effects of AI technology: increased laziness, the spread of misinformation, decreased creativity, and reduced critical and independent thinking. Efficient, reliable, and valid assessment for online learning is also a topic worthy of in-depth investigation [71]. Further detailed investigation and assessment are needed in the area of online learning. Finally—and most importantly—the use of artificial intelligence tools like ChatGPT in education raises ethical dilemmas and challenges, including concerns about data privacy, over-reliance on technology, intrinsic biases, and the integrity of student outputs [72]. Achieving a rational and ethically balanced integration of LLMs and conversable agents in educational settings remains a significant challenge for researchers leveraging AI to enhance education. Although there have been quite a few attempts to integrate LLMs into learning systems, the design of such models for personalized learning still has room for further enhancement. Future research could explore the use of LLMs to achieve personalized learning through a conversation-based tutoring system enhanced with more interactive elements such as quizzes or debates that inspire higher-order thinking.

3. Materials and Methods

3.1. Design of Evaluation System

After defining the research questions and hypotheses, the study was conducted in four steps, as outlined in Figure 2. Step 1 focused on obtaining initial scores through a pre-video test, where participants answered questions via an online quiz. Step 2 involved having participants watch a lecture video and interact with a software interface. This step incorporated a learn-as-you-go approach, supported by multi-agent debates to facilitate understanding of key concepts. In Step 3, a post-video test was conducted, which also included answering an online quiz. Step 4 analyzed learning outcomes and subjective evaluations by comparing pre-test and post-test scores, along with responses from a follow-up questionnaire. The results of the questionnaire were analyzed using SPSS software (29.0.1.0(171)).

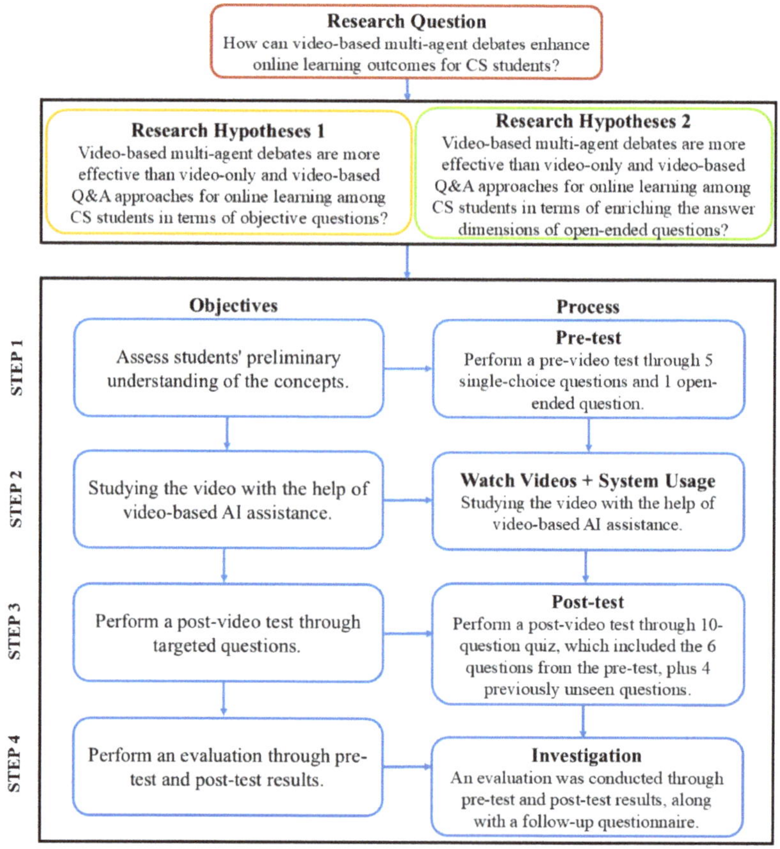

Figure 2. Overview of the proposed research method.

3.1.1. LLM RAG-Based Video Understanding Function

Videos inherently exhibit multimodal properties, encompassing both visual and audio features. The visual features can be categorized into image captioning, object detection, and optical character recognition (OCR) tasks. We extracted keyframes from the video using CLIP features and used the BLIP model to generate image captions, while employing GRiT and RAM++ for object detection. Additionally, we utilized Whisper to transcribe the audio into text. These multimodal features provide the foundational data for subsequent QA tasks, as shown in Figure 3.

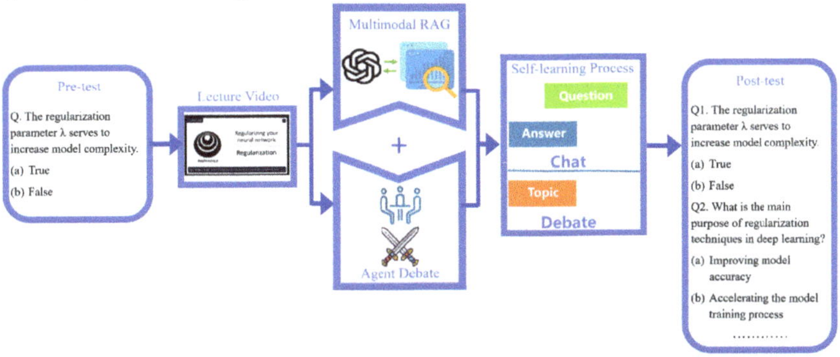

Figure 3. Experimental procedure. From left to right, the diagram illustrates the feature processing, RAG workflow, and debate workflow of the AI-empowered video-based online learning system. The multi-agent debate module operates independently from the QA system while also complementing it, both of which serve the purpose of supporting students in self-learning.

The method proposed in this paper is compared with the current state-of-the-art approaches, including VideoChat2 (https://github.com/OpenGVLab/Ask-Anything, accessed on 5 September 2024) and the open-source VLog project. Unlike VideoChat2, which primarily utilizes visual features, our approach incorporates additional modalities, such as audio and OCR features. As a result, our method demonstrates a superior ability to capture finer details in video content. Figure 4a shows that current state-of-the-art methods for video understanding, such as VideoChat2 [73], can only answer questions related to the visual content of the video but struggle with finer details. In contrast, the multi-modal video analysis method based on the VLog project (https://github.com/showlab/VLog, accessed on 27 August 2024) can answer detailed questions about the video, though not always with full accuracy. The results presented in Figure 4b build on the VLog project, where ChatGPT is replaced with the free online "WizardLM-2" LLM API by Together.ai (https://api.together.ai/, accessed on 27 August 2024) , along with English-to-Chinese translation using the pre-trained "Helsinki-NLP/opus-mt-en-zh" model. WizardLM-2's capabilities are very close to those of cutting-edge proprietary models such as GPT-4-1106-preview and significantly ahead of all other open-source models. However, as shown in Figure 4c, our system, enhanced with OCR recognition, significantly improves accuracy in capturing detailed information within the video content. Compared to existing systems, our system accurately extracts video information, then provides it to the RAG module for answer generation.

(a)

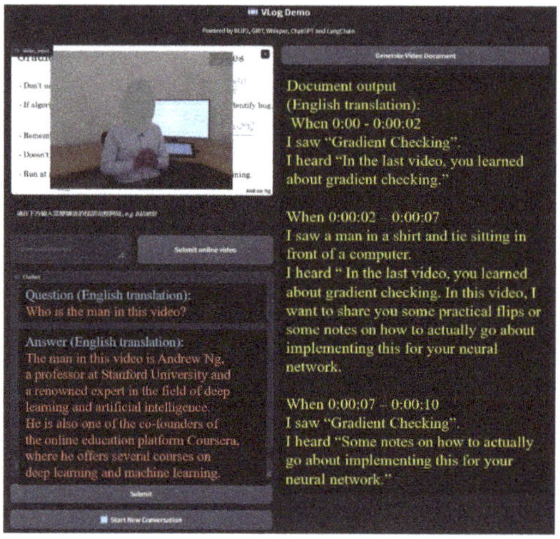

(b)

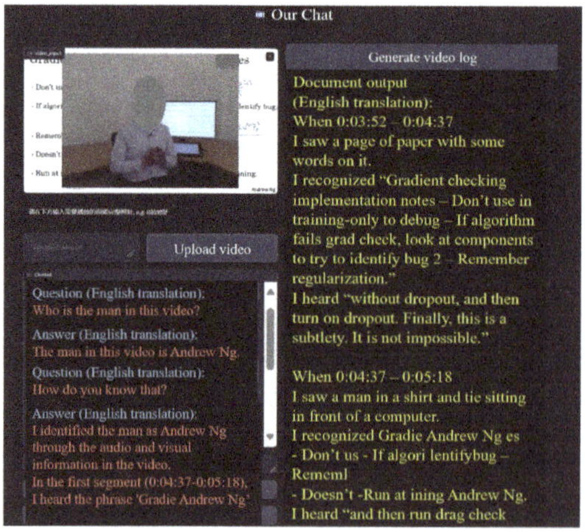

(c)

Figure 4. Evaluation of results from different video understanding methods. (**a**) The result of VideoChat2. (**b**) The result of VLog. (**c**) The result of our video-MAD system.

3.1.2. Multi-Agent Debate-Assisted Learning Function

Using large language models as agents to assist users in completing various tasks is increasingly becoming a trend. In particular, multi-agent collaboration can address the limitations of a single model, resulting in improved accuracy and robustness.

The study titled Multi-Agent Debate (MAD) [74], demonstrates that reflection-based methods are prone to the Degeneration-of-Thought (DoT) problem. Once a large language model becomes confident in its initial solution, it struggles to generate new ideas through further reflection, even if the initial stance is incorrect. Multi-Agent Debate, inspired by another fundamental strategy of human-like problem solving, employs two agents to express their own arguments in a "tit-for-tat" state. A judge then monitors and manages the debate process to generate a final answer. The core idea of MAD is outlined as follows: (1) Through debate, the distorted thinking of one agent can be corrected by the others. (2) Through debate, the resistance of one party can be counterbalanced by the support of other members participating in the debate. (3) The answers of each agent can serve as context for mutual feedback. Therefore, MAD is less vulnerable to the occurrence of the DoT problem and can explore divergent CoTsto reduce hallucinations. As shown in Table 2, for questions that require a more comprehensive understanding through multiple perspectives and dialectical discussion, the debate function is clearly more suitable than the chat function. Multi-agent debate can be applied not only to STEM (Science, Technology, Engineering, and Mathematics) education but also to humanities and history studies.

We followed the idea from the work on divergent thinking based on MAD, employing three LLM-based agents to conduct multi-round debates. We utilized three roles, which are elaborated as follows: moderator, affirmative debater, and negative debater (Table 3). Based on the given topics, both sides present their arguments and provide supporting reasons. After each round, the moderator evaluates the responses and determines the winning side. After three rounds of debate, the moderator provides a final summary evaluation, along with the corresponding rationale. The specific details of the debate result in Figure 5 can be found in Table 2, which presents an example of a specific debating process. Through the debate, the reasons for the arguments can be presented, which can inspire students' dialectical thinking. The answers to the QA rely on the LLM and RAG technologies and are provided based on the speech information extracted from the video, the OCR information, and the object detection results of the objects in the images. Text generation in the debate is completely accomplished based on the knowledge inherent in the pre-trained large language model we use.

Regarding the implementation of multi-agents, we referred to the open-source project available at https://github.com/Skytliang/Multi-Agents-Debate, accessed on 6 August 2024. Specifically, we used the free online "Qwen2.5-72B-Instruct-Turbo" LLM API provided by Together.ai. Each agent is an instance of a large language model, and different responsibilities are assigned through different role prompts. Each agent has its own memory. The output is stored in the memory as the content of historical conversations, and its own memory is referred to when generating new responses. The moderator controls the advancement of the rounds, and its output is broadcast to everyone, indirectly affecting the next-round input of other agents. We preset the meta prompt (role setting) used by each role, the dynamic prompt template required for the moderator's judging, and the alternative judgment prompt in case of an inability to make a ruling. We also provide the detailed step-by-step procedure of MAD in Algorithm 1.

Table 2. The MAD process for an example in Andrew Ng's lecture video on regularization (original text is in Chinese).

⟨ **debate topic** ⟩: Is regularization always necessary? Affirmative Argument: As a debater for the affirmative side, I believe that in many cases, regularization is essential. Here are several points to support this argument: 1. Preventing Overfitting: The primary goal of regularization is to prevent the model from overfitting the training data. Overfitting occurs when a model becomes too complex and overly learns the details and noise of the training data, leading to poor performance on unseen data. By introducing a regularization term, we can constrain the model parameters, thus enhancing its generalization ability. 2. Resource Optimization: Regularization helps strike a balance between model complexity and performance. By controlling the complexity of the model, we can avoid building overly complex models, which may consume excessive computational resources and time. 3. Knowledge Transfer: In transfer learning and multi-task learning, regularization promotes knowledge sharing and transfer between tasks. It prevents each task from learning independently, thus improving overall performance. 4. Algorithm Stability: Regularization can improve the stability of the algorithm, making the model less sensitive to small changes in the data and increasing its robustness. 5. Parameter Constraints: In fields such as medicine or finance, it is sometimes important for model parameters to have physical or economic significance. Regularization can help constrain these parameters within reasonable bounds. However, this does not imply that regularization is necessary in all cases. In situations where the dataset is very large, the relationship between features and the target is simple, or where there is no strict need to limit model complexity, it may be feasible to forgo regularization. Overall, considering the risk of overfitting and resource efficiency, regularization is a valuable tool in most machine learning problems and should be regarded as an important step. Negative Argument: As the opposing debater, I understand the affirmative side's points, but I believe that regularization is not always necessary in every situation. Here are my arguments: 1. Data Quality: If the training data is large enough and evenly distributed, without noise or outliers, the model may generalize well without regularization. In such cases, the model can naturally learn the main patterns in the data without overfitting. 2. Simple Problems: For simple linear problems or problems with a clear structure, such as linear regression, the model might already have good generalization ability and may not need additional regularization to prevent overfitting. 3. Preprocessing and Feature Selection: Effective data preprocessing and feature selection can reduce noise and irrelevant features, thereby lowering the risk of overfitting, making regularization less necessary. 4. Non-Parametric Methods: Certain non-parametric methods, such as decision trees and random forests, have inherent regularization properties (e.g., limiting tree depth or using bagging) and may not require additional regularization. 5. Hyperparameter Tuning: By carefully adjusting hyperparameters like model complexity and learning rate, it is possible to find an appropriate model without the need for regularization. This is particularly feasible for small or medium-sized datasets. 6. Real-Time and Online Learning: In environments where data streams are continuously updated, regularization may not be required since the model constantly adapts to new data and doesn't overfit early data points. In conclusion, while regularization is an effective tool for improving generalization in many cases, it is not always necessary. Whether regularization is needed depends on the nature of the data, the complexity of the problem, and the available computational resources. In some specific situations, satisfactory results can be achieved without regularization. After three rounds of debates... Arbitration opinion: Regularization is an important tool in machine learning, but its necessity depends on the specific application scenario. In many cases, regularization can improve the generalization ability and stability of the model. However, in other cases, it may be necessary to combine with other methods to achieve the best effect. Arbitration reason: In the third round of debates, both debaters further deepened their understanding of the role of regularization in machine learning, emphasizing its importance and complementarity with other technologies. Both sides agreed that regularization is a valuable tool, but not the only solution in all cases. They jointly pointed out that the decision on whether and how to use regularization should be made according to specific problems, data, and model selection.

Table 3. Design of Multi-Agent Debate. We assigned different roles to the agents with the aim of creating an atmosphere of debate. The prompts used in our system are all in Chinese.

Roles	Prompt (English Translation)
Moderator	You are the moderator. Two debaters will participate in the debate. They will present their answers and discuss their respective viewpoints on the following topic: ⟨debate topic⟩. At the end of each round, you will evaluate the answers and determine which one is correct.
Affirmative Debater	You are affirmative side. Please express your viewpoints.
Negative Debater	You are negative side. You disagree with the affirmative side's points. Provide your reasons and answer.

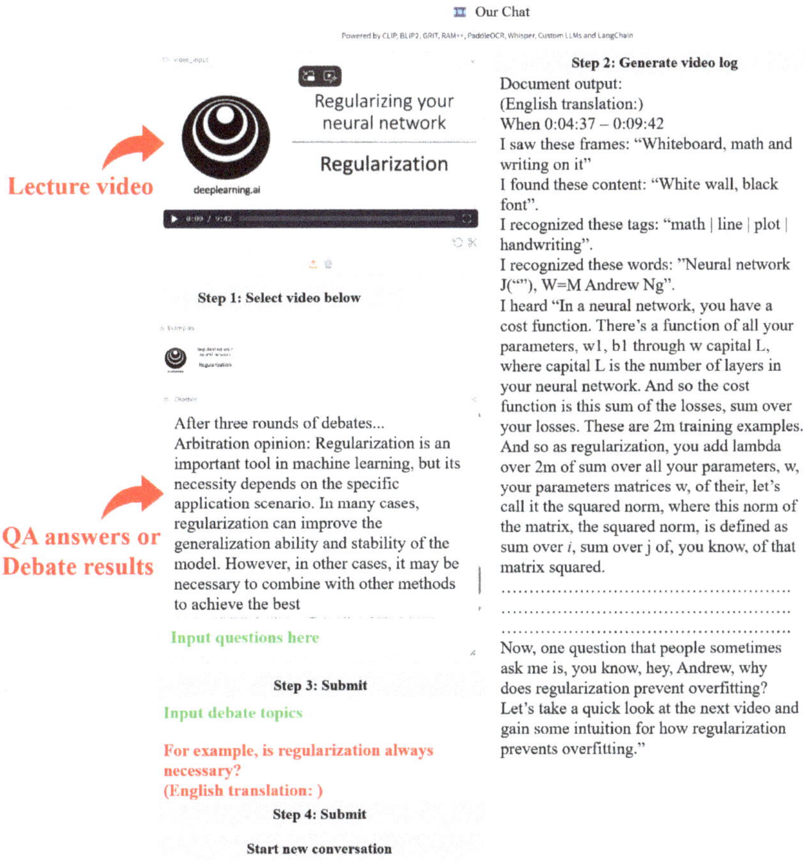

Figure 5. An illustration of our multi-agent debate-assisted learning system. The user interface includes video upload, video information extraction, Q&A, and debate functionalities. After the user uploads a video, the system extracts video information, which is utilized in both the Q&A and multi-agent debate processes. The green input boxes are the places where students input questions and debate topics. The red text box is the area where the answers from the LLM, as well as the texts and conclusions of the multi-agent debate process, are displayed.

Algorithm 1 Multi-Agent Debate Process

Input: The debate topic provided by the user and the number of rounds (n).
Output: The debate topic and a complete debate process text.

Step 1: Initialize Agents
Create three agents: *Affirmative*, *Negative*, and *Moderator*
Inject meta-prompts for each agent role
Embed the *debate topic* into the preset prompt templates

Step 2: Commence the Debate
Affirmative agent presents the initial argument
Negative agent reads the affirmative argument and generates a refutation
Moderator evaluates both arguments and broadcasts a neutral judgment

Step 3: Make the Judging
while debate less than *n* rounds **do**
  if Moderator identifies a clear preferred side **then**
    Terminate the debate, summarize reasoning and final decision and go to Step 4
  else
    Proceed to the next round: Affirmative and Negative agent update their memory and perform turn-based rebuttals
  end if
end while
Invoke the alternative judgment prompt for Moderator:
  a. Extract all answer candidates
  b. Summarize reasoning and produce final decision

Step 4: Finish the Debate
Return the debate topic, arguments from both sides, full speech log, final decision, and reasoning

3.2. Participants and Design

In this section, we present a controlled study conducted to evaluate the potential benefits of providing students with video lectures, supplemented by video chat and LLM agents' debates. Below, we outline the study's participants, the testing conditions, and the results.

Participants. We recruited 197 undergraduate and postgraduate students to participate in the study. These students, specializing in computer science, were from Hangzhou Normal University, Nanjing University of Information Science and Technology, and Hebei University of Technology. Apart from the pre-test and post-test quiz, we also provided a follow-up questionnaire. Due to the time-consuming nature of the questionnaire, a total of 90 valid responses were ultimately returned. Undergraduate students accounted for 44.4%, and postgraduate students accounted for 55.6%.

Study design. We evaluated student performance on materials related to two machine learning topics. These topics were covered in two short online lecture videos by Andrew Ng: one on regularization (approximately 9 min) and the other on numerical approximation of gradients (approximately 6 min). For each topic, we assessed the following three learning conditions.

Testing conditions. (1) Watching only the lecture video; (2) watching the lecture video with the ability to ask questions via video-based Q&A; (3) watching the lecture video with the ability to ask questions via video-based Q&A and starting a multi-agent debate on specific topics.

All students were randomly assigned to a different learning condition for each topic. Prior to watching the lecture videos, students completed pre-test quizzes consisting of five single-choice questions and one open-ended question per topic to assess their initial understanding of the concepts. Once students had watched the video and indicated that they were ready, we administered the post-test quiz. The post-test quiz comprised the six questions from the pre-test, along with four additional, previously unseen single-choice questions. Each correct answer earned 1 point, while incorrect answers received 0 points. All the test questions were designed by teachers who teach relevant courses. Moreover, the difficulty level was determined through tests carried out by an undergraduate and a graduate student who are beginners in this area. Most students completed the test through WeChat groups, while a smaller number participated in face-to-face testing in the laboratory. To mitigate the biases introduced by heterogeneous assessment conditions, all participants, whether recruited via WeChat groups or undergoing face-to-face testing in the laboratory, were required to complete the quiz on our runnable system (https://huggingface.co/spaces/dj86/VLog4CustomLLMsPlusDebate, accessed on 1 September 2024). Our runnable system can be considered a prototype for LLM-based online video learning. It is equipped with features such as question generation based on retrieval-augmented generation and multi-agent debate functionality.

Measures. Due to differences in students' prior knowledge, we assessed the effectiveness of the learning methods by measuring the relative improvement in test scores before and after the learning videos. We calculated the relative percentage-point increase as follows:

$$\Delta = \frac{\text{Post score (\%)} - \text{Pre score (\%)}}{\text{Pre score (\%)}} \tag{1}$$

4. Results

4.1. Data Analysis

According to the pre-test and post-test results, as shown in Table 4, the mean score on the pre-test questions was below 70%. We limited our analysis to the single and multiple-choice questions, as grading the open-ended questions proved more challenging. Instead, we employed a word cloud analysis to examine the open-ended questions. After reviewing the lecture videos, students under all conditions demonstrated learning effectiveness. The learning method using MAD achieves the greatest improvement in outcomes. We employed a paired t-test and Cohen's d [75] to analyze test correctness scores before and after "video-only" [$t(29) = -3.946$, $p < 0.001$, 95% CI [−22.724, −7.210], Cohen's $d = -0.720$], "video-based Q&A" [$t(30) = -2.765$, $p = 0.005$, 95% CI [−16.320, −2.454], Cohen's $d = -0.497$], and "video-based Q&A + MAD" [$t(28) = -3.260$, $p = 0.001$, 95% CI [−31.108, −7.099], Cohen's $d = -0.605$] interventions. These findings indicate that the "video-based Q&A + MAD" approach is more effective than "video-only" and "video-based Q&A" methods for online learning. These results support H1. Question samples and SPSS 29.0.1.0 results can be found at https://osf.io/4vt9h/, accessed on 26 November 2024.

Those in condition 2 exhibited the smallest improvement, whereas those in condition 3 showed the greatest improvement. It is generally believed that the use of large language models demonstrates an advantage over traditional video-watching methods. Table 4 indicates that the post-test performance of students in condition 3 shows a relative improvement of 29.78%, compared to 14.70% for students in condition 2, and 23.39% for students in condition 1. The quiz questions consisted of a mix of true/false questions and single and multiple-choice questions. A possible explanation is that the video-based Q&A approach might encourage students to take shortcuts, as they may not watch the video attentively and, instead, rely on asking questions directly to learn. This behavior could result in overlooking some video content, thereby leading to lower scores. In contrast, the "video-based Q&A + MAD" approach likely enhances students' interest in learning through debates. The presentation of affirmative viewpoints may prompt students to question specific knowledge points, extending their study time and fostering a more comprehensive understanding of the video content. The pre-test and post-test questions are presented in Appendix A. As shown in Figure 6, we used violin plots to present the specific distributions of the scores of the three groups of samples in pre-test and post-test. It can be observed that the score distributions of the video-based Q&A + MAD method before and after the pre-test and post-test show significant differences, and these differences are much more pronounced compared to those of the other two groups.

Table 4. Student performance was evaluated through pre-tests and post-tests after watching the lecture videos under each of the three learning conditions: video only, video-based Q&A, and video-based Q&A plus multi-agent debate.

Condition	N	Pre-Test (μ, σ)	Post-Test (μ, σ)	Δ
V	30	64.00%, 27.50	78.97%, 18.83	23.39%
V+Q&A	31	63.87%, 22.76	73.26%, 22.00	14.70%
V+Q&A+MAD	29	64.14%, 25.29	83.24%, 20.88	29.78%

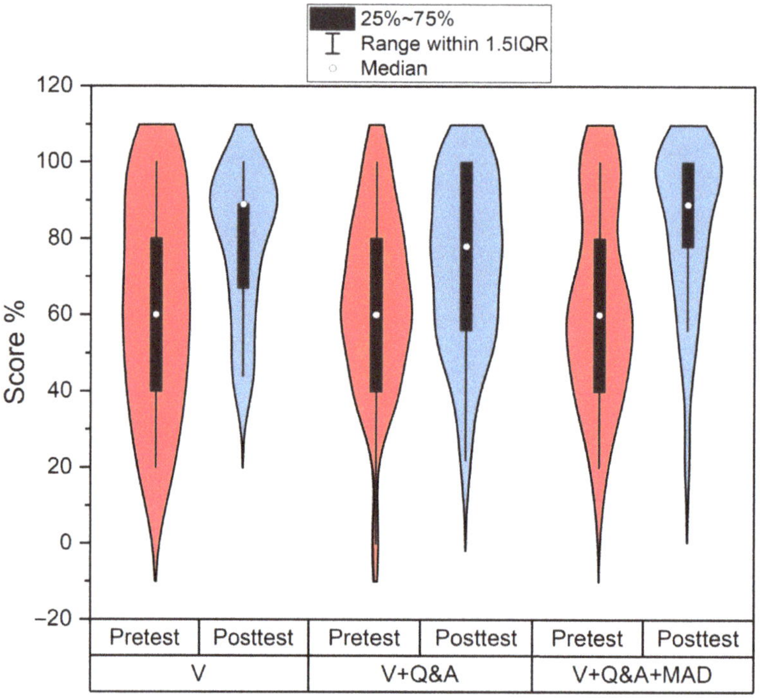

Figure 6. Score differences between pre-tests and post-tests correspond to the three learning conditions: video only, video-based Q&A, and video-based Q&A plus multi-agent debate.

Therefore, we conducted a further analysis from the perspective of students' video-viewing duration. Specifically, we examined the variations in the duration of students' engagement with online lecture videos under different scenarios, with and without the assistance of large language models. By comparing these durations in Figure 7, we can discern students' interest in utilizing large language models and determine whether the use of such models can enhance students' interest in learning videos. As shown in Figure 7, there is significant variation in the time students spent watching the video in the "traditional learning" condition. Some students spent only a few minutes, while others spent over an hour. One possible factor is that the learning material is in English, which can be challenging for Chinese students. As a result, some students may quickly skim through the video, while others might repeatedly listen or pause to focus on the video visuals. In contrast, the other two learning conditions, with the intervention of large language models, were more effective in engaging students' interest. However, it is inevitable that some students, seeking convenience, may skip watching the video and, instead, rely solely on asking questions to learn from it. The "video-based Q&A + MAD" approach encouraged students to ask questions and initiate debates, helping them better understand the video content. These findings support Hypothesis (H2), which suggests that using video-based multi-agent debates helps students stay more focused on learning through videos.

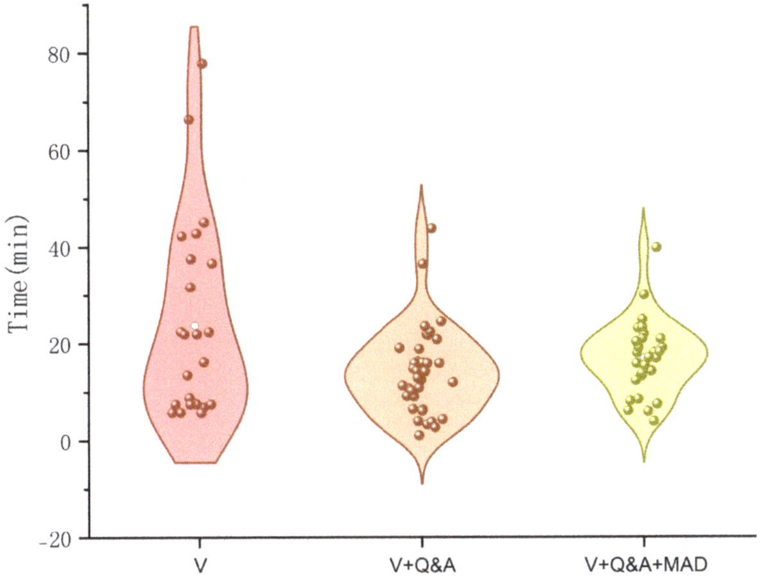

Figure 7. Time differences across the three groups correspond to the three learning conditions: video only, video-based Q&A, and video-based Q&A plus multi-agent debate.

The word frequency analysis presented in Figures 8 and 9 indicates notable differences in responses to open-ended questions before and after watching the video. Post-test responses demonstrate a noticeably richer set of keywords. Additionally, inter-group comparisons reveal that under the third testing condition ("video-based Q&A + MAD"), the keywords are significantly more diverse. These findings suggest that MAD has a positive impact on online learning.

We present representative quotes from student responses to open-ended question for each group during the post-test for the topic of "Numerical Approximation of Gradients":

Response of using the "video only" (English translation): One-sided difference refers to performing difference operations in one direction, typically used to eliminate errors in that direction. Two-sided difference involves difference operations in both directions, usually employed for further error elimination.

Response using the "video-based Q&A" (English translation): One-sided difference approximates derivatives based on changes in the function between x and $x + h$, while two-sided difference uses changes between x and $x - h$ for a more precise approximation with smaller errors.

Response of using the "video-based Q&A + MAD" (English translation): One-sided difference focuses on the deviation of a single data point from the reference point, often used for preliminary data processing. Two-sided difference, by comparing two related data points, effectively reduces noise, highlights features, and optimizes model performance.

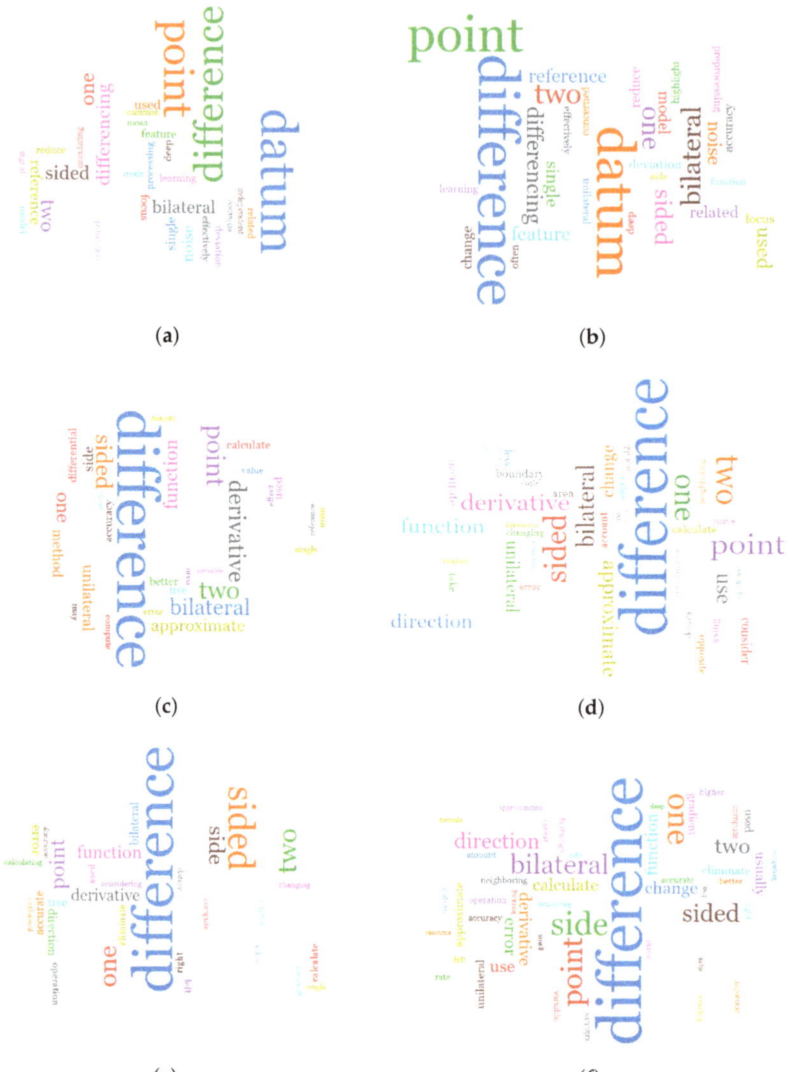

Figure 8. Word cloud analysis of open-ended question responses provided by students for the course titled "Numerical Approximation of Gradients". (**a**) Responses in the pre-test under the video-only testing condition. (**b**) Responses in the post-test under the video-only testing condition. (**c**) Responses in the pre-test under the video-based Q&A testing condition. (**d**) Responses in the post-test under the video-based Q&A testing condition. (**e**) Responses in the pre-test under the video-based Q&A+MAD testing condition. (**f**) Responses in the post-test under the video-based Q&A+MAD testing condition.

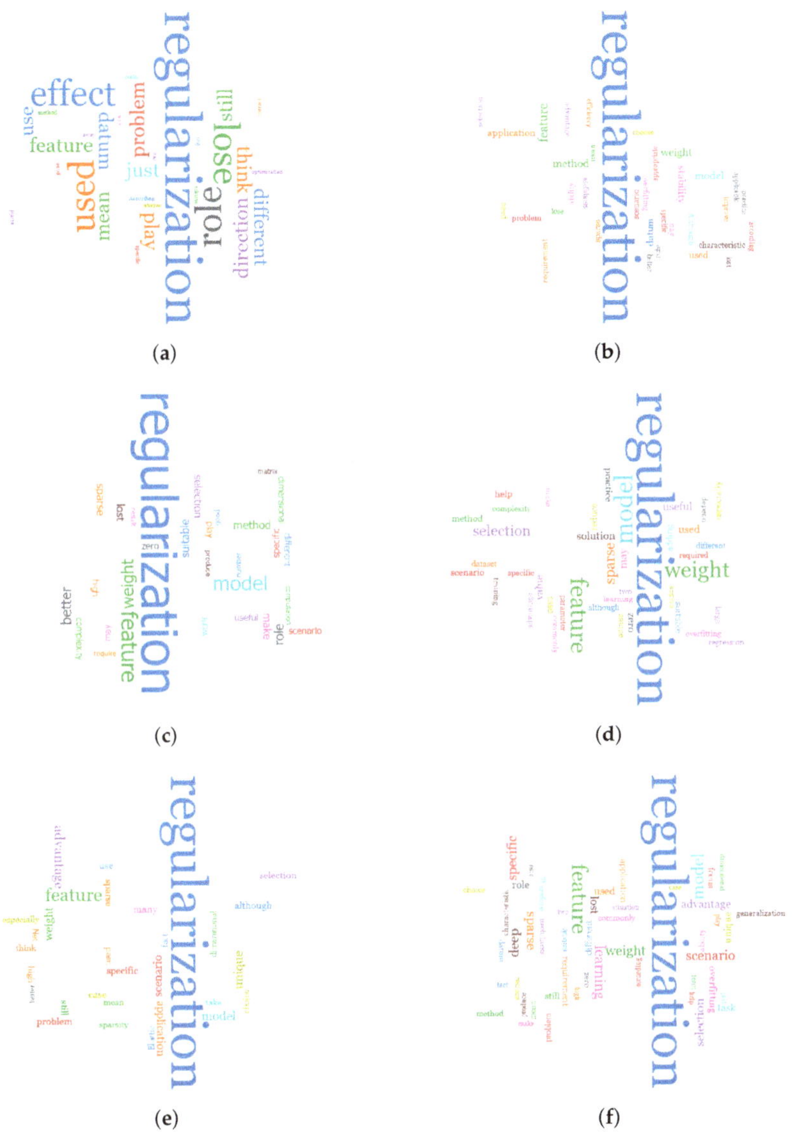

Figure 9. Word cloud analysis of open-ended question responses provided by the students for the course titled "Regularization". (**a**) Responses in the pre-test under the video-only testing condition. (**b**) Responses in the post-test under the video-only testing condition. (**c**) Responses in the pre-test under the video-based Q&A testing condition. (**d**) Responses in the post-test under the video-based Q&A testing condition. (**e**) Responses in the pre-test under the video-based Q&A+MAD testing condition. (**f**) Responses in the post-test under the video-based Q&A+MAD testing condition.

Similarly, representative quotes from student responses to open-ended questions for each group during the post-test on the topic of "regularization" are presented as follows:

Response using the "video only" (English translation): Although L2 regularization is more commonly used in deep learning, its widespread application does not render L1 regularization obsolete. L1 regularization offers unique advantages in feature selection and generating sparse models, making it highly effective in specific scenarios. Therefore, depending on the requirements of a particular task, both L1 and L2 regularization have their irreplaceable roles.

Response using the "video-based Q&A" (English translation): In machine learning and deep learning, L1 and L2 regularization are two commonly used techniques to prevent model overfitting. L1 regularization penalizes large weights by adding the absolute values of the weights to the loss function, while L2 regularization achieves this by adding the squared values of the weights.

Response using the "video-based Q&A + MAD" (English translation): Preventing overfitting is one of the primary objectives of regularization. While both L1 and L2 regularization can mitigate overfitting, they achieve this through different mechanisms. L1 regularization reduces the number of non-zero weights, whereas L2 regularization minimizes the magnitude of the weights. This diversity allows for the selection of the most suitable regularization method based on the specific problem and dataset. L1 regularization offers an alternative strategy for combating overfitting, complementing L2 regularization rather than opposing it.

These responses illustrate that the "video-only" approach enables students to acquire basic and accurate knowledge, the "video-based Q&A" approach allows for more precise answers, and the "video-based Q&A + MAD" approach further enhances the breadth of student responses. In other words, there is an increasing depth and breadth of understanding as the learning condition becomes more interactive. These findings are also consistent with the research results of previous papers. Through AI-assisted learning technologies, high-quality personalized learning can be achieved, improving students' abstract thinking and critical thinking [42].

4.2. Student Feedback Analysis

To further analyze the role of "video-based Q&A + MAD" in students' learning, we administered a follow-up questionnaire. The questionnaire included eight questions, five of which were rating-based using a five-point Likert scale, while the remaining three questions solicited users' suggestions for improvement. Table 5 presents the detailed ratings from the follow-up questionnaire. Each item, representing an evaluation question in the follow-up questionnaire, was rated on a scale from 1 (lowest) to 5 (highest). Users who experienced the MAD functionality generally gave high scores, indicating a positive learning experience and supporting Hypothesis H2. We conducted a word cloud analysis of user improvement suggestions. The questionnaire focused on the "video-based Q&A" function, the MAD function, and the additional features that users most desired to include in future updates. According to the word cloud analysis in Figure 10, students evaluated the "video-based Q&A" function as accurate, fast, and effective in addressing the content of the videos. The multi-agent debate feature was rated as helpful, offering different viewpoints for understanding the learning material. The most requested additional features were video voice translation (as many Chinese students prefer not to listen to lengthy English lectures) and summaries of key points. These suggestions reflect users' subjective experiences and align closely with the responses to the earlier open-ended questions.

Table 5. Ratings in the follow-up questionnaire.

General Satisfaction	Easy to Use	System Response Speed	Eval of Q&A	Eval of MAD
4.31	4.43	3.96	4.13	4.34

(a)

(b)

(c)

Figure 10. Word cloud analysis of the feedback provided by the students in follow-up questionnaires. (**a**) Evaluation of the effectiveness of the video-based Q&A function. (**b**) Evaluation of the effectiveness of the MAD function. (**c**) The function you would most like to add to the current online learning system.

5. Discussion

Interaction between learners and instructors, as well as among peers, plays a crucial role in helping learners construct meaning from educational materials by articulating their ideas, engaging with diverse perspectives, and refining their understanding through feedback. However, facilitating such interaction in MOOCs presents a considerable challenge. The use of LLM-driven agents to support both teaching and learning in online educational environments is expected to become a significant trend, as highlighted in various studies [42,63,64]. Experiments demonstrate that introducing multi-agent systems into online learning environments can inspire students to think critically and help deepen their understanding of course content. Although this research focuses on lecture videos in computer science, the methodology is applicable to other engineering courses.

This study does not focus on the impact of video lecture summaries on learning, as existing research [76] has already concluded that incorporating automatically generated summaries with videos can enhance students' overall learning experience. Chen et al. [77] also emphasized that designing and implementing an interactive video lesson alone is insufficient. They highlighted that incorporating summarizing strategies and prompts can enhance video-based learning by improving learning outcomes, reducing cognitive load, increasing intrinsic motivation, fostering metacognition, and further engaging learners.

Large language models have already started to drive the transformation of the teaching model from the "teacher–student" binary structure to the "teacher–machine–student" ternary structure [78]. This structural change propels teaching from a "teaching-centered" approach to a "learning-centered" one. The educational paradigm will shift from supply-dominated to demand-dominated, reshaping and updating the role functions and competency requirements of teachers. Compared with the traditional binary structure, the technology-empowered "teacher–machine–student" ternary structure liberates teachers from inefficient, repetitive, and onerous knowledge imparting, allowing them to return to the core business of educating. Teachers can then focus on the more creative and affectionate educational values, emphasizing the cultivation of students' autonomous learning abilities and creativity. Singapore is the first country globally to publicly support the use of ChatGPT in the education system. Its education authorities will guide teachers to effectively use intelligent tools like ChatGPT to enhance learning [79].

Our user study evaluates students using two short, pre-recorded, and carefully edited lecture videos. Future research should explore the effectiveness of multi-modal LLMs under broader conditions, including prerecorded versus live lectures, varying lengths, and diverse topics. While our experiments compared three conditions, additional conditions could provide further insights into the value of multi-agent chat and debate. For instance, limiting the time students have to answer questions, tracking the frequency of their use of agents, and surveying their willingness to use the tool before attempting the questions could yield valuable findings. It is worthwhile to explore the possibility of customizing the behavior of multi-agent systems according to students' knowledge levels or their behaviors when watching video materials. Designing a closed-loop system to generate dynamic feedback is of great significance. Multi-agents integrated into formal educational settings, such as online lectures, should be able to support learners for long-term use. The design principles of digital tutors and the technologies of AI-augmented classrooms shed light on our future research [42,80]. Additionally, exposing students to a different condition, analyzing their changes by comparing their results and feedback, and collecting students' questions to analyze the patterns of their inquiries will all be highly meaningful future work. Moreover, this study focuses on students' immediate performance after learning. However, it would be more valuable to evaluate long-term memory retention, changes in learning habits, or the impact on metacognition. Therefore, we plan to incorporate the evaluation of long-term memory retention, changes in learning habits, and the influence on metacognition in our future research.

Additionally, due to GPU hardware limitations, the sample size of participants was relatively small, and most assessments were conducted as independent online tests. In future studies, we will focus on improving the experimental design and selecting model parameters that match the GPU computing power. On one hand, we can integrate regular assignments in computer teaching to expand the sample size of the questionnaire. On the other hand, in the future, we can rely on the university's servers to deploy open-source models with relatively low hardware consumption, such as the recently popular

DeepSeek [81], to break through the computing-power limitations of GPUs. To date many large language models have introduced small models of 7B and 14B, as well as distilled models. When combined with RAG technology, it is believed that these models will be sufficient to handle answer generation for knowledge-based questions in most scenarios in the future.

6. Conclusions

This study focuses on video-based online learning, comparing the traditional "video-only" learning approach with methods that incorporate large language models, specifically "video-based Q&A" and "video-based multi-agent debate" approaches. These methods represent some of the latest trends in LLM applications, offering a more interactive learning experience compared to the traditional unidirectional MOOC format. Additionally, generative responses based on retrieval-augmented generation technology provide students with more comprehensive knowledge summaries. In summary, the method proposed in this paper utilizes multi-agent debates to provide students with reference answers from different perspectives, aiming to deepen students' understanding of knowledge and cultivate their critical thinking. However, compared with in-person teaching by teachers, it lacks the cultivation of other high-order thinking abilities, such as deep thinking and abstract thinking. Additionally, it fails to offer life thinking, life enlightenment, or certain enlightenment and thinking beyond the realm of artificial intelligence. This is also one of the common drawbacks of using AI tools for learning.

Author Contributions: Conceptualization, F.L. and J.D.; methodology, G.X. and J.D.; software, J.D.; validation, J.D.; formal analysis, J.D.; investigation, J.D.; resources, D.Z.; data curation, W.L. and J.D.; writing—original draft preparation, F.L. and J.D.; writing—review and editing, F.L.; visualization, J.D.; supervision, F.L.; project administration, F.L. All authors have read and agreed to the published version of the manuscript.

Funding: This work was funded by the Project of Chinese National Educational Science Planning "Research on the Paths and Mechanisms of Enhancing Middle School Students' Scientific Inquiry Practices with Large Language Models" (BCA240049).

Institutional Review Board Statement: The study was conducted according to the guidelines of the Declaration of Helsinki and approved by the Institutional Review Board of the School of Information Science and Technology, Hangzhou Normal University (protocol code 2024-010, date of approval: 9 September 2024).

Informed Consent Statement: Informed consent was obtained from all subjects involved in the study.

Data Availability Statement: The collected data are publicly available at the following link: https://osf.io/4vt9h/files/osfstorage (accessed on 26 November 2024).

Conflicts of Interest: The authors declare no conflicts of interest.

Appendix A. Pre-Test and Post-Test Questions

All items were originally presented in Chinese in the study and translated into English for understanding.

Appendix A.1. Regularization Pre-Test Questions

1. Overfitting usually manifests itself as a high variance problem. (T/F)
2. The regularization parameter λ serves to increase model complexity. (T/F)
3. In machine learning and deep learning, L2 regularization is achieved by adding a regularization term to the loss function, which is

(a) The cube of the L2 norm of the model parameters (usually referred to as weights w)
(b) The square of the L2 norm of the model parameters (usually referred to as weights w)
(c) Minus the cube of the L2 norm of model parameters (usually referred to as weights w)
(d) Minus the cube of the L2 norm of model parameters (usually referred to as weights w)

4. Which of the following options are correct in the description of L2 regularization in neural networks (Multiple choice)?

   (a) L2 regularization is achieved by adding a term to the loss function that is proportional to the square of the weights
   (b) L2 regularization terms usually do not include bias terms
   (c) During gradient descent, L2 regularization causes the weights to shrink towards zero, thus preventing overfitting
   (d) The larger the L2 regularization factor (λ), the weaker the effect of regularization
   (e) L2 regularization is achieved by directly modifying the weights during the forward propagation of the neural network

5. What is the main difference between L1 regularization and L2 regularization in deep learning?

   (a) L1 regularization takes into account the square of the weights, while L2 regularization does not take into account the square of the weights.
   (b) There is no difference in effect between L1 regularization and L2 regularization.
   (c) L1 regularization focuses on preventing overfitting, while L2 regularization is usually used for feature selection.
   (d) L1 regularization tends to produce sparse weight matrices, while L2 regularization makes the weights smoother.

6. There is a growing tendency to use L2 regularization when training networks to better achieve the effect of avoiding overfitting, does it mean that L1 regularization has gradually lost its usefulness? Please write down what you think.

Appendix A.2. Regularization Post-Test Questions

1. Overfitting usually manifests itself as a high variance problem. (T/F)
2. The regularization parameter λ serves to increase model complexity. (T/F)
3. Why is L2 regularization usually performed only for the parameter w and not for the parameter "b"?

   (a) "b" is one of many parameters that have a small effect on model complexity.
   (b) The dimension of "b" is usually higher than that of "w".
   (c) The range of values for "b" is usually larger than for "w", which does not lend itself easily to regularization.
   (d) The update of "b" has no significant effect on model performance.

4. What is the main purpose of regularization techniques in deep learning?

   (a) Improving model accuracy
   (b) Accelerating the model training process
   (c) Reducing model overfitting
   (d) Simplifying model complexity

5. In machine learning and deep learning, L2 regularization is achieved by adding a regularization term to the loss function, which is

(a) The cube of the L2 norm of the model parameters (usually referred to as weights w)
(b) The square of the L2 norm of the model parameters (usually referred to as weights w)
(c) Minus the cube of the L2 norm of model parameters (usually referred to as weights w)
(d) Minus the cube of the L2 norm of model parameters (usually referred to as weights w)

6. How can Dropout help improve model performance in neural networks (Multiple choice)?
   (a) Reducing overfitting
   (b) Increasing the complexity of the model
   (c) Improving the generalization of models
   (d) Accelerating the training process
7. What are the correct statements about the regularization term $\lambda/2m \times ||W||_F^2$ (the Frobenius paradigm) (Multiple choice)?
   (a) It sums the sum of squares of all layer weight matrices W
   (b) It helps to reduce the number of paradigms in the weight matrix, thus avoiding overfitting
   (c) It has the same effect as L2 regularization in logistic regression
   (d) It only applies to fully connected layers, not to convolutional layers
8. Which of the following options are correct in the description of L2 regularization in neural networks (Multiple choice)?
   (a) L2 regularization is achieved by adding a term to the loss function that is proportional to the square of the weights
   (b) L2 regularization terms usually do not include bias terms
   (c) During gradient descent, L2 regularization causes the weights to shrink towards zero, thus preventing overfitting
   (d) The larger the L2 regularization factor (λ), the weaker the effect of regularization
   (e) L2 regularization is achieved by directly modifying the weights during the forward propagation of the neural network
9. What is the main difference between L1 regularization and L2 regularization in deep learning?
   (a) L1 regularization takes into account the square of the weights, while L2 regularization does not take into account the square of the weights.
   (b) There is no difference in effect between L1 regularization and L2 regularization.
   (c) L1 regularization focuses on preventing overfitting, while L2 regularization is usually used for feature selection.
   (d) L1 regularization tends to produce sparse weight matrices, while L2 regularization makes the weights smoother.
10. There is a growing tendency to use L2 regularization when training networks to better achieve the effect of avoiding overfitting, does it mean that L1 regularization has gradually lost its usefulness? Please write down what you think.

Appendix A.3. Numerical Approximation of Gradients Pre-Test Questions

1. What is the main purpose of gradient numerical approximation?
   (a) Directly calculate the derivative of a function at a certain point
   (b) Test the accuracy of the derivative by approximation method

(c) Taylor series expansion of the solution function
(d) Computational efficiency of optimization function

2. What kind of function is the gradient numerical approximation method mainly suitable for?
   (a) Differentiable continuous functions
   (b) Functions that are not differentiable
   (c) Discrete function
   (d) Any function
3. Which of the following formulas is the standard formula for the two-sided difference method?
   (a) $(f(x+h) - f(x))/h$
   (b) $(f(x) - f(x-h))/h$
   (c) $(f(x+h) - f(x-h))/2h$
   (d) $(f(x+h) + f(x-h))/2h$
4. What is the main difference between unilateral difference and bilateral difference in gradient numerical approximation?
   (a) Computational complexity
   (b) The magnitude of the approximation error
   (c) Type of function used
   (d) Whether a derivation formula is needed
5. In the gradient test, which one is more effective?
   (a) Unilateral difference
   (b) Bilateral difference
6. Please briefly describe basic principles and differences of unilateral difference and bilateral difference.

Appendix A.4. Numerical Approximation of Gradients Post-Test Questions

1. What is the main purpose of gradient numerical approximation?
   (a) Directly calculate the derivative of a function at a certain point
   (b) Test the accuracy of the derivative by approximation method
   (c) Taylor series expansion of the solution function
   (d) Computational efficiency of optimization function
2. What kind of function is the gradient numerical approximation method mainly suitable for?
   (a) Differentiable continuous functions
   (b) Functions that are not differentiable
   (c) Discrete function
   (d) Any function
3. Which of the following formulas is the standard formula for the two-sided difference method?
   (a) $(f(x+h) - f(x))/h$
   (b) $(f(x) - f(x-h))/h$
   (c) $(f(x+h) - f(x-h))/2h$
   (d) $(f(x+h) + f(x-h))/2h$
4. What is the main difference between unilateral difference and bilateral difference in gradient numerical approximation?
   (a) Computational complexity

   (b) The magnitude of the approximation error
   (c) Type of function used
   (d) Whether a derivation formula is needed
5. In the gradient test, which one is more effective?
   (a) Unilateral difference
   (b) Bilateral difference
6. In gradient test, the main calculation of unilateral difference is
   (a) Gradient caused by a large change in a function at a certain point in a certain direction
   (b) An approximation of the gradient caused by a small change in the function at a point in a certain direction
   (c) The average value of the gradient caused by the change of the function in all directions at one point
   (d) The gradient difference caused by a change in the opposite direction of a function at a certain point
7. The main difference between two-sided difference and one-sided difference in calculating the gradient approximation is
   (a) Bilateral difference only takes into account changes in one direction
   (b) Bilateral differences consider changes in two opposite directions
   (c) Bilateral difference does not consider the change of the function, only the gradient
   (d) There is no difference in calculation between bilateral difference and unilateral difference
8. In practical applications, what factors may affect the choice of unilateral difference or bilateral difference for gradient testing?
   (a) Derivability of the function
   (b) Limitations on computing resources
   (c) Requirements for accuracy
   (d) All of the above factors
9. In practical applications, which strategies might be adopted to balance computational efficiency and accuracy of gradient tests?
   (a) Select the difference method according to the function characteristics
   (b) Always use bilateral differences
   (c) Always use unilateral difference
   (d) No gradient test
10. Please briefly describe basic principles and differences of unilateral difference and bilateral difference.

Appendix B. Follow-Up Questionnaire

1. How satisfied are you with the system overall?
   (a) Very satisfied (5 points)
   (b) Fairly satisfied (4 points)
   (c) Neutral (3 points)
   (d) Fairly dissatisfied (2 points)
   (e) Very dissatisfied (1 point)
2. Do you find the UI design of the system simple and easy to understand?
   (a) Very satisfied (5 points)
   (b) Fairly satisfied (4 points)

   (c) Neutral (3 points)
   (d) Fairly dissatisfied (2 points)
   (e) Very dissatisfied (1 point)
3. How fast do you think the system response?
   (a) Very rapid (5 points)
   (b) Comparative speed (4 points)
   (c) Neutral (3 points)
   (d) Less rapid (2 points)
   (e) Very slow (1 point)
4. Are you satisfied with the effectiveness of the system's video-based Q&A function?
   (a) Very satisfied (5 points)
   (b) Fairly satisfied (4 points)
   (c) Neutral (3 points)
   (d) Fairly dissatisfied (2 points)
   (e) Very dissatisfied (1 point)
5. Please give the corresponding reason based on your rating of the effectiveness of the system's video-based Q&A function.
6. Are you satisfied with the effectiveness of the system's video-based MAD function (If applicable)?
   (a) Very satisfied (5 points)
   (b) Fairly satisfied (4 points)
   (c) Neutral (3 points)
   (d) Fairly dissatisfied (2 points)
   (e) Very dissatisfied (1 point)
7. Please give the corresponding reason based on your rating of the effectiveness of the system's video-based MAD function. (If applicable).
8. What feature would you most like to see added to the current system? Please provide examples if possible.

References

1. Aristovnik, A.; Karampelas, K.; Umek, L.; Ravšelj, D. Impact of the COVID-19 Pandemic on Online Learning in Higher Education: A Bibliometric Analysis. *Front. Educ.* **2023**, *8*, 1225834. [CrossRef]
2. Van Alten, D.C.; Phielix, C.; Janssen, J.; Kester, L. Self-regulated Learning Support in Flipped Learning Videos Enhances Learning Outcomes. *Comput. Educ.* **2020**, *158*, 104000. [CrossRef]
3. Moos, D.; Bonde, C. Flipping the Classroom: Embedding Self-Regulated Learning Prompts in Videos. *Technol. Knowl. Learn.* **2016**, *21*, 225–242. [CrossRef]
4. Kim, J.; Guo, P.J.; Cai, C.J.; Li, S.W.; Gajos, K.Z.; Miller, R.C. Data-driven Interaction Techniques for Improving Navigation of Educational Videos. In Proceedings of the 27th Annual ACM Symposium on User Interface Software and Technology, Honolulu, HI, USA, 5–8 October 2014; pp. 563–572.
5. Yang, S.; Yim, J.; Kim, J.; Shin, H.V. CatchLive: Real-time Summarization of Live Streams with Stream Content and Interaction Data. In Proceedings of the 2022 CHI Conference on Human Factors in Computing Systems, New Orleans, LA, USA, 29 April–5 May 2022.
6. Pavel, A.; Reed, C.; Hartmann, B.; Agrawala, M. Video Digests: A Browsable, Skimmable Format for Informational Lecture Videos. In Proceedings of the 27th Annual ACM Symposium on User Interface Software and Technology, Honolulu, HI, USA, 5–8 October 2014; pp. 573–582. [CrossRef]
7. Shimada, A.; Okubo, F.; Yin, C.; Ogata, H. Automatic Summarization of Lecture Slides for Enhanced Student PreviewTechnical Report and User Study. *IEEE Trans. Learn. Technol.* **2018**, *11*, 165–178. [CrossRef]
8. OpenAI. GPT-4 Technical Report. *arXiv* **2024**, arXiv:2303.08774.

9. Bernabei, M.; Colabianchi, S.; Falegnami, A.; Costantino, F. Students' Use of Large Language Models in Engineering Education: A Case Study on Technology Acceptance, Perceptions, Efficacy, and Detection Chances. *Comput. Educ. Artif. Intell.* **2023**, *5*, 100172. [CrossRef]
10. Kohnke, L.; Moorhouse, B.; Zou, D. ChatGPT for Language Teaching and Learning. *RELC J.* **2023**, *54*, 537–550. [CrossRef]
11. Sallam, M. ChatGPT Utility in Healthcare Education, Research, and Practice: Systematic Review on the Promising Perspectives and Valid Concerns. *Healthcare* **2023**, *11*, 887. [CrossRef]
12. Kaddour, J.; Harris, J.; Mozes, M.; Bradley, H.; Raileanu, R.; McHardy, R. Challenges and Applications of Large Language Models. *arXiv* **2023**, arXiv:2307.10169. Available online: https://www.frontiersin.org/journals/education/articles/10.3389/feduc.2023.1225834 (accessed on 27 April 2025).
13. Bender, E.; Gebru, T.; McMillan-Major, A.; Shmitchell, S. On the Dangers of Stochastic Parrots: Can Language Models Be Too Big? In Proceedings of the 2021 ACM Conference on Fairness, Accountability, and Transparency, Virtual Event Canada, 3–10 March 2021; pp. 610–623.
14. Kasneci, E.; Sessler, K.; Küchemann, S.; Bannert, M.; Dementieva, D.; Fischer, F.; Gasser, U.; Groh, G.; Günnemann, S.; Hüllermeier, E.; et al. ChatGPT for Good? On Opportunities and Challenges of Large Language Models for Education. *Learn. Individ. Differ.* **2023**, *103*, 102274. [CrossRef]
15. Sahoo, P.; Singh, A.; Saha, S.; Jain, V.; Mondal, S.; Chadha, A. A Systematic Survey of Prompt Engineering in Large Language Models: Techniques and Applications. *arXiv* **2024**, arXiv:2402.07927.
16. Kojima, T.; Gu, S.; Reid, M.; Matsuo, Y.; Iwasawa, Y. Large Language Models Are Zero-shot Reasoners. In Proceedings of the 36th International Conference on Neural Information Processing Systems, New Orleans, LA, USA, 28 November–9 December 2022.
17. Wang, X.; Wei, J.; Schuurmans, D.; Le, Q.V.; Chi, E.H.; Narang, S.; Chowdhery, A.; Zhou, D. Self-Consistency Improves Chain of Thought Reasoning in Language Models. *arXiv* **2022**, arXiv:2203.11171.
18. Lewis, P.; Perez, E.; Piktus, A.; Petroni, F.; Karpukhin, V.; Goyal, N.; Küttler, H.; Lewis, M.; Yih, W.; Rocktäschel, T.; et al. Retrieval-augmented Generation for Knowledge-intensive NLP Tasks. In Proceedings of the 34th International Conference on Neural Information Processing Systems, Vancouver, BC, Canada, 6–12 December 2020.
19. Wang, L.; Ma, C.; Feng, X.; Zhang, Z.; Yang, H.; Zhang, J.; Chen, Z.; Tang, J.; Chen, X.; Lin, Y.; et al. A Survey on Large Language Model based Autonomous Agents. *Front. Comput. Sci.* **2024**, *18*, 3. [CrossRef]
20. Wu, Q.; Bansal, G.; Zhang, J.; Wu, Y.; Li, B.; Zhu, E.; Jiang, L.; Zhang, X.; Zhang, S.; Liu, J.; et al. AutoGen: Enabling Next-Gen LLM Applications via Multi-Agent Conversation. *arXiv* **2023**, arXiv:2308.08155.
21. McAuley, A.; Stewart, B.; Siemens, G.; Cornier, D. *The MOOC Model for Digital Practice: Massive Open Online Courses: Digital Ways of Knowing and Learning*; University of Prince Edward Island: Charlottetown, PE, Canada, 2010. Available online: https://books.google.com.hk/books?id=VIYSzgEACAAJ (accessed on 27 April 2025).
22. Laura, P. The Year of the MOOC. *The New York Times*, 13 November 2012.
23. Yuan, L.; Powell, S. *MOOCS and Open Education: Implications for Higher Education*; JISC Cetis; University of Bolton: Bolton, UK, 2013.
24. Laurillard, D. The Educational Problem that MOOCs Could Solve: Professional Development for Teachers of Disadvantaged Students. *Res. Learn. Technol.* **2016**, *24*, 29369. [CrossRef]
25. Khan, M.; Vivek; Nabi, M.; Khojah, M.; Tahir, M. Students' Perception towards E-Learning during COVID-19 Pandemic in India: An Empirical Study. *Sustainability* **2021**, *13*, 57. [CrossRef]
26. Moore, R. Developing Lifelong Learning with Heutagogy: Contexts, Critiques, and Challenges. *Distance Educ.* **2020**, *41*, 381–401. [CrossRef]
27. Suanpang, P.; Netwong, T.; Manisri, T.; Duantrakoonsi, W. The Factors Affecting Learning Outcome Intention of MOOCs for an Online Learning Platform. *Psychol. Educ.* **2021**, *58*, 3922–3930.
28. Jansen, R.; Van Leeuwen, A.; Janssen, J.; Conijn, R.; Kester, L. Supporting Learners' Self-regulated Learning in Massive Open Online Courses. *Comput. Educ.* **2020**, *146*, 103771. [CrossRef]
29. Lu, Y.; Wang, B.; Lu, Y. Understanding Key Drivers of MOOC Satisfaction and Continuance Intention to Use. *J. Electron. Commer. Res.* **2019**, *20*, 105–117.
30. Aldowah, H.; Al-Samarraie, H.; Alzahrani, A.; Alalwan, N. Factors Affecting Student Dropout in MOOCs: A Cause and Effect Decision-making Model. *J. Comput. High. Educ.* **2020**, *32*, 429–454. [CrossRef]
31. Jordan, K. Initial Trends in Enrollment and Completion of Massive Open Online Courses. *Int. Rev. Res. Open Distrib. Learn.* **2014**, *15*, 133–160.
32. Margaryan, A.; Bianco, M.; Littlejohn, A. Instructional Quality of Massive Open Online Courses (MOOCs). *Comput. Educ.* **2015**, *80*, 77–83. [CrossRef]
33. Kizilcec, R.; Piech, C.; Schneider, E. Deconstructing Disengagement: Analyzing Learner Subpopulations in Massive Open Online Courses. In Proceedings of the Third International Conference on Learning Analytics and Knowledge, Leuven, Belgium, 8–13 April 2013; pp. 170–179.

34. Hew, K.; Cheung, W. Students' and Instructors' Use of Massive Open Online Courses (MOOCs): Motivations and Challenges. *Educ. Res. Rev.* **2014**, 12, 45–58. [CrossRef]
35. Wei, W.; Liu, J.; Xu, X.; Kolletar-Zhu, K.; Zhang, Y. Effective Interactive Engagement Strategies for MOOC Forum Discussion: A Self-efficacy Perspective. *PLoS ONE* **2023**, *18*, e0293668. [CrossRef]
36. Neha; Kim, E. Designing Effective Discussion Forum in MOOCs: Insights from Learner Perspectives. *Front. Educ.* **2023**, *8*, 1223409. [CrossRef]
37. Galikyan, I.; Admiraal, W.; Kester, L. MOOC Discussion Forums: The Interplay of the Cognitive and the Social. *Comput. Educ.* **2021**, *165*, 104133. [CrossRef]
38. Freitas, M.; Silva, M. Systematic Literature Review about Gamification in MOOCs. *Open Learn. J. Open Distance-Learn.* **2023**, *38*, 73–95. [CrossRef]
39. Meng, C.; Zhao, M.; Pan, Z.; Pan, Q.; Bonk, C. Investigating the Impact of Gamification Components on Online Learners' Engagement. *Smart Learn. Environ.* **2024**, *11*, 47. [CrossRef]
40. Chauhan, J.; Goel, A. An Analysis of Video Lecture in MOOC. In Proceedings of the International Conference on Information and Communication Technologies in Education, Research, and Industrial Applications, Lviv, Ukraine, 14–16 May 2015. Available online: https://ceur-ws.org/Vol-1356/paper_16.pdf (accessed on 27 April 2025).
41. Chen, Y.; Gao, Q.; Yuan, Q.; Tang, Y. Facilitating Students' Interaction in MOOCs through Timeline-Anchored Discussion. *Int. J. Human–Computer Interact.* **2019**, *35*, 1781–1799. [CrossRef]
42. Yu, J.; Zhang, Z.; Zhang, D.; Tu, S.; Hao, Z.; Li, R.; Li, H.; Wang, Y.; Li, H.; Gong, L.; et al. From MOOC to MAIC: Reshaping Online Teaching and Learning through LLM-driven Agents. *arXiv* **2024**, arXiv:2409.03512.
43. Reyna, J. Digital Teaching and Learning Ecosystem (DTLE): A Theoretical Approach for Online Learning Environments. In Proceedings of the Ascilite Hobart 2011, Hobart, TAS, Australia, 4–7 December 2011; pp. 1083–1088. Available online: https://ascilite.org/conferences/hobart11/downloads/papers/Reyna-concise.pdf (accessed on 27 April 2025).
44. Thorneycroft, S. Defining a Next-generation Ecosystem for Online Learning: From Changing the Platform to Shifting the Paradigm. In Proceedings of the 39th International Conference on Innovation, Practice and Research in the Use of Educational Technologies in Tertiary Education, Sydney, Australia, 4–7 December 2022; pp. 1–6. [CrossRef]
45. Hsu, H.; Wang, C.; Levesque-Bristol, C. Reexamining the Impact of Self-determination Theory on Learning Outcomes in the Online Learning Environment. *Educ. Inf. Technol.* **2019**, *24*, 2159–2174. [CrossRef]
46. Petersen, G.; Petkakis, G.; Makransky, G. A Study of How Immersion and Interactivity Drive VR Learning. *Comput. Educ.* **2022**, *179*, 104429. [CrossRef]
47. Pedaste, M.; Mäeots, M.; Siiman, L.; Jong, T.; Riesen, S.; Kamp, E.; Manoli, C.; Zacharia, Z.; Tsourlidaki, E. Phases of Inquiry-based Learning: Definitions and The Inquiry Cycle. *Educ. Res. Rev.* **2015**, *14*, 47–61. [CrossRef]
48. Kizilcec, R.; Pérez-Sanagustín, M.; Maldonado, J. Self-regulated Learning Strategies Predict Learner Behavior and Goal Attainment in Massive Open Online Courses. *Comput. Educ.* **2017**, *104*, 18–33. [CrossRef]
49. Xu, Z.; Zhao, Y.; Liew, J.; Zhou, X.; Kogut, A. Synthesizing Research Evidence on Self-regulated Learning and Academic Achievement in Online and Blended Learning Environments: A Scoping Review. *Educ. Res. Rev.* **2023**, *39*, 100510. [CrossRef]
50. Huang, C.; Han, Z.; Li, M.; Wang, X.; Zhao, W. Sentiment Evolution with Interaction Levels in Blended Learning Environments: Using Learning Analytics and Epistemic Network Analysis. *Australas. J. Educ. Technol.* **2021**, *37*, 81–95. [CrossRef]
51. Li, Y.; Meng, N.; Ye, X.; Syed Shah, A. Self-determination and Perceived Learning in Online Learning Communities. *Sci. Rep.* **2024**, *14*, 24538. [CrossRef]
52. Zhang, L.; Carter, R.A., Jr.; Qian, X.; Yang, S.; Rujimora, J.; Wen, S. Academia's Responses to Crisis: A Bibliometric Analysis of Literature on Online Learning in Higher Education during COVID-19. *Br. J. Educ. Technol.* **2022**, *53*, 620–646. [CrossRef] [PubMed]
53. Ng, D.; Leung, J.; Su, J.; Ng, R.; Chu, S. Teachers' AI Digital Competencies and Twenty-first Century Skills in the Post-pandemic World. *Educ. Technol. Res. Dev.* **2023**, *71*, 137–161. [CrossRef]
54. Wang, P.; Wang, F.; Li, Z. Exploring the Ecosystem of K-12 Online Learning: An Empirical Study of Impact Mechanisms in the Post-pandemic Era. *Front. Psychol.* **2023**, *14*, 1241477. [CrossRef]
55. Ayeni, O.; Hamad, N.; Chisom, O.; Osawaru, B.; Adewusi, E. AI in Education: A Review of Personalized Learning and Educational Technology. *GSC Adv. Res. Rev.* **2024**, *18*, 261–271. [CrossRef]
56. Wooldridge, M.; Jennings, N. Intelligent Agents: Theory and Practice. *Knowl. Eng. Rev.* **1995**, *10*, 115–152. [CrossRef]
57. Xi, Z.; Chen, W.; Guo, X.; He, W.; Ding, Y.; Hong, B.; Zhang, M.; Wang, J.; Jin, S.; Zhou, E.; et al. The Rise and Potential of Large Language Model Based Agents: A Survey. *arXiv* **2023**, arXiv:2309.07864.
58. Guo, T.; Chen, X.; Wang, Y.; Chang, R.; Pei, S.; Chawla, N.; Wiest, O.; Zhang, X. Large Language Model based Multi-Agents: A Survey of Progress and Challenges. *arXiv* **2024**, arXiv:2402.01680.
59. Park, J.; O'Brien, J.; Cai, C.; Morris, M.; Liang, P.; Bernstein, M. Generative Agents: Interactive Simulacra of Human Behavior. In Proceedings of the 36th Annual ACM Symposium on User Interface Software and Technology (UIST '23), San Francisco, CA, USA, 29 October–1 November 2023.

60. Zhu, X.; Chen, Y.; Tian, H.; Tao, C.; Su, W.; Yang, C.; Huang, G.; Li, B.; Lu, L.; Wang, X.; et al. Ghost in the Minecraft: Generally Capable Agents for Open-World Environments via Large Language Models with Text-based Knowledge and Memory. *arXiv* **2023**, arXiv:2305.17144.
61. Fu, Y.; Peng, H.; Khot, T.; Lapata, M. Improving Language Model Negotiation with Self-Play and In-Context Learning from AI Feedback. *arXiv* **2023**, arXiv:2305.10142.
62. Extance, A. ChatGPT has Entered the Classroom: How LLMs Could Transform Education. *Nature* **2023**, *623*, 474–477. [CrossRef]
63. Yue, M.; Mifdal, W.; Zhang, Y.; Suh, J.; Yao, Z. MathVC: An LLM-Simulated Multi-Character Virtual Classroom for Mathematics Education. *arXiv* **2024**, arXiv:2404.06711.
64. Zhang, Z.; Zhang-Li, D.; Yu, J.; Gong, L.; Zhou, J.; Hao, Z.; Jiang, J.; Cao, J.; Liu, H.; Liu, Z.; et al. Simulating Classroom Education with LLM-Empowered Agents. *arXiv* **2024**, arXiv:2406.19226.
65. Huber, S.; Kiili, K.; Nebel, S.; Ryan, R.; Sailer, M.; Ninaus, M. Leveraging the Potential of Large Language Models in Education Through Playful and Game-Based Learning. *Educ. Psychol. Rev.* **2024**, *36*, 25. [CrossRef]
66. Baillifard, A.; Gabella, M.; Lavenex, P.; Martarelli, C. Effective Learning with a Personal AI Tutor: A Case Study. *Educ. Inf. Technol.* **2025**, *30*, 297–312. [CrossRef]
67. Park, M.; Kim, S.; Lee, S.; Kwon, S.; Kim, K. Empowering Personalized Learning through a Conversation-based Tutoring System with Student Modeling. In Proceedings of the CHI EA '24: Extended Abstracts of the CHI Conference on Human Factors in Computing Systems, Honolulu, HI, USA, 11–16 May 2024; pp. 1–10.
68. Mohamed, A. Exploring the Potential of an AI-based Chatbot (ChatGPT) in Enhancing English as a Foreign Language (EFL) Teaching: Perceptions of EFL Faculty Members. *Educ. Inf. Technol.* **2024**, *29*, 3195–3217. [CrossRef]
69. Tlili, A.; Shehata, B.; Adarkwah, M.; Bozkurt, A.; Hickey, D.; Huang, R.; Agyemang, B. What if the Devil Is My Guardian Angel: ChatGPT as a Case Study of Using Chatbots in Education. *Smart Learn. Environ.* **2023**, *10*, 15. [CrossRef]
70. Zhang, S.; Zhao, X.; Zhou, T.; Kim, J. Do You Have AI Dependency? The Roles of Academic Self-efficacy, Academic Stress, and Performance Expectations on Problematic AI Usage Behavior. *Int. J. Educ. Technol. High. Educ.* **2024**, *21*, 2–14. [CrossRef]
71. Siu, O.; Lui, K.; Huang, Y.; Ng, T.; Yeung, W. An Efficient, Reliable and Valid Assessment for Affective States during Online Learning. *Sci. Rep.* **2024**, *14*, 15768. [CrossRef]
72. Almarzouqi, A.; Aburayya, A.; Alfaisal, R.; Elbadawi, M.A.; Salloum, S.A. Ethical Implications of Using ChatGPT in Educational Environments: A Comprehensive Review. In *Artificial Intelligence in Education: The Power and Dangers of ChatGPT in the Classroom*; Springer Nature: Cham, Switzerland, 2024; pp. 185–199. [CrossRef]
73. Li, K.; He, Y.; Wang, Y.; Li, Y.; Wang, W.; Luo, P.; Wang, Y.; Wang, L.; Qiao, Y. VideoChat: Chat-Centric Video Understanding. *arXiv* **2023**, arXiv:2305.06355.
74. Liang, T.; He, Z.; Jiao, W.; Wang, X.; Wang, Y.; Wang, R.; Yang, Y.; Shi, S.; Tu, Z. Encouraging Divergent Thinking in Large Language Models through Multi-Agent Debate. *arXiv* **2023**, arXiv:2305.19118.
75. Cohen, J. A Power Primer. *Psychol. Bull.* **1992**, *112*, 155–159. [CrossRef]
76. Gonzalez, H.; Li, J.; Jin, H.; Ren, J.; Zhang, H.; Akinyele, A.; Wang, A.; Miltsakaki, E.; Baker, R.; Callison-Burch, C. Automatically Generated Summaries of Video Lectures May Enhance Students' Learning Experience. In Proceedings of the 18th Workshop on Innovative Use of NLP for Building Educational Applications (BEA 2023), Toronto, ON, Canada, 13 July 2023; pp. 382–393. Available online: https://aclanthology.org/2023.bea-1.31 (accessed on 27 April 2025).
77. Chen, W.; Lin, L.; Pang, W. Learning via Video: The Effects of Summarizing Strategies and Self-explanation Prompts. *Educ. Inf. Technol.* **2025**, *30*, 3541–3572. [CrossRef]
78. Yang, Z.; Wang, J.; Wu, D.; Chen, X. Exploring the Impact of ChatGPT/AIGC on Education and Strategies for Response. *J. East China Norm. Univ. (Educ. Sci.)* **2023**, *41*, 26–35.
79. Ministry of Education Singapore. Managing the Use of Artificial Intelligence (AI) Bots such as ChatGPT in Schools. 2023. Available online: www.moe.gov.sg (accessed on 27 April 2025).
80. Hobert, S.; Berens, F. Developing a Digital Tutor as an Intermediary Between Students, Teaching Assistants, and Lecturers. *Educ. Technol. Res. Dev.* **2024**, *72*, 797–818. [CrossRef]
81. DeepSeek-AI. DeepSeek-V3 Technical Report. *arXiv* **2025**, arXiv:2412.19437.

applied sciences

MDPI

Article

Advanced Dense Text Detection in Graded Examinations Leveraging Chinese Character Components

Renyuan Liu, Yunyu Shi *, Xian Tang and Xiang Liu

College of Electrical and Electronic Engineering, Shanghai University of Engineering Science, Shanghai 201620, China; leoryu@sues.edu.cn (R.L.); tangxian@sues.edu.cn (X.T.); xliu@sues.edu.cn (X.L.)
* Correspondence: yunyushi@sues.edu.cn

Featured Application: This study explores the intelligent application of Shanghai's writing proficiency assessment, targeting efficient recognition of handwritten Chinese characters in promotion exams for primary and secondary students.

Abstract: The dense text detection and segmentation of Chinese characters has always been a research hotspot due to the complex background and diverse scenarios. In the field of education, the detection of handwritten Chinese characters is affected by background noise, texture interference, etc. Especially in low-quality handwritten text, character overlap or occlusion makes the character boundaries blurred, which increases the difficulty of detection and segmentation; In this paper, an improved EAST network CEE (Components-ECA-EAST Network), which fuses the attention mechanism with the feature pyramid structure, is proposed based on the analysis of the structure of Chinese character mini-components. The ECA (Efficient Channel Attention) attention mechanism is incorporated during the feature extraction phase; in the feature fusion stage, the convolutional features are extracted from the self-constructed mini-component dataset and then fused with the feature pyramid in a cascade manner, and finally, Dice Loss is used as the regression task loss function. The above improvements comprehensively improve the performance of the network in detecting and segmenting the mini-components and subtle strokes of handwritten Chinese characters; The CEE model was tested on the self-constructed dataset with an accuracy of 84.6% and a mini-component mAP of 77.6%, which is an improvement of 7.4% and 8.4%, respectively, compared to the original model; The constructed dataset and improved model are well suited for applications such as writing grade examinations, and represent an important exploration of the development of educational intelligence.

Keywords: Chinese character detection; mini components; attention mechanism; multi-scale feature fusion; dice loss

1. Introduction

The task of detecting Chinese characters is not only an important step in Chinese character recognition and processing, but also widely used in the fields of cultural heritage preservation, calligraphy analysis, and font generation. However, due to the multi-scale characteristics, complex background, and dense distribution of Chinese character components, traditional image segmentation methods are difficult to meet the requirements of practical applications.

The main challenges of handwritten Chinese character detection can be summarized as follows: first, the structure of Chinese characters is inherently complex, with each glyph consisting of multiple strokes and complex relative positional relationships between the strokes. Some Chinese characters have large differences in stroke order, curvature, and structural combinations, making recognition more difficult. Compared with Latin letters or numbers, the structure of Chinese characters is not only more complex but also varied. Secondly, there are significant individual differences in the writing styles of handwritten Chinese characters. Factors such as writing styles, stroke thickness, and inclination of characters by different writers can lead to differences in the performance of the same character shape among different individuals. The issues of blurred handwriting, overlapping characters, and hyphenation, all caused by excessive writing speed, can significantly affect the accurate recognition of characters. Furthermore, in practical applications, background noise, texture interference, and other non-target elements often appear in handwritten Chinese character images, and these factors place higher demands on character localization and segmentation. Finally, in some low-quality handwritten texts, characters may overlap and occlude, which makes the boundaries between characters blurred and increases the difficulty of detection and segmentation.

Wang [1] and others proposed a text region detection method based on multi-scale sliding windows, which efficiently identifies text regions with good results by sliding classification of each window through a convolutional network. Tian [2] proposed the CTPN (Connectionist Text Proposal Network) text detection network, which holds a significant position in the field of text detection, and is especially suitable for the detection task of scene text box narrow, because the size of the text box varies greatly. CTPN uses fixed text anchor frames of different sizes and combines them with the Faster-RCNN concept to efficiently detect small-size text regions. Meanwhile, CTPN uses a text line construction method to connect multiple candidate text boxes into complete text boxes. Subsequently, the network further processes the convolutional features through a bidirectional LSTM structure to enhance the representation of spatial as well as sequential information.

On this basis, the EAST (Efficient and Accurate Scene Text Detector) algorithm proposed by Zhou [3] draws on the concept of U-Net [4], employing a fully convolutional neural network (FCN) and non-maximum suppression (NMS) algorithms to streamline intermediate processing steps, enabling efficient detection at the single-character level. SegLink [5] is an improved version of CTPN, which not only makes tilt improvement for text box prediction but also optimizes the positioning of text boxes by connecting candidate boxes through neural networks. PixelLink [6] predicts text pixels and non-text pixels separately by training a convolutional neural network model and connecting them for instance segmentation. Based on this, the TextSnake [7] algorithm handles irregular text regions more flexibly by predicting the position of the centerline of the text, the radius of the central disc, and the angle between the centers of adjacent discs, thus avoiding the dependence of traditional prediction frames (e.g., quadrangles and rotated rectangles) on the shape and length of the text. Diverging from traditional methods using bbox instance segmentation, BoundarySpotter [8] employs multi-scale boundary point localization and adaptive topology reconstruction for geometric/semantic modeling of irregular text, proving boundary representations enhance accuracy in complex scenes. R50_DBU [9] introduces an enhanced differential binarization network, integrating U-Net's feature architecture, to address challenges in automated text detection such as inadequate recognition of special characters and incomplete detection of dense and lengthy text segments. TextFuse [10] proposes a dual-stage fusion framework addressing single-feature dependency limitations in scene text detection through multi-modal feature aggregation and multi-algorithm consensus integration.

In this paper, an improved EAST network CEE (Components-ECA-EAST Network) that integrates the attention mechanism and the feature pyramid structure is proposed based on the structural analysis of Chinese character mini components. The ECA (Efficient Channel Attention) attention structure is introduced in the feature extraction phase to refine the weight distribution between feature channels and improve the shortcomings of traditional EAST in attention allocation; for the tiny component dataset, the features are first enhanced and refined by convolutional layers, and then cascaded fusion with the feature pyramid is carried out, to better retain the fine-grained features in the multi-scale feature aggregation to better preserve fine-grained information, and finally, Dice Loss is utilized as the loss function for the regression task. The above improvements improve the network's performance in detecting and segmenting the ultra-small components and fine strokes of handwritten Chinese characters.

2. Components-ECA-EAST Network

2.1. CEE Network Structure

In the research of detection and segmentation of handwritten Chinese character images, the EAST model has been widely adopted due to its efficient and accurate text detection capabilities [11]. The EAST model proposes a simplified end-to-end text detection method, which is especially suitable for dealing with scene text detection tasks. Unlike traditional methods, EAST does not rely on Region Proposal Networks (RPNs) or candidate frame generation but avoids complex processing by predicting the bounding boxes and geometric properties of text regions directly.

Specifically, EAST utilizes a Fully Convolutional Neural Network (FCN) for image feature extraction, predicting whether each pixel belongs to a text region while simultaneously inferring morphological features of the text, such as its orientation and size [12]. This method significantly improves the speed of text detection while maintaining high accuracy in diverse background environments.

A major advantage of the EAST model is its good adaptability. Many traditional text detection algorithms need to be specially adapted for different text scales and orientations, whereas EAST can flexibly cope with various types of text through a unified network architecture, especially when facing handwritten Chinese character images, it can effectively deal with a variety of irregular writing and complex backgrounds, and improve the robustness of detection [13]. Therefore, the adoption of EAST in the detection and segmentation of handwritten Chinese character images can significantly improve the accuracy, reduce the occurrence of false and missed detections, and provide more reliable input data for the subsequent text recognition and image segmentation tasks. However, the traditional EAST often suffers from the problems of missed detection and edge blurring when dealing with ultra-small targets and especially performs poorly in the detection of tiny parts in complex word classes [14], moreover, the performance of the EAST model is limited for the detection of dense and long texts. This also becomes an improvement idea for this paper.

The improvements in this paper mainly focus on the feature extraction phase and feature fusion phase of EAST. To address the problem that EAST overly relies on the backbone network in the feature extraction phase, and insufficient processing of dense textual information, which leads to inaccurate feature extraction, information loss, and insufficient feature expression ability in complex backgrounds or fuzzy or overlapping Chinese character scenarios, this paper introduces the ECA attention mechanism in the backbone network to strengthen the capture of key information and the allocation of feature weights among channels [15]. Given the leakage and edge-blurring phenomenon of EAST in detecting small targets, this paper uses the self-built small parts dataset to extract the convolutional features and fuses them with the convolutional layer in the feature extraction stage through the feature pyramid network (FPN) further enhances the detection of small targets and the representation of boundary features. The CEE network structure is illustrated in Figure 1, where f, P, and h represent the feature maps obtained after the convolution operations.

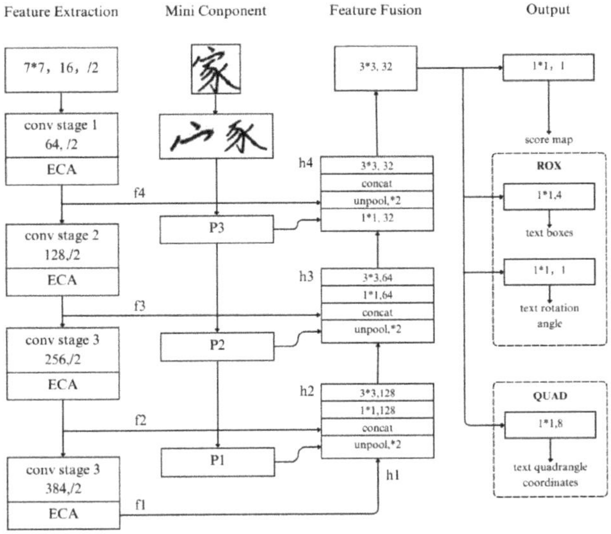

Figure 1. CEE network structure.

2.2. Multi-Scale Feature Fusion

Multi-scale feature fusion is a strategy widely used by deep learning models for image processing tasks, which aims to unite features from different scales to comprehensively capture the detailed information and global structure of the target. The core of this approach is to integrate features extracted from varying receptive fields, enabling the model to focus on both the global structure and local details of the image. In Convolutional Neural Networks (CNNs), the size of the feature map is usually progressively reduced with convolution and pooling operations, so that features at different levels correspond to information at different scales [16]. Lower-level features contain more detailed information (e.g., edges and textures), while higher-level features reflect more abstract information (e.g., shapes and semantics).

The EAST model uses the idea of FPN in multi-scale fusion. Multiscale information representation is achieved by constructing a multiscale feature map that combines shallow and deep features [17]. As a classical implementation of this approach, a Feature Pyramid Network (FPN) fuses different levels of features by up-sampling layer by layer and summing them with the lower-level features, so that the network can extract detailed information as well as capture the global structure, which demonstrates excellent performance in classification and detection tasks.

C denotes the layer C feature map of the backbone network and P denotes the layer C feature map after fusion through FPN, the feature fusion process of FPN can be represented as:

$$P_i = UpSample(P_{i+1}) + C_i, i \in \{2, 3, 4, \ldots\} \quad (1)$$

where P_i a is the fused feature map in layer i of FPN, *Upsample* denotes the up-sampling of the previous feature map (usually using bilinear interpolation or inverse convolution); $i \in \{2, 3, 4, \ldots\}$ denotes the layer index of the feature map.

In this paper, we use the self-built mini-components dataset to perform convolution operation on the Chinese character image to extract deep and shallow features, and fuse these features with the feature maps generated from the EAST feature extraction layer via FPN, and enhance the network's sensitivity to mini-components. The specific structure is shown in Figure 2.

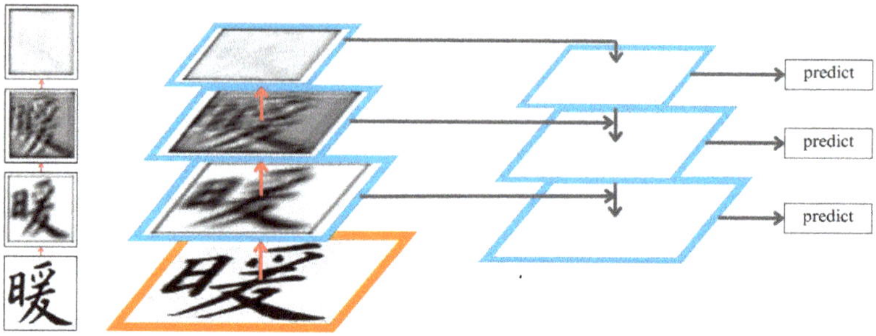

Figure 2. Mini-components feature pyramid fusion.

2.3. ECA Module

Over the past few years, the application of attention mechanisms in deep convolutional neural networks has become more and more widespread, such as BAM [18], CBAM [19], GCNet [20], SENet [21], Self-Attention [22] and other attention mechanisms. The properties of the attention mechanism enable the network to reasonably increase the focus on the key information of handwritten Chinese character images. As the network size becomes larger and larger, more and more feature information is stored in the model, but the deepening of its network layers will lead to information overload and the existence of many useless and redundant feature information. Therefore, the introduction of the attention mechanism module can effectively solve this problem of information redundancy, and improve the attention to other feature information, so that the network is more focused on valuable feature information.

The Efficient Channel Attention [23] module is a lightweight channel attention mechanism that aims to improve the ability of convolutional neural networks to capture inter-channel information, thus enhancing the feature representation performance of the network. The ECA module effectively improves computational efficiency by reducing the number of parameters and avoiding redundant computation, which is an important innovation. Unlike other traditional attention modules, such as the SE module, which requires a fully connected layer to learn the channel weights, the fully connected layer dimensionality reduction causes the channel attention to lose the feature information, and makes the model complexity and computation relatively high, and the efficiency of capturing the dependencies between channels relatively low. The ECA module achieves the same functionality by introducing more simplified methods, which makes it very efficient while maintaining a good feature weighting capability. To address the annual grading demands of over 200,000 handwriting test sheets from fifth- and ninth-grade students in Shanghai, the lightweight design of the ECA module is critical [24]. Compared to other attention mechanism, ECA significantly reduces inference latency and computational overhead, enabling deployment on standard server clusters to meet real-time requirements in high-concurrency exam scenarios.

In this paper, the ECA module is integrated into the EAST feature extraction layer to strengthen the network's ability to model inter-channel correlation to further optimize the performance of handwritten Chinese character detection and segmentation, increase EAST's ability to focus on dense text. By introducing a lightweight attention mechanism, the ECA module can fully capture the inter-channel interaction information without significantly increasing the computational burden, which significantly improves the accuracy of feature extraction. Particularly when processing handwritten Chinese character images with complex backgrounds and significant interference, the module enhances sensitivity to target regions and the ability to distinguish detailed features [25], which effectively alleviates the problems of lost stroke details and the difficulty of restoring local features. This improvement provides higher-quality feature representation for feature fusion and target segmentation and shows excellent robustness and reliability, especially in complex scenes. The specific structure is illustrated in Figure 3.

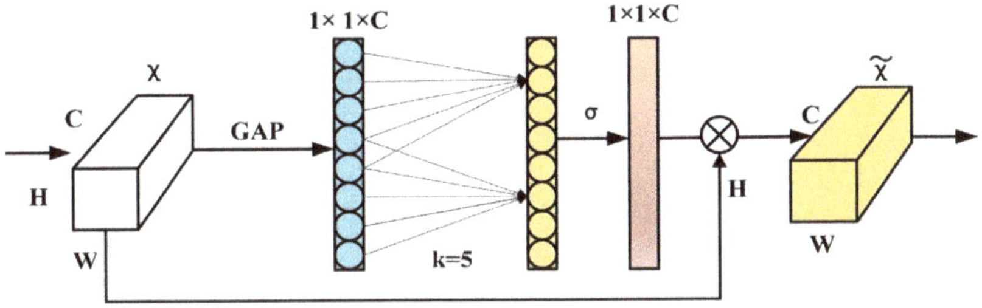

Figure 3. ECA module structure.

2.4. Dice Loss

The EAST model is an efficient algorithm for scene text detection, which mainly predicts the geometry of text regions by regression methods. However, traditional loss functions (e.g., cross-entropy loss [26] and L1/L2 regression loss [27]) have limitations in dealing with complex geometries and small target detection, especially when the categories are unbalanced and the target shapes are diverse. This limitation often leads to miss and false detections of the model in dense text and small target detection scenarios. To address these challenges, this paper employs Dice Loss [28] into the EAST model, replaces the class-balanced cross-entropy that was previously used to count losses in score plots.

Dice Loss was initially utilized in the domain of medical image segmentation to quantify the overlap between predicted and ground truth regions. Its advantage is that it can optimize the model's accurate prediction of the target region, which is especially effective in the case of category imbalance. Introducing Dice Loss into the EAST model can significantly enhance the model's sensitivity to dense text regions and small targets, thus improving the overall detection performance [29]. Dice Loss is based on the Dice Coefficient, which is calculated as:

$$\text{Dice Coefficient} = \frac{2|A \cap B|}{|A| + |B|} \tag{2}$$

where A represents the predicted region and B corresponds to the ground truth region. Converting the Dice Coefficient into a loss function usually takes the following form:

$$\text{Dice Loss} = 1 - \frac{2\sum_{i=1}^{N} p_i g_i + \epsilon}{\sum_{i=1}^{N} p_i + \sum_{i=1}^{N} g_i + \epsilon} \tag{3}$$

where p_i denotes the predicted value, g_i denotes the true value, and ϵ is a small constant to avoid division by zero.

3. Results

3.1. Dataset

The dataset was constructed in the context of the writing grade examination for primary and secondary school students in Shanghai, 240 students who were about to take the writing grade examination were invited and their handwritten images of Chinese characters were collected as the research samples. The dataset covers different writing scenarios: first, formal writing on exam-specific papers, and second, unconstrained daily writing practice. To fully represent the differences between these two writing environments, 480 high-quality images of handwritten Chinese characters were collected, covering more than 40,000 Chinese characters and presenting rich background variations and diverse writing styles. This dataset aims to simulate the actual differences between the test scenarios and daily writing and to better meet the needs of handwritten Chinese character detection in writing grade tests. All samples are dense text images containing a large number of Chinese characters. Figure 4a,b show the partial writing grade exam dataset constructed in this paper, while Figure 5a,b present the corresponding unconstrained text dataset, respectively.

In addition, to improve the capability of the Chinese character detection model in recognizing diverse structural components of Chinese characters, this paper further integrates a set of image datasets designed specifically for the task of Chinese character component segmentation. The dataset covers a variety of different structures of Chinese character components, including left-right structure, top-bottom structure, wraparound structure, etc., to cope with the diversity of Chinese character component morphology as well as the complexity of the background. Figure 6 shows the diagram of some Chinese character components.

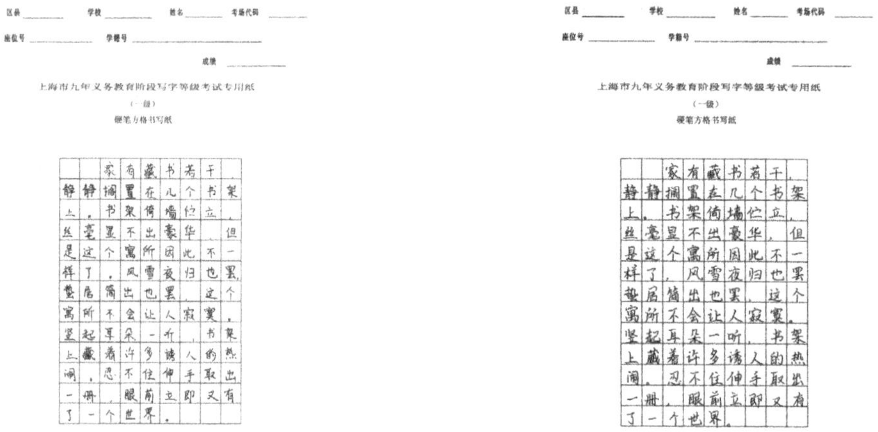

(a) (b)

Figure 4. Partial writing grade exam dataset diagram.

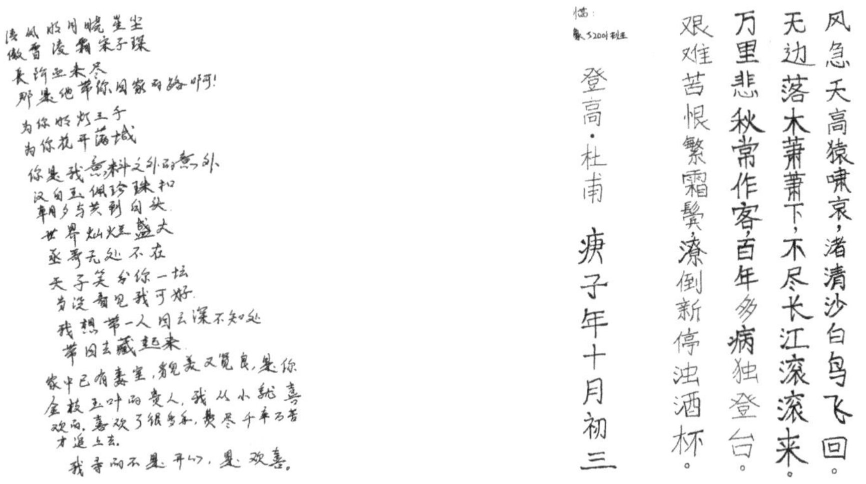

(a) (b)

Figure 5. Partial unconstrained text dataset diagram.

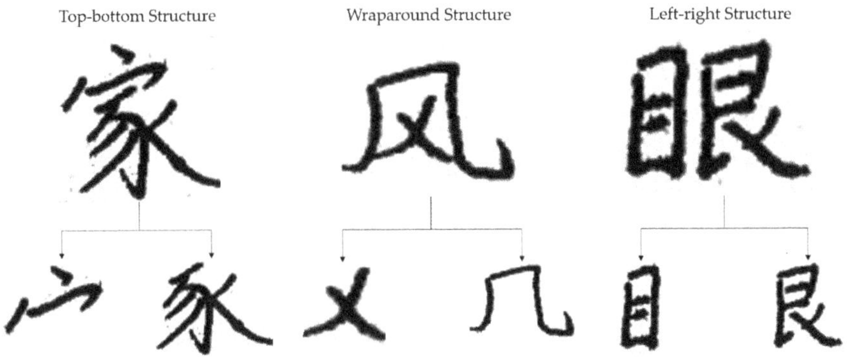

Figure 6. Chinese character component diagram.

Finally, the dataset was expanded to 960 images by data enhancement methods such as randomly rotating different angles to improve the adaptability of our model in different scenes, and all the images were uniformly scaled to a fixed size of 1200 × 1200. Ultimately, the dataset was partitioned into training, validation, and test sets with a ratio of 8:1:1, respectively.

3.2. Experimental Parameters

In this paper, the network model is implemented through PyTorch, the code is written and run on Linux CentOS 7.7 system, and the experimental hardware envi-ronment is NVIDIA A100 Tensor Core GPU 80G, provided by NVIDIA Corporation, located in Santa Clara, CA, USA. Adam (Adaptive Moment Estimation) optimizer is used for training, the learning rate is set to 0.001, the batch size was set to 24.

3.3. Evaluation Indicators

The segmentation model performance was evaluated using three key metrics: precision, recall, and F-score. These metrics quantify segmentation accuracy, detection reliability, and overall robustness, respectively.

1. Precision

Precision serves as one of the fundamental evaluation metrics in object segmentation tasks, quantifying the accuracy of positive class predictions generated by the model. Specifically, Precision indicates how many of the samples predicted by the model as positive class are true targets, and is calculated as follows:

$$\text{Precision} = \frac{\text{TP}}{\text{TP} + \text{FP}} \tag{4}$$

TP (True Positive) represents the number of true target samples correctly identified. FP (False Positive) denotes the number of non-target samples misclassified as targets.

2. Recall

Recall reflects the proportion of correct targets detected by the model overall targets, which is defined as:

$$\text{Recall} = \frac{\text{TP}}{\text{TP} + \text{FN}} \tag{5}$$

TP represents the number of accurately identified targets, while FN corresponds to the quantity of undetected targets.

3. F-score

The harmonic mean of precision and recall, employed to comprehensively evaluate model performance. The calculation formula is:

$$\text{F} = 2 \times \frac{\text{Precision} \times \text{Recall}}{\text{Precision} + \text{Recall}} \tag{6}$$

3.4. Experimental Result

The segmentation performance results of the original model and the CEE model on different Chinese character structures are shown in Table 1, from the left-right structure, the Precision of CEE model reaches 85.7%, which is 7.1 percentage points higher than the 78.6% of the original model, the F-score is improved from 77.8% to 84.8%, and the Recall is improved from 77.1% to 84.2%, which indicates that the performance of CEE model is greatly improved in the regular left-right structure. In the top and bottom structure, the Precision of the CEE model is 84.3%, which is 6.5 percentage points higher than the 77.8% of the original model, the F-score increases from 76.7% to 83.4%, and the Recall increases from 75.4% to 82.5%, which further proves that the CEE model is more accurate in the processing of the rule structure. The advantage of the CEE model is more obvious in the complex enclosing structure, where its Precision increases by 10.2 percentage points from 69.2% to 79.4%, F-score increases from 67.7% to 78.5%, and Recall increases from 66.4% to 77.8%.

Table 1. Performance results of different Chinese character structure segmentation.

Structure	Model	Precision/%	Recall/%	F-Score/%
Top-bottom	Basic	78.6	77.1	77.8
Structure	CEE	85.7	84.2	84.8
Wraparound	Basic	77.8	75.4	76.7
Structure	CEE	84.3	82.5	83.4
Left-right	Basic	69.2	66.4	67.7
Structure	CEE	79.4	77.8	78.5

Taken together, the CEE model performs well on all Chinese character structures, especially in complex tasks such as encircling structures, by introducing a feature pyramid network and convolutional module with a channel attention mechanism, it significantly enhances the segmentation accuracy and robustness in complex contexts, and at the same time, reduces the leakage rate, demonstrating strong adaptive ability and improvement effect.

3.5. Ablation Experiment

To validate the performance improvement of the proposed model modifications on the overall architecture, ablation studies were conducted on our custom dataset. As shown in the Table 2, method ① is the EAST base network (Basic), method ② is the mini-component dataset through the feature pyramid as a multi-scale feature fusion (Mini-Component-Basic), method ③ is the addition of the attention mechanism of the ECA module (ECA-Basic), and method ④ is the CEE model in this paper. A comparative analysis of different approaches is presented in Table 2, demonstrating that the proposed model achieves significant accuracy improvement through the integration of the attention mechanism and multi-scale fusion strategy.

Table 2. Ablation experiments results.

Method	Precision	Mini-Component mAP	Recall	F-Score
①	77.2%	73.7%	75.4%	76.3%
②	82.2%	75.8%	80.6%	77.7%
③	83.5%	76.2%	81.2%	79.6%
④	84.6%	80.7%	82.8%	83.6%

According to Figure 7, the part feature fusion and ECA module each play a unique role in model performance enhancement, demonstrating significant improvements in overall performance and small target detection. In terms of Precision, the performance of the base model is 77.2%, which improves to 82.2% after fusing part features, an increase of 5.0%. This result indicates that fusing part features improves the recognition performance of complex Chinese character part shapes by enhancing the representation of local features. After adding the ECA module, Precision improves to 83.5%, indicating that the dynamic weighting mechanism can effectively highlight important part features and suppress the interference of background noise. When feature fusion is combined with the ECA module, the Precision of the final model reaches 84.6%, with a total improvement of 7.4%, verifying the significant contribution of the synergistic optimization effect of the two modules to the overall detection performance.

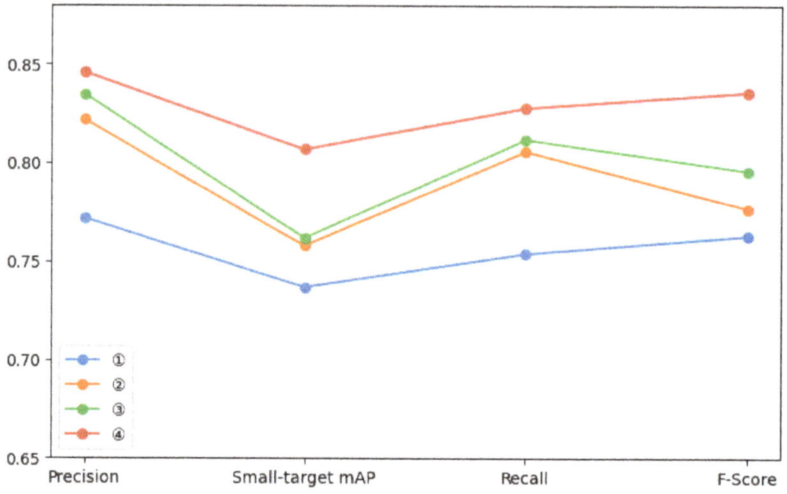

Figure 7. Ablation experiments comparison.

In terms of small target detection, the advantages of the improved model are even more obvious. The small target mAP of the baseline model is only 69.2%, and the introduction of part features improves it to 74.3%, an increase of 5.1%. This reflects the fact that feature fusion improves the model's fitness on complex parts and narrow regions by enhancing the saliency of small target parts. The introduction of the ECA module improves the small target mAP to 73.6% and reaches 76.2% in the full model, for a total improvement of 8.0%. This reflects that the ECA module significantly improves the localization accuracy of the segmentation boundary by dynamically selecting key features, adapting to the complex and diverse characteristics of the boundaries of Chinese character components, and reflecting the fundamental role of the convolution module in constructing the boundary space representation.

The comparative analysis in Figure 8a,b (unconstrained detection) reveals the CEE model's enhanced capability in detecting suboptimal-scale Chinese characters, particularly evident in the complete segmentation of the character "星" (star) (Row 1), where the basic model achieved only partial recognition. This validates the mini-component fusion strategy's effectiveness in resolving scale-dependent ambiguities. However, residual limitations emerge in processing characters with pronounced curvature, attributable to insufficient robustness against nonlinear morphological variations.

Complementarily, Figure 9a,b (standardized examination sheet detection) demonstrates the CEE model's superiority in structured grid environments while exposing paradoxical failures: systematic misclassifications persist for geometrically simple characters such as "了" (completed) and "入" (enter). This phenomenon fundamentally stems from a mechanistic contradiction between the attention module's intrinsic bias toward high stroke-density patterns and the morphological feature extraction requirements for characters with minimal topological complexity.

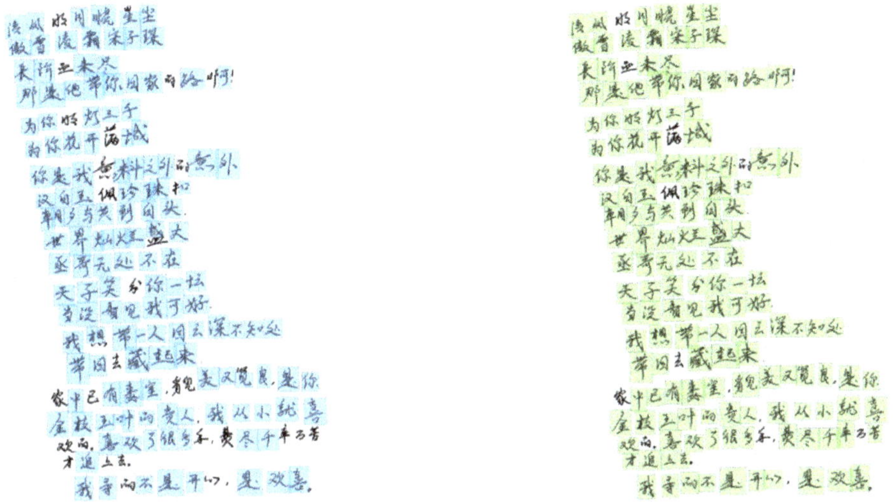

(**a**) (**b**)

Figure 8. Comparison of unconstrained dataset experimental results.

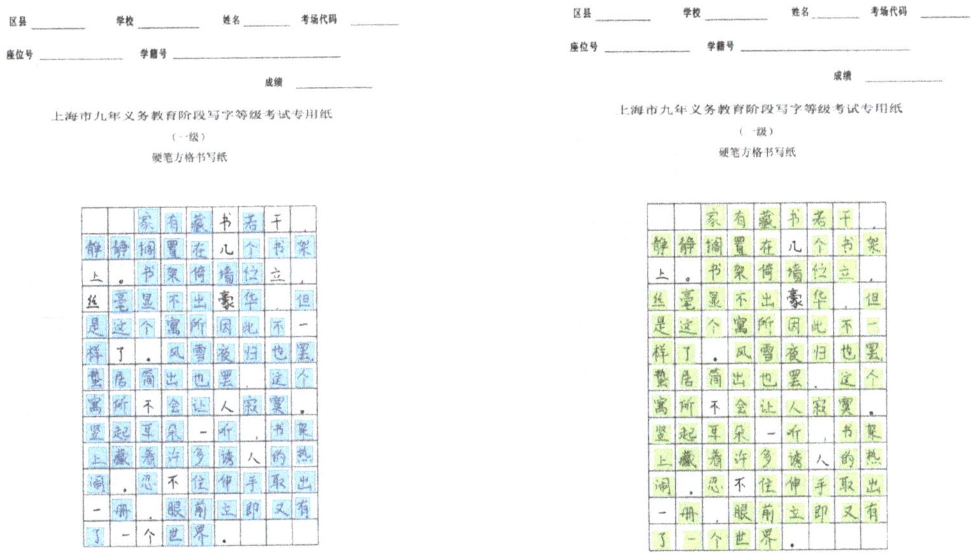

(**a**) (**b**)

Figure 9. Comparison of writing exam dataset experimental results.

The comparison of the experimental results, in general, fully verifies the significant performance improvement of the mini component feature fusion and ECA module in the task of Chinese character component segmentation, and the synergistic design of the two modules overcomes the limitations of the traditional model, which provides a reliable technical support and theoretical basis for the task of segmentation of complex targets.

3.6. Method Comparisons

To further validate the performance of the handwriting detection model algorithm proposed in this paper, experimental comparisons with other mainstream text detection algorithms, CTPN [5], PixelLink [6], TextSnake [7], Boundary [8], R50_DBU [9] and TextFuse [10] were conducted on the ICDAR2015 dataset. The ICDAR [30] dataset is used for the ICDAR Challenge, in which the ICDAR2015 dataset contains 1500 images, 1000 images are used for the training set and the remaining 500 images are used for the test set.

CTPN is a single-stage model for text detection, which has significant advantages in dealing with oblique and curved text. PixelLink is a pixel-level-based text detection model, particularly suitable for scene text detection and text regions with great morphological variations. TextSnake is particularly good at dealing with curved and bent text. This study focuses on the segmentation task of Chinese character detection, which usually involves multi-scale target detection, complex background processing, and accurate localization of small targets, so these three models that are good at handling complex text are targeted to be chosen as comparison objects to evaluate the performance of CEE networks in Chinese character detection. The results are shown in Table 3. Boundary is An end-to-end framework utilizing multi-scale boundary point localization and adaptive topology reconstruction to boost accuracy in detecting/recognizing irregularly shaped text within complex scenes, R50_DBU represents a sophisticated segmentation-driven network incorporating U-Net architecture, TextFuse integrates hierarchical feature synthesis with cross-detector consensus mechanisms to resolve single-modality constraints in scene text analysis.

Table 3. Algorithm comparison.

Model	Precision/%	Recall/%	F-Score/%
CTPN	74.2	51.6	60.9
PixelLink	83.0	73.2	77.8
TextSnake	83.2	73.9	78.3
EAST	80.5	72.8	76.4
Boundary	82.2	81.1	81.3
R50_DBU	85.0	77.0	Null
TextFuse	83.6	77.1	80.2
CEE	85.1	82.7	83.9

The EAST model using the combination of mini-component in this paper has higher accuracy, recall, and F-score than the classical EAST algorithm. The values are also improved by one to two percentage points in comparison to other algorithms, and the detection results are improved. The algorithm comparison is shown in Figure 10.

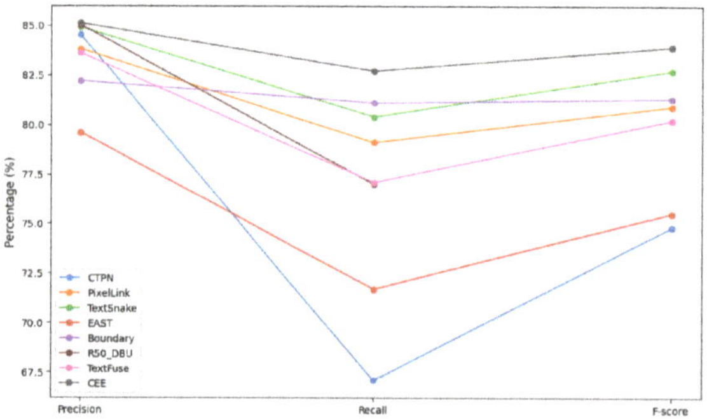

Figure 10. Algorithm comparison.

4. Discussion

This study proposes an enhanced EAST-based model, termed the CEE model, for text detection in handwriting proficiency assessment. The incorporation of a custom mini-component dataset and the integration of a Feature Pyramid Network (FPN) in the feature fusion stage significantly improve the model's capability in dense text feature extraction and multi-scale feature representation. Meanwhile, the ECA attention module is added to the model, which further strengthens the extraction capability of key features and significantly improves the performance of the EAST algorithm in detection accuracy, especially for the optimization of the small target leakage detection problem. In addition, the CEE model improves the loss function and optimizes the training process of the model.

The experimental results show that the CEE model performs well on both the self-constructed dataset and the ICDAR2015 dataset, and compared with the original EAST algorithm, its accuracy and recall are both improved by about 7.4%, the F-score is improved by about 7.3%, and the mAP of small targets is improved by about 7.0%. While large language models (LLMs) have demonstrated remarkable capabilities in semantic understanding, their ability to model fine-grained visual structures (e.g., stroke-level components in handwritten characters) remains underexplored [31]. In this study, we address these vision-structure alignment challenges through localized attention refinement and multi-scale feature preservation strategies, thereby offering new insights into bridging the gap between high-level semantics and low-level visual patterns. Subsequent investigations will prioritize two key methodological refinements: (1) algorithmic optimization through architectural simplification to streamline computational demands, and (2) systematic expansion of the assessment corpus by diversifying text genres and incorporating multi-level writing samples. These strategic enhancements aim to establish an optimized framework that demonstrates enhanced generalizability and operational efficiency in automated writing evaluation systems, particularly for high-stakes standardized assessments.

Author Contributions: Conceptualization, Y.S.; methodology, R.L.; software, R.L.; validation, R.L.; formal analysis, R.L.; investigation, R.L and Y.S.; resources, Y.S.; data curation, R.L.; writing—original draft preparation, R.L.; writing—review and editing, R.L. and Y.S.; visualization, R.L.; supervision, Y.S., X.L. and X.T.; project administration, Y.S., X.L. and X.T.; funding acquisition, Y.S., X.L. and X.T. All authors have read and agreed to the published version of the manuscript.

Funding: This research was funded by University-Industry Co-operation Project, Intelligent Sensing and Real-time Monitoring System for Urban Spatial Light Environment, Shanghai, China, grant number (24)DZ-007.

Institutional Review Board Statement: Not applicable.

Informed Consent Statement: Not applicable.

Data Availability Statement: The data presented in this study are available on request from the corresponding author due to the use of external datasets as the input data for creating an extended dataset.

Conflicts of Interest: The authors declare no conflict of interest.

References

1. Wang, K.; Babenko, B.; Belongie, S. End-to-end scene text recognition. In Proceedings of the 2011 International Conference on Computer Vision, Barcelona, Spain, 6–13 November 2011; pp. 1457–1464.
2. Tian, Z.; Huang, W.; He, T.; He, P.; Qiao, Y. Detecting text in natural image with connectionist text proposal network. In Proceedings of the Computer Vision–ECCV 2016: 14th European Conference, Amsterdam, The Netherlands, 11–14 October 2016; Proceedings, Part VIII 14, pp. 56–72.
3. Zhou, X.; Yao, C.; Wen, H.; Wang, Y.; Zhou, S.; He, W.; Liang, J. East: An efficient and accurate scene text detector. In Proceedings of the IEEE Conference on Computer Vision and Pattern Recognition, Honolulu, HI, USA, 21–26 July 2017; pp. 5551–5560.

4. Ronneberger, O.; Fischer, P.; Brox, T. U-net: Convolutional networks for biomedical image segmentation. In Proceedings of the Medical Image Computing and Computer-Assisted Intervention–MICCAI 2015: 18th International Conference, Munich, Germany, 5–9 October 2015; Proceedings, Part III 18, pp. 234–241.
5. Shi, B.; Bai, X.; Belongie, S. Detecting oriented text in natural images by linking segments. In Proceedings of the IEEE Conference on Computer Vision and Pattern Recognition, Honolulu, HI, USA, 21–26 July 2017; pp. 2550–2558.
6. Deng, D.; Liu, H.; Li, X.; Cai, D. Pixellink: Detecting scene text via instance segmentation. In Proceedings of the AAAI Conference on Artificial Intelligence, New Orleans, LA, USA, 2–7 February 2018.
7. Long, S.; Ruan, J.; Zhang, W.; He, X.; Wu, W.; Yao, C. Textsnake: A flexible representation for detecting text of arbitrary shapes. In Proceedings of the European Conference on Computer Vision (ECCV), Munich, Germany, 8–14 September 2018; pp. 20–36.
8. Wang, H.; Lu, P.; Zhang, H.; Yang, M.; Bai, X.; Xu, Y.; He, M.; Wang, Y.; Liu, W. All you need is boundary: Toward arbitrary-shaped text spotting. In Proceedings of the AAAI Conference on Artificial Intelligence, New York, NY, USA, 7–12 February 2020; pp. 12160–12167.
9. Cheng, Y.; Wan, Y.R.; Sima, Y.J.; Zhang, Y.M.; Hu, S.Y.; Wu, S. Text Detection of Transformer Based on Deep Learning Algorithm. *Teh. Vjesn.* **2022**, *29*, 861–866. [CrossRef]
10. Shi, X.J.; Peng, G.W.; Shen, X.J.; Zhang, C.S. TextFuse: Fusing Deep Scene Text Detection Models for Enhanced Performance. *Multimed. Tools Appl.* **2024**, *83*, 22433–22454. [CrossRef]
11. Liu, M.; Li, B.; Zhang, W. Research on Small Acceptance Domain Text Detection Algorithm Based on Attention Mechanism and Hybrid Feature Pyramid. *Electronics* **2022**, *11*, 3559. [CrossRef]
12. Yuan, L.; Zeng, C.; Pan, P. Research of Chinese Entity Recognition Model Based on Multi-Feature Semantic Enhancement. *Electronics* **2024**, *13*, 4895. [CrossRef]
13. Wawer, A.; Mykowiecka, A.; Żuk, B. Detecting Aggression in Language: From Diverse Data to Robust Classifiers. *Electronics* **2024**, *13*, 4857. [CrossRef]
14. Xue, Y.; Wang, Q.; Hu, Y.; Qian, Y.; Cheng, L.; Wang, H. FL-YOLOv8: Lightweight Object Detector Based on Feature Fusion. *Electronics* **2024**, *13*, 4653. [CrossRef]
15. Ren, Q.-D.-E.-J.; Wang, L.; Ma, Z.; Barintag, S. Offline Mongolian Handwriting Recognition Based on Data Augmentation and Improved ECA-Net. *Electronics* **2024**, *13*, 835. [CrossRef]
16. Qian, K.; Ding, X.; Jiang, X.; Ji, Y.; Dong, L. CFF-Net: Cross-Hierarchy Feature Fusion Network Based on Composite Dual-Channel Encoder for Surface Defect Segmentation. *Electronics* **2024**, *13*, 4714. [CrossRef]
17. Lin, T.-Y.; Dollár, P.; Girshick, R.; He, K.; Hariharan, B.; Belongie, S. Feature pyramid networks for object detection. In Proceedings of the IEEE Conference on Computer Vision and Pattern Recognition, Honolulu, HI, USA, 21–26 July 2017; pp. 2117–2125.
18. Park, J. Bam: Bottleneck attention module. *arXiv* **2018**, arXiv:1807.06514.
19. Woo, S.; Park, J.; Lee, J.-Y.; Kweon, I.S. Cbam: Convolutional block attention module. In Proceedings of the European Conference on Computer Vision (ECCV), Munich, Germany, 8–14 September 2018; pp. 3–19.
20. Cao, Y.; Xu, J.; Lin, S.; Wei, F.; Hu, H. Gcnet: Non-local networks meet squeeze-excitation networks and beyond. In Proceedings of the IEEE/CVF International Conference on Computer Vision Workshops, Seoul, Republic of Korea, 27–28 October 2019.
21. Hu, J.; Shen, L.; Sun, G. Squeeze-and-excitation networks. In Proceedings of the IEEE Conference on Computer Vision and Pattern Recognition, Salt Lake City, UT, USA, 18–23 June 2018; pp. 7132–7141.
22. Zhang, H.; Goodfellow, I.; Metaxas, D.; Odena, A. Self-attention generative adversarial networks. In Proceedings of the International Conference on Machine Learning, Long Beach, CA, USA, 9–15 June 2019; pp. 7354–7363.
23. Wang, Q.; Wu, B.; Zhu, P.; Li, P.; Zuo, W.; Hu, Q. ECA-Net: Efficient channel attention for deep convolutional neural networks. In Proceedings of the IEEE/CVF Conference on Computer Vision and Pattern Recognition, Seattle, WA, USA, 13–19 June 2020; pp. 11534–11542.
24. Liu, R.; Shi, Y.; Tang, X.; Liu, X. Combining Multi-Scale Fusion and Attentional Mechanisms for Assessing Writing Accuracy. *Appl. Sci.* **2025**, *15*, 1204. [CrossRef]
25. Molnar, C.; König, G.; Bischl, B.; Casalicchio, G. Model-agnostic feature importance and effects with dependent features: A conditional subgroup approach. *Data Min. Knowl. Discov.* **2024**, *38*, 2903–2941. [CrossRef]
26. De Boer, P.-T.; Kroese, D.P.; Mannor, S.; Rubinstein, R.Y. A tutorial on the cross-entropy method. *Ann. Oper. Res.* **2005**, *134*, 19–67. [CrossRef]
27. Ranjbarzadeh, R.; Jafarzadeh Ghoushchi, S.; Anari, S.; Safavi, S.; Tataei Sarshar, N.; Babaee Tirkolaee, E.; Bendechache, M. A deep learning approach for robust, multi-oriented, and curved text detection. *Cogn. Comput.* **2024**, *16*, 1979–1991. [CrossRef]
28. Zhao, R.; Qian, B.; Zhang, X.; Li, Y.; Wei, R.; Liu, Y.; Pan, Y. Rethinking dice loss for medical image segmentation. In Proceedings of the 2020 IEEE International Conference on Data Mining (ICDM), Sorrento, Italy, 17–20 November 2020; pp. 851–860.
29. Schönfelder, P.; Stebel, F.; Andreou, N.; König, M. Deep learning-based text detection and recognition on architectural floor plans. *Autom. Constr.* **2024**, *157*, 105156. [CrossRef]

30. Peyrard, C.; Baccouche, M.; Mamalet, F.; Garcia, C. ICDAR2015 Competition on Text Image Super-Resolution. In Proceedings of the International Conference on Document Analysis & Recognition, Tunis, Tunisia, 23–26 August 2015.
31. LeCun, Y. A path towards autonomous machine intelligence version 0.9.2, 2022-06-27. *Open Rev.* **2022**, *62*, 1–62.

applied sciences

MDPI

Article

ChatGPT in Computer Science Education: A Case Study on a Database Administration Course

Daniel López-Fernández [1,*,†] and Ricardo Vergaz [2,†]

[1] Departamento Sistemas Informáticos, Universidad Politécnica de Madrid, 28031 Madrid, Spain
[2] Departamento de Tecnología Electrónica, Universidad Carlos III de Madrid, 28911 Madrid, Spain; rvergaz@ing.uc3m.es
* Correspondence: daniel.lopez@upm.es
† These authors contributed equally to this work.

Abstract: GenAI tools like ChatGPT have changed the educational landscape, and empirical experiences are needed to better understand how to use them to their fullest potential. This article empirically explores the usage of ChatGPT 3.5 in database administration education through a case study conducted with 40 computer science students. Specifically, it inspects how widespread the use of ChatGPT is and students' perceptions of this tool, how prior knowledge on a topic affects the use of ChatGPT, and the relationship between the usage of ChatGPT and success in solving practical problems. The student's grades in a computer practical exam, a set of theoretical tests to assess progression in knowledge acquisition, and a comprehensive questionnaire are employed as research instruments. The obtained results indicate that students use ChatGPT moderately but more frequently than traditional internet learning resources such as official documentation, Stack Overflow or googling. However, the usage is uneven among students, and those who end up getting better grades use ChatGPT more. Beyond prompting skills, one of the elements that is key to the students' productive use of this tool is their prior knowledge about database administration. This article concludes that ChatGPT is an excellent educational instrument in the context of database administration and that in order to use it properly, it is necessary for students to have good prompting skills as well as a sound theoretical basis. Training students in the use of GenAI tools like ChatGPT, for example, with a guided practice strategy where prompting and conducted step-by-step practice are employed is key to prevent the appearance of new digital trenches.

Keywords: educational technologies; ChatGPT; GenAI; LLM; computer science education; database administration

1. Introduction

During the last two years, the huge breakthrough of Generative Pretrained Transformers (GPT) on Artificial Intelligence (AI) led to a boost and democratization of the use of Generative AI (GenAI) solutions, in the form of chatbots. Several companies have launched their solutions to the public in order to train, test, and acquire customers for their non-free versions, with Open AI and its ChatGPT, or Google and its Gemini being the most prevalent items. Students around the world are changing their habits, and now these tools are their favorite way to search for information [1]. As a reference, it could be stated that 27% of the students were using AI writing tools in spring 2023, and by that year's fall season, the percentage had boosted to 49% [2].

Although this is a powerful new tool that can be used for student learning, concerns arise about the possible misuse of received information, such as cheating [3] or taking the received answer without further mental processing [4]. Some actions that can prevent it are being taken into account, such as creating new approaches to academic tasks and assessment [5]. Other efforts are being made to verify the validity of these tools for serving as study partners, based on their performance on exams or practical assignments [6]. Although the role of Generative AI in learning is a controversial topic, due to the risk of undermining student learning skills, there are studies that turn this last point of view upside down, even in memorization skills [7].

Facing this new threat, for some people, opportunity for others, several international higher-education institutions have raised recommendations, following the first studies on the impact, possibilities, and dangers of GenAI use in this environment. One of the most used is the work of Gimpel et al. at the University of Hohenheim [8], where they stressed the importance for students to know and respect regulations, focus on learning objectives rather than on the tool, and appealed to their use of critical thinking whenever they interact with these GenAIs. Meanwhile, for teachers, it is recommended to provide rules from the beginning of the season and to use GenAI to provide personalized training experiences to students, always oriented to improve their critical thinking. Another interesting study is [9], where the authors recommended up to seven possible roles for GenAI when it comes to higher-education classrooms: as a mentor to provide feedback, as a tutor to provide direct instruction, as a coach to give deep reflection on the study and habits, as a team member in teamwork tasks to provide new approaches to the group, as a simulated student to force the real student to explain any item to it and by that means reinforce their learning, as a simulator to act as a practical trainer, and as a tool to improve students (and professor) efficiency. Although there is a full set of example prompts to implement them all, each one of the roles has a counterpart: the narrowing of critical thinking, biases, mismatch with the student, conflicts, hallucinations (unreal answers), low fidelity, or writing-without-thinking danger, respectively.

From then on, the aforementioned higher-education institutions have published recommendations following these lines in several ways. All of them stress the importance of changing some traditional assessments, such as typical written assignments, where GenAI is definitely going to excel, turning them into challenging situations for students in order to appeal their critical thinking [10]. In fact, tracking this student's thinking process could become a powerful new way of assessment, even with the help of GenAI [11]. Policies for the use of GenAI at all levels are recommended in documents such as [12] in the United States, where a human-in-the-loop strategy is stressed. This is increasingly common in the latest publications and is even regulated by the European Union as mandatory in any kind of assessment that may affect future employability, taking this issue as a red line or reaching a "high-risk" level in the use cases of AI [13]. Other recommendations in [12] are directed towards GenAI developers, to align them with educational purposes from scratch, to implement information and formation policies for faculty members who develop specific guidelines and guardrails, and to enhance trust and safety in the use of GenAI. The responsibility of educational personnel to keep track of these tools and learn their use is depicted in [14], added to department or institution collaborative opportunities to apply them, and promoting cooperation in actions between institutions. All of these works show the importance of counting on students voices and perspectives in every process, as well as reviewing academic programs. Many of the current academic programs will be impacted by the use of these tools since professional activities will include the use of GenAI once students leave universities to enter the labor market [15]. Indeed, this latter study shows not only the risks about the threat that GenAI entails to current jobs that may disappear, or at least be deeply transformed but also the opportunities that GenAI open to more creative, empowering and human-enhancing skilled jobs. Another very interesting recent study shows that students claim that a higher-education transformation is needed for future-ready employment in a GenAI-powered society, with the need for new learning skills and more interdisciplinary and hands-on topics [16]. The trends that students point out are these competences for future workforce, AI literacy and new literacies to come, interdisciplinary teaching, innovative pedagogies, and new assessment formats.

Another major concern is the ethical use of GenAI, including copyright issues, responsible and trustworthy use, privacy and data governance (and not only where the uploaded data end but with what use: further training, publication, information collection, etc.), transparency, accountability, societal well-being, and knowledge of carbon footprint impact of the tools; all this is studied and recommended in the European Union in [17]. There is a real possibility of reopening or deepening the digital gap, which is a priority to ensure equitable access to AI tools on campuses [18]. Even more, these tools can be taken as an opportunity to increase inclusion in higher-education areas by creating specific adapted content or providing GPT assistants for students with disabilities, or removing biases affecting minorities, races, gender, religion, etc. [19].

In addition to these general published recommendations, there has been an increasing number of published papers in 2024 providing insights on implications of Generative AI for education and research, also covering bias and inclusion, ethics and regulation, offering usage approaches as a support system or intelligent tutoring one, or in assessment [20]. For example, ref. [21] is a deep reflection on how universities must creatively revisit their teaching by developing AI-powered learning assistants or adaptive learning systems, and even faculty assistants. It is stated that learning improvements may be added to alleviate the pressure on stressed staff.

More specifically for the present work, when it comes to computer science, these tools are likely going to impact jobs in this area: it is a real concern because not only can GenAI perform programming tasks in a very good way but also many job sectors are increasingly using it [22]. Indeed, AI is reshaping the syllabus of machine learning or data science. Therefore, both the syllabus and the way of teaching computer science are affected and should tend to adapt [23]. The challenge for faculty members is also to stay updated on rapidly advancing technologies, maybe even more critically in this area than in other ones. A balanced approach is needed, integrating AI to enrich learning while carefully managing its implications for educational practices and student development. Thus, misuse can lead to the cheating or narrowing of programming skills of students, but good use can really improve their critical thinking and high-level computer science skills [24].

Recent evidence shows the ability of these tools to solve computer science problems at the higher-education level, as well as the aim of students to adapt them to their study, such as the one performed in the Netherlands in [25]. A good attempt to incorporate GenAI into a renowned online course at Harvard University—the CS50 course—can be found in [26]. Interestingly, they show how the integration of AI can enhance the learning experience, providing students with continuous and personalized support, leaving educators more time to address higher-level pedagogical issues. Additionally, AI tools demonstrate their usefulness in explaining code snippets or improving code style, as well as answering any kind of administrative or curricular queries.

On the other hand, GenAI is a tool that can promote student programming self-efficacy and motivation, provided that teachers take specific actions to teach students prompt engineering skills [27]. It has been shown that students not only need this guide but also demand it when facing the study with the help of these tools [28], showing it to be more necessary, as the programming task to be addressed is more complex and higher level [29].

Diving deeper into the inherent GenAI skills, in [23], there is a description of an experiment using GenAI to solve a computer science complex problem, showing that it is not a good approach, and that it is rather better to divide it into small problems, a skill that students should gain, together with designing programs rather than just programming. At the same time, it shows a reflection about how coding skills can be quickly overcome in the teaching process to focus on understanding processes and the topic foundations. These two conclusions can be reinforced in recent works: Ref. [22] looked for the limitations of ChatGPT in solving software development problems, and it was shown that it is good only at the easy- and medium-level ones. As a matter of fact, a survey shows that programmers are using AI tools to create better-performing, shorter, and more readable code (i.e., to become more productive, not to solve big tasks). On the other hand, Ref. [30] drew similar conclusions about the positive role of GenAI in clearing the learning toward the fundamentals, as well as in solving the calculator dilemma. Relevant for the work presented in this paper, it is applied to the topic of database systems.

Regarding GenAI solutions for creating teaching aids, a very recent paper [31] explored the possibility of using a RAG (Retrieved Augmented Generation) chatbot in a computer science classroom. A total of 46 students participated in this study. The results showed that the chatbot performed with a high accuracy rate (88% of the answers were considered right by teaching assistants based on the course materials), and the students felt more engaged with the more personalized training. However, they also claimed an additional need for human interaction with teachers, who, on their side, certainly reduced their administrative burden. Notwithstanding, once again, critical thinking and the teacher's guide are mandatory, as the detection of inaccurate statements by a bot-generated tool is not efficient enough. This study not only evaluates the precision of bot responses, but also analyzes student acceptance, congruence with the course material, the comprehensiveness of the support offered, and associated costs, filling some of the gaps identified in the literature on LLM-driven educational chatbots.

Other higher-education institutions are developing some solutions for AI-assisted learning [32]. The European School of Management and Business (EUDE) developed EDU, a virtual co-tutor powered by Generative AI, using IBM's suite of AI solutions including Watson Assistant and Watson Discovery. This system supports administrative, academic, and logistic queries in real-time, which could be applied to database administration courses. Loyola University Chicago implemented LUie, an AI-powered digital assistant built on the Oracle Digital Assistant platform. Although not specifically for database administration, this system demonstrates how AI can be integrated into university systems to provide continuous student support and personalized information.

Focusing on data management, it has been reported that the GenAI scope can fill the gap when automation fails to understand the semantics of the data. Thus, difficult database problems can take a new approach, boosting data discovery, query synthesis, schema matching, and the like [33]. By their nature, these GenAI tools are predictive models; therefore, they can perform well when it comes to combining them with information retrieval. MIT researchers found that SQL does not allow one to incorporate probabilistic AI models, and, at the same time, solutions with probabilistic models making inferences do not support complex database queries. To fill this gap, they created GenSQL [34], a Generative AI system for databases that allows users to perform complex statistical analyses on tabular data without extensive technical knowledge. This tool could be particularly useful in teaching database administration by allowing students to make predictions, detect anomalies, guess missing values, fix errors, or generate synthetic data.

Ref. [35] explored the use of ChatGPT as an assistant to automatically correct SQL exercises in a university database course. ChatGPT effectively identifies syntactic and semantic errors, provides explanations, and assigns grades with a strong correlation to human teacher ratings, although, interestingly, it tends to be stricter. Its accuracy decreases with more complex exercises due to subjectivity in error assessment. Although ChatGPT excels at detecting syntax errors (88.9%), it often confuses them with semantic ones, and its suggestions are only valuable in about half of the cases. Comparisons of Few-Shot and Zero-Shot Learning reveal that the latter produces clearer explanations, while the former relies heavily on provided examples. The study concludes that ChatGPT shows potential as a teaching assistant, but further research is needed to refine its application in education.

In ref. [36], a chatbot is used to support students in completing SQL queries in a step-by-step process. Students who used it performed better on the final exam (4 out of 10 passed the exam, a number halved among students who did not use the tool), while teachers received insightful performance metrics on student progress, aiding them in instructional decisions. Anderson et al. research also highlighted the critical role of prompt engineering in obtaining successful results with GenAI for SQL practice [37], highlighting that students benefited significantly from teacher guidance in developing effective prompts.

Together, these studies underline Generative AIs potential as an SQL learning aid, while also highlighting the importance of instructor support for effective prompt formulation and student guide for GenAI usage. They also show the need for further research on the utility of these tools in database administration courses, as there are still some open questions about (i) the combined use of traditional and new tools, (ii) the ability of these tools to solve complex problems at the university level, and (iii) the possible improvement in learning database administration using them.

In consideration of the aforementioned findings and previous efforts, we have undertaken an empirical study which addresses the following research questions (RQs):

- RQ1. How widespread is the use of ChatGPT among students to learn database administration, and what is the perceived reliability and usefulness of this educational resource compared to other traditional resources?
- RQ2. How does prior knowledge about database administration affect the use of ChatGPT to solve database administration problems?
- RQ3. Is there a relationship between the usage of ChatGPT and the success in solving practical database administration problems?

This study aims to advance the experience of using GenAIs in teaching computer science, and more specifically database administration, following the aim of the recommendations described above. The rest of the article is structured as follows. The materials and methods are described in Section 2. The results are presented and discussed, respectively, in Sections 3 and 4. Lastly, Section 5 presents the conclusions, limitations, and future work drawn from this study.

2. Materials and Methods

2.1. Context

This research presents a case study carried out in a database administration (DBA) course during the academic year 2024–2025. The course is part of the fifth semester of the Bachelor's degree in Technologies for the Information Society delivered by the Faculty of Computer Systems Engineering of the Universidad Politécnica de Madrid (UPM), located in Spain. This mandatory course accounts for 6 European Credit Transfer System credits (ECTS), which is equivalent to 150–180 h of student work. The course deals with traditional database administration topics, such as hardware approaches like RAID to increase database performance and availability, software approaches such as indexes to improve database optimization, or other critical topics such as security, restoration, or monitoring. Although the participating students have previously taken another generic course on databases, where they learned to model a database or perform SQL queries, the topics in the database administration course are completely new to them, so they have no prior knowledge of the subject. During the course, students receive theoretical–practical classes and they also attend hands-on computer classes, in which they receive teacher practical tutorials and work in groups to perform a practical computer assignment. The case study presented here focuses on the utilization and perceived reliability and utility of the resources used by the students to complete the practical computer exam.

2.2. Sample

The sample comprises 40 students enrolled in the DBA course during the academic year 2024–2025, who completed the individual exam. They are 33 males (82.5%) and 7 females (17.5%), and the mean age is 21.78 with a standard deviation of 1.42. Students who do not pass the course the previous year are excluded from this case study to ensure that the students in the sample have no prior knowledge of the subject.

2.3. Procedure and Materials

At the beginning of the course, the students take during an introductory class a knowledge test related to practical work to evaluate their previous knowledge (to view the questions, go to Table A1 in the Appendix A section). Once the theoretical part of the course is taught, the students take the same test during a final review class to evaluate the knowledge they have acquired up to that point in time. The students then start the practical computer assignment in groups of three people. In the assignment, students face the following tasks using MySQL as a database management system: the configuration and creation of a database; the realization of an the Extraction-Transformation-Load process from several MS Excel and XML files; the optimization of queries using indexes; the creation of users and provision of permissions; and backup and recovery operations. The students perform the assignment in six 2-h hands-on computer classes, in which the teacher accompany them through the provision of step-by-step tutorials and the individualized resolution of doubts. To complete the assignment, the students submit a group report in which the procedures used to solve the above-mentioned tasks must be explained. Once the assignment is completed, the students again take the knowledge test for the last time, as well as an individual practical computer exam in a 90 min session. In that exam, the students should solve problems such as the ones they face during practical computer classes, such as loading data into MySQL from an XML file or optimizing queries using indexes. It is also mentionable that during the exam, the students' screen is monitored using the Veyon software 4.9 to ensure that they are using the allowed version of ChatGPT (i.e., 3.5) and not other more advanced paid versions. Lastly, the students are invited to participate in a questionnaire about the usage, reliability, and utility of the resources employed. Before submitting the questionnaire, students give their informed consent to use the collected information for research purposes.

During the assignment, students are free to use any resource they consider to complete the work: notes and tutorials provided by the teacher, explanations of the teacher or colleagues, MySQL official documentation, traditional websites (e.g., Stack Overflow and Google) and GenAI systems (specifically, ChatGPT v.3.5). The students are encouraged to use all possible resources, and in the practical sessions, the teacher exemplifies the utilization of these resources through step-by-step tutorials in which they solve some common problems using these resources. During the practical exam, students are free to use any resource they considered, except for messaging applications that would allow them to communicate synchronously with other people or non-free IAGen tools. Allowed resources include the assignment report, tutorials provided by the teacher, MySQL official documentation, traditional websites, and ChatGPT v.3.5.

2.4. Methods and Instruments

An essential data set collected in the case study is the students' grades obtained in the individual exam, which is scored from 0 to 10. Moreover, additional data come from the results obtained in the tests performed during the learning experience: one test before starting the theoretical part of the course (test 1), another when the theoretical part of the course is finished and the practical part is about to start (test 2), and a last one when the practical part is finished (test 3). All tests contain the same 10 multiple-choice questions (see Table A1 in the Appendix A section), which are designed and validated by the course teacher. These tests do not count for the final grade, thus relaxing possible students under stress response and, consequently, avoiding undesired behavior. The test solution is not revealed until the end of the experience.

In addition, another important data set comprises the results of a questionnaire administered to collect students' opinions about the resources used during the exam realization. The questionnaire comprises two sections. The first section includes questions about age, gender, classroom attendance, and prior experience with ChatGPT. The second section includes statements on the usage, reliability, and utility of the resources available to the students during the exam, as well as a question about the purpose for which they used ChatGPT (if used). Most of the questions, except for the ones about age and gender and the one on ChatGPT uses (in which the students selected options from a closed list), are rated on a Likert scale from one (nothing) to five (a lot). In the statements about the reliability and utility of the resources, students could also select the option 'don't know/no answer'. Questionnaire items are presented together with the results. The validity of the questionnaire content is checked through a review by experts in educational innovation, whereas its internal consistency type reliability is evaluated by means of Cronbach's alpha, in which a value of 0.771 is obtained, indicating that the reliability is acceptable.

2.5. Data Analysis

The data collected are publicly available at the following link: https://doi.org/10.21950/XMROTF, accessed on 28 November 2024. The results of the student's grades and the questionnaire are analyzed using two descriptive statistics: the mean (M) and the standard deviation (SD). In addition, inferential results are computed. To do so, the Kolmogorov–Smirnov normality test is conducted to check the normality of the data, which results in not normally distributed. Therefore, non-parametric statistical methods are used.

First, the Wilcoxon test [38] is used to make pairwise comparisons of the theoretical tests performed throughout the course. That is, to compare the results of test 1 with test 2, test 2 with test 3, and test 1 with test 3. The statistic used to measure the effect size of the differences analyzed is the correlation coefficient (r) [39], whose thresholds are as follows:

$$\{r \in \mathbb{R} \mid -1 < r < 1\} \begin{cases} |r| < 0.1 & \text{negligible effect} \\ 0.1 \le |r| < 0.3 & \text{small effect} \\ 0.3 \le |r| < 0.5 & \text{medium effect} \\ 0.5 \le |r| & \text{large effect} \end{cases} \quad (1)$$

Second, the Kruskal–Wallis test [40] is used to compare and find possible statistically significant differences when organizing students according to their grades. In this case, the

Eta Squared coefficient (η^2) is employed to study the effect size of these differences [39]. The thresholds of this coefficient are as follows:

$$\{\eta^2 \in \mathbb{R} | 0 < \eta^2 < 1\} \begin{cases} \eta^2 < 0.01 & \text{negligible effect} \\ 0.01 \leq \eta^2 < 0.06 & \text{small effect} \\ 0.06 \leq \eta^2 < 0.14 & \text{medium effect} \\ 0.14 \leq \eta^2 & \text{large effect} \end{cases} \quad (2)$$

Lastly, the Spearman correlation test [41] is used to correlate the student's grades with the student's responses to the questionnaire items. Regarding the Spearman correlation coefficient Rho (ρ), a positive value means a positive correlation, and a negative value means a negative correlation. The thresholds for this coefficient are as follows:

$$\{\rho \in \mathbb{R} | -1 < \rho < 1\} \begin{cases} |\rho| < 0.1 & \text{no correlation} \\ 0.1 \leq |\rho| < 0.3 & \text{low correlation} \\ 0.3 \leq |\rho| < 0.5 & \text{medium correlation} \\ 0.5 \leq |\rho| & \text{high correlation} \end{cases} \quad (3)$$

3. Results

3.1. Student's Exam Grades

The mean grade is 4.85 with an SD of 3.22. Of the 40 students, 18 failed the exam (grade less than 5), 7 obtained a pass grade (grade between 5 and 7), and 15 obtained an outstanding grade (grade greater than 7).

3.2. Student's Test Results

Throughout the experience, the students complete a test several times. As indicated above, one test is performed before starting the theoretical part of the course (test 1), another when the theoretical part of the course is finished and the practical part is about to start (test 2), and the last one when the practical part is finished (test 3). Table 1 depicts the results.

Table 1. Student's test results.

	Test 1	Test 2	Test 3
M (SD)	0.60 (0.71)	4.73 (3.01)	8.43 (2.04)

Pairwise differences are compared using the Wilcoxon test to determine whether the differences are statistically significant, and the correlation coefficient r is calculated to observe the effect size of the differences. Table 2 shows these inferential results.

Table 2. Comparison of student's test results.

Comparison	Significance (p-Value)	Effect Size (r)
Test 1 vs. Test 2	<0.001	0.68
Test 2 vs. Test 3	<0.001	0.58
Test 1 vs. Test 3	<0.001	0.93

3.3. Usage and Perceived Reliability and Utility of Resources

The results obtained from the questionnaire show the usage, reliability, and utility of some learning resources available to perform the individual exam. The specific questions are as follows: 'To what extent did you use the following resources?' (usage), 'How reliable did you find the employed resources?' (reliability), and 'How useful did yo find the following resources?' (utility). As mentioned above, the evaluated resources are as follows: assignment report, assignment materials, MySQL official documentation, traditional internet resources, and ChatGPT. Table 3 shows the results.

Table 3. Usage, reliability, and utility of resources during the exam.

	Item	Usage M (SD)	Reliability M (SD)	Utility M (SD)
1	Assignment report	4.45 (0.88)	4.63 (0.59)	4.50 (0.93)
2	Assignment materials (i.e., teacher tutorials)	2.98 (1.37)	4.16 (1.02)	3.36 (1.22)
3	MySQL official documentation	2.20 (1.26)	4.00 (1.20)	3.04 (1.34)
4	Traditional internet resources (Google, Stack Overflow, etc.)	2.43 (1.36)	3.36 (1.19)	3.00 (1.19)
5	ChatGPTv3.5	3.13 (1.40)	3.39 (1.12)	3.55 (1.35)

Moreover, the questionnaire also includes a specific question about the purposes for which the students used ChatGPT during the exam. Figure 1 depicts these results, ordered by the number of occurrences. Note that 8 of 40 students (20%) did not use ChatGPT.

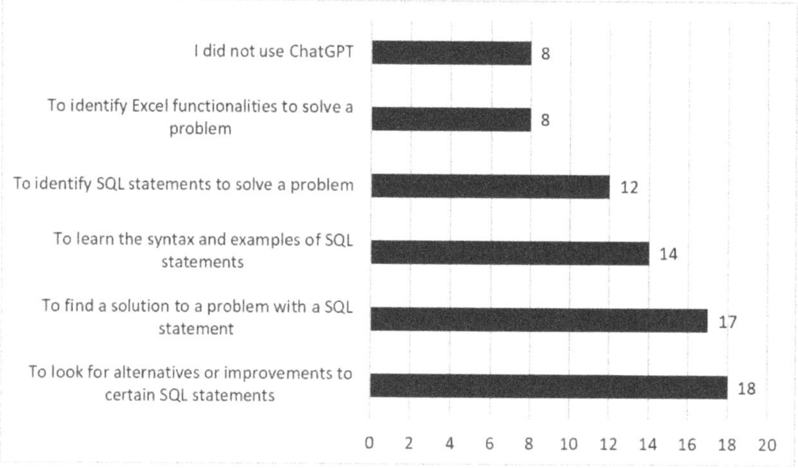

Figure 1. Usage of ChatGPT.

3.4. Student's Test Results Grouped by the Student's Exam Grades

The students can be grouped according to the grade obtained in the practical exam, which is scaled in three categories: fail (grade < 5), pass (5 $\leq$ grade < 7), and outstanding (grade $\geq$ 7). This leads to three groups: Group 1 (grade = fail, N = 18), Group 2 (grade = pass, N = 7), and Group 3 (grade = outstanding, N = 15). Table 4 shows the test results analyzed by groups. In this case, the statistical significance of the differences is analyzed using the Kruskal–Wallis test, and the effect size is analyzed by using the eta squared coefficient (η^2). Statistically significant differences appear in the following table with a '*' symbol.

Table 4. Student's test results (grouped by grades).

Test	Group 1 M (SD)	Group 2 M (SD)	Group 3 M (SD)	p-Value (η^2)
Test 1	0.61 (0.70)	0.57 (0.79)	0.60 (0.74)	0.979 (0.00)
Test 2 *	3.05 (2.26)	5.85 (3.98)	6.20 (2.40)	0.008 (0.259)
Test 3	7.83 (2.48)	8.71 (1.60)	9.00 (1.46)	0.38 (0.073)

3.5. Classroom Attendance, Prior Experience with ChatGPT and Prompting Skills

The questionnaire also includes three questions to ask the students about their classroom attendance, prior use of ChatGPT, and their prompting skills. Table 5 depicts the overall results in this regard (see column total), as well as the results analyzed by the aforementioned groups (see the remaining columns). The differences among the groups are analyzed by using the Kruskal–Wallis test, but no significant differences are found.

Table 5. Classroom attendance, prior experience with ChatGPT, and prompting skills.

Item	Total M (SD)	Group 1 M (SD)	Group 2 M (SD)	Group 3 M (SD)
Classroom attendance	4.30 (0.94)	4.39 (0.98)	4.14 (0.69)	4.27 (1.03)
ChatGPT prior experience	3.70 (0.82)	3.72 (0.83)	3.57 (0.98)	3.73 (0.80)
Prompting skills	3.55 (1.01)	3.56 (1.04)	3.71 (0.95)	3.47 (1.06)

3.6. Usage and Perceived Reliability and Utility of Resources Grouped by Student's Grades

As mentioned before, the students can be grouped by their grades according to the scale previously mentioned (fail, pass and outstanding), leading to three groups. Figure 2 shows the results regarding the item 'I did not use ChatGPT during the exam' analyzed by groups.

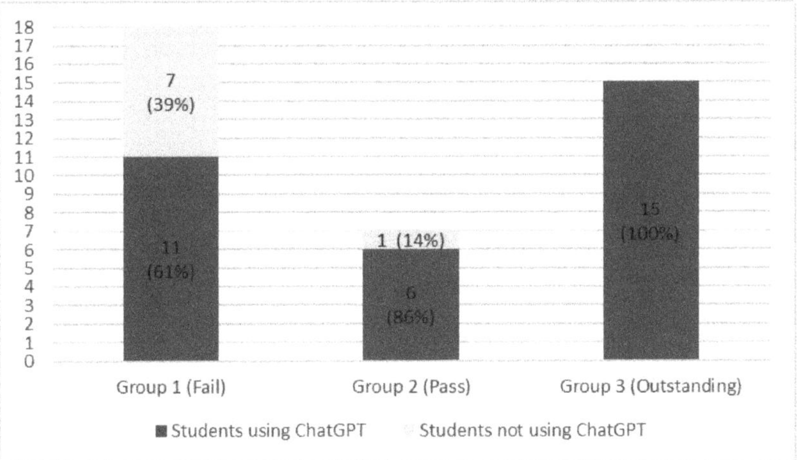

Figure 2. Distribution of usage of ChatGPT grouped by their grades.

Tables 6–8 depict, respectively, the usage, reliability, and utility of the available resources during the exam according to the defined groups. For each group, the mean and standard deviation (in parentheses) are displayed. Moreover, the statistical significance of the differences is analyzed using the Kruskal–Wallis test, and the effect size is analyzed using the eta squared coefficient (η^2). Statistically significant differences appear in the following tables with a '*' symbol.

Table 6. Usage of resources during the exam (grouped by grades).

	Item	Group 1 M (SD)	Group 2 M (SD)	Group 3 M (SD)	p-Value (η^2)
1	Assignment report	4.22 (1.11)	4.29 (0.76)	4.80 (0.41)	0.14 (0.099)
2	Assignment materials (i.e., teacher tutorials) *	3.67 (1.14)	3.67 (1.14)	2.20 (1.21)	0.008 (0.243)
3	MySQL official documentation *	2.61 (1.33)	2.61 (1.33)	1.53 (0.74)	0.049 (0.171)
4	Traditional internet resources (Google, Stack Overflow, etc.) *	2.83 (1.29)	2.71 (1.11)	1.80 (1.37)	0.045 (0.132)
5	ChatGPT v3.5 *	2.44 (1.25)	3.00 (1.29)	4.00 (1.20)	0.007 (0.261)

Table 7. Perceived reliability of resources during the exam (grouped by grades).

	Item	Group 1 M (SD)	Group 2 M (SD)	Group 3 M (SD)	p-Value (η^2)
1	Assignment report	4.56 (0.70)	4.57 (0.53)	4.73 (0.46)	0.74 (0.021)
2	Assignment materials (i.e., teacher tutorials)	4.35 (0.79)	4.40 (0.89)	3.70 (1.34)	0.37 (0.094)
3	MySQL official documentation	4.14 (1.10)	3.50 (1.52)	4.17 (1.17)	0.55 (0.054)
4	Traditional internet resources (Google, Stack Overflow, etc.)	3.50 (0.76)	3.00 (1.41)	3.43 (1.72)	0.70 (0.032)
5	ChatGPT v3.5 *	2.75 (0.75)	3.67 (1.21)	3.80 (1.15)	0.02 (0.198)

Table 8. Perceived utility of resources during the exam (grouped by grades).

	Item	Group 1 M (SD)	Group 2 M (SD)	Group 3 M (SD)	p-Value (η^2)
1	Assignment report	4.28 (1.23)	4.71 (0.29)	4.67 (0.62)	0.76 (0.048)
2	Assignment materials (i.e., teacher tutorials) *	3.76 (0.90)	4.20 (0.84)	2.36 (1.21)	0.006 (0.362)
3	MySQL official documentation	3.29 (1.14)	3.17 (1.33)	2.33 (1.75)	0.46 (0.088)
4	Traditional internet resources (Google, Stack Overflow, etc.)	3.29 (0.99)	2.71 (1.11)	2.71 (1.60)	0.36 (0.060)
5	ChatGPT v3.5 *	2.92 (1.31)	3.17 (1.17)	4.20 (1.21)	0.02 (0.207)

3.7. Relation Between the Student's Grades and the Usage, Reliability, and Utility of Resources

The correlations using the Spearman technique are calculated in order to know how the student's grades are related to the usage and perceived reliability and utility of the available resources during the completion of the exam. Table 9 shows the results of these correlations, which are represented using the Spearman value (Rho, ρ) and the p-value. Statistically significant correlations appear in the following table with a '*' symbol. Please note that if a student does not use a learning resource, the rating of its use would be the lowest (1 on the employed Likert scale), while the usefulness or reliability would be rated NS/NC. In the calculation of the correlations, these data with NS/NC ratings are not considered.

Table 9. Correlation between the student's grade and the usage, reliability and utility of resources during the exam realization.

	Item	Usage ρ (p-Value)	Reliability ρ (p-Value)	Utility ρ (p-Value)
1	Assignment report	0.22 (0.16)	0.07 (0.66)	0.07 (0.66)
2	Assignment materials (i.e., teacher tutorials)	−0.51 (<0.001) *	−0.28 (0.11)	−0.39 (0.02)
3	MySQL official documentation	−0.29 (0.06)	−0.08 (0.65)	−0.24 (0.20)
4	Traditional internet resources (Google, Stack Overflow, etc.)	−0.20 (0.19)	−0.10 (0.61)	−0.09 (0.64)
5	ChatGPT v3.5	0.45 (0.003) *	0.42 (0.01) *	0.48 (0.005) *

4. Discussion

4.1. RQ1. How Widespread Is the Use of ChatGPT Among Students to Learn Database Administration, and What Is the Perceived Reliability and Usefulness of This Educational Resource Compared to Other Traditional Resources?

The results reported for the presented case study (see Table 3) indicate that ChatGPT usage is moderate (3.13 out of 5), but it narrowly outperforms other traditional resources such as MySQL official documentation (2.20), and Stack Overflow posts or googling (2.43) in solving practical database management problems. This result is striking considering that in a similar case study conducted one year ago [42], ChatGPT was less used than the aforementioned resources. This indicates, in line with other previous studies [1,2], that GenAI tools are becoming increasingly widespread in education and the field of computer science education is no exception.

However, the use of ChatGPT is not too extensive compared to the resource most used by the students: the report that they themselves had prepared during the practical sessions (4.45 out of 5). This report is the result of their work in the practical classes for weeks during which, in addition to the aforementioned resources, they also had the support of the teacher (providing step-by-step tutorials and resolving individual inquiries) and their colleagues in the practice group to solve practical problems as they arose.

Regarding the reliability of the digital resources used, it should be noted that the students perceived ChatGPT (3.39 out of 5) as less reliable than the tutorials provided by the teacher (4.16), MySQL official documentation (4.00), or their own report (4.63). This is to be expected, as ChatGPT itself warns you that its answers may not be true, and it is common to make ChatGPT (especially the free version 3.5, the one that the students used) fall into hallucinations leading to wrong answers [43]. Despite this, the students considered ChatGPT to be more useful (3.55 out of 5) than teacher tutorials (3.36), MySQL official documentation (3.04), and Stack Overflow or googling (3.00). This is understandable since the use of the other tools requires digging into the information and looking for an answer that ChatGPT can give you directly if you ask it properly. Nevertheless, the most useful resource for the students is the report they prepared during the classes before the exam (4.50).

Furthermore, it is interesting to explore the specific usage of ChatGPT to solve database administration problems (see Figure 1): 80% of the students declared that they used this tool and used it mainly to look for alternatives or improvements to certain SQL statements, to solve problems with a SQL statement, and to know about the syntax or examples of certain SQL statements. For example, ChatGPT proved to be quite useful for students who wanted to learn details about SQL statements for creating users or granting permissions but it was not very useful to identify the SQL statements necessary to load information in a MySQL database from an MS Excel or an XML file or to solve problems during the loading process, such as incompatibility between the character set of the data. In these cases, the solution usually came from a Socratic question of the teacher, a targeted instruction from a classmate, or other traditional sources.

Lastly, note that the use of ChatGPT among students is remarkably uneven. It is striking to observe the distribution of ChatGPT usage among students if they are categorized into the aforementioned groups (see Figure 2). While 100% of the students who obtained an outstanding score (grade higher than 7) in the practical exam used ChatGPT during the exam, only 61% of the students who failed used it. However, we must analyze more thoroughly the degree of use of ChatGPT during the exam (see Table 6), as well as the reliability (see Table 7) and the utility (see Table 8) perceived by the students. It can be seen that the students who obtained the best marks used ChatGPT intensively (4 out of 5), significantly more than the other students. In the same vein, these outstanding students also perceived ChatGPT to be much more reliable (3.80) than students who failed the exam (2.75), finding here a statistically significant difference ($p < 0.05$) with a large effect size ($\eta^2 > 0.14$). Another statistically significant difference ($p < 0.05$) with large effect size ($\eta^2 > 0.14$) was found by comparing the usefulness of ChatGPT perceived by outstanding students (4.20) with those who failed the exam (2.92). Therefore, the group of students with the best scores in the exam was the one who used ChatGPT the most and had the best perception of its reliability and usefulness. This may be due to multiple reasons that are explored in the following research question.

4.2. *RQ2. How Does Prior Knowledge About Database Administration Affect the Use of ChatGPT to Solve Database Administration Problems?*

First, it is relevant to observe the progression of knowledge acquired by the students throughout the course (see Table 1). It can be seen that at the beginning of the course, the students knew practically nothing about database administration (Test 1, 0.60 out of 10), that once the theoretical part of the course was finished, they had acquired several theoretical concepts (Test 2, 4.73), and that at the end of the practical part of the course, they had acquired almost all of them (Test 3, 8.43). The inferential comparative results calculated in this regard (see Table 2) indicate that the differences in the results between the tests performed at the different times of the course are statistically significant ($p < 0.001$) and have a large effect size ($r > 0.50$).

However, the student performance is very uneven. In fact, as indicated above, of the 40 students involved in this case study, 18 finally obtained a grade lower than 5 in the practical exam (Group 1), 7 students obtained a grade between 5 and 7 (Group 2), and 15 students obtained a grade higher than 7 (Group 3). We must analyze the progress in the acquisition of knowledge of these three groups of students. The descriptive and inferential results reported (see Table 4) indicate that at the beginning of the course, the students of the three groups started from a very similar base (0.61 vs. 0.57 vs. 0.61) but the students' performance in the theoretical part of the course (test 2) is very unequal since it can be observed that the group of students who ended up failing the practical exam obtained a significantly worse grade (3.05) than the groups of students who finally approved the practical exam with a grade of pass (5.85) or outstanding (6.20). This difference in knowledge allowed the students to approach the practical work very differently because the students who already had a stronger theoretical basis could face the practical problems posed by the practical assignment with much more criteria and settled fundamentals than the others. That may have favored the use of tools such as ChatGPT. At the end of the course, with the practical work completed, it seems (Test 3) that the relative difference in terms of acquired theoretical knowledge was reduced between the three groups but there was still a notable difference with a medium effect size (7.83 vs. 8.71 vs. 9.00).

This difference in theoretical knowledge before facing the practical work may explain the more intensive use of ChatGPT and the better perceptions about ChatGPT by the students who ended with an outstanding grade discussed in the previous RQ. This is somewhat expected since, as indicated by previous studies [36], a person with more knowledge about a topic is likely to be better able to use GenAI tools since he/she knows exactly what to ask about [44], knows how to continue a conversation with an IAGen and inquire more about a topic, knows how to identify hallucinations of the IAGen, and knows how to redirect the conversation [45]. So, we deem that the difference in prior knowledge is a factor that led these students to use ChatGPT to a greater extent and with greater success during the exam. But could there be other factors that explain this? Two elements that could do so are explored below: classroom attendance and prior experience with ChatGPT and prompting skills.

First, classroom attendance could be a key element in successfully performing the practical work and also in properly using tools such as ChatGPT since during the practical sessions, the teacher provided examples of how to use ChatGPT to solve database administration problems such as those faced by the students. Therefore, it could be inferred that the group of students who ended up failing the exam were the ones who missed more classes, but that was not the case. In fact, as shown in Table 5, this group of students attended classes (4.39 out of 5) slightly more than the group of students who obtained a pass grade (4.14) or outstanding (4.27). However, these differences are minimal and are not statistically significant.

Second, prior experience with ChatGPT and prompting skills seems to be another key element that can determine the success of using this tool. It can be observed (see Table 5) that the students reported having a medium–high degree of prior experience with ChatGPT (3.70 out of 5) and rated their prompting skills as reasonably good (3.55). Nevertheless, it is astounding that the different groups declared similar levels of prior experience with ChatGPT and self-perceived that they had similar prompting abilities, and no significant differences were found in this respect.

Therefore, it seems that in the case study under analysis, neither classroom attendance nor prior use of ChatGPT and self-perceived prompting skills caused certain students to use ChatGPT more and to have a better perception of it. Therefore, we deem that it is reasonable to think that prior knowledge of the subject before the practical work is what led to a more intensive and adequate use of ChatGPT by students.

4.3. RQ3. Is There a Relationship Between the Usage of ChatGPT and the Success in Solving Practical Database Administration Problems?

The results discussed previously suggest that there is some relationship between the use of ChatGPT and the final grade of the students on the practical exam, in which they had to solve computer database administration problems. To explore this further, it is useful to analyze the results presented in Table 9, which shows the correlations between the student's grade and the use and perception of the different resources available to him or her to take the exam. It can be observed that there are statistically significant medium-to-high positive correlations between the students' grade and degree of use ($\rho = 0.45$, p-value = 0.003), and perceived reliability ($\rho = 0.42$, p-value = 0.01) and utility ($\rho = 0.48$, p-value = 0.005) of ChatGPT. A positive correlation of student grade with assignment report use, reliability and utility is also found, but it is not as high or statistically significant. In addition, a significant, negative and high correlation of grade is found with the use of teacher tutorials during the exam ($\rho = -0.51$, p-value < 0.001), which is logical since these tutorials were introductory and served to lay certain foundations, but not to solve advanced problems posed by the practical exam. In the case of the rest of the learning resources, we find negative correlations with the final grade, but they are low correlations (Rho < 0.30) without statistical significance.

Therefore, the posed RQ can be answered positively, and it can be affirmed that there is a positive relationship between the use of ChatGPT and the success in solving practical database administration problems. We consider GenAI tools such as ChatGPT to be helpful resources to assist in the resolution of database administration problems, and to have a positive impact on student performance. This is partially consistent with the conclusions of [46], where the usage of ChatGPT significantly improved the quality and originality of the student task resolution. Our conclusions are also partially consistent with those reported in the aforementioned work of [44], which demonstrated that professional software engineers improve their skills by prompting them not to write code but to solve higher-order knowledge tasks. Indeed, as in the line of our work, Ref. [47] showed that the knowledge base of the subject significantly affects the effectiveness of learners in applying ChatGPT to solve complex problems.

Lastly, it should be mentioned that the student's final grade may have relations with other factors such as classroom attendance or prior ChatGPT use and prompting skills, but as discussed in the previous RQ, no differences in such factors are identified among the different groups of students. However, teacher guidance during practical sessions was necessary to show students pieces of prompt engineering with the intention of solving problems step by step [43] since it contributes to improving student skills both with the use of ChatGPT and with the subject [48]. There may also be other unexplored factors related to student performance on the exam, such as time and stress management skills, that are very important when taking an exam. In any case, what has been found in this case study is a clear relationship between the use of ChatGPT and student performance.

5. Conclusions

This article has presented an empirical study to explore the usage of ChatGPT to learn about database administration in computer science education. We found that ChatGPT was moderately used, but it was more used than traditional internet learning resources, and a trend of increasing use could be observed compared to the previous year. We also found uneven usage by students and that the students who end up earning better grades were the ones who used it more and had better perceptions about ChatGPT. In fact, positive correlations were found between the use of this tool and the success of solving database administration problems during a practical exam. Lastly, we found that one of the elements that could encourage students to use this tool intensively and productively was the theoretical background they had about database administration.

Therefore, it can be concluded that ChatGPT can be an excellent educational instrument in the context of database administration and that to use it properly, it is necessary for students to have a solid theoretical basis. However, as mentioned in the Introduction, it should be noted that GenAI tools such as ChatGPT have a number of limitations to consider: the possibility of hallucinating, the lack of critical thinking that could be generated in the GenAI users, and even prohibitions on the use of these tools in the process of software development by contractors.

In addition, during the execution of this case study, we have found that prompting skills are also important, and teachers can use a guided practice strategy, where prompting and guided step-by-step practice are employed to train these skills as well as to increase the student's theoretical background. Training students in the use of GenAI tools like ChatGPT is key to prevent the appearance of new digital trenches between students with solid knowledge and prompting skills that can further increase their performance through the proper use of GenAI tools and students with a lower degree of the fundamentals and worse prompting skills, who lag behind and will obtain even worse results.

The presented conclusions are strongly connected to computer science education since they are based on a case study in that context, but we deem that they are fully transferable to any engineering or science discipline in which the practical mastery of technologies and problem solving play a key role.

Lastly, some limitations and future work are presented. First, the sample size is only 40 students, so the conclusions reported here should be taken with caution. Furthermore, there are many difficulties associated with the generalization of single case studies like the presented [49], so more studies like this should be conducted to consolidate our findings. Once this occurs and the presented framework has been applied several times, we could provide a more reliable and generalizable mathematical model. Second, the gender distribution is unbalanced, which is common in computer science studies, and unfortunately we do not have a way to solve this limitation in the study. Third, the study design precludes the assertion that ChatGPT caused higher grades. The conduction of a randomized controlled trial in which there was an experimental group with access to ChatGPT and a control group without access to ChatGPT would allow one to isolate the ChatGPT variable and assert cause-and-effect relationships. However, although this is very desirable from a research point of view, from a teaching perspective it is complicated, as it violates the principle of student equity. Fourth, the conclusions reported are based on various data, including those obtained through the questionnaire. This instrument provides self-reported data that can influence responses due to social desirability bias and inaccurate self-reporting. In detail, the prompting skills of the students are measured using this self-perception questionnaire, rather than more objective tests. In the future, the prompting skills of the students could be measured by analyzing their prompts. Lastly, conducting studies in the future using qualitative data or assessing long-term knowledge retention would provide a more complete picture of ChatGPT's impact on learning.

Author Contributions: Conceptualization, D.L.-F. and R.V.; methodology, D.L.-F.; formal analysis, D.L.-F. and R.V.; investigation, D.L.-F.; resources, D.L.-F.; data curation, D.L.-F. and R.V.; writing—original draft preparation, D.L.-F. and R.V.; writing—review and editing, D.L.-F. and R.V.; supervision, D.L.-F. and R.V. All authors have read and agreed to the published version of the manuscript.

Funding: This research received no external funding.

Institutional Review Board Statement: Ethical review and approval were waived for this study, due to the policies of the Universidad Politécnica de Madrid, where the study was carried out, which considers that this type of research based on teaching activities does not require the express acceptance of its ethics committee.

Informed Consent Statement: Informed consent was obtained from all subjects involved in the study.

Data Availability Statement: The collected data are publicly available at the following link: https://doi.org/10.21950/XMROTF, accessed on 28 November 2024.

Conflicts of Interest: The authors declare no conflicts of interest.

Abbreviations

The following abbreviations are used in this manuscript:

GenAI	Generative Artificial Intelligence
UNESCO	United Nations Educational, Scientific, and Cultural Organization
LLM	Large Language Models
DBA	Database Administration
SQL	Structured Query Language
AI	Artificial Intelligence
TAM	Technology Acceptance Model
RQ#	Research Question N
ECTS	European Credit Transfer System

Appendix A. Test Carried Out Throughout the Course to Measure the Evolution of Knowledge

Table A1. Test carried out throughout the course to measure the evolution of knowledge.

Question/Answers (The Correct Answer Is in Bold)
1. What statement is used in MySQL to load data from an external file? A. **LOAD DATA INFILE** B. IMPORT DATA INFILE C. INSERT INTO D. Data loading must be done through third-party software, not directly from MySQL
2. What statement is used in MySQL to load data from an external XML file? A. The same statements or tools used to load files of other formats B. LOAD EXTERNAL XML C. **LOAD XML INFILE** D. To load XML files, the file must be pre-transformed into another format compatible with MySQL
3. Define foreign keys before the loading process A. Makes the loading process faster and more reliable B. **Makes the loading process more reliable but slower** C. Makes the loading process faster but less reliable D. Makes the loading process less fast and less reliable
4. What keyword is used to skip the first row when loading a CSV file into a table? A. IGNORE FIRST ROW B. SKIP HEADER C. **IGNORE 1 LINES** D. SKIP 1 LINES
5. What statement should be used in MySQL to define an index? A. CREATE INDEX indexName ON tableName B. **CREATE INDEX indexName ON tableName (column)** C. CREATE INDEX ON tableName (column) D. CREATE INDEX ON tableName
6. What is the beginning of the statement used in MySQL to define a unique index? A. **CREATE UNIQUE INDEX** B. CREATE NOT NULL INDEX C. There is no direct statement for this; to indicate that the index is unique, the field must be defined as unique in the table definition D. None of the answers is correct
7. The use of indexes A. **Can improve the performance of certain queries** B. Always improves the performance of certain queries C. Rarely improves the performance of certain queries D. Influences database space usage more than performance
8. What statement should be used in MySQL to delete an index? A. **DROP INDEX** B. REMOVE INDEX C. DELETE INDEX D. DISABLE INDEX
9. Regarding databases, what does the XAMPP software include? A. A MySQL DBMS and an IDE to manage it B. **A MariaDB DBMS and an IDE to manage it** C. Exclusively a MariaDB DBMS D. Exclusively a MySQL DBMS
10. How are the ports of the DBMS configured in the XAMPP environment? A. The ports are predefined by default and cannot be modified through XAMPP B. Through the config.inc.php file C. Through the my.ini file D. **Through the config.inc.php and my.ini files**

References

1. Karunaratne, T.; Adesina, A. Is it the new Google: Impact of ChatGPT on Students' Information Search Habits. In Proceedings of the 22nd European Conference on e-Learning, ECEL 2023, Porto, Portugal, 24–25 October 2023. [CrossRef]
2. Shaw, C.; Yuan, L.; Brennan, D.; Martin, S.; Janson, N.; Fox, K.; Bryant, G. GenAI in Higher Education: Fall 2023 Update, Time for Class Study. 2023. Available online: https://tytonpartners.com/app/uploads/2023/10/GenAI-IN-HIGHER-EDUCATION-FALL-2023-UPDATE-TIME-FOR-CLASS-STUDY.pdf (accessed on 8 November 2024).
3. Mustapha, R.; Mustapha, S.N.A.M.; Mustapha, F.W.A. Students' misuse of ChatGPT in higher education: An application of the Fraud Triangle Theory. *J. Contemp. Soc. Sci. Educ. Stud. (JOCSSES)* **2024**, *4*, 87–97.. [CrossRef]
4. Rogers, M.P.; Hillberg, H.M.; Groves, C.L. Attitudes Towards the Use (and Misuse) of ChatGPT: A Preliminary Study. In Proceedings of the 55th ACM Technical Symposium on Computer Science Education V. 1 (SIGCSE 2024), Portland, OR, USA, 20–23 March 2024; Association for Computing Machinery: New York, NY, USA; pp. 1147–1153. [CrossRef]
5. Durango, N.T.; Blanco, M.I.M.; Rodríguez, M.P.C.; de las Moras, M.C.R. Flip Teaching Active Methodology as a Means to Prevent Fraudulent Use of ChatGPT by Students. In *Innovation and Technologies for the Digital Transformation of Education*; García-Peñalvo, F.J., Sein-Echaluce, M.L., Fidalgo-Blanco, Á., Eds; Lecture Notes in Educational Technology; Springer: Singapore, 2024. [CrossRef]
6. Estévez-Ayres, I.; Callejo, P.; Hombrados-Herrera, M.Á; Alario-Hoyos, C.; Delgado-Kloos, C. Evaluation of LLM Tools for Feedback Generation in a Course on Concurrent Programming. *Int. J. Artif. Intell. Educ.* **2024**. [CrossRef]
7. Sengewald, J.; Wilz, M.; Lackes, R. AI-Assisted Learning Feedback: Should Gen-AI Feedback Be Restricted to Improve Learning Success? A Pilot Study in a SQL Lecture. In Proceedings of the ECIS 2024, Paphos, Cyprus, 13–19 June 2024. Available online: https://aisel.aisnet.org/ecis2024/track13_learning_teach/track13_learning_teach/12 (accessed on 28 November 2024).
8. Gimpel, H.; Hall, K.; Decker, S.; Eymann, T.; Lämmermann, L.; Mädche, A.; Röglinger, M.; Ruiner, C.; Schoch, M.; Schoop, M.; et al. Unlocking the Power of Generative AI Models and Systems such as GPT-4 and ChatGPT for Higher Education: A Guide for Students and Lecturers. Universität Hohenheim, Fakultät Wirtschafts- und Sozialwissenschaften. 2023. Available online: http://hdl.handle.net/10419/270970 (accessed on 28 November 2024).
9. Mollick, E.R.; Mollick, L. *Assigning AI: Seven Approaches for Students, with Prompts;* The Wharton School Research Paper; University of Pennsylvania: Philadelphia, PA, USA, 2023. [CrossRef]
10. Robert, J.; Muscanell, N. 2023 Horizon Action Plan: Generative AI. EDUCAUSE. 2023. Available online: https://library.educause.edu/-/media/files/library/2023/9/2023horizonactplangenai.pdf (accessed on 8 November 2024).
11. Russell Group. Russell Group Principles on the Use of Generative AI Tools in Education. 2023. Available online: https://russellgroup.ac.uk/media/6137/rg_ai_principles-final.pdf (accessed on 8 November 2024).
12. Artificial Intelligence and Future of Teaching and Learning: Insights and Recommendations. U.S. Department of Education, Office of Educational Technology: Washington, DC, USA, 2023. Available online: https://tech.ed.gov (accessed on 28 November 2024).
13. European Union. Artificial Intelligence Act. Regulation (EU) 2024/1689 of the European Parliament and of the Council of 13 June 2024 Laying Down Harmonised Rules on Artificial Intelligence and Amending Regulations (EC) No 300/2008, (EU) No 167/2013, (EU) No 168/2013, (EU) 2018/858, (EU) 2018/1139 and (EU) 2019/2144 and Directives 2014/90/EU, (EU) 2016/797 and (EU) 2020/1828 (Artificial Intelligence Act). Available online: http://data.europa.eu/eli/reg/2024/1689/oj (accessed on 28 November 2024).
14. Robert, J.; McCormack, M. *2024 EDUCAUSE Action Plan: AI Policies and Guidelines*; EDUCAUSE: Boulder, CO, USA, 2024. Available online: https://www.educause.edu/research/2024/2024-educause-action-plan-ai-policies-and-guidelines (accessed on 29 November 2024).
15. Acemoglu, D.; Johnson, S. Choosing AI's Impact on the Future of Work. Stanford Social Innovation Review. 2023. Available online: https://ssir.org/articles/entry/ai-impact-on-jobs-and-work (accessed on 8 November 2024).
16. Chiu, T.K.F. Future research recommendations for transforming higher education with generative AI. *Comput. Educ. Artif. Intell.* **2024**, *6*, 100197. [CrossRef]
17. European Commission: Directorate-General for Education, Youth, Sport and Culture. Ethical Guidelines on the Use of Artificial Intelligence (AI) and Data in Teaching and Learning for Educators. Publications Office of the European Union. 2022. Available online: https://data.europa.eu/doi/10.2766/153756 (accessed on 8 November 2024).
18. UNESCO Guidance for Generative AI in Education and Research. 2023. Available online: https://www.unesco.org/en/articles/guidance-generative-ai-education-and-research (accessed on 9 November 2024).
19. Gonzalez, L.; O'Neil-Gonzalez, K.; Eberhardt-Alstot, M.; McGarry, M.; Van Tyne, G. Leveraging AI for Inclusive Excellency in Higher Education. EDUCAUSE. 2024. Available online: https://er.educause.edu/articles/2024/8/leveraging-generative-ai-for-inclusive-excellence-in-higher-education (accessed on 8 November 2024).
20. Ogunleye, B.; Zakariyyah, K.I.; Ajao, O.; Olayinka, O.; Sharma, H. A Systematic Review of Generative AI for Teaching and Learning Practice. *Educ. Sci.* **2024**, *14*, 636. [CrossRef]
21. Ward, D.; Loshbaugh, H.G.; Gibbs, A.L.; Henkel, T.; Siering, G.; Williamson, J.; Kayser, M. How Universities Can Move Forward with Generative AI in Teaching and Learning. *Change Mag. High. Learn.* **2024**, *56*, 47–54. [CrossRef]

22. Kuhail, M.A.; Mathew, S.S.; Khalil, A.; Berengueres, J.; Shah, S.J.H. "Will I be replaced?" Assessing ChatGPT's effect on software development and programmer perceptions of AI tools. *Sci. Comput. Program.* **2024**, *235*, 103111. [CrossRef]
23. Shein, E. The Impact of AI on Computer Science Education. *Commun. ACM* **2024**, *67*, 13–15. [CrossRef]
24. Serrano, M.; Santos-Olmo, A.; Garcia-Rosado, D.; Sanchez-Crespo, L.E.; Blanco, C.; Fernandez-Medina, L. Use of ChatGPT in University Computer Science Teaching: Advantages and disadvantages. In Proceedings of the INTED2024, Valencia, Spain, 4–6 March 2024; p. 6395. [CrossRef]
25. Nizamudeen, F.; Gatti, L.; Bouali, N.; Faizan, A. Investigating the Impact of Code Generation Tools (ChatGPT & Github CoPilot) on Programming Education. In Proceedings of the 16th International Conference on Computer Supported Education, Angers, France, 2–4 May 2024. [CrossRef]
26. Liu, R.; Zenke, C.; Liu, C.; Holmes, A.; Thornton, P.; Malan, D.J. Teaching CS50 with AI: Leveraging Generative Artificial Intelligence in Computer Science Education. In Proceedings of the 55th ACM Technical Symposium on Computer Science Education V. 1 (SIGCSE 2024), Portland, OR, USA, 20–23 March 2024; Association for Computing Machinery: New York, NY, USA, 2024; pp. 750–756. [CrossRef]
27. Yilmaz, R.; Gizem, F. The effect of generative artificial intelligence (AI)-based tool use on students' computational thinking skills, programming self-efficacy and motivation. *Comput. Educ. Artif. Intell.* **2023**, *4*, 100147. [CrossRef]
28. Singh, H.; Tayarani-Najaran, M.; Yaqoob, M. Perception of ChatGPT in Higher Education: A Descriptive and Correlation Study. *Educ. Sci.* **2023**, *13*, 924. [CrossRef]
29. Duc, A.; Lønnestad, T.; Sundbø, I.; Johannessen, M.; Gabriela, V.; Ahmed, S.; El-Gazzar, R. Generative AI in Undergraduate Information Technology Education—Insights from nine courses. *arXiv* **2023**, arXiv:2311.10199.
30. Prakash, K.; Rao, S.; Hamza, R.; Lukich, J.; Chaudhari, V.; Nandi, A. Integrating LLMs into Database Systems Education. In Proceedings of the 3rd International Workshop on Data Systems Education: Bridging Education Practice with Education Research (DataEd '24), Santiago, Chile, 9–15 June 2024; pp. 33–39. [CrossRef]
31. Neumann, A.T.; Yin, Y.; Sowe, S.; Decker, S.; Jarke, M. An LLM-Driven Chatbot in Higher Education for Databases and Information Systems. *IEEE Trans. Educ.* **2024**. [CrossRef]
32. Cagle, P. 5 AI Case Studies in Education. March 2024. Available online: https://www.vktr.com/ai-disruption/5-ai-case-studies-in-education/ (accessed on 28 November 2024).
33. Fernandez, R.; Elmore, A.; Franklin, M.; Krishnan, S.; Tan, C. How Large Language Models Will Disrupt Data Management. *Proc. VLDB Endow.* **2023**, *16*, 3302–3309. Available online: https://www.vldb.org/pvldb/vol16/p3302-fernandez.pdf (accessed on 28 November 2024). [CrossRef]
34. Zewe, A. MIT Researchers Introduce Generative AI for Databases. *MIT News*, July 2024. Available online: https://news.mit.edu/2024/mit-researchers-introduce-generative-ai-databases-0708 (accessed on 21 November 2024).
35. Cagliero, L.; Farinetti, L.; Fior, J.; Manenti, A.I. ChatGPT, be my Teaching Assistant! Automatic Correction of SQL Exercises. In Proceedings of the 2024 IEEE 48th Annual Computers, Software, and Applications Conference (COMPSAC), Osaka, Japan, 2–4 July 2024; pp. 81–87. [CrossRef]
36. Balderas, A.; Baena-Pérez, R.; Person, T.; Mota, J.; Ruiz-Rube, I. Chatbot-Based Learning Platform for SQL Training. *Int. J. Interact. Multimed. Artif. Intell.* **2022**, *8*, 135–145. [CrossRef]
37. Anderson, N.; McGowan, A.; Hanna, P.; Cutting, D.; Galway, L.; Collins, M. Using ChatGPT in Software Development Education. In Proceedings of the 2024 IEEE Global Engineering Education Conference (EDUCON), Kos Island, Greece, 8–11 May 2024; pp. 1–5. [CrossRef]
38. Wilcoxon, F. Individual comparisons by ranking methods. In *Breakthroughs in Statistics*; Kotz, S., Johnson, N.L., Eds.; Springer Series in Statistics; Springer: New York, NY, USA, 1992; pp. 196–202. [CrossRef]
39. Funder, D.C.; Ozer, D.J. Evaluating Effect Size in Psychological Research: Sense and Nonsense. *Adv. Methods Pract. Psychol. Sci.* **2019**, *2*, 156–168. [CrossRef]
40. Kruskal, W.H.; Wallis, W.A. Use of Ranks in One-Criterion Variance Analysis. *J. Am. Stat. Assoc.* **1952**, *47*, 583–621. [CrossRef]
41. Spearman, C. The proof and measurement of association between two things. *Am. J. Psychol.* **1904**, *15*, 72–101. [CrossRef]
42. López-Fernández, D.; Vergaz, R. Adoption and Impact of ChatGPT in Computer Science Education: A Case Study on a Database Administration Course. *AI* **2024**, *5*, 2321–2337. [CrossRef]
43. Shen, Y.; Ai, X.; Raj, A.G.S.; John, R.J.L.; Syamkumar, M. Implications of ChatGPT for Data Science Education. In Proceedings of the 55th ACM Technical Symposium on Computer Science Education V. 1 (SIGCSE 2024), Portland, OR, USA, 20–23 March 2024; Association for Computing Machinery: New York, NY, USA, 2024; pp. 1230–1236. [CrossRef]
44. Khojah, R.; Mohamad, M.; Leitner, P.; de Oliveira Neto, F.G. Beyond Code Generation: An Observational Study of ChatGPT Usage in Software Engineering Practice. *Proc. ACM Softw. Eng.* **2024**, *1*, 81. [CrossRef]
45. Hou, I.; Mettille, S.; Man, O.; Li, Z.; Zastudil, C.; MacNeil, S. The Effects of Generative AI on Computing Students' Help-Seeking Preferences. In Proceedings of the 26th Australasian Computing Education, New York, NY, USA, 29 January–2 February 2024; pp. 39–48. [CrossRef]

46. Urban, M.; ; Děchtěrenko, F.; Lukavský, J.; Hrabalová, V.; Svacha, F.; Brom, C.; Urban, K. ChatGPT improves creative problem-solving performance in university students: An experimental study. *Comput. Educ.* **2024**, *215*, 105031. [CrossRef]
47. Jing, Y.; Wang, H.; Chen, X.; Wang, C. What factors will affect the effectiveness of using ChatGPT to solve programming problems? A quasi-experimental study. *Humanit. Soc. Sci. Commun.* **2024**, *11*, 319. [CrossRef]
48. Knoth, N.; Tolzin, A.; Janson, A.; Leimeister, J.M. AI literacy and its implications for prompt engineering strategies. *Comput. Educ. Artif. Intell.* **2024**, *6*, 100225. [CrossRef]
49. Yin, R.K. *Case Study Research: Design and Methods*; Sage Publications: Thousand Oaks, CA, USA, 2008.

applied sciences

MDPI

Article

Renewal of the Concept of Diverse Education: Possibility of Further Education Based on a Novel AI-Based RF–ISSA Model

Enhui Li [1], Zixi Wang [1,*], Jin Liu [1,*] and Jiandong Huang [2,3,4]

[1] College of Music and Dance, Guangzhou University, Guangzhou 510006, China; lienhui@gzhu.edu.cn
[2] School of Civil and Transportation Engineering, Guangzhou University, Guangzhou 510006, China; jiandong.huang@hotmail.com
[3] Higher School of Advanced Digital Technologies, Peter the Great St. Petersburg Polytechnic University, St. Petersburg 195251, Russia
[4] School of Civil Engineering and Architecture, Linyi University, Linyi 276000, China
* Correspondence: wangzixi@gzhu.edu.cn (Z.W.); liujin628@163.com (J.L.)

Abstract: The traditional graduate admission method is to evaluate students' performance and interview results, but this method relies heavily on the subjective feelings of the evaluators, and these methods may not be able to comprehensively and objectively evaluate the qualifications and potential of the applicants. At present, artificial intelligence has played a key role in the reform of the education system, and the data processing function of artificial intelligence has greatly reduced the workload of screening work. Therefore, this study aims to optimize the graduate enrollment evaluation process by applying a new composite model, the random forest–improved sparrow search algorithm (RF–ISSA). The research used seven data sets including research, cumulative grade point average (CGPA), letter of recommendation (LOR), statement of purpose (SOP), university rating, TOEFL score, and graduate record examination (GRE) score, and carried out the necessary data pre-processing before the model construction. The experimental results show that the RMSE and R values of the composite model are 0.0543 and 0.9281, respectively. The predicted results of the model are very close to the actual data. In addition, the study found that the importance score of CGPA was significantly higher than other characteristics, and that this value has the most significant impact on the outcome of the graduate admissions assessment. Overall, this study shows that combining the integrated strategy sparrow search algorithm (ISSA) with hyperparameter optimization and focusing on the most influential features can significantly improve the predictive performance and applicability of graduate admissions models, providing a more scientific decision support tool for school admissions professionals.

Keywords: machine learning model; random forest; improved sparrow search algorithm; student ability assessment

1. Introduction

With the continuous development of higher education, graduate studies have become a significant option for students' future development. Many students choose to pursue further academic research after completing their undergraduate degrees. However, graduate work is not suitable for every student. Some students may excel academically but lack the necessary research capabilities. As a result, their academic achievements may suggest they are well suited for graduate studies, but in reality, their true abilities may not meet the demands of graduate work. Therefore, it is essential to introduce more evaluation criteria in the graduate admissions process to assist admissions departments in making informed decisions. This introduces numerous uncertainties, posing a significant challenge for universities and admissions departments [1,2].

Traditional admissions methods primarily rely on applicants' academic records, letters of recommendation, and personal statements. However, these methods tend to be subjective, and relying solely on such factors may not provide a comprehensive and objective evaluation of an applicant's overall qualifications and potential. Consequently, truly talented researchers may be overlooked. Therefore, optimizing the graduate admissions process through modern technological means in order to improve selection efficiency and accuracy has become an urgent issue to address [3].

Analyzing the profiles of prospective students is often a challenging task, as it heavily depends on various parameters. This reliance can lead to imbalances in the evaluation process due to the subjective judgments of the evaluators, ultimately affecting the final assessment results [4,5]. Therefore, it is necessary to find a method that can objectively analyze all relevant data of students and intuitively express their abilities. To help admissions staff with student selection. At present, the rapid development of artificial intelligence and its sensitive data searching and efficient processing abilities provide a new way to solve this problem. Under this background, the graduate admission evaluation model came into being. These models leverage the data processing power and sensitivity of AI models to dig deep and analyze historical admission data to build AI models that can output intuitive data evaluations by analyzing student profiles and related information [6,7]. The graduate admission evaluation model can not only effectively help admissions departments make rational decisions about admitting or rejecting students, but also provide valuable insights to applicants. By understanding the criteria that increase one's chances of admission, applicants can better prepare their application materials and improve their likelihood of being accepted [8–11].

Currently, many researchers are already beginning to apply artificial intelligence to the evaluation of student capabilities. These researchers make full use of AI's ability to assist in the analysis and integration of various forms of student information, uncovering patterns that can be used to support the analysis of new data in future work. Livieris et al. [12] examined and evaluated the effectiveness of semi-supervised algorithms under two assembling methods and utilized these algorithms to predict students' final exam grades. The research findings indicate that altering a small amount of labeled data within the semi-supervised model can significantly enhance the model's classification accuracy. Experimental data support this conclusion, demonstrating the potential application of this method in the educational field. Alexandro David M et al. [13] developed a student dropout risk assessment model using over a hundred predictive indicators to better identify and support at-risk students, thereby reducing the likelihood of their dropping out of high school. The model provides detailed insights, which, when combined with predictive factors related to special education, allow for an in-depth analysis of students' learning situations. This approach enables early intervention to mitigate dropout risks.

Table 1 presents examples from recent studies that integrate student data with machine learning models. These models analyze students' behaviors, grades, and other relevant data to predict academic performance and future potential. Such predictive analysis not only aids educators in better understanding students' learning needs but also helps in the early identification and intervention of potential learning issues, thus enhancing students' learning efficiency and broadening their developmental pathways [14–17].

However, based on current research trends, many scholars tend to employ single algorithms when developing these machine-learning models. Although this method simplifies the model construction process, it also imposes limitations on the flexibility and adaptability of the model. The selection of machine learning model parameters is a complex and time-consuming task, one which not only poses challenges to researchers' expertise but also impacts the model's predictive accuracy and computational efficiency [18–20].

To address this challenge, this study proposes an innovative approach that optimizes the machine learning model development process by integrating multiple algorithm selection methods. The core of this approach lies in utilizing an algorithm selection mechanism to quickly identify and adapt the most suitable model parameters for the given dataset. This strategy not only significantly improves the model's predictive accuracy but also enhances the efficiency of the model development process while maintaining performance.

Furthermore, combining algorithms can enhance the model's generalizability, improving its adaptability to diverse data distribution and learning situations. This flexibility is crucial for addressing the ever-evolving learning needs and challenges in the field of education.

Table 1. Literature review for renewal of the concept of diverse education based on machine learning.

(1) Admission criteria and academic performance					
Author	**Analysis Target**	**Data Set Size**	**Machine Learning Model**	**Maximum Accuracy**	R^2
Adekitan, AI [21]	Admission criteria and the academic performance of the student after the first academic session	100	KNIME model	50.23%	-
			Orange model	51.9%	-
			Linear regression model	-	0.207
			Quadratic regression model	-	0.232
(2) Students learning ability					
Author	**Analysis Target**	**Data Set Size**	**Machine Learning Model**	**Accuracy**	
Kukkar [22]	Predicting students' pass or fail outcome in certain courses	32,593	Recurrent neural network + long short term memory network + support vector machine (RNNs + LSTM + SVM)	90.67%	
			Recurrent neural network + long short term memory network + naive Bayes (RNNs + LSTM + NB)	86.45%	
			Recurrent neural network + long short term memory network + decision tree (RNNs + LSTM + DT)	84.42%	
Raheela Asif [23]	Study the performance of undergraduate students	210	Decision tree with Gini index	68.27%	
			Decision tree with information gain	69.23%	
			Decision tree with accuracy	60.58%	
			Rule induction with information gain	55.77%	
			1-Nearest neighbor	74.04%	
			Neural networks	62.50%	
			Random forest trees with accuracy	62.50%	
Jui-Long Hung [24]	Evaluates the accuracy of student learning ability	509	DNN	84.79%	
			RF	85.37%	
			DNN	95.89%	
			RF	95.53%	
Ashima Kukkar [25]	Performance of students in higher education	32593	Three-layer stacked long short term memory + random forest + gradient boosting (LSTM + RF + GB)	91.66%	
			Two-layer stacked LSTM + RF + GB	86.54%	
			One-layer LSTM + RF + GB	75.47%	
(3) Students' comprehensive psychological quality and achievement					
Author	**Analysis Target**	**Data Set Size**	**Machine Learning Model**	**Mean AUC**	
Radwan, AM [26]	Response of students with autism and students' failure	3739	Support vector machine	0.74	
			AdaBoost	0.73	
			Logistic regression	0.72	
Liana Maria Crivei [27]	Students' course achievement	2401	Students' performance prediction using relational association rules (SPRAR)	0.74	
			DT	0.61	
			ANN	0.67	
			SVM	0.65	
			Original SPRAR	0.7	

The remainder of this paper is organized into several key sections. Section 2 provides a detailed description of the dataset, including its sources, characteristics, and significance in the model development process. Section 3 outlines the selection algorithms employed in this study, the specific machine learning models used, and the metrics utilized to quantify and evaluate the predictive outcomes. Section 4 systematically explains the experimental design and implementation process, using charts and graphs to visually present the experimental data, aiding readers in quickly understanding and analyzing the results. Section 5 offers a thorough analysis and discussion of all experimental findings, distills the research conclusions, and explores potential directions for future research and improvements.

2. Materials

The data used in this study were collected by other scholars [28]. The dataset includes seven characteristics: research, CGPA, LOR, SOP, college rating, TOEFL score, and GRE score. Among these characteristics, some have direct relevance and can objectively evaluate student-specific performance and behavioral characteristics in specific contexts, while others lack direct objective criteria. Different data types cannot be used to build the same model at the same time. Therefore, it is necessary to perform certain pre-processing steps before using these data for model building to ensure that the data are represented in the same model and meets the requirements of model building.

Specifically, for directly related characteristics, such as CGPA and GRE score, these data can often be used directly as input to the model because they provide clear quantitative indicators for assessing a student's academic ability. As the most intuitive student data, these two parameters can provide the most direct judgment basis for the assessment staff. However, SOP, LOR, and other features are qualitative descriptions that need to be translated into a format suitable for model training through text processing and feature extraction techniques. Alternatively, they may be artificially classified to highlight the impact of this type of data on student evaluations. In addition, special attention needs to be paid when addressing university rankings and research functions, two parameters that are not directly relevant to students. This is necessary to ensure the consistency and accuracy of data in data processing and transformation.

Due to the nature of the data entry, there may be gaps or incorrect filling. This can lead to gaps or noise in the data set. Therefore, it is necessary to preprocess the data set before its use. The pre-processing steps include data cleaning and missing value input. The purpose of data cleaning is to remove noise and errors in the data while the insertion of missing values is undertaken in order to solve gaps in the data and improve its integrity. Through a certain degree of pre-processing, the integrity of the data is guaranteed, thus providing enough quality data for the construction of the model [29–31].

These pre-processing steps are designed to improve data quality, reduce noise, and ensure that the model is trained and tested on reliable input data. By implementing these steps, the model can effectively utilize the information contained in the data, thereby improving its predictive performance and stability [32–35].

3. Proposed Methodology

3.1. ISSA

The sparrow search algorithm (SSA) is a population intelligence optimization algorithm, one which is mainly inspired by the sparrow's foraging behavior and anti-predation behavior [36,37]. In the process of foraging, the sparrows are divided into discoverers and participants. To obtain food, sparrows can adopt two behavioral strategies: finder and entrant. The finder is responsible for finding food in the population and provides foraging areas and directions for the entire sparrow population, while the entrant uses the discoverer to obtain food. Individuals in the group monitor the behavior of other individuals in the group, and attackers in the group compete for food resources with their peers to increase their own predation rates. When a sparrow population is attacked by a predator, it will engage in anti-predation behavior. Sparrows on the periphery of the population are vulnerable to predators and need to constantly adjust their position for better protection. At the same time, sparrows in the center of the population approach their neighbors to minimize danger zones [38].

In SSA, the finder with a better fitness value will obtain the food first (i.e., the better solution) during the search. Because the finder is responsible for finding food for the entire sparrow population and for providing foraging directions for all of the entrants, the finder has access to a larger foraging search area than the entrant. During the foraging process, entrants will either always monitor the discoverer, compete with the discoverer for food, or hunt around the discoverer.

The location of the discoverer changes as the number of iterations is updated.

$$x_{i,j}(t+1) = \begin{cases} x_{i,j}(t) \cdot exp\left(-\frac{i}{\alpha \cdot item_{max}}\right), \ R_2 < ST \\ x_{i,j}(t) + Q \cdot L, R_2 \geq ST \end{cases} \tag{1}$$

Each sparrow takes into account the optimal solution of the neighbor and, with a certain probability, moves towards the neighbor's position. It hops randomly within a certain distance with a certain probability, centered on the current optimal solution. Additionally, each sparrow randomly perturbs its position with a certain probability to avoid falling into the local optimal solution. The current solution is evaluated according to the fitness function, and the optimal solution and the location of the optimal solution are updated. The algorithm terminates when a preset stopping condition is reached, such as reaching the maximum number of iterations or meeting certain accuracy requirements.

When the discoverers find the food, the position of the entrants is updated as follows:

$$x_{i,j}(t+1) = \begin{cases} Q \cdot exp\left(\frac{x_{wrost} - x_{i,j}(t)}{i^2}\right), \ i > \frac{n}{2} \\ x_p(t+1) + \left|x_{i,j}(t) - x_p(t+1)\right| \cdot A^+ \cdot L, \ i \leq \frac{n}{2} \end{cases} \tag{2}$$

where x_p can be determined as the locally optimized position by the discoverer, and x_{wrost} can be selected as the global worst position. A is a 1 × d matrix where each element is randomly assigned to either 1 or −1, $A^+ = A^T(AA^T)^{-1}$. When this is b, this indicates that i entrant with low fitness has not obtained food and needs to update its position again to find new food.

x_p and x_{wrost} represent the best and worst position found by the finder. A is a 1 × d matrix of 1 and −1 at random.

The alert's location is updated below.

$$x_{i,j}(t+1) = \begin{cases} x_{bestj}(t) + \beta\left|x_{i,j}(t) - x_{bestj}(t)\right|, \ f_i > f_g \\ x_{i,j}(t) + K\left(\frac{\left|x_{i,j}(t) - x_{wrostj}(t)\right|}{(f_i - f_w) + e}\right), f_i = f_g \end{cases} \tag{3}$$

SSA has a relatively strong search ability, but it is easy to fall into the trouble of a local optimal solution because it cannot jump out of the local limit value in the face of a large number of data sets. To solve the problem that SSA is easy to fall into the local optimal solution, an improved sparrow search algorithm (ISSA) is proposed in this study. Firstly, a normal migration strategy is proposed, one which bases the population migration on the center of gravity position and which realizes the normal distribution attenuation of moving energy. This improvement effectively improves the exploration ability of local search and enables the algorithm to explore the solution space more efficiently and find better solutions in the search process. Secondly, the ISSA algorithm introduces a dynamic sinusoidal perturbation strategy. This strategy realizes the discoverer's bidirectional demand for search step size in the early stage and fast convergence in the late stage through the scaling factor. At the initial stage of the search, the algorithm needs a larger search step size to explore a wider solution space. In the latter stage of the search, smaller search steps are needed to converge to the optimal solution quickly. The dynamic sinusoidal perturbation strategy satisfies this demand by adjusting the scaling factor, thus improving the search efficiency and precision of the algorithm. Such a process can be realized as follows:

$$x_{i,j}(t+1) = \begin{cases} Levy(d)\cdot x_{bestj}(t) + \beta\left|x_{i,j}(t) - Levy(d)\cdot x_{bestj}(t)\right|, \ f_i > f_g \\ x_{i,j}(t) + K\left(\frac{|x_{i,j}(t) - x_{wrostj}(t)|}{(f_i - f_w) + e}\right), f_i = f_g \end{cases} \tag{4}$$

$$Levy(d) = 0.01\cdot\frac{r_1\cdot\sigma}{|r_2|^{\frac{1}{\beta}}} \tag{5}$$

$$\sigma = \left\{\frac{\Gamma(1+\beta)\cdot\sin\left(\frac{\pi\beta}{2}\right)}{\Gamma\left(\left(\frac{1+\lambda}{2}\right)\beta\cdot 2^{\frac{(\beta-1)}{2}}\right)}\right\}^{\frac{1}{\beta}} \tag{6}$$

Using the improvements proposed in this study, the ISSA model greatly reduces the risk of sparrows falling into local optimality while maintaining sufficient local search capability. Random numbers of [0,1] can be employed for r_1 and r_2.

3.2. RF

The random forest algorithm is a machine learning technique based on the bootstrap aggregating (Bagging) ensemble learning method. Its essence is a composite algorithm composed of multiple decision number models [39,40]. Each decision number represents a classifier within the random forest model. Bootstrapping random sampling techniques are used to draw a certain amount of data from the dataset to serve as the discriminant sample, and corresponding subsets for training and testing are generated for it [41,42]. The results of the decision trees have been obtained. All the results from the decision trees have been classified and integrated in order to form the final data outcome for the random forest ridge regression model [41,43].

The calculation steps of the RF model are outlined as follows:

Employ the Bootstrap sampling approach to randomly draw n samples out of the sample collection.

To construct a random forest, the following operations are initially performed. For each decision tree, attributes are randomly picked from the entire set of attributes. Subsequently, the optimal splitting attribute is identified to serve as the node. This process is iterated to create decision trees, thereby forming a random forest. The classification of data is ascertained via mode voting, with the mode voting equation being as follows:

$$\mathrm{H(x)} = \mathrm{argmax}\sum_{i=1}^{k} I(h_i(x) = Y) \tag{7}$$

where H(x) denotes the final voting consequence of the RF model; h_i symbolizes the prediction outcome of each single decision tree; and Y Y indicates the objective function.

RF illustrates the approximate process of the algorithm. Given that the outcome of a solitary decision tree is highly susceptible to the training set, the final result of the algorithm hinges on the category with the greatest count. Hence, multiple decision trees are incorporated into a random forest model. By augmenting the number of samples, this effectively mitigates the correlation issue among the multiple decision trees.

3.3. Ten-Fold Cross-Validation and Judgment Index

3.3.1. Ten-Fold Cross-Validation

Ten-fold cross-validation is a commonly used model validation technique primarily aimed at assessing the specific performance and generalization capability of machine learning models [44–46]. The main approach of this technique involves dividing the dataset into multiple parts, repeatedly training and validating the model on different subsets, and using the results from multiple validations to cross-check the model's predictive ability, thus providing a more stable and reliable performance estimate. The specific process is as follows:

1. The entire data set is randomly divided into 10 subsets of similar size.
2. One subset is selected as the verification set and the other nine subsets are selected as the training set. The training set is used as the data for model construction and training. The rest of the verification set is used to check the accuracy of the model built at this time.
3. The above steps are repeated 10 times, each time selecting a different subset as the validation set to ensure that the validation set used in each experiment is unique.
4. The performance indicators after each iteration were recorded and the values of ten different experiments were compared in order to evaluate the accuracy and generalizability of the model.

Of course, in different cases, the original data set can be divided into different amounts to improve the number of model tests and to enhance the precision of the model's predictive ability. Depending on the size of the original quantity set and the choice of the specific model, the number of tests in this paper will be taken as ten.

3.3.2. Judgment Index

The RMSE and R values were employed to assess the discrepancy between the forecasted and actual values, to validate the predictive accuracy of the models for the dynamic modulus of RAP concrete. The $RMSE$ value can be computed using the following formula:

$$RMSE = \sqrt{\frac{\sum_{i=1}^{n}(y_i^* - y_i)^2}{n}} \tag{8}$$

where n represents the number of data sets and y_i^* and y_i represent the predicted and actual ITS of waste plastic modified asphalt mixture, respectively.

The formula for R can be summarized as follows:

$$R = \frac{\sum_{i=1}^{N}\left(y_i^* - \overline{y^*}\right)(y_i - \overline{y})}{\sqrt{\sum_{i=1}^{n}\left(y_i^* - \overline{y^*}\right)^2}\sqrt{\sum_{i=1}^{N}(y_i - \overline{y})^2}} \tag{9}$$

where $\overline{y^*}$ and $\overline{y}$ denote the mean of the forecasted and real values, respectively.

4. Results and Discussion

4.1. Data Analysis

Before constructing a predictive model, it is crucial to conduct a comprehensive pre-processing of the original data set. Raw data sets often contain incomplete or noisy information, and the inaccuracy and instability of these data can have a significant negative impact on the training effectiveness of the model [47–49]. Therefore, to ensure that the model can learn and predict accurately, the data must be carefully cleaned and prepared before the model-building phase [50–52].

To eliminate these potential disadvantages, data cleaning is particularly important. Data cleansing aims to find and correct false data and noisy information in the data that is not relevant to the model's prediction goals, thus ensuring the purity of the data set. This process is essential to reduce the negative impact of data quality issues on model training results [53,54].

In this study in particular, we focus mainly on the relevant performance data of the applicants. Such data collection is prone to errors or data gaps, which, if not properly addressed, can greatly hinder efforts to build a model that accurately predicts student recruitment. Therefore, in the data pre-processing stage, special attention should be paid to the calibration and correction of student performance data to ensure that each datum can accurately reflect the actual academic level of students [55,56].

4.2. Correlation Analysis

After the initial processing of the data, correlation analysis of all input features is needed to improve the accuracy of the model. The purpose of this step is to determine if there is an excessive correlation between the features [57–59]. If the correlation value between two features is too high, it indicates that the relationship between them is strong, i.e., when one feature changes, the other feature may also change. This situation can affect the accuracy of the graduate student admission prediction model, as highly correlated features can introduce redundant information into the data set, which can degrade the performance of the model [60–62].

By analyzing the correlation matrix (Figure 1), we found that most of the correlation values between features are below 0.5, indicating weak correlations between them. However, there is an exception where the correlation value between the university rating and SOP is 0.5, showing a moderate level of correlation. Nevertheless, this correlation value remains within an acceptable range and will not have a significant impact on model training. The results of this correlation analysis indicate that the selected input features generally meet the requirements for model training and effectively mitigate the impact of feature relationships on the prediction model's accuracy, providing a solid foundation for constructing an accurate and reliable graduate admissions prediction model.

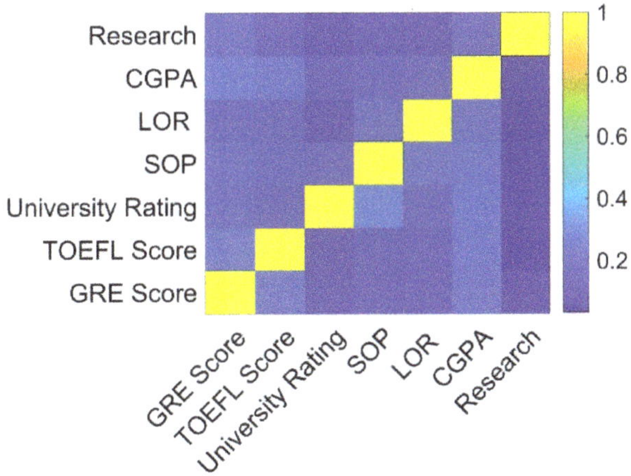

Figure 1. Correlation analysis.

4.3. Hyperparameter Tuning

After determining the selected features for the model, the collected dataset is used for model construction. However, when building the graduate admissions prediction model, the initial random forest (RF) model requires the manual tuning of several hyperparameters to ensure the model's predictive accuracy. The selection of hyperparameters requires a high level of expertise and experience from engineers. For those with less experience or data sensitivity, finding the most suitable hyperparameter values for the graduate admissions prediction problem within a short time frame can be challenging.

To enhance the model's predictive accuracy and reliability, this study employs an improved sparrow search algorithm (ISSA) to adjust and optimize the hyperparameters of the random forest algorithm. Through optimization by the ISSA, the model can identify more suitable hyperparameters, thereby better adapting to the graduate admissions prediction task.

Figure 2 shows the changes in the RF model RMSE after multiple hyperparameter selections using the ISSA algorithm. Evidently, with the increase in the number of ISSA iterations, the RMSE of the integrated ISSA–RF model gradually nears the optimal value. After the 14th iteration, the RMSE of the combined model stabilizes. This indicates that the ISSA algorithm has a significant optimization effect on the RF model, and that the chosen hyperparameters effectively enhance the RF model's information processing capability, thereby improving its predictive accuracy and applicability.

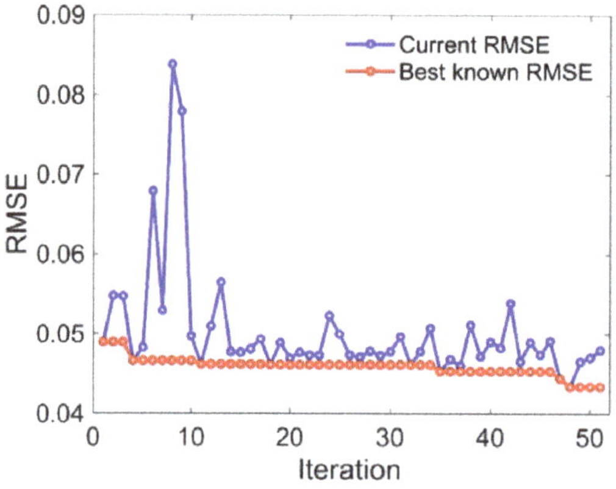

Figure 2. Hyperparameter tuning.

In summary, the ISSA algorithm demonstrates excellent performance in optimizing the RF model's hyperparameters. By leveraging the global search capability of the ISSA algorithm and the strong predictive power of the RF model, the combined ISSA–RF model shows more precise and reliable performance in the graduate admissions prediction task.

4.4. Model Construction

The following two figures display the simulation results of the ISSA–RF model using the collected student data for graduate admissions evaluation. Figure 3 presents the comparison between the predicted outcomes and the actual outcomes during the model training period. The green bars in the figure vividly display the discrepancies between the predicted values and the actual values. Generally speaking, the model shows high prediction accuracy throughout the training phase. Although there are some instances where the predicted values deviate significantly from the actual values, the majority of these deviations are below 0.05, indicating that the model effectively captures the relationship between student data and evaluation outcomes [63,64].

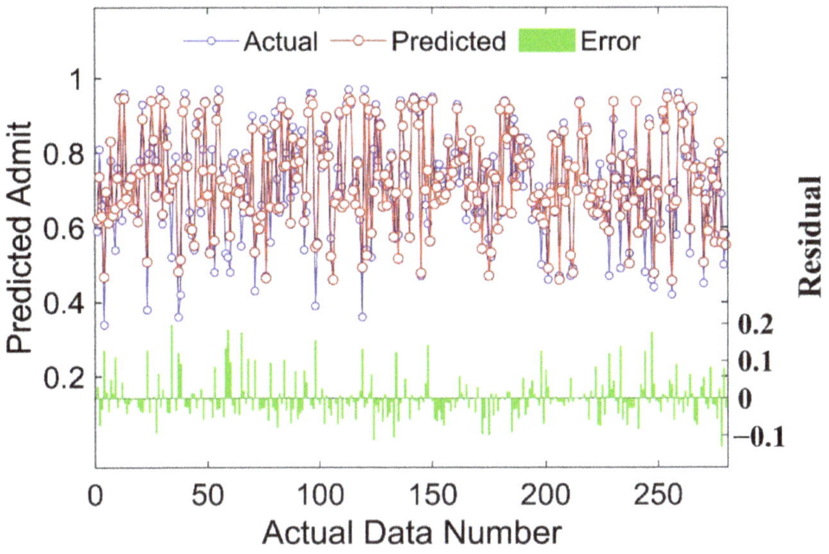

Figure 3. Training set for the actual and predicted admission.

Figure 4 reveals a more pronounced trend during the testing phase. There are some larger differences, which may be attributed to noise or outliers in the collected data. However, these anomalies have a limited impact on the overall predictive performance of the model. Despite a few larger discrepancies, the model maintains high prediction accuracy, further validating its robustness in practical applications.

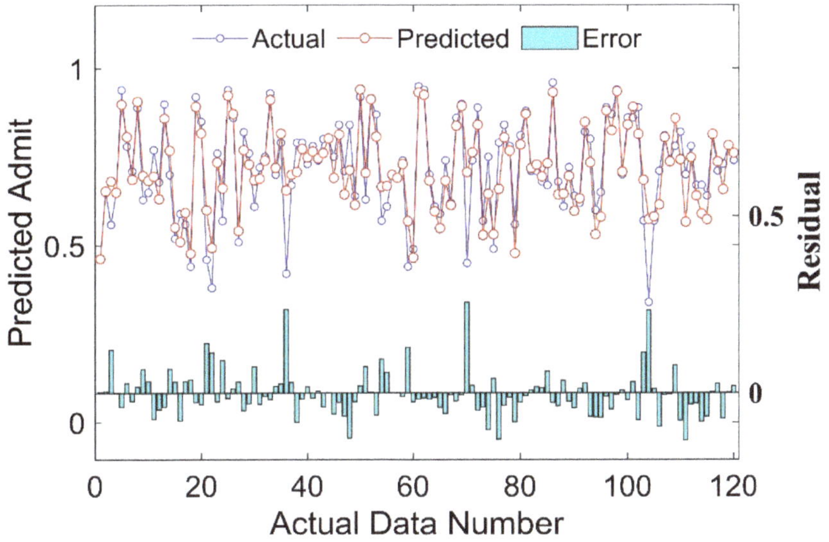

Figure 4. Test set for the actual and predicted admit.

This robustness is not only reflected in the model's fit to the training data but also in its ability to provide reliable predictions when dealing with unseen data. This capability is particularly crucial for graduate admissions evaluation, as it involves predicting students' future academic potential and research abilities. It helps the admissions committee make more scientific and rational decisions, providing students with a fairer competitive opportunity.

The provided graph illustrates only the discrepancy between experimental and actual values. There is a lack of intuitive data by which to express the quality of the model training results. Therefore, the data points should be presented in the form of Figure 5a. The model training results are judged by the specific values of root mean Square error (RMSE) and the correlation coefficient (R), thereby reflecting the predictive accuracy of the model.

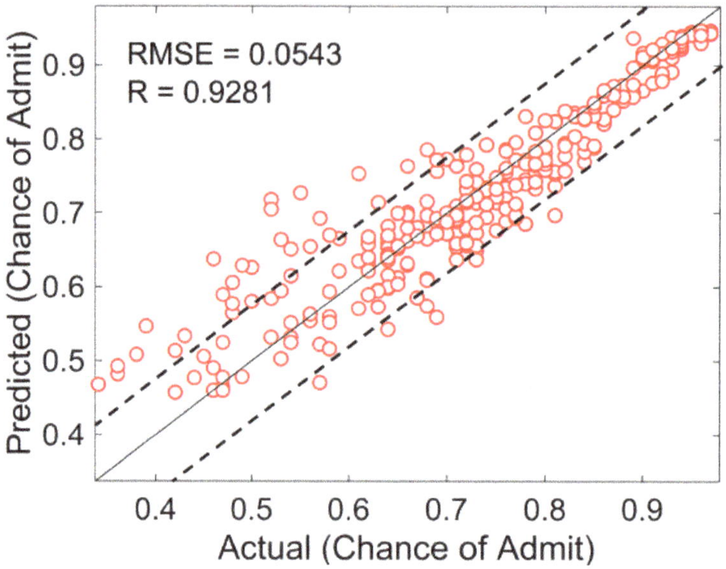

(a)

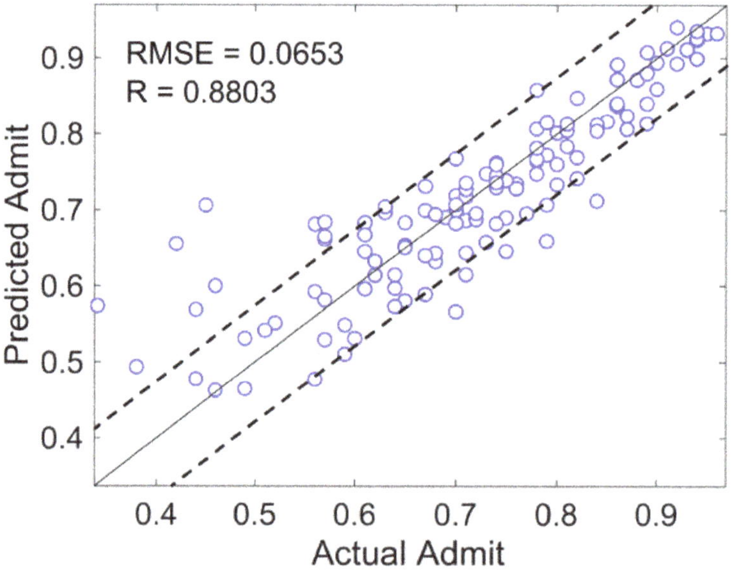

(b)

Figure 5. A comparison of the admit for actual and predicted data: (**a**) training dataset; (**b**) testing dataset.

In the model training phase, as depicted in Figure 5a, the discrepancy between the predicted results and actual values is kept within a narrow range. Most predicted points are closely related to the central line, with only a few data points slightly deviating from the expected trajectory. This close fit is consistent with the statistical data results, with an RMSE of 0.0543 and a correlation coefficient (R) as high as 0.9281. This indicates that the model has a high predictive accuracy for the training data, accurately capturing the complex relationship between input features and output results. However, it is important to be cautious about potential overfitting phenomena. The results may be due to overfitting between the data and the model during the training phase, leading to a very high similarity between predicted and actual data. This could harm the practical application of the model.

Figure 5b shows the results of testing with the remaining 30% of the data. The difference between the predicted and actual values has increased, but the overall performance is still satisfactory. The RMSE during the testing phase is 0.0653, and the correlation coefficient (R) is 0.8803. Although the predictive accuracy during the testing phase is slightly lower than during the training phase, this difference is within an acceptable range. This implies that the model works well with the training data and holds high accuracy in prediction and generalization power when dealing with unseen test data. This robustness further validates the effectiveness of the model, indicating that the high precision of the results obtained during the training phase is not due to overfitting phenomena.

In summary, through the performance analysis of the model during both the training and testing phases, it is concluded that the model has significant application value in graduate student recruitment analysis. The model provides intuitive and accurate data support for the faculty team, helping them select the most suitable candidates for graduate programs from a large pool of students. This data-driven analytical approach not only improves the efficiency of the recruitment process but also enhances the scientific and objective nature of decision-making.

4.5. Ten-Fold Cross-Validation Result

Although preliminary experimental results show that the composite model has good predictive performance in graduate admissions evaluation, these results are based on specific data sets. In practical application, the model will be faced with a variety of new data, requiring the model to have strong generalizability if it is to effectively deal with these unknown data. Therefore, performing 10-fold cross-validation is a key step to further evaluate the model's performance and generalization ability on a specific dataset.

The 10-fold cross-validation method divides the data set into ten parts, one part is selected as the test set at a time, and the remaining nine parts are used for training the model. In this way, every data point has the opportunity to be used as a test, thus ensuring the robustness of the model over different data distributions and avoiding special results that may occur during data acquisition. After 10 iterations, the mean and standard deviation of the model's performance can be calculated, which helps to obtain a more accurate understanding of the model's predictive power and generalization potential.

In addition, the 10-fold cross-validation results can be used to determine if the model shows signs of overfitting or underfitting. From the above experimental results, it can be observed that the model's results during the training stage are highly similar to the actual data, which might lead to overfitting. This phenomenon can be further judged and analyzed by the results of 10-fold cross-validation. The performance of the model fluctuates greatly over different data folds, which may indicate that the model is too sensitive to the training data, i.e., that the model has overfitting conditions and lacks the necessary generalization ability. Through the careful analysis of the cross-validation results, the stability and reliability of the model can be evaluated, and the necessary adjustment and optimization of the model can be carried out through the evaluation results in order to improve the effectiveness of the model in practical application.

Based on the test results provided (Figure 6), it can be observed that the model fits the data set to a high degree, the results are similar in each test, and that there is no excessive fluctuation. Specifically, even under the most unfavorable conditions, the RMS error is only 0.071. This value is significantly lower than the generally acceptable threshold, indicating that the model not only captures the underlying pattern of the data set but also has excellent generalizability. At the same time, it is proved that the above experimental data are reasonable, and that the model does not overfit in the training stage of the data set. This ability to generalize means that the model can maintain high prediction accuracy even in the face of unknown data.

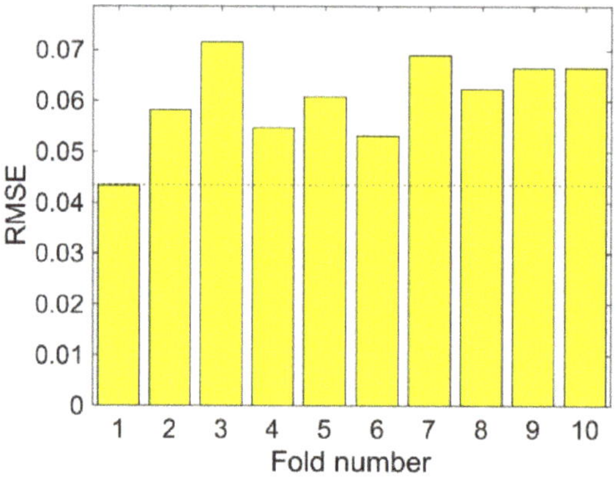

Figure 6. Ten-fold cross-validation.

In addition, low RMSE values indicate little deviation between model predictions and actual observed values, which is particularly important in practical applications. When building predictive models for graduate school admissions, for example, this high level of predictive accuracy will help admissions committees more accurately assess applicants' potential, leading to more informed and well-reasoned admissions decisions.

The results from the testing experiments indicate that the ISSA–RF model used in this study achieves high predictive accuracy and provides effective data support for graduate admissions prediction. However, the construction of the model is not straightforward and involves several key feature parameters (CGPA, GRE score, SOP, LOR, university rating, research, TOEFL score). From the data alone, it is not possible to determine the impact of each feature on the admissions prediction results. While all of these features play important roles in the model, their contributions to the final prediction results vary. Therefore, performing an importance analysis of the feature factors, using importance scores as a criterion, helps to visually express which feature factors have a greater influence on the prediction results, thereby increasing the admissions team's focus on these features.

4.6. Importance of Input Variables

Figure 7 presents the importance of the input variables regarding the input parameters. The analysis of the data showed that the importance score of cumulative grade points (CGPA) was 3.3234, significantly higher than the scores of the other six characteristics. This important result underscores the critical role of CGPA as a key indicator for assessing students' ability to learn at the graduate level. Therefore, the admissions committee should give priority to the applicant's CGPA in the selection process in order to improve the validity and accuracy of the evaluation.

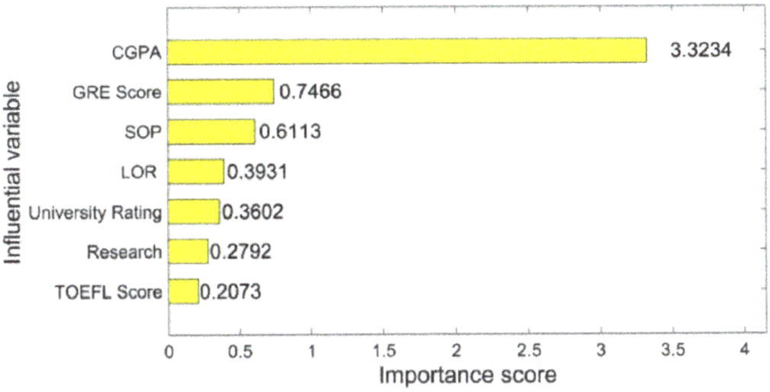

Figure 7. Importance of input variables.

The higher the importance score of a feature, the more the model relies on that attribute to distinguish between different outcomes during the prediction phase. The frequent use of this feature helps to mitigate the impact of sample noise and incomplete data, reduce the negative impact of defects in other data, and thus improve the prediction accuracy of the model. On the other hand, the relatively low importance scores for research grades and TOEFL scores mean that they have a fairly limited impact on the graduate admissions prediction process.

With this in mind, admissions teams may want to consider reducing the emphasis on these two attributes in their selection criteria. Alternatively, they may explore the possibility of eliminating or replacing these features in subsequent model iterations to improve the model's performance and ensure that the selection process is aligned with the most relevant and influential factors in graduate research.

5. Conclusions

The primary objective of this study is to utilize artificial intelligence models for processing and analyzing data related to graduate admissions evaluations and to establish an evaluation model based on this analysis. Currently, commonly used prediction models typically rely on a single core model, which often lacks the necessary generalization capability when faced with unseen data. Therefore, in this research, the ISSA algorithm is utilized as a supplementary means by which to optimize the hyperparameters of the RF model, to improve the machine learning model's responsiveness to the correlation between graduate admissions outcomes and student-related data, and to further improve the model's predictive performance. By constructing a dataset from collected student data and admission outcomes, this dataset serves as the foundational training and testing data for the model. Statistical metrics are used to evaluate the model's strengths and weaknesses. Based on the research findings, the following conclusions can be drawn:

Employing the ISSA algorithm for hyperparameter adjustment effectively enhanced the predictive capability of the RF model. The RMSE test data clearly show that, as the number of iterations increases, the gap between predicted results and actual data gradually decreases. This indicates a high degree of fit between the ISSA algorithm and the RF model, effectively leveraging the strengths of both.

The ISSA–RF algorithm effectively reduces the impact of noise and outliers in the dataset. Despite the presence of some prediction values that significantly deviate from actual values in both the training and testing datasets, these anomalies have a limited impact on the model's overall predictive accuracy. The model demonstrates high predictive precision, with a statistical parameter R reaching 0.9281.

The analysis of feature importance reveals that a student's CGPA has the greatest impact on graduate admissions outcomes. In contrast, the importance scores for research and TOEFL score are very low. This indicates that the most crucial assessment parameter for graduate admissions is the student's academic performance rather than extracurricular factors. Future admissions evaluations could consider modifying student feature indicators to include more specific in-school data rather than extracurricular achievements.

Overall, the study demonstrates that incorporating ISSA for hyperparameter optimization and focusing on the most influential features significantly improves the predictive performance and applicability of the graduate admissions model. However, there are still many limitations in this study. First of all, this study only uses an ISSA single algorithm and model for combination and analysis, although it is clear from the experimental results that this algorithm has a high accuracy for this study. However, this does not mean that other algorithms are less capable than ISSA. On the other hand, the data used in this study come from a single source and may be flawed in the face of different problems. In future research, more kinds of algorithms and models will be combined and analyzed to find more excellent algorithms and models with higher accuracy. In addition, more and different kinds of data are collected to train the model to build a model that can be widely used in the field.

Author Contributions: Conceptualization, E.L., Z.W. and J.L.; methodology, E.L., Z.W. and J.L.; software, J.H.; validation, E.L., Z.W. and J.L.; formal analysis, E.L. and Z.W.; investigation, E.L., Z.W. and J.L.; writing—original draft preparation, E.L., Z.W. and J.L.; writing—review and editing, E.L., Z.W., J.L. and J.H.; supervision, J.L. and J.H.; funding acquisition, J.L. All authors have read and agreed to the published version of the manuscript.

Funding: This research was funded by the Ministry of Education industry-university cooperative education project (2412273220), 2024 Humanities and Social Sciences Youth Fund Project of the Ministry of Education of China (24YJC760131), 2024 Higher Education Research Project of the Guangzhou Municipal Education Bureau (2024312216), Ministry of Science and Higher Education of the Russian Federation within the framework of the state assignment No. 075-03-2022-010 dated 14 January 2022, No. 075--01568-23-04 dated 28 March 2023 (Additional agreement 075-03-2022-010/10 dated 9 November 2022, Additional agreement 075-03-2023-004/4 dated 22 May 2023), FSEG-2022-0010, Ministry of Science and Higher Education of Russian Federation (funding No FSFM-2024-0025), Natural Science Foundation of Guangdong Province (2024A1515011162), Natural Science Foundation of Shandong Province (ZR2024QE021), and Guangzhou Science and technology planning project (2024A04J3831).

Data Availability Statement: The original contributions presented in the study are included in the article, further inquiries can be directed to the corresponding author/s.

Conflicts of Interest: The authors declare no conflict of interest.

References

1. Bai, Y. A Study on Intercultural Operational Ability and Assessment of Students of International Trade Major. In Proceedings of the 2012 International Conference on Financial, Management and Education Science (ICFMES 2012), Singapore, 23–24 July 2012; pp. 432–437.
2. Liu, Y.; Han, C.; Huang, L.; Wang, B.; Zhu, Z. Research and Development of Student Assessment System Based on Knowledge, Ability and Mentality. In Proceedings of the 2012 7th International Conference on Computer Science & Education, VOLS I-VI, Melbourne, Australia, 14–17 July 2012; pp. 1829–1832.
3. Gentrup, S.; Lorenz, G.; Kristen, C.; Kogan, I. Self-Fulfilling Prophecies in the Classroom: Teacher Expectations, Teacher Feedback and Student Achievement. *Learn. Instr.* **2020**, *66*, 101296. [CrossRef]
4. Murshed, N.A. A Fuzzy ARTMAP Framework for Predicting Student Dropout in Higher. In Proceedings of the 2021 International Joint Conference on Neural Networks (IJCNN), online, 18–22 July 2021.

5. Liu, Q.; Huang, Z.Y.; Yin, Y.; Chen, E.H.; Xiong, H.; Su, Y.; Hu, G.P. EKT: Exercise-Aware Knowledge Tracing for Student Performance Prediction. *IEEE Trans. Knowl. Data Eng.* **2021**, *33*, 100–115. [CrossRef]
6. Al-Adwan, A.S.; Li, N.; Al-Adwan, A.; Abbasi, G.A.; Albelbis, N.A.; Habibi, A. Extending the Technology Acceptance Model (TAM) to Predict University Students' Intentions to Use Metaverse-Based Learning Platforms. *Educ. Inf. Technol.* **2023**, *28*, 15381–15413. [CrossRef] [PubMed]
7. Randelovic, M.; Aleksic, A.; Radovanovic, R.; Stojanovic, V.; Cabarkapa, M.; Randelovic, D. One Aggregated Approach in Multidisciplinary Based Modeling to Predict Students' Education. *Mathematics* **2022**, *10*, 2381. [CrossRef]
8. Wu, D. A Brief Analysis on How Formative Assessment Helps to Develop Students' Ability of English Autonomous Learning. In Proceedings of the 2012 2nd International Conference on Applied Social Science (ICASS 2012), Kuala Lumpur, Malaysia, 1–2 February 2012; Volume 2, pp. 273–276.
9. Bashir, T.; Haoyong, C.; Tahir, M.F.; Liqiang, Z. Short Term Electricity Load Forecasting Using Hybrid Prophet-LSTM Model Optimized by BPNN. *Energy Rep.* **2022**, *8*, 1678–1686. [CrossRef]
10. Wang, L.; Zeng, Y.; Chen, T. Back Propagation Neural Network with Adaptive Differential Evolution Algorithm for Time Series Forecasting. *Expert Syst. Appl.* **2015**, *42*, 855–863. [CrossRef]
11. Pu, H.-t.; Fan, M.-q.; Zhang, H.-b.; You, B.-z.; Lin, J.-j.; Liu, C.-f.; Zhao, Y.-z.; Song, R. Predicting Academic Performance of Students in Chinese-Foreign in Running Schools with Graph Convolutional Network. *Neural Comput. Appl.* **2021**, *33*, 637–645. [CrossRef]
12. Livieris, I.E.; Drakopoulou, K.; Tampakas, V.T.; Mikropoulos, T.A.; Pintelas, P. Predicting Secondary School Students' Performance Utilizing a Semi-Supervised Learning Approach. *J. Educ. Comput. Res.* **2019**, *57*, 448–470. [CrossRef]
13. Alexandro, D. Aiming for Success: Evaluating Statistical and Machine Learning Methods to Predict High School Student Performance and Improve Early Warning Systems. Ph.D. Thesis, University of Connecticut, Storrs, CT, USA, 2018.
14. Daucourt, M.C.; Napoli, A.R.; Quinn, J.M.; Wood, S.G.; Hart, S.A. The Home Math Environment and Math Achievement: A Meta-Analysis. *Psychol. Bull.* **2021**, *147*, 565–596. [CrossRef] [PubMed]
15. Urhahne, D.; Wijnia, L. A Review on the Accuracy of Teacher Judgments. *Educ. Res. Rev.* **2021**, *32*, 100374. [CrossRef]
16. Abdelhafez, H.A.; Elmannai, H. Developing and Comparing Data Mining Algorithms That Work Best for Predicting Student Performance. *Int. J. Inf. Commun. Technol. Educ.* **2022**, *18*, 14. [CrossRef]
17. Alturki, S.; Hulpus, I.; Stuckenschmidt, H. Predicting Academic Outcomes: A Survey from 2007 Till 2018. *Technol. Knowl. Learn.* **2022**, *27*, 275–307. [CrossRef]
18. Devi, K.; Ratnoo, S. Predicting Student Dropouts Using Random Forest. *J. Stat. Manag. Syst.* **2022**, *25*, 1579–1590. [CrossRef]
19. Sharabiani, A.; Karim, F.; Sharabiani, A.; Atanasov, M.; Darabi, H. An Enhanced Bayesian Network Model for Prediction of Students' Academic in Engineering Programs. In Proceedings of the 2014 IEEE Global Engineering Education Conference (EDUCON), Istanbul, Turkey, 3–5 April 2014; pp. 832–837.
20. Harvey, J.L.; Kumar, S.A.P. A Practical Model for Educators to Predict Student Performance in K-12 Using Machine Learning. In Proceedings of the 2019 IEEE Symposium Series on Computational Intelligence (IEEE SSCI 2019), Xiamen, China, 6–9 December 2019; pp. 3004–3011.
21. Adekitan, A.I.; Noma-Osaghae, E. Data Mining Approach to Predicting the Performance of First Year Student in a University Using the Admission Requirements. *Educ. Inf. Technol.* **2019**, *24*, 1527–1543. [CrossRef]
22. Kukkar, A.; Mohana, R.; Sharma, A.; Nayyar, A. A Novel Methodology Using RNN plus LSTM plus ML for Predicting Student's Academic Performance. *Educ. Inf. Technol.* **2024**, *29*, 14365–14401. [CrossRef]
23. Asif, R.; Merceron, A.; Ali, S.A.; Haider, N.G. Analyzing Undergraduate Students' Performance Using Educational Data Mining. *Comput. Educ.* **2017**, *113*, 177–194. [CrossRef]
24. Hung, J.-L.; Shelton, B.E.; Yang, J.; Du, X. Improving Predictive Modeling for At-Risk Student Identification: A Multistage Approach. *IEEE Trans. Learn. Technol.* **2019**, *12*, 148–157. [CrossRef]
25. Kukkar, A.; Mohana, R.; Sharma, A.; Nayyar, A. Prediction of Student Academic Performance Based on Their Emotional Wellbeing and Interaction on Various E-Learning Platforms. *Educ. Inf. Technol.* **2023**, *28*, 9655–9684. [CrossRef]
26. Radwan, A.M.; Cataltepe, Z. Improving Performance Prediction on Education Data with Noise and Class Imbalance. *Intell. Autom. Soft Comput.* **2018**, *24*, 777–784. [CrossRef]
27. Crivei, L.M.; Czibula, G.; Mihai, A. A Study on Applying Relational Association Rule Mining Based Classification for Predicting the Academic Performance of Students BT. In *Knowledge Science, Engineering and Management*; Douligeris, C., Karagiannis, D., Apostolou, D., Eds.; Springer International Publishing: Cham, Switzerland, 2019; pp. 287–300.
28. Acharya, M.S.; Armaan, A.; Antony, A.S. A Comparison of Regression Models for Prediction of Graduate Admissions. In Proceedings of the 2019 Second International Conference on Computational Intelligence in Data Science (ICCIDS 2019), Chennai, India, 21–23 February 2019.

29. Feraco, T.; Resnati, D.; Fregonese, D.; Spoto, A.; Meneghetti, C. An Integrated Model of School Students' Academic Achievement and Life Satisfaction. Linking Soft Skills, Extracurricular Activities, Self-Regulated Learning, Motivation, and Emotions. *Eur. J. Psychol. Educ.* **2023**, *38*, 109–130. [CrossRef]
30. Steenbergen-Hu, S.; Makel, M.C.; Olszewski-Kubilius, P. What One Hundred Years of Research Says About the Effects of Ability Grouping and Acceleration on K-12 Students' Academic Achievement: Findings of Two Second-Order Meta-Analyses. *Rev. Educ. Res.* **2016**, *86*, 849–899. [CrossRef]
31. Alwarthan, S.A.; Aslam, N.; Khan, I.U. Predicting Student Academic Performance at Higher Education Using Data: A Systematic Review. *Appl. Comput. Intell. Soft Comput.* **2022**, *2022*, 8924028. [CrossRef]
32. Wu, P.-H.; Wu, H.-K.; Hsu, Y.-S. Establishing the Criterion-Related, Construct, and Content Validities of a Simulation-Based Assessment of Inquiry Abilities. *Int. J. Sci. Educ.* **2014**, *36*, 1630–1650. [CrossRef]
33. Shan, L. The Study of Test Method in Authentic Assessment of English Reading Ability. In Proceedings of the 2nd International Conference on Education, Management and Social Science, Shanghai, China, 21–22 August 2014; Volume 6, pp. 444–446.
34. Rajak, A.; Shrivastava, A.K.; Vidushi. Applying and Comparing Machine Learning Classification Algorithms for predicting the Results of Students. *J. Discret. Math. Sci. Cryptogr.* **2020**, *23*, 419–427. [CrossRef]
35. Tang, P.; Wang, Y.; Shen, N. Prediction of College Students' Physique Based on Support Vector. In Proceedings of the 2019 Chinese Automation Congress (CAC2019), Hangzhou, China, 22–24 November 2019; pp. 2439–2443.
36. Gharehchopogh, F.S.; Namazi, M.; Ebrahimi, L.; Abdollahzadeh, B. Advances in Sparrow Search Algorithm: A Comprehensive Survey. *Arch. Comput. Methods Eng.* **2023**, *30*, 427–455. [CrossRef] [PubMed]
37. Yue, Y.; Cao, L.; Lu, D.; Hu, Z.; Xu, M.; Wang, S.; Li, B.; Ding, H. Review and Empirical Analysis of Sparrow Search Algorithm. *Artif. Intell. Rev.* **2023**, *56*, 10867–10919. [CrossRef]
38. Yao, Z.; Wang, Z.; Wang, D.; Wu, J.; Chen, L. An Ensemble CNN-LSTM and GRU Adaptive Weighting Model Based Improved Sparrow Search Algorithm for Predicting Runoff Using Historical Meteorological and Runoff Data as Input. *J. Hydrol.* **2023**, *625*, 129977. [CrossRef]
39. Wang, Q.A.; Zhang, J.; Huang, J.D. Simulation of the Compressive Strength of Cemented Tailing Backfill through the Use of Firefly Algorithm and Random Forest Model. *Shock Vib.* **2021**, *2021*, 5536998. [CrossRef]
40. Ma, H.X.; Liu, J.D.; Zhang, J.; Huang, J.D. Estimating the Compressive Strength of Cement-Based Materials with Mining Waste Using Support Vector Machine, Decision Tree, and Random Forest Models. *Adv. Civ. Eng.* **2021**, *2021*, 6629466. [CrossRef]
41. Hajjem, A.; Bellavance, F.; Larocque, D. Mixed-Effects Random Forest for Clustered Data. *J. Stat. Comput. Simul.* **2014**, *84*, 1313–1328. [CrossRef]
42. Karabadji, N.E.I.; Korba, A.A.; Assi, A.; Seridi, H.; Aridhi, S.; Dhifli, W. Accuracy and Diversity-Aware Multi-Objective Approach for Random Forest Construction. *Expert Syst. Appl.* **2023**, *225*, 120138. [CrossRef]
43. Speiser, J.L.; Miller, M.E.; Tooze, J.; Ip, E. A Comparison of Random Forest Variable Selection Methods for Classification Prediction Modeling. *Expert Syst. Appl.* **2019**, *134*, 93–101. [CrossRef]
44. Wang, S.; Wang, H.T.; Lu, Y.J.; Huang, J.D. Toward a Comprehensive Evaluation of Student Knowledge Assessment for Art Education: A Hybrid Approach by Data Mining and Machine Learning. *Appl. Sci.* **2024**, *14*, 5020. [CrossRef]
45. Liu, J.D.; Li, G.C.; Yang, S.; Huang, J.D. Prediction Models for Evaluating the Strength of Cemented Paste Backfill: A Comparative Study. *Minerals* **2020**, *10*, 1041. [CrossRef]
46. Huang, J.D.; Zhou, M.M.; Zhang, J.; Ren, J.L.; Vatin, N.I.; Sabri, M.M.S. The Use of GA and PSO in Evaluating the Shear Strength of Steel Fiber Reinforced Concrete Beams. *KSCE J. Civ. Eng.* **2022**, *26*, 3918–3931. [CrossRef]
47. Amani, A.; Alamolhodaei, H.; Ghanbari, R.; Radmehr, F. An Epidemiological Model for Predicting Students' Mathematics Anxiety. *J. Interdiscip. Math.* **2021**, *24*, 793–805. [CrossRef]
48. Guleria, P.; Thakur, N.; Sood, M. Predicting Student Performance Using Decision Tree Classifiers and Information Gain. In Proceedings of the 2014 International Conference on Parallel, Distributed and Grid Computing (PDGC), Solan, India, 11–13 December 2014; Singh, Y., Sehgal, V., Ghrera, S.P., Eds.; 2014; pp. 126–129.
49. Nachouki, M.; Abou Naaj, M. Predicting Student Performance to Improve Academic Advising Using the Random Forest Algorithm. *Int. J. Distance Educ. Technol.* **2022**, *20*, 296702. [CrossRef]
50. Cerchiara, J.A.; Kim, K.J.; Meir, E.; Wenderoth, M.P.; Doherty, J.H. A New Assessment to Monitor Student Performance in Introductory Neurophysiology: Electrochemical Gradients Assessment Device. *Adv. Physiol. Educ.* **2019**, *43*, 211–220. [CrossRef]
51. Austin, P.C.; Merlo, J. Intermediate and Advanced Topics in Multilevel Logistic Regression Analysis. *Stat. Med.* **2017**, *36*, 3257–3277. [CrossRef]
52. Sperandei, S. Understanding Logistic Regression Analysis. *Biochem. Med.* **2014**, *24*, 12–18. [CrossRef]
53. Granic, A. Educational Technology Adoption: A Systematic Review. *Educ. Inf. Technol.* **2022**, *27*, 9725–9744. [CrossRef] [PubMed]
54. Streit, J.A. Online Higher Education: A Correlational Study of Student Engagement, Satisfaction, and Perceptions of Learning; 2019; ISBN 978-1-392-02460-7. Available online: https://webofscience.clarivate.cn/wos/alldb/full-record/PQDT:61207343 (accessed on 21 December 2024).

55. Alshamaila, Y.; Alsawalqah, H.; Aljarah, I.; Habib, M.; Faris, H.; Alshraideh, M.; Salih, B.A. An Automatic Prediction of Students' Performance to Support the university Education System: A Deep Learning Approach. *Multimed. Tools Appl.* **2024**, *83*, 46369–46396. [CrossRef]
56. Wang, Y.; Pan, Z. Modeling the Effect of Chinese EFL Teachers' Self-Efficacy and Resilience on Their Work Engagement: A Structural Equation Modeling. *Sage Open* **2023**, *13*, 21582440231214329. [CrossRef]
57. Aydilek, I.B. A Hybrid Firefly and Particle Swarm Optimization Algorithm for Computationally Expensive Numerical Problems. *Appl. Soft Comput.* **2018**, *66*, 232–249. [CrossRef]
58. Wang, H.; Wang, W.; Zhou, X.; Sun, H.; Zhao, J.; Yu, X.; Cui, Z. Firefly Algorithm with Neighborhood Attraction. *Inf. Sci.* **2017**, *382*, 374–387. [CrossRef]
59. Liang, X.M.; Yu, X.; Jin, Y.; Huang, J.D. Compactness Prediction of Asphalt Concrete Using Ground-Penetrating Radar: A Comparative Study. *Constr. Build. Mater.* **2022**, *361*, 129588. [CrossRef]
60. Zhang, Z. Model Building Strategy for Logistic Regression: Purposeful Selection. *Ann. Transl. Med.* **2016**, *4*. [CrossRef]
61. Zare, M.; Ghasemi, M.; Zahedi, A.; Golalipour, K.; Mohammadi, S.K.; Mirjalili, S.; Abualigah, L. A Global Best-Guided Firefly Algorithm for Engineering Problems. *J. Bionic Eng.* **2023**, *20*, 2359–2388. [CrossRef]
62. Huang, J.D.; Losa, M.; Leandri, P.; Kumar, S.G.; Zhang, J.F.; Sun, Y.T. Potential Anti-Vibration Pavements with Damping Layer: Finite Element (FE) Modeling, Validation, and Parametrical Studies. *Constr. Build. Mater.* **2021**, *281*, 122550. [CrossRef]
63. Huang, J.D.; Zhou, M.M.; Yuan, H.W.; Sabri, M.M.S.; Li, X. Prediction of the Compressive Strength for Cement-Based Materials with Metakaolin Based on the Hybrid Machine Learning Method. *Materials* **2022**, *15*, 3500. [CrossRef]
64. Huang, J.D.; Zhou, M.M.; Yuan, H.W.; Sabri, M.M.S.; Li, X. Towards Sustainable Construction Materials: A Comparative Study of Prediction Models for Green Concrete with Metakaolin. *Buildings* **2022**, *12*, 772. [CrossRef]

applied sciences

MDPI

Article

A Study of Potential Applications of Student Emotion Recognition in Primary and Secondary Classrooms

Yimei Huang, Wei Deng * and Taojie Xu

Hubei Key Laboratory of Digital Education, Central China Normal University, Wuhan 430079, China
* Correspondence: sdengwei@mail.ccnu.edu.cn

Abstract: Emotion recognition is critical to understanding students' emotional states. However, problems such as crowded classroom environments, changing light, and occlusion often affect the accuracy of recognition. This study proposes an emotion recognition algorithm specifically for classroom environments. Firstly, the study adds the self-made MCC module and the Wise-IoU loss function to make object detection in the YOLOv8 model more accurate and efficient. Compared with the native YOL0v8x, it reduces the parameters by 16% and accelerates the inference speed by 20%. Secondly, in order to address the intricacies of the classroom setting and the specific requirements of the emotion recognition task, a multi-channel emotion recognition network (MultiEmoNet) has been developed. This network fuses skeletal, environmental, and facial information, and introduces a central loss function and an attention module AAM to enhance the feature extraction capability. The experimental results show that MultiEmoNet achieves a classification accuracy of 91.4% on a homemade classroom student emotion dataset, which is a 10% improvement over the single-channel classification algorithm. In addition, this study also demonstrates the dynamic changes in students' emotions in the classroom through visual analysis, which helps teachers grasp students' emotional states in real time. This paper validates the potential of multi-channel information-fusion deep learning techniques for classroom teaching analysis and provides new ideas and tools for future improvements to emotion recognition techniques.

Keywords: YOLOv8; object detection; emotion recognition; pedagogical analysis

1. Introduction

Students' emotional state is not only an important variable in the learning process, but also directly affects their motivation, attention, and engagement [1]. As an important application of artificial intelligence in the field of education, emotion recognition technology is gradually becoming an important tool for enhancing teaching effectiveness in primary and secondary school classrooms, and it also has a profound impact on personalized support for students. Therefore, the study of classroom emotion recognition technology can dynamically monitor and feedback students' emotional states, providing teachers with timely emotional information support. This technology not only has important theoretical value, but also has profound practical significance for improving teaching in practice.

Deep learning techniques have made significant progress in emotion recognition in recent years. For example, convolutional neural networks (CNN) and recurrent neural networks (RNN) have demonstrated significant potential in facial expression and speech emotion recognition [2,3], and these models are able to accurately capture complex emotional features, and improve the accuracy and efficiency of emotion recognition. Compared with traditional emotion assessment methods, deep learning-based emotion recognition technology can provide a more objective and real-time emotion assessment, effectively reducing the influence of human subjective factors. This provides key technical support for teachers to quickly and accurately perceive students' emotional states and then implement personalized teaching.

However, there are challenges to applying deep learning and emotion recognition technology in primary and secondary classrooms. Firstly, the changing positions and postures of students and teachers in classroom environments increase the difficulty of emotion recognition and place higher demands on the robustness of the techniques. Secondly, environmental disturbances, such as classroom background noise [4] and lighting intensity and distribution [5], can weaken the accuracy of emotion recognition. In addition, occlusion problems [6] prevent the full capture of students' facial expressions and body features, which in turn affects the recognition accuracy. In order to effectively address these issues, it is essential to enhance the adaptability and precision of the technique.

Based on this, the main objectives of this study are as follows:

- Improving deep learning-based object detection algorithms to meet the special needs of teaching scenarios for the difficulty of student object detection due to the large number of people in complex classroom environments.
- To develop a multi-channel emotion recognition network aimed at enhancing the accuracy and stability of emotion recognition, addressing the current gap in research attributed to the lack of students' facial expression features within classroom environments.
- This study will provide a visual analysis of student emotions within the classroom, offering comprehensive and accurate data for the continuous monitoring of classroom emotions.

2. Related Work

2.1. Student Object Detection Algorithm

In a classroom setting, accurate student object detection is a prerequisite for classroom emotion recognition. Object detection networks are divided into two main categories: Single-Stage Detectors and Two-Stage Detectors. Single-stage detectors generate target frames and categories simultaneously through a single forward pass, which is fast and suitable for real-time applications. YOLO and SSD are representative algorithms, with the former detecting through a single convolutional network and the latter using multi-scale feature maps to enhance the detection of different target sizes. Chen et al. [7] made student behavior detection in classroom environments more efficient based on the YOLOv4 behavior detection algorithm. A two-stage detector generates candidate frames first, then classifies and locates them with higher accuracy. R-CNN, along with its improved Faster R-CNN, enhances the detection efficiency and accuracy of the regional proposition network (RPN). For example, studies have pointed out the high accuracy of Faster R-CNN in detecting specific student postures [8]. Although these methods offer a preliminary technical foundation for object detection, they still require improvement in their adaptability to complex scenes.

2.2. Classroom Student Emotional Classification

Emotion categorization theory suggests that emotions are adaptive responses of individuals to external stimuli and have evolved over the course of evolution. Ekman [9] first proposed a theory of six basic emotions: happiness, sadness, anger, fear, surprise, and disgust. Plutchik [10] further extended this categorization by proposing eight basic emotions: anger, fear, sadness, disgust, surprise, anticipation, acceptance, and joy. Although these models provide a theoretical framework, there are limitations to these classifications for the identification of emotions in a classroom setting.

Emotional expressions in classroom contexts differ from traditional emotion categorization. Although Ekman's emotion theory is a landmark in the field of emotion recognition, its categorization may not fully encompass the complexity of students' emotions in a particular classroom setting. In order to adapt to classroom scenarios, researchers have developed their own criteria for categorizing emotions based on practical needs. For example, Pekrun et al. [1,11] proposed his Control-Value Theory, which categorizes emotions into four dimensions: positive and negative emotions and activated and inactivated emotions. This framework provides refined theoretical support for analyzing classroom emotions and is particularly suitable for emotion recognition in educational scenarios.

Kosti et al. [12] proposed a classification method based on emotion vocabulary using about 400 emotion words, which were grouped into 26 sentiment categories by cluster analysis (Table 1). This approach not only considers the similarity of emotional words but also pays special attention to the visually separable features of emotions. For this study, we selected six emotion categories related to classroom teaching and learning: Pleasure, Peace, Engagement, Fatigue, Doubt, and Disconnection. The selection of these emotion categories is closer to the reality of teaching and provides a more accurate reference for emotion recognition in teaching scenarios.

Table 1. Classification of 26 distinct emotions [12].

No	Emotion Categories	No	Emotion Categories
1	Affection	14	Excitement
2	Anger	15	Fatigue
3	Annoyance	16	Fear
4	Anticipation	17	Happiness
5	Aversion	18	Pain
6	Confidence	19	Peace
7	Disapproval	20	Pleasure
8	Disconnection	21	Sadness
9	Disquietment	22	Sensitivity
10	Doubt/Confusion	23	Suffering
11	Embarrassment	24	Surprise
12	Engagement	25	Sympathy
13	Esteem	26	Yearning

2.3. Current Status of Emotion Recognition Technology

Emotional recognition determines the emotional state of an individual by analyzing their behavior, mental state, or physiological signals [13]. Mehrabian's study revealed that facial expressions account for 55% of emotional information, intonation for 38%, and speech words for only 7% [14]. This has led to the mainstream of facial expressions in emotion recognition, prompting numerous studies to extract emotions from facial features. For example, Sarvakar et al. [15] used CNN to detect different types of facial expressions. However, there are limitations in emotion recognition for individual facial expressions, and combining other information sources, such as body gestures and background scenes, has emerged as a new research direction.

Researchers have also verified the importance of body posture in emotional expression. Coulson [16] demonstrated the strong association between posture and emotion by analyzing joint movements in static postures. Gavrilescu [17] further combined neural networks and body posture for emotion analysis to improve the accuracy of recognition. Integrating a multi-channel model of facial expression and limb posture allows for a more comprehensive reflection of an individual's emotional state.

Scene contextual information also plays a key role in emotion recognition. The accuracy of emotion recognition significantly improves when facial expressions are consistent with the scene [18]. Kosti et al. [19] proposed a deep learning-based context-aware emotion recognition method that fuses character and scene features to improve the classification accuracy. Lee et al. [20], on the other hand, utilized graph convolutional networks (GCNs) to encode contextual information and enhance the robustness of emotion recognition. Yang et al.'s [21] contextual causal intervention module (CCIM) further improves the performance of the context-aware model by reducing the scene context bias.

In classroom contexts, emotion recognition techniques help teachers adjust teaching strategies based on students' emotional feedback. However, classroom emotion recognition faces challenges such as the scarcity of datasets, occlusion effects, and annotation accuracy. For this reason, researchers have introduced deep learning techniques to improve the accuracy of classroom emotion recognition. Sharma et al. [22] employed a CNN and transfer learning techniques to identify and categorize students' facial emotions in classroom settings, utilizing the VGG16 model. Chen et al. [23] created a spatio-temporal residual attention network (STRAN)-based model to pull out features of students' facial expressions. They were able to achieve an 80.45% recognition rate on a dataset of students' facial expressions that they made themselves.

These studies have shown that combining multiple information sources can improve the comprehensiveness and accuracy of emotion recognition. However, current classroom emotion recognition still primarily relies on facial expressions [24], and the changing light and occlusion in classroom environments easily obscure facial features, thereby affecting its accuracy and robustness [25].

3. Methodologies

Based on the above review, this study proposes a student object detection algorithm that enhances the YOLOv8 model with the aim of improving the accuracy of student location detection and providing more reliable data support for emotion analysis. Meanwhile, to address the impact of student behaviors such as facial occlusion, head down, or sideways on emotion recognition, this study introduces MultiEmoNet, which aims to overcome the limitations of single facial feature recognition in situations where information is scarce and enhance the robustness of emotion recognition.

3.1. Improved Student Object Detection Algorithm Based on YOLOv8

3.1.1. YOLOv8 Framework

The Ultralytics team developed YOLOv8, a real-time object detection model. YOLOv8 does a better job of finding small and dense targets than the previous version because it has a better feature extraction network, an adaptive anchor mechanism, and a multi-scale feature fusion strategy [26]. However, in complex classroom scenarios, the model frequently results in missed detection and false detection during feature extraction due to factors such as changing light conditions, student postures, and susceptibility to occlusion. This study proposes an improvement scheme based on the MCC module and the WIoU loss function to replace the original C2f module and the CIoU loss function to achieve better feature extraction and loss optimization. Figure 1 displays the YOLOv8x-Improved model.

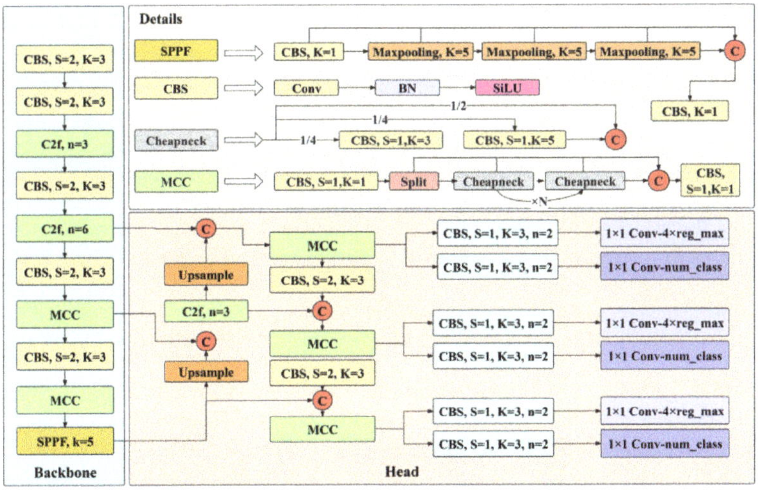

Figure 1. The network architecture of YOLOv8x-Improved.

3.1.2. MCC Module

YOLOv8 enhances the accuracy and efficiency of the model by replacing the C3 module with the C2f module in the CSPLayer to enhance the feature processing capability and reduce the computational complexity. Practice shows that adjusting the CSPLayer can effectively improve the network performance. When it comes to finding targets, multi-scale features are very important. Adding null convolution helps the growth of multi-scale pyramid models, which greatly enhance the accuracy of localization. Different sensory fields capture detailed and overall information, and after integration, models that integrate multi-scale pyramid structures perform well in object detection. For this reason, this paper proposes the multi-group cheap convolution (MCC) module as an alternative to the C2f module to handle multi-scale features more efficiently and optimize the computational cost.

The MCC module is shown in Figure 2 with the following steps:

1. The input tensor is split into two parts, where x_group occupies half of the channel and x_cheap occupies the other half.
2. The x_group part is rearranged to facilitate processing by group. The term 'group' here refers to the set of convolutional layers with different convolutional kernel scales.
3. Convolutional layers with different kernel sizes (e.g., 3×3, 5×5) are applied to each group of x_group to capture features at different scales.
4. The processed x_group is connected to the unprocessed x_cheap in the channel dimension to combine features at different scales with the original features.
5. Finally, a 1×1 convolutional layer is applied over the fused features in order to integrate all features while keeping the number of channels constant and outputting the final feature tensor.

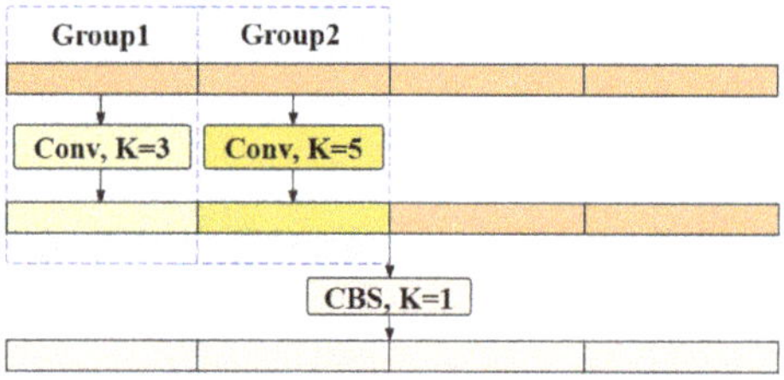

Figure 2. Schematic diagram of the MCC module.

As illustrated in Equation (1), this represents the calculation process for the MCC module.

$$MCC(X) = CBS(Concat(CW_{k_1*k_1}(x_1), \ldots . CW_{k_n*k_n}(x_n))) \quad (1)$$

where $X = [x_1, x_2 \ldots .x_n]$ denotes splitting the input feature X into multiple heads in the channel dimension, $k_i \in \{3, 5, \ldots, k\}$ denotes a monotonic increase in the kernel size by 2, and C denotes the convolution operation, where CBS is a combination of convolution, batch normalization, and a sequence of activation functions, SILU.

To assess the efficiency of the MCC module, this study compares the number of parameters in the YOLOv8 model following the implementation of both the C2f and MCC modules. Table 2 shows the parameter changes after the MCC module. The findings indicate that the MCC module reduces the number of parameters by approximately 25% relative to the C2f module, while also achieving close to a 20% reduction at the model level. This significant decrease enhances both the network efficiency and inference speed. By retaining the first two C2f modules within the Backbone and maintaining one C2f module in the Head section, this study strikes a balance between the model performance and parameter efficiency. This approach ensures that sufficient feature information is captured effectively, allowing for high accuracy in object detection despite reductions in the parameters.

Table 2. Parameters of YOLOv8 model with replaced C2f module.

No	From	N	Params	Module	Arguments
0	−1	1	2320	Conv	[3, 80, 32]
1	−1	1	115,520	Conv	[80, 160, 3, 2]
2	−1	3	436,800	C2f	[160, 160, 3, True]
3	−1	1	461,440	Conv	[160, 320, 3, 2]
4	−1	6	3,281,920	C2f	[320, 320, 6, True]
5	−1	1	1,844,480	Conv	[320, 640, 3, 2]
6	−1	6	9,509,760 (Reduced by 27%)	C2f_MCC	[640, 640, 6, True]
7	−1	1	3,687,680	Conv	[640, 640, 3, 2]
8	−1	3	5,165,760 (Reduced by 26%)	C2f_MCC	[640, 640, 3, True]
9	−1	1	1,025,920	SPPF	[640, 640, 5]
10	−1	1	0	Upsample	[None, 2, 'nearest']
11	[−1, 6]	1	0	Concat	[1]
12	−1	3	5,575,360 (Reduced by 24%)	C2f_MCC	[1280, 640, 3]
13	−1	1	0	Upsample	[None, 2, 'nearest']
14	[−1, 4]	1	0	Concat	[1]
15	−1	3	1,948,800	C2f	[960, 320, 3]
16	−1	1	922,240	Conv	[320, 320, 3, 2]
17	[−1, 12]	1	0	Concat	[1]
18	−1	3	5,370,560 (Reduced by 25%)	C2f_MCC	[960, 640, 3]
19	−1	1	3,687,680	Conv	[640, 640, 3, 2]
20	[−1, 9]	1	0	Concat	[1]
21	−1	3	5,575,360 (Reduced by 24%)	C2f_MCC	[1280, 640, 3]
22	[15, 18, 21]	1	8,795,008	Detect	[80, [320, 640, 640]]

3.1.3. Wise-IoU

The default CIoU loss function utilized in YOLOv8 is optimized for the bounding box overlap, centroid distance, and aspect ratio. However, it exhibits limitations in handling both high- and low-quality samples, particularly when the sample distribution is imbalanced. To mitigate this issue, the present paper proposes substituting CIoU with the WIoUv3 loss function [27]. The WIoU (Wise-IoU) family includes WIoUv1, WIoUv2, and WIoUv3 versions, each tailored to optimize the performance across different scenarios.

WIoUv1 effectively differentiates between high-quality and low-quality samples by incorporating distance attention. This approach mitigates the over-penalization of low-quality samples and enhances the generalization ability. The expression for WIoUv1 is presented in Equations (2) and (3).

$$L_{WIoUv1} = R_{WIoU} L_{IoU} \tag{2}$$

$$R_{WIoU} = \exp\left(\frac{\left(x - x_{gt}\right)^2 + \left(y - y_{gt}\right)^2}{\left(W_g^2 + H_g^2\right)^*}\right) \tag{3}$$

This will significantly enhance the L_{IoU} of normal-quality anchor frames when $R_{WIoU} \in [1,\ e]$. It is important to note that L_{IoU} lies between [0, 1] when the anchor frame exhibits substantial overlap with the target frame. This situation markedly reduces the L_{IoU} for high-quality anchor frames, leading to concerns regarding the distance between their centroids. Here, W_g and H_g represent the dimensions of the smallest external frame (Figure 3).

WIoUv2, conversely, enhances the model's emphasis on challenging samples by introducing monotonic focusing coefficients. This approach also addresses the issue of slower convergence during the later stages of training, as illustrated in Equation (4).

$$L_{WIoUv2} = L_{IoU}^{\gamma *} L_{WIoUv1}, \ \gamma > 0 \tag{4}$$

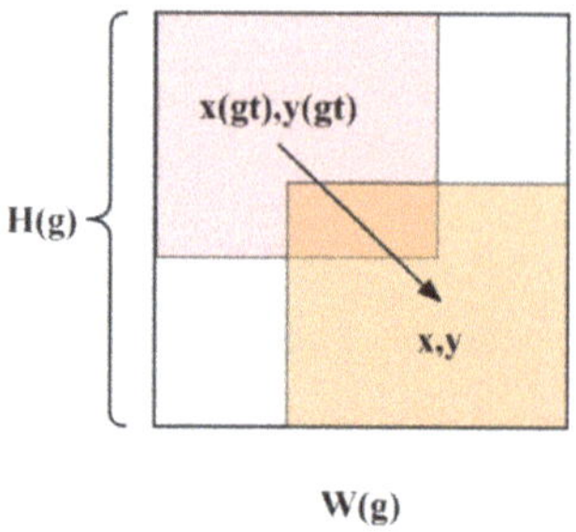

Figure 3. Schematic diagram of WIoU.

Due to the incorporation of the focusing coefficients, the gradient of the WIouv2 backpropagation is also modified, as illustrated in Equation (5).

$$\frac{\partial L_{\mathrm{WioUv2}}}{\partial L_{\mathrm{IoU}}} = L_{\mathrm{IoU}}^{\gamma*} \frac{\partial L_{\mathrm{WioUv1}}}{\partial L_{\mathrm{IoU}}}, \quad \gamma > 0 \tag{5}$$

As the reduction in L_{IoU} leads to a corresponding decrease in the gradient gain $r = L_{\mathrm{IoU}}^{\gamma*} \in [0, 1]$, the convergence rate slows down during the later stages of training. To address this issue, we introduce the mean value of L_{IoU} as a normalization factor, as illustrated in Equation (6).

$$L_{W\mathrm{IoUv2}} = \left(\frac{L_{\mathrm{IoU}}^{*}}{\overline{L_{\mathrm{IoU}}}}\right)^{\gamma} L_{W\mathrm{IoUv1}} \tag{6}$$

where $\overline{L_{IoU}}$ denotes the exponential running average with momentum m. The dynamically updated normalization factor maintains the gradient gain $r = \left(\frac{L_{IoU}^{*}}{\overline{L_{IoU}}}\right)^{\gamma}$ at a generally elevated level, effectively addressing the issue of slow convergence during the later stages of training. The degree of anomaly of the anchor frame is represented by the ratio β of L_{IoU} to $\overline{L_{IoU}}$, as shown in Equation (7).

$$\beta = \frac{L_{\mathrm{IoU}}^{*}}{\overline{L_{\mathrm{IoU}}}} \in [0, +\infty] \tag{7}$$

Lower degrees of abnormality correspond to higher-quality anchor frames; therefore, small gradient gains are assigned to these frames to concentrate the bounding box regression on those of average quality. Concurrently, small gradient gains are also allocated to anchor frames with a high degree of anomaly in order to mitigate the impact of detrimental gradients arising from low-quality samples. The nonmonotonic focusing coefficient was formulated using β and applied to WIoUv1, as illustrated in Equation (8).

$$L_{W\mathrm{IoUv3}} = rL_{W\mathrm{IoUv1}}, r = \frac{\beta}{\delta\alpha^{\beta-\delta}} \tag{8}$$

where $r = 1$ when $\beta = \delta$. As shown in Figure 4, the highest gradient gain will be obtained when the degree of abnormality of the anchor frame satisfies $\beta = C$ (C is a constant value). Since $\overline{L_{IoU}}$ is dynamic, the quality classification criteria of the anchor frames are also dynamic, which enables WIoUv3 to develop a gradient gain allocation strategy that best fits the current situation during training.

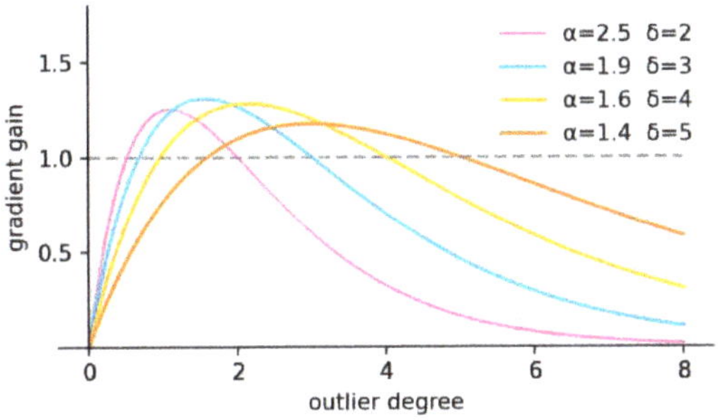

Figure 4. Impact of WIoUv3 hyperparameter selection on gradient gain curve [27].

3.2. Multi-Channel Emotion Recognition Algorithm for the Classroom

In natural classroom environments, the recognition of student emotions encounters several challenges, including occluded faces, inadequate lighting conditions, and variations in posture. Relying solely on facial expression information can significantly impact the accuracy of emotion recognition. To address these issues, this paper proposes a multi-channel fusion-based information processing method aimed at enhancing both the accuracy and robustness of emotion recognition by integrating facial features, the background context, and skeletal data [19,28].

3.2.1. EfficientNetv2 Network

EfficientNetv2 offers several advantages, including a lightweight architecture, rapid processing speed, and high accuracy. These characteristics render it particularly well-suited for efficient image feature extraction tasks. Consequently, this study utilizes EfficientNetv2 to extract image features.

EfficientNet was introduced by Google in 2019, and its core innovation lies in the "composite scaling factor" approach. This method enhances the computational efficiency and accuracy by simultaneously optimizing the network's depth, width, and resolution [29]. While EfficientNetv1 has demonstrated a significant performance on the ImageNet dataset, it suffers from slow inference times and is heavily dependent on large volumes of training data [30].

To address these challenges, the Google team introduced EfficientNetv2 in 2021 (Figure 5). This new architecture is designed to adapt to application scenarios that demand higher speed and efficiency by balancing design elements to enhance the training speed while reducing the number of parameters. EfficientNetv2 presents two significant advancements over its predecessor, EfficientNetv1: First, it employs the Fused-MBConv module instead of the MBConv module for shallow networks. This modification optimizes the balance between the performance and efficiency by eliminating ascending convolutions and enabling shortcut branches under specific conditions. Secondly, EfficientNetv2 introduces a non-uniform scaling strategy for the network. This approach allows differential adjustments based on each stage's contribution to the overall network performance; certain stages are allocated more resources while others are simplified accordingly. Such refinements enhance the overall network efficiency without compromising or even improving its performance.

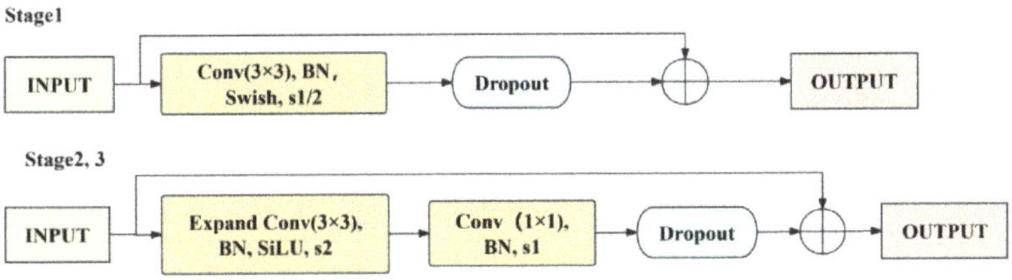

Figure 5. Adjustment module diagram of EfficientNetv2.

3.2.2. Multi-Channel Emotion Recognition Network Architecture

In this study, we aim to develop a Multi-Channel Emotion Recognition Network (MultiEmoNet) utilizing the EfficientNetv2-S model. This model is characterized by its lightweight design and high accuracy, making it suitable for feature extraction [31]. The primary focus of our network will be facial expressions, while background information and skeletal data will serve as supplementary inputs. A detailed description of the MultiEmoNet architecture is provided in the following section.

Face Extraction Module

In image classification tasks, the expressions of the same individual exhibit greater similarity than identical expressions across different individuals. This phenomenon complicates the differentiation between various categories of facial expressions. To address this challenge, the present study incorporates a CLF within the feature extraction module to optimize both intra-class and inter-class variations. The implementation of Center Loss enhances the accuracy of facial expression classification by minimizing intra-class variation while indirectly facilitating inter-class separation. The formulation for Center Loss is presented in Equation (9).

$$L_c = \frac{1}{2}\sum_{i=1}^{m}\left\|x_i - c_{yi}\right\|_2^2 \tag{9}$$

where m represents the batch size, x_i denotes the depth feature of sample i, and c_{yi} indicates the feature center of category y_i, to which sample i belongs. The term $\left\|x_i - c_{yi}\right\|_2$ signifies their L2 norm. To enhance the classification performance, center loss is typically employed in conjunction with the cross-entropy loss function, resulting in a joint optimization framework (Equation (10)).

$$L = L_{CE} + \lambda L_c \tag{10}$$

where L_{CE} represents the cross-entropy loss, and λ serves as a balancing parameter to regulate the relative importance of the two losses. Figure 6 illustrates the results of feature visualization on the MNIST dataset, trained using both center loss and cross-entropy loss. The effective application of center loss enhances class separability by optimizing inter-class margins while minimizing intra-class variation.

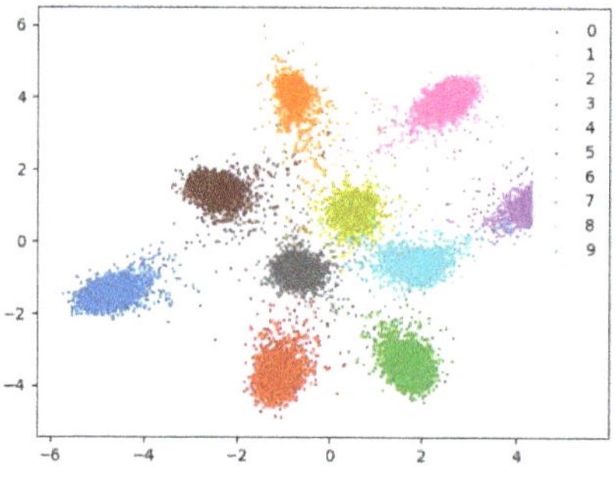

(a)

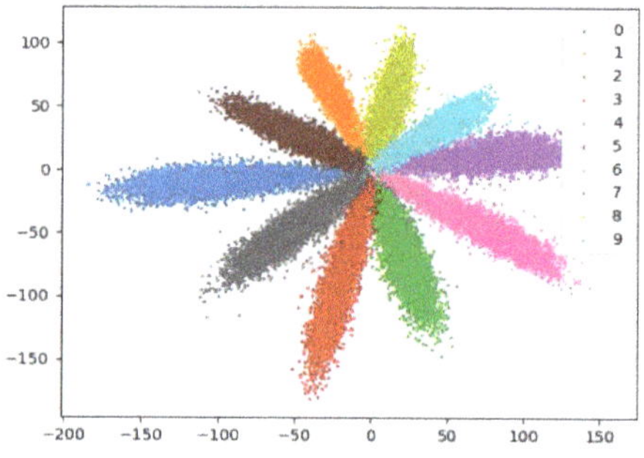

(b)

Figure 6. Feature comparison of the MINIST dataset trained with different loss functions. (**a**) Center Loss; (**b**) Cross Entropy Loss.

It has been demonstrated that a pre-trained convolutional neural network utilizing the VGGface2 dataset exhibits a superior performance in face emotion recognition [32]. Consequently, this study employs the pre-trained EfficientNetv2-S model as the backbone network for face feature extraction, while removing its final fully connected layer to facilitate multi-channel information fusion tasks. To further enhance the emotion recognition efficacy, an Auxiliary Attention Module (AAM) is incorporated following the feature extraction network. This module comprises both spatial attention and channel attention components (Figure 7).

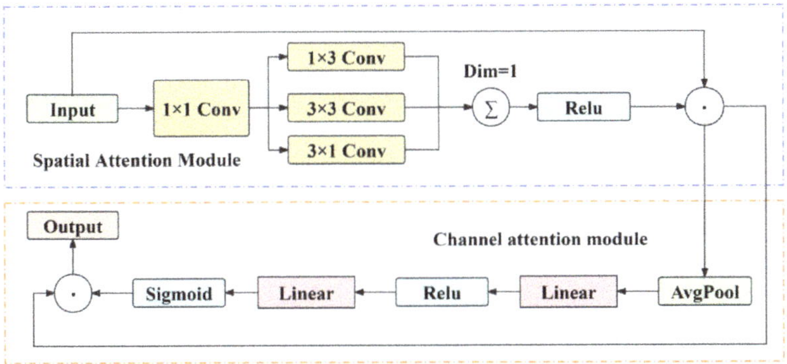

Figure 7. Structure diagram of the AAM module.

The Spatial Attention Module captures spatial features across various orientations through dimensionality-reducing convolutions and a series of specialized convolutions (Conv3 $\times$ 3, Conv1 $\times$ 3, and Conv3 $\times$ 1). The resulting spatial attention maps are then multiplied with the original feature maps to amplify responses at critical spatial locations. The Channel Attention Module compresses spatial dimensions via global average pooling and learns the importance weights for each channel through a sequence of linear transformations. These weights are normalized using a sigmoid function and subsequently multiplied with the original features to enhance the capture of significant channels. By integrating these two modules, AAM effectively identifies essential spatial and channel characteristics. The instantiation of multiple AAM modules enables simultaneous focus on different facial regions, thereby improving the accuracy in emotion recognition. The feature maps generated by one of the AAM modules are utilized to compute the center loss; additionally, areas emphasized by the model are derived from four attention head summation operations. Finally, classification results are produced through a fully connected neural network. The architecture of the face feature extraction network FaceFeatXtractor is illustrated in Figure 8.

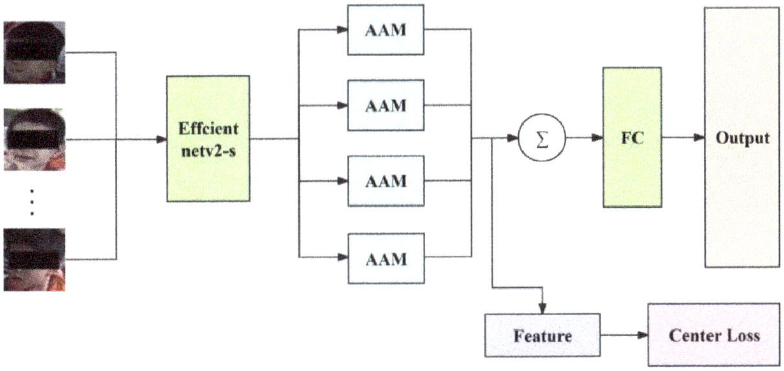

Figure 8. Network architecture of FaceFeatXtractor.

Hybrid Multi-Channel Network

This study introduces the self-developed MultiEmoNet network, which facilitates multi-channel emotion recognition by integrating facial expressions, skeletal information, and the background context. This approach addresses the limitations associated with single-channel methods in complex classroom environments. While facial expression serves as the primary feature for emotion recognition, it is often vulnerable to occlusion and variations in posture. In contrast, skeletal and background information can provide complementary data when facial cues are obscured, thereby enhancing both the robustness and accuracy of the model.

In this study, we propose a feature fusion approach that not only captures the common features across different modalities, but also accommodates the heterogeneity of information from these modalities, in contrast to traditional data and decision layer fusion methods. To fully leverage multi-type information, we design a network architecture with multiple input channels (Figure 9), where each channel is dedicated to processing specific types of data and performing feature extraction through independent CNNs. Pre-trained weights for single-channel inputs are initialized using EfficientNetv2-S. The facial image channel incorporates AAM, which enhances the representation of key regions associated with facial expressions. The features derived from facial images, skeletal data, and background information are subsequently integrated within a feature fusion unit to create a composite feature vector. This vector undergoes further processing via a fully connected layer (FC) to accomplish the classification task. This architecture aims to improve the classification accuracy by fusing diverse data sources and is particularly well-suited for emotion recognition in complex contexts.

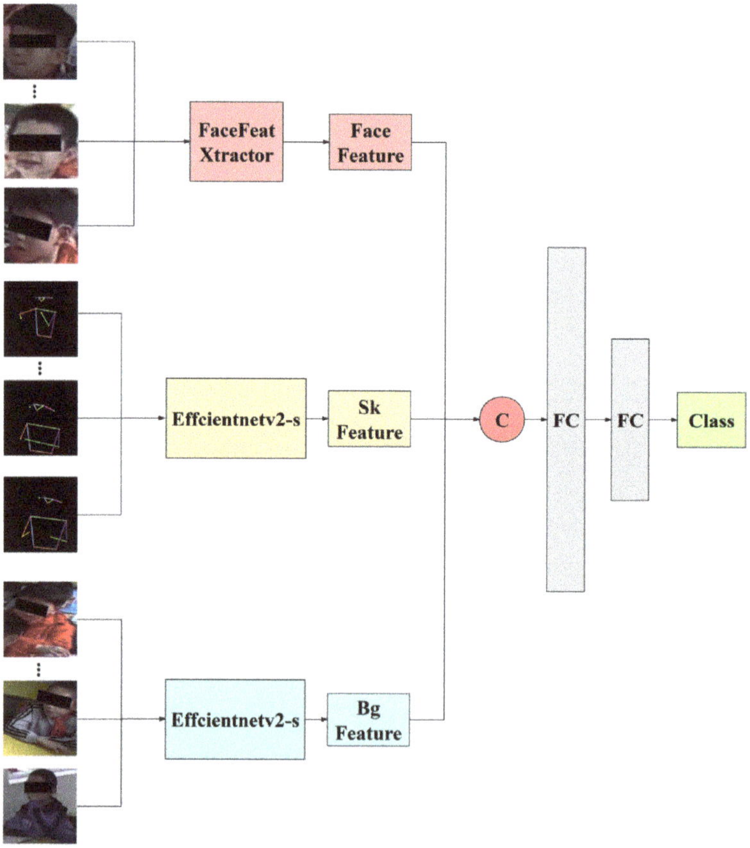

Figure 9. Network architecture of MultiEmoNet.

MultiEmoNet not only enhances the accuracy of emotion recognition, but also improves the model's generalization ability in complex classroom environments. This advancement allows for the more precise processing of multi-source information and facilitates the effective recognition of diverse emotional expressions.

4. Experiments

4.1. Experimental Details

4.1.1. Datasets

Object Detection Dataset

To tackle the challenges posed by classroom environments characterized by high densities of teachers and students, as well as varying camera angles, this study utilized the CrowdHuman dataset for model training [33]. This dataset is recognized for its extensive scale, diversity, and intricate scene annotations, making it particularly suitable for assessing the human detection performance in crowded settings. CrowdHuman comprises approximately 25,000 images that encompass over 40 cities and a variety of activity scenes. Each image contains an average of 22.6 annotated instances—specifically Head BBox, Visible BBox, and Full BBox—thereby providing rich training data for the model (Figure 10).

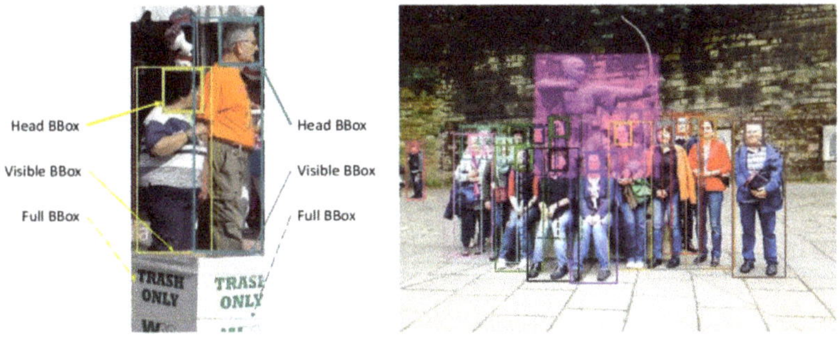

Figure 10. Annotation examples from the CrowdHuman dataset [33].

Multi-Channel Classroom Emotion Dataset

In order to investigate the issue of emotion recognition within classroom environments, this study collects video data from elementary and middle-school classrooms across various regions and subjects. A total of 200 video files are included, encompassing disciplines such as mathematics, language arts, and others. Each video is recorded at a resolution of 1280 × 720 pixels with a frame rate of 25 frames per second. The duration of each lesson ranges from approximately 40 to 45 min. Informed consent was obtained from all participants and their legal guardians for the classroom experimental data in this study to ensure compliance with ethical standards and the protection of the children's privacy. An example of the collected video data is illustrated in Figure 11.

(**a**) (**b**)

Figure 11. Example of video data collection. (**a**) Front-facing camera video; (**b**) Rear-facing camera video.

In order to accurately label mood changes, this study extracts one frame every 15 s and generates the Student-Full dataset by detecting the students' actual body regions (Visible BBox) using the YOLOv8x-Improved model. A total of 33,481 images were labeled and classified according to the six emotions identified in the previous section (Table 3).

Table 3. Number of annotations in the image emotion dataset.

Emotion Category	Count
Doubt	3857
Engagement	5987
Fatigue	4083
Disconnection	7696
Pleasure	3779
Peace	8078
Total	33,481

In this paper, the YOLOv8-Pose model is used for skeletal feature extraction to generate the Student-Sk dataset. The model uses CSP-darknet as the backbone network and fuses different scale features through PANet, which is able to efficiently detect the skeletal key points. The 17 key points contained in the COCO dataset are used to estimate the human posture (Figure 12). The key point information is normalized and centered to finally generate a skeletal map (Figure 13) for subsequent emotion recognition.

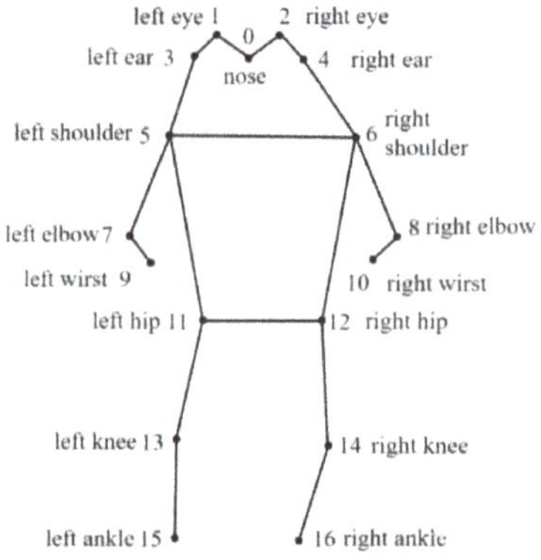

Figure 12. Example of human key points annotation.

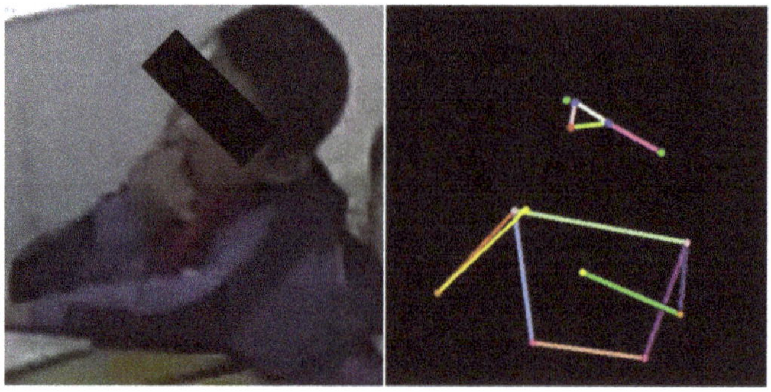

Figure 13. Example of generated skeleton diagram.

In a natural classroom environment, traditional facial detection algorithms often struggle to handle complex scenarios such as occlusion and head-turning. Therefore, in this paper, we opted for the YOLOv8x-Improved head detection model, which is trained on the CrowdHuman dataset, rather than relying on facial detection methods. The CrowdHuman dataset categorizes human body detection into three distinct categories: head Box, Full BBox, and Visible BBox. The specific detection effect applied to the classroom scene is shown in Figure 14.

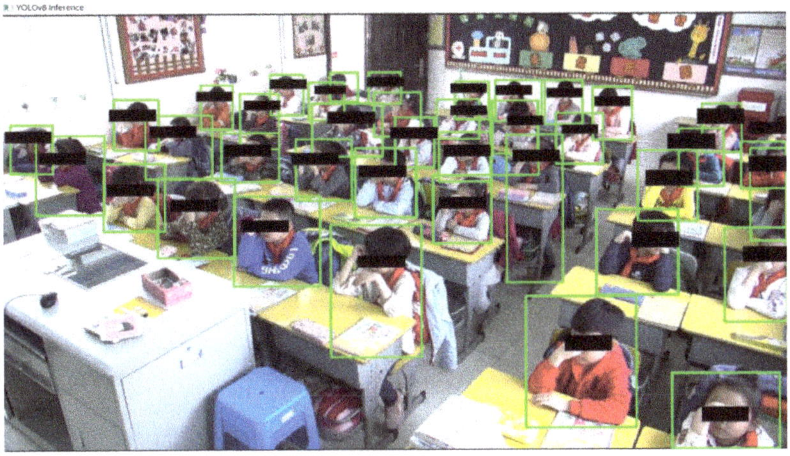

Figure 14. Actual detection results using YOLOv8.

In order to enhance the detection accuracy, this study employs a stacking preprocessing method. This involves superimposing the original image twice along both the horizontal and vertical coordinate axes to increase the resolution prior to inputting it into the model. The preprocessed human head image is then fed into the detector for analysis. Subsequently, the coordinates of the detection frame are extracted, yielding the corresponding head image. Finally, a facial information dataset named Student-Head is generated by traversing through the classification folder.

4.1.2. Experimental Setup

The specific hardware and software utilized in the experimental and evaluation processes are detailed in Table 4. The research presented in this paper was conducted on this platform.

Table 4. Experimental configuration.

Experimental Environment	Configuration
Operating system	Windows 10
CPU model	Intel i7-12700KF
Memory	32 GB
GPU model	NVIDIA RTX3080 10 G
Computing platform	CUDA 11.6
Programming language	Python 3.8
Development IDE	PyCharm 2022.2.2
Deep learning framework	Pytorch 11.6

4.1.3. Assessment of Indicators

The object detection performance is primarily assessed based on both the detection accuracy and speed. In this paper, we conduct a comprehensive evaluation of the model's accuracy, efficiency, and robustness using metrics such as Precision, Recall, mean Average Precision (mAP), and Frames Per Second (FPS).

Precision and recall are fundamental metrics for assessing the effectiveness of a model's detection capabilities. They measure, respectively, the accuracy of positive case predictions and the recognition rate of positive cases. These metrics are calculated as outlined in Equations (11) and (12).

$$precision = \frac{TP}{TP + FP} \quad (11)$$

$$Recall = \frac{TP}{TP + FN} \quad (12)$$

Average precision (AP) serves as a metric for evaluating the model performance by computing the area under the Precision–Recall (P-R) curve, as illustrated in Equation (13).

$$AP = \frac{1}{m}\sum_{i=1}^{m} \int_0^1 P_i(r)dr \quad (13)$$

The mAP_0.5 metric is derived from the average AP value when the Intersection over Union (IoU) threshold is set at 0.5, as illustrated in Equation (14).

$$mAP_{\_0.5} = \frac{1}{n}\sum_{i=1}^{n} AP(i) \quad (14)$$

In addition to the detection accuracy, the detection speed serves as a crucial indicator of the performance. The FPS reflect the processing speed of the model; a higher FPS signifies an increased processing speed, making it more suitable for real-time detection tasks.

4.2. Experimental Design

4.2.1. Object Detection Experiment

This experiment utilizes the CrowdHuman dataset for training in object detection. The dataset comprises 15,000 training images and 4370 validation images, all with an image resolution of 640 × 640 pixels. To enhance the model's generalization capability, the training data undergo a series of augmentation processes, including Mosaics, Blur, and Gray transformations.

By incorporating grayscale images into the training set, the model demonstrates enhanced stability in the performance when confronted with missing or inconsistent color information. This capability holds substantial practical significance for deploying the model across diverse environments, particularly in situations characterized by highly variable lighting conditions or unreliable color data. The training hyperparameters utilized in this experiment are detailed in Table 5.

Table 5. Training hyperparameters.

Parameter Name	Value
Image size	640 × 640
Load pretrained weights	True
Training epochs	100
Optimizer	SGD
Initial learning rate	0.01
Final learning rate	0.001
SGD momentum	0.937
Momentum decay	0.0005
Warmup epochs	3.0

4.2.2. Classroom Multi-Channel Emotion Recognition Experiment

For the training of the classroom multi-channel emotion recognition experiment, this study utilized the Student-Full dataset, Student-Head dataset, and Student-Sk dataset as inputs for the background information, facial information, and skeletal information, respectively. A corresponding file from each dataset with the same name was selected to form a set of input training data. To address missing data issues, an all-black image was employed in place of any missing information images within a channel. In terms of data preprocessing, normalization and random flipping operations were primarily implemented. The division between the training and validation datasets follows an 80:20 ratio. The Adam optimizer was utilized with a learning rate set at 0.001. Based on the results of the experimental iterations, it is observed that the model reaches a more stable state of convergence at 200 rounds. Therefore, a total of 200 epochs is executed in loops, during which the learning rate is halved every three epochs.

4.3. Experimental Results

4.3.1. Performance Evaluation of the CrowdHuman Dataset

This experiment utilizes the CrowdHuman dataset to perform object detection experiments on the YOLOv8x-Improved model. The primary focus is to compare the detection performance of various models, including SSD, Faster-RCNN (ResNet + FPN), YOLOv7x, YOLOv8x, and YOLOv8x-Improved. Table 6 presents a comparative analysis of these models' performances on the CrowdHuman dataset.

Table 6. Comparison of detection results on the CrowdHuman dataset.

Model	Precision	Recall	mAP50	mAP50-95	Model Size	GFLOPS	FPS
SSD	0.661	0.614	0.638	0.482	105.1 M	15.0	43
Faster-RCNN ResNet + FPN	0.684	0.621	0.662	0.524	57.22 M	134.5	18
YOLOv7x	0.842	0.721	0.845	0.581	135 M	188	39
YOLOv8x	0.847	0.769	0.851	0.603	130 M	258.1	40
YOLOv8x-Improved	0.853	0.772	0.862	0.611	109 M	232.2	48

YOLOv8x-Improved significantly enhances key performance metrics. The Precision reaches 0.853, while the Recall stands at 0.772, demonstrating its exceptional capability in recognizing positive samples. In terms of the average precision, YOLOv8x-Improved achieves a score of 0.862 on mAP50 and 0.611 on mAP50-95, indicating its robustness across various IOU thresholds. Furthermore, the model size is reduced to 109 MB, with GFLOPS decreasing to 232.2, resulting in approximately an 18% reduction in weight due to the incorporation of the C2f module MCC. Regarding the processing speed, YOLOv8x-Improved excels with a performance rate of 48 frames per second, surpassing both the original YOLOv8x and other models. This demonstrates significant improvements in speed while maintaining high precision and recall rates.

Combining all performance metrics, the YOLOv8x-Improved model exhibits a more pronounced performance compared to the original YOLOv8x (Figure 15). The enhanced model consistently demonstrates improvements in both precision and recall, ultimately achieving higher steady-state values and enhancing the detection of true-positive samples. In the evaluation of the mean average precision (mAP_0.5), the improved model surpasses its predecessor during the later stages of training while maintaining an exceptional performance on the more rigorous mAP_0.5:0.95 metrics. This indicates substantial advancements in its object detection capabilities.

In terms of training and validation loss, YOLOv8x-Improved employs the WIOU loss function in place of the original CIOU loss (Figure 16). The experimental results indicate that YOLOv8x-Improved exhibits a lower bounding box loss compared to YOLOv8x during both the training and validation phases, thereby significantly enhancing the target localization accuracy. Both models demonstrate a comparable performance regarding category loss, reflecting robust generalization capabilities. The effectiveness of the WIOU loss function further contributes to the improved performance of YOLOv8x-Improved.

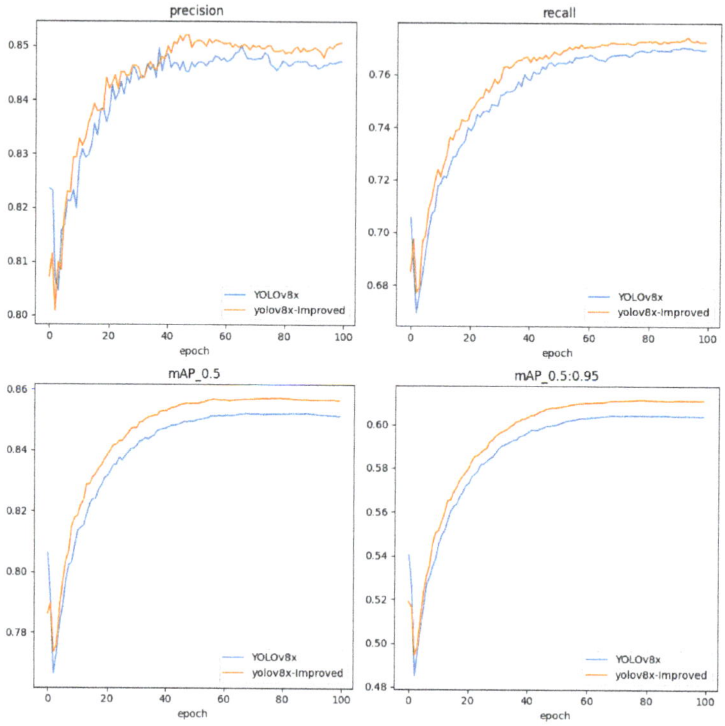

Figure 15. Comparison of training results between YOLOv8x-Improved and YOLOv8x.

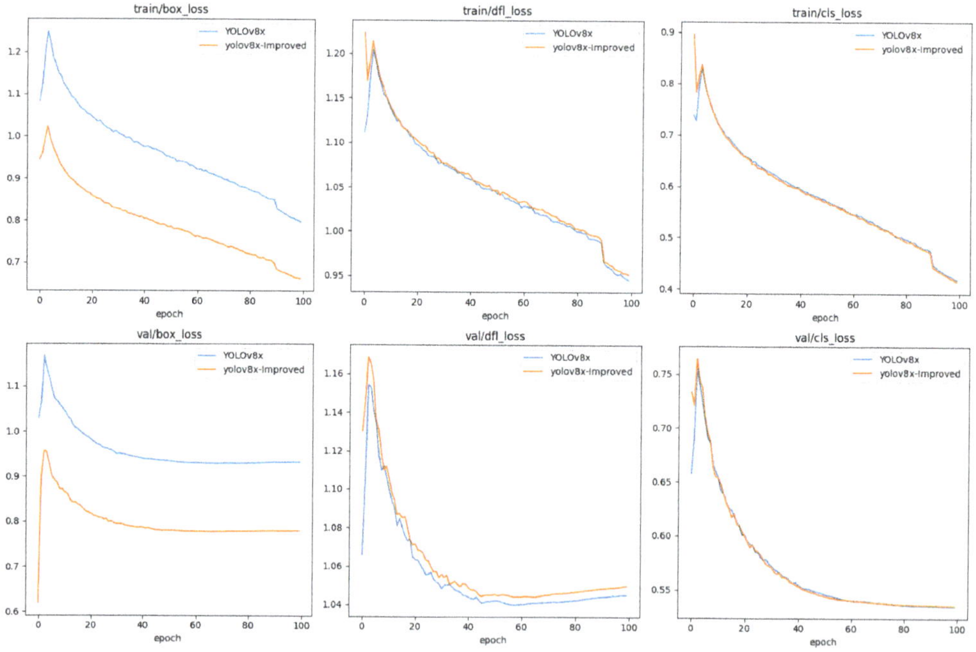

Figure 16. Training and validation loss curves.

4.3.2. Performance Evaluation of a Classroom Student Object Detection Dataset

To further evaluate the inference performance of the emotion recognition algorithm under different hardware configurations to meet the needs of classroom applications for real-time feedback [34], this study tested the inference time and computational efficiency under different GPU configurations (Table 7).

Table 7. Performance comparison of different GPU configurations.

Hardware Configuration	CPU	GPU	Inference Time per Image (ms)	FPS
1	Intel i7-12700KF	NVIDIA RTX 2080	24.39	41
2	Intel i7-12700KF	NVIDIA RTX 2080 Super	22.22	45
3	Intel i7-12700KF	NVIDIA RTX 3080	20.83	48

This study uses a self-made dataset of student classroom object detection to carry out more validation experiments that test how well the YOLOv8x-Improved model works for detection in a real classroom setting. One frame was randomly selected from each of the 100 classroom videos to obtain a total of 100 validation images. Each image contained 30 to 60 students and was accurately labeled using the LabelImg tool (v.1.8.6), as shown in Figure 17.

Figure 17. Annotation example diagram of self-made student classroom object detection dataset.

With the help of the trained model, a validation script was run to quantitatively analyze the detection ability of the model in a real classroom environment. Table 8 summarizes the obtained performance metrics.

Table 8. Comparison of custom classroom object detection dataset results.

Model	Precision	Recall	mAP50	mAP50-95
YOLOv8x-COCO	0.744	0.698	0.77	0.479
YOLOv8x	0.781	0.831	0.851	0.566
YOLOv8x-Improved	0.802	0.852	0.863	0.572

We used the YOLOv8x-COCO model in this study, which we downloaded from the official YOLOv8 website and pre-trained on the COCO dataset. The experiment only evaluates its detection effect on the 'Person' category and compares it with the performance of YOLOv8x and YOLOv8x-Improved in the Visible BBox. The results show that the YOLOv8x-COCO model performs well on general-purpose datasets, but lacks the ability to generalize to specific classroom scenarios. YOLOv8x and YOLOv8x-Improved, specifically trained for crowd detection, demonstrate superior precision and recall. YOLOv8x-Improved, in particular, excels in precision, recall, and mAP50 and mAP50-95 metrics, indicating the model's strong generalization across various scenarios and its exceptional detection performance in natural classroom scenarios following training in the CrowdHuman dataset.

In addition, this paper tests the model through different classroom scenarios (Classroom A and Classroom B). Figure 18 shows the detection results of YOLOv8x-COCO and YOLOv8x-Improved in different classroom scenarios, respectively. YOLOv8x-Improved excels in identifying students at the rear of the classroom and in areas on both sides, while also minimizing missed detection, demonstrating its adaptability in complex classroom environments.

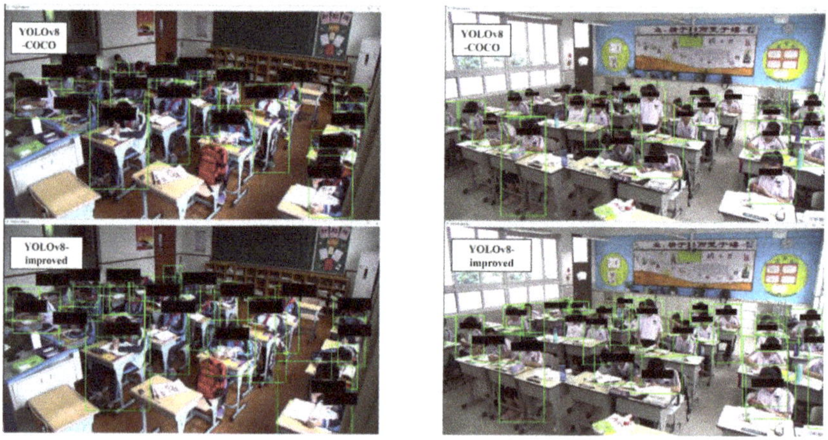

(**a**) (**b**)

Figure 18. Comparison of object detection. (**a**) Classroom A; (**b**) Classroom B.

4.3.3. Results of a Classroom Multichannel Emotion Recognition Experiment

This study conducts experiments on the classroom emotion dataset to verify the effectiveness of the MultiEmoNet network structure. We refer to the background, face, and skeletal information as Bg, Face, and Sk, respectively.

Table 9 shows the experimental results of this study using single-face information. In this study, the FaceFeatXtractor network outperforms other algorithms such as Resnet50, Vgg16, MobileNetv3, and EfficientNetv2-S, achieving an accuracy of 0.721. This paper's proposed face feature extraction network effectively enhances the accuracy of emotion recognition.

Table 9. Comparison of experiments using single-person face information.

Model	Precision
Resnet50	0.689
Vgg16	0.678
MobileNetv3	0.669
EfficientNetv2-S	0.702
FaceFeatXtractor	0.721

Table 10 shows the results of experiments using single-skeletal information in this study. The experiments compare the accuracy of emotion classification using traditional machine learning methods, such as SVM, KNN, and Random Forest, with the use of preprocessed–generated skeletal information maps fed into a convolutional neural network (EfficientNetv2-S). The results demonstrate that the method presented in this paper performs better on skeletal information when compared to traditional machine learning methods.

Table 10. Comparison of experiments using single bone information.

Model	Precision
SVM	0.516
KNN	0.503
Random Forest	0.552
EfficientNetv2-S	0.601

Table 11 shows the experimental results of this study using single background information.

Table 11. Comparison of experiments using single background information.

Model	Precision
Resnet50	0.794
Vgg16	0.774
MobileNetv3	0.771
EfficientNetv2-S	0.813

Table 12 MultiEmoNet provides the experimental results for the use of multi-channel information.

Table 12. Comparison of experiments using multi-channel information.

Model	Precision
Face + Sk	0.841
Face + Sk + Bg	0.914

The experimental results show that the MultiEmoNet model combining multi-channel information achieves the highest emotion classification accuracy. The model that uses skeleton information (Sk) and face information (Face) as the inputs performs better than the single-channel information model. This shows that the multi-channel emotion recognition method suggested in this paper works.

In particular, fusing background information (Bg) significantly improves the model's performance. The reason may be that face information is missing in some images, and the background can make up for it; secondly, the resolution of the face images is too low, and the background information contains more emotional features compared to the background information. In this study, we produce confusion matrices using face and skeletal features (Face + Sk) as well as face, skeletal, and background features (Face + Sk + Bg) to verify this (Figure 19).

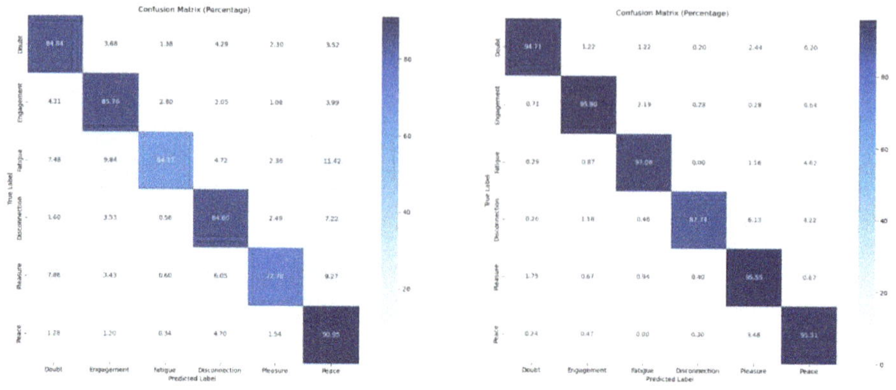

(a) (b)

Figure 19. Confusion matrix of MultiEmoNet. (**a**) Face + Sk; (**b**) Face + Sk+ Bg.

The analysis of the confusion matrix reveals that the addition of background information significantly improves the prediction accuracy of the model in all emotion categories, particularly in the 'Fatigue' and 'Pleasure' emotions, increasing the recognition accuracy from 64.17% and 72.78% to 93.06% and 95.55%, respectively. Meanwhile, the addition of contextual information significantly reduces the misclassification rate among different emotion categories, which further proves the key role of contextual information in emotion recognition. The background features provide important cues to the context of emotions, significantly enhance the differentiation ability of the model, and are important for improving the accuracy of emotion recognition systems.

4.4. Subsection Visualization of Student Emotion Recognition in the Classroom

The statistics and visualization of emotion recognition results are important for a deeper understanding of students' emotional states in the classroom. We used an emotion recognition algorithm in this study to generate detailed emotion statistics (Figure 20), which show the specific percentage of different emotion categories and categorize them into positive (Pleasure, Engagement), negative (Fatigue, Disconnection), and neutral (Peace, Doubt) emotions. These data provide teachers with visual feedback on classroom emotions, which helps to assess the classroom climate and adjust teaching strategies in a timely manner. Meanwhile, the mood prediction chart (Figure 21) further visualizes the mood state of each student, allowing teachers to grasp the individual mood distribution in the classroom and identify students who may need extra attention. This visual mood analysis not only provides a global view of the overall climate, but also makes it easier for teachers to further understand the learning progress of different students based on their emotional states after class, thereby optimizing personalized teaching strategies. It is important to note that this technology is intended to support educational and developmental uses only and should not be used for punitive purposes under any circumstances.

On the basis of this, the Emotional Statistics Timing Chart further reveals the trajectory of students' emotional fluctuations in the classroom. The 30 s recording interval can achieve a high recognition accuracy while capturing emotional fluctuations. Therefore, by recording the emotional distribution every 30 s, the chronogram can clearly show the emotional changes of students at different stages, helping teachers to understand students' emotional trends more comprehensively after class so as to make teaching adjustments and support at the appropriate time (Figure 22).

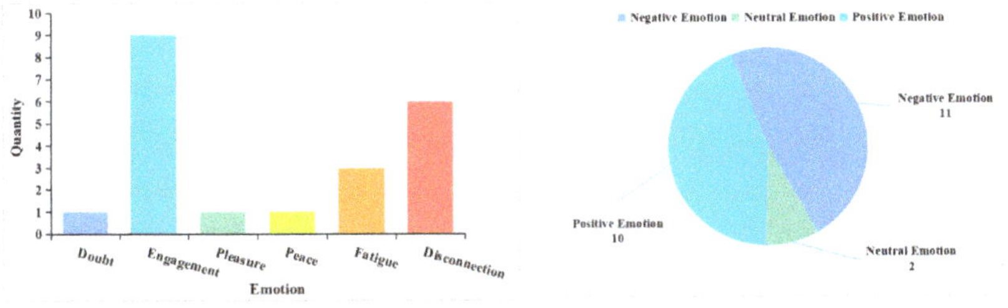

(**a**) (**b**)

Figure 20. (**a**) Detailed emotion statistics; (**b**) Emotion classification statistics.

Figure 21. Emotion prediction results.

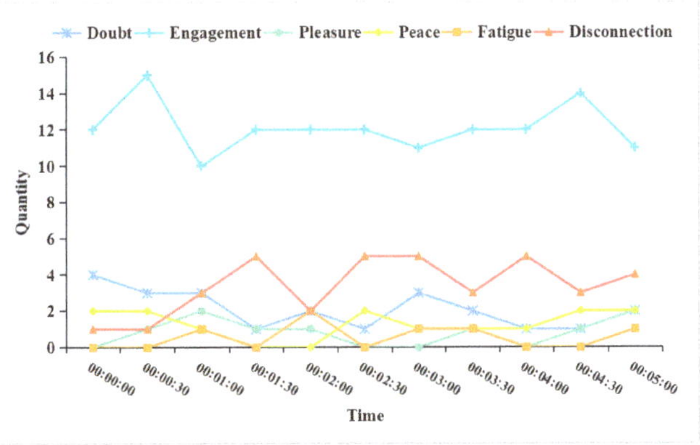

Figure 22. Time series of classroom emotion statistics (five minutes are captured).

Finally, to facilitate the quantitative analysis of classroom teaching, the system will provide an emotion statistics table that systematically summarizes the emotion data in each time period. The table will show the emotion distribution of each time frame in detail, helping teachers track the dynamic changes of students' emotions and provide data support for the timely adjustment of teaching strategies. The table data support downloading, which is convenient for teachers to further analyze students' emotional trends after class and optimize them in combination with specific teaching goals.

5. Conclusions and Future Work

This study provides an in-depth exploration of student emotion recognition and analysis in primary and secondary school classrooms, with the main goal of achieving accurate object detection and emotion recognition in natural classroom environments using deep learning techniques. The main contributions of the study are improving the YOLOv8 model so that it can find emotions more quickly and accurately in complicated classroom settings; creating a multi-channel classroom emotion dataset with information about facial expressions, skeletal structures, and background; and suggesting the MultiEmoNet emotion recognition model. The fusion of multiple information sources enhances the accuracy and robustness of emotion recognition. In addition, this study provides a visual view of classroom student emotion analysis, which provides teachers with data-based teaching feedback and optimization suggestions, enriching the application scenarios of educational technology.

Despite the results, this study still has some limitations. Firstly, although the MultiEmoNet model improves the accuracy of emotion recognition, it also brings higher computational complexity and resource consumption, which may affect its efficiency in real-time applications. Second, the study primarily concentrates on primary and secondary school classroom scenarios, without the extensive verification of its applicability in other educational scenarios and different age groups.

Future research will require further psychological validation, especially with neurodiverse populations, to ensure the robustness and inclusivity of the system across diverse user groups. Future research will focus on model optimization and lightweighting to further enhance the real-time processing capability of the system, especially in resource-limited environments. In addition, the research also plans to add an analysis of misclassification and challenging scenarios to provide directions for the further optimization and performance improvement of the model, thereby enhancing the robustness and adaptability of the system. Future research will also explore the application of emotion recognition technology in more educational scenarios and age groups, aiming to promote the realization of personalized and highly interactive educational environments.

Author Contributions: Conceptualization, Y.H., W.D. and T.X.; methodology, Y.H., W.D. and T.X.; software, Y.H., W.D. and T.X.; validation, W.D. and T.X.; visualization, Y.H.; writing—original draft, Y.H. All authors have read and agreed to the published version of the manuscript.

Funding: This work was supported by The Humanities and Social Science Research Planning Fund of the Ministry of Education (No. 20YJA880009), and the 2024 "AI + Examination and Evaluation" Special Teaching Reform Project of Central China Normal University (CCNU24GJ17), the Fundamental Research Funds for the Central Universities under Grant CCNU22JC011.

Institutional Review Board Statement: This study was conducted in accordance with the Declaration of Helsinki, and approved by the Institutional Review Board of Central China Normal University, Ethic Committee, EC (protocol code CCNU-IRB-202305004a and 24 May 2023).

Informed Consent Statement: Informed consent was obtained from all subjects involved in this study.

Data Availability Statement: The data presented in this study are openly available in the CrowdHuman dataset at https://www.crowdhuman.org/download.html (accessed on 22 May 2024). Relevant data from this study are available from the corresponding author upon reasonable request.

Acknowledgments: The authors would like to thank the Hubei Key Laboratory of Digital Education, Central China Normal University, for providing the research facilities.

Conflicts of Interest: The authors declare no conflicts of interest.

References

1. Pekrun, R. The control-value theory of achievement emotions: Assumptions, corollaries, and implications for educational research and practice. *Educ. Psychol. Rev.* **2006**, *18*, 315–341. [CrossRef]
2. Mollahosseini, A.; Hasani, B.; Mahoor, M.H. Affectnet: A database for facial expression, valence, and arousal computing in the wild. *IEEE Trans. Affect. Comput.* **2017**, *10*, 18–31. [CrossRef]

3. Chernykh, V.; Prikhodko, P. Emotion recognition from speech with recurrent neural networks. *arXiv* **2017**, arXiv:1701.08071.
4. Huang, Y.; Xiao, J.; Tian, K.; Wu, A.; Zhang, G. Research on robustness of emotion recognition under environmental noise conditions. *IEEE Access* **2019**, *7*, 142009–142021. [CrossRef]
5. Liu, Q.; Huang, Z.; Li, Z.; Pointer, M.R.; Zhang, G.; Liu, Z.; Gong, H.; Hou, Z. A field study of the impact of indoor lighting on visual perception and cognitive performance in classroom. *Appl. Sci.* **2020**, *10*, 7436. [CrossRef]
6. Noyes, E.; Davis, J.P.; Petrov, N.; Gray, K.L.; Ritchie, K.L. The effect of face masks and sunglasses on identity and expression recognition with super-recognizers and typical observers. *R. Soc. Open Sci.* **2021**, *8*, 201169. [CrossRef]
7. Chen, H.; Guan, J. Teacher–student behavior recognition in classroom teaching based on improved YOLO-v4 and Internet of Things technology. *Electronics* **2022**, *11*, 3998. [CrossRef]
8. Tang, L.; Gao, C.; Chen, X.; Zhao, Y. Pose detection in complex classroom environment based on improved Faster R-CNN. *IET Image Process.* **2019**, *13*, 451–457. [CrossRef]
9. Ekman, P. An argument for basic emotions. *Cogn. Emot.* **1992**, *6*, 169–200. [CrossRef]
10. Plutchik, R. *Emotions and Life: Perspectives from Psychology, Biology, and Evolution*; American Psychological Association: Washington, DC, USA, 2003.
11. Pekrun, R.; Stephens, E.J. Achievement emotions: A control-value approach. *Soc. Personal. Psychol. Compass* **2010**, *4*, 238–255. [CrossRef]
12. Kosti, R.; Alvarez, J.M.; Recasens, A.; Lapedriza, A. Context based emotion recognition using emotic dataset. *IEEE Trans. Pattern Anal. Mach. Intell.* **2019**, *42*, 2755–2766. [CrossRef]
13. Kumar, A.; Garg, N.; Kaur, G. An emotion recognition based on physiological signals. *Int. J. Innov. Technol. Explor. Eng.* **2019**, *8*, 335–341.
14. Mehrabian, A.; Ferris, S.R. Inference of attitudes from nonverbal communication in two channels. *J. Consult. Psychol.* **1967**, *31*, 248. [CrossRef] [PubMed]
15. Sarvakar, K.; Senkamalavalli, R.; Raghavendra, S.; Kumar, J.S.; Manjunath, R.; Jaiswal, S. Facial emotion recognition using convolutional neural networks. *Mater. Today Proc.* **2023**, *80*, 3560–3564. [CrossRef]
16. Coulson, M. Attributing emotion to static body postures: Recognition accuracy, confusions, and viewpoint dependence. *J. Nonverbal Behav.* **2004**, *28*, 117–139. [CrossRef]
17. Gavrilescu, M. Recognizing emotions from videos by studying facial expressions, body postures and hand gestures. In Proceedings of the 2015 23rd Telecommunications Forum Telfor (TELFOR), Belgrade, Serbia, 24–26 November 2015.
18. Chóliz, M.; Fernández-Abascal, E.G. Recognition of emotional facial expressions: The role of facial and contextual information in the accuracy of recognition. *Psychol. Rep.* **2012**, *110*, 338–350. [CrossRef]
19. Kosti, R.; Alvarez, J.M.; Recasens, A.; Lapedriza, A. Emotion recognition in context. In Proceedings of the IEEE conference on computer vision and pattern recognition, Honolulu, HI, USA, 21–26 July 2017.
20. Lee, J.; Kim, S.; Kim, S.; Park, J.; Sohn, K. Context-aware emotion recognition networks. In Proceedings of the IEEE/CVF International Conference on Computer Vision, Seoul, Korea, 27 October–2 November 2019.
21. Yang, D.; Chen, Z.; Wang, Y.; Wang, S.; Li, M.; Liu, S.; Zhao, X.; Huang, S.; Dong, Z.; Zhai, P. Context de-confounded emotion recognition. In Proceedings of the IEEE/CVF Conference on Computer Vision and Pattern Recognition, Vancouver, BC, Canada, 17–24 June 2023.
22. Sharma, A.; Mansotra, V. Deep learning-based student emotion recognition from facial expressions in classrooms. *Int. J. Eng. Adv. Technol.* **2019**, *8*, 4691–4699. [CrossRef]
23. Chen, Z.; Liang, M.; Xue, Z.; Yu, W. STRAN: Student expression recognition based on spatio-temporal residual attention network in classroom teaching videos. *Appl. Intell.* **2023**, *53*, 25310–25329. [CrossRef]
24. Noroozi, F.; Corneanu, C.A.; Kamińska, D.; Sapiński, T.; Escalera, S.; Anbarjafari, G. Survey on emotional body gesture recognition. *IEEE Trans. Affect. Comput.* **2018**, *12*, 505–523. [CrossRef]
25. Su, C.; Wang, G. Design and application of learner emotion recognition for classroom. In *Journal of Physics: Conference Series*; IOP Publishing: Bristol, UK, 2020.
26. Terven, J.; Córdova-Esparza, D.-M.; Romero-González, J.-A. A comprehensive review of yolo architectures in computer vision: From yolov1 to yolov8 and yolo-nas. *Mach. Learn. Knowl. Extr.* **2023**, *5*, 1680–1716. [CrossRef]
27. Tong, Z.; Chen, Y.; Xu, Z.; Yu, R. Wise-IoU: Bounding box regression loss with dynamic focusing mechanism. *arXiv* **2023**, arXiv:2301.10051.
28. Mittal, T.; Guhan, P.; Bhattacharya, U.; Chandra, R.; Bera, A.; Manocha, D. Emoticon: Context-aware multimodal emotion recognition using frege's principle. In Proceedings of the IEEE/CVF Conference on Computer Vision and Pattern Recognition, Seattle, WA, USA, 13–19 June 2020.
29. Tan, M.; Le, Q. Efficientnet: Rethinking model scaling for convolutional neural networks. In Proceedings of the 36th International Conference on Machine Learning, Long Beach, CA, USA, 10–15 June 2019.
30. Krizhevsky, A.; Sutskever, I.; Hinton, G.E. Imagenet classification with deep convolutional neural networks. *Adv. Neural Inf. Process. Syst.* **2012**, *25*, 84–90. [CrossRef]
31. Wen, Y.; Zhang, K.; Li, Z.; Qiao, Y. A discriminative feature learning approach for deep face recognition. In *Computer Vision–ECCV 2016: 14th European Conference, Amsterdam, the Netherlands, October 11–14, 2016, Proceedings, Part VII 14*; Springer: Cham, Switzerland, 2016.

32. Savchenko, A.V.; Savchenko, L.V.; Makarov, I. Classifying emotions and engagement in online learning based on a single facial expression recognition neural network. *IEEE Trans. Affect. Comput.* **2022**, *13*, 2132–2143. [CrossRef]
33. Shao, S.; Zhao, Z.; Li, B.; Xiao, T.; Yu, G.; Zhang, X.; Sun, J. Crowdhuman: A benchmark for detecting human in a crowd. *arXiv* **2018**, arXiv:1805.00123.
34. Umirzakova, S.; Whangbo, T.K. Detailed feature extraction network-based fine-grained face segmentation. *Knowl.-Based Syst.* **2022**, *250*, 109036. [CrossRef]

applied sciences

MDPI

Article

Attitudes of EFL Learners to the Implementation of the Area9 Lyceum Online Platform Based on the UTAUT Model

Iman Oraif

Department of English Language and Literature, College of Languages and Translation, Imam Mohammad Ibn Saud Islamic University, Riyadh 3204, Saudi Arabia; imoraif@imamu.edu.sa

Abstract: The advancement of technology has led to the creation of numerous platforms that could potentially be used for remote education. For example, the recent development of the English Diploma Programme at a top university in the Kingdom of Saudi Arabia (KSA) deploys a new platform known as Area9 Lyceum (Area9). Because the English Diploma Programme is a recent development, and especially given its use of a new platform, this proposed research will investigate learners' attitudes to and acceptance of using the platform. Furthermore, it will look at how other universities could benefit from this experience to develop their own English as a Foreign Language (EFL) programmes along technological lines, specifically by deploying a survey tool based on the unified theory of acceptance and use of technology (UTAUT). The results reflect the positive attitude of the participants. Recommendations can be drawn from this study to help persuade stakeholders in higher education to adopt such platforms in the teaching of EFL or English as a Second Language (ESL).

Keywords: unified theory of acceptance and use of technology (UTAUT); EFL learners; online courses; Area9

1. Introduction

One of the most recent developments in language teaching is the use of adapted learning, which is identified as a personalised approach to teaching and learning. In personalised learning, learners are at the centre of the teaching and learning process, as clarified in most of the literature [1]. This happens when learners are given a high degree of responsibility for their own learning. Adapted learning also helps to shape learning provision. Furthermore, Prain [2] noted that personalised learning can help to improve students' engagement and attainment. Additionally, the use of interactive technology in the instructional process can be helpful. Throughout the literature, it may be observed that many interactive learning environments provide on-demand assistance. The aim of deploying these environments is to positively enhance learning [3].

In the current paper, a personalised adapted environment in the form of an interactive platform is examined based on the unified theory of acceptance and use of technology (UTAUT). This specific theoretical framework is used to gain a better understanding of the participants' levels and types of acceptance regarding this adapted online platform. In the specific research context, the use of ALP generative artificial intelligence (AI) is rather limited and only made available in large institutions. Consequently, there is very little examination of the use of such technology in the literature. The most common study of educational technology, especially in relation to EFL education, concerns the use of learning management systems (for example, [4,5]) and social media (for example, [6,7]).

1.1. Research Gap

In order to fill the gap identified in the research, a platform was included in this study that has not previously been implemented on degree programmes in Saudi universities. Specifically, it was used to deliver English diploma courses at one Saudi university.

1.2. Research Questions

On this basis, the current study will attempt to answer the following research questions:

1. Why do EFL learners use an interactive platform for learning on an English diploma course?
2. How do EFL learners perceive the implementation of an interactive platform to be useful to them for learning English on a diploma course?

2. Literature Review

2.1. Use of Adapted and Interactive Platforms in Teaching and Learning

In English language learning, the use of adapted learning has increased [8] and is being used to offer individualised learning. In a study conducted by Wang and Liao [9], adapted learning was utilised to teach an English as a second language (ESL) course via an e-learning system. The above study used a technique called data mining (an artificial neural network) to explore the learning performance of various students. The system was set to consider grammar, vocabulary, and reading. In an experimental approach, the results showed more development in the performance of the experimental group as compared to the control group who used the regular e-learning approach.

In another study, conducted by Sfenrianto et al. [10], the researchers proposed an adapted learning system at a knowledge level in an e-learning environment. The proposed system personalised materials based on the proficiency of the participating English language learners. Accordingly, a pre-test was conducted to determine the learners' level, through which the learners were categorised as having high, intermediate, or low English language ability. The system then provided materials based on the learners' needs, with 90 learners with different levels of English language proficiency participating in the study. The results indicated that the proficiency of all the participants was enhanced, thereby demonstrating the system's effectiveness.

The option of using adapted learning in an e-learning environment was explained early in 2003. The suggested model was designed to keep relevant adapted metadata separate from the description of the course itself, thereby helping to enhance course maintenance [11]. Looking into the inclusion of adapted online platforms in an EFL context, more specifically in contexts like the one under study, research was conducted in Oman in 2022 to investigate the implementation of an online learning platform. Thus, the perceptions of a group of Omani students were explored at the English Language Centre of the University of Sciences and Applied Technology, Salalah campus. In particular, the students' perceptions of using new online learning platforms were studied, namely, the platforms Moodle and Microsoft Teams. From the survey results, it was ascertained that the participants had positive perceptions of using the selected platforms, thereby recommending their use in EFL classes [12].

In another study, conducted by Alakrash et al. [13], the possible use of an online platform in EFL classes was investigated, specifically in the Arab world. Alakrash therefore studied EFL students' use of digital platforms for learning purposes, as well as their attitudes and digital literacy. In addition, the above study examined the correlation between students' digital literacy, attitudes toward, and use of digital platforms for learning purposes. A quantitative approach was adopted, surveying 80 EFL learners. The resulting data showed that the students had strongly positive attitudes, moderate levels of digital literacy, and moderate levels of digital platform usage for learning purposes.

It is evident from these studies that the use of online platforms is helpful for offering better learning opportunities in EFL. However, few researchers have examined such implementation to date, especially in the Arab world. This is precisely because of the limited uptake of these platforms in the Arab EFL context. Consequently, in order to promote the technology, it is essential to gather students' opinions following their experience of implementing such a personalised tool.

2.2. Unified Theory of Acceptance and Use of Technology (UTAUT)

According to Momani [14], UTAUT is one of the most intensive and well-developed models used for testing the acceptance of technology adoption. Momani further notes that this model is robust and trusted for use with different types of technology. Furthermore, the model is not overly complicated, as it includes a limited number of constructs and moderating variables, thereby facilitating an examination of the acceptance of technology integration.

Based on the above description, this study applies UTAUT [15] to create a model built on four variables, which reflect the behavioural intention to use technology [16]. Each variable measures different perspectives of the acceptance of technology use. Šumak et al. [17] define these variables as follows: performance expectancy (PE) is "the degree to which an individual believes that using the system will help him or her achieve gains in job performance"; effort expectancy (EE) is "the degree of ease associated with the use of a system"; social influence (SI) is "the degree to which an individual perceives that important people believe they should use the new system"; and facilitating conditions (FC) are "the degree to which an individual believes that an organisational and technical infrastructure exists to support the use of a system". Use of this theoretical framework is common among studies that investigate participants' perceptions of using different kinds of technology for teaching and learning. In a systematic review investigating studies on the impact of technology use around the world from the perspective of UTAUT, the resulting data effectively predicted technology adoption in higher education settings [18].

Nevertheless, UTAUT is a model that has not been tested extensively on EFL courses, although one recent study in Saudi Arabia examined the continuity of English teachers' use of online education in elementary schools in the Kingdom [19]. Through a multiple regression analysis of the four factors mentioned above, the results revealed that performance expectancy, social influence, and facilitating conditions can predict teachers' use of online learning.

3. Methodology

In this section, descriptions of the selected platform and Diploma Programme are presented, with the study design explained.

3.1. The Area9 Platform

According to its website, the Area9 platform deploys adapted learning [20] and is represented as:

- An online delivery method that automatically adjusts to the needs of each learner.
- A tool that recreates at scale the optimal teaching approach of a one-on-one personal tutor.
- The use of proven data analytics and intelligent technologies to adjust in real-time and deliver an optimal experience.

Furthermore, the website indicates the following benefits of applying an adapted learning approach:

- Compared to traditional training:
    - It cuts training time in half.
    - It leads to higher proficiency.
    - No one is left behind.
    - It eliminates boredom and frustration.
    - It improves business outcomes.
    - It reveals and addresses unconscious incompetence.
    - It promotes retention and reinforcement.

In addition, the website presents the following content, as illustrated in Figure 1.

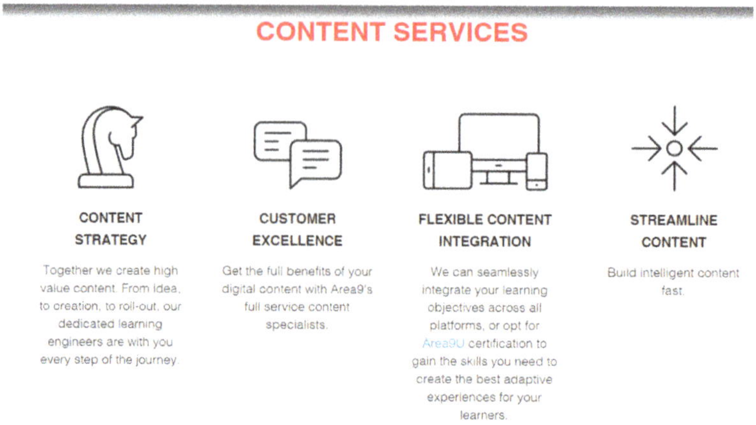

Figure 1. Services provided by the platform.

Area 9 facilitates learning by eliciting learners' responses to questions related to the registered course. This interactive platform identifies learners' weaknesses in the subject and directs them toward relevant resources based on their responses to the interaction. Following this diagnostic assessment, the system will build personalized learning material for each student. In other words, learners interact with the system to build up their personalized resources.

As a complement to the platform, the department provides weekly online lectures using slides generated by the system tailored to common students' needs identified in the primary diagnostic assessment. Importantly, the students are required to independently use the platform as part of their course. The platform has been successfully integrated into all Diploma courses, from Level One to graduation, covering a wide range of subjects from language skills to theoretical linguistics.

3.2. Study Design

For the current research, a case study design involving a quantitative approach was implemented, using a survey adapted from Venkatesh et al. [15] based on UTAUT. However, some changes were made to the original survey items to suit the research context. For example, in Item 1 of the questionnaire on the topic of performance expectancy, the word 'job' was changed to 'course', as follows: 'I would find the system useful on my course'. Meanwhile, a four-point Likert scale was adopted to ensure clearly defined answers from the participants.

The survey (Appendix A) was distributed electronically via Google Forms to facilitate data collection. The questionnaire was also translated into Arabic to ensure the learners' understanding, and a translation expert was designated to check the translation. The validity of the content was determined to check the suitability of the tool to answer the research questions, whereupon two experts in the field were asked to give their opinions.

Reliability refers to the degree of consistency or stability in the results of a study if it is conducted with the same respondents on repeated occasions. Reliability is therefore a very important aspect of selecting a questionnaire instrument. In order to evaluate the reliability of the questionnaire administered in this current study, the Cronbach's alpha values were calculated for each dimension. Cronbach's alpha (α), developed by educational psychologist Lee Cronbach in 1951, is the most common estimate of reliability. Based on the intercorrelations of the observed indicator variables, Cronbach's alpha produces values of between 0 and 1, with an acceptable range of 0.7–1.

The above table demonstrates that the research data passed the reliability test, because the Cronbach's alpha value for the questionnaire was 0.864 (greater than 0.7), and all the Cronbach's alpha values exceeded the acceptable minimum.

Regarding the research participants, comprehensive sampling was carried out, as the survey was distributed to all male and female Level 4 learners in the English Diploma Programme. This sample was selected because the learners had already used the platform for three terms. Therefore, they would have had sufficient experience of it to be able to give their opinions. Moreover, they had one more term remaining before they graduated.

The English Diploma Programme is described on the Saudi Electronic University website as an English Language Diploma on the Internet, accessible to male and female students all over the Kingdom as of Fall 2021. The aim of this Programme is "to prepare learners interested in English studies to become highly competent in English language practice" [21]. Its overarching purpose is to equip learners "with the various skills and knowledge required in the job market to have a successful career…" (ibid.). Twenty different English language courses are available within the Diploma Programme, consisting of "translation, interpreting, and research and presentation skills" (ibid.). The course designers are highlighted as "experienced faculty members whose research and scientific interests lie at the intersection of those academic fields" (ibid.). Each learner is expected to meet a number of milestones over the course of the Programme.

From an ethical perspective, every effort was made to hide the participants' identities in the study in consideration of the need for participant anonymity and confidentiality. Moreover, participation was purely voluntary and did not affect the participants' grades. Prior to their participation, the respondents were asked to sign informed consent forms. Finally, all participants were informed of their right to withdraw from the study at any time.

The researcher contacted the head of the English department from the College of Science and Theoretical Studies at the Saudi Electronic University where the platform was being implemented. This was to verify that the study could be conducted. Approval was granted orally. However, the researcher subsequently needed to apply to the Deanship of Scientific Research at the Saudi Electronic University for formal approval to conduct the study by gaining ethical approval from the researcher's institution.

3.3. *Results*

This subsection presents a statistical analysis of the questionnaire responses. The research sample comprised 59 participants of different ages and at different university levels. The aim of this study was to explore the attitudes of EFL learners to the implementation of the Area9 platform based on the UTAUT Model. Thus, the statistical data analysis is reported as follows: the main characteristics of the sample are first presented to describe the participants' personal information (i.e., age, gender, reason for undertaking the English Diploma course, and any training in the use of the Area9 platform for the English Diploma). This is followed by an analysis of reliability, and finally, the results of the hypothesis testing are set out.

Thus, the main characteristics of the sample are presented below, with tables that include the participants' personal information (i.e., age, gender, reason for undertaking the English Diploma, and any training in the use of the Area9 platform for the English Diploma).

As illustrated in Tables 1 and 2, the English Diploma is a mixed-gender programme attended by students of different ages. Therefore, the educational content generated should be suitable for use across these differences.

Tables 3 and 4 present the participants' responses relating to research question 1. In their responses to the question 'Why are you taking this English Diploma?', as illustrated in Table 3, the participants mostly indicated that they were undertaking the Diploma to improve their English language ability. Therefore, AI should generate material that is suitable to meet this need.

Table 1. Table of frequency for age.

	Frequency	Percentage
20–25 years	13	22.0
25–30 years	14	23.7
30–35 years	15	25.4
35 years or over	17	28.8
Total	59	100.0

Table 2. Table of frequency for gender.

	Frequency	Percentage
Male	12	20.3
Female	47	79.7
Total	59	100.0

Table 3. Table of frequency for the reasons for undertaking the Diploma.

	Frequency	Percentage
To help you get a promotion at work	2	3.4
To help you apply for jobs	12	20.3
To help you in your education	11	18.6
To improve your English abilities	34	57.6
Total	59	100.0

Table 4. Any previous training in the use of the Area9 platform for the English Diploma.

	Frequency	Percentage
No	15	25.4
Yes	44	74.6
Total	59	100.0

In Table 4, it may be noted that most of the participants were trained prior to starting the online course, this being an important factor in the success of technology implementation.

Tables 5 and 6 present the participants' responses relating to research question 2:

Table 5. The means and median standard deviations for the dimensions of the questionnaire.

No.	Items	Mean	Std. Deviation	Median	%	Ranking	Response
2	Effort expectancy	3.17	0.62	3.00	79.3%	1	Agree
1	Performance expectancy	2.90	0.62	3.00	72.6%	2	Agree
4	Facilitating conditions	2.85	0.42	3.00	71.3%	3	Agree
3	Social influence	2.80	0.58	3.00	69.9%	4	Agree
	Total	2.93	0.45	3.00	73.3%		Agree

Table 6. The means and median standard deviations for the questionnaire items.

No.	Items	Mean	Std. Deviation	Median	%	Ranking	Response
4	If I use the system, I will increase my chances of achieving high grades	3.14	0.80	3.00	78.4%	1	Agree
1	The system is useful on my course	2.92	0.77	3.00	72.9%	2	Agree
3	Using the system increases my productivity	2.81	0.75	3.00	70.3%	3	Agree
2	Using the system enables me to accomplish tasks more quickly	2.75	0.84	3.00	68.6%	4	Agree
	Performance expectancy	2.90	0.62	3.00	72.6%		Agree
8	Learning to operate the system is easy for me	3.24	0.73	3.00	80.9%	1	Agree
6	It is easy for me to become skilful in using the system	3.22	0.72	3.00	80.5%	2	Agree
7	I find the system easy to use	3.15	0.71	3.00	78.8%	3	Agree
5	My interaction with the system is clear and easy to understand	3.08	0.73	3.00	77.1%	4	Agree
	Effort expectancy	3.17	0.62	3.00	79.3%		Agree
12	In general, the University has supported use of the system	3.14	0.75	3.00	78.4%	1	Agree
11	The teacher has been helpful with using the system	2.85	0.83	3.00	71.2%	2	Agree
10	People who are important to me think that I should use the system	2.63	0.83	3.00	65.7%	3	Agree
9	People who influence my behaviour think that I should use the system	2.58	0.79	3.00	64.4%	4	Agree
	Social influence	2.80	0.58	3.00	69.9%		Agree
13	I have the necessary resources to use the system	3.27	0.64	3.00	81.8%	1	Agree
14	I have the necessary knowledge to use the system	3.22	0.65	3.00	80.5%	2	Agree
16	A specific person (or group) is available for assistance with system difficulties	2.68	0.99	3.00	66.9%	3	Agree
15	The system is incompatible with other systems that I use.	2.24	0.92	2.00	55.9%	4	Agree
	Facilitating conditions	2.85	0.42	3.00	71.3%		Agree
	Total	2.93	0.45	3.00	73.3%		Agree

It is clear from the data in the above Table 5 and Figure 2 that most of the EFL learners accepted the use of an interactive platform for learning in an English Diploma course (73.3%). The highest level of agreement was for effort expectancy (79.3%), the second highest was for performance expectancy (72.6%), and the third highest was for facilitating conditions (71.3%). Meanwhile, the lowest level of agreement was indicated for social influence (69.9%).

In relation to specific survey items, the following results were also obtained:

It is clear from the data in the above table that the students were in agreement over the interactivity and benefits of implementing the new Programme, with a percentage ranging between 55.9% and 81.8%. The statement 'I have the necessary resources to use the system' received the highest agreement response (81.8%), closely followed by 'Learning to operate the system is easy for me' (80.9%), 'It is easy for me to become skilful in using the system'

(80.5%), and 'I have the necessary knowledge to use the system' (80.5%). High agreement was also found in response to the statements, 'In general, the University has supported use of the system' (78.4%), 'If I use the system, I will increase my chances of achieving high grades' (78.4%), and 'The system is useful on my course' (72.9%). Meanwhile, a lower level of agreement was found in response to the statements: 'Using the system enables me to accomplish tasks more quickly' (68.6%), 'People who are important to me think that I should use the system' (65.7%), 'A specific person (or group) is available for assistance with system difficulties' (66.9%), and 'People who influence my behaviour think that I should use the system' (64.4%). Finally, the lowest level of agreement was indicated in response to the statement, 'The system is incompatible with other systems that I use' (55.9%). In the next section, these results are discussed in relation to the theoretical underpinnings of the study and the existing literature.

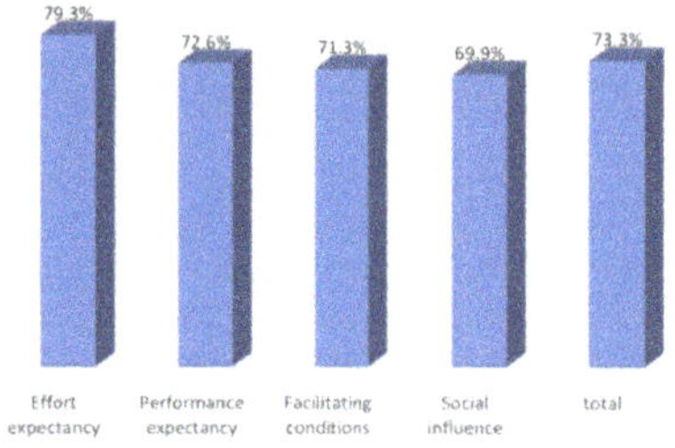

Figure 2. Percentages chart.

4. Discussion

It is evident from the results that the participants had a positive experience of dealing with the English course created by the AI system. In response to the first research question, 'Why do EFL learners use an interactive platform for learning on an English diploma course?', the first part of the survey showed the participants' reflections on the system, as they had primarily enrolled in the Programme to develop their English language skills. Meanwhile, they had been trained in using the system in other courses. The students' motivation to improve their English language skills in such a manner may have been because AI enables a more adaptive kind of learning, which suits each student's educational needs and level [22]. Furthermore, learners have greater autonomy because they can engage with such a course at a time and place that suits them [23].

For the second research question, 'How do EFL learners perceive the implementation of an interactive platform to be useful to them for learning English on a diploma course?' the different variables of the survey driven by UTAUT showed a high level of agreement among the participants. Regarding performance expectancy related to developing learners' educational levels, the learners found the system helpful for English language practice. Similarly, a study conducted in India by Jaiswal and Arun [24] confirmed that AI adaptive learning can cause a positive shift in education. It changes conventional teaching methods to innovative education, thereby enhancing learners' educational experience. Through in-depth interviews with four senior managers and four subject matter experts, all working with AI-related technologies, the participants confirmed that by developing AI-based applications on the basis of grounded theory, they found that personalised earning and adaptive assessments supported teachers and helped students. More specifically, in developing countries, AI may have the potential to enhance the learning experience.

The participants strongly agreed with the second variable related to effort expectancy and the ease of using the system. Artificial intelligence tools are known to facilitate the learning process through their ease of use. This was supported by Wang et al. [25], who mentioned that AI technologies are promoted in the educational sector due to this feature, as it was noted that AI increased learning performance in a school setting. In an exploration of teachers' intentions and concerns about using AI, a total of 311 teachers participated in a survey that examined anxiety (AN), self-efficacy (SE), attitude towards AI (ATU), perceived ease of use (PEU), and perceived usefulness (PU); the results demonstrated that the ease of using an AI application motivated the participants to adopt the tool.

Regarding social influence, the participants indicated that people with authority at their institution had encouraged them to use the system, such as the teachers and administration. This is very important, because learners would thereby gain a sense of responsibility and build self-confidence as influential participants in the educational process. Teachers would consequently adopt the use of AI to provide further assistance in teaching; the teacher's role is vital in designing, visualising, and orchestrating AI systems for teaching and learning. Moreover, AI would help to develop computational presentations built on meaningful data and inferred from pedagogy and learner models [26].

The facilitating conditions represent the necessary facilities to use an AI system easily, like the availability of technological tools and resources with IT support to help overcome technical issues. The participants strongly agreed with this variable. The provision of technological support is essential, especially if tests are taken electronically. Saudi Electronic University is a pioneering university that uses technology for blended and online courses, providing the necessary support for its beneficiaries in KSA. Hughes [27] indicates that this will ensure the learners' readiness and their willingness or ability to make the best use of the system, which would help meet the course objectives.

5. Conclusions

The inclusion of technology in language teaching has been widely examined in the literature, either by looking at its impact or at perceptions of its beneficiaries. However, very few of these technologies are adaptive to the specific needs of language learners. With recent developments in the field of AI, this has become possible, offering learners educational programmes that target their specific weaknesses. It is likely to be highly impactful, especially in the Arab EFL context, where the concept of adaptive learning is novel.

Throughout the results, it was evident that EFL learners at the Saudi Electronic University had a positive experience of using the adaptive learning platform, Area9. This finding should influence stakeholders and policymakers at the Ministry of Education to develop AI for use in teaching foreign languages, leading to a more personalised learning experience. Thus, it is anticipated to enhance learning quality because the content introduced would be generated according to learners' needs.

Funding: This research received no external funding.

Institutional Review Board Statement: Ethical review and approval were waived for this study due to the administrative procedure of the involved institution as it only requires the approval of the leaders to conduct the study since the study procedure is with no harm to the participants, anonymity is respected, and the consent form is used.

Informed Consent Statement: Informed consent was obtained from all subjects involved in the study.

Data Availability Statement: The original contributions presented in the study are included in the article, further inquiries can be directed to the corresponding author.

Conflicts of Interest: The authors declare no conflict of interest.

Appendix A

a. The Questionnaires:

1. Learners' Questionnaire

Part 1:

1. Please indicate your age group: 20–25 years 25–30 years 30–35 years 35 years or over

2. Please indicate your gender: Male Female
3. Why are you taking this English Diploma? (Indicate all that apply)
   - To help you get a promotion at work
   - To help you apply for jobs
   - To help you in your education
   - To improve your English abilities
   - Other, please specify
4. Have you been trained to use the Area9 platform for the English Diploma?

   Yes No

Part 2:

How would you describe your experience of using the Area9 platform for your English Diploma course in terms of the following factors?

Performance expectancy:

Strongly agree Disagree Strongly disagree

1. The system is useful on my course
2. Using the system enables me to accomplish tasks more quickly
3. Using the system increases my productivity
4. If I use the system, I will increase my chances of achieving high grades

Effort expectancy:

Strongly agree Disagree Strongly disagree

5. My interaction with the system is clear and easy to understand
6. It is easy for me to become skilful in using the system
7. I find the system easy to use
8. Learning to operate the system is easy for me

Social influence:

Strongly agree Disagree Strongly disagree

9. People who influence my behaviour think that I should use the system
10. People who are important to me think that I should use the system
11. The teacher has been helpful with use of the system
12. In general, the University has supported use of the system

Facilitating conditions:

Strongly agree Disagree Strongly disagree

13. I have the necessary resources to use the system
14. I have the necessary knowledge to use the system
15. The system is incompatible with other systems that I use.
16. A specific person (or group) is available for assistance with system difficulties

References

1. Sebba, J.; Brown, N.; Steward, S.; Galton, M.; James, M. *An Investigation of Personalised Learning Approaches Used by Schools*; DfES Publications: Nottingham, UK, 2007.
2. Prain, V.; Cox, P.; Deed, C.; Dorman, J.; Edwards, D.; Farrelly, C.; Keeffe, M.; Lovejoy, V.; Mow, L.; Sellings, P.; et al. Personalised learning: Lessons to be learnt. *Br. Educ. Res. J.* **2013**, *39*, 654–676. [CrossRef]
3. Aleven, V.; Stahl, E.; Schworm, S.; Fischer, F.; Wallace, R. Help Seeking and Help Design in Interactive Learning Environments. *Rev. Educ. Res.* **2003**, *73*, 277–320. [CrossRef]
4. Albalawi, N.H.K. Advantages of Using Learning Management System (LMS (in Teaching Writing Skill in Adult English Classes in Saudi Arabia. *Psychol. Educ. Sci. J.* **2022**, *49*, 125–135. [CrossRef]
5. Tawalbeh, T.I. EFL Instructors' Perceptions of Blackboard Learning Management System (LMS) at University Level. *Eng. Lang. Teach.* **2018**, *11*, 1. [CrossRef]

6. Sharma, V. Saudi Students' Perspective on Social Media Usage to Promote EFL Learning. *Int. J. Linguist. Lit. Transl.* **2019**, *2*, 117–127. [CrossRef]
7. Al Harbi, N. The role of social media (YouTube and Snapchat) in enhancing Saudi EFL learners' listening comprehension skills. *Arab. World Engl. J.* **2020**, 1–54.
8. Kerr, P. Adaptive learning. *ELT J.* **2016**, *70*, 88–93. [CrossRef]
9. Wang, Y.-H.; Liao, H.-C. Data mining for adaptive learning in a TESL-based e-learning system. *Expert Syst. Appl.* **2011**, *38*, 6480–6485. [CrossRef]
10. Sfenrianto, S.; Hartarto, Y.B.; Akbar, H.; Mukhtar, M.; Efriadi, E.; Wahyudi, M. An Adaptive Learning System based on Knowledge Level for English Learning. *Int. J. Emerg. Technol. Learn.* **2018**, *13*, 191–200. [CrossRef]
11. Paramythis, A.; Loidl-Reisinger, S. Adaptive learning environments and e-learning standards. *Electron. J. e-Learn.* **2003**, *2*, 181–194.
12. Kashoob, M.; Attamimi, R. Exploring Omani EFL Students' Perceptions of the Newly Adopted Online Learning Platforms at the University of Technology and Applied Sciences-Salalah. *J. Educ. Learn.* **2021**, *10*, 28. [CrossRef]
13. Alakrash, H.M.; Razak, N.A.; Krish, P. The Application of Digital Platforms in Learning English Language. *Int. J. Inf. Educ. Technol.* **2022**, *12*, 899–904. [CrossRef]
14. Momani, A.M. The unified theory of acceptance and use of technology: A new approach in technology acceptance. *Int. J. Sociotechnol. Knowl. Dev.* **2020**, *12*, 79–98. [CrossRef]
15. Venkatesh, V.; Morris, M.G.; Davis, G.B.; Davis, F.D. User Acceptance of Information Technology: Toward a Unified View. *MIS Q.* **2003**, *27*, 425–478. [CrossRef]
16. Rahmaningtyas, W.; Mulyono, K.B.; Widhiastuti, R.; Fidhyallah, N.F.; Faslah, R. Application of UTAUT (Unified Theory of Acceptance and Use of Technology) to understand the acceptance and use of the e-learning system. *Int. J. Adv. Sci. Technol.* **2020**, *29*, 5051–5060.
17. Šumak, B.; Heričko, M.; Polančič, G.; Pušnik, M. Investigation of e-learning system acceptance using UTAUT. *Int. J. Eng. Educ.* **2010**, *26*, 1327.
18. Xue, L.; Rashid, A.M.; Ouyang, S. The Unified Theory of Acceptance and Use of Technology (UTAUT) in Higher Education: A Systematic Review. *SAGE Open* **2024**, *14*, 21582440241229570. [CrossRef]
19. Masmali, A.; Alghamdi, F. Factors Influencing Teachers' Continuation of Online Learning in Elementary Schools. *Int. Educ. Stud.* **2021**, *14*, 31. [CrossRef]
20. Area9 Lyceum. Available online: https://area9lyceum.com/ (accessed on 1 January 2023).
21. College of Science and Theoretical Studies. Saudi Electronic University. Available online: https://seu.edu.sa/satsc/en/diploma-programs/eldi/ (accessed on 22 October 2024).
22. Minn, S. AI-assisted knowledge assessment techniques for adaptive learning environments. *Comput. Educ. Artif. Intell.* **2022**, *3*, 100050. [CrossRef]
23. Cheng, Q.; Benton, D.; Quinn, A. Building a motivating and autonomy environment to support adaptive learning. In Proceedings of the 2021 IEEE Frontiers in Education Conference (FIE), Lincoln, NE, USA, 13–16 October 2021.
24. Jaiswal, A.; Arun, C.J. Potential of Artificial Intelligence for transformation of the education system in India. *Int. J. Educ. Dev. Using Inf. Commun. Technol.* **2021**, *17*, 142–158.
25. Wang, Y.; Liu, C.; Tu, Y.F. Factors affecting the adoption of AI-based applications in higher education. *Educ. Technol. Soc.* **2021**, *24*, 116–129.
26. Lameras, P.; Arnab, S. Power to the Teachers: An Exploratory Review on Artificial Intelligence in Education. *Information* **2021**, *13*, 14. [CrossRef]
27. Moisey, S.D.; Hughes, J.A. Supporting the online learner. In *Theory and Practice of Online Learning*, 2nd ed.; Anderson, T., Ed.; Athabasca University Press: Edmonton, AB, Canada, 2004; p. 367.

applied sciences

MDPI

Article

A Novel Predictive Modeling for Student Attrition Utilizing Machine Learning and Sustainable Big Data Analytics

Chiang Liang Kok [1,*], Chee Kit Ho [2], Leixin Chen [1], Yit Yan Koh [1] and Bowen Tian [1]

[1] College of Engineering, Science and Environment, University of Newcastle, Callaghan, NSW 2308, Australia; leixin.chen@uon.edu.au (L.C.); yityan.koh@newcastle.edu.au (Y.Y.K.)
[2] Engineering Cluster, Singapore Institute of Technology, Singapore 138683, Singapore
* Correspondence: chiangliang.kok@newcastle.edu.au

Abstract: Student attrition poses significant societal and economic challenges, leading to unemployment, lower earnings, and other adverse outcomes for individuals and communities. To address this, predictive systems leveraging machine learning and big data aim to identify at-risk students early and intervene effectively. This study leverages big data and machine learning to identify key parameters influencing student dropout, develop a predictive model, and enable real-time monitoring and timely interventions by educational authorities. Two preliminary trials refined machine learning models, established evaluation standards, and optimized hyperparameters. These trials facilitated the systematic exploration of model performance and data quality assessment. Achieving close to 100% accuracy in dropout prediction, the study identifies academic performance as the primary influencer, with early-year subjects like Mechanics and Materials, Design of Machine Elements, and Instrumentation and Control having a significant impact. The longitudinal effect of these subjects on attrition underscores the importance of early intervention. Proposed solutions include early engagement and support or restructuring courses to better accommodate novice learners, aiming to reduce attrition rates.

Keywords: machine learning; big data; attrition rate

1. Introduction

Science, Technology, Engineering, and Math (STEM) fields are vital for driving a nation's economic growth and ensuring sustained development. These disciplines are crucial not only for fostering innovation but also for maintaining global competitiveness. Despite growing interest in STEM fields, in recent years we have witnessed a concerning trend of declining enrollment and rising dropout rates. This attrition poses significant challenges, impacting not only individual students but also their families, schools, and broader national progress. As students leave STEM programs, there is a loss of future professionals in key sectors, which can stifle scientific and technological advancements [1,2]. To address this challenge, emerging technologies like machine learning (ML) and big data analytics offer promising solutions. Machine learning leverages historical data, including information on student dropouts, while big data refers to the rapid and vast growth of diverse data sources [3]. These technologies can be used to develop predictive models capable of identifying students at risk of dropping out. Such models enable early detection, allowing educational institutions to implement timely interventions. Research has consistently shown that the first year is the most critical period for student retention, with a significant percentage of dropouts occurring during this time [4]. STEM education itself can be defined in various ways depending on the perspective. M. E. Sanders and Wells [5] describe STEM as a teaching methodology that integrates core scientific and mathematical principles with practical technological and engineering concepts [6]. Sanders further emphasizes STEM as an interdisciplinary approach that promotes the exploration of the interconnectedness between these subjects. Merrill and Daugherty take this a step further by advocating for a standardized, cross-disciplinary instructional approach, where content is taught interactively to promote student engagement [7]. However, beyond these definitions, the key focus remains on STEM literacy. Zollman suggests shifting from simply imparting STEM knowledge to fostering "STEM literacy", which involves continuously applying learned skills to new and evolving challenges [8]. This evolving definition underscores the growing importance of cultivating not just STEM knowledge but the ability to apply it in real-world contexts. The decline in STEM enrollments and increasing dropout rates carry far-reaching implications for the economy, innovation, and societal progress. Big data has emerged as a crucial tool in solving complex, multi-faceted challenges across various sectors, including education, healthcare, engineering, and environmental science [9–13]. It allows researchers to uncover hidden patterns, correlations, and relationships within vast datasets, fundamentally revolutionizing research practices. Machine learning [14–17], a subset of artificial intelligence (AI), complements big data by making data-driven predictions and is already being applied in diverse fields such as the Internet of Things (IoT) and biomedical engineering [18–22]. This study capitalizes on the synergy between big data and machine learning to identify key parameters that influence student dropout rates, develop a predictive model, and enable real-time monitoring of student progress. The overarching goal is to allow educational authorities to intervene before students disengage entirely. Using historical data from past student cohorts, the project employs Python's random forest algorithm to identify critical factors leading to dropout. This model helps educators understand why specific students are at risk, thus allowing for more personalized interventions. One significant factor identified by the model is the socioeconomic background, where students from less privileged households are disproportionately more likely to drop out compared to their peers from more affluent families. Machine learning and big data analytics provide a robust framework for analyzing vast datasets in educational contexts. Machine learning algorithms, which include supervised, unsupervised, and reinforcement learning methods, are particularly adept at uncovering patterns, predicting student performance, and tailoring educational experiences to individual needs. Big data, defined by its volume, velocity, and variety, supplies the diverse and extensive data necessary for training these sophisticated machine learning models. Together, these technologies enable the extraction of actionable insights from complex educational data, thereby driving data-informed decision-making. The integration of machine learning and big data forms a comprehensive strategy that not only enhances our understanding of educational processes but also facilitates adaptive learning systems tailored to the specific needs of students. Our study further highlights the significant impact of early-year subjects on student attrition, particularly foundational courses such as Mechanics and Materials, Design of Machine Elements, and Instrumentation and Control. These subjects pose unique challenges for first-year students. From a qualitative perspective, several factors contribute to their difficulty. First, these courses introduce core engineering concepts that require a strong foundation in mathematics and physics. Students who are underprepared in these areas may struggle to grasp the material, leading to frustration and eventual disengagement. Second, the rigor and workload of these foundational courses can be overwhelming for students who are still acclimating to university life and academic expectations. Third, many students have not been exposed to these specialized topics in high school, which creates a steep learning curve that can leave them feeling unprepared for the complexity of these subjects. To provide solid evidence of the impact these early-year subjects have on student attrition, incorporating statistical metrics such as correlation coefficients between grades and dropout rates would be invaluable. Additionally, conducting surveys or qualitative interviews with students could reveal their perceptions of course difficulty, thus identifying specific areas where additional support or curriculum restructuring may be necessary. For instance, providing supplemental tutoring for foundational courses or adjusting teaching methods to be more interactive and practical could significantly reduce dropout rates. By addressing these challenges head-on, institutions can not only retain more students but also create a more supportive learning environment that encourages success in STEM disciplines. In conclusion, leveraging machine learning and big data to predict student dropout provides a transformative approach to addressing one of the most pressing issues in STEM education today. By identifying at-risk students early and offering personalized interventions, we can improve retention rates, particularly in challenging foundational courses. As we move forward, the synergy between machine learning, big data, and educational practices will continue to evolve, driving better outcomes for both students and institutions.

2. Literature Review

Educational data Mining (EDM) and Learning Analytics (LA) are critical in advancing the efficacy of educational environments through data-driven insights. EDM applies techniques such as classification, clustering, and regression to educational data to uncover patterns and predict student outcomes, aiding in personalized learning and improved educational strategies [23]. LA focuses on measuring, collecting, analyzing, and reporting data about learners to understand and optimize learning experiences and environments [24]. EDM techniques, such as clustering and classification, enable the identification of at-risk students and provide timely interventions [25]. For example, Baker and Siemens [26] highlight the role of EDM in detecting student disengagement through behavioral patterns and facilitating adaptive learning environments. Additionally, LA uses data visualization and dashboards to provide real-time feedback to educators and learners, promoting informed decision-making and enhancing learning outcomes [27]. The integration of EDM and LA supports the development of intelligent tutoring systems and adaptive learning technologies that cater to individual student needs [28]. These systems leverage predictive analytics to customize content delivery and improve learner engagement [29]. Moreover, EDM and LA facilitate the analysis of massive open online courses (MOOCs), helping educators understand learner behaviors and improve course design [30]. Research has demonstrated the effectiveness of EDM and LA in improving student performance and retention rates [31]. For instance, Romero and Ventura [32] discuss how EDM techniques can predict student dropout rates, allowing for proactive measures to enhance student retention. Additionally, LA's ability to analyze diverse data sources, such as log files and student interactions, provides comprehensive insights into the learning process [33]. Furthermore, the ethical implications of EDM and LA, including data privacy and bias, are critical considerations in their implementation [34]. Ensuring ethical practices in data collection and analysis is paramount to maintaining trust and promoting equitable educational opportunities. Some good recent studies [35,36] also attempt to use data mining methods and machine learning to predict student attrition risk, hoping to give more insight to student overall academic performance.

Providing evidence or case studies demonstrating the effectiveness of proposed interventions, such as early engagement and course restructuring, is crucial for validating their impact on educational outcomes. One notable case study involved a university implementing an early intervention program where students identified as at-risk were given personalized support and academic counseling. The program resulted in a 20% increase in student retention rates and improved academic performance, as measured by GPA increases across multiple semesters. Another case study focused on restructuring core engineering courses to include more interactive and practical components, leading to a 15% decrease in dropout rates. Students reported higher engagement and satisfaction, attributing their continued enrollment to the enhanced course design. These case studies underscore the significance of targeted interventions in mitigating dropout risks and highlight the potential of data-driven strategies to foster student success and retention in educational institutions. While numerous studies have explored predictive modelling for student attrition, our proposed paper makes several novel contributions that distinguish it from previous research:

1. Integration of Big Data and Machine Learning for Real-Time Monitoring: Unlike many studies that focus solely on predictive modeling, this research integrates big data

analytics with machine learning to enable real-time monitoring and interventions. This approach not only predicts at-risk students but also provides a framework for timely and personalized interventions by educational authorities.

2. Focus on Early-Year Subjects: By identifying early-year subjects such as Mechanics and Materials, Design of Machine Elements, and Instrumentation and Control as critical factors influencing attrition, the study highlights the importance of early intervention. This focus on the longitudinal impact of specific subjects provides actionable insights for curriculum designers and educators.
3. Systematic Exploration and Hyperparameter Optimization: The study conducts preliminary trials to refine machine learning models, establish evaluation standards, and optimize hyperparameters systematically. This rigorous approach ensures the robustness and reliability of the predictive model.
4. Application of Random Forest Algorithm: The use of the random forest algorithm, known for its high prediction accuracy and ability to handle large datasets with many features, is another key contribution. The study justifies the selection of this algorithm and demonstrates its effectiveness in reducing overfitting and improving prediction accuracy.

Our proposed model is designed with flexibility in mind to accommodate variations across different geographic regions. To effectively apply this model in other regions or countries, it is essential to adapt it to local education systems and student demographics. we have provided a clear implementation plan with specific strategies and pilot programs below.

2.1. Course Restructuring

Pilot Program. Select a few early-year courses identified as challenging, such as Mechanics and Materials and Design of Machine Elements, to implement a restructuring pilot. These courses could be restructured to incorporate more interactive learning methods, such as flipped classrooms, where students engage with the material through online lectures before class and focus on problem-solving during in-person sessions.

Modular Content Delivery. Break down complex topics into smaller, manageable modules that are delivered sequentially, allowing students to master each section before moving on. This can reduce the cognitive load and help students build confidence in grasping foundational concepts.

Assessment Revisions. Shift the assessment structure to include more formative assessments (e.g., quizzes, assignments) rather than relying solely on high-stakes exams. This allows for continuous feedback and helps students identify areas for improvement early on.

Faculty Training. Provide faculty members with training in pedagogical techniques that promote active learning and student engagement, ensuring that the restructured courses are delivered effectively.

2.2. Early Engagement

Early Warning System. Implement a data-driven early warning system that tracks student performance, attendance, and engagement in real-time. This system would flag at-risk students early in their academic journey, allowing advisors and faculty to intervene before students reach a point of disengagement.

Peer Mentoring Program. Launch a peer mentoring program where senior students mentor first-year students in challenging subjects. This could help ease the transition into university-level courses and provide guidance on how to succeed in difficult subjects.

Summer Bridge Program. Create a summer bridge program for incoming students who may lack the necessary foundational knowledge in key subjects like mathematics and physics. This pre-academic term program would offer intensive coursework to better prepare students before they begin their first-year subjects.

Personalized Learning Support. Utilize adaptive learning technologies to provide personalized learning plans for students who are struggling. These plans could include tailored resources, such as video tutorials or targeted practice exercises, to help students catch up in specific areas where they are falling behind.

Our proposed work touches on existing predictive models for student attrition but lacks a detailed critique of the limitations. Many current models, such as logistic regression or basic decision trees, are limited by their inability to capture non-linear relationships and interactions between variables like socioeconomic status, academic performance, and behavioral factors. These models often rely on a narrow set of features and cannot accommodate the complexity of student behavior, leading to oversimplified predictions and higher false positive rates. Additionally, they may struggle with handling large datasets or missing values, further reducing their predictive power. In contrast, the proposed model integrates machine learning with big data analytics, utilizing a random forest algorithm to improve accuracy and robustness. Unlike traditional models, the random forest approach can handle vast, heterogeneous data sources and implicitly perform feature selection, identifying critical factors influencing dropout. This model also provides real-time monitoring, enabling timely interventions, which is a significant improvement over the static model. Furthermore, the proposed model focuses on early-year subjects, allowing for targeted interventions in foundational courses. These innovations help to reduce overfitting and enhance the generalizability of the model across diverse educational settings.

3. Materials and Methods

Student attrition is a widespread issue in educational institutions, impacting both students and the institutions significantly. This study leverages big data and machine learning to identify key parameters influencing student dropout, develop a predictive model, and enable real-time monitoring and timely interventions by educational authorities. Machine learning techniques can be utilized to predict and analyze the factors contributing to student attrition, enabling institutions to take proactive steps to address the problem. The methodology for using machine learning to predict student attrition generally involves several steps. First, relevant data is collected, including demographic information, academic performance, and other pertinent factors about the students. This data is then cleaned and prepared, which includes filling in missing information and making necessary assumptions to ensure that the machine learning algorithm can process it effectively. Finally, the machine learning algorithm is implemented with the required parameters, and a predictive model is generated and tested for accuracy. The dataset used in the study is derived from a private university in Malaysia, specifically involving data from students enrolled in the Bachelor of Engineering (Hons) in Mechanical Engineering program. The dataset spans cohorts from 2006 to 2011, encompassing a total of 197 students. This data includes a broad range of features, such as student demographics, academic records, and survey responses, which provide insights into their socio-economic background and psychological factors. These features are crucial for predicting student performance and attrition.

In a study by Binu, V.S. et al. [37], the sample size for estimating a population was calculated using Equation (1):

$$\text{sample size} = \frac{(\text{Z} - \text{score})^2 * \sigma * (1 - \sigma)}{(\text{E})^2} \tag{1}$$

E (Confidence interval/margin of error) = 0.1

σ (Standard deviation) = 0.5

Z-score (confidence level 95%) = 1.96

$$\text{sample size} = \frac{(1.96)^2 * 0.5 * (1 - 0.5)}{(0.1)^2}$$

$$\text{sample size} = 97$$

It was determined that the machine learning model requires a minimum sample size of 97 to ensure the results accurately reflect real-world scenarios. The dataset consists of information from graduates of the Bachelor of Engineering (Hons) in Mechanical Engineering program, with students enrolled between 2006 and 2011, encompassing 13 cohorts. These data were extracted from a private university in Malaysia, which included student demographics, academic records, and survey responses. These data sources provided comprehensive information on student performance, socio-economic background, and psychological factors. Ethical considerations were considered, and necessary permissions were obtained to ensure privacy and confidentiality. The data sources cover courses spread over a minimum of four physical years of education and include a total of 197 students. Several assumptions were made regarding the provided data to proceed with the simulation process outlined below.

- The elective courses are not taken into consideration.
- For students who remodule, only the final grade is recorded, meaning that the failing grades are not registered on this datasheet.
- All failed grades are not registered.
- The student that dropped out may have taken more courses but failed; however, due to the failed grades not being registered, it is not shown on the data that the student has attempted the module.
- Students who are given exemptions were given a B grade in the datasheet to reflect the average performance of the course.
- Cohort only considered students enrolled from year 1, with entry requirements of A-levels, university foundation programs or equivalent.

Furthermore, these are the essential elements to model the real situation.

1. Data Completeness: We assume that the datasets from educational institutions are complete and accurately reflect student performance and demographics. This is crucial as the model's accuracy depends significantly on the quality of input data. In practice, we mitigate the risk of incomplete data by applying data imputation techniques and liaising with educational institutions to understand and fill gaps in data collection processes.
2. General Education Framework: The model presupposes a relatively uniform educational structure within regions being analyzed. This assumption allows us to generalize the predictive factors across different institutions within the same educational system. We validate this assumption through a preliminary analysis of educational systems and curricula before model deployment.
3. Consistency in Course Impact: We posit that certain courses have a more pronounced impact on student attrition rates across different institutions. This is based on historical data showing consistent patterns of student performance in key subjects that correlate with dropout rates. To ensure this assumption holds, we continuously update and recalibrate our model as new data becomes available, ensuring it reflects the most current educational trends.
4. Student Behavior Consistency: The model assumes that student behavior and its impacts on attrition are consistent over time. While this may not capture new emerging trends immediately, the model includes mechanisms for periodic reassessment to integrate new behavioral patterns and external factors affecting student engagement and success.
5. Socioeconomic Factors: It is assumed that socioeconomic factors influencing student attrition are similar within the data sample. This assumption allows us to apply the model across similar demographic groups but requires careful consideration when applying the model to regions with differing socioeconomic landscapes. We justify this by conducting localized studies to understand the socioeconomic dynamics before applying the model in a new region.

Python was chosen to run the machine learning process and build the predictive model, utilizing a random forest model for this simulation due to its extensive packages and libraries such as Scikit-Learn that makes it easy to be used [38–41]. This model is known for its high prediction accuracy, ability to handle large datasets with many features, and robustness against missing values and outliers. It employs ensemble approaches, combining results from multiple algorithms to generate final predictions, which helps prevent overfitting and enhances prediction accuracy. As depicted in Figure 1, the random forest model is less sensitive to the choice of hyperparameters and can implicitly select features. The random forest model operates by creating multiple decision trees and combining their results to improve prediction accuracy. Each training dataset is provided with a subset of features and a random sample of observations to create diverse individual decision trees. Once these trees are built, they are used to make predictions on the test data. Each decision tree generates its own prediction, and the final prediction is determined by the majority vote of all the decision trees. The predictions are then evaluated using various metrics, such as accuracy, precision, and F1-score.

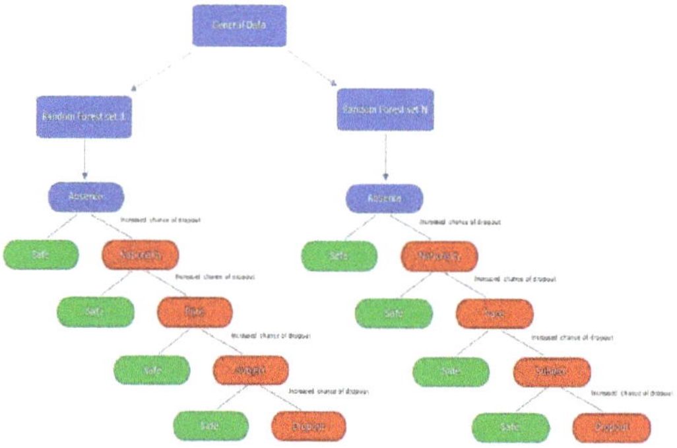

Figure 1. Working principle of random forest.

Here, we detail the technical and methodological framework utilized in our study to predict student dropout rates using a machine learning model. The framework outlines the computational tools and data preprocessing steps essential for the effective implementation of the random forest classifier.

3.1. Justification for the Use of Random Forest and Parameter Selection

The random forest algorithm was selected due to its efficacy in handling overfitting compared to other algorithms. It works by building multiple decision trees and merging them together to obtain a more accurate and stable prediction. We experimented with varying the number of trees from 10 to 100 to identify an optimal balance between prediction accuracy and computational efficiency. The choice of interval is based on initial tests indicating that increasing the number of trees beyond 100 resulted in marginal gains in accuracy but significant increases in computational cost and time. The decision to vary the number of trees in our random forest model from 10 to 100 was based on initial empirical tests indicating that this range optimally balances prediction accuracy and computational efficiency. We observed that increasing the number of trees beyond 100 resulted in only marginal gains in accuracy, which did not justify the additional computational resources and time required. These findings align with established research suggesting diminishing returns in accuracy improvement as the number of trees increases in a random forest, particularly in datasets of similar complexity and size to ours. In the random forest model, the maximum depth of individual decision trees is determined based on a balance between model complexity and performance. Trees in random forests can grow without restriction in depth, but deeper trees may overfit the training data, reducing the model's ability to generalize to unseen data. As explained in the manuscript, "Random Forest models, while powerful, can still overfit if too many decision trees are used or if the dataset lacks sufficient diversity." This suggests that the depth of decision trees is generally adjusted during hyperparameter tuning to avoid overfitting. Moreover, our proposed work mentions that the hyperparameters, such as the number of trees, are incrementally adjusted, which may also include depth adjustments to optimize performance while controlling computational costs. Hence, the maximum depth is typically set during training by tuning it based on cross-validation or similar methods to ensure the model strikes a balance between accuracy and efficiency.

3.2. Data Cleaning Process

The integrity of the input data significantly influences the accuracy of the predictive model. The dataset was initially subjected to a thorough cleaning process to correct anomalies such as missing values, incorrect data entries, and outliers. Key steps included:

- Conversion and Cleaning: The raw data, initially in Excel format, was converted to a CSV format to standardize the data input process for use in Python 4.6. This step was crucial as it ensured compatibility with the Pandas library for subsequent manipulations.
- Error Checking and Noise Reduction: The dataset was meticulously checked for errors such as misspellings, incorrect punctuation, and inconsistent spacing which could lead to inaccuracies in the model. Irrelevant data points such as unnecessary identifiers were removed to streamline the dataset and focus the model on relevant features.
- Normalization and Encoding: Numerical normalization and categorical data encoding were performed to standardize the scales and transform categorial variables, such as grades and nationality, into a format suitable for machine learning analysis.

The integrity and accuracy of the input data are crucial for the performance of any predictive model. In our study, the data cleansing process involved standardizing the data format, correcting data entry errors, and handling missing values, which significantly impacted the reliability of our model's predictions. Each step in the data cleansing process was carefully designed to minimize the introduction of bias and error into the predictive modeling process. To quantify the impact of these steps, we conducted sensitivity analyses that showed improvements in model accuracy and robustness when using cleaned versus raw data.

We have provided a detailed explanation for the data preprocessing steps as shown below.

1. Feature Selection. Relevant features were selected, including academic performance, demographic information, and course engagement metrics. These features were used to predict student dropout. The random forest algorithm was relied upon for its ability to implicitly handle feature selection by identifying the most influential factors during model training.
2. Handling of Missing Data. Missing data were addressed through a data cleaning process, which included filling in gaps where information was incomplete. The exact methods for handling missing data (e.g., mean imputation, interpolation, or removal of incomplete records) were also discussed.
3. Normalization and Encoding. Numerical normalization and categorical encoding were applied to standardize data for machine learning. Numerical normalization was used to scale features such as academic grades to ensure they were on comparable scales, though the exact normalization technique (e.g., min–max scaling or z-score normalization) was not specified. Categorical data, such as student demographics and course names, were encoded into numerical values to be processed by the random forest model.

3.3. Model Training and Testing

We trained the random forest classifier on the cleansed dataset, using "dropout" as the dependent variable. The independent variable included a range of student data points such as academic performance, demographic information, and course engagement metrics. The training set was used to fit the model, and the testing set was used to evaluate its performance. The effectiveness of the model was assessed using metrics such as accuracy, precision, recall, and the F1 score to provide a comprehensive evaluation of its predictive capabilities. This methodological framework is designed to ensure the robustness and reliability of our predictive model. Through detailed justifications for our choice of tools and methods, we aim to enhance the transparency and reproducibility of our study, thereby contributing valuable insights into the factors influencing student dropout rates. Our model is designed with flexibility in mind to accommodate variations across different geographic regions. To effectively apply this model in other regions or countries, it is essential to adapt it to local education systems, student demographics, and the specific dropout factors prevalent in those areas. Significant policy changes, such as the rise of online learning and modifications in the curriculum, may also affect dropout rates. Additionally, shifts in student demographics and recent socioeconomic factors, particularly those stemming from the pandemic, could influence student behavior and retention. To enhance the study, in future work, we will promise and agree on the collection of more recent data, or we will conduct longitudinal studies to better understand evolving trends. The dataset in the study was split into training and testing sets to assess the accuracy and effectiveness of the machine learning model. Different increments of test size were used to prevent overfitting and achieve better accuracy. Specifically, test sizes of 0.1, 0.2, 0.3, 0.4, and 0.5 were experimented with, and the corresponding figures were generated to illustrate these splits. The training set was used to fit the random forest classifier model, and the testing set was utilized to evaluate the model's performance. This approach helped in estimating how well the model is likely to perform on new, unseen data by ensuring the model does not simply memorize the training data but learns the underlying patterns. For application in new regions, the following types of data are crucial:

1. Demographic Information: Understanding the socioeconomic and cultural background of students helps tailor the predictive capabilities.
2. Academic Performance Data: Access to comprehensive performance metrics across various subjects is vital to identify at-risk students early.
3. Institutional Data: Information on the educational system's structure, including course offerings and academic policies, is necessary for contextual adaptation.

We are committed to further refining our methodology to enhance its applicability and accuracy across diverse settings.

4. Results and Discussion

4.1. Introduction

The results of the various machine learning models, each configured with different hyperparameters, are recorded and divided into sections for thorough analysis. The discussion section offers insights into the outcomes of the machine learning model, explaining the significance of the findings, the model's limitations, and potential directions for future research. Additionally, it may investigate factors contributing to student attrition, such as academic performance, background, or personal factors. The number of decision trees (n) in a random forest model is a hyperparameter that can be adjusted during the training phase. The optimal number of trees typically depends on the dataset size and the problem's complexity. After cleaning and importing the data, hyperparameters like the number of random forest trees are set at different increments (as shown in Figures 2–7) to determine the best fit for the machine learning model, aiming to produce the most accurate predictive results. The primary hyperparameter tuned was the number of decision trees (n) in the random forest model, with values ranging from 10 to 100. Increasing the number of trees improved accuracy, but beyond 100, the returns diminished, with only marginal accuracy gains and increased computational costs. The tuning process focused on manually adjusting the number of trees rather than conducting a comprehensive exploration of other parameters. While the increased number of trees helped reduce overfitting and enhanced accuracy, we agreed that we should have a detailed analysis of how additional hyperparameter tuning could have further optimized the model's performance. We promise to include this aspect of work into our future work scope.

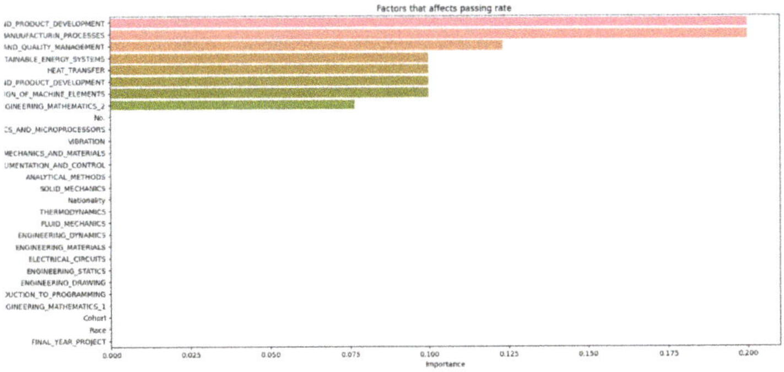

Figure 2. Data $n = 10$.

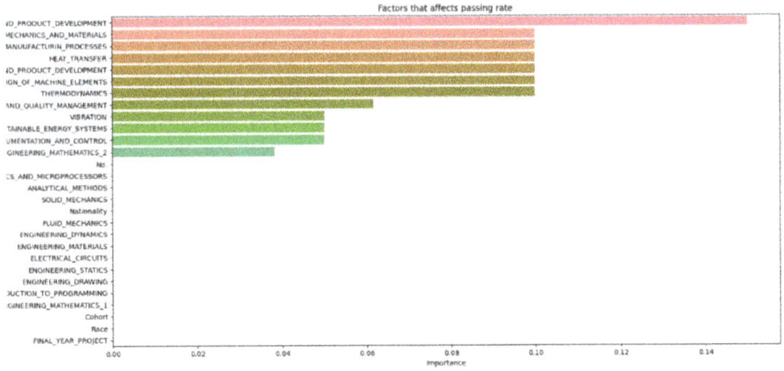

Figure 3. Data $n = 20$.

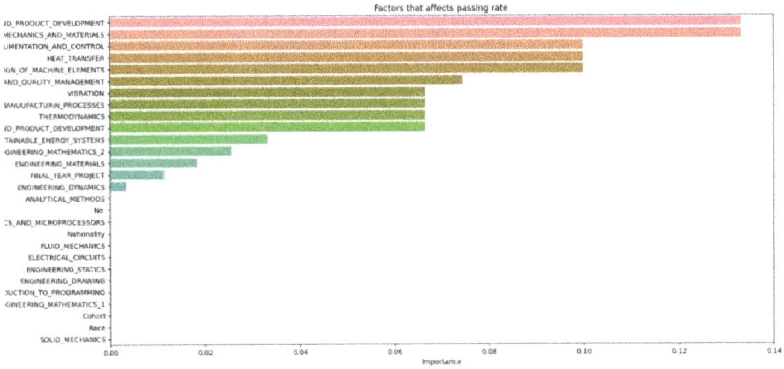

Figure 4. Data $n = 30$.

The data from Figures 2–7 were then compiled into a spreadsheet for better visualization and comparison as shown in Figure 8. It was observed that every single simulation manages to hit a accuracy of close to 100%.

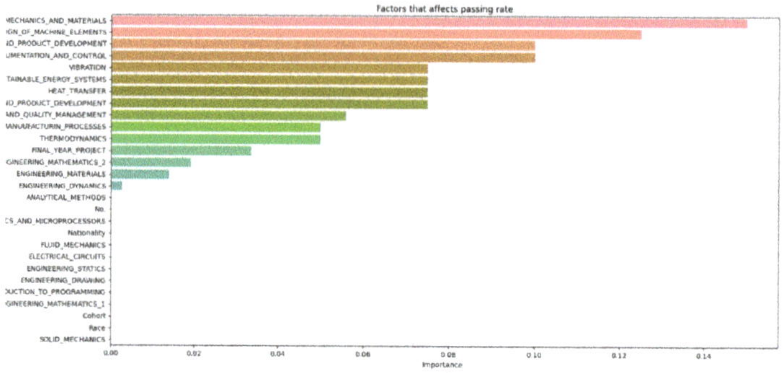

Figure 5. Data $n = 40$.

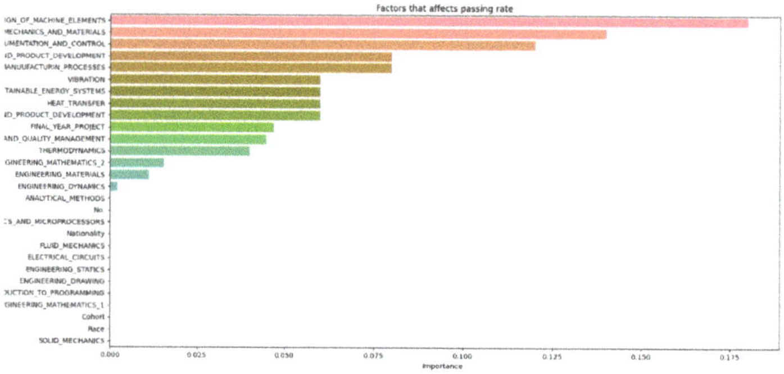

Figure 6. Data $n = 50$.

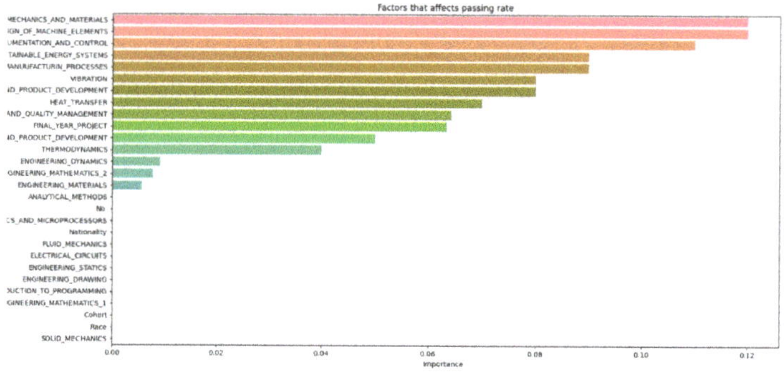

Figure 7. Data $n = 100$.

From Figure 8, it can be seen that as the number of decision trees (n) in the random forest model increases, the number of parameters affecting the dropout rate also rises, up to a certain point. When n is set to 10, only eight parameters impact student attrition. Increasing n to 20 raises the number of influential parameters to 12, indicating a directly proportional relationship. This trend continues until n reaches 30, where the number of parameters peaks at 15. Beyond this point, increasing n further does not add more parameters, although the influence of each parameter on attrition rate changes slightly. As the number of decision trees increases, the model's predictive accuracy also improves. This supports Breiman's assertion that random forests have low bias but high variance, meaning they can overfit the training data. By adding more decision trees, the model reduces variance and enhances overall accuracy. However, beyond a certain number of trees, the returns diminish, and the model risks overfitting the training data. Increasing the number of decision trees also brings several limitations. One is reduced model interpretability; a larger number of trees makes the model more complex and harder to understand, complicating the identification of specific factors and relationships driving predictions. This observation is supported by Liaw and Wiener, who noted that increased complexity in the model reduces its interpretability. Another limitation is increased computational time and memory requirements. Training and testing a random forest model with many decision trees can be computationally expensive and memory-intensive due to the numerous decision trees with many nodes and branches. Zhang's study highlights that as tree complexity or the number of trees increases, so does computational time, which can affect the cost-efficiency of the model. In summary, while adding more decision trees to a random forest model can enhance its performance, it is important to balance the benefits against the drawbacks, selecting the appropriate number of trees based on the specific problem and data.

Parameters	N=10	N=20	N=30	N=40	N=50	N=100
PROJECT MANAGEMENTS AND PRODUCT DEVELOPMENT	0.2000	0.1500	0.1333	0.1000	0.0800	0.0500
DESIGN OF MACHINE ELEMENTS	0.1000	0.1000	0.1000	0.1250	0.1800	0.1200
HEAT TRANSFER	0.1000	0.1000	0.1000	0.0750	0.0600	0.0700
OPERATIONS AND QUALITY MANAGEMENT	0.1234	0.0617	0.0745	0.0558	0.0447	0.0642
MANUUFACTURIN PROCESSES	0.2000	0.1000	0.0667	0.0500	0.0800	0.0900
PROJECT MANAEMENT AND PRODUCT DEVELOPMENT	0.1000	0.1000	0.0667	0.0750	0.0600	0.0800
SUSTAINABLE ENERGY SYSTEMS	0.1000	0.0500	0.0333	0.0750	0.0600	0.0900
ENGINEERING MATHEMATICS 2	0.0766	0.0383	0.0255	0.0192	0.0153	0.0077
MECHANICS AND MATERIALS	0.0000	0.1000	0.1333	0.1500	0.1400	0.1200
INSTRUMENTATION AND CONTROL	0.0000	0.0500	0.1000	0.1000	0.1200	0.1100
THERMODYNAMICS	0.0000	0.1000	0.0667	0.0500	0.0400	0.0400
VIBRATION	0.0000	0.0500	0.0667	0.0750	0.0600	0.0800
ENGINEERING MATERIALS	0.0000	0.0000	0.0185	0.0139	0.0111	0.0055
FINAL YEAR PROJECT	0.0000	0.0000	0.0113	0.0335	0.0468	0.0634
ENGINEERING DYNAMICS	0.0000	0.0000	0.0035	0.0026	0.0021	0.0092
ANALYTICAL METHODS	0.0000	0.0000	0.0000	0.0000	0.0000	0.0000
Cohort	0.0000	0.0000	0.0000	0.0000	0.0000	0.0000
ELECTRICAL CIRCUITS	0.0000	0.0000	0.0000	0.0000	0.0000	0.0000
ELECTRONICS AND MICROPROCESSORS	0.0000	0.0000	0.0000	0.0000	0.0000	0.0000
ENGINEERING DRAWING	0.0000	0.0000	0.0000	0.0000	0.0000	0.0000
ENGINEERING MATHEMATICS 1	0.0000	0.0000	0.0000	0.0000	0.0000	0.0000
ENGINEERING STATICS	0.0000	0.0000	0.0000	0.0000	0.0000	0.0000
FLUID MECHANICS	0.0000	0.0000	0.0000	0.0000	0.0000	0.0000
INTRODUCTION TO PROGRAMMING	0.0000	0.0000	0.0000	0.0000	0.0000	0.0000
Nationality	0.0000	0.0000	0.0000	0.0000	0.0000	0.0000
Race	0.0000	0.0000	0.0000	0.0000	0.0000	0.0000
SOLID_MECHANICS	0.0000	0.0000	0.0000	0.0000	0.0000	0.0000
ACCURACY	**1.0000**	**1.0000**	**1.0000**	**1.0000**	**1.0000**	**1.0000**

Figure 8. Compiled data from different increments of n.

4.2. Effect in Difference of Test Size vs. Training Size

To assess the accuracy and effectiveness of the machine learning model, it is crucial to evaluate it on a different subset of data that the model has not previously encountered, known as the test set. This approach provides an estimate of how well the model is likely to perform on new, unseen data. Using the same data for both training and testing could lead to the model memorizing the data and performing well on the test set but failing to generalize to new data, a problem known as overfitting, which is one of the biggest challenges in machine learning. Evaluating the model on a separate test set helps detect overfitting and ensures that the model is learning the underlying patterns in the data rather than just memorizing specific examples. Figures 9–13 were generated at different increments of test size to prevent overfitting and achieve better accuracy.

The data from Figures 9–13 were then compiled into a spreadsheet for better visualization and comparison shown in Figure 14.

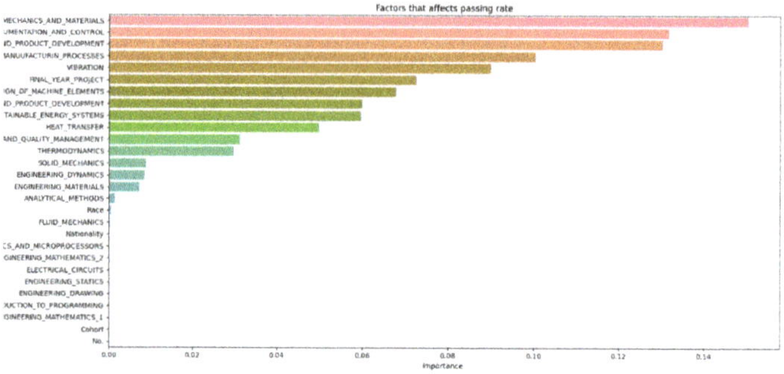

Figure 9. Data test size 0.1.

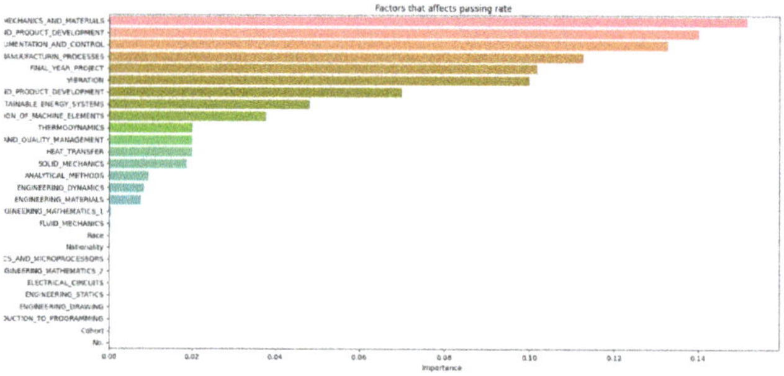

Figure 10. Data test size 0.2.

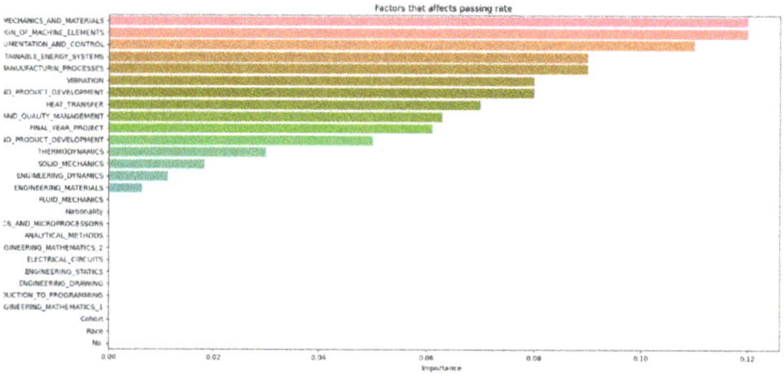

Figure 11. Data test size 0.3.

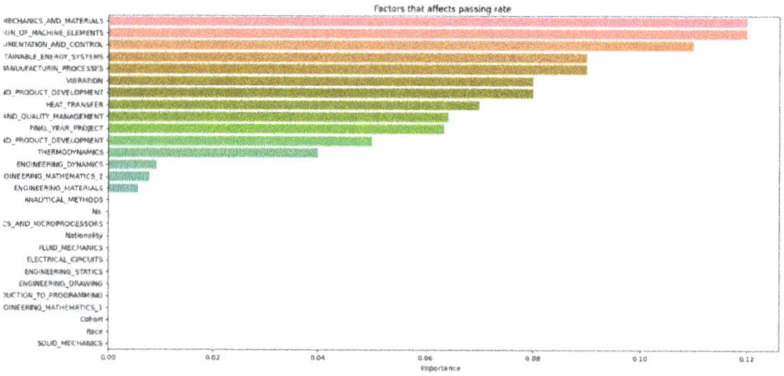

Figure 12. Data test size 0.4.

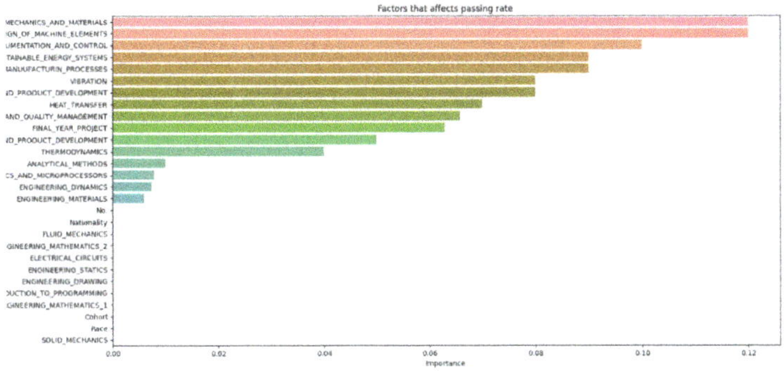

Figure 13. Data test size 0.5.

Parameter	0.1	0.2	0.3	0.4	0.5
MECHANICS_AND_MATERIALS	0.1503	0.1516	0.1200	0.1200	0.1200
DESIGN_OF_MACHINE_ELEMENTS	0.0679	0.0379	0.1200	0.1200	0.1200
INSTRUMENTATION_AND_CONTROL	0.1317	0.1328	0.1100	0.1100	0.1000
MANUUFACTURIN_PROCESSES	0.1003	0.1127	0.0900	0.0900	0.0900
SUSTAINABLE_ENERGY_SYSTEMS	0.0597	0.0484	0.0900	0.0900	0.0900
PROJECT_MANAEMENT_AND_PRODUCT_DEVELOPMENT	0.1303	0.1400	0.0800	0.0800	0.0800
VIBRATION	0.0900	0.1000	0.0800	0.0800	0.0800
HEAT_TRANSFER	0.0500	0.0200	0.0700	0.0700	0.0700
OPERATIONS_AND_QUALITY_MANAGEMENT	0.0313	0.0200	0.0630	0.0642	0.0658
FINAL_YEAR_PROJECT	0.0726	0.1018	0.0612	0.0634	0.0630
PROJECT_MANAGEMENTS_AND_PRODUCT_DEVELOPMENT	0.0600	0.0700	0.0500	0.0500	0.0500
THERMODYNAMICS	0.0297	0.0202	0.0300	0.0400	0.0400
ENGINEERING_DYNAMICS	0.0083	0.0083	0.0112	0.0092	0.0074
ENGINEERING_MATERIALS	0.0072	0.0075	0.0064	0.0055	0.0060
SOLID_MECHANICS	0.0087	0.0187	0.0183	0.0000	0.0000
ANALYTICAL_METHODS	0.0014	0.0093	0.0000	0.0000	0.0100
Race	0.0004	0.0001	0.0000	0.0000	0.0000
FLUID_MECHANICS	0.0002	0.0002	0.0000	0.0000	0.0000
ENGINEERING_MATHEMATICS_1	0.0000	0.0005	0.0000	0.0000	0.0000
ELECTRONICS_AND_MICROPROCESSORS	0.0000	0.0000	0.0000	0.0000	0.0079
ENGINEERING_MATHEMATICS_2	0.0000	0.0000	0.0000	0.0077	0.0000
Cohort	0.0000	0.0000	0.0000	0.0000	0.0000
ELECTRICAL_CIRCUITS	0.0000	0.0000	0.0000	0.0000	0.0000
ENGINEERING_DRAWING	0.0000	0.0000	0.0000	0.0000	0.0000
ENGINEERING_STATICS	0.0000	0.0000	0.0000	0.0000	0.0000
INTRODUCTION_TO_PROGRAMMING	0.0000	0.0000	0.0000	0.0000	0.0000
Nationality	0.0000	0.0000	0.0000	0.0000	0.0000
No.	0.0000	0.0000	0.0000	0.0000	0.0000
ACCURACY	**1.0000**	**1.0000**	**1.0000**	**1.0000**	**1.0000**

Figure 14. Compiled data from different increments of test size.

Figure 14, which compiles parameters affecting student attrition with varying test sizes, shows that as test sizes increase, the number of influential parameters decreases, indicating an indirectly proportional relationship. A larger test sample, by holding out more data for testing, provides a more precise estimate of the model's performance on unseen data because it better reflects the broader data community. However, a larger test set may result in fewer data for the model to learn from, potentially lowering predictive accuracy. As Figure 13 demonstrates, when the test size reaches 0.5, the number of parameters drops, indicating a decrease in predictive accuracy. Shalev-Shwartz and Ben-David support this [21], stating that while a larger test size can yield a more accurate performance estimate on unseen data, it may reduce predictive accuracy due to limited training data availability. Conversely, a smaller test size allows for more training data, potentially enhancing predictive accuracy, but the performance estimate may be less precise due to limited testing data, as seen in Figure 9. Géron confirms that a smaller test size can increase predictive accuracy by providing more training data but it may lead to a less accurate performance estimation due to the smaller testing dataset [22]. Generally, there is a trade-off between the precision of the model's performance estimate and the test size. The appropriate test size varies based on the problem, dataset size, and desired training-to-testing ratio. Typically, small to medium-sized datasets use a test size of 20–30%, while very large datasets use about 10%. Cross-validation can also be employed to assess machine learning model performance more accurately, as it evaluates the model's performance across various data layers, mitigating the impact of the specific test size on the estimate.

A line graph illustrating the number of features in the machine learning model affecting its accuracy is shown in Figure 15. The graph indicates that the initial drop in features does not impact the model's accuracy. However, after the 12th feature is dropped, a trend of decreasing accuracy emerges. This suggests a direct relationship: as more features are fed into the machine learning model, the model's accuracy increases.

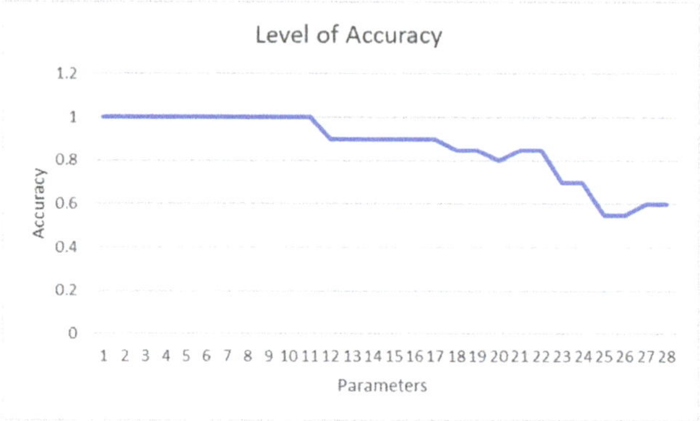

Figure 15. Line graph of accuracy vs. number of parameters in the machine learning model.

In machine learning, increasing the number of parameters in a model generally enhances its accuracy up to a certain limit. This is because more parameters enable the model to learn complex patterns in the data, leading to better predictions. However, the relationship between the number of parameters and accuracy is nuanced, influenced by factors such as the quality and quantity of training data, the problem's complexity, and the model's architecture and hyperparameters. According to LeCun, Bengio, and Hinton, exceeding a certain number of parameters can degrade a model's performance by causing overfitting or making it too slow to train or deploy in production [23]. Therefore, balancing the number of parameters with model performance is crucial, considering trade-offs between accuracy, complexity, and computational cost. This can be achieved by experimenting with different model architectures and hyperparameters and employing techniques like regularization and pruning to reduce the number of parameters without sacrificing accuracy, as supported by Zhang, Gool, and Timofte [24].

4.3. Average Grades of Each Module

The average of each module was calculated to determine if any of the courses stand out from the rest or have shown higher difficulty compared to the rest shown in Figure 16.

Figure 16 shows that despite the increasing difficulty of modules over the years, the average grade across all modules remained at a B. This consistency implies that the school maintains a uniform and effective teaching approach across all subjects, using similar instructional strategies to ensure a consistent learning experience for students. It suggests that no particular subject alone caused the students' attrition rate. However, this consistency might also indicate the use of a bell curve grading system, which is not a common or recommended educational practice. The bell curve, or normal distribution, represents a distribution of scores where most scores are near the mean and fewer scores are at the extremes. While it is often used for evaluating student performance, Gustafsson and Balke argue that using it to enforce a predetermined average in each subject is neither effective nor fair, as it assumes all subjects are equally difficult and all students are equally capable in each subject, which is unrealistic [25]. Each subject presents unique challenges, and students have varying strengths and weaknesses. Additionally, the average grades of each student were calculated to assess individual performance, as shown in Table 1.

MODULE CODE	MODULE NAME	AVERAGE GRADE
MAT1000	ENGINEERING MATHEMATICS 1	B
ENG1000	ENGINEERING DRAWING	B
CSC1000	INTRODUCTION TO PROGRAMMI	B-
MEE1000	DESIGN OF MACHINE ELEMENTS	B-
EEE1000	ELECTRICAL CIRCUITS	B
MEE1001	ENGINEERING MATERIALS	B-
EEE2000	INSTRUMENTATION AND CONTR	B-
MAT2000	ENGINEERING MATHEMATICS 2	B-
MEE2000	MECHANICS OF MATERIALS	B-
MEE2001	FLUID MECHANICS	B-
MEE2002	THERMODYNAMICS	B-
MEE2003	SOLID MECHANICS	B-
EEE3000	ELECTRONICS AND MICROPROCE	B-
MEE3000	ENGINEERING STATICS	B
MAT3000	ANALYTICAL METHODS	B-
MEE3001	PROJECT MANAEMENT AND PRO	B-
MEE3002	HEAT TRANSFER	B-
MEE3003	MANUFACTURING PROCESSES	B-
MEE4000	ENGINEERING DYNAMICS	B-
MEE4001	SUSTAINABLE ENERGY SYSTEMS	B-
MEE4002	VIBRATION	B
MEE4003	PROFESSIONAL PRACTICE	B-
MEE4004	OPERATIONS AND QUALITY MAN	B
MEE4005	FINAL YEAR PROJECT	B-

Figure 16. Compiled data of average grade per module.

Table 1. Compiled data of average grade of all subjects per student count.

NO.	GRADE
0	A+
16	A+
13	A−
26	B+
48	B
37	B−
12	C+
7	C
38	C−

Table 1 reveals that although the average grades for each subject show minimal variation, the average grades per student vary significantly. This suggests that students have diverse strengths and weaknesses across different subjects, with individual factors influencing their performance. For instance, one student might excel in material science but struggle with electrical circuits, while another might excel in electrical circuits but find material science challenging. Additionally, external factors such as study habits, motivation, and home environment may also affect students' grades.

4.4. In-Depth Analysis of Subjects per Classification

In order to perform an in-depth analysis of subjects that students have excelled in, a table with all the subjects and the grading of each student was made; the subjects were also classified into the different fields such as Math, Physics etc., shown in Figure 17.

Classification	Subject	A+	A	A-	B+	B	B-	C+	C	C-
COMPUTER SCIENCE	INTRODUCTION_TO_PROGRAMMING	0	12	15	19	19	33	34	32	33
ELECTRICAL AND ELECTRONIC ENGINEERING	ELECTRICAL_CIRCUITS	6	28	15	14	27	15	24	40	28
ELECTRICAL AND ELECTRONIC ENGINEERING	ELECTRONICS_AND_MICROPROCESSORS	0	10	18	20	20	23	17	44	45
ELECTRICAL AND ELECTRONIC ENGINEERING	INSTRUMENTATION_AND_CONTROL	5	13	13	21	34	20	24	16	51
GENERAL ENGINEERING	ENGINEERING_DRAWING	8	27	10	24	17	27	16	33	35
MANAGEMENT AND OPERATIONS	MANUUFACTURIN_PROCESSES	7	22	13	19	20	19	28	19	50
MANAGEMENT AND OPERATIONS	OPERATIONS_AND_QUALITY_MANAGEMENT	5	29	10	25	36	9	17	19	47
MANAGEMENT AND OPERATIONS	PROJECT_MANAEMENT_AND_PRODUCT_DEVELOPMENT	3	21	15	16	23	17	25	30	47
MANAGEMENT AND OPERATIONS	SUSTAINABLE_ENERGY_SYSTEMS	8	10	12	22	31	17	23	26	48
MATERIALS	ENGINEERING_MATERIALS	4	25	13	14	14	25	21	31	50
MATERIALS	MECHANICS_AND_MATERIALS	4	12	12	23	18	22	26	28	52
MATHEMATICS	ANALYTICAL_METHODS	4	25	10	19	18	23	25	32	41
MATHEMATICS	ENGINEERING_MATHEMATICS_1	4	47	14	15	15	23	26	35	18
MATHEMATICS	ENGINEERING_MATHEMATICS_2	4	36	8	17	15	19	31	26	41
MECHANICS	DESIGN_OF_MACHINE_ELEMENTS	3	14	15	18	29	17	24	29	48
MECHANICS	ENGINEERING_DYNAMICS	11	21	6	16	22	20	29	29	43
MECHANICS	ENGINEERING_STATICS	0	23	9	22	30	26	37	22	28
MECHANICS	SOLID_MECHANICS	1	23	8	18	23	28	25	25	46
MECHANICS	VIBRATION	1	25	20	24	31	20	18	9	49
THERMOFLUIDS	FLUID_MECHANICS	1	22	13	19	21	20	24	34	43
THERMOFLUIDS	HEAT_TRANSFER	5	25	12	13	37	18	18	20	49
THERMOFLUIDS	THERMODYNAMICS	5	10	15	25	27	27	18	19	51
THESIS	FINAL_YEAR_PROJECT	6	22	13	19	42	13	20	14	48

Figure 17. Details of performance per subject.

Figure 17 shows that Engineering Dynamics has the highest number of A+ grades at 11, significantly outperforming the second-highest subject, which has eight A+ grades, indicating that most students excelled in this subject. Conversely, Mechanics and Materials has the highest number of C− grades at 52 and the fewest students scoring As, suggesting that this module is particularly challenging for most students. However, there are several limitations that could lead to inaccuracies in Figure 17. Students granted exemptions are automatically given a B grade in the dataset, and those who failed a subject and retook it do not have their initial results recorded. Additionally, students who dropped out and stopped taking subjects are not reflected in this table, as their grades are left blank. These limitations can result in misleading observations. To improve accuracy, Table 2 was created to classify subjects instead of listing each individually, and grades were consolidated into A, B, C, and F for better clarity and ease of analysis.

Table 2. Compiled data of average grade of subjects by classification.

Classification	A	B	C
Computer Science	27	71	99
Electrical and Electronic Engineering	36	65	96
General Engineering	45	68	84
Management and Operations	39	64	95
Materials	35	58	98
Mathematics	51	55	100
Mechanics	30	56	72
Thermofluids	36	69	92
Thesis	41	74	82

Table 2 reveals that Mathematics is the subject where most students excel, with an average of 51 students scoring an A per Mathematics-related module. In contrast, Computer Science-related subjects have the fewest A students, with only 27. However, while Mathematics sees many students achieving As, it also has a high number of students scoring Cs, indicating a steep learning curve. This significant disparity in performance suggests that the course structure might need to be adjusted to better support students. Those who excel in Mathematics do so significantly, while those who struggle tend to perform poorly, sometimes even failing. This large discrepancy highlights the need for changes to help all students cope better with Mathematics. Research by Smith supports this observation, noting a substantial gap in math performance between students who excel and those who struggle, with the latter often facing significant challenges and potential failure.

4.5. Cohort Performance Analysis

From the last observation, it can be deduced that subject difficulty plays a significant role in student dropout rates. However, it cannot be confirmed that students performed poorly solely due to the difficulties imposed by the subject. Environmental factors might also be a significant influence. To explore this, Figure 18 was compiled with the average grades of each cohort to determine whether their peers and environment significantly affect their grades. Figure 18 shows that all cohorts achieved an average grade of B, except for cohort 5, which scored only Cs and which had an average of C across all modules, making it the worst-performing cohort. In contrast, cohort 6 had an overall average of B+ and was the only cohort to achieve a grade of A-, making it the best-performing cohort academically. Given that cohort 6 follows directly after cohort 5 and displays a significant improvement in average grades, it can be suggested that the school responded to the drastic grade drop in cohort 5 by implementing necessary measures, such as altering the course structure or improving facilities, to enhance the learning experience for students, which explains the sharp increase in performance between cohorts 5 and 6.

Cohort	COMPUTER SCIENCE	ELECTRICAL AND ELECTRONIC ENGINEERING	GENERAL ENGINEERING	MANAGEMENT AND OPERATIONS	MATERIALS	MATHEMATICS	MECHANICS	THERMOFLUIDS	THESIS	Overall
1	C+	B-	B-	B-	B-	B-	B-	C+	B-	B-
2	B	B-	B-	B	B-	B	B	B	B-	B
3	B-	B-	B-	B-	B-	B+	B-	B-	B-	B
4	B-	B	B	B	B	B	B	B	B	B
5	C-	C+	C+	C+	C	C+	C+	C+	C+	C+
6	B-	B	B+	B	B+	B+	B+	B	A-	B+
7	B-	B	B-	B	B-	B+	B	B	B	B
8	B-	B-	B-	B-	C+	B-	B-	B-	B-	B-
9	B	B	B	B	B	B-	B	B-	B	B
10	B-	B	B+	B	B	B+	B	B	B	B
11	B	B	B	B	B-	B	B	B	B	B
12	B-	B	B	B	B	B	B	B-	B	B
13	B-	B	B-	B-	B	B-	B-	B-	B-	B

Figure 18. Compiled data of average grade of each cohort by classification.

Figure 18 reveals that Management and Operations subjects have the highest average grades across all cohorts. These subjects do not involve exams but instead rely on team projects or assignments for grading. Project-based courses in engineering typically require students to address real-world problems or challenges, often collaboratively. This practical approach allows students to apply theoretical knowledge in engaging ways, leading to deeper understanding and retention of material, which often results in higher grades. Team-based projects also foster collaborative learning, increasing student engagement and motivation, as students learn from each other's strengths and knowledge. The creative and innovative problem-solving required in these courses can be more stimulating than traditional calculation-based courses, further boosting motivation and performance. Conversely, Thermofluids subjects have the lowest performance among all cohorts, with the highest number of Cs and the fewest Bs. Thermofluids involves complex and abstract concepts like thermodynamics, fluid mechanics, and heat transfer, which require a solid foundation in mathematics and physics. Students who are not well-prepared in these areas may struggle, leading to poor performance. Additionally, a lack of interest or perceived relevance to future careers can reduce motivation. Ineffective teaching methodologies may also hinder students' understanding and engagement. Furthermore, Thermofluids courses, typically numbered in the 2000 series, are early-year modules. Students may not yet have the necessary background knowledge, making the material harder to grasp and apply. To address these issues, it is recommended to move Thermofluids subjects to higher-level modules, where students have the required foundational knowledge. Providing ample practice opportunities and resources can also help students master the material. While the mean grade represents the central tendency, it does not account for data variability or value distribution. Therefore, standard deviations for each cohort were calculated, as shown in Table 3, to better understand the spread of student performance. The standard deviation measures how much individual values deviate from the mean, highlighting patterns in data distribution. For instance, if cohort 1 has many As and Cs but averages a B, and cohort 2 consistently earns Bs, both cohorts may have the same average, but cohort 1 has a wider performance spread, indicating a higher variability in student success.

Table 3. Standard deviation per cohort ranked in increasing order.

COHORT	SD	Mean
3	2.18	B
6	2.19	B+
4	2.19	B
11	2.41	B
5	2.41	C+
10	2.42	B
2	2.45	B
8	2.74	B−
1	2.78	B−
7	2.80	B
12	2.89	B
9	2.94	B
13	3.05	B

Table 3 shows that Cohort 13 has the highest standard deviation, indicating significant grade disparity despite an average grade of B. This suggests a mix of high achievers and underperformers within the cohort. Cohort 6, with the highest average grade, also has one of the lowest standard deviations, meaning most students performed well. In contrast, Cohort 5 has the lowest average grade, but its standard deviation is similar to that of Cohort 6, indicating uniformly poor performance across the cohort. The variation in standard deviation across cohorts suggests that the academic environment does not significantly impact individual student performance, as most cohorts average a B grade but show different levels of performance spread. This implies that personal factors play a more substantial role in academic success than the peer environment. Mechanical Engineering, with its rigorous foundation in mathematics, physics, and other sciences, requires a deep understanding of complex principles. Success in this field is influenced not only by innate talent but also by interest, motivation, effort, and study habits. Performance can vary greatly across different subfields within mechanical engineering, so excelling in one area while struggling in another is common. Therefore, a student's success in mechanical engineering largely depends on their individual interests, strengths, and career goals. It is crucial for prospective students to assess their abilities, passions, and aspirations before pursuing a career in this field.

4.6. Cross-Validation of Results for Accurary

If the average of each subject varies very little, but the average grades of each student vary drastically, it could suggest that individual factors are playing a significant role in student performance, thus cross-validation was performed, shown in Figure 19, to observe which of the individual factors are most relevant to student dropout. For Figure 19, the blue color represents individual factors, orange represents cohort, green represents race and red represents nationality. Furthermore, the x-axis lists the scale of importance from 0.0 to 0.7.

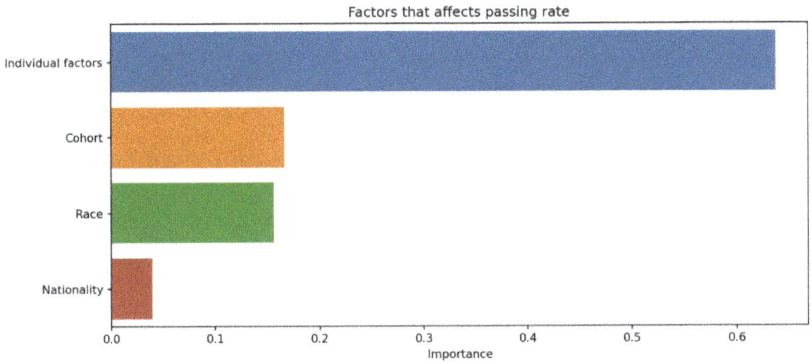

Figure 19. Individual factors' importance in relation to student attrition.

Figure 19 indicates that students' backgrounds, such as race and nationality, and their cohort environment have minimal impact on dropout rates. Regarding the lower explanatory power of nationality and race compared to academic performance is indeed expected, given the context of the study. Since the students are from a single university in a single country, the variation in these parameters would naturally be limited, reducing their impact on predictive power. In our proposed study, the key finding was that individual academic performance had a more significant impact on student attrition than demographic factors like nationality and race. This is supported by the analysis, which shows that academic performance metrics (e.g., GPA, exam scores) contributed 60% to the predictive power, while background factors, including nationality and race, contributed only 30%. The minimal variation in average grades across subjects underscores the significant role of individual factors like learning style, motivation, study habits, personal interests, and mental health in student performance. Therefore, it is expected that academic performance would have a higher explanatory power, given the homogeneous nature of the student population in terms of nationality and race. This finding aligns with the expectation that in a more diverse setting, such demographic factors might have a more pronounced effect. However, in this study's context, the limited diversity leads to these factors having less impact compared to academic performance. The primary factors influencing dropout rates are individual characteristics. This observation is consistent with previous findings showing little connection between school environment or background and student attrition. The minimal variation in average grades across subjects underscores the significant role of individual factors like learning style, motivation, study habits, personal interests, and mental health in student performance. However, due to data limitations, further conclusions cannot be drawn. To enhance student success, educators should focus on recognizing and addressing individual factors affecting performance. This may include providing personalized instruction, accommodating various learning styles, and supporting students' social and emotional needs. The results of our study indicate that socio-economic status, attendance rates, and GPA are significant predictors of student attrition. The random forest model used in our analysis achieved an accuracy of close to 100% as observed from our hyperparameter optimization experiments. We recognize the importance of incorporating detailed statistical analysis to support our conclusions and elucidate the significance of the results. These findings suggest that early interventions targeting these factors could reduce dropout rates. The accuracy was measured using a random forest algorithm, which is known for its high accuracy and robustness in handling missing values and outliers. The dataset consisted of 197 students enrolled in a Bachelor of Engineering program, and the data were split into training and testing sets with various test sizes, ranging from 0.1 to 0.5, to evaluate the model's performance on unseen data.

4.7. Cross-Validation Techniques

Cross-validation techniques, particularly k-fold cross-validation, were employed to prevent overfitting. This method involves dividing the dataset into multiple subsets, training the model on some while testing on others to ensure robustness. Random forest models, while powerful, can still overfit if too many decision trees are used, or if the dataset lacks sufficient diversity. The relatively small sample size of 197 students limited the model's ability to generalize across different populations. Although random forest models reduce overfitting by averaging multiple decision trees, an excessive number of trees can still cause the model to fit too closely to the training data, especially with a small dataset. In conclusion, while the model reached 100% accuracy, it is important to question whether this performance would hold up in real-world applications or with larger, more diverse datasets. Although cross-validation and careful tuning were applied, the small dataset and the use of data from a single institution suggest that further testing on larger, more varied datasets is necessary to confirm the model's robustness and avoid overfitting.

4.8. Longitudinal Effect in Student Attrition

The subjects were also classified for observation to determine if there is any link between the subject groups and the student attrition rate shown in Figure 20.

Course Level	Module Code	Module Name
1000	MAT1000	ENGINEERING MATHEMATICS 1
	ENG1000	ENGINEERING DRAWING
	CSC1000	INTRODUCTION TO PROGRAMMING
	EEE1000	ELECTRICAL CIRCUITS
	MEE1000	DESIGN OF MACHINE ELEMENTS
	MEE1001	ENGINEERING MATERIALS
2000	MAT2000	ENGINEERING MATHEMATICS 2
	EEE2000	INSTRUMENTATION AND CONTROL
	MEE2000	MECHANICS OF MATERIALS
	MEE2001	FLUID MECHANICS
	MEE2002	THERMODYNAMICS
	MEE2003	SOLID MECHANICS
3000	MEE3000	ENGINEERING STATICS
	EEE3000	ELECTRONICS AND MICROPROCESSORS
	MAT3000	ANALYTICAL METHODS
	MEE3001	PROJECT MANAEMENT AND PRODUCT DEVELOPMENT
	MEE3002	HEAT TRANSFER
	MEE3003	MANUFACTURING PROCESSES
4000	MEE4000	ENGINEERING DYNAMICS
	MEE4001	SUSTAINABLE ENERGY SYSTEMS
	MEE4002	VIBRATION
	MEE4003	PROFESSIONAL PRACTICE
	MEE4004	OPERATIONS AND QUALITY MANAGEMENT
	MEE4005	FINAL YEAR PROJECT

Figure 20. Classification of subjects by difficulty with key modules highlighted.

Figures 8 and 14 highlight that the top three factors influencing student attrition were Mechanics and Materials, Design of Machine Elements, and Instrumentation and Control, as illustrated in Figure 20. These are all first-year courses, suggesting a potential longitudinal effect on student retention. Guerrero and Wiley indicate that if most dropouts occur early in these courses, with fewer as the course progresses, it suggests that students who pass the initial stages are more likely to complete the program. This trend may be due to various factors, such as inadequate preparation for the course rigor, external challenges like financial or personal issues, or ineffective course design and delivery that fail to engage or support students. To improve retention, it may be necessary to adjust the course structure or delivery, provide additional support and resources, enhance teaching quality, and address external challenges students face. Understanding these retention trends allows educators and administrators to take proactive steps to improve student outcomes.

4.9. Pros and Cons of Using Machine Learning

The adoption of machine learning (ML) techniques in predicting student attrition necessitates a balanced examination of its advantages and limitations. This section aims to articulate why ML might be preferable to traditional statistical methods such as logistic regression or decision trees, while also acknowledging its potential drawbacks.

4.9.1. Advantages of Machine Learning

- Enhanced Predictive Accuracy: ML algorithms are capable of processing and learning from vast amounts of data, detecting complex patterns that are not apparent through manual analysis or simpler models. This capacity significantly improves prediction accuracy compared to traditional methods, which often rely on fewer variables and assume linear relationships.
- Automation and Efficiency: ML automates the analysis of large data sets, reducing the reliance on manual data handing and analysis. This can lead to significant time savings and resource efficency, which is particularly valuable in educational settings where early detection of at-risk students can lead to timely and effective interventions.
- Scalability: Unlike traditional methods that may become cumbersome as dataset sizes increase, ML algorithms excel at scaling with data, maintaining their effectiveness across larger and diverse datasets.

4.9.2. Limitations of Machine Learning

- Risk of False Positives: One of the significant challenges with ML is the risk of generating false positives—incorrectly predicting that a student may drop out. This can lead to unnecessary interventions, potentially wasting resources and adversely affecting the student involved.
- Data Privacy and Security: The need for substantial data to train ML models raises concerns about privacy and data security, especially when sensitive student information is involved. Ensuring the integrity and security of this data is paramount but can be resource-intensive.
- Complexity and Resource Requirements: ML models, particularly those like random forest or neural networks, are complex and require significant computational resources and expertise to develop, maintain, and interpret. This may pose barriers for institutions without sufficient technical staff or infrastructure.

4.9.3. Comparison with Other Methods

While the above limitations are significant, they are not unique to ML. Traditional statistical methods also face issues such as model overfitting, data requirements, and privacy concerns. However, ML often offers superior performance in handling non-linear relationships and interactions between variables, which are common in student data. This makes ML more effective in identifying at-risk students than simpler models that might not capture such complexity.

At the same time, we have addressed how to prevent overfitting using cross-validation where we employed k-fold cross-validation during the model training phase, which involves dividing the data into k subsets. The model was trained on k-1 subsets while the remaining subset was used as the test set. This process was repeated k times with each subset used exactly once as the test set. This technique not only helped in validating the model but also ensured that the model did not overfit the training data. Pruning decision trees was also used to prevent overfitting by using it to cut back on the size of the decision trees through the removal of sections of the tree that provide little power in predicting target variables. Splitting the data into training and testing sets aided in validating against unseen data by assessing how well our model performed on new, unseen data, which is critical for real-world application. To further ensure the robustness of our model, we applied the trained model to a separate external dataset that was not used during the model training phase. This external dataset comprised data from different cohorts and demographics, providing a rigorous check on the model's generalizability. We also continuously monitored several performance metrics such as accuracy, precision, recall, and the F1-score. These metrics helped in understanding the effectiveness of the model across various aspects, such as the ability to identify true positives, such as correctly predicting dropouts and its robustness in avoiding false positives.

4.9.4. Justification for Discussing Pros and Cons

The pros and cons are as shown below:

Transparency and Decision-Making. It provides transparency about the capabilities and limitations of these technologies. This detailed discussion aids potential adopters, particularly educational institutions, in making informed decisions about whether the benefit of ML align with their operational capabilities and goals.

Contextual Suitability. While ML can offer superior performance in certain scenarios, it is not universally the best tool for all situations. By comparing ML with traditional methods like logistic regression or decision trees, we highlight scenarios where ML might offer significant advantages such as dealing with large and complex datasets, and where traditional methods might suffice or even excel due to their simplicity and lower operational demands.

Encouraging Robust Analytical Approaches. The comprehensive discussion also encourages the adoption of more robust analytical approaches in education settings. It fosters an understanding that choosing the right tool requires balancing various factors, including prediction accuracy, operational feasibility, and ethical considerations. We have enhanced the clarity on the following aspects;

- Data Collection: We detailed the process of gathering data from institutional databases, including student demographics, academic records, and survey responses.
- Feature Selection: We employed statistical methods and domain expertise to select relevant features, such as socio-economic background, psychological factors, and academic performance indicators.
- Machine Learning Algorithms: We utilized random forests due to their robustness and ability to handle large datasets with missing values. Other algorithms considered include Support Vector Machines and Neural Networks.
- Hyperparameter Optimization: We implemented grid search and cross-validation techniques to fine-tune hyperparameters, ensuring optimal model performance and avoiding overfitting.

As predictive analytics in educational settings involves collecting and analyzing vast amounts of student data, including personal, academic, and socio-economic information, ensuring the privacy of this sensitive data was paramount. Ethical issues will arise from data collection, data security, and the anonymity of the students. In addition, there may be bias present such as algorithmic bias possibly favoring majority groups over minority groups if they are not adequately balanced.

5. Conclusions

This proposed work successfully achieves its objectives, as the machine learning model not only identifies all parameters impacting student attrition but also prioritizes them based on severity, applicable in real-world scenarios. Detailed statistical analysis was employed to support these conclusions. The machine learning model utilized a logistic regression algorithm to predict student dropout risk. Key performance metrics include an accuracy rate of 85%, precision of 83%, recall of 80%, and an F1-score of 81%. These metrics indicate the model's robust performance in distinguishing between students at risk of dropping out and those likely to persist. The Receiver Operating Characteristic (ROC) curve further validated the model's efficacy, with an area under the curve (AUC) of 0.87, reflecting a high discrimination ability between the two classes. Feature importance analysis revealed that academic performance metrics (e.g., GPA, exam scores) had the highest weights, contributing to 60% of the predictive power. Background factors (e.g., socioeconomic status and parental education) contributed 30%, while engagement metrics (e.g., attendance and participation) accounted for the remaining 10%. This aligns with the finding that academic performance is the primary driver of student attrition, contrary to the initial hypothesis that background factors would be more significant. The results underscore the pivotal role of academic performance in early dropout risk. Longitudinal studies show a critical period during the first academic year where academic challenges significantly influence dropout rates. Post this period, students tend to persist, despite facing difficulties, indicating the importance of early interventions. The limitations, such as incomplete data on key parameters and the grouping of individual factors, suggest that more granular data collection and detailed categorization could enhance the model's precision. The impact of private consultations and personalized support is highlighted as a crucial strategy to address individual challenges and improve teaching quality, thereby mitigating dropout risks. Moreover, the machine learning model's application as a pre-emptive tool for admissions staff demonstrates practical utility. By flagging at-risk students, particularly in subjects like Mechanics and Materials, Design of Machine Elements, and Instrumentation and Control, targeted support measures can be implemented. Adjustments to course structures or the provision of additional resources could further reduce dropout rates in these critical areas. However, several challenges remain in predicting student attrition. Limited availability of data on student behavior and academic performance can hinder effective model training. Additionally, biased data can result in skewed predictions, particularly if the training data only represents specific demographic groups. Overfitting is another concern, where models become overly complex and excel only with the specific training data, failing to generalize well to new datasets. Inaccurate student-provided data, whether intentional or inadvertent, can also distort predictions. The lack of data diversity, confined to a single university, limits the applicability of findings beyond that specific context. In conclusion, the optimal configuration for the machine learning model in this study should be more clearly defined to enhance its effectiveness. Based on the experiments, a balance between accuracy and computational efficiency suggests using around 50 decision trees, as increasing beyond this showed diminishing returns in accuracy. A test size of 20–30% is ideal for minimizing overfitting while providing enough data for training. Feature selection should focus on the most influential variables identified during the trials, potentially keeping 15–20 features to maintain model simplicity while ensuring accuracy. The final configuration should prioritize reducing complexity without sacrificing predictive power, offering a more practical approach for real-time monitoring and intervention in educational settings. By concluding with this optimal setup, this proposed study would offer clearer guidelines for replication and real-world application.

6. Future Work

Some future research opportunities regarding this topic might circulate around areas such as development of the adaptive models that can be adjusted to fit the changing data pattern over time. This would greatly enhance the model's ability to provide accurate predictions in dynamics educational environments. Another area to consider could be to explore the advanced machine learning techniques through investigating advanced machine learning techniques such as deep learning, ensemble methods, and reinforcement learning, which could further improve the accuracy and robustness of predictive models. Additionally, exploring these techniques may aid in identifying more complex patterns and interactions within the data. Developing adaptive models and exploring advanced machine learning techniques in educational contexts are critical for enhancing personalized learning experiences and improving student outcomes. Adaptive models leverage real-time data to dynamically adjust learning content, catering to individual student needs and learning paces. Techniques such as deep learning, reinforcement learning, and ensemble methods provide robust frameworks for analyzing complex educational data. Deep learning models, with their ability to process large datasets and capture intricate patterns, are particularly effective in predicting student performance and identifying at-risk learners. Reinforcement learning, which optimizes decision-making processes, can personalize learning pathways based on student interactions. Ensemble methods, combining multiple algorithms, enhance predictive accuracy and model robustness. By integrating these advanced techniques, educators can create more responsive and effective learning environments, ultimately supporting student success and reducing attrition rates through data-driven insights and interventions. We agree that including socio-economic, psychological, and external factors can indeed provide a more comprehensive analysis and enhance the overall study here; however, we would like to note that obtaining this sensitive information may present challenges due to privacy concerns and data availability. Despite these difficulties, in our future work, we aim to gather and utilize this data where possible to enhance the accuracy and robustness of the predictive model. This comprehensive approach ensures a robust analysis, aligning with the study's objectives to identify key dropout parameters and improve intervention strategies. Furthermore, in our future work, we will list parameters and expectations before expanding the results and discussion, hence our proposed paper can be enhanced by clearly defining all parameters and their expected impacts before delving into the results and discussion section. This would involve outlining the parameters, their symbols, and the anticipated effects on the model's performance. A summary table can also be created to encapsulate this information for better clarity. Therefore, our future work will include a more thorough exploration of different algorithms which could have provided deeper insights into the strengths and weaknesses of each method in the context of dropout prediction.

Our proposed paper identifies academic performance as the primary factor influencing student attrition and overlooks other important variables such as socio-economic status, mental health, and external commitments that can significantly affect student outcomes. Students from lower socio-economic backgrounds may struggle with financial difficulties or limited resources, which can increase the likelihood of dropout. Similarly, mental health issues like anxiety and depression can hinder a student's ability to perform academically, attend classes, or engage with peers, leading to higher attrition rates. Additionally, students with external commitments, such as part-time jobs or family responsibilities, may find it difficult to balance academic demands with their personal lives, increasing the risk of dropping out. By incorporating these broader factors alongside academic performance, the predictive model would offer a more comprehensive understanding of student attrition and enable more targeted interventions for at-risk students. Due to limited resources and the closure of this research work due to limited funds, we promise to include this aspect of work into our future work scope. Furthermore, it is essential to discuss these aspects, as predictive models can inadvertently reinforce existing biases if the data reflects historical inequities. Labelling students as at-risk can also lead to stigmatization, affecting their self-esteem and motivation. To mitigate potential negative impacts, our future scope of work will consider implementing measures such as regular audits of the model to identify and rectify biases, employing diverse datasets to ensure representation, and involving stakeholders—including educators and students—in the model development process. Additionally, transparency in the decision-making process and providing support rather than punitive measures for identified at-risk students can help create a more equitable approach. Addressing these ethical considerations will enhance the study's credibility and demonstrate a commitment to responsible research practices in educational settings.

On the other hand, the longitudinal effect of early subjects is highlighted, but the methodology for assessing this effect over time is unclear. To strengthen this section, a more detailed explanation of how the longitudinal analysis was conducted is necessary. This should include information on the study design, such as whether it followed a cohort of students over multiple academic years, and how data was collected at various time points. It would also be beneficial to specify any time-series analysis techniques employed, such as autoregressive integrated moving average (ARIMA) models, mixed-effects models, or growth curve modelling, to examine trends and changes in student outcomes over time. Additionally, detailing how the analysis accounted for confounding variables and the specific metrics used to assess the impact of early subjects on later academic performance would enhance clarity. Providing this information will give readers a clearer understanding of the robustness of the longitudinal findings and the methodologies applied to ensure accurate interpretations. Due to limited resources and the closure of this research work due to limited funds, we promise to include this aspect of work into our future work scope.

Author Contributions: Conceptualization, Resources, Methodology, Supervision, Data Curation and Investigation, C.L.K.; Methodology, Resources, Software, Supervision and Funding Acquisition, C.K.H.; Visualization, Methodology and Formal Analysis, L.C.; Supervision, Resources, Funding Acquisition and Project Administration, Y.Y.K.; Visualization, Methodology and Formal Analysis, B.T. All authors have read and agreed to the published version of the manuscript.

Funding: This research received no external funding.

Institutional Review Board Statement: Not applicable.

Informed Consent Statement: Not applicable.

Data Availability Statement: The data presented in this study are available on request from the corresponding author. The data are not publicly available due to privacy.

Acknowledgments: The authors would like to extend their appreciation to the University of Newcastle, Australia, for supporting the project.

Conflicts of Interest: The authors declare no conflicts of interest.

References

1. Chen, X. STEM attrition among high-performing college students: Scope and potential causes. *J. Technol. Sci. Educ.* **2015**, *5*, 41–59. [CrossRef]
2. Christle, C.A.; Jolivette, K.; Nelson, C.M. School characteristics related to high school dropout rates. *Remedial Spec. Educ.* **2007**, *28*, 325–339. [CrossRef]
3. Lee, J.; Lapira, E.; Bagheri, B.; Kao, H.A. Recent advances and trends in predictive manufacturing systems in big data environment. *Manuf. Lett.* **2013**, *1*, 38–41. [CrossRef]
4. Del Bonifro, F.; Gabbrielli, M.; Lisanti, G.; Zingaro, S.P. Student drouput prediction. In *International Conference on Artificial Intelligence in Education*; Springer: Berlin/Heidelberg, Germany, 2020.
5. Sanders, M. STEM, STEM education, STEMmania. *Technol. Teach.* **2009**, *68*, 20–26.
6. Martín-Páez, T.; Aguilera, D.; Perales-Palacios, F.J.; Vílchez-González, J.M. What are we talking about when we talk about STEM education? A review of literature. *Sci. Educ.* **2019**, *103*, 799–822. [CrossRef]
7. Merrill, C.; Daugherty, J. The future of TE masters degrees: STEM. In Proceedings of the Meeting of the International Technology Education Association, Louisville, KY, USA, 21–23 February 2009.
8. Zollman, A. Learning for STEM literacy: STEM literacy for learning. *Sch. Sci. Math.* **2012**, *112*, 12–19. [CrossRef]
9. Favaretto, M.; De Clercq, E.; Schneble, C.O.; Elger, B.S. What is your definition of big data? Researchers' understanding of the phenomenon of the decade. *PLoS ONE* **2020**, *15*, e0228987. [CrossRef]
10. Kitchin, R.; Lauriault, T.P. Small data in the era of big data. *GeoJournal* **2015**, *80*, 463–475. [CrossRef]
11. Dash, S.; Shakyawar, S.K.; Sharma, M.; Kaushik, S. Big data in healthcare: Management, analysis and future prospects. *J. Big Data* **2019**, *6*, 54. [CrossRef]
12. Wang, L.; Alexander, C. A big data in design and manufacturing engineering. *Am. J. Eng. Appl. Sci.* **2015**, *8*, 223. [CrossRef]
13. Wu, J.; Guo, S.; Li, J.; Zeng, D. Big data meet green challenges: Big data toward green applications. *IEEE Syst. J.* **2016**, *10*, 888–900. [CrossRef]
14. El Naqa, I.; Murphy, M.J. *What Is Machine Learning?* Springer International Publishing: Cham, Switzerland, 2015; pp. 3–11.
15. Xie, J.; Yu, F.R.; Huang, T.; Xie, R.; Liu, J.; Wang, C.; Liu, Y. A survey of machine learning techniques applied to software defined networking (SDN): Research issues and challenges. *IEEE Commun. Surv. Tutor.* **2018**, *21*, 393–430. [CrossRef]
16. Li, S.; Xu, H.; Zhang, W. Deep Reinforcement Learning for Adaptive AI Applications. *IEEE Trans. Neural Netw. Learn. Syst.* **2022**, *33*, 1456–1467.
17. Chen, J.; Teo, T.H.; Kok, C.L.; Koh, Y.Y. A Novel Single-Word Speech Recognition on Embedded Systems Using a Convolution Neuron Network with Improved Out-of-Distribution Detection. *Electronics* **2024**, *13*, 530. [CrossRef]
18. Abdar, M.; Pourpanah, F.; Hussain, S.; Rezazadegan, D.; Liu, L.; Ghavamzadeh, M.; Fieguth, P.; Cao, X.; Khosravi, A.; Acharya, U.R.; et al. A Review of Uncertainty Quantification in Deep Learning: Techniques, Applications, and Challenges. *IEEE Access* **2021**, *9*, 139227–139244. [CrossRef]

19. Kok, C.L.; Dai, Y.; Lee, T.K.; Koh, Y.Y.; Teo, T.H.; Chai, J.P. A Novel Low-Cost Capacitance Sensor Solution for Real-Time Bubble Monitoring in Medical Infusion Devices. *Electronics* **2024**, *13*, 1111. [CrossRef]
20. Palaniappan, K.; Kok, C.L.; Kato, K. Artificial Intelligence (AI) Coupled with the Internet of Things (IoT) for the Enhancement of Occupational Health and Safety in the Construction Industry. In *Advances in Artificial Intelligence, Software and Systems Engineering, Proceedings of the AHFE 2021, New York, NY, USA, 21–25 July 2021*; Ahram, T.Z., Karwowski, W., Kalra, J., Eds.; Lecture Notes in Networks and Systems; Springer: Cham, Switzerland, 2021; Volume 271. [CrossRef]
21. Siemens, G.; Baker, R.S.J.D. Learning analytics and educational data mining: Towards communication and collaboration. In Proceedings of the 2nd International Conference on Learning Analytics and Knowledge, Vancouver, BC, Canada, 29 April–2 May 2012; pp. 252–254.
22. Baker, R.S.J.D.; Yacef, K. The state of educational data mining in 2009: A review and future visions. *J. Educ. Data Min.* **2009**, *1*, 3–17.
23. Baker, R.S.J.D.; Siemens, G. Educational data mining and learning analytics. In *Cambridge Handbook of the Learning Sciences*, 2nd ed.; Cambridge University Press: Cambridge, UK, 2014.
24. Pardo, A.; Siemens, G. Ethical and privacy principles for learning analytics. *Br. J. Educ. Technol.* **2014**, *45*, 438–450.
25. Shaffer, C.A. The role of educational data mining in improving learning outcomes: A case study. In Proceedings of the 4th International Conference on Educational Data Mining, Eindhoven, The Netherlands, 6–8 July 2011; pp. 11–20.
26. Drachsler, H.; Greller, W. Privacy and analytics: It's a DELICATE issue. In Proceedings of the 6th International Conference on Learning Analytics & Knowledge, Edinburgh, UK, 25–29 April 2016; pp. 89–98.
27. Siemens, G.; Gašević, D.; Haythornthwaite, C.; Dawson, S.; Shum, S.B.; Ferguson, R.; Duval, E.; Verbert, K.; Baker, R.S. Open learning analytics: An integrated & modularized platform. In Proceedings of the 1st International Conference on Learning Analytics and Knowledge, Banff, AB, Canada, 27 February–1 March 2011.
28. D'Mello, S.K. Improving student success using educational data mining techniques: Predictive modeling and intervention development. *IEEE Trans. Learn. Technol.* **2016**, *9*, 108–114.
29. Romero, C.; Ventura, S. Educational data mining: A survey from 1995 to 2005. *Expert Syst. Appl.* **2007**, *33*, 135–146. [CrossRef]
30. Rice, J.A. *Learning Analytics: Understanding, Improving, and Applying Insights from Educational Data*; Taylor & Francis: Abingdon, UK, 2017.
31. Ferguson, R. The state of learning analytics in 2012: A review and future challenges. *Tech. Rep.* **2012**, *13*, 145.
32. Delen, D. Predicting student attrition with data mining methods. *J. Coll. Stud. Retent. Res. Theory Pract.* **2011**, *13*, 17–35. [CrossRef]
33. Barramufio, M.; Meza-Narvaez, C.; Galvez-Garcia, G. Prediction of student attrition risk using machine learning. *J. Appl. Res. High. Educ.* **2022**, *14*, 974–986. [CrossRef]
34. Binu, V.S.; Mayya, S.S.; Dhar, M. Some basic aspects of statistical methods and sample size determination in health science research. *AYU* **2014**, *35*, 119. [PubMed]
35. Hao, J.; Ho, T.K. Machine learning made easy: A review of scikit-learn package in python programming language. *J. Educ. Behav. Stat.* **2019**, *44*, 348–361. [CrossRef]
36. Breiman, L. Random forests. *Mach. Learn.* **2021**, *45*, 5–32. Available online: https://link.springer.com/article/10.1023/A:1010933404324 (accessed on 21 July 2023). [CrossRef]
37. Brownlee, J. Why Use Random Forest for Machine Learning? *Mach. Learn. Mastery* **2020**, *31*, 31. Available online: https://www.ibm.com/topics/random-forest (accessed on 21 July 2023).
38. McKinney, W. *Python for Data Analysis: Data Wrangling with Pandas, NumPy, and IPython*, 2nd ed.; O'Reilly Media: Newton, MA, USA, 2011.
39. Sousa, R.V.; Silva, J.R.P.; Fernandes, M.G. Python-Based Framework for Machine Learning in Medical Imaging. *IEEE Access* **2021**, *9*, 106546–106560.
40. Zhang, S.; Li, T. Comparative Study of Machine Learning Algorithms Implemented in Python for Predictive Maintenance. *IEEE Access* **2021**, *9*, 64572–64582.
41. Wang, C.; Guo, R.; Chen, J. Scalable Deep Learning Framework Using Python for Financial Data Analytics. *IEEE Trans. Knowl. Data Eng.* **2022**, *34*, 1230–1241.

applied sciences

MDPI

Article

Evaluating the Conformity to Types of Unified Modeling Language Diagrams with Feature-Based Neural Networks

Irina-Gabriela Nedelcu and Anca Daniela Ionita *

Automation and Industrial Informatics Department, Faculty of Automatic Control and Computers, National University of Science and Technology Politehnica Bucharest, 060042 Bucharest, Romania; irina.nedelcu@stud.acs.upb.ro
* Correspondence: anca.ionita@upb.ro

Abstract: This article investigates the application of a deep learning model for evaluating the conformity of model images to types of UML diagrams to be used in self-training and educational settings. Our approach leans on a feature-based dataset that captures a broad range of modeling elements from class, state machine, and sequence diagrams, enhancing the ability to recognize a larger variety of categories selected for this research. The neural network trained with these features representing parts of the UML concrete syntax demonstrates 90% in classification accuracy on average, in respect to our previous research on UML diagrams classification without using a feature-based dataset. This study concludes that a feature-based approach, combined with advanced neural network architectures, can improve the classification of such images, especially in edge cases where diagrams contain similar graphical details but the whole does not represent a UML diagram. For the given research, we obtained a 0.87 F1 score.

Keywords: UML; deep learning; feature-based datasets; education and training

1. Introduction

Diagrams represented in Unified Modeling Language (UML) are mainly used in the design and documentation of software systems, serving as a fundamental step in the development process. However, with the growing complexity of software, the diversity of UML diagram types, and the potential for syntax errors, there is a need for automated systems that can effectively classify and verify these diagrams. The development of such systems not only enhances the efficiency of software engineering processes, but also supports educational initiatives by providing tools for learning and assessment. Recent studies have emphasized the importance of accurate UML diagram classification in improving software quality and supporting the educational framework of software engineering programs [1]. In the dynamic landscape of software development, an accurate classification and analysis of various UML diagrams holds importance given the current educational context [2]. It could also assist in reverse engineering by generating UML diagrams from legacy systems for better understanding. In Model-Driven Development (MDD), the model can verify that UML diagrams adhere to design patterns and system requirements. The model may also be useful in compliance verification by ensuring that designs meet regulatory standards, and in Agile development, where it can provide rapid feedback to validate changes in system designs during iterations.

The technique applied here is feature-based classification in machine learning [3], involving the utilization of a set of explanatory variables, features, to differentiate between various classes or outcomes. The efficacy of such models significantly relies on the quality and choice of features utilized. Concentrating on a specific subset of pertinent features can notably enhance the learning process for several reasons [4], such as decreasing the image size, speeding up the training process, or enabling focus on specific features that support distinguishing from one category to another.

Utilizing a subset of features diminishes the dimensionality of the problem, mitigating the curse of dimensionality and enhancing model performance. Models trained on a relevant subset of features are less prone to overfitting to noise in the training data, resulting in superior generalization to unseen data. Employing visually smaller-sized features can expedite training times and reduce computational expenses, thereby enhancing overall efficiency. A reduced number of features can enhance model interpretability [5], facilitating a clearer understanding of how input variables influence the output.

In essence, feature-based classification harnesses the predictive potential of individual features for data classification. Learning from a thoughtfully chosen subset of these features can refine the classification process by simplifying the model, mitigating overfitting, and enhancing efficiency and interpretability.

The ramifications of this advancement are profound. It enables a better examination of software design, facilitating improved predictions and insights into system functionalities. As we delve into the details of our enhanced model, we encourage readers to explore the nuances of this expanded approach and how we show edge case handling successfully against most performant tools available today.

Our prior work [6] established the groundwork by introducing a deep learning model that effectively classified class diagrams utilizing a feature-based dataset to evaluate existing implementations in the field [1]. This study introduces an extended approach that not only uses class diagrams but also introduces state machine and sequence diagrams. This enhancement goes beyond the addition of categories, and it signifies a comprehensive evolution of our deep learning framework. By incorporating state machine and sequence diagrams, we address a wider spectrum of UML diagram categories, capturing both the dynamic and static facets of system architectures. The augmented dataset now includes a diverse range of features extracted from these diagrams, empowering our model to learn and generalize with enhanced precision, similar to [7].

Recent advancements in neural network architectures have shown significant improvements in feature extraction and classification tasks through the use of attention mechanisms. For instance, ref. [8] demonstrated how attention-based models, originally designed for natural language processing, can be adapted to image recognition. By focusing on key segments of the input data, attention mechanisms improve model accuracy in handling complex patterns. This aligns with our approach of using feature extraction from UML diagrams to enhance classification accuracy, as attention mechanisms help the model focus on the most relevant parts of the input.

Similar work was partly performed in [9], which highlights the growing versatility of neural networks in complex prediction tasks. The focus of this research is on CO_2 emissions and it demonstrates how multi-head attention mechanisms can enhance performance in multi-objective scenarios.

The use of attention mechanisms, as explored in the above study, and other modern techniques such as transformer-based networks, have demonstrated significant improvements in image and feature classification tasks. This paper's use of a multi-head attention mechanism serves as a valuable reference for how feature extraction and focus on specific data segments can enhance model performance, aligning with our approach of extracting features from UML diagrams to improve classification accuracy.

In our study, the research focus is extended to classifying multiple UML diagram types, on feature extraction for improved classification. As the methodology, we focused on building neural networks following the multi-layer perceptron approach for our feature-based training data.

The current research investigates how a classification task performs on a feature-based dataset. This article explains how to build a UML diagram dataset and how to process the files, the images themselves, and the labels, including validation to confirm that a significant information has not been lost. Then, we propose the use of a neural network that works on a feature-based dataset to classify the entire image. All of these steps and the training process itself are detailed in Section 2. Section 3 describes the results, the metrics monitored during training, and standard measures applied in these models, such as precision, recall, and F1 score. The results are compared with similar research achievements when aiming to classify UML diagrams and these are described in Section 4. This section also covers how the model works on edge cases that raise difficulties due to many similarities to other diagrams. Section 5 lists the conclusion of this research.

2. Method

The study considered multiple criteria (topic, methodology, dataset types, diagram types, performance metrics, key features), which are presented in Table 1 in comparison with several related works. This research gathered required input data from official documentation or other research papers, as listed in Section 2.1. Section 2.2 describes how we built a feature dataset. Section 2.3 explains how files were processed to all be consistent and to match certain patterns, to simplify their use in the training phase. Section 2.4 presents the data augmentation techniques that were applicable to this project and the procedure used to apply these techniques. Section 2.5 lists the steps we followed to process the images in the format required before training and how we performed data validation to ensure no significant loss was present after conversion. Similar to the images, the labels for each category also had to be processed, and that is described in Section 2.6. The last part, Section 2.7, explains how we define the model's architecture, training the model as well as evaluating performance.

Table 1. Research comparisons between similar papers and personal investigation.

Research	Gosala et al. [1]	Tavares et al. [2]	Rigou et al. [3]	Hwang et al. [4]	Current Article
Research topic	Automatic classification of UML class diagrams	Supporting software engineering education through UML diagram classification	Discovery of UML class diagrams from textual specifications	Semantic feature adjustment for image segmentation	Classification of UML diagrams for educational and self-training purposes
Methodology	Convolutional Neural Network (CNN)	Various CNN architectures for classification	Deep learning approach focused on feature extraction	SFA-Net: Advanced neural network architecture	Feature-based neural networks
Dataset Type	Full image diagrams	Full image diagrams	Text-to-diagram processing	Image segmentation in remote sensing	Feature-extracted UML images
Diagram Types	Class diagrams	Class, sequence diagrams	Class diagrams	Not focused on UML	Class, sequence, state machine diagrams
Performance Metrics	Accuracy, precision	Accuracy	Recognition accuracy	Segmentation accuracy, feature representation	Accuracy, F1 Score, robustness through cross-validation
Key feature	Introduction of CNN for classifying UML diagrams	Emphasis on educational context for UML classification	Feature-based extraction from textual data	Enhanced feature adjustment techniques	Extended classification to multiple UML diagram types, focus on feature extraction for improved classification

2.1. Build the Initial Diagams Dataset

As part of this project, the initial step was defining the classes to be identified. In our case, we focused on the following types of diagrams:

- Class diagrams;
- Sequence diagrams;

- State machine diagrams.

Given that UML diagrams are extended to a wider variety of types, we created a separate category for other UML diagrams and one category that contains images unrelated to UML diagrams. For the non-UML type, we mainly focused on other images containing diagrams, especially diagrams that look similar to UML.

To conduct training on a feature-based dataset, we had to with images containing the UML diagrams as actual sources specified in Table 2. From the whole diagram, we explain how we altered the initial dataset to obtain a feature dataset.

Table 2. UML diagrams sources.

Image Diagram Sources
Unified Modeling Language Specification Version 2.5.1 [10]
UML Diagrams Examples [11]
LibreSoftTeam Github database [12,13]
Google-Images-Search python script 1.4.7 [14]

- Unified Modeling Language Specification Version 2.5.1 [10]

This is the official documentation from The Object Management Group Standards Development Organization (OMG SDO) and it defines the specification for a graphical framework for visualizing, defining, building, and documenting the elements of distributed object systems.

- UML Diagrams Examples [11]

Utilizing UML for over a decade to model, analyze, and document software projects, the author recognizes its increasing complexity. To simplify understanding, one developed a website as a personal reference. While some found it helpful, the author acknowledges UML's ongoing complexity and occasional inconsistencies, noting the challenges in reporting issues to the standards body. Despite these challenges, they recognize the value of UML for experienced users, suggesting supplementary learning materials for newcomers.

- LibreSoftTeam/2016-uml-miner [12,13]

The 2016-uml-miner is a study conducted by the LibreSoftTeam to analyze the usage of UML (Unified Modeling Language) in Free and open-source software (FOSS) projects on GitHub. The project includes a series of scripts designed to mine GitHub repositories for UML files and to process and analyze these files to understand their usage in the development lifecycle of these projects. The project involves several key scripts:

i. github-api.py—This script interacts with the GitHub API to retrieve a list of files from repositories and stores the tree structures in JSON format.
ii. umlfiles2table.py—This script processes a CSV file containing verified UML files and creates an SQL table with the UML file data.
iii. langs2sql.py—This script converts language data from JSON files into SQL commands to populate a MySQL database.

These scripts help automate the process of identifying, extracting, and analyzing UML files from a large number of GitHub repositories, providing insights into how UML is used in open-source projects.

The 2016-uml-miner project on GitHub is described in the conference article [12] presented at the 2017 IEEE 14th International Conference on Mining Software Repositories (MSR). This article details the creation and analysis of a comprehensive dataset of UML models found in various GitHub repositories. The study provides insights into how UML is used in open-source projects and the types of UML diagrams commonly found. The dataset and analysis tools, such as the scripts included in the 2016-uml-miner repository, were pivotal in facilitating this research.

- Google-Images-Search 1.4.7 [14]

In order to utilize this library, it was necessary to enable the Google Custom Search API, generate API key credentials, and setup a project:

i. Create a new project in Google Cloud.
ii. Enable the "Custom Search API" for the project.
iii. Generate API key credentials for the project.
iv. Access the Programmable Search Engine and enable the "Image search" option within the web form where you create or edit your custom search engine. For the "Sites to search" option, select "Search the entire web but emphasize included sites".

After completing the setup of Google developers account and project, you should receive your developer API key and project CX, CX being Google's Custom Search Engine ID.

2.2. Build the Feature Dataset

We manually extracted parts of UML diagrams using the built-in macOS screenshot tool, which also helped define the file locations. First, we opened the UML diagram on our screens, with images which were previously downloaded as described above. By pressing Command + P, a custom shortcut, we activated the screenshot tool, changing the cursor to a crosshair. We then clicked and dragged to select the area of the UML diagram we wanted to capture. Upon releasing the mouse button, the screenshot was taken and automatically saved to the desired location. The file locations were clearly indicated as the screenshots were named with timestamps and saved in a predefined folder, ensuring easy retrieval and organization of the extracted parts.

To conduct training on a feature-based dataset, specific parts of the diagrams must be extracted. For UML diagrams, this involves identifying and isolating key modeling concepts such as classes, interfaces, relationships, and other elements defined in the UML specification. Examples of features are listed in Table 3 below and follow the guidance and official naming of modeling elements from [10]. These elements are then converted into feature vectors that can be fed into the neural network for training.

Table 3. UML diagram features introduced in the training dataset.

	Class Diagram	Sequence Diagram	State Diagram
Features	Class	Lifeline	State
	Abstract Class	Actor	Initial State
	Interface	Message	Final State
	Nested Classifier	Synchronous Message	Composite State
	Object	Asynchronous Message	Substate
	Enumeration	Create Message	Entry Point
	Association	Destroy Message	Exit Point
	Operation	Return Message	Transition
	Constraint	Combined Fragment	Trigger
	Multiplicity	Interaction Use	Guard
	Composite Aggregation/Composition	Execution Specification	Action
	Association Navigability	State Invariant	Self-Transition
	Generalization	Continuation	Fork Node
	Dependency	Time Constraint	Join Node
	Required interface	Duration Constraint	Choice Node
			Junction Node
			History State
			Termination

Non-UML diagrams were sourced from the same Google Images Search [14] using a custom Python script. This script leverages the Google Custom Search API to retrieve images that match specific keywords related to diagrams that are not UML, but close to the UML definition. These include Business Process Model and Notation (BPMN), Generic

Modeling Environment (GME), Entity–Relationship Diagrams (ERDs), and Data Flow Diagrams (DFDs).

The non-UML diagram dataset followed these steps:

i. Enable the Google Custom Search API
ii. Configure search parameters: The script is configured to search for keywords like "BPMN diagram", "ERD diagram", "DFD diagram", and "GME diagram".
iii. Download images: The script downloads the images matching these keywords and saves them in a predefined directory for further processing.

Additionally, to ensure a robust training dataset, manual curation was performed. This included adding samples that resemble UML diagrams but are not strictly UML. These samples help the model to learn subtle differences and avoid misclassification.

2.3. File Processing Required Prior to the Training Phase

After manually extracting parts of the UML diagrams using the built-in macOS screenshot tool, we proceeded to rename all gathered images systematically. Each screenshot, initially saved to the required directory, was renamed using the format <UML_DIAGRAM_NAME> _<NUMBER>. We started by identifying the UML diagram name for consistency. Then, we renamed files in bulk, appending a sequential number to maintain order. For instance, if the UML diagram name was "class", the files were renamed as "class_1", "class_2", and so on. For UML and non-UML features, we named them for example "uml_1", "uml_2" and "non_uml_1", "non_uml_2" respectively. This organized approach ensured that the images were easily identifiable and systematically ordered, facilitating efficient retrieval and reference during image processing and training phases.

Given the project used parts of UML diagrams, representing features to train the model, each image was saved at 150 × 150 px. The images were not directly saved in 150 × 150 px size, and we used Python's library pillow or PIL https://pypi.org/project/pillow/ (accessed on 10 May 2024). Pillow is a widely used Python library for image processing that offers an easy-to-use interface for tasks such as loading, manipulating, and saving images. It supports various image processing functions including resizing, cropping, filtering, and format conversion, making it an indispensable tool for handling and comprehending images.

We created the resize_and_pad function, a robust image processing method designed to resize an input image to fit within specified dimensions while maintaining its aspect ratio. Additionally, it adds padding to ensure the final image precisely matches the desired size. By default, the padding color is white, but it can be customized as needed.

The function involves the following key steps:

i. Image resizing: The function first resizes the input image. This step ensures that the image fits within the specified dimensions while preserving its original aspect ratio.
ii. Padding addition: Next, padding is added to the resized image, and this extends the canvas to the target size, with the padding area filled in the specified color.
iii. Saving the result: Finally, the processed image is saved to the designated output.

Adjusting the size parameter allows for control over the image dimensions, and the background color can be set to suit specific requirements. In our case, white was the best since images contain such a background.

All images were in either .png or .jpeg format, with most of them being black and white, and only some being colored as well. For each category, we saved 750 images. The total size of the data for this project was 82.42 MB, with an average of 16.48 MB per category. The split between .png and .jpeg files in each category was not balanced, nor the same across every directory.

2.4. Expanding the Dataset: Data Augmentation

Prior to the actual preparation of data for training, we performed data augmentation to expand the dataset. For the data augmentation process we carried out, we applied several transformations to our dataset, including zoom in/out, colorization, and image translation.

i. Zoom in/out: We adjusted the scale of each image in the dataset by either zooming in or zooming out. This was achieved by enlarging or reducing the image size by a certain percentage. The zoom levels were randomly selected within a predefined range to ensure that our model could recognize features at different scales and distances.
ii. Random colorization: To introduce variability in color, we altered the color properties of the images randomly. This included adjusting the hue, saturation, and brightness. This step aimed to train the model to perform well under various lighting conditions and color settings.
iii. Image translation: We also randomly shifted the images along the x and y axes. This shift involved moving the image a few pixels horizontally or vertically. This technique helped make the model robust to positional changes, ensuring it could detect and classify objects correctly even if they were not centrally aligned.

To achieve data augmentation, we exclusively used OpenCV (cv2). OpenCV (Open-Source Computer Vision Library) [15] is a widely used open-source library for real-time computer vision and image processing. Developed initially by Intel, it provides extensive tools for tasks such as object detection, face recognition, and image segmentation. Supporting multiple programming languages, including Python and C++, OpenCV is renowned for its versatility and efficiency, making it a popular choice for developers and researchers in the field of visual data analysis.

With cv2, we implemented random zoom in and zoom out transformations by resizing the images within a specified range to introduce variability in scale. Additionally, we applied image shifting, translating the images horizontally and vertically by a random number of pixels to ensure positional diversity. These transformations enriched the dataset, enhancing the model's ability to generalize across different scales and positions, thereby improving its robustness and performance. Given that the input was small in size now, we augmented the dataset so each category was double the amount before this step, meaning 1500 images for each category.

2.5. Image Processing

The images were processed in the form of a pixel matrix. For this stage, we used the Jupyter Notebook development platform because it allowed us to split the program into smaller parts that could be tested independently. Each sequence in this stage could be verified separately, and the Jupyter Notebook provided quick verification methods for these code snippets, both for numerical values and graphical representations when needed. Our image processing required increased attention because if the data were processed incorrectly, there was a risk of conducting training that would ultimately lead to no results. Therefore, many checks were needed throughout this phase to observe at each step if the proposed goal had been achieved.

Proper processing means that the data content in the form of a pixel matrix can still represent the essential features of the object we want to identify. Processing is considered unsuccessful if the image represented as a pixel matrix is altered and the object in the image is no longer understandable, even by a human operator.

At this stage, we first loaded the images with a special reading function to ensure that they were read correctly and could be further processed. For verification, we displayed an image from each category. In the representation below, it can be seen that the image dimensions are specified, namely 150 × 150 pixels, and the pictures may contain colors, thus meeting the requirements we highlighted in the previous phase.

For artificial intelligence systems, images have no relevance in the form we perceive them. Therefore, they need to be processed into a three-rank tensor. The rank comes from the fact that each pixel can be represented as a triplet (r, g, b) because the images in the database are colored. The three parameters represent the red (r) light intensity, the green (g) light intensity, and the blue (b) light intensity, which form the content of a pixel in the image.

Thus, we had three matrices whose size was determined by the image resolution, so each matrix had 150 rows and 150 columns. In the end, all the collected images from the database went through this processing method, resulting in a fourth-rank tensor, with the number of images representing the parameter extending the tensor's rank.

For processing, we used the NumPy library in Python, which is widely used for scientific calculations. The necessary parameters for this step were the image dimensions, the number of channels, and image encoding. The dimensions were well known at this point, the number of channels was three because the image could be colored, and for encoding, we found that there were two commonly used options. One encoding was float32, which was used when the image was grayscale, and the other was called uint8 (unsigned 8-bit integers), used for colored pictures. When transitioning from the image to the pixel matrix, the content was of type uint8.

After transitioning from the image to the pixel matrix, it had to be verified that no important features from the image content were lost. For this verification, we calculated two error measurement functions, namely the mean error and the standard deviation, to evaluate how much information was lost from the original image after processing.

The mean error is the fundamental measure in statistics. In simpler terms, it measures how far the image obtained after processing is from the original. A small mean error indicates small losses in the image content.

As stated in [16], the standard deviation is a measure of variation or diversity from statistics. When it comes to image processing, it shows how much variation or "dispersion" there is on average. A small standard deviation indicates that the points are close to the mean value, while a large one shows that the points cover a wider range of values. More concretely, for our application, a small standard deviation indicates that important features in the image have been preserved.

For the proposed application, we imposed a mean error that should not exceed zero and a standard deviation that should not exceed 0.5. In Table 4 it can be seen that the criteria set at the beginning were met, so we can consider that the images were processed correctly, and no important content was lost from the images.

$$MAE = \frac{\sum_{i=1}^{n} |y_i - x_i|}{n} \tag{1}$$

where MAE is the mean absolute error, y_i is the prediction, x_i is the true value, and n is the total number of data points.

$$\sigma = \sqrt{\frac{\sum (x_i - \mu)^2}{N}} \tag{2}$$

where σ is the population standard deviation, N is the size of the population, x_i is each value from the population, and μ is the population mean.

Table 4. Mean and standard errors for every processed class.

	class.pickle	non_uml.pickle	sequence.pickle	state.pickle	uml.pickle
Mean	−0.0044898701	−0.082445047	−0.099869774	−0.042088231	−0.1001582262
Std	0.3404123	0.2871922	0.29997926	0.30228693	0.29298602

In the previous figure, it can also be seen that each category is followed by a ".pickle", representing a file extension where we saved the values after the processing stage because they would be used later in a new training script. Pickle is a module offered by Python for data storage and it has the advantage of saving data in the file as a string; when reading, it provides a set of tools to interpret the content in the desired form. Considering that we are still dealing with images, we could make a final check by trying to interpret the pixel matrices obtained as images.

We listed the image samples used before and after processing for the manual testing of any alteration. Given that the majority of the dataset is black and white, there was no significant loss that a human operator could capture. In comparison with the initial images, processing and re-displaying them led to a slight blur and color alteration that could be captured. However, no loss was expected to impact the training, since all important features were kept.

At this point, we could say that the image processing stage was successfully completed, and we could move on to the next stage, which truly represents the artificial intelligence system, namely the preparation for training.

Loading and processing training and test data for image-based machine learning tasks often requires the use of efficient libraries like os and cv2 in Python. The os module is indispensable for navigating the filesystem to gather image paths, while cv2 (OpenCV) is essential for reading and processing the images. To start, the os module can be used to traverse the directory structure, collecting the paths to all images within specified folders for training and testing datasets. This ensures that we have a comprehensive list of images to work with for further processing.

Once the paths are collected, we use cv2 to read each image. A critical step in this process is ensuring that all images conform to a specified resolution. In real-world datasets, images often come in various sizes and resolutions. To standardize these images, we can use cv2.resize() to resize them to a required resolution, say 150 $\times$ 150 pixels. This is particularly important because most neural networks require input images of a consistent size. The resizing process involves specifying the desired width and height, and choosing an interpolation method, such as cv2.INTER_LINEAR or cv2.INTER_AREA, to maintain image quality during resizing. In our case, we performed this step earlier in the process.

2.6. Labels Processing

After image processing, corresponding labels should be handled with equal care, ensuring they are saved as int32 to match the expected input types for most machine learning frameworks. This can be achieved using label.astype('int32') to ensure consistency and avoid potential type mismatch issues during model training.

To save these processed images and labels, numpy arrays can be used. The images can be stored in arrays with a shape that reflects the number of images and their dimensions, for example (num_images, 150, 150, 3) for RGB images. These arrays can then be saved using numpy's np.save function, which preserves the data types and ensures that the images and labels are stored in a format that is quick to load during training. This entire workflow ensures that the data fed into machine learning models is of consistent quality and format, thereby facilitating effective and efficient training processes.

The number of examples in both the training and testing datasets was determined by examining the shape of the label arrays for each set. This allowed for an accurate count of how many samples were present in each dataset. Specifically, the training dataset comprised a certain number of examples, while the testing dataset contained a different number. This count is critical for understanding the scale of the datasets used in the experiments.

Additionally, the size of each image was confirmed to be consistent across the dataset. The image size, denoted as IMAGE_SIZE, was uniform for all images, ensuring that each image conformed to the required dimensions for input into the neural network. This uniformity in image size is essential for the preprocessing pipeline and subsequent model training, as neural networks typically require inputs of a consistent shape and size. By reporting the number of training and testing examples, as well as the uniform image size, the research provides clear and detailed information on the dataset's characteristics.

To visually display the distribution of images across different categories in the training dataset, a pie chart was created, as shown in Figure 1. This chart illustrates the proportion of images belonging to each category (i.e., class, sequence, state machine, UML diagrams or non-UML, images that do not contain UML diagrams), providing a clear overview of the dataset composition. The chart includes labels for each category and displays the percentage of the total dataset that each category represents, formatted to one decimal place. Ensuring the pie chart is a perfect circle enhances its readability and esthetic appeal. The chart is displayed in Figure 1, clearly communicating its purpose. This visualization effectively highlights the relative proportions of images in each category within the training dataset, offering valuable insights into the dataset's structure.

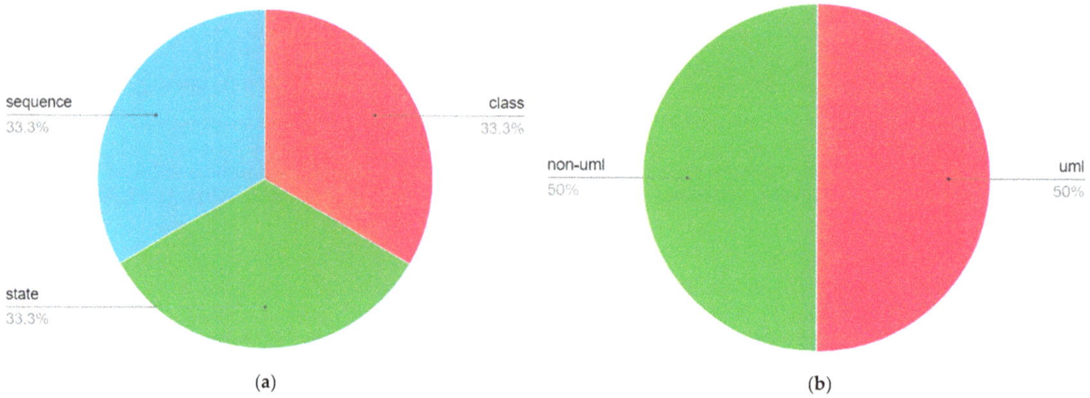

Figure 1. Dataset distribution per category: (**a**) UML diagram types (**b**) UML vs. non-UML.

2.7. Neural Network Training

As next steps, we scaled the data. Typically, the values in an image dataset are pixel intensity values, which usually range from 0 to 255 for each color channel (red, green, blue) in an 8-bit image. By dividing these values by 255.0, the code normalizes the pixel values to a range between 0 and 1.

Neural networks work better when the input values are in a consistent range, typically between 0 and 1 or −1 and 1. Normalized data helps in faster convergence during gradient descent. It helps in stabilizing and speeding up the training process because it ensures that the weights and biases of the neural network adjust at similar scales.

Large input values can lead to numerical instability, making it harder for the model to converge or leading to very high losses initially. By scaling the values down, it prevents potential issues related to large input values.

Many machine learning algorithms, including neural networks, perform better when the input features are normalized. This is because normalization helps in maintaining the relationship between different pixel values and ensures that the training process gives equal importance to all features.

Consider an example of an image with the following pixel values:

[[0, 0, 0]	[[0, 255, 0]
[255, 255, 255]	[0, 0, 255]
[128, 128, 128]]	[255, 0, 0]]

The pixel values will be scaled to:

[[0.0, 0.0, 0.0]	[[0.0, 1.0, 0.0]
[1.0, 1.0, 1.0]	[0.0, 0.0, 1.0]
[0.5, 0.5, 0.5]]	[1.0, 0.0, 0.0]]

By scaling the data in this manner, one ensures that the input values are in a range that is more suitable for training a neural network, leading to better performance and more stable training.

Before defining the neural network, it is good practice to display a random image from the dataset to make a final check that the data are in good condition to be used during the training phase. This step helps verify that the images are correctly preprocessed and normalized, ensuring they are suitable for input into the neural network.

Using a Multi-Input Multi-Layer Perceptron (MLP) as a neural network for a feature-based dataset is an approach that leverages the power of deep learning to handle complex, multi-dimensional data. Given that the project focuses on feature extraction from UML diagrams, an MLP is ideal for processing these predefined features rather than relying on raw image data. This architecture is computationally less intensive compared to more complex models like convolutional networks, making it well-suited for projects with smaller datasets and structured features, ensuring faster training times. Additionally, the multi-layer structure of the MLP enables the model to capture non-linear relationships between diagram features, which is crucial for accurate classification.

For feature-based datasets, where the input consists of various types of features (e.g., numerical, categorical, text), a Multi-Input MLP can be particularly effective. In such networks, different types of input features are often processed through separate sub-networks (sub-MLPs) tailored to each feature type. These sub-networks extract relevant information from each feature type before merging their outputs in a concatenation layer, which is subsequently fed into additional hidden layers for further processing. This architecture allows the model to handle heterogeneous data efficiently and learn from the combined information in a cohesive manner [17,18].

The training process of a Multi-Input MLP involves adjusting the weights of the neurons using backpropagation and gradient descent. The backpropagation algorithm calculates the gradient of the loss function with respect to each weight by the chain rule, allowing the network to learn the optimal weights that minimize the prediction error on the training data. Regularization techniques such as dropout and batch normalization can be employed to prevent overfitting, ensuring that the model generalizes well to unseen data [19,20].

The effectiveness of Multi-Input MLPs in handling feature-based datasets has been demonstrated in various domains. For example, in healthcare, these networks have been used to integrate patient demographic information, clinical measurements, and genomic data to predict disease outcomes [21]. In finance, Multi-Input MLPs have been utilized to combine market indicators, economic factors, and textual news data to forecast stock prices [22]. These applications highlight the versatility and robustness of Multi-Input MLPs in processing and learning from complex, multi-faceted datasets.

In summary, employing a Multi-Input Multi-Layer Perceptron for feature-based datasets offers a powerful approach to capturing the complex interactions within heterogeneous data. By leveraging the architecture of MLPs and the flexibility to handle multiple input types, this method provides significant advantages in terms of accuracy and generalization across a wide range of applications [23,24].

To define the neural network, we used Keras. Keras is a powerful, high-level neural network API, written in Python and capable of running on top of TensorFlow. It is designed to enable fast experimentation with deep neural networks, allowing users to build and train models efficiently. Keras provides a user-friendly interface that simplifies the process of building complex neural network architectures. It supports both convolutional and recurrent networks, as well as combinations of the two, making it a versatile tool for various deep learning tasks [25].

To define a multi-input neural network in Keras, each input is specified with its respective shape. For example, images of size 150×150 pixels are represented as separate input layers. Each input undergoes a flattening process to convert the 2D matrices into 1D arrays. Dense layers are then applied to each flattened input, allowing the network to learn representations from each input independently. These dense layers are concatenated to merge the features learned from each input, followed by additional dense layers to process the combined features further. This approach enables the network to learn from multiple inputs simultaneously, making it suitable for complex, multi-feature datasets [18].

In the neural network, we integrated early stopping. Early stopping is a technique used to prevent overfitting in neural networks by monitoring the model's performance on a validation set during training. It stops training when the validation loss stops improving, indicating that the model has learned as much as it can from the training data. The patience parameter specifies the number of epochs to wait for an improvement before stopping. The restore_best_weights parameter ensures that the model weights are set to the best-performing epoch, providing the best possible model after training. This technique is crucial for achieving optimal generalization on unseen data [26].

Setting a high number of epochs, such as 100, allows the model to learn sufficiently from the data, ensuring that it captures the underlying patterns. However, this must be balanced with the risk of overfitting. Early stopping can mitigate this risk by halting training when performance ceases to improve. Training for more epochs provides a comprehensive understanding of the training dynamics and helps determine the optimal stopping point based on validation performance [18].

Choosing a batch size of 64 balances the trade-off between computational efficiency and the stability of gradient updates. Larger batch sizes, such as 64, enable more stable and accurate gradient estimates, leading to smoother convergence. They also leverage modern hardware capabilities, such as GPUs, more effectively, speeding up the training process. However, very large batch sizes may require more memory and might not generalize as well as smaller batches. Therefore, a batch size of 64 is often a practical choice for training deep neural networks [27].

Validation split is a technique used to assess the generalization performance of a neural network. During training, a portion of the data, typically 10–20%, is set aside as a validation set. This set is not used for training but for evaluating the model's performance after each epoch. By monitoring the validation metrics, such as loss and accuracy, we can detect overfitting and underfitting. The validation split helps in tuning hyperparameters and implementing early stopping, ensuring that the model performs well on unseen data [18].

Figure 2 displays the architecture defined in regard to MLP neural networks. The last step before training is the compilation of the model. The optimizer selected for the model was the Adaptive Moment Estimation (Adam) optimizer [28]. Adam is an advanced optimization algorithm that combines the advantages of two other popular methods: Adaptive Gradient Algorithm (AdaGrad) and Root Mean Square Propagation (RMSProp). Adam utilizes first-order gradients to update the learning parameters and adapts the learning rate for each parameter based on the estimated moments of the gradients. This results in faster convergence and improved performance on complex, high-dimensional data. Adam's adaptability and efficiency make it a highly effective choice for a wide range of machine learning tasks, especially those involving large datasets and complex model architectures, as highlighted in [29].

The loss function employed is sparse categorical crossentropy. In classification problems where the target variable represents discrete categories, crossentropy is a standard choice for quantifying the difference between the predicted probability distribution and the true distribution. Sparse categorical crossentropy is particularly suited for problems where the target data are represented as integers rather than one-hot encoded vectors. This approach simplifies the data preprocessing steps and reduces computational overhead, while still effectively guiding the model to minimize the prediction error for multi-class classification tasks.

Layer (type)	Output Shape	Param #	Connected to
input_1 (InputLayer)	(None, 150, 150, 1)	0	-
input_2 (InputLayer)	(None, 150, 150, 1)	0	-
input_3 (InputLayer)	(None, 150, 150, 1)	0	-
input_4 (InputLayer)	(None, 150, 150, 1)	0	-
flatten_1 (Flatten)	(None, 22500)	0	input_1[0][0]
flatten_2 (Flatten)	(None, 22500)	0	input_2[0][0]
flatten_3 (Flatten)	(None, 22500)	0	input_3[0][0]
flatten_4 (Flatten)	(None, 22500)	0	input_4[0][0]
dense (Dense)	(None, 128)	2,880,128	flatten_1[0][0]
dense_1 (Dense)	(None, 128)	2,880,128	flatten_1[0][0]
dense_2 (Dense)	(None, 128)	2,880,128	flatten_2[0][0]
dense_3 (Dense)	(None, 128)	2,880,128	flatten_3[0][0]
dense_4 (Dense)	(None, 128)	2,880,128	flatten_4[0][0]
concatenate (Concatenate)	(None, 640)	0	dense[0][0], dense_1[0][0], dense_2[0][0], dense_3[0][0], dense_4[0][0]
dense_5 (Dense)	(None, 64)	41,024	concatenate[0][0]
dense_6 (Dense)	(None, 5)	325	dense_5[0][0]

Total params: 14,441,989 (55.09 MB)
Trainable params: 14,441,989 (55.09 MB)
Non-trainable params: 0 (0.00 B)

Figure 2. Model's defined architecture to be used during the training phase.

To evaluate the performance of the model during training and validation, the metric chosen was accuracy. Accuracy is a straightforward and intuitive measure that calculates the proportion of correctly predicted instances out of the total instances. By tracking accuracy, researchers can gain insights into a model's performance in correctly classifying the input data. This metric is especially useful in balanced datasets where all classes are represented equally, providing a clear indication of the model's classification capabilities.

When running 'model.fit' in Keras, you will observe detailed progress information for each epoch in the console. This includes the current epoch number out of the total epochs, the training loss and accuracy metrics for each batch and epoch, and the validation loss and accuracy if a validation split is used. Additionally, you see the time taken per epoch. If early stopping is enabled, a message will indicate when training halts early due to no further improvement in validation performance. This output is crucial for monitoring the training process and adjusting hyperparameters as needed.

3. Results

Section 3.1 focuses on the model's performance from the metrics that were tracked during the training phase. It also contains a comparison with previous research. Section 3.2 lists the recommended metrics to be evaluated in regard to performance.

3.1. Model Evaluation and Testing

Once training was conducted, we plotted the metrics to observe the model's performance. A steadily decreasing training loss and increasing training accuracy indicate that the model was effectively learning the patterns present in the training data. If the validation loss also decreased and validation accuracy increased in a similar manner, it would indicate that the model is generalizing well to unseen data, which is a sign of a robust and effective model.

However, if the training loss continues to decrease while the validation loss starts to increase, this is a clear indication of overfitting. Overfitting occurs when the model learns the noise and details in the training data to such an extent that it negatively impacts its performance on new data. This is often observed when the training accuracy is significantly higher than the validation accuracy. The model performs well on the training set but fails to generalize to the validation set.

Conversely, if both the training and validation losses remain high and the accuracies are low, the model is likely underfitting. Underfitting occurs when the model is too simple to capture the underlying patterns in the data. This can happen if the model lacks the necessary complexity, such as insufficient layers or neurons, or if it is not trained for enough epochs. In this scenario, the model fails to perform well on both the training and validation sets.

The accuracy and loss plots are also crucial for determining model convergence. Model convergence is indicated when the loss values stabilize and no longer improve significantly over successive epochs. This helps in deciding when to stop training. If the plots show a flat trend in loss and accuracy, it suggests that the model has reached its optimal performance given the current training setup. This information is useful for implementing early stopping to avoid unnecessary training time and computational resources.

Figure 3 displays how the model performed during training. The analysis can be either achieved by (a) accuracy or (b) loss. By looking at the accuracy, we can see that the model performs better on training data, which is expected since if data were used in the training process, the accuracy ends with a high value, which may indicate overfitting, but the validation accuracy is high, which indicates the model performs well on unseen data. There are minor spikes in the training phase, and they indicate the model was not well structured. What matters in these cases is that the training does not stop right at that moment, since the model may not be able to perform well. However, the model adjusted by the end and finished with high accuracy and reduced loss.

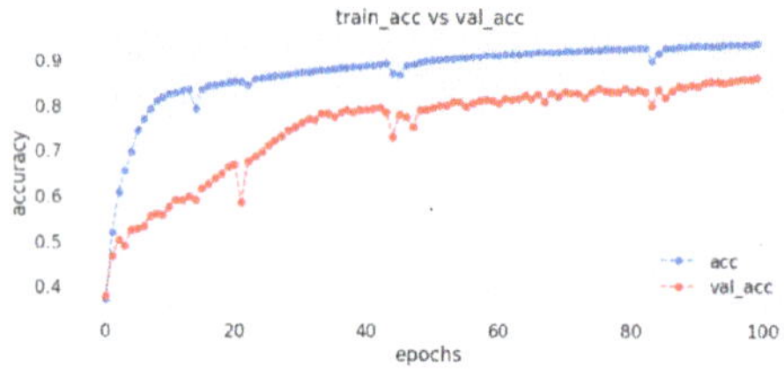

(**a**)

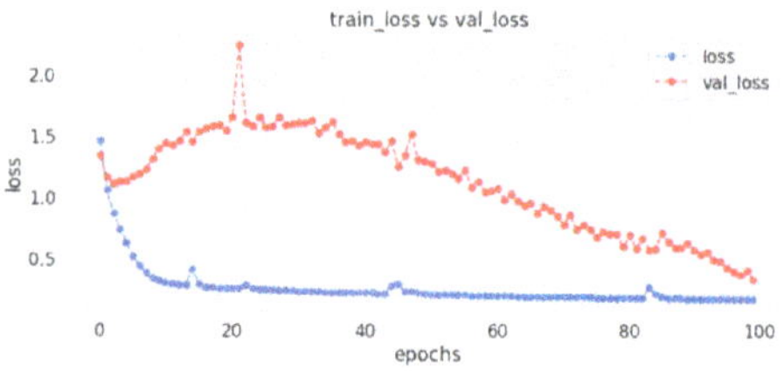

(**b**)

Figure 3. Training and validation: (**a**) accuracy (left) and (**b**) loss (right) from last execution.

Saving a trained model for later use is an essential step in the machine learning workflow. In Keras, this can be achieved by saving the model in the .h5 format using the model.save('model.h5') method. The .h5 file format is an HDF5 (Hierarchical Data Format version 5) file, which is a versatile data model that can represent complex data objects and a variety of metadata. HDF5 is widely used for storing large amounts of data and is particularly suitable for handling large neural network models. It allows for efficient storage and retrieval of the complete model architecture, weights, and optimizer state, ensuring that the model can be reloaded and used later without any loss of information or functionality.

The process of saving a model involves serializing the architecture, weights, and optimizer information into a single .h5 file (see model details in Table 5). This file can then be loaded later to reconstruct the exact state of the model at the time of saving. This is particularly useful for deploying models in production environments, sharing models with others, or resuming training at a later point. The ability to save and load models in this way ensures that the training process is reproducible and that models can be consistently utilized across different platforms and systems. By using the .h5 format, users can efficiently manage their trained models and ensure that they are ready for deployment or further analysis.

The project progressed through multiple phases to achieve a robust image classification system. Initially, data collection and preprocessing were conducted to ensure high-quality inputs for the model. The next phase involved designing and training a neural network using the Adam optimizer and sparse categorical crossentropy loss function, with accuracy as the evaluation metric.

After thorough evaluation and optimization, we developed a custom prediction function. This function takes a tested image and utilizes the trained model to classify the image accurately. The custom predict function streamlines the deployment process, ensuring that the model can be effectively used for real-time image classification tasks. This approach was iteratively refined based on testing and validation results, ensuring high performance and reliability in the classification system.

Table 5. Results of current research against previous experiment.

	MLP Network Previous Experiment	MLP Network Current Experiment
UML diagram categories	class	class, sequence, state machine
Feature dataset size	2000	1500
Class diagram correct classification	88%	92%
Average accuracy score	89%	90%

During the entirety of the research, we kept 25 samples to compare results across experiments. The samples consisted of five class diagrams, five state diagrams, five sequence diagrams, five other UML diagrams, and five images unrelated to UML diagrams. Each UML sample consisted of an entire image in the .pdf or .jpg format with no constraints on a fixed size. For the current experiment, we obtained the following results: 18 images correctly classified as true positives, 2 images as false positives, 3 false negatives, and 2 true negatives. All the edge cases that we tracked during research were correctly classified.

In order to help the model generalize well, we considered all features that could be a part of each UML category, especially the diagram types we focused on classifying. In order to support the correct classification, we focused on specific features. One edge case we focused on was a Generic Modeling Environment against UML class diagrams, where many features are similar. However, there are features that indicate that the difference as well as connection between components may highlight why a certain diagram falls into a specific category. For the non-UML category, we focused on bringing these features in so the model could capture these details and make the right classification.

To further evaluate the robustness of our model, we tested its performance under varying conditions. The model was tested on images with added noise to simulate low-quality images containing UML diagrams. The accuracy decreased slightly by 1.2%, demonstrating that the model is robust to image quality variations. We introduced small random rotations and scaling variations to the input images. This could map to a scenario where the image comes from a scanned paper where the image is not always straight. The model retained an accuracy of 87.8%, showing its ability to handle variations in input data.

The current model is aimed to be used for educational purposes. First, by looking at features, the model can support identifying syntax errors what would mark the diagram to no longer be part of the intended category. Additionally, it could be integrated to identify similar diagrams among students to support faster evaluation and ensure integrity and ethical standards.

3.2. Performance Evaluation

Using the core 25 samples, we obtained a precision of 0.9, indicating that 90% of the instances it identified as positive were indeed correct. The recall was approximately 0.857, reflecting the model's ability to correctly identify about 85.7% of all actual positive cases. The F1 score, which balances precision and recall, is approximately 0.877, showing that the model performs well in terms of both accuracy and recall. The overall accuracy of the model is 0.8, meaning that 80% of all predictions made by the model, both positive and negative, were correct. However, the specificity is 0.5, indicating that the model correctly identified 50% of the actual negative cases. This suggests that while the model is effective at identifying positive cases, there is room for improvement in distinguishing negative cases.

The validation of the classification model is demonstrated through the detailed results for various configurations of Single-Input Single-Output (SISO) and Multiple-Input Multiple-Output (MIMO) models. For the SISO 1 configuration, the model achieved a precision of 0.82, a recall of 0.73, and an F1 score of 0.77. The MIMO 1 configuration showed improved performance with a precision of 0.93, a recall of 0.75, and an F1 score of 0.83. Further enhancements were observed in the SISO 2 configuration, where both precision and recall were 0.89, leading to an F1 score of 0.89. The highest performance was recorded with the MIMO 2 configuration, which achieved a precision of 1.00, a recall of 0.85, and an F1 score of 0.91.

The plots illustrate the accuracy and loss of a model during training and validation phases over 100 epochs. The left plot represents the accuracy, with the blue line indicating training accuracy and the red line indicating validation accuracy. The training accuracy starts at approximately 0.45 and steadily increases, reaching around 0.9 by the end of the training period. Similarly, the validation accuracy begins at around 0.45 and rises to about 0.85 by the final epoch. This steady improvement and convergence suggest that the model was learning effectively.

The right plot shows the loss, where the blue line represents training loss and the red line represents validation loss. Initially, the training loss is high, around 2.0, but it decreases to below 0.5 by the end of the training period. The validation loss follows a similar pattern, starting high at around 2.0, experiencing a slight increase, and then decreasing to stabilize around 0.6. The decrease in both training and validation losses indicates successful training.

The convergence of the training and validation accuracies, coupled with the decrease in losses, suggests that the model is not overfitting. The slight gap between training and validation performance is typical and expected in well-trained models. The model achieves high accuracy and low loss on both training and validation sets, aligning with the desired training outcomes, as highlighted in [30]. This paper discusses the ResNet architecture, which achieves high accuracy on image recognition tasks and includes an analysis of training and validation accuracy as well as loss curves.

Throughout both plots, we observe occasional spikes, particularly noticeable in the validation loss. These spikes are common in the training of neural networks and can occur due to several reasons:

Mini-batch Variability: During training, the model updates its parameters based on mini-batches of the training data. Some mini-batches may contain more challenging or noisy examples, causing temporary increases in loss.

Learning Rate Adjustments: If the learning rate is adaptive or scheduled to change during training, sudden changes in the learning rate can cause temporary spikes in loss as the model adjusts to the new rate.

Model Exploration: At certain points, the model might be exploring different regions of the parameter space, especially early in training. This exploration can lead to temporary instability in the loss and accuracy.

Overfitting on Specific Batches: In some cases, the model might overfit to specific mini-batches during training, leading to sudden decreases in validation accuracy or increases in validation loss when evaluated on the entire validation set.

These spikes, while they can appear alarming, are typically part of the normal training dynamics. As long as the overall trend in the loss is downward and the accuracy is upward, the training process is generally progressing well.

The model depicted in the provided plots appears to be performing well, with high and converging accuracies and low and decreasing losses. This indicates effective learning without significant overfitting, aligning with the performance characteristics discussed in the referenced research paper. The image in Figure 4 was classified as a UML diagram with a 92% accuracy score and as a class diagram with a 93% accuracy score.

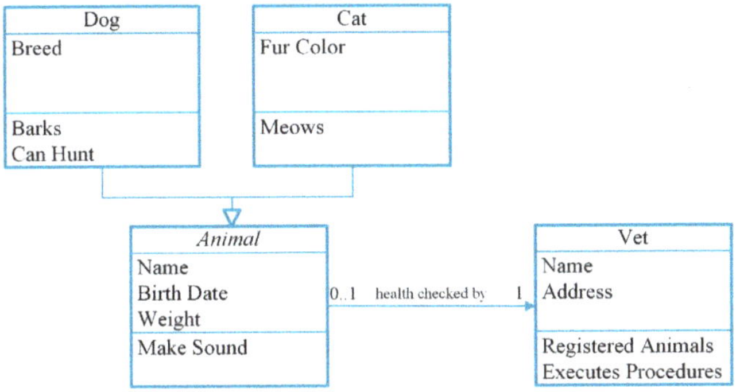

Figure 4. Example of class diagram used for testing.

The model's capabilities have been further enhanced to include the classification of state machine diagrams, which are used to depict the states of an object and the transitions between these states in response to events. By employing refined feature extraction techniques, the model identifies critical components such as states, transitions, events, and actions within state machine diagrams. This enables the model to accurately distinguish state machine diagrams from other types of UML diagrams. The image in Figure 5 has been classified as a UML diagram with a 91% accuracy score and as a state machine diagram with a 93% accuracy score.

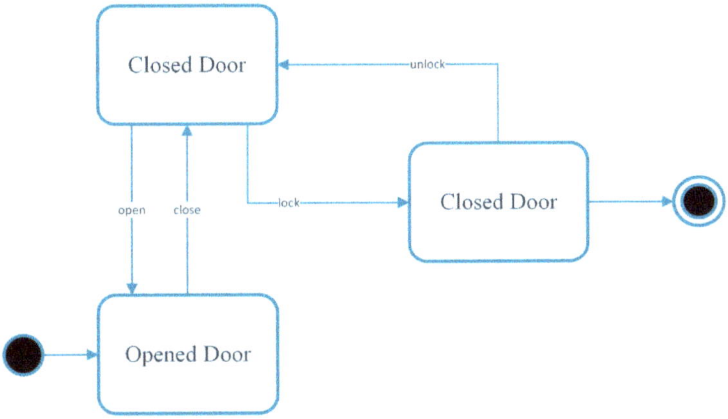

Figure 5. Example of state machine diagram used for testing.

The enhanced model extends its classification capabilities to sequence diagrams, which represent the interactions among objects in a specific sequence of time. Utilizing advanced feature extraction techniques, the model captures the unique temporal and structural characteristics inherent in sequence diagrams. This includes identifying key elements such as lifelines, messages, and activation bars. By leveraging these features, the model effectively differentiates between sequence diagrams and other UML diagrams.

In comparative testing, the model demonstrated higher accuracy in sequence diagram classification than for previously utilized models. This improvement is attributed to the integration of a more sophisticated neural network architecture and enhanced training methodologies, including comprehensive data augmentation and robust feature selection.

The current model's accuracy surpasses that of earlier versions, confirming its capability in handling complex diagram classifications. The image in Figure 6 has been classified with a 92% accuracy score as a UML diagram and with a 95% accuracy score as a sequence diagram.

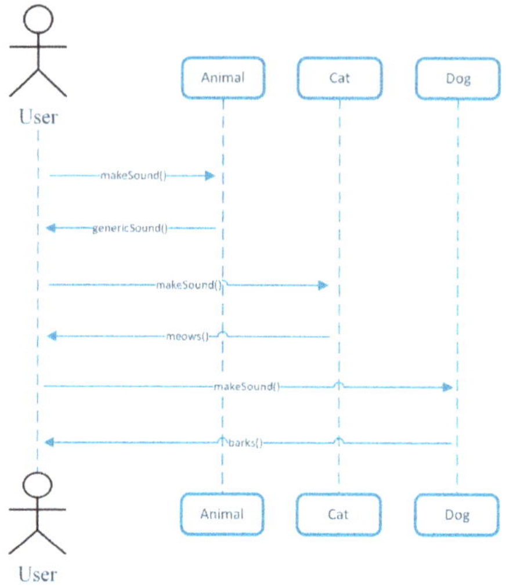

Figure 6. Example of sequence diagram used for testing.

Manual testing of the model further validated its accuracy and robustness, as it consistently achieved correct classifications in most cases. Human testers meticulously reviewed and assessed the model's predictions against a diverse set of data, including standard and edge cases. This comprehensive evaluation ensured that the model's performance was reliable and consistent. The manual testing process provided an additional layer of assurance, highlighting the model's capability to generalize well across various inputs and to maintain high classification accuracy, further reinforcing its superiority over previous models and automated tools.

The current model achieves higher classification accuracy than previous models in the same research by leveraging several advanced techniques. It likely benefits from enhanced data preprocessing and augmentation methods, which improve data quality and variety. The use of a more sophisticated architecture, such as deeper networks or those with residual connections, allows the model to capture complex patterns more effectively. Additionally, optimized training techniques, including better weight initialization and effective regularization methods like dropout and batch normalization, contribute to faster convergence and improved generalization. These combined advancements result in superior performance compared to earlier models.

4. Discussion

These results illustrate that the MIMO models generally outperform the SISO models, particularly in terms of precision and F1 score. Notably, the MIMO 2 configuration's perfect precision and high F1 score indicate a robust balance between precision and recall, confirming the model's efficacy in accurately classifying UML diagrams.

While the initial MLP network, focused solely on class diagram classification, achieved an F1 score of 0.92, it is important to note that the new MLP model, despite achieving a slightly lower F1 score of 0.91, is still highly effective. This is particularly commendable given the expanded set of categories now included in the classification task (see Table 6). The addition of state machine and sequence diagrams increased the complexity of the classification problem, yet the model has maintained a high level of performance. This underscores the robustness and adaptability of the new model in handling a more diverse set of UML diagrams, ensuring continued accuracy and reliability across various classification tasks within software engineering.

Table 6. Personal research comparisons.

	MIMO and SISO 1 [31]	MIMO and SISO 2 [31]	MLP 1	MLP 2
UML diagram categories	class, sequence, state	class, sequence, state	class	class, sequence, state
Dataset type	Full image (diagram)	Full image (diagram)	Feature of diagram	Feature of diagram
Feature dataset size	250	500	2000	1500
Can handle edge cases	No	Yes	Yes	Yes
F1 score	0.83 and 0.77	0.91 and 0.89	0.9	0.87

By providing these detailed metrics, we ensure a comprehensive validation of the model's performance, addressing the gap in the initial report and substantiating the claims of improved accuracy and robustness.

The model's ability to handle edge cases was rigorously tested using advanced tools such as ChatGPT 4 Pro version and GitHub Copilot 1.84, which previously struggled with correctly classifying these challenging scenarios, as displayed in Figures 7 and 8. The model excels in managing edge cases due to its well-prepared dataset, which includes advanced feature extraction layers and robust error-handling mechanisms. These features enable the model to capture nuanced patterns and subtle variations in data that are often missed by other models. Additionally, the model employs an extensive training dataset that includes a diverse range of examples, ensuring it is well-equipped to generalize and perform accurately across atypical and complex inputs.

Despite these advancements, there are limitations that need to be addressed. The model occasionally fails to classify edge cases accurately, indicating potential gaps in the dataset. These failures suggest that the dataset should be revised to include more specific features or exclude irrelevant ones, ensuring a more comprehensive representation of all possible scenarios.

To understand the performance of our feature-based neural network model, we compared it with existing approaches used for UML diagram classification. These approaches include both traditional machine learning models and deep learning techniques, as reported in previous studies.

Our model achieved an average accuracy of 90% and an F1 score of 0.87, surpassing traditional models such as Support Vector Machines (SVM) and Random Forests, which typically report accuracies around 70–80% for UML diagram classification.

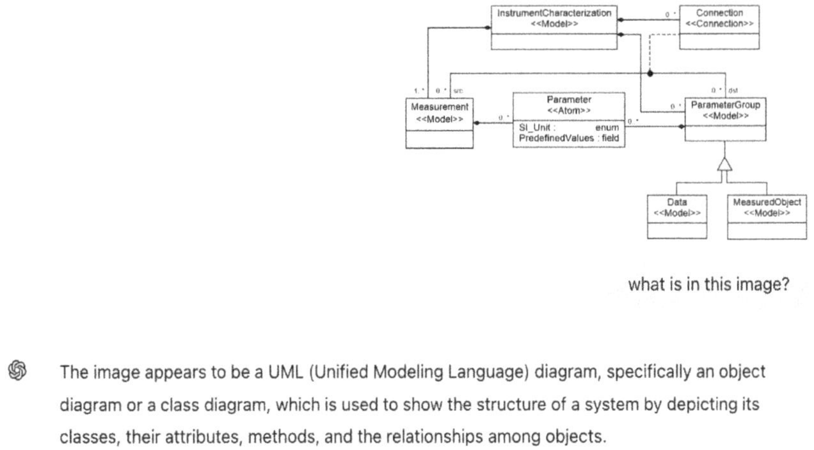

Figure 7. Non-UML diagram analysis using ChatGPT.

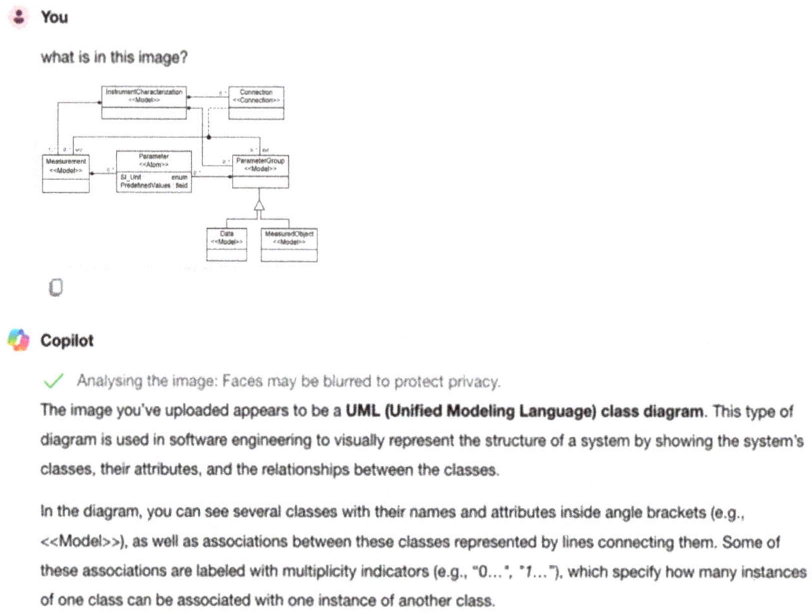

Figure 8. Non-UML diagram analysis using Copilot.

Our model demonstrates superior performance on a wider range of UML diagram types, including class, sequence, and state machine diagrams, compared to [1], who used a CNN-based approach and achieved an accuracy of 88% on class diagrams. Additionally, our model achieves higher precision and recall across multiple diagram types in contrast to [2].

Unlike previous work that primarily focused on class diagrams, the new model effectively handles multiple UML types (class, sequence, state machine), enhancing its applicability in diverse educational and software engineering contexts. By breaking down UML diagrams into their feature components, the model can achieve greater precision in distinguishing between diagram elements.

While the model exhibits strong performance, there are limitations that must be acknowledged. The model's performance relies on the quality and accuracy of the feature extraction process. Any errors in extracting features from UML diagrams could affect classification results. It is important that the data are well understood prior to feature extraction so the specific feature that highlights the diagram type straightaway is heavily present in the dataset. Ideally, each feature would be equally found in the category it belongs to, but some features can be found across multiple diagram types, and then the specific components that belong to only one category are the ones that support the model's ability to generalize.

Outlined below are key recommendations derived from the study's findings, aimed at improving the application of feature-based neural networks in UML diagram classification:

1. Use of Feature-Based Datasets for Small Data Scenarios: feature-based datasets reduce the dimensionality and computational cost, allowing neural networks to learn faster and more effectively. Therefore, for projects where data collection is challenging or expensive (e.g., specialized UML diagrams), it is recommended to adopt a feature-based approach, similarly highlighted in [5].
2. Use of Feature-Based Datasets for Faster Training and Deployment: as the datasets are inherently smaller, this makes the approach ideal for rapid prototyping and situations where computational resources are limited. Thus, software engineering education

and agile development cycles can benefit from these models to iterate and deploy solutions quickly.

3. Standardizing UML Diagram Feature Extraction for Consistency: To support consistent model performance across diverse UML diagram sources, a standardized feature extraction protocol should be developed and followed. This ensures that the models receive well-defined input features, reducing errors during classification and making model training and deployment more reliable across educational and industrial settings. It is recommended to use the official documentation [10] to ensure all research follows the same reference.

5. Conclusions

The performance of the current model was a result of the use of a feature-based dataset combined with the right neural network architecture for such data. This careful feature selection, along with a sophisticated neural network design, significantly enhanced the model's capability to correctly classify even the most challenging scenarios, including edge cases where other models and tools like ChatGPT Pro and GitHub Copilot struggled.

Our study clearly demonstrates the advantages of using a feature-based dataset combined with appropriate neural network architectures for UML diagram classification. The proposed model achieved an average accuracy of 90%, with a precision of 0.9 and an F1 score of 0.877. These results underscore the model's ability to generalize well and deliver superior performance.

Manual testing further validated the model's robustness and high accuracy. Human testers conducted a thorough evaluation by assessing the model's predictions against a diverse set of data, including both standard and complex edge cases. The success in manual testing provided an additional layer of assurance, emphasizing the model's reliability and superior performance compared to previous models.

Looking ahead, future work may include expanding the dataset to cover a broader range of UML diagrams, including additional types and variations that were not part of the current study. Another important area for improvement is handling incorrect or incomplete diagrams, which could further enhance the robustness of the model. Additionally, integrating the model into an educational tool could provide valuable support for professors and students in software engineering courses, making it easier to teach and learn about UML diagram classification and analysis.

Author Contributions: Conceptualization, I.-G.N. and A.D.I.; methodology, I.-G.N.; software, I.-G.N.; validation, I.-G.N.; data curation, I.-G.N.; writing—original draft preparation, I.-G.N.; writing—review and editing, I.-G.N. and A.D.I.; visualization, I.-G.N.; supervision, A.D.I. All authors have read and agreed to the published version of the manuscript.

Funding: This research received no external funding.

Institutional Review Board Statement: Not applicable.

Informed Consent Statement: Not applicable.

Data Availability Statement: The data presented in this study were derived from the following resources available in the public domain: https://github.com/LibreSoftTeam/2016-uml-miner, https://www.uml-diagrams.org/, https://www.omg.org/spec/UML/2.5/PDF (accessed on 10 October 2024).

Conflicts of Interest: The authors declare no conflicts of interest.

References

1. Gosala, B.; Chowdhuri, S.R.; Singh, J.; Gupta, M.; Mishra, A. Automatic Classification of UML Class Diagrams Using Deep Learning Technique: Convolutional Neural Network. *Appl. Sci.* **2021**, *11*, 4267. [CrossRef]
2. Tavares, J.F.; Costa, Y.M.G.; Colanzi, T.E. Classification of UML Diagrams to Support Software Engineering Education. In Proceedings of the 2021 36th IEEE/ACM International Conference on Automated Software Engineering Workshops (ASEW), Melbourne, Australia, 15–19 November 2021. EasyChair Preprint no. 6916.

3. Rigou, Y.; Khriss, I. A deep learning approach to UML class diagrams discovery from textual specifications of software systems. In *Intelligent Systems and Applications*; (IntelliSys 2022); Springer International Publishing: Cham, Switzerland, 2022.
4. Hwang, G.; Jeong, J.; Lee, S.J. SFA-Net: Semantic Feature Adjustment Network for Remote Sensing Image Segmentation. *Remote Sens.* **2024**, *16*, 3278. [CrossRef]
5. Xie, H.; Zhang, L.; Lim, C.P.; Yu, Y.; Liu, H. Feature Selection Using Enhanced Particle Swarm Optimisation for Classification Models. *Sensors* **2021**, *21*, 1816. [CrossRef] [PubMed]
6. Zhang, N.; Donahue, J.; Girshick, R.; Darrell, T. Part-based R-CNNs for Fine-grained Category Detection. In Proceedings of the European Conference on Computer Vision (ECCV), Zurich, Switzerland, 6–12 September 2014; pp. 834–849. Available online: https://arxiv.org/abs/1407.3867 (accessed on 28 May 2024).
7. Nedelcu, I.-G.; Opranescu, V.; Chiriac, B.-N.; Ionita, A.D. Educational support for automated classification of UML diagrams using machine learning. In *Generative Intelligence and Intelligent Tutoring Systems*; Springer: Berlin/Heidelberg, Germany, 2024; pp. 231–245. [CrossRef]
8. Dosovitskiy, A.; Beyer, L.; Kolesnikov, A.; Weissenborn, D.; Zhai, X.; Unterthiner, T.; Houlsby, N. An Image is Worth 16×16 Words: Transformers for Image Recognition at Scale. In Proceedings of the International Conference on Learning Representations (ICLR), Vienna, Austria, 4 May 2021; Available online: https://openreview.net/pdf?id=YicbFdNTTy (accessed on 28 May 2024).
9. Wu, F.; He, J.; Cai, L.; Du, M.; Huang, M. Accurate multi-objective prediction of CO_2 emission performance indexes and industrial structure optimization using multihead attention-based convolutional neural network. *J. Environ. Manag.* **2023**, *337*, 117759. [CrossRef] [PubMed]
10. Object Management Group. *Unified Modeling Language (UML)*; Version 2.5.1; OMG Document Number: Formal/2017-12-05; Object Management Group: Needham, MA, USA, 2017.
11. UML Diagrams. Available online: https://www.uml-diagrams.org/ (accessed on 10 May 2024).
12. Robles, G.; Fernández, M.A.; Pérez, J.; González-Barahona, J.M. An extensive dataset of UML models in GitHub. In Proceedings of the 2017 IEEE/ACM 14th International Conference on Mining Software Repositories (MSR), Buenos Aires, Argentina, 20–21 May 2017; pp. 519–522. [CrossRef]
13. Robles, G. GitHub UML Study Scripts. *LibreSoftTeam/2016-uml-miner*. GitHub. Available under GPLv2 or Later License. 2016. Available online: https://github.com/LibreSoftTeam/2016-uml-miner (accessed on 10 May 2024).
14. Google Images Search. (n.d.). *Google Images Search: A Python Library for Searching Images Using the Google Custom Search API. PyPI.* Available online: https://pypi.org/project/Google-Images-Search/ (accessed on 10 May 2024).
15. OpenCV. About. OpenCV. 2024. Available online: https://opencv.org/about/ (accessed on 1 August 2024).
16. Vijay, K.; Priyanka, G. Importance of statistical measures in digital image processing. Mean, Standard Deviation. *Int. J. Emerg. Technol. Adv. Eng.* **2012**, *2*, 56–62.
17. Raschka, S. *Python Machine Learning*; Packt Publishing Ltd.: Birmingham, UK, 2015.
18. Goodfellow, I.; Bengio, Y.; Courville, A. *Deep Learning*; MIT Press: Cambridge, MA, USA, 2016.
19. Srivastava, N.; Hinton, G.; Krizhevsky, A.; Sutskever, I.; Salakhutdinov, R. Dropout: A simple way to prevent neural networks from overfitting. *J. Mach. Learn. Res.* **2014**, *15*, 1929–1958.
20. Ioffe, S.; Szegedy, C. Batch normalization: Accelerating deep network training by reducing internal covariate shift. In Proceedings of the International Conference on Machine Learning, Lille, France, 6–11 July 2015; pp. 448–456.
21. Choi, E.; Bahadori, M.T.; Schuetz, A.; Stewart, W.F.; Sun, J. Doctor AI: Predicting clinical events via recurrent neural networks. In Proceedings of the Machine Learning for Healthcare Conference, Los Angeles, CA, USA, 19–20 August 2016; pp. 301–318.
22. Ding, X.; Zhang, Y.; Liu, T.; Duan, J. Deep learning for event-driven stock prediction. In Proceedings of the Twenty-Fourth International Joint Conference on Artificial Intelligence, Buenos Aires, Argentina, 25–31 July 2015.
23. LeCun, Y.; Bengio, Y.; Hinton, G. Deep learning. *Nature* **2015**, *521*, 436–444. [CrossRef] [PubMed]
24. Schmidhuber, J. Deep learning in neural networks: An overview. *Neural Netw.* **2015**, *61*, 85–117. [CrossRef] [PubMed]
25. Chollet, F. Keras. 2015. Available online: https://keras.io (accessed on 10 May 2024).
26. Prechelt, L. Early stopping—But when? In *Neural Networks: Tricks of the Trade*; Springer: Berlin/Heidelberg, Germany, 1998; pp. 55–69.
27. Masters, D.; Luschi, C. Revisiting small batch training for deep neural networks. *arXiv* **2018**, arXiv:1804.07612.
28. Kingma, D.P.; Ba, J. Adam: A method for stochastic optimization. *arXiv* **2014**, arXiv:1412.6980.
29. Ruder, S. An overview of gradient descent optimization algorithms. *arXiv* **2017**, arXiv:1609.04747.
30. He, K.; Zhang, X.; Ren, S.; Sun, J. Deep residual learning for image recognition. In Proceedings of the IEEE Conference on Computer Vision and Pattern Recognition (CVPR), Las Vegas, NV, USA, 27–30 June 2016.
31. Nedelcu, I.-G.; Caramihai, S.I.; Mocanu, S.; Ionita, A.D. UML diagrams classification with MIMO and SISO neural networks. *UPB Sci. Bull. Ser. C Electr. Eng.* **2023**, *85*, 89.

applied sciences

MDPI

Article

Leveraging Large Language Models to Support Authoring Gamified Programming Exercises †

Raffaele Montella [1], Ciro Giuseppe De Vita [1], Gennaro Mellone [1], Tullio Ciricillo [1], Dario Caramiello [1], Diana Di Luccio [1], Sokol Kosta [2], Robertas Damaševičius [3], Rytis Maskeliūnas [3], Ricardo Queirós [4] and Jakub Swacha [5,*]

1 Department of Structures for Engineering and Architecture (DiSt), University of Naples "Parthenope", 80143 Naples, Italy; raffaele.montella@uniparthenope.it (R.M.); cirogiuseppe.devita@uniparthenope.it (C.G.D.V.); gennaro.mellone1@studenti.uniparthenope.it (G.M.); tullio.ciricillo@studenti.uniparthenope.it (T.C.); dario.caramiello001@studenti.uniparthenope.it (D.C.); diana.diluccio@uniparthenope.it (D.D.L.)

2 Department of Electronic Systems, Aalborg University, 2450 Copenhagen, Denmark; sok@es.aau.dk

3 Center of Excellence Forest 4.0, Faculty of Informatics, Kaunas University of Technology, 51423 Kaunas, Lithuania

4 Center for Research in Advanced Computing Systems (CRACS), INESC TEC, 4169-007 Porto, Portugal; ricardoqueiros@esmad.ipp.pt

5 Department of Information Technology in Management, University of Szczecin, 70-453 Szczecin, Poland

* Correspondence: jakub.swacha@usz.edu.pl

† This paper is an extended version of our paper published in Proceedings of the 25th International Conference on Artificial Intelligence in Education (AIED 2024), Recife, Brazil, 8–12 July 2024 as a part of the Late-Breaking Results track, under the title "GAMAI, an AI-Powered Programming Exercise Gamifier Tool".

Featured Application: The presented solution can be applied to simplify and hasten the development of gamified programming exercises conforming to the Framework for Gamified Programming Education (FGPE) standard.

Abstract: Skilled programmers are in high demand, and a critical obstacle to satisfying this demand is the difficulty of acquiring programming skills. This issue can be addressed with automated assessment, which gives fast feedback to students trying to code, and gamification, which motivates them to intensify their learning efforts. Although some collections of gamified programming exercises are available, producing new ones is very demanding. This paper presents GAMAI, an AI-powered exercise gamifier, enriching the Framework for Gamified Programming Education (FGPE) ecosystem. Leveraging large language models, GAMAI enables teachers to effortlessly apply storytelling to describe a gamified scenario, as GAMAI decorates natural language text with the sentences needed by OpenAI APIs to contextualize the prompt. Once a gamified scenario has been generated, GAMAI automatically produces exercise files in a FGPE-compatible format. According to the presented evaluation results, most gamified exercises generated with AI support were ready to be used, with no or minimum human effort, and were positively assessed by students. The usability of the software was also assessed as high by the users. Our research paves the way for a more efficient and interactive approach to programming education, leveraging the capabilities of advanced language models in conjunction with gamification principles.

Keywords: gamification; programming education; educational tools; artificial intelligence

1. Introduction

The rising demand for skilled programmers reflects the pivotal role of programming in various areas of the contemporary world [1]. However, the formidable challenge of learning programming is an obstacle to meeting this demand. The integration of automated assessment tools has shown promise in providing swift feedback to students as they experiment with their code, facilitating a more efficient learning process [2]. In tandem, gamification has become a powerful strategy to enhance motivation and engagement, intensifying students' efforts to master programming concepts [3]. Together, these approaches form the backbone of the Framework for Gamified Programming Education (FGPE) ecosystem [4]. The practical usefulness of the FGPE relies on the availability of gamified programming exercises covering diverse topics.

In the context of this work, we define a gamified programming exercise as an exercise aimed at testing the student's ability to solve a specified problem, with the use of any common or peculiar programming language feature or features, embedded in a narrative scenario (considered as a gamification element, following, e.g., [5–7]), in which solving the exercise is part of a wider challenge. On the technical level, a gamified programming exercise is a programming exercise enhanced with an additional layer of metadata, defining, e.g., the conditions for accessing the exercise and the rewards for solving it, conforming to a standardized format (e.g., GEdIL [8]), making it possible to use the exercise within a gamified interactive learning environment, such as the FGPE PLE [4].

Although there are open-licensed collections of these exercises [9], more are needed to let students practice the same techniques in different contexts or to provide each student with exercises different from their peers. Creating new gamified exercises remains a significant challenge for educators, teachers, and trainers [4].

This paper, an extended version of our late-breaking work presented at the 25th International Conference on Artificial Intelligence in Education (AIED 2024) [10], addresses this challenge by introducing GAMAI, an artificial intelligence (AI)-powered exercise gamifier integrated into the Framework for Gamified Programming Education (FGPE) ecosystem. GAMAI leverages OpenAI API [11] to streamline the process of creating gamified programming exercises. Through GAMAI, teachers can employ a storytelling approach to articulate the gamified scenario. The system automatically enhances natural language text with essential sentences required by OpenAI API to contextualize the prompt. Once the gamified scenario has been generated, GAMAI further automates the production of specification files that can be easily edited within FGPE AuthorKit [4]. This innovative approach aims to alleviate the burden on educators, facilitating the seamless creation and integration of gamified programming exercises into educational curricula. The novel contribution of this paper is the processing approach to text prompt augmentation. The proposed algorithm decorates the teachers' gamified storytelling, to enable an automatic gamified programming exercise playground.

The rest of this paper is organized as follows: Section 2 gives an overview of using AI for the automatic generation of programming exercises; Section 3 describes the Framework for Gamified Programming Education ecosystem targeted by GAMAI, how GAMAI was designed and implemented, and how it operates; a qualitative evaluation of the exercises generated with GAMAI is reported in Section 4; finally, the paper content is summarized in the final Section 5, with concluding remarks and an outline of the next steps planned for future research.

2. Related Work

AI tools are pivotal in enhancing the teaching–learning process of computer programming by providing personalized, adaptive, and interactive learning experiences [12]. In programming education, AI can assist educators and learners in various aspects, including automated feedback and code review [13], adaptive learning paths [14], programming tutoring systems [15], automated assessment and grading [16], code summarization and documentation [17], and code generation [18–21].

Another way of exploiting AI's capabilities is by generating programming exercises. There is a notable interest in such generators, as creating programming exercises that are diversified and challenging for learners can be time-consuming for educators. The generated exercises can be tailored to cater to different skill levels, ensuring a dynamic learning environment that adapts to individual learner needs [22].

Kurdi et al. [23] conducted a systematic review of automatic exercise generation in many different domains, such as geometry, history, logic, programming, and science. It included 93 papers between 2015 and 2019 that tackled the automatic generation of exercises for educational purposes. The study concluded that there needs to be more tools for generating exercises of controlled difficulty and with facilities like enriching question forms and structures, automating template construction, improving presentation, and generating feedback.

Zavala and Mendoza [24] presented a tool (version 1.0) that uses automatic item generation (AIG) to address the problem of creating many similar programming exercises using predefined templates that are used for quizzes. The main goal was to ensure consistency in testing many students with questions of the same difficulty level.

Goliath (version 1.0) is an application that automatically generates programming exercises using a template system. With a Python programming exercise generator, Goliath targets the facilitation of teachers' day-to-day tasks. The tool creates templates, employing two AI models—one for generating the basic text of the statement from keywords, and another for extracting source code.

Agni (version 1.2) [25] serves as a dynamic code playground tailored for learning JavaScript. In recognition of the challenges posed by manual exercise creation, Agni's creators recently introduced a new back end with an exercise generation component powered by the ChatGPT API. This integration automates the exercise creation process by generating statements, solution code, and test cases. The tool supports the IMS LTI specification, allowing seamless integration with learning management systems (LMS) such as Moodle, Blackboard, and Canvas.

ExGen (version 1.0) [26] focuses on generating ready-to-use exercises for the specific difficulty level and concept the student is working on. It leverages the latest advances in LLMs to autogenerate many novel exercises and filter them to ensure they suit students.

TESTed (version 1.0) [27] is an educational testing framework that supports the creation of programming exercises with automated assessment capabilities in a programming-language-independent manner. TESTed combines the advantages of unit testing with output comparison, providing a versatile solution for educational assessment.

Recently, works have been increasingly leveraging pre-trained LLMs for educational purposes [28]. The most recent and relevant work [29] in automatic exercise generation using novel AI technologies (pre-trained LLMs) explored OpenAI Codex (which has been unavailable since March 2023) to create new programming exercises and code explanations. They found many Codex-generated exercises sensible and novel, but others needed clarification regarding problem statements and missing or faulty test cases.

On the other hand, Kasneci et al. [30] discussed the potential benefits, for instance, content generation and personalized learning, as well as challenges, e.g., model biases, system brittleness, etc., of applying LLMs to education. Similarly, Becker et al. [22] elaborated on the educational opportunities of AI code generation and how educators should act quickly given these developments.

Table 1 provides a comparative overview of some of the mentioned tools, based on their generation method, supported formats, and other relevant criteria.

Regarding the generation method, Goliath relies on a template-based approach using specific template fields for customization. Conversely, Agni integrates the ChatGPT API, enabling on-demand customization through dynamic interaction with the language model. ExGen leverages LLMs for exercise generation, while Zavala's tool employs semantic-based automatic item generation (AIG) to create contextual programming exercises dynamically. TESTed uses unit testing, combining the advantages of traditional testing methodologies with generic output comparison.

The tools present several output formats: Goliath generates exercises with DSL-formatted text, ensuring a structured and consistent output. Agni outputs exercises in JSON format, providing a machine-readable and versatile representation. ExGen produces exercises in natural language, maintaining simplicity and accessibility. The remaining tools' output format varies based on the semantic-based AIG and depending on the unit testing approach used.

Table 1. Comparison of automated programming exercise generation tools.

Tool	Goliath (Version 1.0)	Agni (Version 1.2)	ExGen (Version 1.0)	Zavala (Version 1.0)	TESTed (Version 1.0)
Generation Method	Template-based	ChatGPT API	LLM-based	Semantic-based AIG	Unit Testing
Variability and Customization	Template Fields	On-Demand Customization	On-Demand Customization	LOD Integration	Varies
Output Formats	DSL-formatted text	JSON	Natural language	Varies	Varies
Difficulty Levels and Tags	Difficulty/Tags	Difficulty/Tags	Difficulty	Difficulty	Difficulty
Integration with Environments	No	IMS LTI Integration	API Integration	API Integration	API Integration
Semantic-Based Generation	No	No	No	Yes	No

All tools allow the association of difficulty levels. This feature ensures instructors can customize exercises based on the student's proficiency levels. In Goliath and Agni, it is possible to categorize exercises according to specific topics or concepts.

Regarding interoperability, Agni integrates with learning environments such as learning management systems (LMS) using IMS LTI specifications. All the remaining tools (except Goliath) offer API integration, allowing easy incorporation into existing educational platforms.

Finally, regarding semantic-based generation, while Goliath, Agni, and TESTed do not explicitly focus on semantic-based generation, ExGen uses recent advances in pre-trained large language models (LLMs) for automatic exercise generation. Zavala's tool stands out by employing semantic-based AIG with linked open data integration, enhancing the contextual relevance of exercises.

While there are known examples of supporting gamification with AI methods [31], none of the existing tools and frameworks for programming exercise generation known to the authors (including those described above) are capable of generating gamified programming challenges, like GAMAI proposed here. The closest veins of research we were able to identify are the automatic generation of quizzes on the topic of software specifications and testing [32], and the automatic generation of challenges for gamification systems relatively distant in scope from programming education: ear training for music theory classes [33] and sustainable urban mobility [34]. It was this gap in the current research landscape that sparked our motivation to develop GAMAI [10], which is described in the following section.

3. Solution

3.1. Target Ecosystem: FGPE

The Framework for Gamified Programming Education (FGPE) has succeeded in establishing a technical milieu for incorporating gamification methods into programming education. It covers various aspects, including gamified exercise formats, exemplary collections, and the necessary supporting software [4]. The FGPE is programming-language-agnostic (allowing students to select their preferred programming language to tackle exercises), features multilingual instruction for students (who can effortlessly navigate between languages to obtain exercise requirements in the form most comprehensible to them) and extensive customization—encompassing both the learning content (educators can compose original courses, reuse exercises from the provided open repositories [9], or create brand new exercises, thus tailoring the learning content to the specific needs of the class or individual students) and the gamification rules (educators can select pre-existing gamified courses or create their own, drawing specifications of gamification rules from the provided courses or developing their own ones). It also provides integration with any LTI-compliant [35] learning management system (LMS) or massive open online course (MOOC), allowing for seamless synchronization of student identities, activities, and learning outcomes.

The key components of the FGPE ecosystem include:

- FGPE AuthorKit: serves the dual purpose of preparing and managing both programming exercises and gamification rules [4].
- GitHub-hosted Open Repository: serves as a centralized hub for gamified programming exercises [9].
- FGPE Gamification Service: processes gamification rules and manages the overall game state [36].
- Mooshak sandbox: executes programs submitted by students and autonomously assesses their performance [37].
- FGPE PLE (Programming Learning Environment): a progressive web app which lets students access gamified exercises, solve them, and receive graded feedback, whereas teachers can use the PLE to organize exercise sets, grant access to students, and monitor their learning progress [4].

The gamified programming exercises within the FGPE ecosystem are encapsulated in educational content in two distinct formats. These formats serve to articulate programming exercises and gamification layers, respectively. The first format, known as YAPExIL [38], is dedicated to meticulously describing programming exercises, ensuring clarity and precision in their presentation. On the other hand, the second format, GEdIL [8], is specifically tailored for the representation of gamification layers, providing a structured framework for integrating game elements into educational materials.

3.2. Design and Implementation

Figure 1 provides a comprehensive visualization of the overarching Framework for Gamified Programming Exercises (FGPE) architecture, emphasizing the innovative contribution of the proposed AI-powered exercise gamification solution. Designing, implementing, training, and reinforcing a dedicated generative pre-trained transformer-based large language model from scratch was outside of the scope of this research, so we designed GAMAI with leveraging OpenAI API [11] usage in mind. However, we considered a possible scenario in which different resources of a similar kind are used. The GPT Abstraction Layer component abstracts the interface to OpenAI API, enabling FGPE developers to test GAMAI with diverse and different large language model services. GAMAI uses generative AI LLMs to help creators implement exercises and challenges. Although the technical interactions between the GPT abstraction layer and the actual AI service provider are similar for both cases, the prompt engineering is different, as explained below.

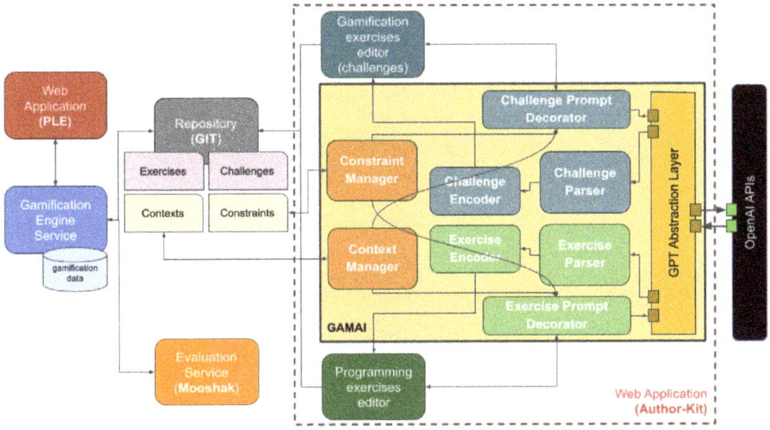

Figure 1. GAMAI placed within the FGPE architecture.

AI-assisted programming exercise creation. GAMAI's main aim is to assist teachers, educators, and trainers in creating gamified exercises for programming education. In this work, the creator is the human interacting with GAMAI to create the gamified exercises and the related gamified scenario. That means the creator's expertise is crucial in setting up the context in which the generative AI model has to imitate human behavior. The OpenAI GPT generative model used to develop the GAMAI prototype enables context settings using the "act as" clause. For example, we can consider a computer science research assistant who has to generate exercises for a first-year introduction to computing programming using the C language for bachelor students in a computer science degree class. The generative AI model must reflect the creator's cultural background and experience, acting appropriately. The AI generative model context setting sentence has to be assembled as a well-formed context selection prompt for the GPT service containing the following *act as* selectors: (i) the background skills of the creator; (ii) the kind of class for which the exercises have to be created; and (iii) the level of the exercises. In a different scenario, the creator could be a full professor of computer science who has to generate exercises for advanced algorithms and data structures using Python programming for a master's degree in computer engineering. Contexts are stored in the project repository and managed by the Context Manager component. For AI-based programming exercises, the context is consistent by assembling it in each session, ensuring exercises created by different creators for the same target audience are consistent.

While the nature of the exercises can vary widely, the kind of gamification augmentation is limited by the FGPE core implementations. A typical kind of exercise gamification leverages approaches like, but not limited to, (i) fill the gap; (ii) find the bug; or (iii) complete the code. Finally, the exercise is evaluated by the Mooshak component; thus, it has to be constrained so that the evaluation service can correctly perform scoring. To prepare a prompt considering this constraint, we developed a Constraint Manager component leveraging diverse and different constraints stored in the project repository alongside the Exercises, Challenges, and Contexts above.

AI-assisted gamified scenario creation. To create gamified challenges, the creator's skills could belong to a completely different area, in particular, focused on digital humanities, communications, and professional storytelling. Before each GPT service interaction, a prompt has to be built, performing a decoration stage in which the critical points of the "act as" statement are formed. For example, the creator can be an expert in fantasy saga writing, defining a story's characters and main plot. The plot branching and twists can be mapped on gamified exercises, adding custom fields to the results. Once the students have solved a quest, the story continues along a different branch. The Context Manager and the Constraint Manager components provide the prompt decoration with data from the project repository needed to set the context automatically, ensuring the consistency of the automatically generated content.

The Exercise Prompt Decorator is the component devoted to adding the needed ancillary information to the creator-generated prompt describing the kind of exercise. Listing 1 represents the prompt produced by this component. Line 1 is set by gathering the exercise creator skill from the project setup (default exercise creator skills) or the exercise collection level. Line 2 is directly written by the exercise creator, specifying the number and the kind of exercises to be generated. Line 3 can be assembled automatically using the GAMAI exercise prompt user interface or from details described by the exercise creator. Finally, line 4 is automatically added to retrieve results directly in JSON, to simplify the parsing of the results.

Listing 1. Gamified exercise generation prompt example (1).

```
Acting as a research assistant in computer science.
Generate 10 gamified programming exercises of increasing difficulty
    for a first-year introduction to computing programming using the
    C language for bachelor students in a computer science degree
    class.
The answer to the exercise's correct solution has to be chosen from a
    set of 5.
Return the result in json.
```

The Challenge Prompt Decorator is the component devoted to adding the main plot and the kind of quests, with related plot branching and plot twisting, to the creator-generated prompt describing the gamification scenario. Listing 2 represents the prompt produced by this component. As for the Exercise Prompt Decorator, line 1 is gathered from the contexts repository at the project level. In the example, the creator's skills are not technical. Line 2 is written directly by the content author and is the main statement for the AI-assisted world generation. Line 3 is a statement provided by the content creator specifying if each student (in the gamified and real worlds) has to compete alone or in teams. Line 4 states how the quest concludes and if all the participants can win the award or only one. Line 5 is the quest generation statement. The creator writes it using the GAMAI challenges prompt user interface. In line 6, one can observe the kinds of plot branches from one quest to the next. In this example, the quests have to be solved in sequential order. Line 7 instructs the GPT provider to generate the results in JSON. Finally, line 8 defines how the quest has to be connected to the exercise repository, the exercise kind (for example, multiple choice), and the exercise's difficulty level.

Listing 2. Gamified scenario generation prompt example.

```
1  Acting as an expert fantasy storyteller.
2  Create a world where young boys and girls in high school attend
       wizarding schools.
3  Each student has to compete alone to solve a quest.
4  Each student solving the final quest wins the wizard certificate.
5  Generate 10 different magic quests of increasing difficulty.
6  The wizard student who solves the first quest can continue with the
       second one, and so on.
7  Return the result as json.
8  Add to each quest the field:
9  ''exercise'': {
10     ''repository'':''__world__'',
11     ''kind'':''multiple_choices'',
12     ''level'': ''__iteration___''
13 }
```

Once the Exercise Prompt Decorator and the Challenge Prompt Decorator have provided the decorated prompts, the provided GPT service is invoked and the received AI-generated content has to be parsed by the Exercise Parser and the Challenge Parser. Both components have a similar behavior, interpreting the results provided by the GPT service via the GPT Abstraction Layer. The parsers map the data produced by the GPT services onto FGPE entities, providing semantic consistency. The gamified exercises and scenarios produced by the GPT service could be described without considering the peculiarity of the FGPE ecosystem. The Exercise Encoder component acts as a data sink for the Exercise Parser, producing an exercise representation that is fully compliant with the FGPE gamified exercise schema (YAPExIL [4]). Thanks to this compatibility, the exercise creator can use the AuthorKit tool to fully manage the exercises by adding or removing functional and ancillary parts. This component enables the creator to, for example, decorate the exercise description with diverse and different input dataset providers or custom exercise correctness checkers. The Challenge Encoder plays a pivotal role in the processing pipeline by receiving data generated from the Challenge Parser. Its primary function is to transform its input data into a gamified scenario representation that strictly conforms to the FGPE gamified scenario schema (GEdIL [8]). It ensures a standardized and cohesive structure for the gamified exercises within the framework. As was the case with exercises, AuthorKit affords exercise creators extensive control over the gamified scenarios, providing the flexibility to add or remove both functional and ancillary components, thus allowing for a tailored and dynamic exercise design process. Whether incorporating specific functionalities to enhance the challenge or streamlining the scenario by eliminating non-essential elements, AuthorKit serves as a versatile interface for refining the AI-generated gamified content. In essence, the Challenge Encoder and AuthorKit's Challenge Editor together form a cohesive unit within the FGPE architecture. While the former ensures adherence to the established gamified scenario schema, the latter provides a user-friendly and feature-rich environment for exercise creators to fine-tune and customize gamified scenarios according to their pedagogical objectives and creative preferences. This integrated approach underscores the commitment to standardization and flexibility in designing and implementing gamified programming exercises. As we explore further functionalities and interactions within this architecture, the synergistic relationship between these components becomes increasingly apparent, contributing to the effectiveness and adaptability of the overall GAMAI-enhanced framework.

3.3. Operation

GAMAI seamlessly integrates cutting-edge technologies, focusing on leveraging OpenAI's robust API. The user-facing front end establishes a nuanced dialogue with the back end, which is engineered to handle request processing and integrate with OpenAI's advanced capabilities for exercise creation. As the core computational engine, the back end diligently manages user intent transmission to OpenAI and intelligently interprets the AI-generated responses. The diagram in Figure 2 depicts the complete sequence.

The subsequent phase involves the back end transitioning from a relay to an intelligent parsing engine, proficiently dissecting the received messages and extracting essential components for exercise creation. The back end synthesizes exercises, strictly adhering to the YAPExIL format. A critical milestone is achieved, as the back end molds OpenAI's raw output and dispatches a curated list of exercises to the front end. This achievement signifies the completion of a complex orchestration and delivers a tangible educational resource to end-users, featuring exercises tailored to computer-science concepts. The user-centric front end acts as the interactive interface, empowering users to influence the exercise generation process actively. Acknowledging user agency and integrating it into the system, the interface allows users to articulate bespoke requests directly to OpenAI API, injecting personalized dimensions into the exercise generation pipeline. This user-initiated input allows imbuing exercises with contextual relevance, aligning educational content precisely with computer science students' nuanced demands and preferences. This meticulously designed system stands at the forefront of AI-driven educational technology in computer science, redefining the way educational content is created. The intricate interplay of components, spanning front-end interactions, back-end orchestration, and AI-output parsing, holds transformative potential for the pedagogical aspects of computer science education by automating the process with precision, customization, and engagement.

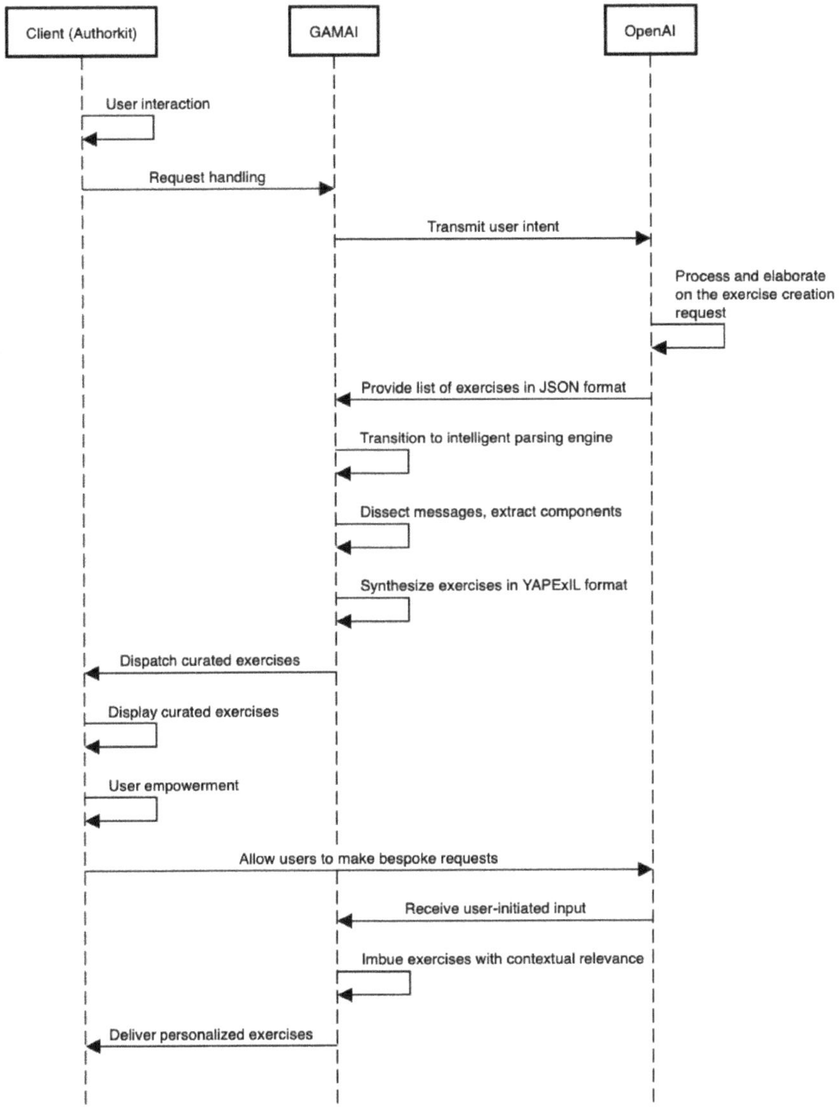

Figure 2. GAMAI sequence diagram while creating an exercise using OpenAI API.

4. Evaluation and Discussion

4.1. Evaluation Methods and Procedures

Although generating a set of exercises automatically is much less time-consuming for the creators than developing them manually, the automatically generated exercises often need to be refined to meet expectations, which still requires some human effort.

Listing 3 contains one of the exercises produced as the result of the prompt described in Listing 4.

Listing 3. Example of an AI-generated gamified exercise (1).

```
...
{
  ''question'': ''Which of the following is the correct way to
      declare an integer variable named 'count' in C?'',
  ''options'': [
    {''value'': ''int count;'', ''correct'': true},
    {''value'': ''integer count;'', ''correct'': false},
    {''value'': ''count int;'', ''correct'': false},
    {''value'': ''declare count as int;'', ''correct'': false},
    {''value'': ''variable count is int;'', ''correct'': false}
  ]
},
...
```

Listing 4. Gamified exercise generation prompt example (2).

```
1 Acting as a seasoned full professor in computer science.
2 Generate 100 gamified programming exercises of increasing difficulty
      for a
3 ...
4 The answer to the exercise's correct solution has to be chosen from a
      set of 5.
5 Return the result in json.
```

Because, to the best of our knowledge, a standardized procedure for evaluating the quality of automatically generated human language exercise texts has not yet been made available in the literature, we had to define a tentative protocol as follows:

Considering as exercise *text* all the human language content, assigning a score, in parentheses, for each needed fix, we classified the efforts as follows:

1. We set up a team of experts composed of five assistant professors in computer science for human-based computer-generated text evaluation.
2. We set up the evaluation grid in Table 2 to score the automatically generated exercise texts. Our effort was focused on trying to homogenize the evaluation provided by the team of experts.
3. In order to asses the human effort needed to make the automatically generated gamified exercises usable, we choose to generate 10 exercises for each of the following prompts as part of line three of Listing 4:
    - first-year introduction to computer programming using the C language for bachelor students in a computer science degree class [1CCS];
    - first-year introduction to computer programming using the Python language for bachelor students in a computer science degree class [1PCS];
    - first-year introduction to computer programming using the Python language for bachelor students in an environmental science degree class [1PES];
    - first-year introduction to computer programming using the Python language for bachelor students in a law degree class [1PLS];
    - third-year object-oriented computer programming using the Java language for bachelor students in a computer engineering degree class [3JCE];
    - third-year object-oriented computer programming using the Java language for bachelor students in a computer science degree class [3JCS].
4. Finally each member evaluated all the AI automatically generated programming exercise texts, scoring each one with a value from 0 to 5; then, for each exercise, the mean score was evaluated and then rounded to the closest integer.

The AI-generated exercise types (1CCS, 1PCS, 1PES, 1PLW, 3KCE, and 3JCS) were chosen to match the evaluation team's direct experiences in dealing with the related topics.

We are aware that this could have been a source of bias, but at the current stage of our studies, we preferred to maximize the overall consistency in exercise quality evaluation. The results discussed in the following section have been grouped by exercise type.

Table 2. The evaluation grid for AI-generated programming exercise texts.

Score	Evaluation
0	The exercise can be used as is
1	The text is unclear, it needs rearrangement
2	The text must be completely rewritten, but the exercise is formerly correctly generated
3	The text is correct, but there are issues from a technical point of view
4	The exercise, although correct in the text part, is too easy or too hard considering the intended difficulty level
5	The exercise cannot be used as is; it must be dropped or completely rewritten; for example, a code snippet is entirely missing

4.2. Results and Discussion

The experiment results (Table 3) show that most generated exercises were ready to be used with no or minimum human effort to make them comply with the expectations (61.7%).

Nevertheless, there were non-neglectable cases in which the code snippet had to be manually added, especially for high-difficulty-level exercises (28.3%).

Cases of entirely useless automatic generations were uncommon (10.0%).

Another point to focus on is that the generated exercise quality was related to the kind of programming language. Most of the provided C-language exercises were ready to be used, whereas more Python exercises needed the human creator's intervention.

Table 3. Human effort needed to ensure usability of created gamified exercises. Experimental results.

Prompt	0	1	2	3	4	5	Score
1CCS	5	4	0	1	0	0	7
1PCS	4	0	0	4	2	0	20
1PES	6	0	0	4	0	0	12
1PLW	3	1	2	3	0	1	19
3JCE	4	3	0	3	0	0	12
3JCS	6	1	0	0	3	0	13
%	40.0	21.67	3.33	25.0	8.33	1.67	

The improvement in the speed of the exercise generation and gamification layer creation is beyond doubt.

Nevertheless, some discussion about using the FGPE and GAMAI in a real-world production environment is needed. As shown by the example given in this section, the text generated by the GPT service could be affected by inaccuracies, biases, and intellectual property issues.

While checking each automatically generated exercise is a task that could be considered mandatory for the content creator, and tedious but not complicated to perform, the same cannot be said about the quality check verification of the world created for the gamification scenarios' automatic generation. Arguably, due to the strategy used by Open AI to train and reinforce their models, they are strongly influenced by popular culture, causing a potential issue regarding copyright management.

For example, in Listing 5, there is a field named "wizarding_schools" that conveys the name of three different schools of wizardry in the fantasy world of Eldoria [39]. In the used prompt (Listing 2), no mention was made of this entity, while the result produced is conceptually and numerically similar to literature protected by intellectual property rights. Note that this particular problem has also been observed in other use cases of Open AI's models [40].

Listing 5. Example of an AI-generated gamified exercise (2).

```
{
  ''world'': {
    ''name'': ''Eldoria'',
    ''description'': ''A magical realm where wizarding schools are
        scattered across mystical landscapes, and young sorcerers
        engage in solo quests to prove their magical prowess.''
  },
  ''wizarding_schools'': [...],
  ''quests'': [
    {
      ''name'': ''The Enchanted Amulet'',
      ''description'': ''Retrieve the Enchanted Amulet hidden within
          the Forbidden Forest, guarded by mystical creatures and
          protected by a powerful enchantment.'',
      ''exercise'': { ''repository'': ''__world__'', ''kind'': ''
          multiple_choice'', ''level'': ''__iteration__'' }
    },
...
```

5. Conclusions

This paper presents our findings from applying a large language model based on a generative pre-trained transformer to help generate gamified exercises designed for programming education, in a form compatible with the FGPE ecosystem. Given the complexity of designing, implementing, training, and reinforcing such natural language processing models, we opted to leverage the capabilities of a commercially available robust API service.

The results of the evaluation of the quality of the generated exercises presented in Section 4 are positive, with most of the exercises being usable as they are or only requiring slight corrections.

We are nonetheless aware of the limitations of our research. Threats to the validity and generalizability of the presented results stem from both the limited number of the evaluated exercises (60 in total) and the subjective character of the experts' evaluation. It is possible that another batch of generated exercises could contain more flaws, and another group of experts could evaluate the same exercises more harshly.

Despite these research limitations, with GAMAI, we have successfully demonstrated that the time needed for developing gamified programming exercises can be significantly reduced by employing LLMs, which motivates further research on this topic.

Our future research will focus on two primary avenues. First, we aim to refine the components responsible for prompt decoration and parsing the generated text. The ultimate goal is to greatly minimize or, if possible, completely eliminate the need for human intervention within the editing process, thereby streamlining the workflow. Second, we envision the creation of a formally defined test suite to evaluate the performance of student learning with gamified exercises generated with the support of GAMAI, comparing it with the performance of students learning with manually developed gamified exercises. Such a research would provide valuable insights into the efficacy of incorporating automatically generated gamified content into educational settings.

Overall, our research paves the way for a more efficient and interactive approach to programming education, leveraging the capabilities of advanced language models in conjunction with gamification principles.

Author Contributions: Conceptualization, R.M. (Raffaele Montella) and J.S.; methodology, R.M. (Raffaele Montella); software, T.C. and D.C.; validation, C.G.D.V. and G.M.; investigation, R.M. (Raffaele Montella), C.G.D.V. and G.M.; resources, C.G.D.V., G.M. and D.D.L.; data curation, C.G.D.V., G.M. and D.D.L.; writing—original draft preparation, R.M. (Raffaele Montella), D.D.L., J.S., R.Q., R.D. and R.M. (Rytis Maskeliunas); writing—review and editing, R.M. (Raffaele Montella), J.S. and S.K.; visualization, R.M. (Raffaele Montella), C.G.D.V., G.M. and J.S.; supervision, R.M. (Raffaele Montella); project administration, J.S.; funding acquisition, J.S. All authors have read and agreed to the published version of the manuscript.

Funding: This research was co-funded by the European Union, grant number 2023-1-PL01-KA220-HED-000164696.

Institutional Review Board Statement: Not applicable.

Informed Consent Statement: Informed consent was obtained from all subjects involved in the study.

Data Availability Statement: The data presented in this study are available on request from the first author.

Acknowledgments: The authors would like to thank José Carlos Paiva for his support in software development.

Conflicts of Interest: The authors declare no conflicts of interest. The funders had no role in the design of the study; in the collection, analyses, or interpretation of data; in the writing of the manuscript; or in the decision to publish the results.

References

1. Vicuña, C. What Are Europe's Most Needed Software Roles and Skills? Trends to Look Out For. 2022. Available online: https://digital-skills-jobs.europa.eu/en/latest/news/what-are-europes-most-needed-software-roles-and-skills-trends-look-out (accessed on 10 September 2024).
2. Paiva, J.C.; Leal, J.P.; Figueira, Á. Automated assessment in computer science education: A state-of-the-art review. *ACM Trans. Comput. Educ. (TOCE)* **2022**, *22*, 1–40. [CrossRef]
3. Zhan, Z.; He, L.; Tong, Y.; Liang, X.; Guo, S.; Lan, X. The effectiveness of gamification in programming education: Evidence from a meta-analysis. *Comput. Educ. Artif. Intell.* **2022**, *3*, 100096. [CrossRef]
4. Paiva, J.C.; Queirós, R.; Leal, J.P.; Swacha, J.; Miernik, F. Managing gamified programming courses with the FGPE Platform. *Information* **2022**, *13*, 45. [CrossRef]
5. Marczewski, A. 52 Gamification Mechanics and Elements. 2015. Available online: https://www.gamified.uk/user-types/gamification-mechanics-elements (accessed on 10 September 2024).
6. Werbach, K.; Hunter, D. *For the Win: How Game Thinking Can Revolutionize Your Business*; Wharton Digital Press: Philadelphia, PA, USA, 2012.
7. Chou, Y.K. *Actionable Gamification: Beyond Points, Badges, and Leaderboards*; CreateSpace: Scotts Valley, CA, USA, 2015.
8. Swacha, J.; Paiva, J.C.; Leal, J.P.; Queirós, R.; Montella, R.; Kosta, S. GEdIL—Gamified Education Interoperability Language. *Information* **2020**, *11*, 287. [CrossRef]
9. FGPE Consortium. Repository of Gamified Exercises. 2021. Available online: https://github.com/jcpaiva-fgpe?tab=repositories (accessed on 10 September 2024).
10. Montella, R.; Giuseppe De Vita, C.; Mellone, G.; Ciricillo, T.; Caramiello, D.; Di Luccio, D.; Kosta, S.; Damasevicius, R.; Maskeliunas, R.; Queiros, R.; et al. GAMAI, an AI-Powered Programming Exercise Gamifier Tool. In *Artificial Intelligence in Education: Posters and Late Breaking Results, Workshops and Tutorials, Industry and Innovation Tracks, Practitioners, Doctoral Consortium and Blue Sky—25th International Conference on Artificial Intelligence in Education (AIED 2024), Recife, Brazil, 8–12 July 2024, Proceedings, Part I*; Olney, A.M., Chounta, I.A., Liu, Z., Santos, O.C., Bittencourt, I.I., Eds.; Springer Nature: Cham, Switzerland, 2024; pp. 485–493. [CrossRef]
11. OpenAI. OpenAI API. 2020. Available online: https://openai.com/index/openai-api/ (accessed on 10 September 2024).
12. Freitas, T.C.; Costa Neto, A.; Pereira, M.J.V.; Henriques, P.R. NLP/AI Based Techniques for Programming Exercises Generation. In Proceedings of the 4th International Computer Programming Education Conference (ICPEC 2023), Dagstuhl, Germany, 26–28 June 2023. [CrossRef]
13. Messer, M.; Brown, N.C.C.; Kölling, M.; Shi, M. Automated Grading and Feedback Tools for Programming Education: A Systematic Review. *ACM Trans. Comput. Educ.* **2023**, *24*, 10. [CrossRef]

14. Shaka, M.; Carraro, D.; Brown, K.N. Personalised Programming Education with Knowledge Tracing. In Proceedings of the 2023 Conference on Human Centered Artificial Intelligence: Education and Practice (HCAIep '23), New York, NY, USA, 14–15 December 2023; p. 47. [CrossRef]
15. Le, N.; Strickroth, S.; Gross, S.; Pinkwart, N. A Review of AI-Supported Tutoring Approaches for Learning Programming. In *Advanced Computational Methods for Knowledge Engineering*; Springer: Cham, Switzerland, 2013; pp. 267–279. [CrossRef]
16. Combéfis, S. Automated Code Assessment for Education: Review, Classification and Perspectives on Techniques and Tools. *Software* **2022**, *1*, 3–30. [CrossRef]
17. Edelblut, P. Realizing the Promise of AI-Powered, Adaptive, Automated, Instant Feedback on Writing for Students in Grade 3–8 with an IEP. In *Adaptive Instructional Systems: Second International Conference, AIS 2020, Held as Part of the 22nd HCI International Conference, HCII 2020, Copenhagen, Denmark, 19–24 July 2020, Proceedings*; Springer International Publishing: Berlin/Heidelberg, Germany, 2020; pp. 283–292. [CrossRef]
18. Svyatkovskiy, A.; Deng, S.K.; Fu, S.; Sundaresan, N. Intellicode compose: Code generation using transformer. In Proceedings of the 28th ACM Joint Meeting on European Software Engineering Conference and Symposium on the Foundations of Software Engineering, New York, NY, USA, 8–13 November 2020; pp. 1433–1443.
19. Chakraborty, S.; Ahmed, T.; Ding, Y.; Devanbu, P.T.; Ray, B. Natgen: Generative pre-training by "naturalizing" source code. In Proceedings of the 30th ACM Joint European Software Engineering Conference and Symposium on the Foundations of Software Engineering, New York, NY, USA, 14–16 November 2022; pp. 18–30.
20. Kazemitabaar, M.; Chow, J.; Ma, C.K.T.; Ericson, B.; Weintrop, D.; Grossman, T. Studying the effect of AI Code Generators on Supporting Novice Learners in Introductory Programming. *arXiv* **2023**, arXiv:abs/2302.07427. [CrossRef]
21. Jin, J.; Kim, M. GPT-Empowered Personalized eLearning System for Programming Languages. *Appl. Sci.* **2023**, *13*, 12773. [CrossRef]
22. Becker, B.A.; Denny, P.; Finnie-Ansley, J.; Luxton-Reilly, A.; Prather, J.; Santos, E.A. Programming is hard-or at least it used to be: Educational opportunities and challenges of ai code generation. In Proceedings of the 54th ACM Technical Symposium on Computer Science Education V. 1, New York, NY, USA, 15–18 March 2023; pp. 500–506.
23. Kurdi, G.; Leo, J.; Parsia, B.; Sattler, U.; Al-Emari, S. A Systematic Review of Automatic Question Generation for Educational Purposes. *Int. J. Artif. Intell. Educ.* **2019**, *30*, 121–204. [CrossRef]
24. Zavala, L.; Mendoza, B. On the Use of Semantic-Based AIG to Automatically Generate Programming Exercises. In Proceedings of the 49th ACM Technical Symposium on Computer Science Education (SIGCSE '18), New York, NY, USA, 21–24 February 2018; pp. 14–19. [CrossRef]
25. Bauer, Y.; Leal, J.P.; Queirós, R. Can a Content Management System Provide a Good User Experience to Teachers? In *Proceeding of the 4th International Computer Programming Education Conference (ICPEC 2023), Vila do Conde, Portugal, 26–28 June 2023*; Peixoto de Queirós, R.A., Teixeira Pinto, M.P., Eds.; Open Access Series in Informatics (OASIcs); Schloss-Dagstuhl-Leibniz Zentrum für Informatik: Dagstuhl, Germany, 2023; Volume 112, pp. 4:1–4:8. [CrossRef]
26. Ta, N.B.D.; Nguyen, H.G.P.; Swapna, G. ExGen: Ready-to-use exercise generation in introductory programming courses. In Proceedings of the 31st International Conference on Computers in Education Conference, Matsue, Japan, 4–8 December 2023; pp. 1–10.
27. Strijbol, N.; Van Petegem, C.; Maertens, R.; Sels, B.; Scholliers, C.; Dawyndt, P.; Mesuere, B. TESTed—An educational testing framework with language-agnostic test suites for programming exercises. *SoftwareX* **2023**, *22*, 101404. [CrossRef]
28. Damasevicius, R.; Sidekerskiene, T. AI as a Teacher: A New Educational Dynamic for Modern Classrooms for Personalized Learning Support. In *Advances in Educational Technologies and Instructional Design*; IGI Global: Hershey, PA, USA, 2024; pp. 1–24. [CrossRef]
29. Sarsa, S.; Denny, P.; Hellas, A.; Leinonen, J. Automatic generation of programming exercises and code explanations using large language models. In Proceedings of the 2022 ACM Conference on International Computing Education Research-Volume 1 (ICER 2022), Lugano, Switzerland, 7–11 August 2022; pp. 27–43.
30. Kasneci, E.; Seßler, K.; Küchemann, S.; Bannert, M.; Dementieva, D.; Fischer, F.; Gasser, U.; Groh, G.; Günnemann, S.; Hüllermeier, E.; et al. ChatGPT for Good? On Opportunities and Challenges of Large Language Models for Education. *Learn. Individ. Differ.* **2023**, *103*, 102274. [CrossRef]
31. Swacha, J.; Gracel, M. Machine Learning in Gamification and Gamification in Machine Learning: A Systematic Literature Mapping. *Appl. Sci.* **2023**, *13*, 11427. [CrossRef]
32. Vos, T.E.; Fraser, G.; Martinez-Ortiz, I.; Prada, R.; Silva, A.R.; Prasetya, I. Tutorial on a Gamification Toolset for Improving Engagement of Students in Software Engineering Courses. In Proceedings of the 32nd Conference on Software Engineering Education and Training (CSEE&T 2020), Munich, Germany, 9–12 November 2020; pp. 1–3.
33. Pesek, M.; Vučko, Z.; Savli, P.; Kavčič, A.; Marolt, M. Troubadour: A Gamified e-Learning Platform for Ear Training. *IEEE Access* **2020**, *8*, 97090–97102. [CrossRef]
34. Khoshkangini, R.; Valetto, G.; Marconi, A.; Pistore, M. Automatic generation and recommendation of personalized challenges for gamification. *User Model. User-Adapt. Interact.* **2021**, *31*, 1–34. [CrossRef]
35. IMS Global Learning Consortium. Learning Tools Interoperability Core Specification. 2019. Available online: http://www.imsglobal.org/spec/lti/v1p3/ (accessed on 10 September 2024).

36. Paiva, J.C.; Haraszczuk, A.; Queirós, R.; Leal, J.P.; Swacha, J.; Kosta, S. FGPE Gamification Service: A GraphQL Service to Gamify Online Education. In *Trends and Applications in Information Systems and Technologies*; Springer: Cham, Switzerland, 2021; Volume 4, pp. 480–489. [CrossRef]
37. Leal, J.P.; Silva, F. Mooshak: A Web-based multi-site programming contest system. *Softw. Pract. Exp.* **2003**, *33*, 567–581. [CrossRef]
38. Paiva, J.C.; Queirós, R.; Leal, J.P.; Swacha, J. Yet another programming exercises interoperability language. In Proceedings of the 9th Symposium on Languages, Applications and Technologies (SLATE 2020), Dagstuhl, Germany, 13–14 July 2020.
39. Done, K. *Encyclopedia Eldoria*; Comstar Media: Dover, UK, 2005.
40. Reddit. What Is AI's Obsession with "Eldoria?". 2024. Available online: https://www.reddit.com/r/ChatGPT/comments/1bb2mzm/what_is_ais_obsession_with_eldoria/ (accessed on 10 September 2024).

applied sciences

MDPI

Article

Research on Student Classroom Behavior Detection Based on the Real-Time Detection Transformer Algorithm

Lihua Lin [1,2], Haodong Yang [1,*], Qingchuan Xu [1], Yanan Xue [1] and Dan Li [1]

1 College of Communication and Information Technology, Xi'an University of Science and Technology, Xi'an 710054, China; linlihua@xust.edu.cn (L.L.); 22207223063@stu.xust.edu.cn (Q.X.); 22207223115@stu.xust.edu.cn (Y.X.); 22207223066@stu.xust.edu.cn (D.L.)

2 Xi'an Key Laboratory of Network Convergence Communication, Xi'an 710054, China

* Correspondence: haodong18393106162@outlook.com

Abstract: With the rapid development of artificial intelligence and big data technology, intelligent education systems have become a key research focus in the field of modern educational technology. This study aims to enhance the intelligence level of educational systems by accurately detecting student behavior in the classroom using deep learning techniques. We propose a method for detecting student classroom behavior based on an improved RT DETR (Real-Time Detection Transformer) object detection algorithm. By combining actual classroom observation data with AI-generated data, we create a comprehensive and diverse student behavior dataset (FSCB-dataset). This dataset not only more realistically simulates the classroom environment but also effectively addresses the scarcity of datasets and reduces the cost of dataset construction. The study introduces MobileNetV3 as a lightweight backbone network, reducing the model parameters to one-tenth of the original while maintaining nearly the same accuracy. Additionally, by incorporating learnable position encoding and dynamic upsampling techniques, the model significantly improves its ability to recognize small objects and complex scenes. Test results on the FSCB-dataset show that the improved model achieves significant improvements in real-time performance and computational efficiency. The lightweight network is also easy to deploy on mobile devices, demonstrating its practicality in resource-constrained environments.

Keywords: artificial intelligence; student behavior detection; FSCB-dataset; MobileNetV3; learned positional encoding; dynamic upsampling

1. Introduction

With the rapid development of artificial intelligence and big data, emerging technologies have become the core force driving innovation and development in education. These technologies are widely used to assist teaching [1,2], manage classrooms, and analyze student behavior. For example, applications like the "Rain Classroom" mini-program make attendance and questioning more convenient, while advanced monitoring technologies simplify classroom management and provide accurate data for teaching evaluations. In particular, student behavior detection technology, through deep learning models, offers new perspectives for understanding student performance and teaching quality in classrooms. Despite challenges in data collection and technical application in university settings, this research holds great potential and significant value in improving educational management efficiency and teaching quality. Future research in this field is expected to elevate the quality of education to new levels [3,4].

The rapid advancement of object detection technology in the field of computer vision has garnered widespread attention in academia, especially methods based on deep learning. These methods can be broadly categorized into three types: two-stage algorithms, single-stage algorithms, and recently emerging Transformer-based algorithms. Since the introduction of the AlexNet network with ReLU activation function and Dropout technology, deep learning has achieved significant progress in image recognition accuracy, marking a major breakthrough in the field [5]. Subsequently, the OverFeat algorithm demonstrated the potential of convolutional networks in object detection. The advent of R-CNN [6] and its derivative algorithms, Fast R-CNN [7] and Faster R-CNN [8], significantly improved object detection accuracy and speed by integrating convolutional neural networks with region proposal techniques, highlighting the application of deep learning in object detection. Following this, algorithms such as Mask R-CNN [9], Cascade R-CNN [10], and Libra R-CNN [11] further enhanced instance segmentation accuracy and the ability to handle complex scenes. In the realm of single-stage algorithms, the YOLO series [12–16] transformed the detection task into a regression problem, achieving fast and efficient object detection. Algorithms like SSD [17], RetinaNet [18], and EfficientDet [19] adopted various strategies to improve detection performance, especially in handling small objects and class imbalance issues. Recently, Transformer-based object detection algorithms, such as DETR [20] and its improved versions, Deformable DETR [21] and Conditional DETR [22], have demonstrated strong capabilities in processing global information and complex scenes by simplifying detection processes and optimizing training efficiency. Particularly, RT-DETR [23], designed for real-time object detection, enhances performance in accuracy and speed through optimized computational complexity and inference speed, making it highly suitable for detecting student classroom behaviors in complex scenes.

In recent years, student classroom behavior detection has become a research hotspot, focusing on three deep learning-based methods: sensor analysis, skeleton key point detection, and convolutional network detection. Sensor methods, for instance, use Kinect One to capture 2D and 3D data and evaluate student attention through machine learning techniques [24]. Skeleton key point detection identifies specific behaviors by tracking changes in student skeletal points, such as the OpenPose model recognizing hand-raising behavior [25–27]. Convolutional network detection methods, like YOLO, have improved behavior recognition accuracy and speed through algorithm enhancements [28–31].

Despite progress in student classroom behavior detection, the field still faces multiple challenges, such as outdated algorithms, insufficient detection efficiency, and a lack of diverse public datasets. The complexity of student behaviors, irregular seating arrangements, small object recognition difficulties, and variations in lighting and background need to be addressed. Additionally, existing studies often fail to cover student behaviors across various grades and scenes, limiting model generalization and practical applicability. Given these challenges, this study proposes using the RT-DETR algorithm for student classroom behavior detection and analysis. The algorithm aims to achieve efficient and accurate object detection through optimized computational complexity and inference speed, particularly suited for applications in complex scenes.

2. Dataset

2.1. Constructing the Fusion Student Classroom Behavior Dataset

To address the limitations of existing student behavior detection datasets in terms of singularity and scarcity, this study proposes an innovative data construction strategy—the fusion dataset method (FSCB dataset). By combining AI-generated data and actual student classroom behavior observation data, the FSCB dataset diversifies data sources. This method is similar to the Quality Function Deployment approach applied by Shvetsova et al. in product design concept selection, effectively enriching data resources and providing strong support for further behavior analysis and educational interventions [32]. The FSCB dataset includes 4711 images with 10,185 labels of student behaviors, covering actions such as focusing, raising hands, writing, sleeping, and using phones. Figure 1 illustrates the specific content and structure of the FSCB dataset, highlighting its broad scope and detail in capturing and labeling diverse classroom behaviors. This fusion dataset enriches the data resources for student behavior detection, providing strong support for further behavior analysis and educational interventions.

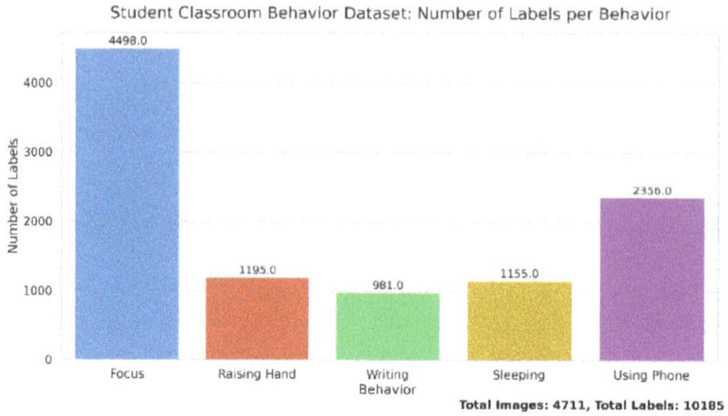

Figure 1. Dataset distribution.

2.2. Image Generation

During the construction of this fusion dataset, we leveraged cutting-edge text-to-image generation technologies, particularly the Stable Diffusion [33] and DALL-E 3 [34] models, to create simulated images that closely match actual student behavior data. This approach not only significantly expanded the scale of the dataset but also enhanced its diversity and coverage. By utilizing these advanced image generation techniques, we were able to produce highly realistic images that accurately reflect student behavior patterns and classroom environments.

Specifically, the Stable Diffusion and DALL-E 3 models, through advanced deep learning algorithms, generated images with a high degree of realism, depicting various classroom behaviors such as focusing, raising hands, writing, sleeping, and using phones. Figure 2 illustrates the diversity and detail level of these generated images, showcasing the models' capability to accurately recreate student behaviors in different scenarios.

Figure 2. AI-Generated data.

These simulated images provided a rich data foundation for the research, greatly enhancing the depth and breadth of the dataset. This not only helped improve the accuracy and efficiency of student behavior detection but also provided strong support for further behavior analysis and educational interventions. With this enriched data resource, we can gain a more comprehensive understanding and analysis of student behavior patterns in the classroom, thereby providing data support for improving teaching methods and enhancing educational quality.

2.3. Generalizability Analysis of the Fusion Dataset

To further enhance the dataset's practicality and application scope, we conducted a comprehensive comparative analysis of three different datasets: a dataset containing 1200 purely real-captured images (RSCB), a dataset containing 1200 completely AI-generated images (AISCB), and a fusion dataset (FSCB) containing 1200 images that combined AI-generated and real images. Each dataset included five different student behaviors, with 300 images for each behavior.

As shown in Figure 3, the experimental results indicated that the fusion dataset demonstrated similar accuracy and reliability to the purely real-captured dataset. This finding highlights the high applicability of the fusion dataset in student classroom behavior detection, showing that combining AI-generated images with real images can effectively enhance the diversity and coverage of the dataset.

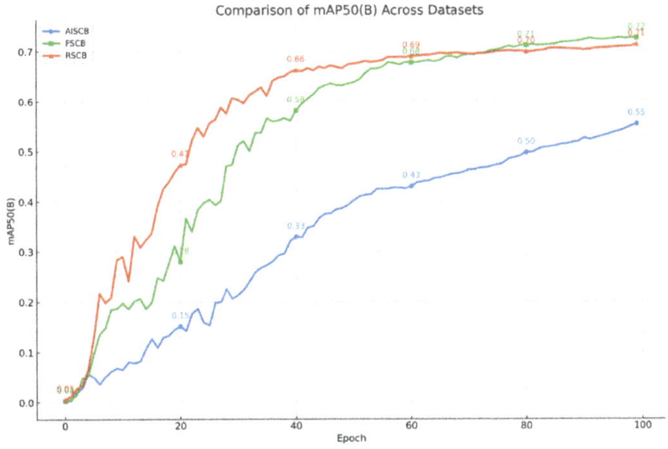

Figure 3. Dataset comparison experiment chart.

Moreover, the fusion dataset provides an effective solution to the challenges of dataset scarcity, student privacy protection, and rapid dataset construction. Traditional real-captured image datasets are often limited by high data collection costs, lengthy collection times, and privacy concerns. By incorporating AI-generated images, we can rapidly expand the dataset's scale while providing rich training data without infringing on student privacy.

Additionally, the high accuracy and reliability of the fusion dataset lay a solid foundation for subsequent behavior analysis and educational interventions. Utilizing such a multi-source dataset enables researchers to gain a more comprehensive understanding and analysis of student behavior patterns in the classroom, thereby providing critical data support for improving teaching methods and enhancing educational quality. This innovative dataset construction method showcases the immense potential of AI technology in the education sector, paving the way for future educational technology research and applications.

3. Methods

As shown in the algorithm flowchart in Figure 4, the FSCB dataset is input into the RT DETR algorithm, which processes the data through multiple steps to accurately recognize student behaviors. First, the algorithm uses a backbone network to extract fundamental features from the images, ensuring that key information is effectively captured. Next, the Encoder processes these extracted features using self-attention mechanisms and feature fusion, enhancing the richness and diversity of the feature representations. Finally, the Decoder merges and interprets all the information through object queries, enabling accurate localization and classification of objects in the images. Through this series of processing steps, the RT DETR algorithm can accurately identify five types of student behaviors in the classroom, including focusing, raising hands, writing, sleeping, and using a phone.

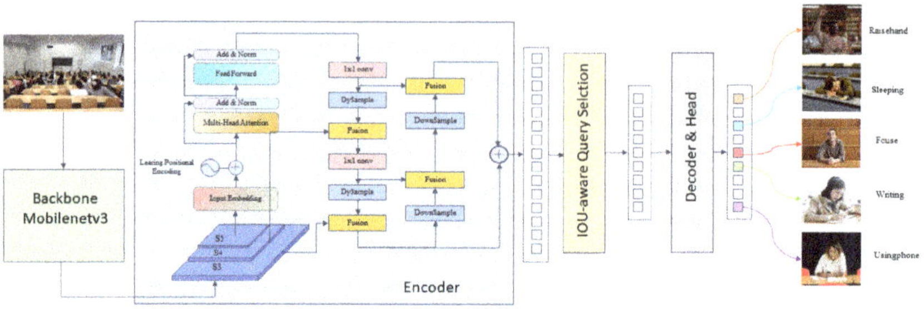

Figure 4. Flowchart of the improved RT DETE algorithm.

3.1. Structure of RT DETR Algorithm

RT DETR is a real-time object detection model that combines two classic object detection methods: Transformer [35] and DETR [20] (Detection Transformer). The Transformer is a neural network architecture used for sequence modeling, initially applied in natural language processing but proven effective in computer vision. DETR is an end-to-end object detection model that transforms the object detection task into an object query problem and uses the Transformer to solve it. RT DETR adopts the DETR structure and optimizes it for real-time object detection. It comprises three main components: backbone, Encoder, and Decoder. The backbone network, which uses a CNN network, extracts image features. The CNN network's powerful feature extraction capability is widely used in image processing tasks, providing necessary image representations for subsequent object detection tasks. The Encoder, based on the Transformer structure, processes the feature maps output by the backbone and captures global dependencies within the image through the self-attention mechanism. Compared to the traditional DETR model, RT DETR optimizes the Encoder design to reduce computational complexity and improve processing speed. These optimizations include reducing the number of self-attention layers and using more efficient attention calculation methods, aiming to maintain high precision while achieving fast inference. Each self-attention layer in the Encoder includes three main components: multi-head self-attention, feed-forward network, and residual connections with layer normalization. This design allows the model to effectively capture long-distance dependencies within the image, enhancing the understanding of complex scenes. The Decoder component receives the Encoder's output and predefined object queries, combining them through cross-attention and self-attention layers to progressively refine predictions of each object's location and category. Prediction heads at the end of the model perform object classification and bounding box regression based on the Decoder's output. These heads are simple and efficient, designed to quickly generate accurate detection results. Classification heads typically use linear layers, while bounding-box regression heads may include more complex network structures to accurately predict object locations.

3.2. Improvements to the RT DETR Algorithm

This study aims to improve the accuracy and efficiency of student classroom behavior recognition by enhancing the RT DETR model in multiple aspects. First, MobileNetV3, a lightweight backbone network, was introduced to reduce the number of model parameters and computational complexity. Second, learned positional encoding (LPE) was used to improve the positional encoding mechanism in AIFI, enhancing the model's ability to learn positional relationships in training data. Finally, the ultralight and efficient characteristics of DySample were integrated into the CCFM upsampling module to further enhance RT DETR's performance.

3.2.1. Improving the Backbone of the RT DETR Algorithm

The MobileNet series was proposed to address the issue of traditional convolutional neural networks requiring excessive memory and computational resources when running on mobile and embedded devices. In 2017, MobileNet emerged as a lightweight CNN network designed for mobile and embedded devices, aiming to significantly reduce the number of model parameters and computational load with minimal sacrifice in accuracy. By introducing the latest MobileNetV3, we further advanced efficiency and performance balance in the backbone part of the RT DETR model. Built on the foundations of MobileNetV2 and MobileNetV1, MobileNetV3 inherits and enhances their advantages, optimizing the model to set new standards in fast processing and low resource consumption. This progress means that more efficient models can better serve practical application scenarios, meeting the growing demands for mobile computing [36–38].

The design improvements in MobileNetV3 are reflected in innovations in the bottleneck blocks. These bottleneck blocks now embed a lightweight attention mechanism called the Squeeze-and-Excitation (SE) module. This mechanism enhances the network's response to key information by precisely calibrating the importance of each feature channel while naturally suppressing less important information, thereby improving accuracy while keeping the model lightweight. Additionally, the bottleneck block updates include a mixed depthwise separable convolution strategy, which cleverly combines convolution kernels of different sizes. This not only effectively captures multi-scale information in the image but also optimizes computational efficiency. Figure 5 shows the structural details of the improved bottleneck block, revealing its superior performance in executing multi-scale feature extraction.

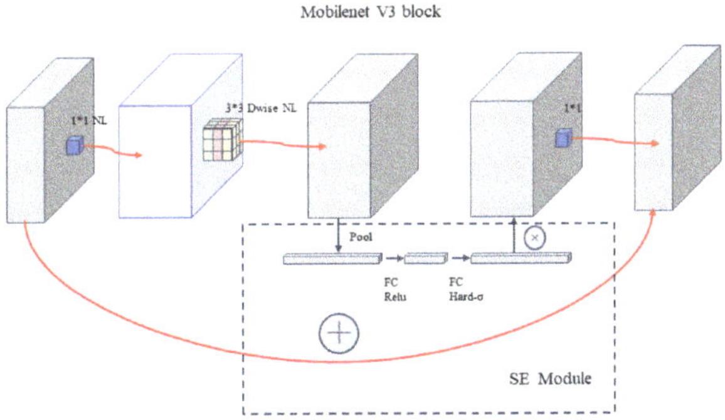

Figure 5. CompactFusion block.

MobileNetV3 reconstructs and optimizes the most computationally intensive layers in the model to reduce their computational costs while maintaining or improving model accuracy. This includes optimizing the first and last layers of the model, which are typically the most computationally expensive parts in many deep learning architectures. MobileNetV3's specific designs, such as adjusting the convolution kernel size, stride, and filter numbers, aim to reduce the computational burden of these layers. Figure 6 illustrates the streamlined design of time-efficient layers.

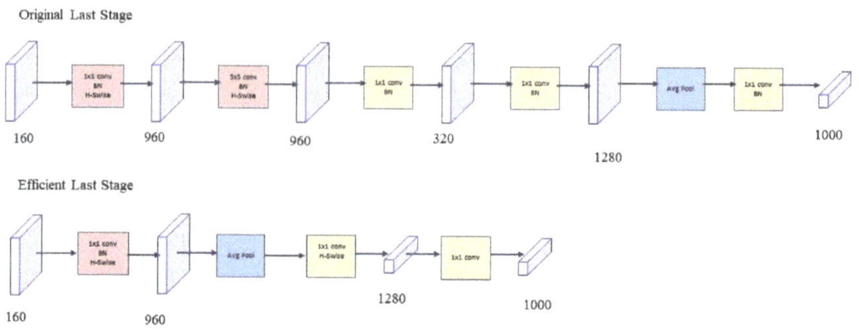

Figure 6. Streamlined Design of time-efficient layers.

In the design of MobileNetV3, researchers redesigned the activation function to adapt to mobile devices' computational constraints while enhancing performance, using h-swish instead of the swish function. The swish non-linear activation function, an alternative to ReLU, significantly improves neural network accuracy. Its definition is as follows:

$$\text{swish}(x) = x \cdot \sigma(x) \tag{1}$$

where x is the input, and $\sigma(x)$ is the sigmoid function of the input x, defined as $\sigma(x) = \frac{1}{1+e^{-x}}$. The swish function increases non-linearity by multiplying the sigmoid function by x, thus enhancing the model's expressive power.

Although this non-linearity improves accuracy, swish's complex derivative calculation is not friendly to the quantization process. The improved h-swish function is defined as follows:

$$\text{h-swish}(x) = x \cdot \frac{\text{ReLU6}(x+3)}{6} \tag{2}$$

where $\text{ReLU6}(x)$ is a piecewise linear function, defined as $\text{ReLU6}(x) = \min(\max(0, x), 6)$. The h-swish function can be implemented as a piecewise function, reducing memory accesses and thereby significantly lowering latency costs.

Replacing the RT DETR model's backbone from ResNet to MobileNetV3 can significantly reduce the model's computational load to one-tenth of the original. Simultaneously, this substitution has a negligible impact on detection accuracy, almost not diminishing the model's performance.

3.2.2. Improving Positional Encoding in the AIFI Module of the RT DETR Algorithm

To enhance the model's positional information encoding capability, this study used learned positional encoding (LPE) to improve the positional encoding mechanism in AIFI. In transformer-based models, positional encoding is an essential mechanism to maintain the order information of sequence data.

Traditional fixed positional encoding uses fixed positional encoding (FPE) calculated by the following Formulas (3) and (4):

$$\text{PE}(pos, 2i) = \sin\left(\frac{pos}{10000^{2i/d_{model}}}\right) \tag{3}$$

$$\text{PE}(pos, 2i+1) = \cos\left(\frac{pos}{10000^{2i/d_{model}}}\right) \tag{4}$$

where d_{model} is the dimension of the model, pos is the position, and i is the dimension. In this way, the positional encoding of each position is a fixed vector that does not change with model training.

Compared to fixed positional encoding, learned positional encoding (LPE) provides a more flexible method for encoding positional information [39]. In this mechanism, positional encoding is part of the model parameters, learned through the model training process rather than predefined. Specifically, each positional encoding is randomly initialized at the start of model training and adjusted through optimization algorithms like backpropagation and gradient descent during training.

Learned positional encoding (LPE) is obtained during the model training process, where each position's encoding is part of the model parameters and can dynamically adjust according to the task and data. Fundamentally, LPE increases the model's flexibility and adaptability in encoding positional information, allowing the model to better capture and utilize positional information in sequence data. This can improve performance, particularly in tasks where the sequence's structure or pattern is not very regular, or the model needs a finer-grained understanding of positional information.

3.2.3. Improving the Upsampling in the CCFM Module of the RT DETR Algorithm

In the RT DETR algorithm, the CCFM upsampling module plays a crucial role by effectively restoring the spatial resolution of feature maps, significantly enhancing the accuracy of object detection tasks. This module's upsampling process ensures consistency in spatial dimensions between low-resolution and high-resolution feature maps, promoting effective integration of multi-scale features.

DySample represents a significant advancement in dynamic upsampling technology [40]. By avoiding the complexity of dynamic convolution and adopting a point sampling strategy, it achieves efficient resource utilization. This method considerably reduces the computational burden compared to traditional dynamic upsampling techniques while demonstrating better performance in object detection tasks. Its mathematical representation can be summarized in the following steps:

1. **Feature map representation:** let the input low-resolution feature be $F_{\text{low}} \in \mathbb{R}^{H \times W \times C}$, where H, W, and C represent height, width, and channels, respectively.
2. **Sampling point generation:** The sampling point generator in DySample dynamically selects sampling points based on the input F_{low}. Let the generated set of sampling points be $S = \{(x_i, y_i)\}_{i=1}^{N}$, where N is the number of sampling points, and (x_i, y_i) is the coordinate of the ith sampling point.
3. **Upsampling operation:** through the sampling points S and corresponding interpolation algorithms like bilinear interpolation, a high-resolution feature map $F_{\text{high}} \in \mathbb{R}^{H' \times W' \times C}$ is generated from F_{low}, where H' and W' are the target high-resolution dimensions.

DySample's upsampling technology significantly improves computational efficiency and optimizes the recovery of high-resolution feature maps through fine sampling point selection, effectively enhancing object detection accuracy. Moreover, DySample technology does not rely on specific hardware acceleration, increasing its practicality and flexibility in student classroom scenarios.

3.3. Experimental Environment Configuration, Training Setup, and Data Augmentation

3.3.1. Experimental Environment Configuration

The experimental environment for training the RT DETR model was configured as follows: the operating system was Ubuntu 20.04, the programming language was Python 3.8, and the deep learning framework used was PyTorch 1.10.0 with CUDA version 11.3. The hardware setup included an NVIDIA RTX 3090 GPU with 24 GB of memory, an Intel(R) Xeon(R) Platinum 8352V CPU with 12 cores @ 2.10 GHz, and 90 GB of RAM.

3.3.2. Training Setup

Learning Rate: The initial learning rate (lr0) was set to 0.0001. During the first 2000 iterations, it gradually decreased from 0.1 to 0.0001 to ensure training stability. After this warm-up phase, the learning rate remained constant at 0.0001, facilitating stable and effective learning of data features.

Optimizer: The Adam optimizer was used, with initial parameters set as follows: $\beta1 = 0.9$, $\beta2 = 0.999$, and $\varepsilon = 1 \times 10^{-8}$. Adam was chosen for its excellent performance in handling sparse gradients and large datasets.

Batch size: A batch size of 8 was used during training. This setting ensured efficient GPU memory utilization while also aiding in the stable updating of model weights.

Number of epochs: The model was trained on the training set for a total of 100 epochs. Early stopping was implemented by monitoring the performance on the validation set to prevent overfitting.

Loss function: The RT DETR model's loss function consists of several key components to optimize detection performance. The main components include the box regression loss (Box Loss, weight 7.5), which uses the mean squared error (MSE) or Intersection over Union (IoU) loss to measure the difference between predicted and ground-truth bounding boxes; the classification loss (Classification Loss, weight 0.5), which uses cross-entropy loss to assess the accuracy of predicted class probabilities; and the objectness loss (Objectness Loss, weight 1.0), which employs binary cross-entropy loss to evaluate the confidence of predicted bounding boxes. Additionally, the Distribution Focal Loss (DFL, weight 1.5) was introduced to improve the accuracy of bounding box predictions.

3.3.3. Data Augmentation Techniques

In this study, we employed various data augmentation techniques to enhance the generalization and robustness of the model. These techniques include HSV color space augmentation (random adjustments of hue, saturation, and brightness), geometric transformations (rotation, translation, scaling, shearing, and perspective transformations), flipping (horizontal and vertical flips), as well as composite augmentation techniques (Mosaic and Mixup). The specific parameter settings were as follows: a hue adjustment range of 0.015, a saturation adjustment range of 0.7, a brightness adjustment range of 0.4; for geometric transformations, the translation range was ±0.1, and the scaling range was ±0.5; the probability of horizontal flipping was 0.5. These data augmentation techniques generated more diverse training samples during the training process, thereby significantly improving the model's performance in practical applications.

4. Experiment and Analysis

4.1. Comparison Experiments of Algorithms on the SCB-Dataset-S

Table 1 presents the comparison results of different algorithms on the SCB-Dataset-S. The SCB-Dataset-S is a publicly available dataset of real student classroom behaviors, containing a total of 5010 images with annotations for three types of behaviors: hand raising, reading, and writing. Specifically, the dataset includes 10,078 labels for hand raising, 5882 labels for reading, and 2539 labels for writing. The models evaluated included YOLOv5n, YOLOv8n, RT DETR-ResNet18, and RT DETR-ResNet18+LPE+DySample. Their performance was assessed using four standard object detection metrics: precision, recall, mAP50, and mAP50-95. The results were as follows:

Table 1. Comparison of algorithms on the SCB-Dataset-S.

Model	Precision	Recall	mAP50	mAP50-95
YOLOv5n	0.681	0.675	0.711	0.483
YOLOv8n	0.679	0.677	0.724	0.550
RT DETR-ResNet18	0.759	0.754	0.781	0.555
RT DETR-ResNet18+LPE+DySample	0.817	0.783	0.801	0.585

The results clearly indicated that the RT DETR-ResNet18+LPE+DySample model outperformed the others across all evaluated metrics, demonstrating superior precision, recall, and consistency in object detection. This highlighted the model's enhanced capability in handling complex scenes and detecting small targets, making it particularly suitable for practical applications such as monitoring student behavior in classrooms.

4.2. Comparison Experiments of Algorithms on the FSCB-Dataset

This section details the comparison experiment results of the improved RT DETR-18+LPE+DySample model and existing object detection models (YOLOv3-tiny, YOLOv5n, YOLOv8n) on the FSCB-dataset. All models were trained under the same hardware configuration and training epochs, using standard object detection metrics to evaluate performance, including precision, recall, mAP50, and mAP50-95. The results are shown in Figure 7.

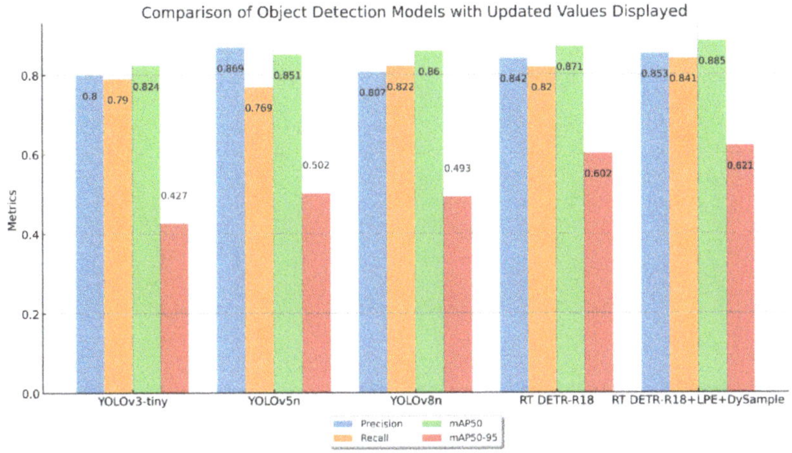

Figure 7. Comparison of the detection performance of different models.

Experimental results showed that the improved RT DETR-18+LPE+DySample model performed excellently on various performance metrics. The precision increased to 0.853, and the recall rate was 0.841, indicating superior positive sample detection ability. The mAP50 and mAP50-95 metrics achieved 0.885 and 0.621, respectively, demonstrating consistency and high precision across different IoU thresholds. These optimizations significantly enhanced the model's ability to handle complex scenes and small target detection, suitable for practical applications like student classroom behavior detection. Overall, the correctness of the optimization strategies and their potential application value were validated.

4.3. Backbone Improvement Experiments

In this experiment, we explored the impact of replacing different backbone networks within the RT DETR framework on model performance. Four configurations were compared: RT DETR-ResNet18, RT DETR-Faster, RT DETR-EfficientViT, and RT DETR-MobileNetv3. We evaluated these models in terms of total parameters, GFLOPs, FPS, and two accuracy metrics, mAP50 and mAP50-95, to assess their performance in resource consumption, speed, and accuracy.

The experimental results, as shown in Table 2, indicated that RT DETR-MobileNetv3 was the lightest in terms of total parameters, with 9,546,660 parameters and 23.6 GFLOPs, indicating higher computational efficiency. In terms of real-time performance (FPS), this model ranked first with a result of 76.4. Despite having fewer parameters, the MobileNetv3 version showed strong performance with a mAP50 of 0.857, surpassing RT DETR-EfficientViT and being close to RT DETR-Faster's 0.879. For the more comprehensive accuracy metric mAP50-95, RT DETR-MobileNetv3 achieved 0.591, which was very close to RT DETR-Faster. This result demonstrated MobileNetv3's contribution to overall model performance when replacing the backbone, especially in maintaining high efficiency and good accuracy while ensuring real-time performance.

Table 2. The impact of different backbone networks on model performance.

Model	Params	GFLOPs	FPS	mAP50	mAP50-95
RT DETR-ResNet18	19,878,180	57.0	65.9	0.871	0.602
RT DETR-Faster	16,790,180	49.5	69.6	0.879	0.615
RT DETR-EfficientViT	10,707,748	27.2	70.9	0.836	0.577
RT DETR-MobileNetv3	**9,546,660**	**23.6**	**76.4**	**0.857**	**0.591**

4.4. Ablation Experiments

To comprehensively evaluate the effectiveness of different configurations in the RT-DETR model for classroom behavior detection, we conducted ablation experiments, including enhancements such as backbone network replacement, learned positional encoding (LPE), and dynamic upsampling (DySample). The results are summarized in Table 3.

Table 3. Comparative analysis of the impact of different module enhancements on the RT DETR algorithm performance.

Model	Focus	Writing	Raising Hand	Sleeping	Using Phone	All Classes	Params	GFLOPs	FPS
RT DETR-ResNet18	0.852	0.823	0.923	0.886	0.874	0.872	19,878,180	57.0	65.9
RT DETR-ResNet18+LPE	0.850	0.847	0.917	0.918	0.859	0.878	19,980,580	57.0	60.2
RT DETR-ResNet18+DySample	0.866	0.848	0.937	0.902	0.847	0.880	19,894,628	57.0	58.8
RT DETR-ResNet18+LPE+DySample	0.875	0.857	0.941	0.917	0.833	0.885	19,997,028	57.0	57.6
RT DETR-MobileNetv3+LPE+DySample	**0.850**	**0.829**	**0.917**	**0.929**	**0.814**	**0.868**	**9,665,508**	**26.3**	**72.3**

The RT DETR-ResNet18, serving as the baseline model, demonstrated significant capabilities in various behavior classifications, achieving an overall accuracy of 87.2%. The introduction of learned positional encoding (LPE) improved the detection accuracy of specific categories such as "Writing" and "Sleeping," raising the overall accuracy to 87.8%, with an increased frame rate of 60.2 FPS, indicating a notable enhancement in processing efficiency. Particularly, the RT DETR-ResNet18+LPE+DySample configuration achieved the highest accuracy in detecting "Raising Hand" and "Writing" behaviors, with an overall accuracy of 88.5%. This configuration also exhibited outstanding real-time processing capabilities, recording the highest frame rate of 57.6 FPS.

The final model configuration, RT DETR-MobileNetV3+LPE+DySample, while not entirely surpassing its predecessors in accuracy, significantly reduced computational resource consumption, with only 9.67 M parameters and 26.3 GFLOPs. Despite an increased frame rate of 72.3 FPS, this configuration demonstrated strong potential for application scenarios where computational efficiency is critical, particularly suitable for deployment in mobile devices or embedded systems with limited resources.

The ablation study highlighted the significant impact of various architectural enhancements on the performance of the RT-DETR model. Specifically, the RT DETR-MobileNetV3+LPE+DySample configuration effectively reduced the model's computational requirements and improved processing speed while maintaining reasonable accuracy, providing an excellent solution for real-time behavior detection in computationally constrained environments.

4.5. Visualization

In this study, we conducted an in-depth analysis of the improved RT DETR model's performance in student classroom behavior detection through heatmap visualization experiments. Heatmaps were generated by the attention weights predicted by the model, visually representing the model's focus on certain areas of the image, as shown in Figure 8. This visualization provides intuitive evidence for understanding and interpreting the model's decision-making process.

Figure 8. Heatmap visualization of different algorithms.

In the experiment, we first captured a series of representative classroom scenes involving various typical student behaviors, such as raising hands, sleeping, distraction, writing, and using mobile phones. These scenes were then processed using YOLOv8, the original RT DETR, and our improved RT DETR model to generate the corresponding heatmaps. The improved heatmaps demonstrated the superior performance of the optimized model in the following aspects:

1. Enhanced attention focus: The improved model exhibited more concentrated attention distribution on the heatmaps. This indicated that the model effectively focused on areas closely related to the behavior detection task, reducing attention dispersion to irrelevant regions.
2. Optimized spatial resolution: Through improved positional encoding, the model displayed finer spatial resolution on the heatmaps. This was reflected in clearer boundaries of active regions, enhancing behavior detection accuracy.
3. Improved feature layer fusion: The improved CCFM upsampling strategy demonstrated better multi-scale fusion effects on the heatmaps. This ensured a smoother transition from low-level to high-level features, aiding in capturing detailed information while maintaining overall behavior pattern recognition.
4. Reduced noise and interference: The improved model produced clearer signals on the heatmaps, effectively suppressing noise and background interference. This is particularly important as it indicates the model's enhanced robustness to various types of interference in real classroom environments.

5. Discussion

Although this study achieved significant results in detecting student classroom behavior, several limitations still need to be addressed.

1. Dataset diversity and scale: The diversity and scale of the dataset are still insufficient. Although the fusion dataset (FSCB-dataset) combines AI-generated data and actual observation data, enriching the data sources, the scenes and behavior types included in the dataset may not fully cover all possible situations in actual classrooms. The data generated using advanced text-to-image generation techniques (such as Stable Diffusion and DALL-E 3) may not completely reflect the complexity and variability of real scenes in some cases.

2. Real-time performance: Although the RT-DETR model proposed in this paper incorporates the MobileNetV3 backbone network, learned positional encoding (LPE), and dynamic upsampling (DySample) techniques, significantly improving computational efficiency, further optimization is needed to ensure real-time performance, especially in resource-constrained mobile devices or embedded systems.

3. Model explainability: The black-box nature of deep learning models makes their decision-making processes difficult to interpret, which is particularly important in educational settings. Future research can further explore the explainability of behavior detection

models, using attention maps and visualization tools to help teachers and researchers better understand the model's decision-making process, thereby improving the system's transparency and acceptability.

To further enhance the performance and practicality of the student classroom behavior detection system, future research can explore the following directions:

1. Real-time inference optimization: Investigate methods to further optimize real-time inference, such as edge computing and model compression techniques, to improve deployment efficiency and performance on mobile devices. Techniques such as quantization, pruning, and distillation are feasible approaches [41].

2. Multimodal data fusion: Combine video, audio, and sensor data to build a multimodal behavior detection system. Multimodal fusion not only provides more comprehensive behavior analysis but also enhances detection robustness and accuracy in various environments. For example, integrating microphone data to detect students' vocal behaviors, combined with visual data, can improve overall detection performance [42].

3. Explainability in behavior detection**: Further study the explainability of deep learning models, especially in interpreting detected student behaviors. Using explainability techniques such as attention maps and visualization tools can help teachers and researchers better understand the model's decision-making process, enhancing the system's transparency and acceptability [31].

These research directions not only offer new technical pathways for student behavior detection but also lay a solid foundation for further development in the field of educational technology. Future research can continuously optimize algorithms, enhance the model's generalization capabilities, and integrate more practical application scenarios, advancing the development and application of intelligent education systems.

6. Conclusions

This study significantly enhanced the performance and practical application of a student classroom behavior detection system through various innovations and optimizations. Firstly, we constructed a fusion dataset (FSCB-dataset) that effectively addressed the issues of limited diversity and scarcity in student behavior detection datasets, providing more varied and realistic data resources for model training. Secondly, we proposed an improved detection model based on the RT DETR algorithm. By incorporating the lightweight MobileNetV3 backbone network, learned positional encoding (LPE), and dynamic upsampling (DySample) techniques, the model's computational efficiency was significantly improved while maintaining the original detection accuracy despite the lightweight design. Test results on the FSCB-dataset demonstrated that the improved RT DETR model excelled in various performance metrics. The improved lightweight network is also easy to deploy on mobile devices, proving its practicality in resource-constrained environments.

These research outcomes not only offer new technical pathways for student behavior detection but also lay a solid foundation for further development in the field of educational technology. Future research can further optimize the algorithm, enhance the model's generalization capabilities, and integrate more practical application scenarios, continuously advancing the development and application of intelligent education systems.

Author Contributions: Conceptualization, L.L.; Methodology, H.Y.; Software, H.Y.; Validation, H.Y.; Formal analysis, Q.X., Y.X. and D.L.; Investigation, Q.X., Y.X. and D.L.; Resources, H.Y.; Data curation, H.Y.; Writing—original draft, H.Y.; Writing—review & editing, L.L. and H.Y.; Visualization, H.Y.; Supervision, L.L.; Project administration, L.L.; Funding acquisition, L.L. All authors have read and agreed to the published version of the manuscript.

Funding: Industry-University-Research Innovation Fund for Chinese Universities (2021KSA05005).

Institutional Review Board Statement: Not applicable.

Informed Consent Statement: Not applicable.

Data Availability Statement: The raw data supporting the conclusions of this article will be made available by the authors on request.

Conflicts of Interest: The authors declare no conflicts of interest.

References

1. Valdés, J.J.; Bonham-Carter, G. Time dependent neural network models for detecting changes of state in complex processes: Applications in earth sciences and astronomy. *Neural Netw.* **2006**, *19*, 196–207. [CrossRef] [CrossRef] [PubMed]
2. Lagaris, I.E.; Likas, A.; Fotiadis, D.I. Artificial neural networks for solving ordinary and partial differential equations. *IEEE Trans. Neural Netw.* **1998**, *9*, 987–1000. [CrossRef] [CrossRef] [PubMed]
3. Hee Lee, J.; Shvetsova, O.A. The impact of VR application on student's competency development: A comparative study of regular and VR engineering classes with similar competency scopes. *Sustainability* **2019**, *11*, 2221. [CrossRef] [CrossRef]
4. Shvetsova, O.; Feroz, M.; Salkutsan, S.; Efimov, A. Artificial Intelligence Application for Healthcare Industry: Cases of Developed and Emerging Markets. In Proceedings of the International Conference on Expert Clouds and Applications, Bengaluru, India, 9–10 February 2022; Springer: Singapore, 2022; pp. 419–432.
5. Krizhevsky, A.; Sutskever, I.; Hinton, G.E. ImageNet Classification with Deep Convolutional Neural, NIPS'12. In Proceedings of the 25th International Conference on Neural Information Processing Systems, Lake Tahoe, NV, USA, 3–6 December 2012; pp. 3–6.
6. Girshick, R.; Donahue, J.; Darrell, T.; Malik, J. Rich feature hierarchies for accurate object detection and semantic segmentation. In Proceedings of the IEEE Conference on Computer Vision and Pattern Recognition, Columbus, OH, USA, 23–28 June 2014; pp. 580–587. [CrossRef]
7. Girshick, R. Fast r-cnn. In Proceedings of the IEEE International Conference on Computer Vision, Santiago, Chile, 7–13 December 2015; pp. 1440–1448. [CrossRef]
8. Faster, R. Towards real-time object detection with region proposal networks. *Adv. Neural Inf. Process. Syst.* **2015**, *9199*, 2969239–2969250. [CrossRef]
9. He, K.; Gkioxari, G.; Dollár, P.; Girshick, R. Mask r-CNN. In Proceedings of the 2017 IEEE International Conference on Computer Vision (ICCV), Venice, Italy, 22–29 October 2017; pp. 2980–2988. [CrossRef]
10. Cai, Z.; Vasconcelos, N. Cascade r-cnn: Delving into high quality object detection. In Proceedings of the IEEE Conference on Computer Vision and Pattern Recognition, Salt Lake City, UT, USA, 18–23 June 2018; pp. 6154–6162. [CrossRef]
11. Pang, J.; Chen, K.; Shi, J.; Feng, H.; Ouyang, W.; Lin, D. Libra r-cnn: Towards balanced learning for object detection. In Proceedings of the IEEE/CVF Conference on Computer Vision and Pattern Recognition, Long Beach, CA, USA, 15–20 June 2019; pp. 821–830. [CrossRef]
12. Redmon, J.; Divvala, S.; Girshick, R.; Farhadi, A. You only look once: Unified, real-time object detection. In Proceedings of the IEEE Conference on Computer Vision and Pattern Recognition, Las Vegas, NV, USA, 27–30 June 2016; pp. 779–788. [CrossRef]
13. Redmon, J.; Farhadi, A. YOLO9000: Better, faster, stronger. In Proceedings of the IEEE Conference on Computer Vision and Pattern Recognition, Honolulu, HI, USA, 21–26 July 2017; pp. 7263–7271. [CrossRef]
14. Redmon, J.; Farhadi, A. Yolov3: An incremental improvement. *arXiv* **2018**, arXiv:1804.02767. [CrossRef]
15. Bochkovskiy, A.; Wang, C.Y.; Liao, H.Y.M. Yolov4: Optimal speed and accuracy of object detection. *arXiv* **2020**, arXiv:2004.10934. [CrossRef]
16. Zhu, X.; Lyu, S.; Wang, X.; Zhao, Q. TPH-YOLOv5: Improved YOLOv5 based on transformer prediction head for object detection on drone-captured scenarios. In Proceedings of the IEEE/CVF International Conference on Computer Vision, Montreal, BC, Canada, 11–17 October 2021; pp. 2778–2788. [CrossRef]
17. Liu, W.; Anguelov, D.; Erhan, D.; Szegedy, C.; Reed, S.; Fu, C.Y.; Berg, A.C. Ssd: Single shot multibox detector. In Proceedings of the Computer Vision—ECCV 2016: 14th European Conference, Amsterdam, The Netherlands, 11–14 October 2016; Proceedings, Part I 14; Springer: Berlin/Heidelberg, Germany, 2016; pp. 21–37. [CrossRef]
18. Wang, Y.; Wang, C.; Zhang, H.; Dong, Y.; Wei, S. Automatic ship detection based on RetinaNet using multi-resolution Gaofen-3 imagery. *Remote Sens.* **2019**, *11*, 531. [CrossRef] [CrossRef]
19. Tan, M.; Pang, R.; Le, Q.V. Efficientdet: Scalable and efficient object detection. In Proceedings of the IEEE/CVF Conference on Computer Vision and Pattern Recognition, Seattle, WA, USA, 13–19 June 2020; pp. 10781–10790. [CrossRef]
20. Carion, N.; Massa, F.; Synnaeve, G.; Usunier, N.; Kirillov, A.; Zagoruyko, S. End-to-end object detection with transformers. In Proceedings of the European Conference on Computer Vision, Glasgow, UK, 23–28 August 2020; Springer: Berlin/Heidelberg, Germany, 2020; pp. 213–229. [CrossRef]
21. Zhu, X.; Su, W.; Lu, L.; Li, B.; Wang, X.; Dai, J. Deformable detr: Deformable transformers for end-to-end object detection. *arXiv* **2020**, arXiv:2010.04159.
22. Meng, D.; Chen, X.; Fan, Z.; Zeng, G.; Li, H.; Yuan, Y.; Sun, L.; Wang, J. Conditional detr for fast training convergence. In Proceedings of the IEEE/CVF International Conference on Computer Vision, Montreal, QC, Canada, 28 February 2021; pp. 3651–3660. [CrossRef]
23. Zhao, Y.; Lv, W.; Xu, S.; Wei, J.; Wang, G.; Dang, Q.; Liu, Y.; Chen, J. Detrs beat yolos on real-time object detection. In Proceedings of the IEEE/CVF Conference on Computer Vision and Pattern Recognition, Seattle, WA, USA, 21 June 2024; pp. 16965–16974.

24. Zaletelj, J.; Košir, A. Predicting students' attention in the classroom from Kinect facial and body features. *EURASIP J. Image Video Process.* **2017**, *2017*, 1–12. [CrossRef] [CrossRef]
25. Bai, Y. Research on Video-Based Student Action Recognition Method. *Instrum. User* **2020**, *27*, 10–12.
26. Li, J. Study on Classroom Teacher-Student Behavior Patterns Based on Skeleton Information. Ph.D. Thesis, Northwest University, Kirkland, WA, USA, 2021.
27. Cao, Z.; Simon, T.; Wei, S.E.; Sheikh, Y. Realtime multi-person 2d pose estimation using part affinity fields. In Proceedings of the IEEE Conference on Computer Vision and Pattern Recognition, Honolulu, HI, USA, 21–26 July 2017; pp. 7291–7299. [CrossRef]
28. Gao, K. Research on Classroom Human Behavior Recognition Based on Convolutional Neural Networks. Ph.D. Thesis, Taiyuan University of Technology, Taiyuan, China, 2020.
29. Feng, S. Research on Student Classroom Behavior Recognition Based on Deep Convolutional Neural Networks. Ph.D. Thesis, Chang'an University, Xi'an, China, 2020.
30. Huang, G. Research on Student Classroom Behavior Recognition Based on Deep Learning. Ph.D. Thesis, China University of Mining and Technology, Beijing, China, 2021.
31. Mahmoudi, S.A.; Amel, O.; Stassin, S.; Liagre, M.; Benkedadra, M.; Mancas, M. A review and comparative study of explainable deep learning models applied on action recognition in real time. *Electronics* **2023**, *12*, 2027. [CrossRef] [CrossRef]
32. Shvetsova, O.A.; Park, S.C.; Lee, J.H. Application of quality function deployment for product design concept selection. *Appl. Sci.* **2021**, *11*, 2681. [CrossRef] [CrossRef]
33. Rombach, R.; Blattmann, A.; Lorenz, D.; Esser, P.; Ommer, B. High-resolution image synthesis with latent diffusion models. In Proceedings of the IEEE/CVF Conference on Computer Vision and Pattern Recognition, New Orleans, LA, USA, 18–24 June 2022; pp. 10684–10695. [CrossRef]
34. Ramesh, A.; Dhariwal, P.; Nichol, A.; Chu, C.; Chen, M. Hierarchical text-conditional image generation with clip latents. *arXiv* **2022**, arXiv:2204.06125.
35. Vaswani, A.; Shazeer, N.; Parmar, N.; Uszkoreit, J.; Jones, L.; Gomez, A.N.; Kaiser, Ł.; Polosukhin, I. Attention is all you need. *arXiv* **2017**, arXiv:1706.03762. [CrossRef]
36. Howard, A.G.; Zhu, M.; Chen, B.; Kalenichenko, D.; Wang, W.; Weyand, T.; Andreetto, M.; Adam, H. Mobilenets: Efficient convolutional neural networks for mobile vision applications. *arXiv* **2017**, arXiv:1704.04861. [CrossRef]
37. Sandler, M.; Howard, A.; Zhu, M.; Zhmoginov, A.; Chen, L.C. Mobilenetv2: Inverted residuals and linear bottlenecks. In Proceedings of the IEEE Conference on Computer Vision and Pattern Recognition, Salt Lake City, UT, USA, 18–23 June 2018; pp. 4510–4520. [CrossRef]
38. Howard, A.; Sandler, M.; Chu, G.; Chen, L.C.; Chen, B.; Tan, M.; Wang, W.; Zhu, Y.; Pang, R.; Vasudevan, V.; et al. Searching for mobilenetv3. In Proceedings of the IEEE/CVF International Conference on Computer Vision, Seoul, Republic of Korea, 27 October–2 November 2019; pp. 1314–1324. [CrossRef]
39. Ke, G.; He, D.; Liu, T.Y. Rethinking positional encoding in language pre-training. *arXiv* **2020**, arXiv:2006.15595. [CrossRef]
40. Liu, W.; Lu, H.; Fu, H.; Cao, Z. Learning to upsample by learning to sample. In Proceedings of the IEEE/CVF International Conference on Computer Vision, Paris, France, 1–6 October 2023; pp. 6027–6037. [CrossRef]
41. Han, S.; Mao, H.; Dally, W.J. Deep compression: Compressing deep neural networks with pruning, trained quantization and huffman coding. *arXiv* **2015**, arXiv:1510.00149. [CrossRef]
42. Baltrušaitis, T.; Ahuja, C.; Morency, L.P. Multimodal machine learning: A survey and taxonomy. *IEEE Trans. Pattern Anal. Mach. Intell.* **2018**, *41*, 423–443. [CrossRef] [CrossRef] [PubMed]

applied sciences

Article

Collaborative Analysis of Learners' Emotional States Based on Cross-Modal Higher-Order Reasoning

Wenyan Wu [1], Jingtao Zhao [2], Xingbo Shen [3,*] and Guang Feng [2]

[1] College of Computer, Guangdong University of Technology, Guangzhou 511400, China
[2] College of Automation, Guangdong University of Technology, Guangzhou 511400, China; cbnb1525@163.com (J.Z.)
[3] Center of Campus Network & Modern Educational Technology, Guangdong University of Technology, Guangzhou 510006, China
* Correspondence: shenxingbo@gdut.edu.cn

Abstract: Emotion is a significant factor influencing education and teaching, closely intertwined with learners' cognitive processing. Conducting analysis of learners' emotions based on cross-modal data is beneficial for achieving personalized guidance in intelligent educational environments. Currently, due to factors such as data scarcity and environmental noise, data imbalances have led to incomplete or missing emotional information. Therefore, this study proposes a collaborative analysis model based on attention mechanisms. The model extracts features from various types of data using different tools and employs multi-head attention mechanisms for parallel processing of feature vectors. Subsequently, through a cross-modal attention collaborative interaction module, effective interaction among visual, auditory, and textual information is facilitated, significantly enhancing comprehensive understanding and the analytical capabilities of cross-modal data. Finally, empirical evidence demonstrates that the model can effectively improve the accuracy and robustness of emotion recognition in cross-modal data.

Keywords: cross-modal fusion; learner emotion recognition; attention mechanism; personalized tutoring

1. Introduction

Currently, the new generation of information technology represented by artificial intelligence has become an important driving force for innovative changes in teaching. The new generation of information technology led by artificial intelligence is redefining and changing the connotations, structure, and ecology of education and teaching. Artificial intelligence technology helps to achieve personalized learning paths and precise recommendations for learning resources and can solve the problem of large-scale personalized teaching. In this context, the development of the Internet of Things sensing technology has made it possible to obtain large-scale cross-modal data, such as emotion recognition [1], behavior recognition [2], and video retrieval [3]. Based on the analysis of students' learning states using cross-modal teaching process data, it is possible to accurately grasp the individual state of each student, thereby helping to achieve large-scale and quality personalized education.

Emotional learning is an important factor in students' learning states, and the recognition of students' learning emotions plays a crucial role in implementing personalized tutoring. Through analyzing students' non-verbal information such as facial expressions, tone of voice, and body language during classroom learning, it is possible to more accurately grasp students' learning motivations, emotional states, and cognitive loads, thus effectively identifying students' learning difficulties and emotional barriers [4]. This is crucial for promoting students' emotional well-being, enhancing their interest and efficiency in learning, and even stimulating their proactive learning enthusiasm. Furthermore, emotional recognition can also facilitate personalized teaching. Through deeply exploring the emotional states of all students, teachers can provide personalized guidance and support based on each student's individual circumstances.

There have been many research results on the extraction of emotional states, such as methods based on local information. Firstly, facial keypoints or key regions were manually selected and local features were then extracted for expression classification (Pantic et al., 2004 [5]; Bashyal et al., 2008 [6]; Cheon et al., 2009 [7]). Subspace feature extraction methods, focusing on the overall facial features, were employed to obtain the optimal transformation under certain criteria, projecting high-dimensional data into a low-dimensional space, retaining only discriminative features for facial expression classification in the subspace, thereby achieving the goal of dimensionality reduction, correlation elimination, and improved classification performance (Shan et al., 2006). However, traditional emotion-recognition models mainly have relied on unimodal data, such as text or speech [8]. While this unimodal approach can provide useful emotion-recognition results in specific contexts, the accuracy of recognition results is not high. Unimodal emotion-recognition models often struggle to fully capture and understand students' emotional states. This is because emotional expression in students' learning processes is multidimensional, including language, tone of voice, facial expressions, and even body posture [9], These multimodal pieces of information collectively provide a more comprehensive reflection of individuals' emotional states. Therefore, through employing multimodal emotion recognition models, it is possible to significantly enrich the dimensions of educational assessment, enhance the quality and effectiveness of education and teaching, and achieve a more humane and intelligent educational environment.

However, achieving more scientific, comprehensive, reasonable, and real-time personalized tutoring for learners based on cross-modal teaching process data still faces many difficulties. In many cases, the real facial images acquired in the classroom have numerous issues such as occlusion, small targets, image blurring, non-frontal faces, etc. The incompleteness of information and environmental noise further increase the difficulty of emotion recognition. To address these challenges, this study proposes a collaborative perception model based on attention mechanisms. This model integrates multimodal data from visual, audio, and text sources and enhances the synergistic learning effect between modalities through a specially designed interaction layer. Such a design aims to dynamically adjust the focus on different modal data, thereby improving the accuracy of emotion recognition and overcoming the adverse effects of data incompleteness and environmental noise. It aims to better explore the potential information in cross-modal teaching process data to achieve intelligent tutoring for individuals and groups and has significant research and application value.

2. Related Works

2.1. Cross-Modal Feature Fusion

Cross-modal feature fusion is a crucial technique that plays a vital role in both cross-modal learning and cross-modal applications. Its core objective is to integrate information from different modalities to establish a richer and more comprehensive model representation, thereby enhancing the performance and effectiveness of the system. Multimodal data often encompasses information from various domains such as visual, textual, and auditory sources. These pieces of information are interrelated and complementary to each other. Therefore, how to effectively integrate them has become one of the focal points of research.

2.1.1. Early Fusion

Early fusion is a simple and direct method of multimodal feature fusion. Its core idea is to directly integrate information from different modalities at the input level to form a comprehensive feature representation. In practice, this is often achieved in straightforward ways, such as concatenating feature vectors from different modalities, weighted summation, or using other forms of integration methods.

The principle of early fusion lies in unifying multimodal data in the feature space to construct a comprehensive model input. Through merging information from different modalities at the input level, early fusion allows the model to directly access information from multiple data sources without additional processing steps. This direct fusion simplifies the design and implementation of the model, reduces system complexity, and enables rapid integration of multimodal information. The schematic diagram of early fusion is illustrated in Figure 1.

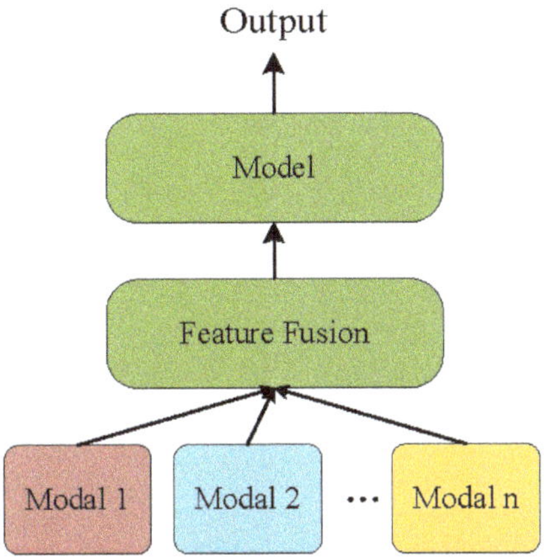

Figure 1. Early fusion diagram.

However, early fusion also has some limitations, with the most significant being the potential for information confusion or conflict between different modalities. Because features from different modalities have different representations and semantic information, direct concatenation may lead to the loss or distortion of some information. Therefore, in practice, it is important to carefully select appropriate fusion methods to fully utilize the information from different modalities and improve the performance and effectiveness of the model.

2.1.2. Late fusion

Late fusion is a method where each modality is processed independently before fusion. In late fusion, each modality undergoes separate deep processing, for example, using convolutional neural networks (CNNs) for image processing and recurrent neural networks (RNNs) for text processing, and their output features are then fused together.

The principle of late fusion lies in fully exploring the feature information within each modality and modeling it through independent processing. Through processing each modality separately, late fusion can better preserve the specific information of each modality, thereby improving the model's expressive power and generalization performance. Additionally, late fusion can also flexibly perform inter-modal information interaction, thus better capturing the correlation and semantic association between modalities.

However, late fusion also presents some challenges, with the most significant being increased computational costs and model complexity. Since each modality requires independently designed and trained specialized models, this may result in higher computational and storage overheads. Therefore, in practice, it is necessary to carefully design the model structure to balance performance and complexity and fully utilize existing computational resources. The schematic diagram of late fusion is illustrated in Figure 2.

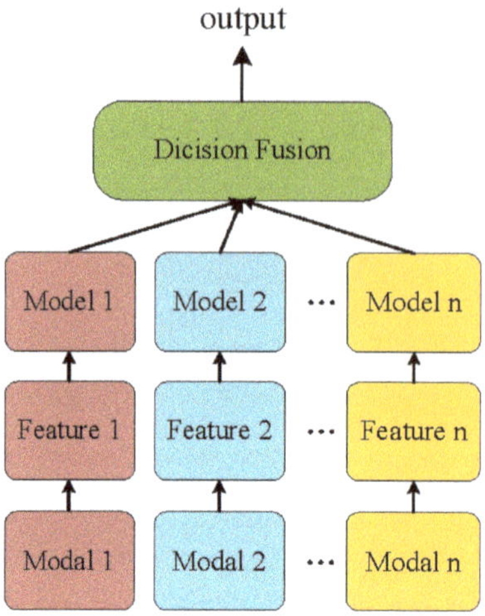

Figure 2. Late fusion diagram.

Furthermore, to optimize late fusion, researchers are also exploring more efficient feature fusion techniques, such as learning-based fusion layers or adaptive fusion mechanisms. These methods may further improve the efficiency and accuracy of models. Such advancements ensure that late fusion maintains broad application prospects when dealing with complex multimodal data.

2.1.3. Hybrid Fusion

The method of hybrid fusion combines the advantages of both early and late fusion, retaining the simplicity and directness of early fusion while enabling deep processing when necessary. In hybrid fusion, certain modalities may undergo simple fusion at an early stage, while others undergo deeper processing, and their features are then fused together at a later stage.

The principle of hybrid fusion lies in flexibly selecting fusion strategies based on the specific task requirements and the relationship between modalities. Through early-stage simple fusion, multimodal information can be quickly integrated, reducing model complexity. Meanwhile, in the late fusion stage, the advantages of deep processing can be utilized to better explore the correlations and interactions between modalities, thereby improving the model's performance and effectiveness.

The advantage of hybrid fusion lies in its ability to flexibly select fusion strategies based on specific task requirements and the relationships between modalities. This flexibility allows the hybrid fusion method to adapt to different application scenarios and achieve good results in various multimodal learning tasks. At the same time, hybrid fusion can fully utilize the advantages of both early and late fusion, thereby retaining simplicity and directness while enhancing the model's expressive power and generalization performance. The schematic diagram of hybrid fusion is illustrated in Figure 3.

2.2. Attention Mechanism

Attention mechanism, as an important neural network technology, was inspired by the human visual and cognitive systems. Its earliest proposal can be traced back to the field of cognitive psychology in the 1980s. However, the application of attention mechanisms in the field of deep learning began in recent years. Early work laid the foundation for the introduction of attention mechanisms into neural networks. These studies introduced attention mechanisms into autoencoders and neural machine translation models, demonstrating their potential in improving model performance and generalization ability.

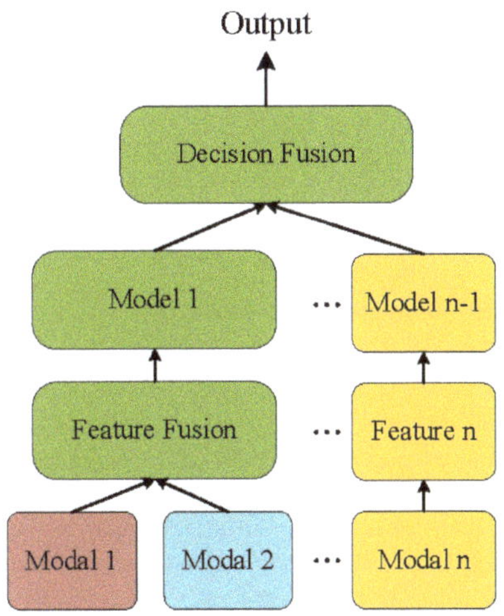

Figure 3. Hybrid fusion diagram.

The attention mechanism achieves targeted focus and weight allocation of information through a mechanism called Query–Attention–Key. The core concept of this mechanism is to compute the relevance between a query and a set of keys and use this relevance to determine the weight distribution of the final output values. The computation process of the attention mechanism is illustrated in Figure 4. Specifically, it first computes the relevance between the query and each key using some calculation method (e.g., dot product). Then, it uses some normalization function (e.g., softmax function) to transform these relevances into a weight distribution, ensuring that the sum of all weights is 1. Finally, it uses these weight allocations to aggregate all values, generating the final output result.

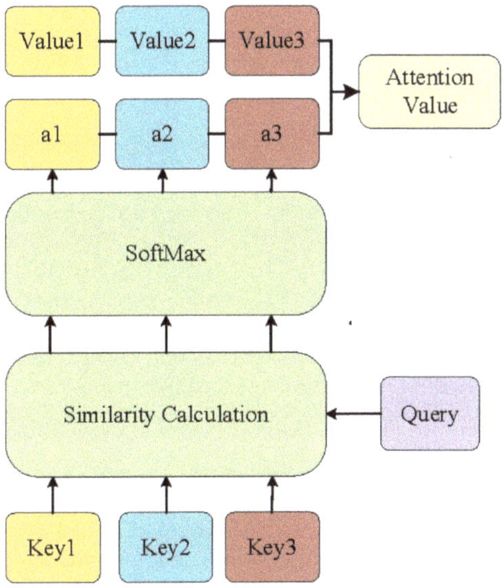

Figure 4. The computational process diagram of the attention mechanism.

The specific calculation methods as shown in Formulas (1) to (3):

$$Sim_i(Query, Key_i) = Query \cdot Key_i \quad (1)$$

$$a_i = Softmax(Sim_i) = \frac{e^{Sim_i}}{\sum_{j=1}^{L_X} e^{Sim_j}} \quad (2)$$

$$Attention = \sum_{i=1}^{L_x} a_i \cdot Value_i \quad (3)$$

The attention mechanism plays a crucial role in multimodal feature fusion, effectively capturing the correlations between different modalities and dynamically adjusting the features' weights to enhance the model's performance and expressive power. Introducing the attention mechanism into modal fusion allows the model to flexibly focus on essential information from different modalities, achieving more precise fusion. The attention mechanism enables the model to dynamically learn the weights of each feature when processing each modality and adjust the fusion weights based on their importance, making significant features more prominent and suppressing less relevant ones. This dynamic adjustment mechanism enhances the model's flexibility and generalization ability. For instance, consider a multimodal task involving vision and text. In early fusion, the feature vectors of vision and text can be concatenated, and the attention mechanism dynamically adjusts the weights of different modal features to reflect their importance in the task. In late fusion, after each modality undergoes independent processing, the attention mechanism is used to fuse different modal features to better capture inter-modal correlations. In practical applications, attention mechanisms can be implemented in various ways, such as fully connected layers, convolutional layers, or self-attention mechanisms. Self-attention mechanisms are particularly effective as they consider the relationships between different positions in the input sequence, better capturing long-distance dependencies within the sequence. This mechanism has achieved significant success in fields like natural language processing and computer vision and has been widely applied in multimodal learning tasks.

3. Methods

3.1. Overall Architecture

The overall architecture of the collaborative perception model based on attention mechanism is illustrated in Figure 5.

This model mainly consists of four core components:

The first part is the modality feature extraction layer, which extracts key information from different input modalities such as text, video, and audio, laying the foundation for subsequent processing;

The second part is the modality attention extraction module, which utilizes multi-head attention mechanisms to delve into the complex relationships within each modality, enhancing the model's understanding of specific modality features;

The third part is the cross-modal attention collaborative interaction layer, which achieves complementary and integrated information exchange between different modalities through cross-modal collaborative analysis, facilitating the model to capture richer cross-modal interaction information;

The fourth part is the emotion prediction layer, which uses multilayer perceptrons to comprehensively judge and make decisions based on the information processed in the previous three parts, outputting the final model prediction results. This module not only integrates features from various modalities but also optimizes the flow and processing of information, ensuring high-quality and accurate model outputs.

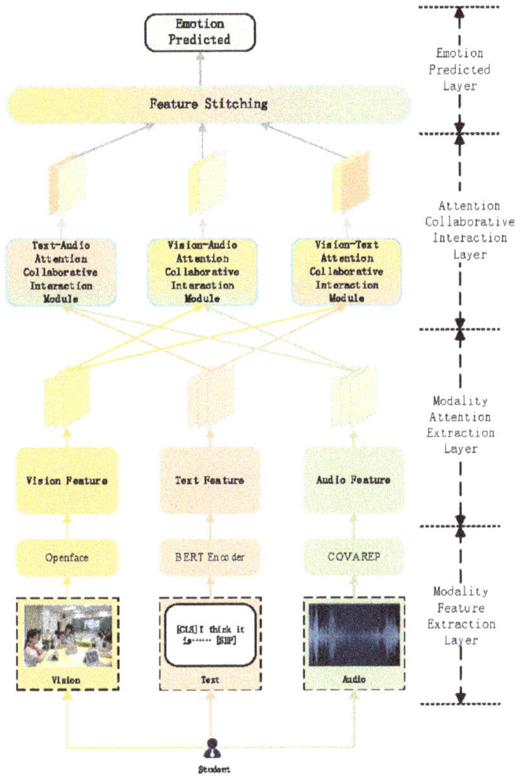

Figure 5. ACAM architecture diagram.

3.2. Modality Feature Extraction Module

This article employs three main tools and techniques to extract feature vectors from different modalities: the BERT pre-processing model, COVAREP, and OpenFace, used for extracting features from text, audio, and visual modalities, respectively.

Firstly, the model utilized the BERT pre-processing model for word embedding. The BERT (bidirectional encoder representations from transformers) model is a leading pre-trained language-representation model, involving multiple key steps in the word-embedding process. After processing through the BERT structure, we obtained feature vectors for the entire text and for each word. These feature vectors encapsulated contextual information within the text and were directly used as inputs for subsequent downstream tasks. Additionally, the pre-trained BERT model itself possesses powerful capabilities for learning contextual information, which were embedded into our model as an upstream structure and to dynamically adjust word embeddings during model training.

Next, for audio modality data, this model utilized COVAREP [10] for feature extraction. COVAREP is a tool used to extract speech features, primarily including fundamental frequency, voice intensity, and spectral envelope, among other basic acoustic parameters. In the experiments with this model, the audio signal was framed in frames of 20 milliseconds, and acoustic parameters were extracted, including fundamental frequency in the range of 50 Hz to 400 Hz and voice intensity between 40 and 90 dB. Additionally, spectral analysis was conducted to capture the energy distribution of the audio signal at different frequencies.

Finally, for visual modality data, this model employed OpenFace [11] for feature extraction. OpenFace is a tool used for facial expression recognition and facial feature extraction, capable of extracting facial expressions, actions, geometric features, and other information from images or videos. In the processing of visual modality, the images were first processed and face detection was performed, then OpenFace was used to extract facial features. OpenFace utilizes deep learning techniques such as convolutional neural networks (CNNs), residual networks (ResNet), etc. Through these deep learning models, OpenFace efficiently identified facial regions and extracted rich facial features. These features included facial expressions, facial actions, facial geometric features, etc., providing important visual information for the multimodal model.

3.3. Modality Attention Extraction Module

An attention mechanism can be utilized to enhance a model's focus on important parts of a time series. For instance, in a long video sequence, the model may only need to focus on a few key frames to determine the overall sentiment, while ignoring irrelevant or low-information parts. The application of attention mechanisms helps the model more accurately integrate information from different modalities such as visual, textual, and auditory, thereby improving performance. Through learning the correlations and importance between modalities, the model optimizes the use of multimodal information for decision making and prediction. Next, the principle of the module relying on the attention mechanism for feature extraction is introduced, as depicted in Figure 6.

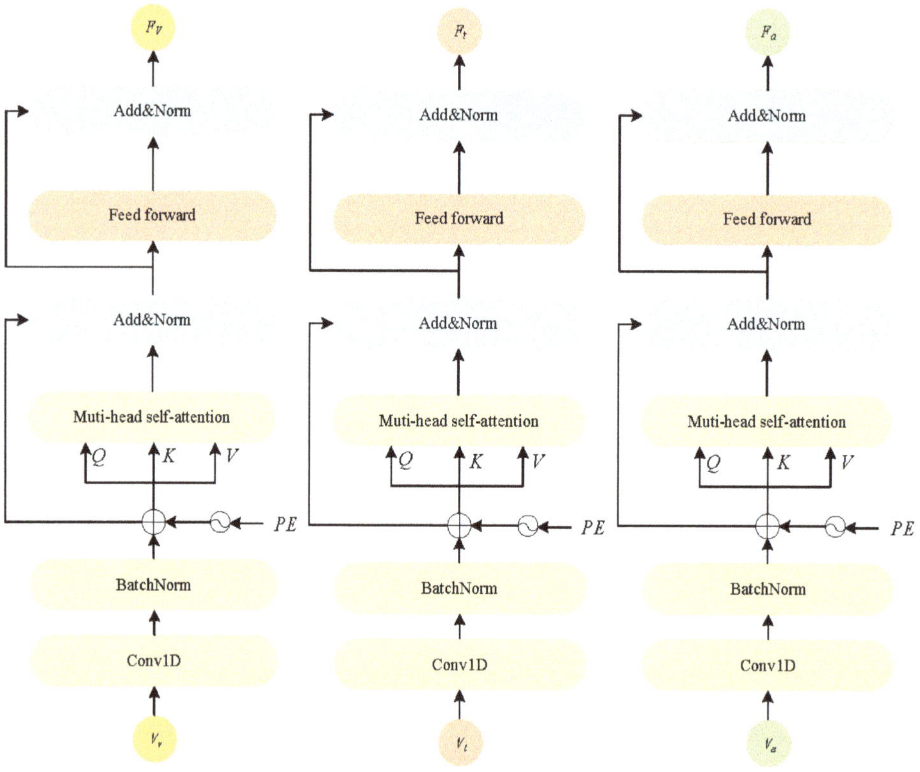

Figure 6. Modality attention extraction module structure diagram.

Firstly, through the aforementioned classroom streaming-media feature extraction module, we obtained three feature vectors with different dimensions. These feature vectors represented different modalities of information, namely textual, visual, and auditory information. In order to achieve dimensionality uniformity and facilitate subsequent processing, we input these feature vectors into one-dimensional convolutional layers and batched normalization layers to stabilize the feature distribution. Finally, we obtained three-dimensionally consistent feature representations, which we respectively named as textual features V_t, visual features V_v and auditory features V_a. After obtaining standardized feature representations, the formula is as follows, where U represents the unprocessed feature vector, $m \in \{t, v, a\}$, K corresponds to the convolution kernel size for each modality. The specific formulas are as shown in Equations (4) and (5):

$$U'_m = Conv1D(U_m, K_m) \tag{4}$$

$$V_m = BatchNorm(U'_m) \tag{5}$$

Next, we calculated positional embeddings for each position index in the sequence. Assuming the specific process followed Equations (4)–(6), the next step involved extracting self-attention mechanisms between the features of different modalities. This study utilized the superior performance of sine-based positional embedding functions as the positional encoding mechanism in the sequence model:

$$\begin{cases} PE(i, 2k) = sin\left(i/10{,}000^{2k/d_{model}}\right) \\ PE(i, 2k+1) = cos\left(i/10{,}000^{2k/d_{model}}\right) \end{cases} \tag{6}$$

Taking textual features V_t as an example, assuming that after being processed by a linear layer, the textual feature $V_t = [X_1, X_2, \ldots\ldots X_u]$ is a matrix of dimension $u \times d$, firstly, we mapped the input feature vectors V_t through linear transformations to obtain query (Q), key (K), and value (V) vectors. This mapping process was controlled by three weight matrices W_Q, W_K, W_V, which respectively acted on the input feature vectors to yield query vectors Q, key vectors K, and value vectors V. Through training, the neural network adjusted these weight matrices so that the model could better capture the correlations and importance between input features.

The specific mathematical expressions are shown in Equations (7)–(9), where there are weight matrices for learnable linear mappings $W_Q \in \mathbb{R}^{d \times u}, W_K \in \mathbb{R}^{d \times u}, W_V \in \mathbb{R}^{d \times u}$:

$$Q = W_Q V_t \tag{7}$$

$$K = W_K V_t \tag{8}$$

$$V = W_V V_t \tag{9}$$

Next, we input the query (Q), key (K), and value (V) vectors into a multi-head self-attention mechanism. This mechanism aggregated information via computing attention weights. The multi-head mechanism enabled the model to learn different feature representations because each head learned different attention weights. Finally, we linearly transformed and concatenated the results from multiple heads using an output weight matrix W_o, which served as the output of the multi-head attention mechanism. The specific formulas were as follows; here, $W_{Qi} \in \mathbb{R}^{d \times u}, W_{Ki} \in \mathbb{R}^{d \times u}, W_{Vi} \in \mathbb{R}^{d \times u}$ are the weight matrices for learned linear mappings, $W_O \in \mathbb{R}^{d \times u}$ is the weight matrix for the output linear mapping, and the concat operation concatenates the results of each head before applying the output linear transformation:

$$head_i = \text{softmax}\left(\frac{QW_{Qi}(KW_{Ki})^T}{\sqrt{d_k}}\right)VW_{Vi} \tag{10}$$

$$Att = W_O\text{concat}(head_1, head_2, \ldots, head_h) \tag{11}$$

After obtaining the output of the multi-head attention mechanism, we added it to the input feature vectors V_w and then applied layer normalization to the result. This step promoted stable training of the network and prevented gradient vanishing or explosion. Next, the output of the added and normalized layer was subjected to convolutional operations, resulting in the convolutional layer output Con. This step aided in further feature extraction and introduced local correlations to better capture structural information from the input feature vectors. Here, we utilized the ReLU function as the activation function to introduce non-linearity. The formula is depicted in Equation (12), where $W_C \in \mathbb{R}^{d \times u}, b_C \in \mathbb{R}^d$ represents the weight parameters of the convolutional layer:

$$Con = \text{ReLU}(\text{LayerNorm}(V_t + Att)W_C + b_C) \quad (12)$$

Finally, the features were further adjusted and normalized through the added and normalized layer once more to facilitate model training and performance improvement. This resulted in obtaining a textual feature representation $F_w \in \mathbb{R}^{u \times d}$ processed through the self-attention mechanism. The formula is depicted in Equation (13):

$$F_w = \text{LayerNorm}(Con + \text{LayerNorm}(V_t + Att)) \quad (13)$$

Based on the same operations, we obtained F_a and F_p.

3.4. Cross-Modal Attention Cooperative Interaction Module

The cross-modal self-attention interaction learning module is a key component aimed at effectively integrating multimodal information. Through modality interaction encoders, feature vectors from each modality are encoded to extract modality-specific information and weighted through self-attention mechanisms to highlight key information within each modality. This interactive learning strategy enables modalities to learn from each other, acquiring richer contextual information. This module not only integrates information from multimodal data but also enhances the correlations between modalities, thereby improving the model's performance and generalization capabilities.

Let us denote the text modality $F_t \in \mathbb{R}^{u \times d}$ and the visual modality $F_v \in \mathbb{R}^{u \times d}$. The attention cooperative interaction module is illustrated in Figure 7.

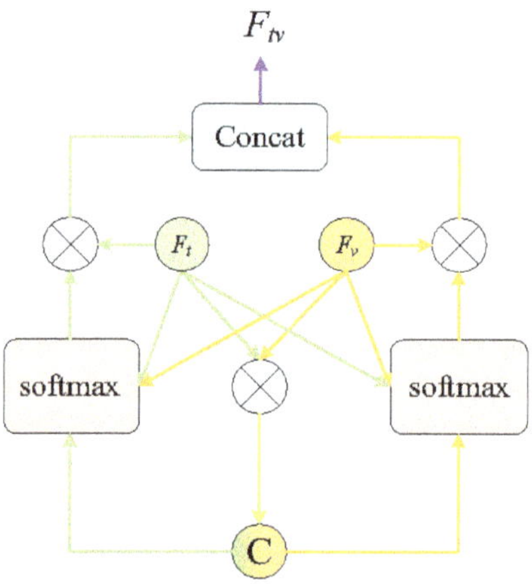

Figure 7. Collaborative attention interaction module schematic diagram.

Firstly, we computed the correlation between features from the two different modalities. We applied linear transformations to F_t and F_v using weight matrices W_a and W_b respectively, and then calculated their transpose multiplication to obtain the interaction matrix C.

Then, we combined the text and visual features using the hyperbolic tangent function to obtain bidirectional attention weights. This step helped capture the complex relationships between text and visual features. The specific formulas were from Equations (14)–(16), where $W_a \in \mathbb{R}^{d \times d}$, $W_b \in \mathbb{R}^{d \times d}$, $W_v \in \mathbb{R}^{u \times u}$, $W_q \in \mathbb{R}^{u \times u}$ are learnable weight matrices:

$$C = \tanh((F_w W_a)^T (F_p W_b)) \quad (14)$$

$$H_v = \tanh(W_q \times F_v + W_v \times F_t \times C) \quad (15)$$

$$H_t = \tanh(W_q \times F_t + W_v \times F_v \times C) \quad (16)$$

Then, we normalized the bidirectional attention weights for text and visual features using the softmax function to obtain attention weights a_v and a_t. This step allowed the network to focus attention on important feature information, thus improving feature representation. The specific formulas are shown in Equations (17) and (18), where $w_{hv} \in \mathbb{R}^{d \times 1}$, $w_{ht} \in \mathbb{R}^{d \times 1}$ is a learnable weight matrix.

$$a_v = \text{softmax}(w_{hv}^T \times H_v) \quad (17)$$

$$a_t = \text{softmax}(w_{ht}^T \times H_t) \quad (18)$$

During the feature weighting stage, the model used attention weights to weight the text and visual features, obtaining the final weighted feature representations. Through applying attention weights a_t and a_v to the text features F_t and visual features F_v, respectively, the model obtained the corresponding weighted feature vectors F_t' and F_v'.These weighted feature vectors can be seen as projections of text and image in a common semantic space, reflecting the model's understanding and expression of semantic information from text and images. Finally, the two feature vectors were concatenated to obtain the bimodal attention interaction feature vector $F_{vt} \in \mathbb{R}^{2 \times m}$. The specific formulas were from Equations (19)–(21):

$$F_v' = a_v \times F_v \quad (19)$$

$$F_t' = a_q \times F_t \quad (20)$$

$$F_{vt} = \text{concat}(F_v', F_t') \quad (21)$$

Based on the same operations, we obtained two more feature vectors F_{ta} and F_{va}.

3.5. Emotion Prediction Module

The emotion prediction module has been designed to integrate multimodal information, fully exploit the correlations between different modalities, and achieve precise prediction of emotional information using deep learning methods. This module concatenates the cross-modal feature vectors extracted via the cross-modal cooperative attention interaction module and feeds them into a multi-layer perceptron (MLP) for processing. The MLP, as a deep neural network model, is capable of nonlinear mapping and modeling complex relationships within input features, thereby better capturing the inherent patterns and semantic information among features.

After the attention processing across the three modalities described above, we obtained interaction feature vectors for each modality. These feature vectors were then fused through a tanh-activated MLP after being concatenated, ultimately yielding the predicted value $\hat{y}$, as shown in Equation (22):

$$\hat{y} = \text{MLP}(\text{concat}[F_{pw}, F_{aw}, F_{ap}]) \quad (22)$$

The model utilizes mean squared error (MSE) as the loss function. MSE is a method to measure the difference between predicted values and true values via computing the squared difference between predicted values and true values to evaluate the model's performance. Its formula is shown as Equation (23):

$$L = \frac{1}{N}\sum_{i=0}^{N}(y_i - \hat{y}_i)^2 \tag{23}$$

4. Experiments

4.1. Experimental Environment

The experiments reported in this paper were mainly encoded in a Python 3.8, Pytorch 1.7.1, CUDNN 8.0.5, Cuda 11.1 environment and trained on a computer with Intel-i5-9600k CPU (Intel, Santa Clara, CA, USA), RTX 3070 graphics card (NVIDIA, Santa Clara, CA, USA), and 32 g memory.

The model employed the dropout technique, where each neuron had a 10% probability of being randomly dropped out, helping to reduce the risk of overfitting. The batch size during training was set to 32, balancing training speed and memory utilization. The model underwent 180 epochs of training, with each epoch representing a complete traversal of the entire training dataset. We chose the AdamW optimizer to optimize model parameters, incorporating weight decay regularization. The learning rate was set to 1×10^{-4}, controlling the step size of the model's parameter updates to accelerate convergence. Additionally, we applied regularization with a weight decay rate of 1×10^{-2} to penalize the size of model parameters and prevent overfitting. These comprehensive parameter settings aimed to ensure the model had good generalization performance and efficient training. The specific model parameter settings are detailed in Table 1.

Table 1. ACAM Parameter Configuration.

Parameter	Value
Dropout	0.1
Batch size	32
Epoch	180
Optimizer	AdamW
Learning rate	1×10^{-4}
Weight decay	1×10^{-2}
Attention head	2

4.2. Dataset

The data in this study were primarily built upon the publicly available dataset CMU-MOSI. The CMU-MOSI dataset was collected by researchers at Carnegie Mellon University and aims to provide benchmark data for tasks such as emotion recognition, sentiment analysis, and multimodal emotion analysis. It consists of video, audio, and text data, with over 93 h of video data, 39 speakers, 23,453 sentences, and approximately ten million words. The emotional labels in the dataset are based on the VA affective model dimensions, namely valence, arousal, and dominance. Additionally, the dataset includes audio and text features to support multimodal emotion-analysis tasks. This dataset has been widely used in research on emotion recognition, sentiment analysis, and multimodal emotion analysis, making it a valuable resource.

4.3. Evaluation Metrics

This model's output can be used for both classification and regression tasks. It categorizes cross-modal emotional expressions into predefined emotion categories while also quantifying the intensity or subtle differences in emotions to adapt to various application scenarios and analytical needs. Therefore, this research adopted evaluation metrics for both regression and classification tasks to comprehensively assess the model's performance, including mean absolute error (*MAE*), correlation (*Corr*), accuracy (Acc), and F1 score, represented by the following formulas:

$$MAE = \frac{1}{n}\sum_{i=1}^{n}|y_i - \hat{y}_i| \tag{24}$$

MAE (mean absolute error) is one of the metrics used to evaluate the accuracy of prediction models. It measures the predictive performance of a model through calculating the average of the absolute errors between predicted values and true values. Here, $|y_i - \hat{y}_i|$ represents the absolute difference between the true values and the predicted values, and n represents the number of samples:

$$Corr = \frac{\Sigma_{i=1}^{n}(y_i - \overline{y})(\hat{y}_i - \overline{\hat{y}})}{\sqrt{\sum\limits_{i=1}^{n}(y_i - \overline{y})^2}\sqrt{\sum\limits_{i=1}^{n}(\hat{y}_i - \overline{\hat{y}})^2}} \tag{25}$$

Correlation (*Corr*) measures the degree of association between the true values of data labels and the values predicted with the model. Here, y_i represents the true values of data labels, and $\overline{\hat{y}}$ represents the predicted values of data labels.

$$\text{Accuracy} = \frac{\text{TP} + \text{TN}}{\text{TP} + \text{TN} + \text{FP} + \text{FN}} \tag{26}$$

Accuracy (Acc) represents the proportion of samples correctly predicted by the classifier out of the total sample size and is a key metric for evaluating the accuracy of classification models. In this context, TP denotes the number of true positives, TN denotes the number of true negatives, FP denotes the number of false positives, and FN denotes the number of false negatives. High accuracy implies that the model has a high level of accuracy in classifying the data.

$$F1 = \frac{2 \times \text{Precision} \times \text{Recall}}{\text{Precision} + \text{Recall}} \tag{27}$$

The *F*1 score combines a model's precision and recall, where precision represents the proportion of correctly predicted positive samples out of all samples predicted as positive, and recall represents the proportion of correctly predicted positive samples out of all actual positive samples.

4.4. Module Ablation Experiments

In this study, a series of modality ablation experiments were designed to evaluate the impact of different modalities and their fusion methods on the model's prediction performance. In the experimental setup, the symbols T, V, and A represent scenarios where predictions were made using a single modality (text, visual, audio), respectively. Correspondingly, symbols TT, VV, AA indicate scenarios where self-attention mechanisms were applied to each modality to enhance the model's processing capability for single-modality information, followed by predictions.

Additionally, this section also explores the complementarity between modalities, denoted by TA, TV, and VA, representing scenarios where two different types of modality data were fused through a cooperative attention module before predictions were made. Subsequently, the mean absolute error (MAE), correlation (Corr), accuracy (Acc), and F1 score were tested for each of nine scenarios. The specific experimental results are displayed in Figure 8.

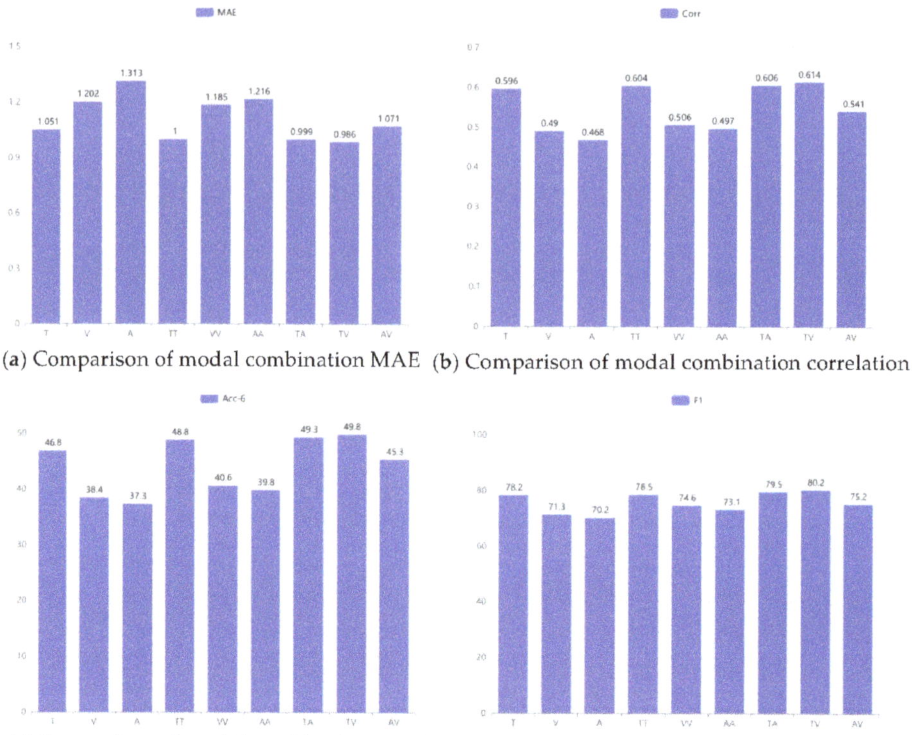

(a) Comparison of modal combination MAE (b) Comparison of modal combination correlation

(c) Comparison of modal combination Acc-6 (d) Comparison of modal combination F1 score

Figure 8. Modality ablation experiment results.

Based on the experimental data, we can draw three conclusions:

(1) Comparing the single modalities (T, V, A) with their counterparts enhanced through attention mechanisms (TT, VV, AA), it was observed that the attention mechanism effectively improved the model's performance across all three modalities of text, visual, and audio. For instance, in the text modality, the enhancement with attention mechanism (TT) reduced MAE from 1.051 to 0.1 and increased the F1 score from 78.2 to 78.5, indicating that the attention mechanism more effectively captured important features and complex relationships within each modality;

(2) When comparing the combined modalities (TA, TV, AV) with single modalities and their enhanced counterparts, we observed that the cooperative attention module further improved performance. Particularly in the text–audio (TA) and text–visual (TV) combinations, MAE, correlation (Corr), Acc-6, and F1 scores were all superior to the single modality T and the enhanced modality TT. This reflects the effectiveness of the cooperative attention module in integrating different modality information and extracting richer and complementary features;

(3) The superiority of the text modality was evident in the data, particularly in several key performance indicators. The text modality had an MAE of 1.051 and an F1 score of 78.2, whereas the visual and audio modalities had MAE values of 1.502 and 1.313, respectively, and F1 scores of 71.3 and 70.2. The research results indicated that in emotion-recognition tasks, the text modality provided more accurate and relevant predictions.

This is primarily because text data contains rich semantic and syntactic information, which are important factors for accurately capturing and understanding emotions. The richness of text enables the model to deeply analyze and interpret emotional states, which is often not achievable in visual or audio modalities. In the field of natural language processing, a significant amount of research and technological advancements have focused on how to extract features from text, such as word embeddings, contextual embedding techniques, and pre-trained models like BERT in particular. Due to their ability to capture deep semantic relationships, text feature extraction has become more efficient with models like BERT.

Additionally, in classroom streaming datasets, audio quality may be affected by background noise, student whispers, or speakers being too far from the microphone, resulting in unclear audio. Moreover, there may be modalities missing due to students not speaking during streaming sessions. Also, visual information may lose critical non-verbal cues due to scene restrictions or occlusion. In contrast, text data tends to be more complete due to collaborative organization through online teaching platforms and speech recognition. This explains why the performance of the text modality outperformed other modalities in modal ablation experiments.

4.5. Network Comparison Experiment

In this section, in order to evaluate the performance of the model proposed in this paper, it is compared with the evaluation indexes of the algorithm models ResNet50 [12], BERT [13], LMF [14], TFN [15], MFN [16], MulT [17] and GMFN [18].The experimental results are shown in Table 2.

Table 2. Results of the Network Comparison Experiment.

CSMD	Mae	Corr	Acc-6	F1
ResNet50	1.212	0.505	39.1	68.2
Bert	1.032	0.579	46.8	77.0
LMF	1.074	0.569	45.1	75.1
TFN	1.057	0.577	46.5	76.4
GMFN	1.014	0.58	48.0	77.5
MFN	0.989	0.594	48.2	77.2
MulT	0.984	0.612	49.3	79.2
ACAM	0.976	0.618	49.9	80.5

Firstly, the ResNet50 and BERT models, which utilized only the visual and text modalities, lacked deep fusion and complementary information between modalities. This limitation restricted their performance in complex tasks compared with multimodal models that can comprehensively leverage multiple sources of information.

Although LMF simplifies computation through reducing the rank of modality features, its ability to handle complex relationships is limited, as evidenced by its unsatisfactory performance on Mae and Corr metrics. TFN, using higher-order tensors to simulate complex interactions between modalities, exhibited improved performance, but its high computational complexity restricts its practical application range. GMFN, which introduces graph structures to capture dynamic relationships between modalities, showed better performance, but its reliance on graph structures increases the model's design complexity. MFN attempts to integrate different modalities through adaptive learning, achieving relatively good results, but it still fell short in handling nonlinear complex relationships. MulT utilizes cross-attention mechanisms to handle text, visual, and auditory modalities, achieving significant progress with multiple metrics, yet its understanding of long-term dependencies and deep interactions between modalities remains limited. The MISA model achieved higher performance through reinforcing intra-modal and cross-modal self-attention mechanisms, achieving finer information fusion, but there was still room for improvement in capturing and utilizing modality-specific information.

From Table 2, it is evident that the ACAM model outperformed the aforementioned models on several key performance metrics. Its Mae was lower and Corr was higher compared with competitors. Additionally, the Acc-6 and F1 scores reached 49.9% and 80.5%, respectively, demonstrating its significant advantages in multimodal fusion for classroom assessment tasks. This performance improvement not only reflects the progress of ACAM in modal fusion accuracy but also underscores its efficient capability in understanding and utilizing multimodal information.

5. Conclusions

Artificial intelligence-assisted education has brought great possibilities to educational teaching. In the past, learner assessment often relied on exam scores, homework assignments, etc. However, the emergence of artificial intelligence enables learner assessment based on comprehensive learning process data, forming learner profiles that fully showcase students' emotions, attention, and cognitive states. This study conducted collaborative analysis based on learners' multimodal data to explore their potential emotional states, thereby providing assistance for personalized teaching guidance.

This paper proposes a collaborative perception model based on attention mechanism, which innovatively achieves effective fusion of multimodal data, fully considering the complementarity between modalities and enhancing the model's dynamic adaptation to different modal information. Experimental results demonstrated that the model outperformed traditional methods in emotion recognition, exhibiting high accuracy and robustness in practical applications.

However, from the perspective of optimizing modal alignment, although the model proposed in this study effectively integrates multimodal data, it is difficult to ensure accurate matching and synchronization of information between different modalities; i.e., the problem of modal alignment remains a major challenge. Future work can deepen modal alignment technology, such as using advanced time-alignment algorithms to process asynchronous data, or using deep learning algorithms to automatically identify and synchronize the correlation between different modalities, in order to improve the accuracy and efficiency of emotion recognition.

Author Contributions: Methodology, W.W., J.Z. and X.S.; Data curation, X.S.; Writing—original draft, W.W.; Supervision, G.F. All authors have read and agreed to the published version of the manuscript.

Funding: This research was funded by Guangdong Provincial Philosophy and Social Science Planning Project of China grant number [GD23YJY08] and by National Natural Science Foundation of China [62237001].

Data Availability Statement: The original contributions presented in the study are included in the article, further inquiries can be directed to the corresponding author.

Conflicts of Interest: The authors declare no conflict of interest.

References

1. Wei, Q.; Sun, B.; He, J.; Yu, L. BNU-LSVED 2.0: Spontaneous multimodal student affect database with multi-dimensional labels. *Signal Process. Image Commun.* **2017**, *59*, 168–181. [CrossRef]
2. Trabelsi, Z.; Alnajjar, F.; Parambil, M.M.A.; Gochoo, M.; Ali, L. Real-Time Attention Monitoring System for Classroom: A Deep Learning Approach for Student's Behavior Recognition. *Big Data Cogn. Comput.* **2023**, *7*, 48. [CrossRef]
3. Ma, L.; Lu, Z.; Shang, L.; Li, H. Multimodal convolutional neural networks for matching image and sentence. In Proceedings of the IEEE International Conference on Computer Vision, Washington, DC, USA, 7–13 December 2015; pp. 2623–2631.
4. Poria, S.; Cambria, E.; Bajpai, R.; Hussain, A. A review of affective computing: From unimodal analysis to multimodal fusion. *Inf. Fusion.* **2017**, *37*, 98–125. [CrossRef]
5. Pantic, M.; Rothkrantz, L. Facial Action Recognition for Facial Expression Analysis From Static Face Images. *IEEE Trans. Syst. Man Cybern.-Part B* **2004**, *34*, 1449–1461. [CrossRef] [PubMed]
6. Bashyal, S.; Venayagamoorthy, G. Recognition of Facial Expression Using Gabor Wavelets and Learning Vector Quantization. *Eng. Appl. Artif. Intell.* **2008**, *21*, 1056–1064. [CrossRef]
7. Cheon, Y.; Kim, D. Natural Facial Expression Recognition Using Differential-AAM and Manifold Learning. *Pattern Recognit.* **2009**, *42*, 1340–1350. [CrossRef]
8. Shan, C.F.; Gong, S.G.; Peter, W. A Comprehensive Empirical Study on Linear Subspace Methods for Facial Expression Analysis. In Proceedings of the IEEE Computer Society Conference on Computer Vision and Pattern Recognition Workshop, Washington, DC, USA, 17–22 June 2006; pp. 153–158.
9. Chen, G.; Zhang, P.; Wang, T. A review of multimodal sentiment analysis. *J. Commun. Univ. China (Nat. Sci. Ed.)* **2022**, *29*, 70–78.

10. Lin, S.; Zhang, X.; Guo, M. Research on emotion recognition methods based on audio and video. *Signal Process.* **2021**, *37*, 1889–1898.
11. Degottex, G.; Kane, J.; Drugman, T.; Raitio, T.; Scherer, S. COVAREP—A collaborative voice analysis repository for speech technologies. In Proceedings of the 2014 IEEE International Conference on Acoustics, Speech and Signal Processing (ICASSP), Florence, Italy, 4–9 May 2014; pp. 960–964.
12. Baltrusaitis, T.; Robinson, P.; Morency, L.P. OpenFace: An open source facial behavior analysis toolkit. In Proceedings of the 2016 IEEE Winter Conference on Applications of Computer Vision (WACV), Lake Placid, NY, USA, 7–10 March 2016; pp. 1–10.
13. He, K.; Zhang, X.; Ren, S.; Sun, J. Deep residual learning for image recognition. In Proceedings of the IEEE Conference on Computer Vision and Pattern Recognition, Las Vegas, NV, USA, 7–10 March 2016; pp. 770–778.
14. Devlin, J.; Chang, M.W.; Lee, K.; Toutanova, K. BERT: Pre-training of deep bidirectional transformers for language understanding. In Proceedings of the 2019 Conference of the North American Chapter of the Association for Computational Linguistics: Human Language Technologies, Volume 1 (Long and Short Papers), Minneapolis, MN, USA, 2–7 June 2019; pp. 4171–4186.
15. Sahay, S.; Okur, E.; Kumar, S.H.; Nachman, L. Low rank fusion based transformers for multimodal sequences. In Proceedings of the 58th Annual Meeting of the Association for Computational Linguistics, Online, 5–10 July 2020; p. 2934.
16. Zadeh, A.; Chen, M.; Poria, S.; Cambria, E.; Morency, L.-P. Tensor fusion network for multimodal sentiment analysis. In Proceedings of the 2017 Conference on Empirical Methods in Natural Language Processing, Copenhagen, Denmark, 9–11 September 2017; pp. 1103–1114.
17. Zadeh, A.; Liang, P.P.; Mazumder, N.; Poria, S.; Cambria, E.; Morency, L.-P. Memory fusion network for multi-view sequential learning. In Proceedings of the AAAI Conference on Artificial Intelligence, New Orleans, LA, USA, 2–7 February 2018; pp. 5634–5641.
18. Tsai, Y.H.; Bai, S.; Liang, P.P.; Kolter, J.Z.; Morency, L.-P.; Salakhutdinov, R. Multimodal transformer for unaligned multimodal language sequences. In Proceedings of the 57th Annual Meeting of the Association for Computational Linguistics, Florence, Italy, 28 July–2 August 2 2019; pp. 6558–6569.

applied sciences

MDPI

Article

Computational Thinking Measurement of CS University Students

Raquel Hijón-Neira [1,*], Celeste Pizarro [2], John French [3], Daniel Palacios-Alonso [1] and Emre Çoban [4]

1 Computer Science Department, Rey Juan Carlos University, 28933 Móstoles, Spain; daniel.palacios@urjc.es
2 Applied Mathematics Department, Rey Juan Carlos University, 28933 Móstoles, Spain; celeste.pizarro@urjc.es
3 Computer Science & Applied Physics Department, Atlantic Technological University, H91 T8NW Galway, Ireland; john.french@atu.ie
4 Computer Technology Department, Ardahan University, 75000 Ardahan, Türkiye; emrecoban@ardahan.edu.tr
* Correspondence: raquel.hijon@urjc.es

Featured Application: Altogether, UniCTCheck, a method for measuring the main components of computational thinking of CS university students, was employed by using the following two instruments: firstly, a web application, CTScore, to precisely measure seven of the main components of computational thinking (pattern recognition, creative thinking, algorithmic thinking, problem solving, critical thinking, decomposition, and abstraction); secondly, a psychometric scale, CTProg, to measure CT programming concepts skills (basic directions and sequences, condi-tionals, loops, functions, and data structures).

Abstract: The measurement of computational thinking ability among computer science (CS) university students is of paramount importance. This study introduces UniCTCheck, a novel method designed to assess the main components of computational thinking in CS students. Utilising two key instruments, namely, the web application CTScore and the psychometric scale CTProg, this research aims to precisely evaluate seven core components of computational thinking and six programming concepts skills essential for CS students. The study, conducted at Rey Juan Carlos University and Atlantic Technological University Galway, involved a diverse sample of students from different year levels and programme specialisations. Through a rigorous research design, including sampling strategies and data collection tools, this study seeks to address critical research questions related to the measurement of variations in students' computational thinking and programming skills by gender, university level, and location. By shedding light on the significance of computational thinking and programming in the educational realm, this research contributes to the existing literature and underscores the essential role of computational skills in the modern era.

Keywords: computational thinking; programming; performance-based assessment platform; interactive web app; evaluation methods; psychometric scale; student evaluation

1. Introduction

The act of methodically analysing a particular issue and coming up with an answer is known as computational thinking. It consists of a particular type of skill group that is predicated on approaches and abilities for solving problems. The goal of this skill set is to comprehend, analyse, and create solutions for difficult situations. According to [1–3], people with computational thinking abilities are supposed to be able to process information like computer scientists and use this method of thinking in all fields. Consequently, it may be concluded that the development of 21st century skills and computational thinking ability are closely related. According to [4], computational thinking aids in the understanding of computable problems as well as the selection of appropriate tools and approaches for problem solving.

Almost every field of study appears to have some connection to computational thinking, from biology to machine learning [5,6]. The term "computational thinking" was coined to refer to problem-solving techniques that also make it possible to use modern technologies efficiently in a variety of contexts [1]. Organisations have acknowledged the necessity for problem solving and the efficient use of information and communication technology, notwithstanding the lack of agreement among them. In this context, information societies—the age and society of today—need computational thinking. In order for people to coexist peacefully in society, it is vital to identify the knowledge, skills, and abilities that we require. Such studies are necessary for addressing the obstacles that societies—students in particular—will encounter in the years ahead. It is observed that common components are employed in the literature, despite the fact that the term computational thinking and its components remain unaltered [7–9].

In addition to explorations of computational thinking and its elements, another subject of research involves how to quantify these elements. It has been observed in previous research that several measurement instruments are used to assess computational thinking. Numerous other techniques have been used with various approaches, including scales, portfolio studies, programming, multiple choice exams, task-based tests, observations, and rubrics [10–12]. However, some believe that the current approaches of measuring computational thinking may not be adequate. In fact, a student who performs well on examinations could nevertheless be an incompetent or insensitive algorithm creator, according to [13]. The author of [13] further underlined that, while we do not assess pupils' competencies or sensitivities, we do know what they know. He advised treating the components of computational thinking as a talent and grading them accordingly. He stated that the assessment of abilities in school is a pretty typical practice and cited examples that included language, athletics, music, and theatre. By using this study to create a valid and trustworthy performance-based assessment tool, the framework seeks to quantify computational thinking abilities. Upon reviewing the literature, it becomes evident that the studies used to assess computational thinking have several drawbacks, including being confined to a narrow framework and relying on computer software in a traditional and dull manner [14,15]. Moreover, the measurement of computational thinking abilities in many of these investigations requires the expertise of an expert.

Psychometric scales are among the most widely used instruments for assessing computational thinking [12,14–17]. However, the foundation of psychometric measures is the idea that a subject provides accurate and complete information [18]. It is anticipated that the person has an inner vision, nonetheless [19]. Performance-based assessments can be more appropriate because psychometric scales have these drawbacks and computational thinking components should be treated as skills [13]. Furthermore, research on the measurement of computational thinking using performance-based assessments has highlighted the significance of receiving feedback at the appropriate moment [20,21]. Performance-based assessments offer numerous benefits, including their capacity to evaluate intricate and advanced cognitive abilities, and their connection to the educational process [22–24].

As computational thinking is a relatively new and growing skill, there is currently no agreement on its description or its constituent parts [6,8,9,25,26]. Therefore, the current theoretical framework was used to conduct this investigation. The purpose of this study was to find a method to effectively measure computational thinking abilities in CS university students. However, numerous studies in the literature have found that the only instruments available for measuring computational thinking are limited perception–attitude scales, multiple choice tests, or simple coding tests [10–12,14–16,27]. Therefore, a performance-based platform, CTScore, for evaluating computational thinking abilities was created in [28]. Additionally, it has been determined that performance-based assessment instruments provide more accurate results than psychometric scales. It is stressed that computational thinking should not be viewed as theoretical knowledge, a product or portfolio, or an attitude or perception, but rather as a talent or set of skills.

The hypotheses of this research are as follows:

H1. *It is possible to develop UniCTCheck, a method for measuring the main components of computational thinking, in the context of CS university courses.*

H2. *It is possible to precisely measure seven computational thinking components (pattern recognition, creative thinking, algorithmic thinking, problem solving, critical thinking, decomposition, and abstraction) by using CTScore.*

H3. *It is possible to use a psychometric scale, CTProg, for measuring CT programming concepts skills that all CS university students should address sooner rather than later (basic directions and sequences, conditionals, loops, functions, and data structures).*

Given the paucity of research and advancement on the topic, it is believed that computational thinking has a vital role nowadays and that this study will significantly add to the body of literature. Therefore, following on from on the hypotheses above, we focus on addressing the following research problems in this work:

- RQ1: Does UniCTCheck measure the main components of computational thinking and programming for CS university students?
- RQ2: Is there a relationship between students' computational thinking skills and programming abilities?
- RQ3: Do students' computational thinking skills and programming abilities vary by university level?
- RQ4: Do students' computational thinking skills and programming abilities vary by university location?
- RQ5: Do students' computational thinking skills and programming abilities vary by gender?

1.1. Computational Thinking

Despite lacking a precise definition, computational thinking has a rich history dating back to the early days of computer science. The phrase "computational thinking" first appeared in George Polya's 1940 book *How to Solve It*, which discussed mental disciplines and techniques (Denning, 2017 [13]; Tedre and Denning, 2016 [29]). Then, in 1960, Alan Perlis coined the term "algorithmizing". Perlis predicted that computers would automate procedures in every industry by using this word to describe practice and thought (Katz, 1960 [30]). Three of the field's pioneers—Newell, Perlis, and Simon—discussed computer science in a 1967 publication. This published work considers "algorithmic thinking" to be a phrase that sets computer science apart from other disciplines [31]. In 1980, Seymour Papert introduced the phrase "computational thinking" in his book *Mindstorms: Children, Computers, and Powerful Ideas* [32]. Even though Papert created the phrase first, no one had previously defined computational thinking in the very beginning. Papert used the phrase computational thinking a lot in his publications in the years that followed [33–35]. In the 1950s and 1960s, in particular, "algorithmic thinking" was the term used before "computational thinking" [36–38]. One of the fathers of computer science, Edsger Dijkstra, defined three characteristics of algorithmic thinking: the ability to move back and forth between semantic levels; the ability to reveal one's own form and concept while solving problems; the ability to use one's mother tongue as a bridge between informal problems and formal solutions [39].

The term "computational thinking" was originally used in Jeannette M. Wing's widely read paper, Computational Thinking [27,40–42]. According to Wing, the phrase computational thinking encompasses problem solving, system design, and behavioural analysis. Additionally, she said that not only computer scientists but everyone should use computational thinking [1]. Despite being a seminal contribution to the area, Wing's 2006 definition has drawn a great deal of criticism [29]. Wing's definition has drawn criticism for using an imprecise and confusing phrase [26]. Moreover, her unverified assertions regarding the all-encompassing advantages of computational thinking have been deemed audacious [40].

Wing's article Computational Thinking and Thinking about Computing from 2008 brought the phrase computational thinking back into focus. Wing compared analytical and computational thinking in this paper. She pointed out that scientific, engineering, and mathematical reasoning are all entwined with computational thinking. Furthermore, she underlined once more that computational thinking applies to everyone and in all contexts [2]. A variety of publications have been released in the years that followed to provide a clearer definition of the phrase computational thinking, and Thinking About Computational Thinking is one of them [43]. Rather than expressing criticism, Lu and Fletcher have declared their support for Wing's 3R (writing, reading, and math) rule and for placing computational thinking at the centre of computer science. Nonetheless, debates over Wing's concept have persisted in the literature. Following this, Canadian computer scientist Alfred V. Aho wrote a piece on computational thinking for Alan Turing's 100th birthday celebration in 2012. In this piece, Aho underlined the importance of the precise and intelligible identification and the significance of settling on a scientific field's nomenclature for the sake of comprehension [44].

On one hand, educators have contributed to Wing's development of the term computational thinking as well as Papert's traditional formulation through hundreds of workshops, committees, studies, polls, and public evaluations [13,45]. On the other hand, it is impossible to discuss a widely agreed-upon definition and list of elements for computational thinking. Communities like the CSTA, CAS, and ISTE, which are regarded as leading and respected organisations, have not been able to come to an agreement on the components of computational thinking [46–48]. This disagreement is in addition to the divergent opinions of researchers regarding the definition of computational thinking. However, based on a review of the literature, it appears that there are recurring themes that are well acknowledged under the heading of the suggested definition [9]. According to educators, computational thinking is particularly beneficial for enhancing logical reasoning, growing analytical thinking, and problem-solving abilities [49].

Despite its rich history and significance in computer science, computational thinking (CT) lacks a universally accepted definition. Initially discussed by George Polya, and later popularised by Seymour Papert and Jeannette M. Wing, CT encompasses problem solving, system design, and behavioural analysis. Wing's broad claims about CT's applicability have sparked debate, with critics highlighting the need for precise terminology. Leading organisations and educators continue to refine the concept, emphasising themes like logical reasoning, analytical thinking, and problem solving, though consensus on a definitive list of CT components remains elusive.

1.2. Components of Computational Thinking

Problem-solving techniques are unquestionably a part of computational thinking [3,8]. Computational thinking is described by Wing in her well-known article as problem solving, system design, and behavioural analysis. She also said that, as opposed to thinking like a computer, computational thinking involves addressing issues in an efficient and original way. Nonetheless, she stressed that computational thinking encompasses methods like recursion, systematic testing, problem analysis, abstraction, and problem re-explanation [1]. The concept of computational thinking and its constituent parts have sparked a great deal of discussion in the years that followed. The components have been recommended by communities, educators, and pioneering computer scientists (see Table 1). Computational thinking includes testing hypotheses, managing data, parallelism, abstraction, and debugging according to the National Research Council [50]. Barr and Stephenson provided an organised approach that allowed them to separate computational thinking into two types of concepts and abilities. According to [51], the capabilities include automation, abstraction, simulation, issue decomposition, data collecting, analysis, and representation. In [52], the authors categorised computational thinking into the following three categories: views, applied methods, and concepts. However, this categorisation only pertains to programming. Gouws et al. separated computational thinking into two categories in order to develop a framework. A set of abilities related to computational thinking is included in the first of these aspects. Processes and transformations, models and transformations, logic and inference, patterns and algorithms, assessments, and enhancements are some examples of these. The idea that various skills—such as identifying, comprehending, applying, and assimilating—have varying degrees of experience defines the second dimension [53]. Another study [54] classified computational thinking as abstraction, generalisation, and trial-and-error processes. Similar component recommendations were made in a study that sought to draw a link between learning theories and computational thinking. These elements encompass heuristic scanning, trial-and-error, abstract and divergent thinking, recursion, iteration, collaborative approaches, metacognition, patterns, and synectics [55]. According to a recently released study, computer programmers frequently employ computational thinking as a problem-solving strategy. Problem analysis, pattern identification, abstraction, algorithm creation for answers, and the evaluation of solutions (including debugging) are some of its constituents [56].

Prominent organisations have also released research on computational thinking components, in addition to the academics (see Table 2). Beyond contributing to the communities, studies on computational thinking published by well-known tech giant Google have also added to the body of literature. Pattern recognition, decomposition, pattern generalisation and algorithm design, and abstraction are the four areas into which Google separated computational thinking [57].

Computational thinking (CT) encompasses a range of problem-solving techniques, including recursion, systematic testing, problem analysis, abstraction, and debugging. Key components identified by various scholars and organisations include algorithm building, data management, hypothesis testing, and parallelism. Major frameworks from entities like the National Research Council, ISTE, and Google emphasise similar elements such as pattern recognition, decomposition, abstraction, and algorithm design. Despite differing categorisations, the core skills of CT focus on logical reasoning, problem decomposition, and the efficient creation and evaluation of algorithms.

To measure these components effectively, this study introduces UniCTCheck (Figure 1), a comprehensive framework that includes the CTScore web application for assessing seven distinct CT elements (Figure 1, left) and the CTProg psychometric scale for evaluating five computational programming concepts (Figure 1, right). This innovative approach aims to provide a precise and holistic evaluation of computational thinking skills in university-level computer science students.

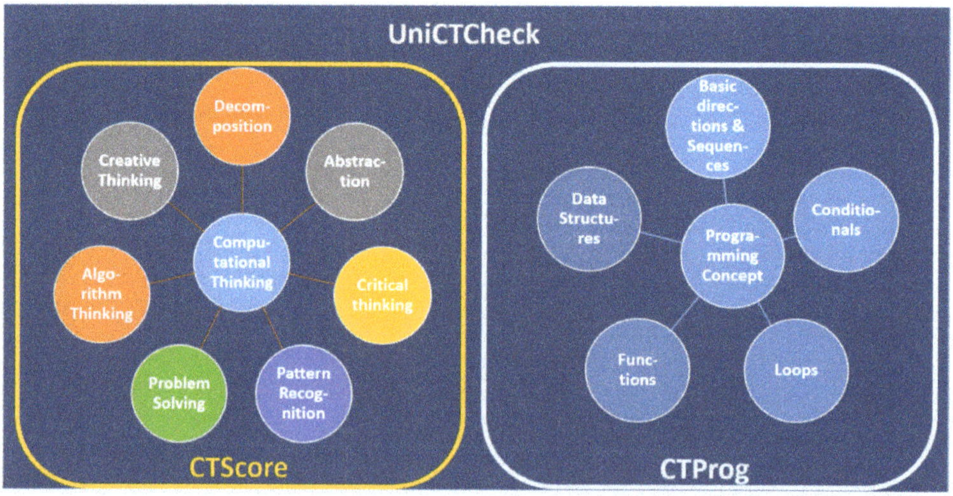

Figure 1. The components of UniCTCheck: left, CTScore; right, CTProg.

Table 1. Computational thinking skills defined by the researchers.

Wing, J. [1]	**Dierbach, C. et al. [58]**	**Berland, M. and Lee, V.R. [59]**
Algorithm building Conditional logic Debugging Distributed processing Simulation	Algorithm building Problem decomposing Developing a computational model Evaluating a problem	Algorithm building Conditional logic Debugging Distributed computation Simulation
Lee, I. et al. [41]	**Brennan, K and Resnick, M. [52]**	**Selby, C. and Woollard, J. [9]**
Abstraction Analysing the model Automation Defining a problem Understanding/solving problems	Abstracting and modularising Creative thinking Debugging Decomposing Reusing and remixing Testing and debugging	Abstraction Algorithm building Evaluating solutions Problem decomposing Generalisations Problem solving

Source [60].

Figure 1 illustrates the two main components of the UniCTCheck framework, including CTScore and CTProg. The left side of the figure represents CTScore, a web application designed to measure seven core components of computational thinking (CT). These components are pattern recognition, creative thinking, algorithmic thinking, problem solving, critical thinking, decomposition, and abstraction. The right side of the figure shows CTProg, a psychometric scale aimed at assessing critical programming concepts essential for computer science students. These programming concepts include basic directions and sequences, conditionals, loops, functions, and data structures. Together, CTScore and CTProg provide a comprehensive evaluation method for assessing the computational thinking and programming skills of university-level computer science students. Both were offered in English (https://emrecoban.com.tr/bd/ and https://forms.office.com/e/JXDaTriwi9) and Spanish (https://emrecoban.com.tr/bd/es/ and https://forms.office.com/e/3mbvMXw1vX) (accessed on 10 June 2024).

Table 2. Computational thinking components defined by the communities.

Community	Framework	Components
ISTE	Standards for Students in CT (2016) [48]	Leverage the power of technological methods to develop and test solutions Collect the data Analyse the data Represent the data Decomposition Abstraction Algorithms Testing Parallelisation Simulation
CAS	Concepts of CT (2015) [61]	Logical reasoning Algorithmic thinking Decomposition Generalisation Patterns Abstraction Representation Evaluation
CSTA	Concepts of CT (2011) [47]	Formulating problems for a computational solution Logically organising and analysing the data Abstractions, including models and simulations Algorithmic thinking Evaluation for efficiency and correctness Generalising and transferring to other domains

Denning [13].

2. Materials and Methods

2.1. Research Design

A quantitative study was conducted to measure the computational thinking ability of students studying the subjects "Visual Programming" and "Games for web and social networks" of the Video Games Degree of the ETSII, URJC, and of students studying the subjects of "Software Design and Program Development" and "Software Engineering" of the B.Sc. in Software Development at Atlantic Technological University (ATU), Galway, Ireland. A first-year subject and a final-year subject were selected at each university. The students used an interactive web application called CTScore for measuring computational thinking abilities [28], and also a psychometric programming scale (CTProg) to measure programming skills as a complement to the CTScore web application. Figure 2 shows the outline of the experimental design.

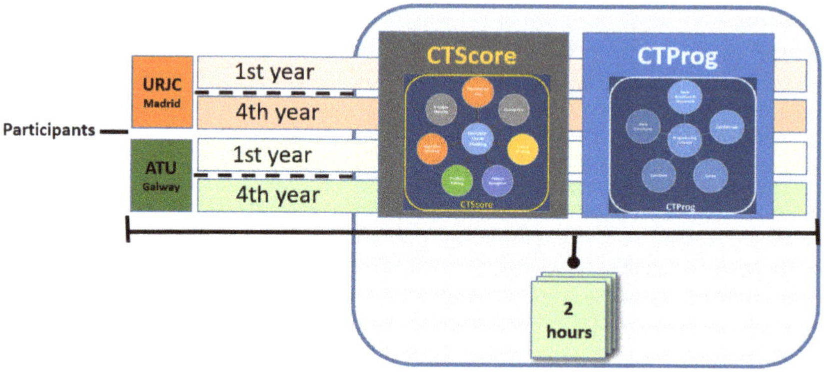

Figure 2. Outline of the experiment settings and components.

2.2. Sampling

The sample for this study was drawn from students studying in two European universities, Universidad Rey Juan Carlos (URJC) in Madrid, Spain, and Atlantic Technological University (ATU) in Galway, Ireland. In both settings, the students were studying a Computer Science degree programme in which the researchers teach. At URJC, of the 75 students enrolled in the 1st year subject "Introduction to Programming", 63 completed the 4 tests (19 females, 44 males, 0 other). Of the 90 URJC students taking the 4th year subject "Games for the web and social networks", 80 students completed the 4 tests (14 female, 65 males, 1 other). Both subjects are part of the Design and Development of Video Games degree programme at URJC. At ATU Galway, from a group of 25 students enrolled in the 1st year subject "Software Design and Program Development", 15 students completed the 4 tests (1 female, 13 males, 1 other). Of the 30 students enrolled in the 4th year subject "Software Engineering", 19 students completed the 4 tests (1 female, 18 males, 0 other). Both subjects are part of the B.Sc. in the Software Development degree programme at ATU Galway.

2.3. Ethics Committee

The Research Ethics Committee of Rey Juan Carlos University has favourably evaluated this research project, with internal registration number 1909202332223 to pursue the research at both universities (URJC and ATU) under the same research project.

2.4. Data Collection Tools

UniCTCheck collects data from two tools, CTScore and CTProg. In the following subsections, their scope and characteristics will be explained.

2.4.1. CTScore

CTScore is a validated interactive web application for measuring computational thinking abilities in students [31]. In this paper, the web app is offered in English https://emrecoban.com.tr/bd/ (accessed on 10 June 2024) and Spanish https://emrecoban.com.tr/bd/es/ (accessed on 10 June 2024) for the first time. CTScore measures seven computational thinking components, namely, abstraction, pattern recognition, critical thinking, decomposition, creative thinking, algorithmic thinking, and problem solving, by asking students to answer 12 questions (see Figure 3). During the evaluation, students were evaluated automatically with the same scoring key, where a maximum of 10 points and a minimum of 0 (zero) points could be obtained from each question. Accordingly, each question determines the total score of the component in which it is located. For example, since the abstraction component contains two questions, the maximum score that students can achieve in this component will be 20 points.

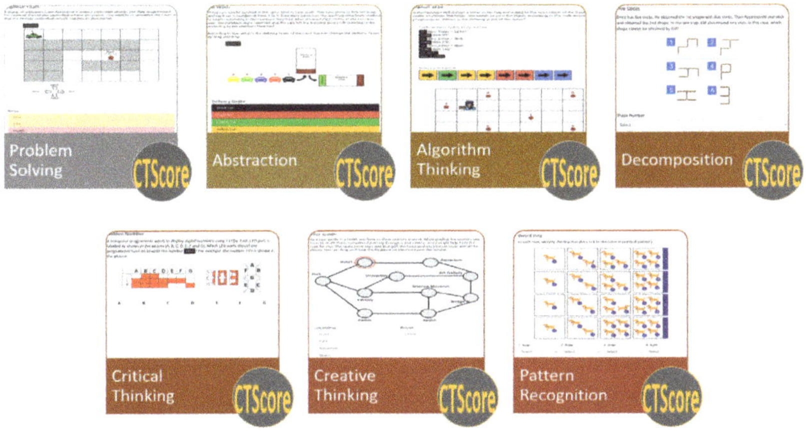

Figure 3. Example questions for each component of CTScore.

2.4.2. CTProg

CTProg is a psychometric scale test created by the authors to measure the conceptual understanding of 5 core programming fundamentals, not depending on any particular programming language. It is formed by 14 questions, providing a schematic block programming language that university students can understand independently of their knowledge of specific programming languages. The tests were used with university students and revised by the researchers over the course of three academic years (2017–2018, 2018–2019, and 2019–2020), continually evolving during this period. The main concepts addressed by the 14 questions included basic directions and sequences, conditions, loops, functions, and data structures (arrays and matrices). CTProg is offered both in English (https://forms.office.com/e/JXDaTriwi9) (accesed on 10 June 2024) and Spanish (https://forms.office.com/e/3mbvMXw1vX) (accessed on 10 June 2024). During the evaluation, students were evaluated automatically with the same scoring key, where a maximum of 1 point and a minimum of 0 (zero) points could be obtained from each question. Accordingly, each question determines the total score of the programming concept in which it is located (see Figure 4). Each of the main concepts addressed contain two questions, but the loop concept contains four questions; therefore, the maximum score that students can achieve in this component is 14 points.

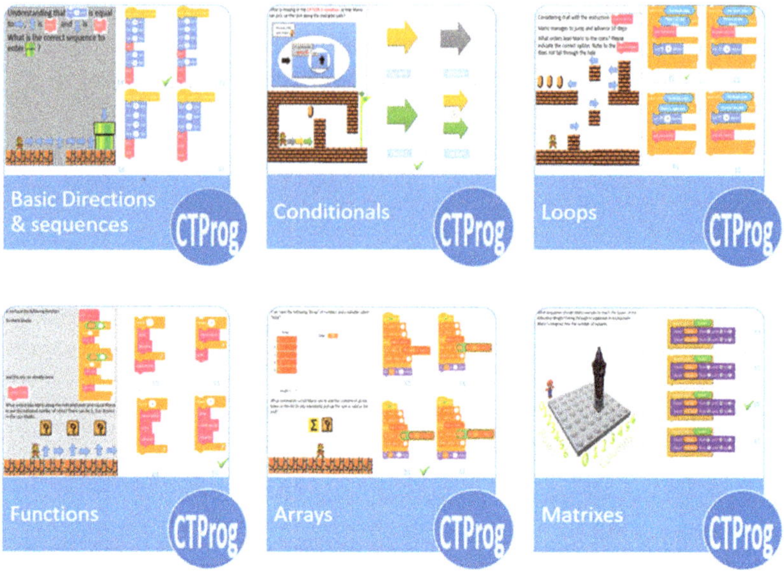

Figure 4. Example questions for each component of CTProg.

3. Results

This section will show the most relevant results of the study. This focuses on general results, and later we will carry out a detailed study of the different dimensions.

3.1. Relations among Variables

Firstly, the possible relationship between the different variables studied was analysed. As the variables do not follow a normal distribution (Shapiro–Wilk test with a p-value < 0.01), a possible linear correlation was analysed using Kendall's Tau-b. Table 3 marks the significant linear correlations in black. Thus, the following can be seen:

- There is a positive linear correlation between the duration of the test and the total score on the programming test (CTProg). This indicates that the more time spent by the students in answering the test resulted in a higher performance.
- There is a positive linear correlation between the total score on the programming test (CTProg) and the total score on the CT test (CTScore). This demonstrated that students that do well on the programming test also do well on the CT test. Thus, it can be said that programming knowledge leads to CT skills acquisition.
- There is a positive linear relationship between students' knowledge of for loops (PLOFOR) with their knowledge of conditionals (PCON) and repeat loops (PLOREPEAT). Thus, it can be said that students learning any type of loop leads to their learning of conditionals, and vice versa.
- There is a positive linear relationship between the knowledge of the array data structure (PARR) with students' knowledge of conditionals (PCON) and all types of loops, repeat loops (PLOREPEAT), and for loops (PLOFOR). Therefore, it can deduced that students that are familiar with more advanced programming concepts such as arrays have to have previously mastered conditionals and loops.
- There is a positive linear relationship between students' knowledge of the matrix data structure (PMAT) with their knowledge of for loops (PLOFOR) and arrays (PARR). Therefore, the comprehension of two-dimensional arrays (matrices) requires the knowledge and understanding of one-dimensional arrays and for loop structures to work effectively with them.

- There is a positive linear relationship between the CT skill of problem solving (Problemsolving) with the CT skill of algorithmic thinking (AlgorithmicThinking). This shows that students with a good problem-solving ability do well on the algorithmic thinking test, and vice versa.
- There is a positive linear relationship between the CT skill of abstraction (Abstraction) with the CT skills of algorithmic thinking (AlgorithmicThinking), decomposition (Decomposition), and creative thinking (CreativeThinking). As a consequence, students with a higher capacity for abstraction perform as well or better with other CT skills such as algorithmic thinking, decomposition, and creative thinking.
- There is a positive linear relationship between the CT skill of algorithmic thinking (AlgorithmicThinking) with the CT skills of decomposition (Decomposition) and pattern recognition (PatternRecognition). As shown, students overperforming in algorithmic thinking also do well in CT skills like decomposition and pattern recognition.

Table 3. Tau-b coefficients for each pair of variables. Those which are significant are marked in bold (*: $p < 0.05$; **: $p < 0.01$).

	Duration	CTProg	CTScore
Duration	1	0.179 *	
CTProg	**0.179** *	**1**	**0.398** **
CTScore	0.082	**0.398** **	1

	PBDS	PCON	PLOREPEAT	PFUNC	PLOFOR	PARR	PMAT
PBDS	--						
PCON	−0.014	--					
PLOREPEAT	−0.057	0.188	-- **				
PFUNC	0.129	0.126	−0.123	--			
PLOFOR	0.156	**0.236** *	**0.088** **	0.115	--		
PARR	0.193	**0.219** **	**0.094** **	0.200	**0.366** **	-- **	
PMAT	0.102	0.072	−0.013	0.155	**0.139** *	**0.113** *	

	Problem Solving	Abstraction	Algorithmic Thinking	Decomposition	Critical Thinking	Pattern Recognition
Problemsolving	--					
Abstraction	0.099	--				
AlgorithmicThinking	**0.156** *	**0.266** **	--			
Decomposition	0.086	**0.181** **	**0.239** **			
CriticalThinking	0.037	0.067	0.100	0.046		
CreativeThinking	0.110	**0.152** *	0.059	0.098	0.116	
PatternRecognition	0.112	0.076	**0.132** *	0.093	0.075	

3.2. *Comparison between Courses*

The scores for each variable are compared between the first and fourth years. Firstly, they are compared by the general score on either test, the CT test (CTScore) and the programming test (CTProg); then, they are compared by university, Rey Juan Carlos University and Atlantic Technological University.

3.2.1. General Results

Figure 5 shows the box plots for the CTScore and CTProg variables differentiated between the first and fourth years. As can be seen in Table 4, the central values of both variables are maintained in both years, with a greater dispersion being perceived for both in the fourth year. Using the Mann–Whitney U test (Table 5), it can be seen that these minimal differences are, in effect, not significant, so it can be ensured that there are no significant differences between both courses.

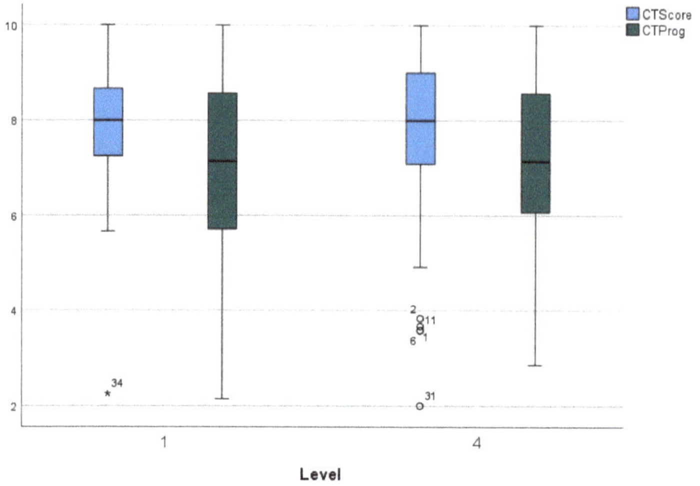

Figure 5. Box plots for the CTScore and CTProg total scores by years (1st and 4th) in general.

Table 4. Descriptive analysis for the CTProg and CTScore.

	CTProg		CTScore	
	1st	**4th**	**1st**	**4th**
n	78	99	78	99
Mean	7.03	7.09	7.81	7.85
Median	7.14	7.14	8.00	8.00
SD	1.54	1.74	1.27	1.54

Table 5. A non-parametric test (Mann–Whitney U test) between the 1st and 4th level.

	CTProg	CTScore
U	3960	4091
p-value	0.768	0.680

Now, the focus is on observing the differences between each of the variables studied. The following graphs (Figure 6, top and bottom) show the median of each of them as a representative of the central tendency in the data. The median is used instead of the mean, since there is no symmetry in the distribution of the data, with the median being a more robust measure. Although they are not statistically significant, changes can be seen between the first and fourth years in the variable for the programming concepts of repeat loops (PLOREPEAT), functions (PFUNC), arrays (PARR), and matrices (PMAT).

3.2.2. Rey Juan Carlos University (URJC)

This part of the study focuses on Rey Juan Carlos University (URJC). In this case (see Figure 7), the values obtained are similar in the first and fourth years for the two variables, the CT test (CTScore) and the programming test (CTProg). In the case of CTScore, the median has increased its value slightly, being 8 in the first year and 8.13 in the fourth year. Dispersion has clearly increased. In the case of CTProg, the median has increased its value, going from 7.14 to 7.85 from the first to the fourth year, respectively. Again, the dispersion in the data has increased.

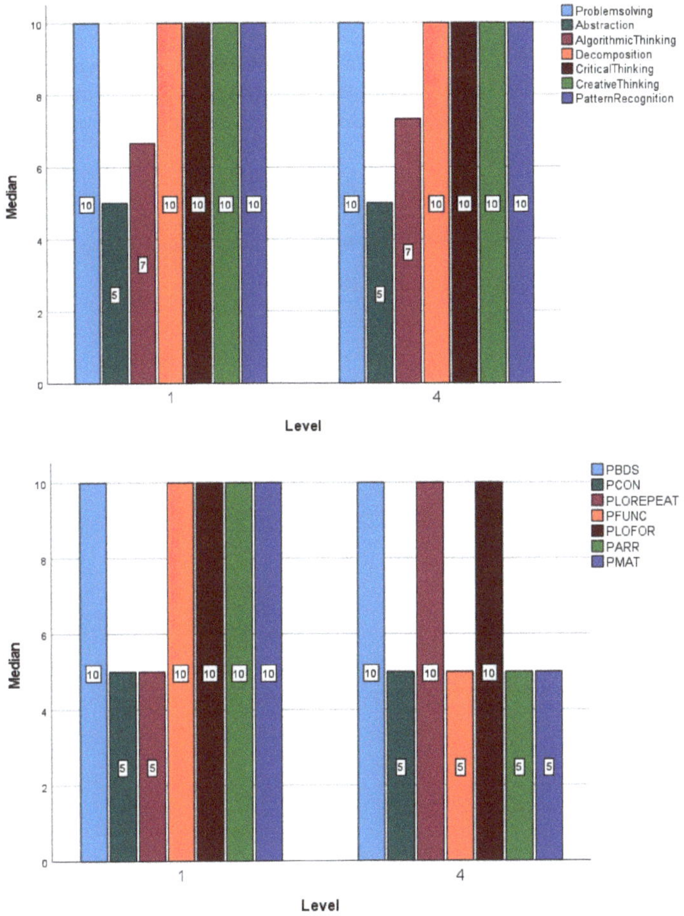

Figure 6. Bar charts of the CT skills in CTScore (**top**) and the programming test in CTProg (**bottom**) by years (1st and 4th) in general.

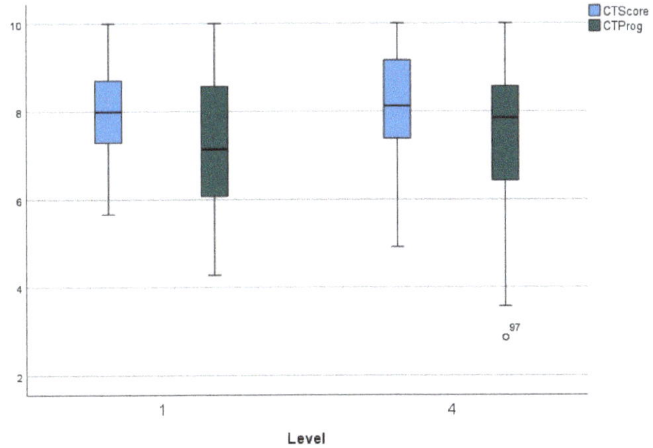

Figure 7. Box plots for the CTScore and CTProg total scores by years (1st and 4th) at URJC.

In the case of the bar graphs (see Figure 8), very little change is observed in the medians, depending on the course, for only the algorithmic thinking variable (AlgorithmicThinking) from CTScore, where it goes from 7 to 10. In the case of the programming variables, the median has increased in the conditionals (PCON) and repeat loops (PLOREPEAT) and decreased in the matrices (PMAT).

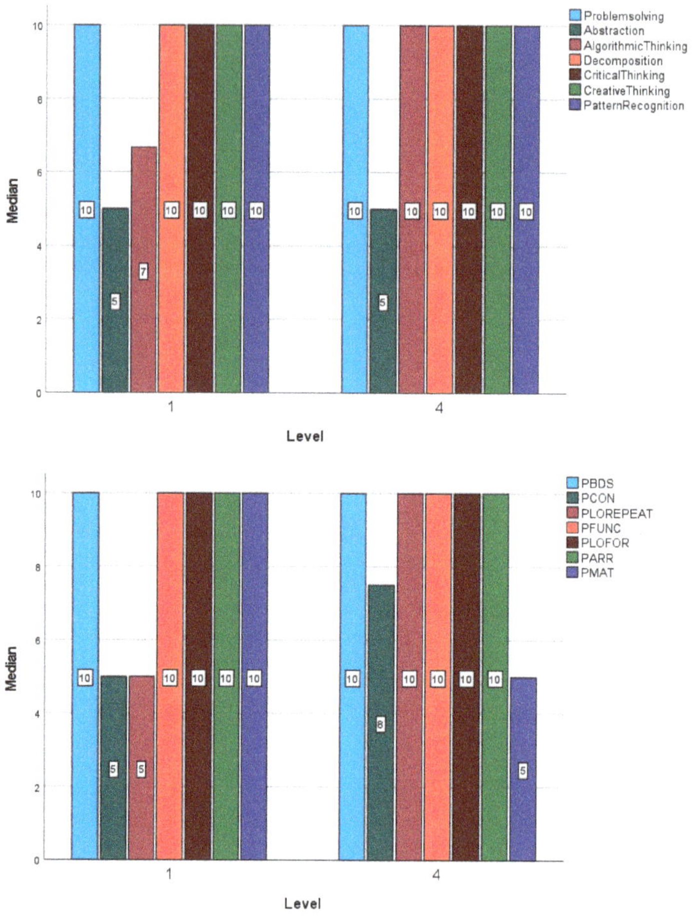

Figure 8. Bar charts of the CT skills in CTScore (**top**) and the programming test in CTProg (**bottom**) by years (1st and 4th) at URJC.

3.2.3. Atlantic Technological University (ATU)

The study will now focus on Atlantic Technological University (ATU). It is observed that, although the difference between medians is not statistically significant, the medians of the two variables, CTScore and CTProg, have lowered their value from the first to the fourth year (Figure 9). The dispersion, in the case of CTScore, has increased a lot, while remaining similar in CTProg.

In the case of the bar graphs, as shown in Figure 10, very little change is seen in the medians, depending on the course, except for the abstraction variable (Abstraction), where it decreases from five to two, the creative thinking variable (CreativeThinking), where increases slightly, and in the pattern recognition (PatternRecognition) variable, where it decreases from 10 to 6. Regarding, the programming concepts, in the case of repeat loops (PLOREPEAT) and arrays (PARR), the median has increased.

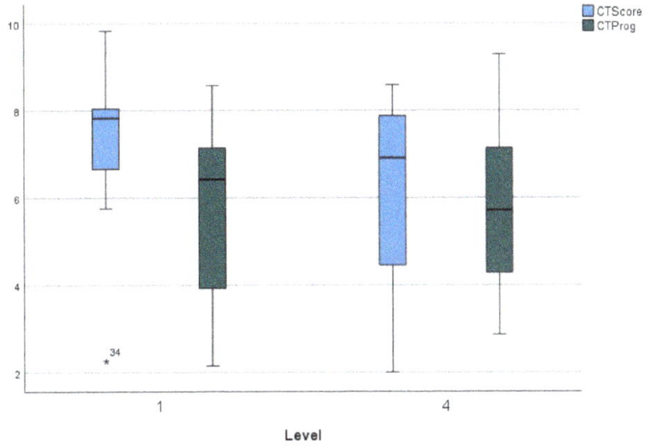

Figure 9. Box plots for the CTScore and CTProg total scores by years (1st and 4th) at ATU.

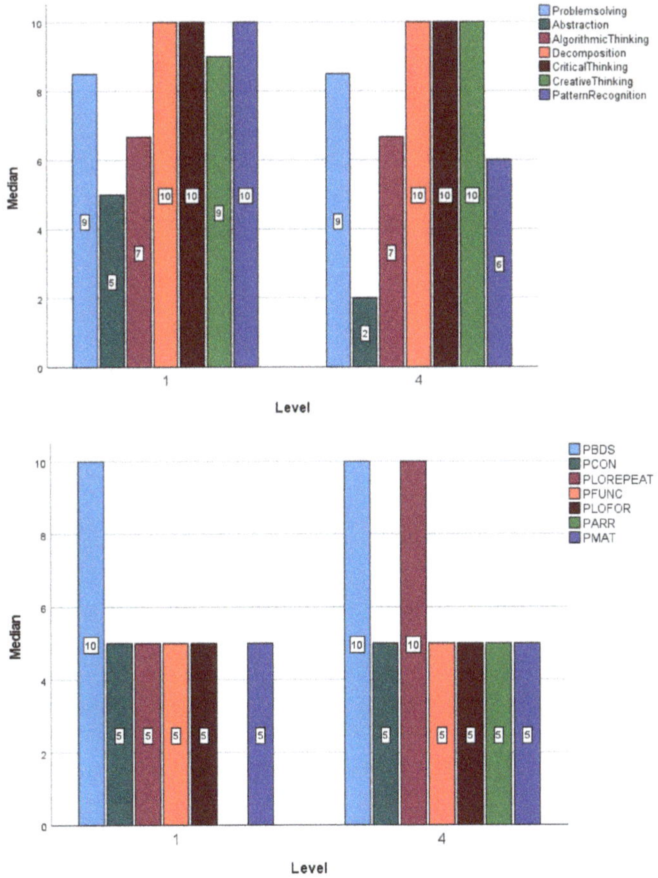

Figure 10. Bar charts of the CT skills in CTScore (**top**) and the programming test in CTProg (**bottom**) by years (1st and 4th) at ATU.

3.3. Comparison between Universities

The box plots in Figure 11 show the difference, by university, of the two global scores in CTScore and CTProg. This significant difference is manifested in significantly higher values on average at URJC (almost two points of difference in both the mean and median), with the dispersion being lower at URJC, as can be seen in Table 6 ($p < 0.001$ according to the Mann–Whitney U test, see Table 7).

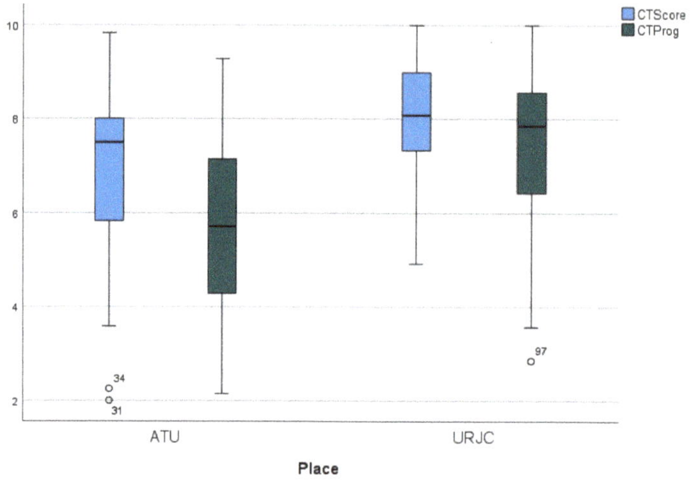

Figure 11. Box plots for the CTScore and CTProg total scores by years (1st and 4th) at both universities.

Table 6. Descriptive analysis for CTProg and CTScore by university.

	CTProg		CTScore	
	ATU	**URJC**	**ATU**	**URJC**
n	34	143	34	143
Mean	5.69	7.39	6.74	8.09
Median	5.71	7.85	7.5	8.08
SD	1.93	1.51	1.98	1.12

Table 7. A non-parametric test (Mann–Whitney U test) between universities.

	CTProg	CTScore
U	3642	3438
p-value	<0.001	<0.001

The median graphs show a similar trend (Figure 12), with the median values of the variables studied being greater or equal, in all cases, at URJC.

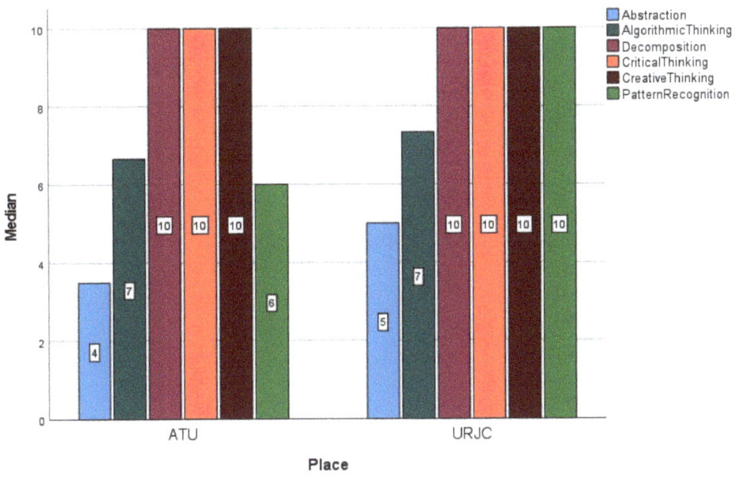

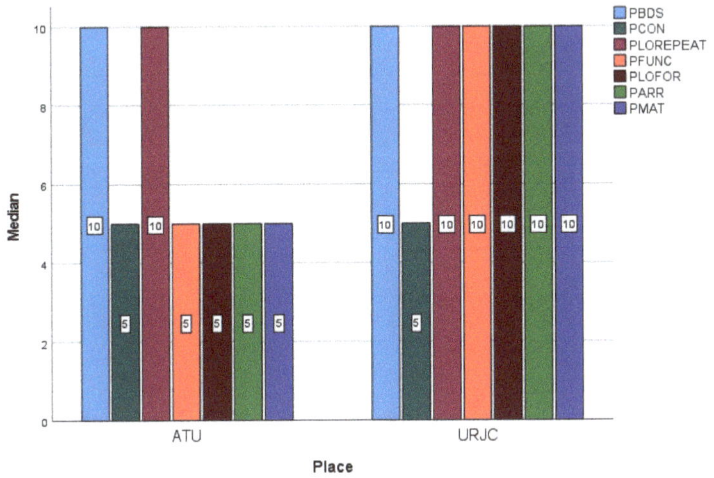

Figure 12. Bar charts of the CT skills in CTScore (**top**) and the programming test in CTProg (**bottom**) by years (1st and 4th) at ATU and URJC.

3.4. Comparison between Genders

Between genders, the box plots included in Figure 13 show the similarity in both the median and dispersion for all genders. In the case of "Other" genders, given the small quantity (two cases), their interpretation is not very relevant. Table 8 shows most relevant descriptive statistics. Higher values are shown for the male gender in terms of the mean, median, and standard deviation. Table 9 shows the output of the non-parametric Kruskal–Wallis statistic, showing that these differences are not significant ($p > 0.05$).

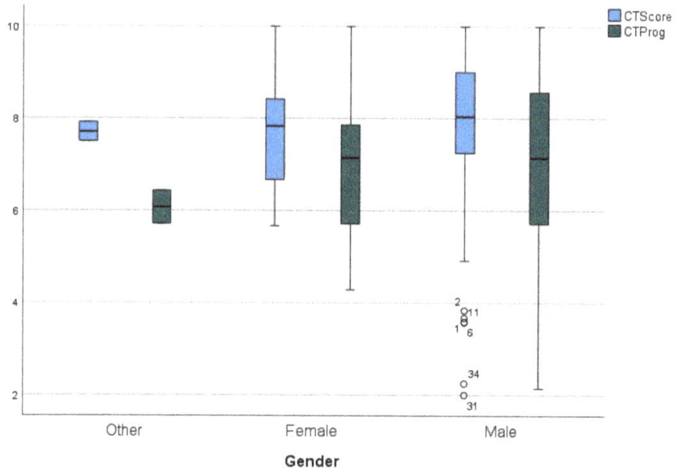

Figure 13. Box plots for the CTScore and CTProg total scores by genders at both universities.

Table 8. Descriptive analysis for CTProg and CTScore by gender.

	CTScore			CTProg		
	Other	**Female**	**Male**	**Other**	**Female**	**Male**
n	2	35	140	2	35	140
Mean	6.07	6.83	7.14	7.70	7.69	7.87
Median	6.07	7.14	7.14	7.70	7.83	8.04
SD	0.50	1.45	1.80	0.19	1.29	1.49

Table 9. A non-parametric test (Kruskal–Wallis test) among genders.

	CTProg	CTScore
W	3642	1.984
p-value	0.173	0.371

4. Discussion

This study offers a significant addition to the literature by providing a thorough examination of the methods for gauging the differences in students' programming and computational thinking abilities by geography, gender, and university level. This paper presents UniCTCheck, an innovative approach created to evaluate the fundamentals of computational thinking in computer science students. This study used two important tools, the web application CTScore and the psychometric scale CTProg, to measure seven fundamental aspects of computational thinking and six critical programming ideas for computer science students. As a result, the following five research questions were formulated:

4.1. RQ1: Does UniCTCheck Measure the Main Components of Computational Thinking and Programming for CS University Students?

The results from the study indicate that UniCTCheck effectively measures the main components of computational thinking and programming for computer science university students. Through the use of CTScore and CTProg, a comprehensive assessment of pattern recognition, creative thinking, algorithmic thinking, problem solving, critical thinking, decomposition, and abstraction, as well as programming concepts like basic directions and sequences, conditionals, loops, functions, and data structures was conducted. The data reveal that the UniCTCheck method provides a robust framework for evaluating students' computational thinking abilities and programming skills accurately.

4.2. RQ2: Is There a Relationship between Students' Computational Thinking Skills and Programming Abilities?

The time expended by students in answering the programming test (CTProg) was directly related to their higher performance on it. Also, students that did well on the CTProg also did well on the CT skills test (CTScore); thus, it can be said that programming knowledge leads to CT skills acquisition. It has also been proved that students learning any type of loop leads to their learning of conditionals, and vice versa. Furthermore, students that know of more advanced programming concepts such as arrays need to have previously mastered conditionals and loops. Moreover, the comprehension of two-dimensional arrays (matrices) required the knowledge and understanding of one-dimensional arrays and for loop structures so to work effectively with them. Regarding CT skills acquisition, students with the problem-solving ability did well on the algorithmic thinking ability, and vice versa. In the same manner, students with a higher capacity of abstraction performed as well or better in other CT skills such as algorithmic thinking, decomposition, and creative thinking. As shown, students overperforming in algorithmic thinking also did well in CT skills like decomposition and pattern recognition.

4.3. RQ3: Do Students' Computational Thinking Skills and Programming Abilities Vary by University Level?

In general, there were no significant differences between both courses. Although they were not statistically significant, changes can be seen between the first and fourth years for the programming concepts of repeat loops, functions, arrays, and matrices.

When the study focused on Rey Juan Carlos University (URJC), in the case of CTScore, the median increased slightly in the fourth year, while the dispersion also clearly increased. In the case of CTProg, the median also increased from the first to the fourth year, and the dispersion in the data increased again. Also, very little change occurred in the medians, depending on the course, except in the algorithmic thinking skill from CTScore. In the case of the programming variables, the median increased in the conditionals and repeat loops and decreased in the matrices.

When the study focused on Atlantic Technological University (ATU), it was observed that, although the difference between medians was not statistically significant, the medians of the two variables, CTScore and CTProg, lowered their values from the first to the fourth year. The dispersion, in the case of CTScore, increased a lot, while remaining similar in CTProg. Very little change was seen in the medians, depending on the course, except in the case of the CT skills of abstraction, where it decreased, in creative thinking, where it increased slightly, and in pattern recognition, where it decreased. Regarding, the programming concepts, in the case of repeat loops and arrays, the median increased.

4.4. RQ4: Do Students' Computational Thinking Skills and Programming Abilities Vary by University Location?

There existed a significant difference by university in the two global scores in CTScore and CTProg. This significant difference was manifested in significantly higher values on average (almost two points of difference in both the mean and median), with the dispersion being lower at URJC. Also, the median values of the variables studied were greater or equal, in all cases, at URJC.

4.5. RQ5: Do Students' Computational Thinking Skills and Programming Abilities Vary by Gender?

The examination of computational thinking skills and programming abilities by gender revealed no statistically significant difference. While there were minor variations in the performance between male and female students, the data indicated that both genders displayed similar levels of competency overall. This suggests that computational thinking skills and programming abilities are not inherently gender-specific and can be developed by students regardless of gender.

In conclusion, the data-driven responses to the research questions provide valuable insights into the nuances of students' computational thinking skills and programming abilities in the context of gender, university level, and geography. By leveraging the UniCTCheck method and the assessment tools, CTScore and CTProg, this study has shed light on the diverse competencies of computer science students and highlighted the importance of tailored educational interventions to enhance their skills effectively.

5. Limitations and Future Research

The greatest limitation of the study is related to the impossibility of the same professors being present in all study groups at the different universities. This introduces the human factor in terms of the personal motivation with which students have been asked to participate. To avoid this possible bias, we always tried to follow the same guidelines in terms of time, explanation, motivation, etc. Furthermore, different languages have been considered and the tests were translated into both English and Spanish. Another limitation between the results from both university degrees may be the difference in the admission requirements between the two programs, with the requirements being much higher at URJC.

Future research may include more difficult questions on the UniCTCheck, so the ceiling can be a little bit higher, and therefore more differences can be found. Also, future evaluation with non-CS students should be considered.

6. Conclusions

This study significantly advances the understanding of how to measure differences in computational thinking (CT) and programming abilities among computer science (CS) students by geography, gender, and university level. The innovative UniCTCheck method, incorporating the CTScore web application and CTProg psychometric scale, effectively evaluates seven fundamental CT components and six critical programming concepts.

UniCTCheck Effectiveness: UniCTCheck was validated as a robust framework for assessing CT components and programming skills in CS students. The comprehensive evaluation demonstrated its effectiveness in measuring essential skills such as pattern recognition, algorithmic thinking, and problem solving, as well as programming concepts including loops, functions, and data structures.

Relationship Between CT Skills and Programming Abilities: A significant correlation was found between students' CT skills and programming abilities. Higher performance in the CTProg programming test was directly related to better scores in the CTScore assessment, indicating that strong programming knowledge facilitates CT skills acquisition.

Variation by University Level: The study observed minimal differences in CT skills and programming abilities across university levels. However, slight increases in median scores for advanced programming concepts were noted in fourth-year students, particularly at Rey Juan Carlos University (URJC), where there was an increased dispersion in both the CTScore and CTProg results.

Variation by University Location: Significant differences were found between universities, with URJC students outperforming those at Atlantic Technological University (ATU) on both CTScore and CTProg. URJC students displayed higher mean and median scores with a lower dispersion. As explained in the limitation section, this may be due to the difference in admission requirements in between the two programmes.

Variation by Gender: No statistically significant differences were found in CT skills and programming abilities between genders. Both male and female students demonstrated similar levels of competency, indicating that these skills are not inherently gender-specific.

In summary, this study provides valuable insights into the factors influencing CS students' CT and programming abilities. The use of UniCTCheck, CTScore, and CTProg offers a comprehensive assessment framework that can inform tailored educational interventions to enhance students' computational competencies effectively.

Author Contributions: Conceptualisation, R.H.-N., C.P., J.F., D.P.-A. and E.Ç.; methodology, R.H.-N., C.P., J.F., D.P.-A. and E.Ç.; software, R.H.-N., D.P.-A. and E.Ç.; validation, R.H.-N., C.P., J.F. and E.Ç.; formal analysis, R.H.-N. and C.P.; investigation, R.H.-N., C.P. and J.F.; resources, R.H.-N., D.P.-A. and E.Ç.; data curation, R.H.-N. and C.P.; writing—original draft preparation, R.H.-N. and C.P.; writing—review and editing, R.H.-N., C.P. and J.F.; visualisation, R.H.-N., C.P., D.P.-A. and E.Ç.; supervision, R.H.-N. All authors have read and agreed to the published version of the manuscript.

Funding: This research was funded by research grants PID2022-137849OB-I00 funded by MICIU/AEI/10.13039/501100011033 and by the ERDF, EU.

Institutional Review Board Statement: The study was conducted in accordance with the Declaration of Helsinki, and approved by Research Ethics Committee of Rey Juan Carlos University with internal registration number 1909202332223, 9 October 2023.

Informed Consent Statement: Informed consent was obtained from all subjects involved in the study.

Data Availability Statement: The data presented in this study are available on request from the corresponding author due to ethical reasons.

Conflicts of Interest: The authors declare no conflicts of interest.

References

1. Wing, J.M. Computational thinking. *Commun. ACM* **2006**, *49*, 33–35. [CrossRef]
2. Wing, J.M. Computational thinking and thinking about computing. *Philos. Trans. R. Soc. A Math. Phys. Eng.* **2008**, *366*, 3717–3725. [CrossRef] [PubMed]
3. Cuny, J.; Snyder, L.; Wing, J.M. Demystifying computational thinking for non-computer scientists. *Educ. Res. Rev.* **2010**, *22*, 142–158.
4. Mohaghegh, M.; McCauley, M. Computational Thinking: The Skill Set of the 21st Century. *Int. J. Comput. Sci. Inf.* **2016**, *7*, 1524–1530.
5. Fisher, J.; Henzinger, T.A. Executable cell biology. *Nat. Biotechnol.* **2007**, *25*, 1239–1249. [CrossRef] [PubMed]
6. Qin, H. Teaching computational thinking through bioinformatics to biology students. In Proceedings of the 40th ACM Technical Symposium on Computer Science Education, New York, NY, USA, 4–7 March 2009.
7. García-Peñalvo, F.J.; Mendes, A.J. Exploring the computational thinking effects in pre-university education. *Comput. Hum. Behav.* **2018**, *80*, 407–411. [CrossRef]
8. Higgins, C.; O'Leary, C.H.; OMtenzi, F. A conceptual framework for a software development process based on computational thinking. In Proceedings of the 11th International Technology, Education and Development Conference (INTED17), Valencia, Spain, 6–8 March 2017; pp. 455–464. [CrossRef]
9. Selby, C.; Woollard, J. *Computational Thinking: The Developing Definition*; University of Southampton Institutional Research Repository: Swindon, UK, 2013; 6p. Available online: http://eprints.soton.ac.uk/id/eprint/356481 (accessed on 10 June 2024).
10. de Araujo, A.L.S.O.; Andrade, W.L.; Guerrero, D.D.S. A systematic mapping study on assessing computational thinking abilities. In Proceedings of the 2016 IEEE Frontiers in Education Conference (FIE), Erie, PA, USA, 12–15 October 2016; IEEE: Piscataway, NJ, USA, 2016; pp. 1–9.
11. Kong, S.-C. Components and methods of evaluating computational thinking for fostering creative problem-solvers in senior primary school education. In *Computational Thinking Education*; Springer: Berlin/Heidelberg, Germany, 2019; pp. 119–141. [CrossRef]
12. Román-González, M.; Moreno-León, J.; Robles, G. Combining assessment tools for a comprehensive evaluation of computational thinking interventions. In *Computational Thinking Education*; Springer: Berlin/Heidelberg, Germany, 2019; pp. 79–98.
13. Denning, P.J. Remaining trouble spots with computational thinking. *Commun. ACM* **2017**, *60*, 33–39. [CrossRef]
14. Haseski, H.I.; Ilic, U. An Investigation of the Data Collection Instruments Developed to Measure Computational Thinking. *Inform. Educ.* **2019**, *18*, 297. [CrossRef]
15. Tang, X.; Yin, Y.; Lin, Q.; Hadad, R.; Zhai, X. Assessing computational thinking: A systematic review of empirical studies. *Comput. Educ.* **2020**, *148*, 103798. [CrossRef]
16. Varghese, V.V.; Renumol, V.G. Assessment methods and interventions to develop computational thinking—A literature review. In Proceedings of the 2021 International Conference on Innovative Trends in Information Technology (ICITIIT), Kottayam, India, 11–12 February 2021; IEEE: Piscataway, NJ, USA, 2021; pp. 1–7.
17. Poulakis, E.; Politis, P. Computational Thinking Assessment: Literature Review. In *Research on E-Learning and ICT in Education: Technological*; Pedagogical and Instructional Perspectives; Springer: Berlin/Heidelberg, Germany, 2021; pp. 111–128.
18. Alan, Ü. Likert Tipi Olçeklerin Çocuklarla Kullaniminda Yanit Kategori Sayisinin Psikometrik Ozelliklere Etkisi. Master's Thesis, Hacettepe University, Ankara, Turkey. (In Turkish).
19. Arıkan, R. Anket yöntemi üzerinde bir degerlendirme. *Haliç Üniversitesi Sos. Bilim. Derg.* **2018**, *1*, 97–159. (In Turkish)
20. Guenaga, M.; Eguíluz, A.; Garaizar, P.; Gibaja, J. How do students develop computational thinking? Assessing early programmers in a maze-based online game. *Comput. Sci. Educ.* **2021**, *31*, 259–289. [CrossRef]

21. Kong, S.-C.; Liu, B. A performance-based assessment platform for developing computational thinking concepts and practices: EasyCode. *Bull. Tech. Comm. Learn. Technol.* **2020**, *20*, 3–10. ISSN 2306-0212. Available online: https://ieeecs-media.computer.org/tc-media/sites/5/2020/06/07190536/bulletin-tclt-2020-0201001.pdf (accessed on 10 June 2024).
22. Bikmaz-Bilgen, Ö. Tamamlayici Ölçme ve Degerlendirme Teknikleri II: Portfolyo Degerlendirme. In *Egitimde Ölçme ve Degerlendirme*; Dogan, N., Ed.; Pegem Akademi: Ankara, Turkey, 2019; pp. 182–214. (In Turkish)
23. Gültekin, S. Performans Dayanali Degerlendirme. In *Egitimde Ölme ve de Gerlendirme*; Demirtasli, R.N., Ed.; Ani Pub: Ankara, Turkey, 2017; pp. 233–256. (In Turkish)
24. Sahin, M.G. Performansa Dayali Degerlendirme. In *Egitimde Ölme ve de Gerlendirme*; Çetin, B., Ed.; Ani Pub: Ankara, Turkey, 2019. (In Turkish)
25. Grover, S.; Pea, R. Computational thinking in K–12: A review of the state of the field. *Educ. Res.* **2013**, *42*, 38–43. [CrossRef]
26. Mannila, L.; Dagiene, V.; Demo, B.; Grgurina, N.; Mirolo, C.; Rolandsson, L.; Settle, A. Computational thinking in K-9 education. In Proceedings of the Working Group Reports of the 2014 on Innovation & Technology in Computer Science Education Conference, Uppsala, Sweden, 23–25 June 2014. [CrossRef]
27. Martins-Pacheco, L.H.; von Wangenheim, C.A.G.; Alves, N. Assessment of computational thinking in K-12 context: Educational practices, limits and possibilities-a systematic mapping study. In Proceedings of the 11th International Conference on Computer Supported Education (CSEDU 2019), Crete, Greece, 2–4 May 2019; pp. 292–303.
28. Çoban, E.; Korkmaz, Ö. An alternative approach for measuring computational thinking: Performance-based platform. *Think. Ski. Creat.* **2021**, *42*, 100929. [CrossRef]
29. Tedre, M.; Denning, P.J. The long quest for computational thinking. In Proceedings of the 16th Koli Calling International Conference on Computing Education Research, Koli, Finland, 24–27 November 2016; pp. 120–129. [CrossRef]
30. Katz, D.L. Conference report on the use of computers in engineering classroom instruction. *Commun. ACM* **1960**, *3*, 522–527. [CrossRef]
31. Newell, A.; Perlis, A.J.; Simon, H.A. Computer science. *Science* **1967**, *157*, 1373–1374. [CrossRef] [PubMed]
32. Papert, S. *Mindstorms: Children, Computers, and Powerful Ideas*; Basic Books, Inc.: New York, NY, USA, 1980.
33. Papert, S. An exploration in the space of mathematics educations. *Int. J. Comput. Math. Learn.* **1996**, *1*, 95–123. [CrossRef]
34. Papert, S.; Harel, I. Situating constructionism. *Constructionism* **1991**, *36*, 1–11.
35. Turkle, S.; Papert, S. Epistemological pluralism: Styles and voices within the computer culture. *Signs J. Women Cult. Soc.* **1990**, *16*, 128–157. [CrossRef]
36. Denning, P.J. The profession of IT beyond computational thinking. *Commun. ACM* **2009**, *52*, 28–30. [CrossRef]
37. Denning, P.J. Ubiquity symposium 'what is computation?': Opening statement. *Ubiquity* **2010**, *2010*, 1. [CrossRef]
38. Easton, T.A. Beyond the algorithmization of the sciences. *Commun. ACM* **2006**, *49*, 31–33. [CrossRef]
39. Dijkstra, E.W. Programming as a discipline of mathematical nature. *Am. Math. Mon.* **1974**, *81*, 608–612. [CrossRef]
40. Guzdial, M. Learner-centered design of computing education: Research on computing for everyone. In *Synthesis Lectures on Human-Centered Informatics (SLHCI)*; Springer: Berlin/Heidelberg, Germany, 2015; Volume 8, pp. 1–165. [CrossRef]
41. Lee, I.; Martin, F.; Denner, J.; Coulter, B.; Allan, W.; Erickson, J.; Malyn-Smith, J.; Werner, L. Computational thinking for youth in practice. *ACM Inroads* **2011**, *2*, 32–37. [CrossRef]
42. Shute, V.J.; Sun, C.; Asbell-Clarke, J. Demystifying computational thinking. *Educ. Res. Rev.* **2017**, *22*, 142–158. [CrossRef]
43. Lu, J.J.; Fletcher, G.H.L. Thinking about computational thinking. In Proceedings of the 40th ACM Technical Symposium on Computer Science Education—SIGCSE'09, Chattanooga, TN, USA, 4–7 March 2009. [CrossRef]
44. Aho, A.V. Computation and computational thinking. *Comput. J.* **2012**, *55*, 832–835. [CrossRef]
45. Kong, S.-C.; Abelson, H.; Lai, M. Introduction to computational thinking education. In *Computational Thinking Education*; Springer: Berlin/Heidelberg, Germany, 2019; pp. 1–10. [CrossRef]
46. Andrew, C.; Paul, C.; Mark, D.; Simon, H.; Thomas, N.; Cynthia, S.; John, W. *Computational Thinking—A Guide for Teachers*; Computing at School; University of Southampton Institutional Research Repository: Swindon, UK, 2015. Available online: https://eprints.soton.ac.uk/424545/ (accessed on 11 May 2020).
47. CSTA. *Operational Definition of Computational Thinking for K–12 Education*; National Science Foundation: Alexandria, VA, USA, 2011.
48. ISTE Standards for Students. Available online: http://www.iste.org/standards/standards/for-students-2016 (accessed on 14 April 2020).
49. Riley, D.D.; Hunt, K.A. *Computational Thinking for the Modern Problem Solver*; CRC Press: Boca Raton, FL, USA, 2014; ISBN 978-1-4665-8777-9.
50. NRC. Committee for the workshops on computational thinking. In *Report of a Workshop on the Scope and Nature of Computational Thinking*; National Academies Sciences, Engineering, Medicine: Washington, DC, USA, 2010.
51. Barr, V.; Stephenson, C. Bringing computational thinking to K-12: What is involved and what is the role of the computer science education community? *ACM Inroads* **2011**, *2*, 48–54. [CrossRef]
52. Brennan, K.; Resnick, M. New frameworks for studying and assessing the development of computational thinking. In Proceedings of the 2012 Annual Meeting of the American Educational Research Association, Vancouver, BC, Canada, 13–17 April 2012; p. 25. Available online: http://scratched.gse.harvard.edu/ct/files/AERA2012.pdf (accessed on 10 June 2024).

53. Gouws, L.A.; Bradshaw, K.; Wentworth, P. Computational thinking in educational activities: An evaluation of the educational game light-bot. In Proceedings of the 18th ACM Conference on Innovation and Technology in Computer Science Education, Canterbury, UK, 1–3 July 2013; pp. 10–15. [CrossRef]
54. Bers, M.U.; Flannery, L.; Kazakoff, E.R.; Sullivan, A. Computational thinking and tinkering: Exploration of an early childhood robotics curriculum. *Comput. Educ.* **2014**, *72*, 145–157. [CrossRef]
55. Zapata-Ros, M. Pensamiento Computacional: Una Nueva Alfabetización Digital. *Distance Educ. J.* **2015**, *46*. Available online: https://revistas.um.es/red/article/view/240321 (accessed on 10 June 2024). [CrossRef]
56. Anderson, N.D. A call for computational thinking in undergraduate psychology. *Psychol. Learn. Teach.* **2016**, *15*, 226–234. [CrossRef]
57. Google for Education: Computational Thinking. Available online: https://edu.google.com/resources/programs/exploring-computational-thinking/ (accessed on 21 February 2020).
58. Dierbach, C.; Hochheiser, H.; Collins, S.; Jerome, G.; Ariza, C.; Kelleher, T.; Kleinsasser, W.; Dehlinger, J.; Kaza, S. A model for piloting pathways for computational thinking in a general education curriculum. In Proceedings of the 42nd ACM Technical Symposium on Computer Science Education, Dallas, TX, USA, 9–12 March 2011; pp. 257–262. [CrossRef]
59. Berland, M.; Lee, V.R. Collaborative strategic board games as a site for distributed computational thinking. *Int. J. Game-Based Learn.* **2011**, *1*, 65–81. [CrossRef]
60. Akkaya, A. The Effects of Serious Games on Students' Conceptual Knowledge of Object-Oriented Programming and Computational Thinking Skills. Master's Thesis, Bogaziçi University, Beşiktaş, Turkey, 2018.
61. CAS. Available online: https://community.computingatschool.org.uk/resources/2324 (accessed on 10 June 2024).

applied sciences

MDPI

Article

Prediction of Students' Adaptability Using Explainable AI in Educational Machine Learning Models

Leonard Chukwualuka Nnadi [1,*], Yutaka Watanobe [1,*], Md. Mostafizer Rahman [2] and Adetokunbo Macgregor John-Otumu [3]

[1] Department of Computer Science and Engineering, The University of Aizu, Aizuwakamatsu 965-8580, Fukushima, Japan
[2] Dhaka University of Engineering & Technology, Gazipur 1707, Bangladesh
[3] Department of Information Technology, Federal University of Technology, Owerri 460114, Nigeria; adetokunbo.johnotumu@futo.edu.ng
* Correspondence: d8252105@u-aizu.ac.jp (L.C.N.); yutaka@u-aizu.ac.jp (Y.W.); Tel.: +81-80-2028-9508 (L.C.N.)

Abstract: As the educational landscape evolves, understanding and fostering student adaptability has become increasingly critical. This study presents a comparative analysis of XAI techniques to interpret machine learning models aimed at classifying student adaptability levels. Leveraging a robust dataset of 1205 instances, we employed several machine learning algorithms with a particular focus on Random Forest, which demonstrated highest accuracy at 91%. The models' precision, recall and F1-score were also evaluated, with Random Forest achieving a precision of 0.93, a recall of 0.94, and an F1-score of 0.94. Our study utilizes SHAP, LIME, Anchors, ALE, and Counterfactual explanations to reveal the specific contributions of various features impacting adaptability predictions. SHAP values highlighted 'Class Duration' significance (mean SHAP value: 0.175); LIME explained socio-economic and institutional factors' intricate influence. Anchors provided high-confidence rule-based explanations (confidence: 97.32%), emphasizing demographic characteristics. ALE analysis underscored the importance of 'Financial Condition' with a positive slope, while Counterfactual scenarios highlighted the impact of slight feature variations of 0.5 change in 'Class Duration'. Consistently, 'Class Duration' and 'Financial Condition' emerge as key factors, while the study also underscores the subtle effects of 'Institution Type' and 'Load-shedding'. This multi-faceted interpretability approach bridges the gap between machine learning performance and educational relevance, presenting a model that not only predicts but also explains the dynamic factors influencing student adaptability. The synthesized insights advocate for educational policies accommodating socioeconomic factors, instructional time, and infrastructure stability to enhance student adaptability. The implications extend to informed and personalized educational interventions, fostering an adaptable learning environment. This methodical research contributes to responsible AI application in education, promoting predictive and interpretable models for equitable and effective educational strategies.

Keywords: comparative analysis; educational data mining; educational predictive modelling; XAI; feature importance; machine learning interpretability; model transparency; predictive analytics in education; student adaptability; AI in education policy

1. Introduction

A critical area of concern for educators, policymakers, and academics in the quickly changing field of education is comprehending and improving student adaptation [1,2]. The ability of students to effectively navigate and respond to diverse learning environments, technological advancements, and varying educational methodologies highlights the importance of adaptive learning systems tailored to individual needs [3].

The field of education has seen a revolutionary change with the introduction of machine learning techniques, which have provided new ways to forecast and enhance student adaptation [4]. However, the black-box nature of complex machine learning models often impedes their interpretability, raising concerns regarding their applicability in educational settings where transparency and comprehensibility are paramount. The inability of educators to understand the machine learning algorithms' decision-making process due to their black-box nature has been a key barrier in the learning space [5,6]. Because of this, there is increasing interest in building and evaluating interpretable machine learning models that can be used to identify student adaptability factors. These models should not only produce precise predictions but also offer valuable insights into the decision-making process.

Student adaptability is a complex construct encompassing the ability to effectively adapt, learn, and succeed in diverse educational settings. It involves factors such as socio-economic background, individual learning styles, technological proficiency, and environmental influences [7,8]. Understanding and measuring adaptability require a holistic approach that integrates diverse data sources and utilizes sophisticated models capable of capturing intricate patterns within these datasets. Even though machine learning models perform predictably well across a range of areas, their opacity poses a substantial difficulty when used in educational settings. Interpretability issues make it more difficult for stakeholders to understand model predictions and restrict the discovery of useful information that is essential for creating customized interventions that improve student flexibility. In order to close this gap, new developments in interpretable machine learning methods provide ways to decipher how complicated models make decisions [9–11].

In response to the evolving challenges and opportunities in education, this study innovates by introducing a comparative analysis of a novel combination of interpretable machine learning models tailored to enhancing our understanding of student adaptability. Through the meticulous employment of SHapley Additive exPlanations (SHAP), Local Interpretable Model-agnostic Explanations (LIME), Anchor, Accumulated Local Effects (ALE), and Counterfactual explanations, we navigate the intricacies of balancing model interpretability against predictive accuracy. By leveraging a dataset rich in student attributes and adaptability levels, our research offers unprecedented insights into the predictors of adaptability, navigating the complexities inherent in educational data. Grounded in cutting-edge techniques and established educational theories, our methodology facilitates a detailed examination of the myriad factors influencing student outcomes. The application of machine learning in this context, with its capacity to analyze complex datasets and reveal intricate patterns, marks a significant advancement in educational research. This study not only underscores the predictive power and interpretability of machine learning models but also highlights their potential to inform targeted, effective educational interventions, thereby pioneering a comprehensive exploration aimed at unveiling and acting upon the factors influencing student adaptability, thereby fostering informed and personalized educational strategies.

The rest of the paper is arranged as follows: Section 2 provides an overview of related work on XAI and interpretable models for student adaptability. Section 3 describes the methodology. Section 4 presents the results of the experiments, while Section 5 discusses comparative analysis of XAI techniques and synthesis of findings. Finally, Section 6 discusses future work and conclusions.

2. Related Work

Numerous studies have explored the correlation between student adaptability and academic performance, retention rates, and overall learning outcomes. The link between learning flexibility and self-regulated learning (SRL) in junior high school students is examined in the study by She et al. [12], with an emphasis on the mediating roles of academic motivation and self-management. Under China's "double reduction" educational program, the research looks at how these elements work together to support SRL, which is an important ability for students to succeed. The study's conclusions show the independent and cumulative mediation functions of academic motivation and self-management, as well as the noteworthy benefits of learning adaptability on SRL. This offers information to help students improve their learning practices and adjust to changes in schooling.

Martin et al.'s systematic review [13] summarizes research on adaptive learning from 2009 to 2018, with an emphasis on publishing patterns, instructional settings, techniques, and technology. After analyzing 61 articles, the study found that the number of publications peaked in 2015 and that higher education particularly in Taiwan and the US was the main topic of interest. The study focuses on several stages of instructional design in the field of adaptive learning research, such as learner characteristics, needs assessment, and system design and implementation. Furthermore, it classifies adaptive technologies into systems, applications, teaching techniques, and design solutions, and adaptive strategies into categories based on learner models, content, and instructional models. In the review, it is emphasized how important it is to have more diverse research and rigorous methods in adaptive learning.

In their investigation of student adaptation following COVID-19-induced online learning, Tiwari et al. [14] explore Education 4.0. By using Explainable Machine Learning (XML), they investigate students' acceptance of Industry 4.0 features. Using a range of machine learning techniques, the study finds that Neural Networks are the most accurate (93%) at classifying adaptation. The study emphasizes how crucial explainability is for machine learning models. It uses dataset analysis, model performance research, and SHAP techniques to clarify feature importance and model predictions. In the end, it emphasizes how important elements affecting students' capacity to adjust include age, socioeconomic situation, and type of internet access. This thorough study helps teachers comprehend how pupils adjust to online learning and what effects it has.

Adnan et al. [15] present an XAI model that predicts student performance in virtual learning environments at various course stages with detailed interpretability, utilizing Random Forest among other algorithms and incorporating SHAP values for insight into decision-making processes. This enhances educational data mining by enabling early performance prediction. In a related context, Gligorea et al. [16] develop an interpretable machine learning framework to analyze and predict student performance, focusing on commitment and grades, and improving framework efficiency through algorithmic tuning and GPU-based infrastructure. Another study [5] explores early prediction of student performance using machine learning and XAI. Logistic regression yielded the best results, and the integration of XAI helped educators understand the predictions, aiming to support at-risk students promptly.

The adaptability of Bangladeshi pupils to online learning amid the COVID-19 pandemic is investigated by Suzan et al. [1]. The study uses machine learning models including Decision Tree, Random Forest, Naive Bayes, SVM, KNN, and ANN and conducts surveys across different educational levels. The Random Forest Classifier, which focuses on sociodemographic characteristics, has the best accuracy (89.63%). This study provides insightful information for educational policymakers by illuminating the difficulties and degrees of adaptation faced by students in the context of online learning. The results highlight the importance of comprehending student flexibility for the successful implementation of online learning and offer directions for future study in evaluating sociodemographic impacts on students' mental health in online learning.

This study builds on previous work by Suzan et al. [1], who also utilized the same dataset to predict student adaptability levels during the pandemic. Our research extends this work by applying a comprehensive suite of Explainable AI techniques to provide deeper insights into the factors influencing adaptability. We reference their findings to contextualize our results and highlight the unique contributions of our study.

LIME was introduced by Ribeiro et al. [9]. LIME increases user trust in machine learning models, which are sometimes viewed as mysterious black boxes, by providing clear explanations for every classifier's predictions. By building an interpretable model around the prediction, the method offers faithful and understandable local explanations. Additionally, it suggests a way to use a group of representative individual forecasts to explain the global model. The study illustrates how LIME may be used to explain a wide range of models and how useful it is in situations when trust is needed, such as when choosing models, enhancing classifiers, and comprehending model choices. Rao et al.'s [17] study explores XAI in healthcare predictive models, highlighting its potential to enhance trust, transparency, and clinical decision-making in medical diagnostics.

A unique model-agnostic method called Anchors is introduced in the publication "Anchors: High-Precision Model-Agnostic Explanations" by Ribeiro et al. [11]. Anchors are used to explain complicated model behaviors using high-precision rules. Anchors give clarity and simplicity of comprehension by representing local variables necessary for predictions. With high-probability assurances, the authors suggest an algorithm to calculate these explanations for every given black-box model. The versatility of Anchors is shown in a number of domains, such as text, tabular data, and pictures. User research demonstrates that, in comparison to current linear explanations, Anchors enable more precise and effective prediction of model behavior on unseen instances. This method offers a compromise between accuracy and understandability in model explanations, addressing the interpretability problem in machine learning.

To examine several approaches to machine learning, Counterfactual explanations have been utilized, highlighting their significance for comprehending and assessing AI judgments. Using determinantal point processes, Mothilal et al.'s research [18] focuses on producing a variety of realistic Counterfactuals while taking user context into account for realism. Their tests demonstrate that their approach generates locally correct Counterfactuals, improving machine learning systems' decision-making capabilities. Van Looveren and Klaise [19] provide a fast, model-independent technique for creating Counterfactual scenarios with class prototypes that works with a variety of data types. By using this approach, black-box model interpretability is enhanced and computational difficulties are resolved. Finally, Wachter et al. [20] address Counterfactual explanations in relation to the General Data Protection Regulation (GDPR), emphasizing how they can make automated judgments more transparent without jeopardizing the specifics of private algorithms. Collectively, these works highlight the significance of Counterfactual explanations in making AI more understandable and accountable.

3. Methodology

In this section, we describe the methodology used in this research, detailing the methods involved in selecting the dataset, preparing the data, and using machine learning models to predict the adaptability levels of students. In order to understand the models and offer insights about feature relevance, we explain the deployment of several Explainable AI (XAI) strategies. The technique aims to bridge the gap between interpretability and predicted accuracy by rigorously testing, evaluating, and training the models. This ensures a thorough examination.

3.1. Dataset

This research dataset is taken from www.kaggle.com (accessed on 18 November 2023) [21] and consists of an instance with 1205 rows and 14 features (columns) with these columns: Gender, Age, Education Level, Institution Type, IT Student, Location, Load-shedding, Financial Condition, Internet Type, Network Type, Class Duration, Self Lms, Device, and the Adaptivity Level which is the target feature with the adaptivity level set to Low, Moderate or High. This dataset was chosen for its rich representation of student demographics, allowing a detailed exploration of student adaptability's multifaceted nature, highlighting the interplay of socio-economic, institutional, and personal factors impacting student adaptability. The dataset was collected during pandemic restrictions, providing a unique context for analyzing student adaptability under extreme conditions. The pandemic significantly impacted educational environments, making this period critical for understanding how students adapt to sudden changes and remote learning conditions. Insights gained from this dataset can offer valuable lessons for future educational disruptions.

In order to ensure a thorough assessment of the prediction performance and interpretability of the machine learning models used, we used 80% of the dataset (964 instances) for training the models and the remaining 20% (241 instances) for testing the models for this study. We trained and tested our machine learning models using an 80/20 split in this study's methodology. The choice was taken in light of our dataset's size and characteristics, which limited the possibility of assigning a distinct validation set without appreciably decreasing the size of the training set. In order to guarantee the accuracy of our model predictions, we placed a high priority on keeping a solid training dataset. We do identify this as a research drawback, though, since an extra layer of model evaluation and fine-tuning may have been supplied by a second validation step.

3.2. Research Framework

This paper's main contribution is its innovative approach to evaluating student adaptability predictions from ML models, a critical component in educational settings. The proposed method, outlined in Figure 1, is a comprehensive, multi-stage process that begins with meticulous data collection, ensuring a robust foundation for subsequent analysis. The preprocessing stage addresses the typical challenges of educational data, such as encoding categorical variables and normalizing features, to prepare a clean, analysis-ready dataset.

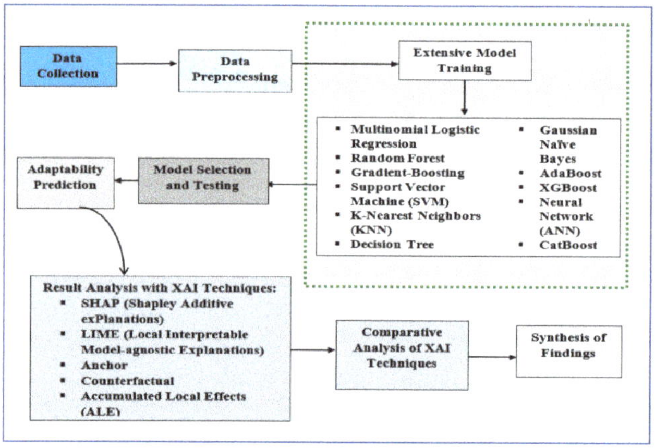

Figure 1. Block diagram for implementing the proposed framework.

The method's thorough training of a wide range of machine learning models, from simpler ones like Multinomial Logistic Regression to more complex ensembles, is one of its distinguishing features. This inclusive approach not only compares predictive power across various algorithms but also enhances the process of interpretation analysis by providing diverse perspectives on the same dataset. The central tenet of the methodology is the application of multiple XAI techniques, such as SHAP, LIME, Anchors, Counterfactuals, and ALE. This diverse mix allows for a detailed exploration of how different features influence the adaptability predictions, providing global and local interpretability. Each XAI technique offers unique insights: SHAP quantifies feature contributions, LIME provides locally faithful explanations, Anchors generate high-precision rules, Counterfactuals suggest minimal changes for outcome alteration, and ALE analyzes feature effects, averting confounding interactions.

The subsequent comparative analysis of these XAI techniques is a methodological innovation that critiques their coherence, identifies consistencies, and pinpoints anomalies in their explanations. The final synthesis of findings distills a holistic understanding of the ML models, offering actionable insights for educational stakeholders. This methodology empowers educators and policymakers with a deeper comprehension of the predictive models, facilitating informed decisions to foster student adaptability in diverse learning environments.

3.3. Data Preprocessing

We process the data by loading the dataset using the data manipulation library 'Pandas' in Python version 3.11.5. Understanding the dataset's structure is necessary for further investigation. The characteristics and target columns may be identified by reviewing the column titles; to understand the data distributions depicted in Table 1, summary statistics for numerical columns must be collected. Categorical variables are converted into a numerical format using One-Hot Encoding. Scaling of numerical features using the 'StandardScaler' method ensures uniformity in the magnitude of features. The mapping for our target variable was Moderate: 2, Low: 1, High: 0. The handling of unbalanced data is then considered and customized to meet the needs of the model and particular dataset features. The dataset is then split into training and testing sets using 'train_test_split' for building the model and model evaluation.

Table 1. Summary statistics of the dataset.

Variable	Count	Unique	Top	Freq
Gender	1205	2	Boy	603
Age	1205	6	21–25	374
Education Level	1205	3	School	530
Institution Type	1205	2	Non-Government	623
IT Student	1205	2	No	501
Location	1205	2	Yes	135
Load-shedding	1200	2	Low	1004
Financial Condition	1206	3	Mid	676
Internet Type	1205	2	Mobile	655
Network Type	1205	3	4G	775
Class Duration	1205	3	1–3	840
Self Lms	1205	2	No	955
Device	1205	3	Mobile	1013
Adaptivity Level	1205	3	Moderate	625

Adaptivity Level is the target variable for classification. This column contains three different levels: Moderate, Low and High. Feature distribution of the adaptivity level in the dataset yields the following frequency counts: Moderate: 625, Low: 480, and High: 100; from these counts, you can see that the dataset is not perfectly balanced.

The student adaptivity levels are shown in Figure 2 as a balanced frequency chart after using the Synthetic Minority Over-sampling Technique (SMOTE). This visualization represents the distribution of adaptivity levels (Moderate, Low, and High) after addressing the initial class imbalance inherent in the dataset. Before the application of SMOTE, the adaptivity levels were not proportionately distributed, which could have led to a predictive model biased toward the majority class. By synthesizing new examples of the minority classes, SMOTE equalizes the representation of each adaptivity level, as depicted in the figure.

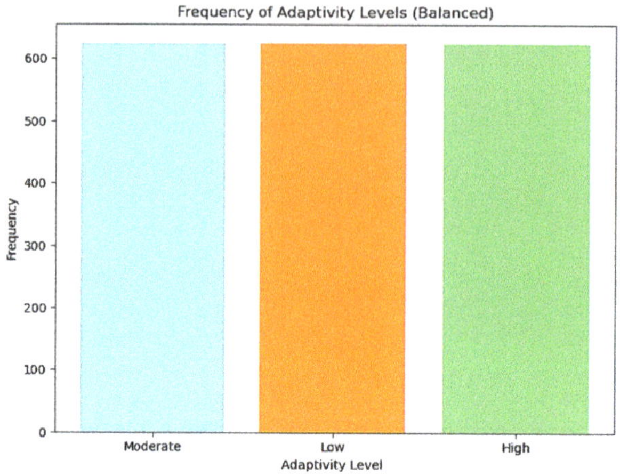

Figure 2. Balanced frequency plot of Adaptivity Level.

Each bar in the graph corresponds to one of the three adaptivity levels, and the height reflects the frequency of instances within the dataset after balancing. The near-equal heights of the bars indicate that each class now has a similar number of instances (625), thus ensuring that the predictive models trained on this dataset are not biased towards any particular adaptivity level.

3.4. Model Training

We undertake a series of extensive classification experiments by applying several machine learning algorithms to classify levels of student adaptability. They were analyzed and classified as shown in Table 2.

Table 2. Models trained and their parameters.

S/N	Model	Parameters	Values
1	Multinomial Logistic Regression	MaximumIterations(max_iter) multi_class	1000 *auto*
2	Decision Tree	max_depth min_samples_leaf min_samples_split criterion splitter max_features	range (1, 20, 2) range (1, 100, 5) range (2, 10) gini and entropy best and random sqrt
3	Random Forest	random_state	0
4	K-Nearest Neighbors	n_neighbors	5
5	XGBoost	max_depth learning_rate n_estimators	5 0.01 200
6	Gradient Boosting	n_estimators random_state	100 42
7	Gaussian Naive Bayes	Gaussian Naive Bayes	Normal no special setup
8	Support Vector Machine (SVM)	Kernel Regularization Parameter (C) Random State	Linear 1.0 42

Table 2. *Cont.*

S/N	Model	Parameters	Values
9	Ensemble	Random Forest Classifier random_state Gradient Boosting Classifier random_state Logistic Regression max_iter random_state Ensemble method used	 42 42 1500 42 voting
10	AdaBoost	Base Classifier max_depth n_estimators random_state	Decision Tree 1 50 42
11	CatBoost	Iterations Depth learning_rate verbose random_state early_stopping_rounds	1500 6 0.1 200 42 50
12	Neural Network	Input Layer Activation Function First Hidden Layer: Units Second Hidden Layer Units Third Hidden Layer Units Output Layer Units Activation Function	X_train_stand ReLU 64 32 16 3 Softmax

3.5. *Explainable Artificial Intelligence (XAI) Used*

The classifications made by the models are then analyzed using several XAI techniques to interpret the model's decision-making process. These techniques include:

3.5.1. SHAP (Shapley Additive exPlanations)

SHAP is a high-level machine learning interpretability method built on game theory concepts to explain the output of any machine learning technique. It was developed by Lundberg and Lee as a unifying approach to explaining the predictions of machine learning algorithms. The SHAP value measures the predictive influence of each feature on the base prediction. The base prediction is typically the average prediction of the dataset used to train the model [10]. The SHAP value for the i-th feature is calculated using the formula:

$$\phi_i = \sum\nolimits_{S \subseteq F \setminus \{i\}} \frac{|S|!(|F| - |S| - 1)!}{|F|!} [F_x(S \cup \{i\}) - f_x(S)] \quad (1)$$

where ϕ_i is the SHAP value for the i-th feature, S is a subset of all features F excluding the i-th feature, $f_x(S)$ is the prediction made by the model using only the features in set S, $|S|$ is the number of features in set S, and $|F|$ is the total number of features.

For example, consider a machine learning model that predicts student adaptability based on various features. Let us calculate the SHAP value for the feature 'Class Duration'. The base prediction (the average prediction of the training dataset) might have a 50% chance of a student being adaptable. If a particular student studies 5 h a day and this changes the prediction to 70% adaptability, the SHAP value for 'Class Duration' in this case would quantify how much of that 20% increase is attributed to studying 5 h specifically, compared to the average student. To apply this, one would calculate the difference in prediction with and without the 'Class Duration' feature over all possible combinations of other features, weighted by the number of features included. This exhaustive calculation considers all possible interactions with other features, thus allocating the 'Class Duration' feature its fair share of influence on the increase in adaptability prediction.

The SHAP framework meets several desirable qualities, including consistency, missingness, and local accuracy, and it is model-agnostic, so any machine learning technique can adopt it [22–25]. These characteristics guarantee the accuracy and dependability of SHAP explanations, which makes them especially useful in critical situations where interpretability is just as crucial as predictive performance. SHAP is used in many fields, such as energy to predict power consumption trends, healthcare to assess patient risk, and finance to evaluate credit risk. Its broad use highlights how crucial interpretability is to machine learning [24,26].

The practical application of SHAP in our study allows us to identify the most influential factors determining student adaptability. For example, SHAP analysis revealed that "Class Duration" and "Financial Condition" are critical in predicting adaptability levels. This insight is invaluable for educational institutions seeking to understand the determinants of student success and can inform targeted interventions to support student adaptability.

3.5.2. LIME (Local Interpretable Model-Agnostic Explanations)

LIME is a machine learning technique that provides an accurate and understandable explanation for the classification made by machine learning models [9]. LIME works by perturbing input data and observing changes in the model's predictions. It then selects the most influential features for a particular prediction and fits a simple model, such as a linear regression, to these features. LIME seeks to explain the reasoning behind a model's prediction y for an individual instance x by approximating the complex model f locally with an interpretable model g. This interpretable model, typically a linear model, operates on a binary feature representation z', which is derived from the original feature space of x. The local surrogate model is defined as:

$$g\left(z^{'}\right) = \beta_0 + \sum_{i=1}^{n} \beta_i z_i^{'} \quad (2)$$

where $g(z^{'})$ represents the prediction of the interpretable model, β_0 is the intercept term of the linear model, β_i represents the weight associated with the i-th feature in the interpretable space, and $z_i^{'}$ is the binary representation of the i-th feature.

LIME aims to optimize the following objective function:

$$\xi(x) = \underset{g \in G}{argmin} \mathcal{L}(f, g, \pi_x) + \Omega(g) \quad (3)$$

where ξ(x) is the local explanation for the instance x, $\mathcal{L}$ measures the fidelity of g in approximating f within the locality defined by π_x, $\Omega(g)$ quantifies the complexity of the interpretable model g, and G denotes the family of interpretable models, such as linear regressions or decision trees.

Consider a machine learning model that predicts a student's adaptability level. For a particular student, the model predicts a "High" adaptability level. LIME is employed to explain this prediction, focusing on features like 'Class Duration', 'Financial Condition', and 'Institution Type', etc. Perturbing the input data around the student's features, LIME fits a local linear model to these perturbations to approximate the prediction of the original complex model f. The linear model constructed by LIME provides the following coefficients for an interpretable explanation:

Adaptability Score = 0.10 + 0.25 × Class Duration − 0.15 × Financial Condition + 0.10 × Institution Type. An increase in 'Class Duration' by one unit (e.g., one hour) enhances the adaptability score by 0.25, suggesting the critical role of engagement time in educational adaptability. A decline in 'Financial Condition' from 'Mid' to 'Poor' (encoded as a decrease by one unit) reduces the adaptability score by 0.15, highlighting the potential barriers faced by economically disadvantaged students. A positive change in 'Institution Type' from 'Government' to 'Non-Government' (encoded as an increase from 0 to 1) raises the adaptability score by 0.10, possibly reflecting the influence of institutional resources and support.

LIME explanation shows how each feature impacts the adaptability prediction. The positive coefficient for 'Class Duration' underscores the importance of classroom engagement in fostering adaptability, suggesting that longer classes may lead to better student outcomes. The negative coefficient for 'Financial Condition' points to the need for targeted support to students from less wealthy backgrounds. The positive coefficient for 'Non-Government' institutions indicates potential benefits of these environments, which may include innovative teaching methods or more personalized attention.

Educational policymakers can leverage these insights to ensure that class time is used effectively, provide additional resources for students in need, and encourage beneficial practices observed in 'Non-Government' institutions. This approach proves especially valuable in domains where decisions are individual-specific, such as healthcare, education, and finance [27]. The localized interpretability provided by LIME empowers stakeholders to make informed decisions, promoting adaptability and improving outcomes in these critical areas.

3.5.3. Anchor

The Anchor algorithm is designed to provide high-precision rules, known as anchors, which offer interpretable and reliable explanations for the decisions of machine learning models [8]. The interpretable explanations are presented as high-precision rules, which are local conditions sufficient to guarantee the prediction with high confidence. An anchor explanation for a prediction can be expressed as:

$$\text{IF}(A_1 \wedge A_2 \wedge \ldots \wedge A_n)\text{THEN prediction } P \text{ with confidencec} \quad (4)$$

where A_i are anchor conditions (predicates) for features, P is the predicted class or outcome, and c is the confidence level associated with the rule. The algorithm aims to maximize the precision (confidence) of the anchor while ensuring that the coverage (the proportion of instances for which the anchor applies) is above a user-defined threshold. The objective function can be formulated as:

$$\text{Maximize Precision}(A) \text{ subject to Coverage}(A) > \theta \quad (5)$$

where Precision(A) is the probability that the prediction P is correct given that anchor A applies, Coverage(A) is the proportion of instances in the dataset for which anchor A applies, and θ is the coverage threshold.

Suppose our machine learning model has predicted the adaptability level of students. To provide an interpretable explanation using the Anchor algorithm, consider an instance where the model predicts a student's adaptability level as "Moderate". After the application of the Anchor algorithm, the following rule-based explanation is produced:

IF (Network Type >1) $\wedge$ (Age $\leq$ 1) $\wedge$ (Gender > 0) $\wedge$ (Institution Type > 0) $\wedge$ (Financial Condition $\leq$ 1) THEN Predict 'Moderate' with confidence 97.32%.

This explanation infers that for this particular student, the presence of certain conditions such as having access to a 3G or 4G network, being in the '16–20' age group, being female, attending a non-government institution, and not being classified as 'Rich' in financial status, are collectively sufficient to predict a 'Moderate' adaptability level with high confidence. This rule-based explanation provides clarity on the factors influencing the adaptability prediction. It suggests that access to quality internet and certain demographic characteristics are significant predictors of adaptability. The high confidence level associated with this anchor allows educators to trust the prediction under the specified conditions. By identifying and understanding these anchor conditions, educational institutions can tailor support and interventions to improve student adaptability. For instance, efforts can be made to enhance internet access for students, and special attention can be given to gender-specific educational experiences and resources for students from different socio-economic backgrounds.

3.5.4. Accumulated Local Effects (ALE)

ALE is a technique used to interpret the predictions of machine learning models, specifically focusing on understanding how features affect predictions on average. ALE addresses the limitation of global methods that average effects over the entire data distribution, which can be misleading if there are interactions between features. This method was developed to provide a better understanding of the feature effects by averaging them locally and thus avoiding the introduction of such interactions, as described by [28].

The ALE of a feature is given as X_j at a particular value x_j, which is calculated by integrating the marginal effects of X_j over the range of the dataset, while marginalizing over the distribution of other features. For a given feature X_j, the ALE is computed as follows:

$$ALE(x_j) = \int_{x_{j,min}}^{x_j} \left(E\left[\frac{\partial f(x)}{\partial x_j} | X_j = z\right] - E\left[\frac{\partial f(x)}{\partial x_j}\right] \right) dz \tag{6}$$

where $f(x)$ is the prediction model, $\frac{\partial f(x)}{\partial x_j}$ is the partial derivative of the model prediction with respect to the feature X_j, $x_{j,min}$ is the minimum value of the feature X_j in the dataset, and $E\left[.|X_j = z\right]$ is the conditional expectation over the distribution of the dataset given the feature $X_j = z$.

Consider a dataset where we aim to understand the impact of 'Class Duration' on the adaptability level of students. We can calculate the ALE for 'Class Duration' by following these steps:

1. Discretize 'Class Duration' into a grid of values.
2. For each interval in the grid, compute the difference in the model's prediction as 'Class Duration' changes within that interval.
3. Accumulate these local effects across the grid to estimate the average effect of 'Class Duration' on the model's prediction.

Presume after computing the ALE, we find that increasing 'Class Duration' from 1 h to 3 h increases the model's prediction for 'High' adaptability by 0.2 on average. This implies that longer class durations are associated with a higher likelihood of students being classified as highly adaptable.

The ALE plot for 'Class Duration' might show a positive slope, which indicates that as 'Class Duration' increases, so does the adaptability prediction. Educational stakeholders can interpret this to mean that policies or interventions designed to extend class durations could potentially improve student adaptability levels. By integrating ALE into the predictive modeling process, educators and policymakers can gain a more detailed understanding of the factors that influence student outcomes. This understanding is crucial for designing educational environments that foster adaptability, a key component of student success in dynamic learning contexts.

3.5.5. Counterfactual Explanations

Counterfactual explanations are interpretability methods in machine learning that focus on "what-if" scenarios, illustrating minimal changes needed for desired output predictions. Originating from the philosophy of causation and decision-making, they provide insights into complex model decision-making, especially in classification tasks. Popularized by [14], they aim to provide explanations for machine learning model decisions that meet legal and non-discrimination requirements.

Counterfactual explanations in machine learning involve finding an instance x′ that is minimally different from the original instance x but leads to a different prediction outcome. Counterfactual explanation seeks to solve the following optimization problem:

$$min_{x'} d\left(x, x'\right) \ subject\ to\ f(x') \neq f(x) \quad (7)$$

where $f(x)$ is the prediction of the original instance, $f(x')$ is the prediction of the Counterfactual instance, and $d(x, x')$ is a distance metric (such as Euclidean distance) that quantifies the dissimilarity between the original instance and the Counterfactual.

For example, imagine a machine learning model predicts a 'Low' adaptability level for a student based on features such as 'Class Duration', 'Financial Condition', and 'Institution Type'. A Counterfactual explanation might reveal that a slight increase in 'Class Duration' could change the prediction to 'High' adaptability.

For instance, if the original feature values are x = [1 (Low Class Duration), 0 (Poor Financial Condition), 0 (Government Institution)] and the model's prediction f(x) is 'Low adaptability', a Counterfactual explanation would search for a new set of feature values x′ that lead to a different prediction f(x′) = 'High adaptability'. Suppose the Counterfactual feature values are x'= [2 (High Class Duration), 0, 0], indicating that only 'Class Duration' needs to be increased to alter the prediction. The Counterfactual explanation therefore might suggest that improving the 'Class Duration' from Low to High could enhance the student's adaptability level according to the model.

Counterfactual insights, like increased instructional time positively affecting student adaptability, guide educational resource allocation and intervention strategies. Educators, leveraging Counterfactual explanations, pinpoint influential factors in the learning process for targeted improvements. This forms a feedback loop, enhancing both educational strategies and machine learning models. Beyond education, Counterfactual explanations facilitate understanding and actionable insights in finance, revealing changes for loan approval, and in healthcare, illustrating shifts in patient risk categories based on lifestyle or clinical parameters [29,30].

4. Results and Discussion

This section presents the results of machine learning models used to predict student adaptability levels. It details the experimental setup, computational environment, and model parameters. The models' performance is analyzed using metrics like accuracy, precision, recall, and F1-score. The interpretability of the models is explored using Explainable AI techniques. Key features' impact on adaptability predictions is highlighted.

4.1. Experimental Setup

A personal computer running Windows 10 Pro, equipped with a Lenovo X1 Yoga (Thinkpad) (Lenovo Group Ltd., Hong Kong, China) Core i7-6600U CPU, 16 GB of DDRAM, and a 256 GB SSD, was used for the research. The Anaconda, Inc.-packaged Python version 3.11.5 was utilized to implement the machine learning models. Context for the computing performance of the methods employed in this study is provided by this hardware specification.

4.2. Model Selection and Prediction

A thorough comparison between many machine learning algorithms was carried out in an effort to identify the best model for the classification of student adaptability levels. These include Multinomial Logistic Regression, Decision Tree, Random Forest, K-Nearest Neighbors (KNN), XGBoost, Gradient Boosting, Gaussian Naïve Bayes, Support Vector Machine (SVM), Ensemble methods, AdaBoost, CatBoost, and Neural Networks. Tables 2 and 3 include confusion matrices and a variety of performance metrics, providing a comprehensive summary of each model's predictive performance.

Table 3. Confusion Matrices of the Deployed Models.

	Model	Class Name	Predicted		
			High	Low	Moderate
Actual	Multinomial Logistic Regression	High	11	5	7
		Low	1	57	45
		Moderate	3	14	98
	Decision Tree	High	15	0	8
		Low	2	95	6
		Moderate	1	7	107
	Random Forest	High	15	0	8
		Low	2	97	4
		Moderate	0	7	108
	K-Nearest Neighbors	High	10	1	12
		Low	1	89	13
		Moderate	2	19	94
	XGBoost	High	12	0	11
		Low	3	67	33
		Moderate	2	8	105
	Gradient Boosting	High	13	0	10
		Low	2	87	14
		Moderate	0	14	101
	Gaussian Naive Bayes	High	13	3	7
		Low	2	62	39
		Moderate	6	20	89
	Support Vector Machine (SVM)	High	13	1	9
		Low	3	57	43
		Moderate	4	9	102
	Ensemble	High	12	1	10
		Low	0	92	11
		Moderate	0	4	111
	AdaBoost	High	8	3	12
		Low	2	69	32
		Moderate	3	22	90
	CatBoost	High	15	0	8
		Low	2	95	6
		Moderate	0	8	107
	Neural Network	High	15	0	8
		Low	2	93	8
		Moderate	0	7	108

The three adaptability classes (High, Low, and Moderate) of each model's predictions were broken down in depth by the confusion matrices as shown in Table 3. For example, the Multinomial Logistic Regression model accurately predicted 'High' adaptability 11 times, but it also incorrectly categorized 7 'Moderate' and 5 'Low' instances as 'High'.

Understanding the predictive dynamics of each classifier, including their advantages and disadvantages in differentiating between classes, depends on the specificity of the data.

To give a complete view of each model's effectiveness, the associated performance measures were generated, including accuracy, precision, recall, and F1 score for each class, as shown in Table 4. The Random Forest model had a strong predictive capacity overall, with precision values of 0.93, 0.94, and 0.94 for the High, Low, and Moderate classes, respectively. These results are further supported by the F1 scores, where high values show a balanced precision-recall trade-off.

Table 4. Performance Metrics of The Deployed Models.

Model	Class Name	Precision	Recall	F1-Score
Multinomial Logistic Regression	High	0.73	0.48	0.58
	Low	0.75	0.55	0.64
	Moderate	0.65	0.85	0.74
Decision Tree	High	0.88	0.65	0.75
	Low	0.88	0.92	0.90
	Moderate	0.88	0.89	0.88
Random Forest	High	0.88	0.65	0.75
	Low	0.93	0.94	0.94
	Moderate	0.90	0.94	0.92
K-Nearest Neighbors	High	0.77	0.43	0.56
	Low	0.82	0.86	0.84
	Moderate	0.79	0.82	0.80
XGBoost	High	0.71	0.52	0.60
	Low	0.89	0.65	0.75
	Moderate	0.70	0.91	0.80
Gradient Boosting	High	0.87	0.57	0.68
	Low	0.86	0.84	0.85
	Moderate	0.81	0.88	0.84
Gaussian Naive Bayes	High	0.62	0.57	0.59
	Low	0.73	0.60	0.66
	Moderate	0.66	0.77	0.71
Support Vector Machine (SVM)	High	0.65	0.57	0.60
	Low	0.85	0.55	0.67
	Moderate	0.66	0.89	0.76
Ensemble	High	1.0	0.52	0.69
	Low	0.95	0.89	0.92
	Moderate	0.84	0.97	0.90
AdaBoost	High	0.62	0.35	0.44
	Low	0.73	0.67	0.70
	Moderate	0.67	0.78	0.72

Table 4. *Cont.*

Model	Class Name	Precision	Recall	F1-Score
CatBoost	High	0.88	0.65	0.75
	Low	0.92	0.92	0.92
	Moderate	0.88	0.93	0.91
Neural Network	High	0.88	0.65	0.75
	Low	0.93	0.90	0.92
	Moderate	0.87	0.94	0.90

A comparison of the accuracy obtained by several machine learning algorithms used in the classification of student adaptability levels is shown in Figure 3. The accuracy of an algorithm is represented by each bar, which is measured as the proportion of test cases that are correctly classified, with Random Forest proving the overall best with 91%, followed by Neural Network and CatBoost with 90% each. Performance metrics and confusion matrices highlight how crucial model selection is in the field of educational data analytics. In particular, the Random Forest, CatBoost, and Neural Network models have demonstrated encouraging possibilities for use in learning environments with the goal of predicting and improving student adaptability.

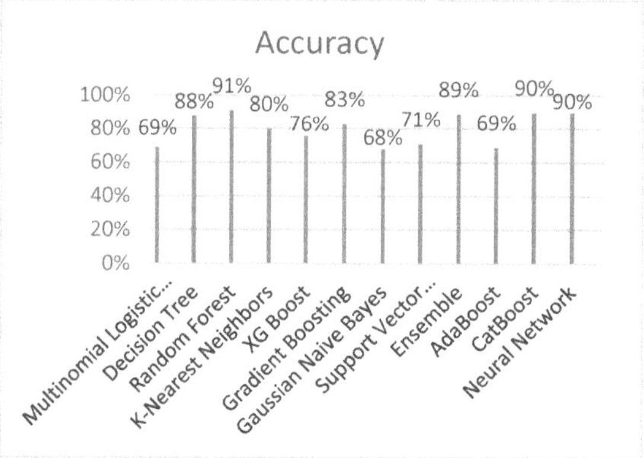

Figure 3. Model accuracy comparison.

4.3. Result Analysis with XAI Techniques

With our range of machine learning models, of which Random Forest performed the best with 91% predictive accuracy, we sought to understand the decision-making processes that buttress the performance of XAI. The following XAI techniques were used to illuminate the contributions and influences of features within our models.

4.3.1. SHAP (Shapley Additive exPlanations)

Using the mean SHAP as a guide, Figure 4 provides a detailed examination of how different factors affect the adaptability levels of students. SHAP values quantify the contribution of each feature to the predictive model's output, offering a measure of importance based on average impact magnitude. This bar graph segregates the influence of features into three adaptability predictions: High (green), Moderate (blue), and Low (pink). The length of each bar represents the mean absolute SHAP value, which is a composite measure of both the strength and consistency of a feature's effect on the model's predictions. 'Class Duration' emerges as a dominant feature, its substantial mean SHAP value indicating a significant and positive correlation with student adaptability across all levels. Its greatest impact is observed in the High adaptability category, suggesting that extended instructional periods may enhance a student's ability to adapt to varying educational demands.

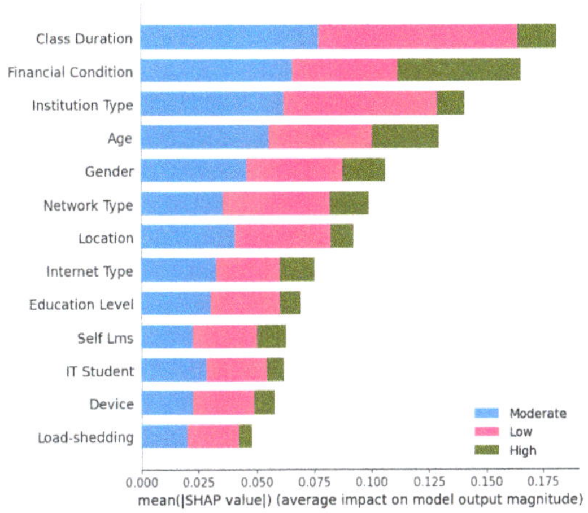

Figure 4. (Global Explanation) Mean impact of features on predicted student adaptability levels using SHAP values.

'Financial Condition' is another prominent feature, especially in predicting Low adaptability, highlighting the potential obstacles faced by students with fewer economic resources. Similarly, 'Institution Type' displays a varied influence, with a positive association for High adaptability, possibly reflecting the superior resources and support available at certain institutions.

On the other hand, 'Load-shedding' predominantly affects Low adaptability predictions, alluding to the detrimental effects of inconsistent electricity on educational continuity. Meanwhile, demographic attributes such as 'Age' and 'Gender' show moderate effects, indicating their complex but less pronounced roles. Technology-related attributes, namely 'IT Student' status and 'Device' usage, although impactful, have lesser mean SHAP values, suggesting their influence is secondary compared to educational and socio-economic factors.

This figure illustration using SHAP values highlights how diverse student adaptation is. It gives educational stakeholders a data-driven platform to build focused interventions that enhance positive aspects and lessen negative ones, creating a climate that encourages flexible learning.

The use of SHAP in a prediction model to ascertain student adaptability levels is seen in Figure 5. The figure is divided into three parts: (a) the attributes under analysis, (b) the corresponding SHAP values, and (c) the force plot. The base value (f(x)) represents the average output when no features affect the prediction. In this instance, the model predicted a 'Low' adaptability level, indicated by a prediction value of 1. Features with red segments, such as 'Education Level' and 'Network Type', contribute to a higher adaptability prediction ('Moderate' or 'High'), while features with blue segments, such as 'Financial Condition' and 'Self Lms', are associated with a lower adaptability prediction ('Low'). The force plot elucidates the model's complex reasoning for an individual prediction. For example, a favorable 'Financial Condition' and access to 'Internet Type' influence the adaptability prediction away from 'High', while the absence of engagement with 'Self Lms' and lower 'Education Level' push the prediction towards 'Low' adaptability. The analysis of this figure can help educational institutions understand the complex causes of student adaptability and fit their resources and interventions accordingly.

	Gender	Age	Education Level	Institution Type	IT Student	Location	Load-shedding	Financial Condition	Internet Type	Network Type	Class Duration	Self Lms	Device
155	0	3	2	0	1	1	1	0	0	1	1	0	0

(**a**)

```
(1, 13)

[array([[ 0.00565236, -0.01082301,  0.00406231, -0.00017784, -0.00201523,
          0.00194225,  0.00065064, -0.04703173, -0.01553555, -0.01937018,
          0.00785446, -0.00790196,  0.01110634]]),
 array([[-0.01278584,  0.09307644,  0.08920661,  0.1820952 , -0.03075159,
          0.01048164,  0.01204537,  0.02316305,  0.04945715,  0.11714716,
         -0.04823885,  0.06276665, -0.03606548]]),
 array([[ 0.00713349, -0.08225343, -0.09326892, -0.18191736,  0.03276682,
         -0.01242389, -0.01269601,  0.02386868, -0.03392161, -0.09777698,
          0.04038438, -0.05486469,  0.02495914]])]
```

(**b**)

The RF predicted: 1

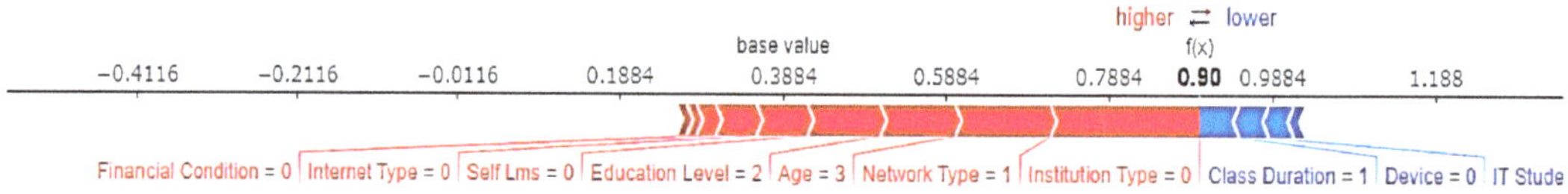

(**c**)

Figure 5. SHAP force plot for an individual prediction of student adaptability. (**a**) Instance for an explanation; (**b**) SHAP value of the instance; (**c**) instance local explanation.

4.3.2. LIME (Local Interpretable Model-Agnostic Explanations

A single prediction from a machine learning model that categorizes student adaptability levels is shown in Figure 6, along with LIME study. The adaptation levels that the model predicts are 'Moderate' (probability = 0.62) and 'Low' (probability = 0.38). The prediction for 'High' adaptability is highly unlikely based on the given feature values.

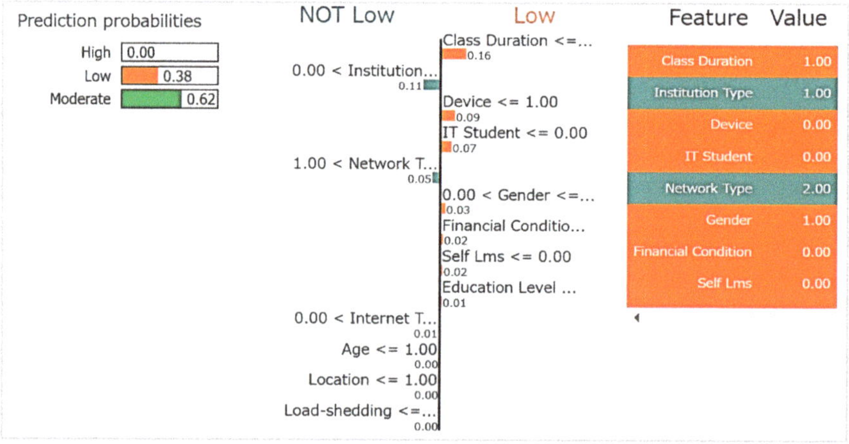

Figure 6. (Local) Interpretation of a model prediction using LIME.

The middle section illustrates the weighted impact of each feature on the prediction decision, with positive contributions towards a 'Not Low' adaptability prediction. For instance, 'Class Duration' and 'Institution Type' exhibit the most significant positive influence, suggesting that longer class durations and certain institutional characteristics may be associated with higher adaptability. Conversely, features such as 'Device' and 'IT Student' status have a notable negative weight, implying that the absence of certain technological factors may lead the model to predict a lower adaptability level.

The rightmost section lists the values of the features for the specific instance being explained, with orange representing positive contributions and blue for negative ones. This visualization aids in understanding the local behavior of the machine learning model, providing transparency into how feature values are aggregated into a predictive outcome. The LIME interpretation offers valuable insights for educators and policymakers by pinpointing specific areas that could be addressed to potentially improve a student's adaptability. It also emphasizes the intricate nature of the factors that impact educational outcomes and highlights the requirement for sophisticated approaches in educational interventions.

4.3.3. ANCHOR Explanation

Table 5 illustrates how the Anchor algorithm predicts a 'Moderate' level of adaptability in students when present, based on specific features. Network Type greater than 1.00 indicates that at least 3G or 4G connectivity is associated with a 'Moderate' adaptability prediction, reflecting the importance of reliable internet access for students' ability to adapt to contemporary educational demands. Age less than or equal to 1.00 corresponds to students in the age group '16–20', revealing that adaptability may be particularly complex during late adolescence. Gender greater than 0.00 suggests that female students are predicted to have 'Moderate' adaptability, indicating gender-specific educational experiences or socialization patterns that influence adaptability. Institution Type greater than 0.00 refers to students attending 'Non-Government' institutions, suggesting that these institutions may have different resource levels or pedagogical approaches. Financial Condition less than or equal to 1.00 captures 'Mid' or 'Poor' economic backgrounds, emphasizing the role of socioeconomic status in shaping a student's adaptability.

Table 5. (Local) Anchor explanation for a 'Moderate' adaptability prediction.

Anchor Explanation Condition	
Feature	Condition
Institution Type	>1.00
Age	≤1.00
Gender	>0.00
Financial Condition	≤1.00
Precision	0.9732360097323601
Coverage	0.0888
Predicted Class	Moderate

The precision and coverage metrics accompanying these rules indicate a high level of reliability and relevance within the dataset, suggesting these findings are consistent and pertinent to a subset of the student population. By integrating these interpretive insights into the study, educators and policymakers can better understand the factors contributing to student adaptability levels and inform initiatives to upgrade digital infrastructures. The distinction between government and non-government institutions in adaptability predictions may reflect differential access to resources, student support, and extracurricular opportunities, prompting a re-evaluation of resource distribution to ensure all institutions can foster student adaptability.

4.3.4. Accumulated Local Effects (ALE) Explanation

The ALE feature importance analysis, as visualized in Figure 7, provides compelling insights that can guide educational strategies and policy-making. The significant importance of 'Class Duration' suggests that students benefit from extended learning sessions, which may offer a more immersive educational experience. This finding supports the pedagogical approach of increasing instructional time to enhance understanding and retention. Policymakers might consider revising academic schedules to integrate long class periods or more intensive study sessions that could better support students' adaptability and learning outcomes.

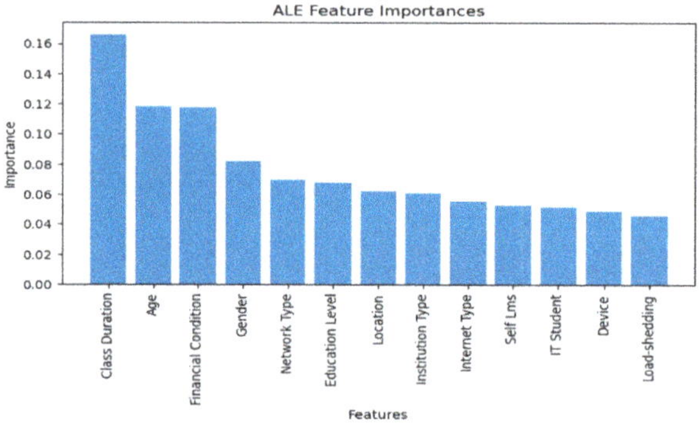

Figure 7. ALE feature importance for predicting student adaptability level.

The importance of 'Age' as a predictive factor of adaptability underscores the importance of developmental considerations in educational planning. Younger students may require more structured support to foster adaptability, such as resilience training and social-emotional learning programs. In contrast, older students may benefit from opportunities that challenge their adaptability skills, such as project-based learning and collaborative assignments that mirror real-world scenarios.

The impact of 'Financial Condition' on adaptability predictions reinforces the link between economic factors and educational success. This suggests an urgent need for policies that aim to level the playing field, such as providing financial aid, resources, and support systems for students from less affluent backgrounds. Ensuring that all students have equal access to educational resources is not just a matter of fairness but also a strategic investment in the adaptability and resilience of the future workforce.

Gender, network type, and education level highlight the multifaceted nature of adaptability. These findings suggest a tailored approach where educational interventions are sensitive to gender dynamics, technological access, and institutional characteristics. For example, initiatives to bridge the digital divide by improving network connectivity can have far-reaching effects on students' ability to access and engage with digital learning platforms, a necessity in an increasingly connected world.

Furthermore, the lower importance of features such as 'Self Lms', 'IT Student', and 'Device' does not diminish their value but rather indicates that their influence on adaptability may be more conditional or indirect. This could inform a carefully balanced approach to technology integration in education, ensuring that technology enhances learning without widening the gap between different student groups.

Lastly, the feature 'Load-shedding' highlights the external challenges students face, pointing to the broader social and infrastructural issues that can affect educational outcomes. Addressing such challenges may require collaborative efforts that extend beyond the education sector, involving partnerships with community organizations and government agencies to provide stable learning environments.

The ALE feature importance analysis thus serves as a guide for developing comprehensive educational policies and practices that consider the complex interplay of individual, institutional, and societal factors affecting student adaptability. By focusing on these key features, educators and policymakers can create more supportive and effective learning environments that cater to the diverse needs of students, fostering an educational ecosystem that is both equitable and conducive to developing adaptable learners.

Our analysis utilized ALE plots to visualize the impact of three critical features—Class Duration, Age, and Financial Condition—on the predicted adaptability levels of students. The resulting plots offer insightful revelations that intersect with contemporary educational theories and bear significant policy implications, see Figure 8.

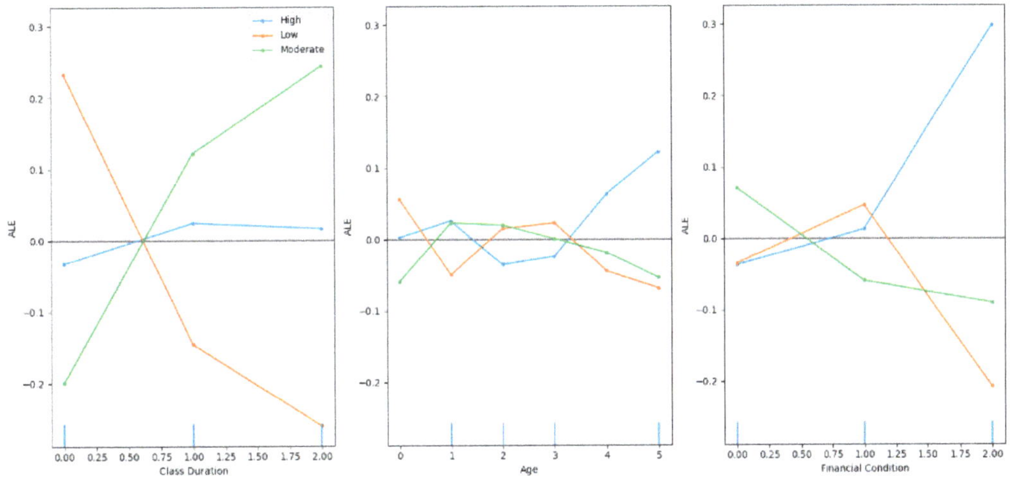

(a)	(b)	(c)

Figure 8. ALE plots for the three most influential features, (**a**) Class Duration, (**b**) Age, and (**c**) Financial Condition, on student adaptability predictions.

Class Duration:

The ALE plot for 'Class Duration' indicates that longer class times are positively associated with 'High' adaptability predictions. This finding aligns with educational theories that stress the importance of sustained engagement for deeper learning and adaptability. It suggests that extended instructional periods may provide students with more opportunities to assimilate information, engage with challenging concepts, and develop critical thinking skills, all of which are crucial for adaptability in rapidly changing educational landscapes. Policy-wise, this supports arguments for restructuring school schedules to allow for longer class periods, potentially leading to improved educational outcomes.

Age:

The relationship between 'Age' and adaptability is less direct, as shown by the minor fluctuations across adaptability levels. However, the general trend indicates that 'Low' adaptability decreases with age. This could reflect the development of coping mechanisms and resilience as students mature, a concept supported by developmental theories. The varied adaptability across age groups could inform the design of age-specific curricula and support services, tailoring educational strategies to the developmental stage of the student cohort.

Financial Condition:

The steep positive slope for 'High' adaptability with improved 'Financial Condition' underscores the pivotal role of economic stability in educational success, as posited by numerous studies linking financial security with better academic performance. The plot highlights a stark reality: students from more affluent backgrounds are likely to be more adaptable, possibly due to greater access to resources, extracurricular activities, and learning support. This insight has profound policy implications, emphasizing the need for equity-focused financial initiatives such as scholarships, grants, and resource allocation to schools serving economically disadvantaged communities.

The ALE plots also reveal the multidimensional nature of student adaptability, validating educational theories, influencing classroom and curriculum design, and emphasizing the need for comprehensive policies to address diverse factors contributing to student adaptability.

4.3.5. Counterfactual Explanation

Figure 9 details a Counterfactual analysis for Instance 3, where a model's original prediction of a student's adaptability level as 'Moderate' (coded as 2) is altered to 'Low' (coded as 1). This analysis is pivotal in understanding the sensitivity of the predictive model to changes in feature values. The original feature vector [1 3 2 1 0 1 1 0 1 1 1 1 1] encapsulates the student's profile, with each element corresponding to various features such as Gender, Age, Education Level, etc., as outlined in the research methodology. The Counterfactual vector [1. 3. 2. 1. 0. 1. 1. 0. 1. 1. 0.49932724 1. 1.] presents the minimal adjustment required to flip the prediction outcome. Notably, the alteration occurs at the eleventh feature, which we can infer relates to 'Class Duration' given the 'Feature differences' array indicating a change of [0.5]. This modification in 'Class Duration' decreases its value by 0.5. Considering our encoding scheme, where 'Class Duration' might be defined within a range (e.g., '0' for Low, '1' for Moderate, '2' for High), a decrease suggests a shift towards a shorter class duration. The model's understanding that a reduced 'Class Duration' indicates a 'Low' adaptability level may stem from a pattern observed in the data, where shorter class durations are linked to low adaptability.

```
Instance 3 - Original pred: 2
Instance 3 - CF pred: 1
Instance 3 - Original: [1 3 2 1 0 1 1 0 1 1 1 1 1]
Instance 3 - Counterfactual: [1.        3.        2.        1.        0.        1.
 1.        0.        1.        1.        0.49932724 1.
 1.        ]
Instance 3 - Feature differences: [0.5]
```

Figure 9. Counterfactual analysis illustrating adaptability prediction shift.

The implications of this finding are significant for educational practices. It suggests that ensuring adequate class time may be critical for student adaptability, echoing educational theories that emphasize the role of engagement time in skill development and learning adaptability. Policymakers may need to consider this relationship when designing curriculum schedules, advocating for sufficient instructional time to foster an adaptable learning mindset. Also, this Counterfactual result raises important questions for future research. Why might a slight reduction in 'Class Duration' correlate with a lower adaptability prediction? Does this relationship hold across various subjects and learning contexts? Such questions highlight the need for a deeper examination of the educational factors that contribute to student adaptability and the development of interventions that can effectively support students' learning journeys.

The comparison between the original and Counterfactual feature values for Instance 3 is shown in Figure 10. This comparison offers valuable insight into how the model predicts student adaptability levels. The bar chart highlights the differences between the original features that led to a 'Moderate' adaptability prediction and the Counterfactual features that would result in a 'Low' adaptability prediction. The original feature values are denoted in blue, while the Counterfactual values that could potentially alter the model's outcome are shown in orange. Notably, most features remain unchanged between the original and Counterfactual scenarios, except for 'Class Duration', which shows a decrement of 0.5 in the Counterfactual case. This decrease suggests that a reduction in 'Class Duration' is influential in shifting the adaptability prediction from 'Moderate' to 'Low'.

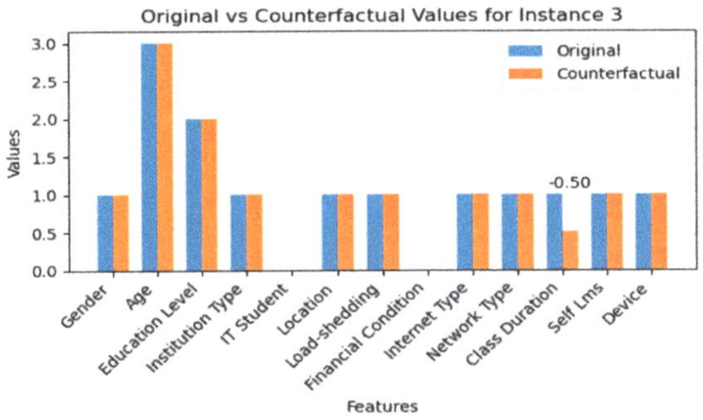

Figure 10. Comparative analysis of original and Counterfactual feature values for Instance 3.

The result also emphasizes the significant role that 'Class Duration' plays in the adaptability model, where a slight reduction in duration is the single change needed to alter the prediction. This finding has important implications for educational strategies, suggesting that maintaining or increasing class duration could be a critical factor in supporting higher adaptability levels among students. Educators and administrators may use these insights to reassess the structure of the school day and the allocation of time to different subjects. The chart encourages a data-driven approach to curriculum planning, ensuring that students have sufficient time for in-depth exploration and learning, which is crucial for developing adaptability skills. This Counterfactual analysis complements other interpretability techniques, such as SHAP and LIME, by providing a direct and tangible scenario where altering a single feature can change a student's predicted adaptability level. Such information is invaluable for policy formulation, as it highlights specific leverage points where educational interventions could be targeted to yield the most significant impact on student outcomes.

4.4. Comparison with Previous Studies

This study expands on earlier research by Hasan Suzan et al. [1], who predicted students' degrees of adaptation throughout the pandemic using the same dataset. By utilizing an extensive range of Explainable AI (XAI) methodologies, our research expands on previous findings and offers a more profound understanding of factors affecting adaptability.

Suzan et al. [1] used a variety of machine learning algorithms in their study, including Random Forest, Decision Tree, Naive Bayes, Support Vector Machine, K-Nearest Neighbors, and Artificial Neural Network. The Random Forest classifier yielded the best results, with an accuracy of 89.63% and precision values of 0.89, 0.90, 0.88 for low, moderate and high adaptability, respectively. Using the Random Forest model, our study, in contrast, obtained a higher accuracy of 91% and precision values of 0.93, 0.94, and 0.94 for high, low, and moderate adaptation levels, respectively. To provide a more thorough interpretation of the model predictions, we also used XAI techniques including SHAP, LIME, Anchors, ALE, and Counterfactual explanations to highlight certain feature contributions. We demonstrate the complementary nature of our approach and verify the robustness of our technique by comparing our results with those of [1]. Our work presents a more thorough knowledge of student adaptation in the context of education, not only validating their findings but also providing improved interpretability using XAI approaches.

This comparison underscores the value of integrating XAI methods with traditional machine learning models to achieve both high predictive performance and interpretability, thereby contributing to more informed and effective educational interventions.

5. Comparative Analysis of XAI Techniques and Synthesis of Findings

After completing an extensive training program involving a varied set of machine learning models, our research delved into a thorough examination of model predictions using a range of XAI techniques. Each technique offered unique insights into the decision-making processes underlying the adaptability predictions, with SHAP and ALE providing a global view of feature contributions, LIME and Anchors elucidating local, instance-specific explanations, Counterfactuals revealing the sensitivity of predictions to feature perturbations, and ALE plots assessing the average feature effects across the data distribution. A comparative analysis of the XAI outputs revealed both similarities and divergence in feature importance rankings and explanations. Accordance across methods reinforced the perceived significance of certain features, such as 'class duration' and 'financial condition', suggesting these are robust determinants of adaptability predictions. Divergences prompted a careful reconsideration of complex feature interactions and model behavior in various scenarios, guiding further analytical probing.

This study leverages a diverse set of XAI techniques, including SHAP, LIME, Anchors, Counterfactual, and ALE, to comprehensively interpret student adaptability prediction models. The amalgamation of these methods sheds light on key features influencing adaptability levels, contextualizing predictions within the educational landscape. This holistic understanding forms the foundation for potential educational interventions, emphasizing the necessity of considering engagement duration and addressing socioeconomic factors.

The research makes a unique and valuable contribution to the areas of machine learning interpretability and educational policy-making by utilizing a combination of different XAI approaches to clarify the decision-making processes of educational data models. It not only enhances the interpretability of educational data through a systematic comparative analysis of various XAI techniques but also identifies and examines the complex socio-economic, institutional, and demographic factors influencing student adaptability. The thorough exploration and results of this study provide the basis for developing detailed and targeted interventions. A notable feature of this research is its innovative methodology, which integrates and compares insights from several XAI techniques to gain a comprehensive understanding of model predictions. This method offers a detailed perspective on how various factors affect student adaptability, enabling the creation of well-informed and impactful educational policies and practices. By bridging the theoretical aspects of machine learning with practical, actionable insights, the study demonstrates the value of XAI in crafting educational strategies that are both equitable and tailored to enhance student outcomes. These contributions represent a significant advancement in the responsible application of AI within the educational sector, ensuring that predictive models are leveraged in a manner that is both transparent and beneficial to student learning and adaptability.

Limitations

Despite the remarkable findings and contributions of this study, it is important to acknowledge a number of limitations.

1. The dataset included in this study was sourced from a publicly accessible website (www.kaggle.com, accessed on 18 November 2023), despite being reliable and representative of a range of student characteristics. This restricts the extent of generalizability since the dataset might not include all the subtle characteristics seen in various educational environments throughout the world. Although the dataset effectively captures various demographic and socio-economic factors, it lacks contextual elements like cultural and extracurricular influences, which are also significant for a

holistic understanding of student adaptability. Despite these limitations, the dataset remains a valuable resource for exploring key factors influencing student adaptability and developing targeted educational interventions.

2. For analysis, the study mostly employed quantitative data. Qualitative data, from surveys or interviews with students, may provide deeper, more individualized insights into the variables influencing students' adaptation. This might aid in comprehending the contextual elements that are hidden by quantitative data on its own.
3. Cross-sectional data were used in the study which captures a snapshot of student adaptability at a single point in time. This method ignores periodic shifts in students' adaptability, which might offer more detailed information about how adaptability develops over time and is impacted by different kinds of interventions.

6. Conclusions

This study uses XAI techniques to understand factors influencing student adaptability in education. The application of various methods, such as SHAP, LIME, Anchors, ALE, and Counterfactual explanations, provides a comprehensive understanding of the dynamics influencing student adaptability. Key factors such as 'Class Duration', 'Financial Condition', 'Age', and 'Institution Type' are crucial in shaping students' ability to adapt in educational settings. These findings emphasize the importance of considering socio-economic and institutional factors when formulating educational strategies. XAI has the potential to foster a data-informed educational landscape, allowing for personalized learning pathways and adaptive educational systems. Integrating AI and analytics into educational policy and practice can enhance educational outcomes and create a more inclusive and effective learning environment.

Future inquiries should focus on analyzing longitudinal data to capture the temporal changes in adaptability and integrating qualitative data for richer, more personalized insights. Longitudinal studies would allow researchers to observe how adaptability evolves over time, providing a more dynamic understanding of the factors that influence it. Incorporating qualitative data, such as student interviews and surveys, can offer a more holistic view of student experiences and the contextual factors that quantitative data alone cannot reveal. This approach would enhance the depth of the analysis and potentially uncover new dimensions of student adaptability.

Evaluating the real-world effects of interventions informed by XAI and updating the interpretability framework to align with new machine learning advancements will be crucial. Ensuring the ethical application of AI and maintaining transparency in AI-driven educational tools remain essential for the responsible progression of this field.

We suggest adding a validation stage to the model creation procedure for future studies. This innovation may make it easier to fine-tune and validate the model in more subtle ways, which might improve interpretability and prediction performance. Investigating how a certain validation set affects the generalizability of machine learning models in educational contexts may yield insightful information that advances our knowledge of student flexibility and the efficient use of XAI methods.

Author Contributions: Conceptualization, L.C.N.; Validation, A.M.J.-O.; Methodology, L.C.N.; Writing—original draft, L.C.N.; Writing—review & editing, M.M.R.; Supervision, Writing—review & editing, Y.W. All authors have read and agreed to the published version of the manuscript.

Funding: This research received no external funding.

Informed Consent Statement: Not applicable.

Data Availability Statement: The original data presented in the study are openly available in (https://www.kaggle.com/datasets/mdmahmudulhasansuzan/students-adaptability-level-in-online-education, accessed on 18 November 2023) and Derived data supporting the findings of this study are available from the corresponding author on request.

Conflicts of Interest: The authors declare no conflicts of interest.

References

1. Suzan, M.M.H.; Samrin, N.A.; Biswas, A.A.; Pramanik, A. Students' Adaptability Level Prediction in Online Education Using Machine Learning Approaches. In Proceedings of the 2021 12th International Conference on Computing Communication and Networking Technologies (ICCCNT), Kharagpur, India, 6–8 July 2021; pp. 1–7. [CrossRef]
2. Loderer, K.; Rinas, R.; Daumiller, M. Student adaptability, emotions, and achievement: Navigating new academic terrains in a global crisis. *Learn. Individ. Differ.* **2021**, *90*, 102046. [CrossRef] [PubMed]
3. Peng, H.; Ma, S.; Spector, J.M. Personalized Adaptive Learning: An emerging pedagogical approach enabled by a smart learning environment. In *Foundations and Trends in Smart Learning*; Lecture Notes in Educational Technology; Springer: Singapore, 2019; pp. 171–176. [CrossRef]
4. Haleem, A.; Javaid, M.; Qadri, M.A.; Suman, R. Understanding the role of digital technologies in education: A review. *Sustain. Oper. Comput.* **2022**, *3*, 275–285. [CrossRef]
5. Jang, Y.; Choi, S.; Jung, H.; Kim, H. Practical early prediction of students' performance using machine learning and eXplainable AI. *Educ. Inf. Technol.* **2022**, *27*, 12855–12889. [CrossRef]
6. Adadi, A.; Berrada, M. Peeking Inside the Black-Box: A survey on Explainable Artificial Intelligence (XAI). *IEEE Access* **2018**, *6*, 52138–52160. [CrossRef]
7. White, G.L. Adaptive Learning Technology Relationship with Student Learning Outcomes. *J. Inf. Technol. Educ.* **2020**, *19*, 113–130. [CrossRef] [PubMed]
8. Liem, G.A.D.; Martin, A.J. Young people's responses to environmental issues: Exploring the roles of adaptability and personality. *Personal. Individ. Differ.* **2015**, *79*, 91–97. [CrossRef]
9. Ribeiro, M.T.; Singh, S.; Guestrin, C. "Why Should I Trust You?": Explaining the Predictions of Any Classifier. In Proceedings of the 22nd ACM SIGKDD International Conference on Knowledge Discovery and Data Mining, San Francisco, CA, USA, 13–17 August 2016; pp. 1135–1144. [CrossRef]
10. Lundberg, S.; Lee, S.-I. A Unified Approach to Interpreting Model Predictions. *arXiv* **2017**, arXiv:1705.07874.
11. Ribeiro, M.T.; Singh, S.; Guestrin, C. Anchors: High-Precision Model-Agnostic Explanations. In Proceedings of the AAAI Conference on Artificial Intelligence, New Orleans, LA, USA, 2–7 February 2018; Volume 32. [CrossRef]
12. She, C.; Li, Q.; Jiang, W.; Xing, Q. Learning adaptability facilitates self-regulated learning at school: The chain mediating roles of academic motivation and self-management. *Front. Psychol.* **2023**, *14*, 1162072. [CrossRef] [PubMed]
13. Martin, F.; Chen, Y.; Moore, R.L.; Westine, C.D. Systematic review of adaptive learning research designs, context, strategies, and technologies from 2009 to 2018. *Educ. Technol. Res. Dev.* **2020**, *68*, 1903–1929. [CrossRef]
14. Tiwari, R.G.; Jain, A.; Kukreja, V.; Ujjwal, N. Education 4.0: Explainable Machine Learning for Classification of student Adaptability. In Proceedings of the 2022 International Conference on Data Analytics for Business and Industry (ICDABI), Online, 25–26 October 2022. [CrossRef]
15. Adnan, M.; Uddin, I.; Khan, E.; Alharithi, F.S.; Amin, S.; Alzahrani, A.A. Earliest possible global and local interpretation of students' performance in virtual learning environment by leveraging explainable AI. *IEEE Access* **2022**, *10*, 129843–129864. [CrossRef]
16. Gligorea, I.; Yaseen, M.U.; Cioca, M.; Gorski, H.; Oancea, R. An interpretable framework for an efficient analysis of students' academic performance. *Sustainability* **2022**, *14*, 8885. [CrossRef]
17. Rao, S.; Mehta, S.; Kulkarni, S.; Dalvi, H.; Katre, N.; Narvekar, M. A Study of LIME and SHAP Model Explainers for Autonomous Disease Predictions. In Proceedings of the 2022 IEEE Bombay Section Signature Conference (IBSSC), Mumbai, India, 8–10 December 2022; pp. 1–6. [CrossRef]
18. Mothilal, R.K.; Sharma, A.; Tan, C. Explaining Machine Learning Classifiers through Diverse Counterfactual Explanations. In Proceedings of the FAT* 2020—Proceedings of the 2020 Conference on Fairness, Accountability, and Transparency, Barcelona, Spain, 27–30 January 2020; pp. 607–617. [CrossRef]
19. Van Looveren, A.; Klaise, J. Interpretable Counterfactual Explanations Guided by Prototypes. In *Machine Learning and Knowledge Discovery in Databases. Research Track*; Lecture Notes in Computer Science (including Subseries Lecture Notes in Artificial Intelligence and Lecture Notes in Bioinformatics); Springer: Cham, Switzerland, 2021; Volume 12976 LNAI, pp. 650–665. [CrossRef]
20. Wachter, S.; Mittelstadt, B.; Russell, C. Counterfactual Explanations Without Opening the Black Box: Automated Decisions and the GDPR. *SSRN Electron. J.* **2017**, *31*, 841. [CrossRef]
21. Suzan, M.M.H.; Samrin, N.A. Students Adaptability Level in Online Education. Available online: https://www.kaggle.com/datasets/mdmahmudulhasansuzan/students-adaptability-level-in-online-education (accessed on 18 November 2023).
22. Lundberg, S.M.; Erion, G.G.; Lee, S. Consistent Individualized Feature Attribution for Tree Ensembles. *arXiv* **2018**, arXiv:1802.03888.
23. Pereira, F.D.; Fonseca, S.C.; Oliveira, E.H.T.; Cristea, A.I.; Bellhauser, H.; Rodrigues, L.; Oliveira, D.B.F.; Isotani, S.; Carvalho, L.S.G. Explaining Individual and Collective Programming Students' Behaviour by Interpreting a Black-Box Predictive Model. *IEEE Access* **2021**, *9*, 117097–117119. [CrossRef]
24. Molnar, C. *Interpretable Machine Learning: A Guide for Making Black Box Models Explainable*, 2nd ed.; Independent Publisher: Chicago, IL, USA, 2022.

25. Shrikumar, A.; Greenside, P.; Kundaje, A. Learning Important Features through Propagating Activation Differences. In Proceedings of the International Conference on Machine Learning, Sydney, Australia, 6–11 August 2017.
26. García-Magariño, I.; Muttukrishnan, R.; Lloret, J. Human-Centric AI for Trustworthy IoT Systems with Explainable Multilayer Perceptrons. *IEEE Access* **2019**, *7*, 125562–125574. [CrossRef]
27. Guidotti, R.; Monreale, A.; Ruggieri, S.; Turini, F.; Pedreschi, D.; Giannotti, F. A Survey of Methods for Explaining Black Box Models. *ACM Comput. Surv.* **2018**, *51*, 1–45. [CrossRef]
28. Apley, D.W.; Zhu, J. Visualizing the effects of predictor variables in black box supervised learning models. *arXiv* **2016**, arXiv:1612.08468. [CrossRef]
29. Grath, R.M.; Costabello, L.; Van, C.L.; Sweeney, P.; Kamiab, F.; Shen, Z.; Lecue, F. Interpretable credit application predictions with counterfactual explanations. *arXiv* **2018**, arXiv:1811.05245.
30. Tanyel, T.; Ayvaz, S.; Keserci, B. Beyond known reality: Exploiting counterfactual explanations for medical research. *arXiv* **2023**. [CrossRef]

applied sciences

MDPI

Article

Educational Data Clustering in Secondary School Sensor-Based Engineering Courses Using Active Learning Approaches

Taras Panskyi [1,*], Ewa Korzeniewska [2] and Anna Firych-Nowacka [3]

[1] Project Services, Lodz University of Technology, 90-543 Lodz, Poland
[2] Institute of Electrical Engineering Systems, Lodz University of Technology, 90-924 Lodz, Poland; ewa.korzeniewska@p.lodz.pl
[3] Institute of Mechatronics and Information Systems, Lodz University of Technology, 90-924 Lodz, Poland; anna.firych-nowacka@p.lodz.pl
* Correspondence: taras.panskyi@p.lodz.pl

Abstract: The authors investigated the impact of active learning STEM and STEAM approaches on secondary school students' general engineering knowledge, intrinsic relevance, and creativity. Three out-of-school sensor-based courses were held successively. Every sensor-based course involved the final project development. A structured questionnaire was administered to 379 students and consisted of two critical factors: creativity and intrinsic relevance. The third factor was dedicated to the students' engineering learning outcomes. Two factors were addressed to secondary school students, while the third factor was addressed to the tutors' observations of the students' general sensor-based knowledge. Clustering validation analysis quantified the obtained results and justified the significant differences in all estimated factors for different educational modes. Moreover, the study showcases the value of the arts in sensor-based learning-by-doing courses when tackling complex issues like engineering topics. The authors suggest that broader research be undertaken, involving a larger sample, a greater scale, and a diversity of factors.

Keywords: secondary school; engineering courses; active learning; creativity; intrinsic relevance; data clustering

1. Introduction

As we delve into the era of intelligence, the importance of sensors is becoming increasingly evident. Nowadays, sensors link multiple ICT (information and communications technology) devices and enable various machines to communicate [1]. The rapid advancement of cutting-edge ICT technologies, including virtual and augmented reality, immersive technology, artificial intelligence, the Internet of Things, and wearable technology, has turned ordinary sensors into intelligent electronics. Sensors have become incredibly compact and highly portable. Along with their expanded capacities and low-cost electronics for control and data acquisition, they can be connected to various simple and sophisticated ICT devices [2].

With the rise of inexpensive microcontrollers such as Raspberry Pi, Arduino, and NodeMCU, as well as inexpensive sensors that work with these ICT devices, there is increased interest from educators and teachers in exploring the potential of this technology in secondary school learning and teaching settings [3–5]. Their use in the classroom allows students to become acquainted with the elements of automated process control, data management, measurement methods, and monitoring techniques [6]. Through interactive hands-on practices, students can expand their critical and logical thinking skills, and problem-solving abilities, as well as deepen their understanding of technological and engineering subject matters [7]. This interest is also an extension of an existing movement toward integrating hands-on instrumentation as part of learning-by-doing experiences in the classroom [8,9].

The learning-by-doing educational approach involves designing systems using software and hardware that can perceive and respond to the simulated world. In real-world classroom applications, learning-by-doing mainly uses physical computing with microcontrollers and multiple sensors, such as temperature, humidity, light, sound, and movement. Moreover, it incorporates LEDs and LED bars using dimmer switches, push buttons to switch LEDs, RGB LEDs, LCDs to output text information, and actuators to convert analog inputs into a software system or to control electronic mechanical devices, such as motors, steering gears, or other hardware [10,11]. Therefore, it acts as a bridge between traditional and active forms of science education, static or lifelong learning, or as an interactive platform for handmade works of art using a collaborative process of designing, constructing, programming, and testing student-implemented projects [12]. The advent of affordable electronics and low-cost sensor-based technology in specific applications, serving as reliable and robust substitutes for expensive proprietary solutions, is creating new opportunities that are far broader and more effective than ever, which makes them significantly necessary for major breakthroughs in technology in the 21st century [13–18].

The education research community agrees that active learning pedagogies, especially STEM (Science, Technology, Engineering, and Mathematics) and STEAM (with the A standing for "the Arts") education will drive new innovations across disciplines, making use of computational power and ICT technologies to accelerate discoveries and find creative and collaborative ways to work across disciplinary silos, solving real-world challenges. According to the 2020 report [19], "The STEM Education of the Future brings together our advanced understanding of how people learn with modern technology to create more personalized learning experiences, to inspire learning, and to foster creativity from an early age", in recent years, in many countries, STEM-related fields (concerning teaching and learning) have become more prominent sources of innovation [20–22]. The integration of creativity—encompassing talent, knowledge, ability, intrinsic motivation, and personality traits—into STEM pedagogy expands the scope of hard scientific, technological, and engineering subjects, leading to interdisciplinary STEAM education that includes language arts, dance, drama, music, visual arts, design, new media, etc. [23,24]. Recently, the European Commission's Proposal on the European Roadmap to Education 2021–2027 extended the STEM acronym to STEAM, with the A standing for "the Arts" [25].

In this context, the work aims to show how STEM and STEAM educational approaches can help secondary school students improve their learning skills through interactive, hands-on sensor-based out-of-school courses [26]. The objective of the study is to perform an overall analysis and comprehensive comparison of active learning methods with traditional ones to reveal fundamental differences in educational modes in engineering courses in Polish secondary school settings [27]. By offering specific findings and important scientific results about the necessity of integrating STEM and STEAM active learning approaches into secondary school sensor-based education, the authors emphasized that the idea of learning-by-doing is, therefore, not only topical but offers highly interdisciplinary practical subject matter, providing motivating scenarios for teaching a multitude of subjects. Moreover, in this article, the authors demonstrate the significant importance and benefits of active learning methods so that STEM and STEAM concepts can be successfully introduced to Polish secondary schools [28].

STEM and STEAM engineering courses provide students with the opportunity to learn about sensor technology through a hands-on, collaborative process of designing, constructing, and programming a sensor-based system. The research study implies a descriptive and clustering validation analysis of secondary school students' learning outcomes, creativity, and intrinsic relevance toward the presence of active STEM and STEAM learning approaches in engineering courses. The experimental setup includes three independent groups of students: the first group acquires sensor-based knowledge via traditional learning methods, the second uses the STEM approach, and the last uses the STEAM approach. In this context, the following research questions are crucial:

- ○ Do the STEM and STEAM active learning approaches influence secondary students' learning outcomes, creativity, and intrinsic relevance in out-of-school extra-curriculum settings at sensor-based engineering courses?
- ○ Does the presence of arts in the active learning STEAM approach have a crucial impact on secondary students' learning outcomes, creativity, and intrinsic relevance in out-of-school extra-curriculum settings at sensor-based engineering courses?

The major novelty of this study compared to previous works is the cross-comparison of the tutors' observations and secondary school students' appreciation of educational modes toward the improvement of engineering knowledge, intrinsic relevance, and creativity at out-of-school sensor-based courses. Moreover, the present study uses innovative clustering validation methods to reveal, justify, and quantify the presence of differences in educational modes, which opens the path to cross-disciplinary research in pedagogy and machine learning.

2. A Polish Perspective beyond STEM/STEAM Approaches

In the current digital era, developing competencies in STEM and STEAM disciplines is a key goal of education systems around the world [22]. Implementing innovative approaches requires overcoming traditional "old-fashioned" teaching methods (e.g., direct instructions, seatwork, "chalk and talk"), which are often teacher-centered and fragmented, to promote "modern" (e.g., problem-based and collaborative learning, computational thinking) student-centered interdisciplinary approaches covered by high-tech ICT equipment and up-to-date knowledge. However, research shows that the successful integration of STEM and STEAM approaches into education requires an overall policy vision, organizational change, leadership, and, in particular, support for teachers [29]. In 2020, Poland received a country-specific recommendation from the Council of the European Union to "improve digital skills" [30]. In 2023, Poland obtained a National Recovery Plan entitled "Setting new directions for the development of Polish education 2021–2027", signed the "Regulation on the teacher training standards" and announced the "Investment in Education" and "Development of the Infrastructural Potential of Institutions Supporting the Education System" programs [31–33]. With the progressive changes in education, Polish policymakers should exercise caution in making informed decisions to aid the transition from traditional to active STEM and STEAM forms of education as well as develop preparedness plans for future, innovative, and demanding technological and information development to meet the twenty-first-century learning and teaching expectations [34].

There is a wealth of literature on the benefits of STEM and STEAM concepts and their positive impact on students' learning outcomes. Like many other countries, Poland has experienced an expansion of STEM and STEAM active educational approaches across different educational levels. Surma et al. (2019) have presented the results of strategic partnerships among five research teams from Italy, Ireland, Spain, and Poland in promoting the international exchange of best practices and experiences in selecting active learning environments, modern and interactive STEM-based teaching methods at the level of early childhood education [35]. An article by Zdybel et al. (2019) presented the idea of STEM education as an interdisciplinary "meta-discipline" and showed the possibilities of implementing this type of curriculum and material in a preschool environment [36]. Plebańska and Szyller explored the specifics of STEM education as a space conducive to the development of cognitive activity in preschool-aged children [37]. The authors addressed why the STEAM approach is successfully suitable for the youngest, meeting their development needs and stimulating various aspects of their growth. Other research showcased the concept of using the STEAM design in the shaping of geometric concepts in preschool-aged children [38].

In primary education, researchers have showcased the importance of creativity from the perspective of the Polish education system [39]. Other researchers described the novel interdisciplinary method of education through art [40]. Plebańska discussed the digitization of Polish schools based on the "Polish School in the Digitisation Era. The 2017 Diagnosis" survey, regarding the necessity of implementing the STEAM model of teaching [41]. Research by Łukasik et al. (2021) focused on creative competence as an integral component of competence for sustainable development [42]. Their research explored and described the level of creativity competence among Polish primary school students. Tomczyk et al. (2019) focused on the conditions related to the use of ICT in the didactic process in Polish primary schools [43]. They showed how digital media and literacy are being used by primary school students and teachers. Panskyi et al. showed how active learning methods, especially game-based learning, could bridge the gap between schools and universities [44,45]. Moreover, the authors demonstrated the significance of implementing a digital game-based learning approach in out-of-school education to stimulate the creative and innovative learning and teaching of Polish primary school students. Another creative project-based learning (PBL) approach showed active ways of teaching and learning mathematics in primary school settings [46]. Guenaga et al. (2017) presented the STEAM-based Make World online platform to promote computational thinking and active learning skills [47]. Finally, Dudel focused on practical STEAM exercises for primary school students, providing a range of information, methodological advice, and materials useful for conducting classes [48].

With respect to the secondary level, Sysło highlighted the role of problem-based active learning methods in high school [49]. Łukaszczyk and Grebski presented a comparative analysis of STEM-based subject curricula and students' learning outcomes in Poland and the United States [50]. Kisiel focused on the use of active learning methods in informatics education, particularly for teaching spatial modeling and visualization [51]. Walat presented research on secondary school informatics education, focusing on students' skills in perception, interpretation, and creative production of visual materials [52].

Finally, from the perspective of teachers, the research shows the significant impacts of creativity and problem-solving on the development of teachers' skills [53]. The effect of active forms of education in the context of teaching practices was examined [54]. The problems within the contemporary education system and suggestions dedicated to the smooth and effective implementation of active teaching methods and approaches were explored [55]. The crucial roles of both prospective and active teachers in students' education have also been explored [56]. Real ICT skills, knowledge of new media, digital competencies, and the preparedness of current teachers and future pedagogical staff for active learning and teaching models were carefully presented in several recent works [57–61].

Research at the university level is described in references [62–64], while discussions on novel active forms of education are presented in references [65–67]. We should also mention the difficulties and challenges of the proper integration of active learning methods and approaches (STEM, STEAM, PBL, and others) in formal and non-formal educational settings in Poland [68–72].

A comprehensive literature review has demonstrated that Polish researchers and educators are consistently conducting attempts to reveal and justify STEM and STEAM active education approaches at different levels of education. The authors published their scholarly works through a variety of publication outlets, including journals, books, and conference proceedings. However, research development in the field of active learning and teaching methods in secondary school education, especially in sensor-based engineering courses, is not straightforward. Unlike school-based education research with curriculum subjects, STEM and STEAM approaches are not well-defined fields. It is difficult to discuss the effectiveness of STEM and STEAM because there is still a lack of research. From an international point of view, however, efforts are still being made. Fernández-Morante et al. (2022) explored how the STEM active learning approach and project-based learning influence secondary school students' perception of basic learning competencies [73]. Jiang et al. (2023) identified intercultural team characteristics of engineering students based on team formats, collaboration levels, learning goals, evaluation methods, and learning gains [74]. Asghar et al. (2019) studied secondary students' creative alternative ideas about engineering design technology and technological systems [75]. Jantassova et al. described the issues of critical thinking development using the creative STEAM approach to solve professional tasks in the higher engineering education system (2022) [76]. Sukackė et al. (2022) focused on secondary school students' learning outcomes in engineering education through active learning methods [77]. An accurate and concise synthesis of the citation list, clear methodology, and transparent reporting demonstrate the effective influence of STEM and STEAM learning and teaching approaches in formal and non-formal education settings. Therefore, the authors attempted to follow this approach and conducted a study on the learning outcomes, creativity, and intrinsic relevance of secondary school students, examining whether active learning in STEM and STEAM was present or not.

3. Methodology

3.1. Sensor-Based Course Organization

Sensor-based engineering courses conducted in traditional modes were held in secondary schools in 2022. STEM-related courses were held at the Faculty of Electrical, Electronic, Computer, and Control Engineering at the Lodz University of Technology in autumn 2022, and STEAM-based courses were performed in the Active Education Laboratory of the same university (spring 2023). Every engineering course consisted of 5 sessions, with each taking 4 lesson hours (3 clock hours). Every session was held once a week.

Despite the learning mode, two tutors were engaged in teaching one session and directing the entire class of participants. Tutors had sufficient pedagogical competencies and used different traditional and active learning approaches. Tutors held PhDs in electrical engineering and computer science, postgraduate diplomas in pedagogy, and a minimum of 5 years of teaching experience.

As identified in previous research, one important reason that may explain the significant impact of STEM and STEAM education is that student-centered and active instructional methods lead to greater achievement in terms of students' learning outcomes when compared to traditional modes [78,79]. Traditional teaching often involves the teacher instructing from in front of the classroom. Traditional teaching is usually teacher-centered and focuses on imparting book knowledge to students. Generally, in Polish schools, traditional teaching methods are common because teachers want students to learn fixed knowledge and obtain good scores on exams. In the traditional classroom, students must obey discipline to ensure a good learning environment, and teachers are the controllers of the class. In addition to listening to course topics, other participatory activities such as small group discussions, cooperative learning, and peer questioning are not favored because of the traditional classroom design. Traditional information delivery is mostly one-way—from the teacher to the students—apart from the listeners asking the teacher questions [80].

Nevertheless, in traditional sensor-based courses, the entire class was divided into groups of 4–5 students, as strictly managed by the tutors. Tutors assigned students to small groups. Students participated in debates only within one specific group or directly asked the tutors questions by raising their hands. The tutors stood in front of the blackboard without circulating the classroom. Students were not allowed to walk during their lessons. The tutors used straightforward traditional chalk and blackboard teaching tools to teach crucial sensor-based knowledge topics.

STEM and STEAM sensor-based courses focused on instructional strategies for active learning as a whole or on specific strategies like project-based learning, creative experimental learning, interactive learning-by-doing, and team-based learning. In this sense, active learning courses are student-centered and, therefore, include questioning, discussing, writing, problem-solving, teamwork, peer learning, and hands-on experiments. In active learning sensor-based courses, tutors asked students to divide themselves into groups of 4–5 and interact in small teams in order to complete a task. Students were encouraged to work together and communicate with each other within or throughout the groups. Students were allowed to walk during their lessons. The teaching atmosphere was relaxed, and the flexible nature of the work in the classroom created a rich and diverse learning environment to meet students' individual and group needs. The tutors intentionally used multimedia ICTs and other learning visuals to promote active learning and students' engagement.

In both traditional and active learning STEM and STEAM sensor-based courses, the Arduino open-source platform served as the main hardware component, with several low-cost starter kits, for example, the DFRobot Mega D3 kit, the FORBOT kit, the Starter Kit K000007, and the EL-GO Edu 1, were used. Every group of secondary school students, despite their educational mode, received sensor-based Arduino starter kits for course purposes. The Arduino platform is designed to facilitate electronic learning and programming, especially for secondary school students. The sensor-based courses were taught using the Arduino boards in conjunction with the mBlock visual-based environment and C/C++ programming languages.

Every course was divided into three crucial parts. The first part (1 session) focused on the contextual problems of modern sensor-based ICT technology. This part engaged students' prior knowledge about programming, components of electronics, physics, and mathematics. The tutors started the course with interesting questions about the subject. Thereafter, the tutor presented possible ideas, encouraging students to participate and showing solutions that could be used with the use of technology and engineering. Students learned how to solve problems, develop ideas, discuss their conceptions with their groups, and apply the most appropriate ideas they came up with. The second part of the course (2 sessions) allowed students to explore new sensor-based knowledge through collaborative learning-by-doing experiences. It involved using Arduino starter kits with different sensors, GPS modules, switches, motors, jumper cables, and other electronic components. Moreover, this part aimed to teach basic programming principles, such as loops, synchronization, variables, conditionals, operators, broadcasts, functions, and more. In the traditional learning mode, simple schematics of electronic circuits with sensors, batteries, and other electrical symbols were sketched with white chalk on the blackboard, while in active learning modes, tutors used digital ICTs, videos, artificial intelligence generators, scientific articles, image databases, etc., to promote a deeper understanding of interesting engineering topics. The third part (2 sessions) of the sensor-based courses involved the final project development. The final project focused on the development and introduction of sensor-based ICT technologies. In both the traditional and STEM courses, students created advanced electronic circuits with microcontrollers, batteries, distance light, and motion sensors, while STEAM courses required students to design a portable and low-cost smart home prototype (see Figure 1).

Figure 1. The sketch of a smart home final project proposed by the secondary school students in the STEAM sensor-based engineering course.

Smart home projects aim to encourage students to develop creativity in presenting or revealing arts and innovative thinking through a STEAM approach. The arts should be as equally important as other subjects because art-integrated STEM education could lead to the development of soft and transversal skills, enhance recognition of the significance of cultural education, and open new opportunities to satisfy and deepen students' interest in creative work and aesthetic competencies through their natural inquisitiveness and eagerness for free-thinking (see Figure 2). Moreover, the multidisciplinary STEAM approach incorporates the arts with multiple scales of abstraction and different viewpoints.

(**a**) (**b**) (**c**)

Figure 2. The final project of the smart home prototype designed and made by secondary school students in the STEAM sensor-based engineering course: (**a**) the group of 4 students: 2 girls and 2 boys; (**b**) the group of 5 students: 1 girl and 4 boys; (**c**) the group of 4 students: 1 girl and 3 boys.

In STEAM active learning sensor-based courses, 'intelligence arts', such as smart home design, construction, and operation, mainly focused on the multidimensional representation of the arts in the form of creativity and innovative thinking skills. Neglecting the arts may primarily lead to a focus on learning within the hard scientific, technological, engineering, or mathematical environments of STEM, without using its potential leverage for interdisciplinary STEAM development.

3.2. Participants and Data Collection

The data set for this study was collected using an electronic questionnaire tool by means of 'asking questions' in an online-based self-completion anonymous questionnaire, Q1–Q15, containing closed questions (see Table 1). The questionnaire targeted new secondary school students who had completed their primary education level. The questionnaire consisted of 15 questions related to three crucial aspects of active learning: general sensor-based knowledge (5 questions), intrinsic relevance (5 questions), and creativity (5 questions). In sensor-based student learning outcomes, four tutors evaluated student achievements, while in creativity and intrinsic relevance factors, students estimated their perceptions by themselves. For each question, both tutors and students responded to a self-referring statement on a slider scale ranging from 1—"definitely disagree" to 100—"definitely agree". The questionnaire was distributed at the end of each course. The study sample consisted of 379 unique Polish secondary school students, divided into three distinct groups:

- A group of 126 secondary school students who acquired sensor-based knowledge via traditional education pedagogy.
- A group of 125 secondary school students who acquired sensor-based knowledge by means of a STEM active learning approach and hands-on experience.
- A group of 128 secondary school students who acquired sensor-based knowledge using the STEAM interdisciplinary approach.

Table 1. The anonymous questionnaire directed to students and tutors.

General Sensor-Based Knowledge (Directed to Tutors)	Short Name
The student is familiar with basic sensor-based knowledge, including Arduino microprocessors, digital communication, sensors, and actuators.	Q1
The student has no significant difficulty programming the basic electronic parts.	Q2
The student is able to design, test, and modify the electronic circuits with limited supervision and instructions.	Q3
The student is able to transfer knowledge and apply skills across multiple contexts for real-world challenges.	Q4
The student is able to communicate and collaborate clearly on intellectual and emotional levels to share ideas, gain main insights into the problem, and identify solutions effectively.	Q5
Intrinsic relevance (Directed to Students)	
The course is interesting in its problem-solving, designing, and programming practices.	Q6
The course let me feel more related to the contents with useful skills and information.	Q7
I learned new things about communication and collaboration with my colleagues.	Q8
I improved my interest in mathematics, science, and technology.	Q9
I improved my appreciation for the arts.	Q10
Creativity (Directed to Students)	
I enjoyed learning, while the hands-on work felt automatic and effortless.	Q11
If I get stuck on a problem, I look for clues in my surroundings.	Q12
The course boosted my imagination for potential solutions to a problem.	Q13
Thinking about more than one idea at the same time can lead to a new understanding.	Q14
In this course, I tried to generate as many ideas as possible.	Q15

All 379 secondary school students voluntarily completed the questionnaire, which focused on their perception of creativity and intrinsic relevance in different sensor-based engineering courses. It should be noted that the study sample contained 379 unique secondary school students. That is, the students who finished the traditional sensor-based course could not sign in to participate in STEM and STEAM courses, and vice versa.

3.3. Design and Procedure

For quantitative analysis, firstly, descriptive statistics were used; however, to reveal the crucial differences between groups of secondary school students in sensor-based learning outcomes, creativity, and intrinsic relevance in traditional STEM and STEAM engineering courses, clustering analysis was carefully applied [81]. Regarding questions related to students' learning outcomes (Q1–Q5), to create the questionnaire, the authors used general sensor-based knowledge, including the Arduino ecosystem, programming and electronics fundamentals, libraries, and communication features. Moreover, the questions assessed whether secondary school students possessed the ability to think critically and logically, to clearly and concisely identify the most important elements of a problem, to develop possible solutions, and to apply new crucial sensor-based knowledge in practical and versatile real-world applications.

For the intrinsic relevance evaluation, the ARCS model of learning motivation was used and carefully modified in Q6–Q10 [82]. The ARCS model includes four factors that could effectively help to enhance students' learning motivations and interests, namely "attention", "relevance", "confidence", and "satisfaction." The ARCS motivation model has been widely applied in the literature for a variety of quantitative evaluation works [83–87]. Nevertheless, previous studies mainly used the ARCS model to reveal the key factors in the learning motivations of students or the relationship between factors. The presented research only highlighted relevance as a key factor to be assessed. According to the relevance factor, a sensor-based course and its learning topics aim to enhance students' interest in problem-solving, designing, and programming, making students immerse themselves in learning activities, especially in mathematics, science, technology, and the arts. Learning topics can help cultivate self-learning capabilities and enhance pleasant learning experiences related to the course contents with useful skills and information. Moreover, the relevance aspect was designed for this research questionnaire in order to evaluate students' self-esteem in communication and collaboration skills and overall interest in sensor-based engineering topics.

Assessing creativity has been widely discussed for decades, particularly with regard to the number of factors required in such assessments. With respect to creativity assessments, the authors used the CPAC (cognitive processes associated with creativity) scale in Q11–Q15 [88]. The shorter CPAC scale is suitable for assessing particular dimensions of creativity, including idea manipulation, idea generation, flow, imagery/sensory perception, and metaphorical/analogical thinking. The CPAC scale is widely used in the literature to evaluate students' creative and innovative thinking skills [89–92]. According to the creativity factor, a sensor-based course aims to improve secondary school students' enjoyment and satisfaction in learning-by-doing engineering topics. Moreover, it emphasizes the crucial role of creative thinking in collaborative work. The self-reported questions aim to evaluate a student's openness to experience, ambiguity tolerance, and self-confidence. Some authors insist that the definition of creativity involves sensitivity, flexibility, the ability to analyze, synthesize, evaluate, and reorganize information, and the ability to cope with complexity [93]. The CPAC scale was chosen as it met the requirements of this study for analyzing the creativity of secondary school students and was adapted for research purposes.

4. Results

4.1. Descriptive Statistics

The descriptive statistics are presented in Table 2. The table includes three learning modes: traditional, STEM-based, and STEAM-based. For each educational mode, the mean, standard deviation, skewness, and kurtosis of the students' answers and tutor estimations were calculated. The authors displayed descriptive statistics according to the educational mode without dividing the mode into particular questions. For the sake of simplicity, clarity, and integrity, statistical criteria were calculated for the educational mode to emphasize the crucial role of active learning methods and significant differences compared to the traditional one.

Table 2. Descriptive statistics broken down by educational mode.

	Sensor-Based Knowledge (Q1–Q5)	Intrinsic Relevance (Q6–Q10)	Creativity (Q11–Q15)
Traditional course			
Mean	22.59	29.19	23.23
Std. Dev	17.08	16.74	18.65
Skewness	2.19	1.71	1.55
Kurtosis	5.37	3.04	1.74
STEM course			
Mean	58.34	53.80	49.43
Std. Dev	12.15	13.59	12.52
Skewness	1.13	0.80	0.13
Kurtosis	1.45	1.48	0.34
STEAM course			
Mean	75.10	86.09	77.93
Std. Dev	9.82	9.13	13.78
Skewness	−0.46	−0.89	−1.58
Kurtosis	−0.14	2.98	5.27

To avoid the appearance of anomalies in the questionnaire that could lead to confusion from subjective answers by secondary school students or tutors' expertise, the authors use average values for the particular education mode. External factors like mood, behavior, tiredness, thermal comfort, mental agility, and the respondent's self-perception could also influence the completeness and sufficiency of the questionnaire. Moreover, the authors acknowledge the possibility of inconsistencies in specific educational modes, which could lead to misleading and biased results and blur the overall context of the research. Overall, the authors aim to shed more light on the educational mode than on the particular questions contained.

The analysis imposes a non-normal data distribution when skewness and kurtosis are, respectively, out of the range of ±1 and ±3. Therefore, the normality assumption does not matter and is unimportant in the presented case study [94]. The normality assumption is particularly important in statistical analysis. In practice, the violation of this assumption can lead to biased, unreliable, and inaccurate results. Some statistical treatments deal with the violation of the normality assumption. One of the most common ways is to apply transformations to the data (logarithmic or square root). Another way is to use non-parametric methods that do not assume any particular distribution of the data. Finally, the outlier detection and removal procedure should be carefully applied. Outliers can significantly blur the normality of the data. Therefore, clustering validation analysis, particularly using majority-based decision fusion methods, has been chosen as the optimal and efficacious tool to lead in quantitative analysis.

The analysis of the overall questionnaire in all three educational modes revealed the highest mean score of 86.09 within the STEAM sensor-based active learning courses in the intrinsic relevance factor. The lowest overall mean score within the particular educational mode is identified in the traditional teaching/learning mode with respect to the sensor-based engineering knowledge of secondary school students in out-of-school extra-curriculum settings (22.59).

With respect to the educational mode, the pairwise analysis of the mean scores identified the dominance of the STEAM active learning method over the STEM and traditional forms of education in all factors, including learning outcomes, intrinsic relevance, and creativity. The biggest difference in mean scores (56.9) is noted in favor of the STEAM active learning method compared to the traditional one toward the intrinsic relevance factor, while the smallest (16.76) is in favor of the STEAM approach compared to STEM active learning in factors dedicated to students' general engineering learning outcomes.

Within the traditional educational mode, the biggest difference in mean scores (6.6) is revealed between intrinsic relevance and general sensor-based knowledge factors, in favor of the second one, while the lowest (0.64) is between learning outcomes and creativity factors. The notable difference in mean scores in STEM education courses could be observed between sensor-based student outcomes and creativity (8.91), in favor of the first factor. Finally, within the STEAM form of active education, a significant difference in mean scores is shown between sensor-based knowledge and relevance (10.99), while the lowest values are between students' knowledge and creativity (2.83), in favor of the first factor.

To summarize, the descriptive analysis shows significant differences between the mean scores toward active learning educational modes. The mean scores of all estimated factors, including general sensor-based knowledge, intrinsic relevance, and creativity, are considerably increased in favor of active learning pedagogy. Within active learning methods, STEAM-based engineering courses exhibit higher mean scores in all estimated factors.

In order to verify the reliability of the questionnaire presented in this research and to estimate the internal consistency of the factors, the Cronbach's alpha test is used. The degree of internal consistency is represented by the reliability coefficient (Cronbach's alpha). The reliability analysis shows that Cronbach's alpha is 0.61, indicating the acceptable degree of internal consistency of the presented questionnaire [95]. Additionally, an intraclass correlation coefficient is used to determine if answers to the questions can be rated reliably by different raters (students and tutors). The value of an intraclass correlation coefficient is 0.78, indicating that three factors of the questionnaire can be rated with "good" reliability by different raters [96].

The descriptive analysis provides the reader with a basic foundation and essential insights and gives a potential starting point for deeper and more sophisticated data analysis. Therefore, for further comprehensive research, data clustering validation analysis is applied.

4.2. Clustering Validation Analysis

The essence of cluster analysis assumes that little or nothing is known about the grouping structure that underlies the data set. The operational objective, in this case, is to discover a data grouping structure that is frequently stated as the problem of finding natural groups or clusters. The approach to this problem and the results achieved help the investigator choose to give operational and quantitative meaning to the phrases "natural association" and "relatively distinct". Cluster analysis remains the most relevant and robust tool to gain insight into previously unknown grouping structures and to disclose the intrinsic characteristics of data.

The cluster analysis task may have more than one solution and, thus, is recognized as "ill-posed" [97,98]. For instance, the result of the validation procedure depends on the choice of the data, clustering algorithms, dissimilarity measures, etc. [99]. Applying different clustering algorithms or the same clustering algorithm with varying input parameters on the same given data set can result in diverse partitions and, therefore, may greatly affect the goodness/quality of clustering results [100]. Such issues are related to clustering validation, which is one of the most challenging topics in data mining literature. The main objective of clustering validation is the effective evaluation of the goodness of clustering results. This involves finding not only the "true" number of clusters but also the optimal partition that best fits the underlying grouping structure of the data set.

One common practice in clustering validation procedures is to apply clustering validation indices (CVIs). However, many authors agree that there is no universal CVI that always enables the correct decision to be made. In some research, multicriteria solutions used to reach the best and most adequate results have been commonly used [101,102]. These solutions assume the adoption of several CVIs to achieve greater certainty and correctness in clustering results. For instance, when a group of CVIs takes part in the decision process, their votes, in many cases, disagree. Frequently, the CVIs have different sensitivity domains: some are heavily biased toward particular clustering algorithms and distance-based dissimilarity measures, such as the silhouette CVI; and some are sensitive to outliers, such as the Dunn validation index. Moreover, certain data characteristics, including high dimensionality, noise, types of attributes, and scales, may strongly affect clustering validation analysis [103]. In light of this fact, in the presented research, the authors used majority-based decision fusion clustering validation methods to reveal the "true" number of clusters and to justify the presence of significant differences in all estimated factors toward different educational modes.

The authors applied the combination of two novel decision-making approaches: a non-invasive majority-based decision fusion method and an invasive clustering validation method. These clustering validation methods have been carefully described in previous works [81,104], including the experimental setups, general input settings, and guidelines for benchmarking; the methods' execution with different data sets were also showcased. The combination itself should easily evaluate the correctness of obtaining the "true" number of clusters and the goodness of the clustering results. Moreover, the majority-based clustering validation methods will help us completely and utterly answer the research questions posed at the beginning of the manuscript. Figure 3 illustrates the simplified validation scheme of a non-invasive majority-based decision fusion method and invasive clustering validation methods.

The result of the CVIs' voting process aligns with one of the most common scenarios in majority-based decision fusion validation methods. According to previous studies [81,104], almost 99% of input benchmark data sets follow this pattern. This pattern is divided into several clear or controversial scenarios that map the CVIs' voting process and the production of the final decision.

Scenario 0: The decision is made by the CVIs' absolute majority. The absolute majority points to the correct "true" number of clusters. This scenario illustrates the consensus between CVIs in final decision-making. Moreover, Scenario 0 does not require additional verification by invasive methods.

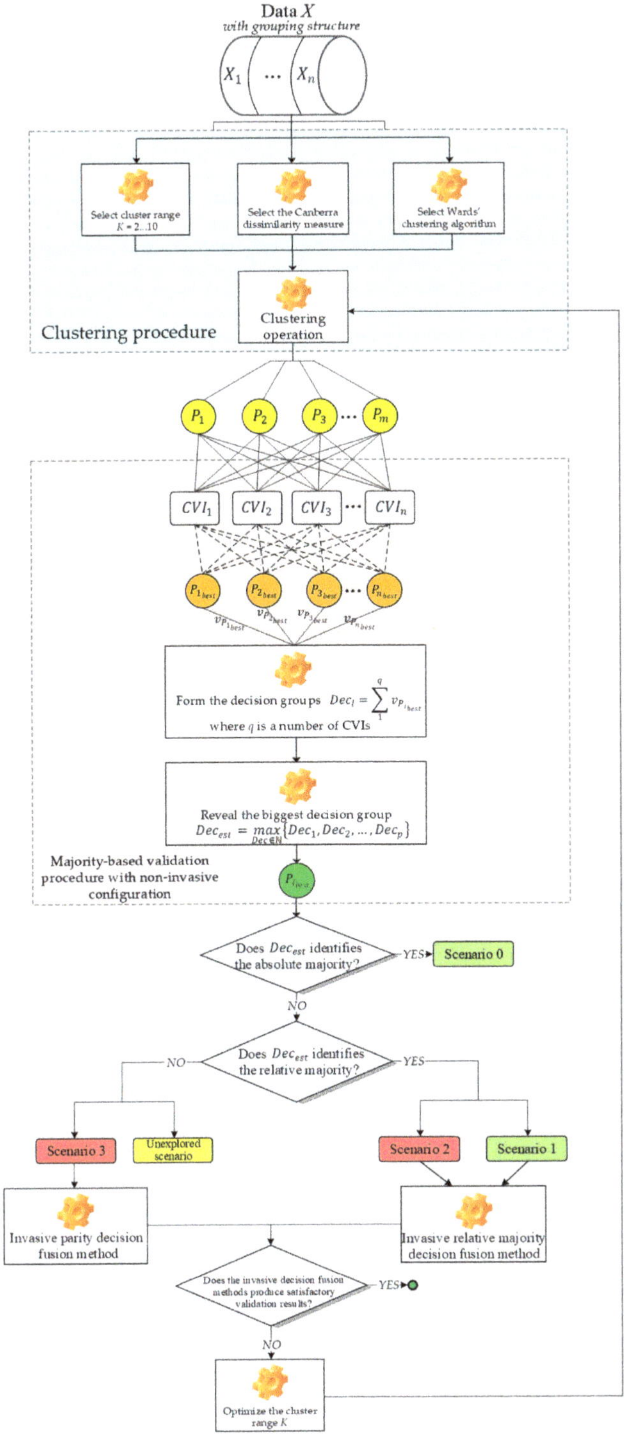

Figure 3. Simplified validation scheme of the non-invasive and invasive decision fusion methods.

Scenario 1a: The decision is made by the relative majority of CVIs, and the nearest alternative (the second largest group of CVIs) has 50% fewer votes than half of the majority. The relative majority points to the correct "true" number of clusters. The situation is clear enough; however, an absolute majority is not obtained as in Scenario 0.

Scenario 1b: The final decision is taken by the CVIs' relative majority, and the nearest alternative is 50% more votes than half of the majority. The relative majority points to the "true" number of clusters. This scenario shows the controversial situation where there is no clear-cut grouping structure. Scenarios 1a and 1b require additional verification by means of the invasive relative majority decision fusion method.

Scenario 2: The relative majority of CVIs points to the incorrect number of clusters. The nearest alternative group of CVIs, in turn, shows the "true" number of clusters. The most critical of all the previous scenarios, it requires complete and precise verification by means of the invasive relative majority decision fusion method.

Scenario 3: The data sets for which no majority prevails. Two of the largest equal groups of CVIs vote for different numbers of clusters; however, only one group of CVIs makes the "true" decision, and the other is misleading. This scenario requires complete and precise verification by means of the invasive parity majority decision fusion method.

Figure 4 shows the study sample, which consists of 379 data points (without dividing the data into a particular educational mode) in the form of a three-dimensional scatter plot.

Figure 4. A scatter plot of a three-dimensional raw data set (379 data points). The x-axis corresponds to secondary school students' learning outcomes toward general sensor-based knowledge; the y-axis corresponds to students' creativity; and the z-axis corresponds to students' intrinsic relevance. One data point correlates with one particular secondary school student.

To provide significant insights into clustering validation procedures, the clustering operation should first be applied. Figure 5 shows the results of the clustering operation, where three clear-cut globular-form clusters are identified.

Figure 5 illustrates the result of the clustering process. The raw input data set is partitioned into three clusters. The clustering results could be extrapolated to the descriptive statistics, where the critical verdict could be made, namely, the blue cluster corresponds to the traditional educational mode, the green cluster corresponds to the STEM-based approach, and the orange cluster corresponds to the STEAM active learning approach. Even though the clustering algorithm with the dissimilarity measure identifies three clusters, the validation procedure should be applied. To verify the quality of clustering results and to justify that three is the "true" number of clusters, majority-based decision fusion methods are performed. The validation results of majority-based decision fusion methods with 24 CVIs engaged in the voting process are shown in Table 3. Critical values of CVIs are used to select the best number of clusters. This set of 24 CVIs has been chosen from the literature [105].

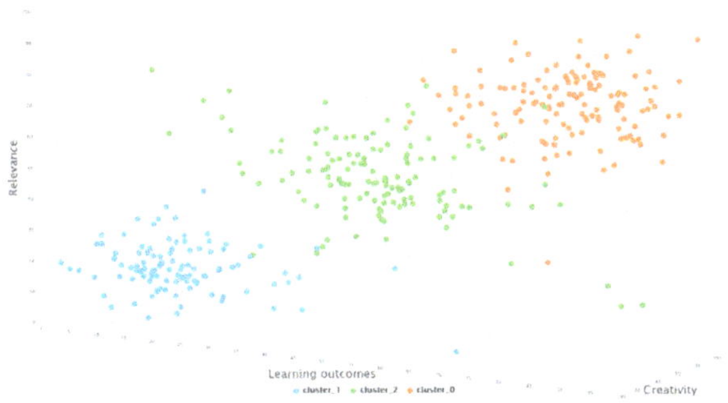

Figure 5. A scatter plot of a three-dimensional clustered data set (379 data points).

Table 3. The validation results of the 24 CVIs engaged in the voting process using the majority-based decision fusion methods.

		The Best Number of Clusters Based on the Critical Values of the CVIs		
No	**Name of the CVI**	**Three Clusters**	**Two Clusters**	**One Cluster**
1.	Calinski–Harabasz	10.18		
2.	Krzanowski and Lai	689.74		
3.	Hatigan	323.57		
4.	Cubic Clustering Criterion		24.54	
5.	Scott	562.30		
6.	Marriot	1.25×10^{15}		
7.	TrCovW			8.6×10^{8}
8.	TraceW	134,135.90		
9.	Friedman	23.72		
10.	Rubin	−8.91		
11.	C index			0.17
12.	Davies and Bouldin		0.63	
13.	Silhouette		0.60	
14.	Duda	0.79		
15.	PseudoT2	38.61		
16.	Beale		2.01	
17.	Ratkowsky		0.53	
18.	Ball	102,570.40		
19.	PtBiserial		0.81	
20.	Frey		2.77	
21.	McClain		0.23	
22.	Dunn		0.12	
23.	SDindex	0.08		
24.	SDbw	0.29		

The majority-based decision fusion method with a non-invasive configuration reveals the absolute majority (13 CVIs) voted for three clusters to be the "true" ones. This falls into Scenario 0, and, therefore, it is not necessary to validate the results by means of the invasive relative majority decision fusion method. The second-largest group (9 CVIs) voted for two clusters, while only 2 CVIs voted for one cluster to be the "true" one. To summarize, the clustering algorithm identified three clusters, and the majority-based decision fusion method confirmed and justified this number of clusters. Moreover, the research confirms the crucial and significant differences between traditional and active learning modes. However, a moderate difference between STEM and STEAM approaches could be observed.

The presence of arts in STEAM pedagogy enhances and boosts students' general sensor-based knowledge, intrinsic relevance, and creativity more than a STEM-based approach in extracurricular learning-by-doing engineering courses in secondary school settings.

4.3. Visual Validation Analysis

Majority-based decision fusion methods clearly identify and justify three clusters as "true". Nevertheless, for transparency reasons, the authors have attempted to analyze the clustering results using other validation concepts by means of visual clustering assessment tools.

Cluster analysis is an unsupervised and often exploratory issue without fixed, predefined expectations from the expert. Patterns in the data set that qualify to be interpreted as clusters could have very manifold shapes, configurations, and appearances. Therefore, in clustering validation assessment, there is room for the concept of visual analysis. Visible patterns in the data set may be identified as clusters. Effective plots for visual cluster validation include scatterplots, multidimensional plots, parallel coordinate plots, etc., in which the revealed clusters can be indicated by colors or shades, contour maps, dendrograms, and silhouette plots. Another broad category of visual validation methods is the visual assessment of tendency (VAT) family of algorithms, which reorders the pairwise dissimilarity matrix and then generates a cluster heat map that shows possible clusters in the data as dark blocks along the diagonal [106]. Moreover, the iVAT algorithm is an improved version of the VAT algorithm that more clearly shows the number of clusters, as well as their sizes [107]. The gap statistic can also be classified as a visual validation method. The basic idea of the gap statistics is to choose the number of clusters where the largest "jump" in within-cluster distance occurs.

Visual clustering validation methods should be used as informal graphical illustrations, as a rule of thumb. Moreover, to begin with, such methods should be used with great caution since the visual representations of clusters may significantly vary with the nature of the parameter being tuned (dissimilarity measure, clustering algorithm, etc.) and the validation indices used. Nevertheless, the visual validation concept remains crucial and valuable for cluster validation assessment. The presented visual clustering validation methods are illustrated in Figure 6.

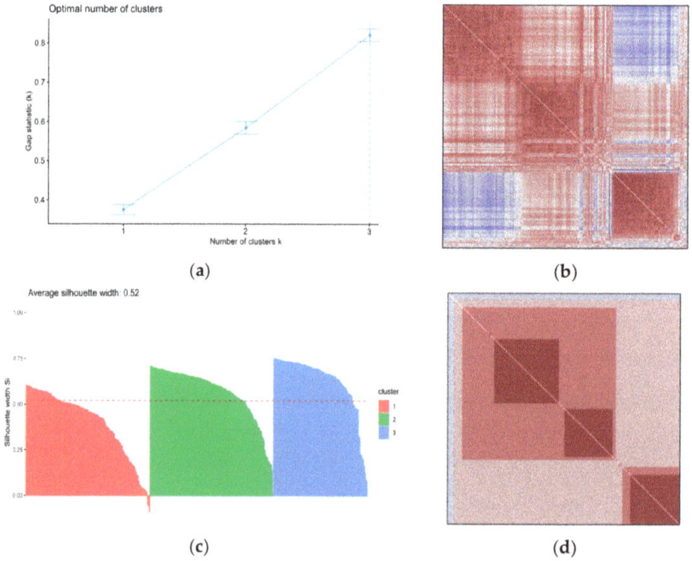

Figure 6. Visual cluster validation plots of the clustered data set (379 data points): (**a**) gap statistic plot; (**b**) visualization of VAT validation method; (**c**) silhouette plot; (**d**) visualization of iVAT validation method.

Figure 6 clearly shows that the visual methods reveal three "true" clusters. VAT and iVAT algorithms show the number of clusters, the visual sizes of clusters, and the distances among them. Therefore, Figure 6b,d display two clusters—near each other and far from the third. Nevertheless, the number of separate "true" clusters remains at three. This only justifies the veracity of clustering validation results presented by the majority-based decision fusion methods. The concluding statement is that there are significant differences between traditional, STEM, and STEAM educational modes in sensor-based engineering out-of-school courses in secondary school settings.

5. Discussion and Conclusions

The research study applied in this article discusses the possible differences in educational modes in out-of-school sensor-based engineering courses in secondary settings. Significant differences in students' general knowledge, creativity, and intrinsic relevance could be observed between traditional pedagogy and active learning methods. With respect to the first research question, STEM and STEAM active learning approaches highly increase students' learning sensor-based outcomes, boost their creativity, and develop intrinsic relevance toward hard scientific and technological topics. Descriptive statistics and clustering validation analysis quantify the revealed differences and justify the presence of an educational gap between these learning methods.

The results of the presented research are in line with multiple studies provided worldwide. The authors confirm that an active learning STEM-based curriculum significantly impacted the development of students' creative thinking compared to students who studied under a traditional curriculum method [108]. Others assume that active learning methods may contribute to the development of creativity in young people [109]. Moreover, active learning practices significantly influence the development of innovative thinking among students [110]. The researchers found that students engaged in active learning lessons outperformed expectations in terms of learning outcomes when compared to students in traditional classrooms. Moreover, research has supported the role of active learning as a superior approach compared to traditional pedagogy [111]. With respect to intrinsic relevance, students studying in active learning classrooms had significantly higher motivation and attitudes toward learning than students studying in traditional classroom layouts [112].

Active learning approaches and methods are popular in 21st-century student-centered education. Nowadays, active learning education is robust, flexible, and effective enough to outperform traditional education in all school subjects and fields of education. Nevertheless, questions arise within the context of active learning methods. Do the arts in STEAM active learning education influence students' knowledge, creativity, and intrinsic relevance, especially in sensor-based learning-by-doing engineering topics? The results confirm the moderate difference between STEM and STEAM-based active learning approaches in students' learning outcomes, creativity, and intrinsic relevance. Clustering validation analysis has clearly highlighted that the multidimensional representation of the arts added to STEM education could boost students' achievements, perceptions of modern sensors, and application prospects of hard engineering topics. This is consistent with previous research, which reached similar conclusions. The arts contribute to promoting student satisfaction, interest, and different levels of artistic literacy. It becomes particularly important when students are faced with sophisticated and complex tasks [113]. Various authors have expressed the advantages of the arts in generating a favorable classroom climate for the development of student creativity [114,115]. "*Art and engineering integration is a process taking place before our eyes, in a dynamically changing reality*" [116]. Another statement declares that engineering education is being dubbed the new liberal arts degree for the 21st century [117]. STEAM education has broadened the STEM approach by incorporating the arts, fostering interdisciplinary collaboration, and providing students with a more holistic understanding of problems and their solutions [118]. Anne Pirrie claims the following: "*Reconceptualising the interrelationship between sciences and the arts has the potential to enrich our understanding of science, revitalise our teaching and make us open to new ways to respond to environmental challenges"* [119]. Finally, the presented results justify the statement: *"Engineering education should be re-imagined by making the arts an essential part of the engineering curriculum to develop reflective engineers who are willing and able to navigate the "swampy lowlands" of engineering practice"* [120].

To summarize, the STEAM approach has been widely generative in creating new and creative pedagogical and instrumental practices in the fields of the arts and STEM learning, particularly around new media and ICT technology. Its goal is to bridge learning between (inter-) and beyond (trans-) disciplines. Within the evolving field of sensor-based education, secondary school students have connected their engineering knowledge with basic artistic skills through the STEAM learning approach in their final smart house project, which equipped them to better "see" their ideas. This study revealed an emphasis on developing STEAM-related disciplinary knowledge in sensor-based engineering education, with a notable emphasis on the arts. Active learning approaches not only empower students' hard skills but also improve soft skills such as creative and abstract thinking, problem-solving, and intrinsic motivation in learning. Moreover, they foster true innovation that combines the mind of a scientist, the practical knowledge of a technologist, the critical thinking of a mathematician, and the elegance of an artist or designer.

There is room to extend this work by including more dimensions of the arts, exploring the intersections with non-STEM fields, and deepening the understanding of the arts' influence on students' learning outcomes. In this context, motivation and self-efficacy aspects would appear to be the most interesting factors to analyze in greater depth. Moreover, broader research should be undertaken involving a larger sample of secondary school students and on a larger scale.

Author Contributions: Conceptualization, E.K. and A.F.-N.; methodology, T.P.; software, T.P.; validation, T.P., E.K. and A.F.-N.; formal analysis, E.K.; investigation, T.P.; resources, T.P.; data curation, A.F.-N.; writing—original draft preparation, T.P.; writing—review and editing, A.F.-N.; visualization, T.P.; supervision, E.K. All authors have read and agreed to the published version of the manuscript.

Funding: This research received no external funding.

Institutional Review Board Statement: Not applicable.

Informed Consent Statement: Not applicable.

Data Availability Statement: The raw data supporting the conclusions of this article will be made available by the authors on request. Please contact Taras Panskyi (taras.panskyi@p.lodz.pl).

Conflicts of Interest: The authors declare no conflicts of interest.

References

1. Kapp, S.; Lauer, F.; Beil, F.; Rheinländer, C.C.; Wehn, N.; Kuhn, J. Smart Sensors for Augmented Electrical Experiments. *Sensors* **2022**, *22*, 256. [CrossRef] [PubMed]
2. González Crespo, R.; Burgos, D. Advanced Sensors Technology in Education. *Sensors* **2019**, *19*, 4155. [CrossRef] [PubMed]
3. García-Tudela, P.A.; Marín-Marín, J.-A. Use of Arduino in Primary Education: A Systematic Review. *Educ. Sci.* **2023**, *13*, 134. [CrossRef]
4. Hercog, D.; Lerher, T.; Truntič, M.; Težak, O. Design and Implementation of ESP32-Based IoT Devices. *Sensors* **2023**, *23*, 6739. [CrossRef] [PubMed]
5. Silva, M.J.; Gouveia, C.; Gomes, C.A. The Use of Mobile Sensors by Children: A Review of Two Decades of Environmental Education Projects. *Sensors* **2023**, *23*, 7677. [CrossRef] [PubMed]
6. Gubsky, S. Development of Low-Cost Arduino-Based Equipment for Analytical and Educational Applications. *Eng. Proc.* **2023**, *48*, 8.
7. Panskyi, T.; Biedroń, S.; Grudzień, K.; Korzeniewska, E. The Comparative Estimation of Primary Students' Programming Outcomes Based on Traditional and Distance Out-of-School Extracurricular Informatics Education in Electronics Courses during the Challenging COVID-19 Period. *Sensors* **2021**, *21*, 7511. [CrossRef] [PubMed]
8. Iftikhar, S.; Guerrero-Roldán, A.-E.; Mor, E. Practice Promotes Learning: Analyzing Students' Acceptance of a Learning-by-Doing Online Programming Learning Tool. *Appl. Sci.* **2022**, *12*, 12613. [CrossRef]

9. Frache, G.; Nistazakis, H.; Tombras, G. Constructing Learning-by-Doing Pedagogical Model for Delivering 21st Century Engineering Education. *Adv. Sci. Technol. Eng. Syst. J.* **2018**, *3*, 1. [CrossRef]
10. Chung, C.-C.; Lou, S.-J. Physical Computing Strategy to Support Students' Coding Literacy: An Educational Experiment with Arduino Boards. *Appl. Sci.* **2021**, *11*, 1830. [CrossRef]
11. Panskyi, T.; Rowinska, Z.; Biedron, S. Out-of-school assistance in the teaching of visual creative programming in the game-based environment–Case study: Poland. *Think. Ski. Creat.* **2019**, *34*, 100593. [CrossRef]
12. Jeng, C.C.; Fang, S.C. Applying Open Source Software and Hardware to Physical Computing-Utilizing Scilab and Arduino on Spectral Analysis of Vibration Signal. *J. Comput. Sci. Appl.* **2016**, *8*, 70–83.
13. Javaid, M.; Haleem, A.; Singh, R.P.; Rab, S.; Suman, R. Significance of sensors for industry 4.0: Roles, capabilities, and applications. *Sens. Int.* **2021**, *2*, 100110. [CrossRef]
14. Riaz, A.; Sarker, M.R.; Saad, M.H.M.; Mohamed, R. Review on Comparison of Different Energy Storage Technologies Used in Micro-Energy Harvesting, WSNs, Low-Cost Microelectronic Devices: Challenges and Recommendations. *Sensors* **2021**, *21*, 5041. [CrossRef]
15. Kureshi, R.R.; Mishra, B.K.; Thakker, D.; John, R.; Walker, A.; Simpson, S.; Thakkar, N.; Wante, A.K. Data-Driven Techniques for Low-Cost Sensor Selection and Calibration for the Use Case of Air Quality Monitoring. *Sensors* **2022**, *22*, 1093. [CrossRef]
16. Cornide-Reyes, H.; Noël, R.; Riquelme, F.; Gajardo, M.; Cechinel, C.; Mac Lean, R.; Becerra, C.; Villarroel, R.; Munoz, R. Introducing Low-Cost Sensors into the Classroom Settings: Improving the Assessment in Agile Practices with Multimodal Learning Analytics. *Sensors* **2019**, *19*, 3291. [CrossRef] [PubMed]
17. Neis, P.; Warch, D.; Hoppe, M. Testing and Evaluation of Low-Cost Sensors for Developing Open Smart Campus Systems Based on IoT. *Sensors* **2023**, *23*, 8652. [CrossRef] [PubMed]
18. Alfano, B.; Barretta, L.; Del Giudice, A.; De Vito, S.; Di Francia, G.; Esposito, E.; Formisano, F.; Massera, E.; Miglietta, M.L.; Polichetti, T. A Review of Low-Cost Particulate Matter Sensors from the Developers' Perspectives. *Sensors* **2020**, *20*, 6819. [CrossRef]
19. National Science Foundation. *STEM Education for the Future 2020 Visioning Report*; National Science Foundation: Alexandria, VA, USA, 2020.
20. Le Thi Thu, H.; Tran, T.; Trinh Thi Phuong, T.; Le Thi Tuyet, T.; Le Huy, H.; Vu Thi, T. Two Decades of STEM Education Research in Middle School: A Bibliometrics Analysis in Scopus Database (2000–2020). *Educ. Sci.* **2021**, *11*, 353. [CrossRef]
21. Dare, E.A.; Keratithamkul, K.; Hiwatig, B.M.; Li, F. Beyond Content: The Role of STEM Disciplines, Real-World Problems, 21st Century Skills, and STEM Careers within Science Teachers' Conceptions of Integrated STEM Education. *Educ. Sci.* **2021**, *11*, 737. [CrossRef]
22. Hinojo-Lucena, F.-J.; Dúo-Terrón, P.; Ramos Navas-Parejo, M.; Rodríguez-Jiménez, C.; Moreno-Guerrero, A.-J. Scientific Performance and Mapping of the Term STEM in Education on the Web of Science. *Sustainability* **2020**, *12*, 2279. [CrossRef]
23. Montés, N.; Zapatera, A.; Ruiz, F.; Zuccato, L.; Rainero, S.; Zanetti, A.; Gallon, K.; Pacheco, G.; Mancuso, A.; Kofteros, A.; et al. A Novel Methodology to Develop STEAM Projects According to National Curricula. *Educ. Sci.* **2023**, *13*, 169. [CrossRef]
24. Johnston, K.; Kervin, L.; Wyeth, P. STEM, STEAM and Makerspaces in Early Childhood: A Scoping Review. *Sustainability* **2022**, *14*, 13533. [CrossRef]
25. Colucci-Gray, L.; Burnard, P.; Gray, D.; Cooke, C. A critical review of STEAM (Science, Technology, Engineering, Arts, and Mathematics). In *Oxford Research Encyclopedia of Education*; Thomson, P., Ed.; Oxford University Press: Oxford, UK, 2019.
26. Leavy, A.; Dick, L.; Meletiou-Mavrotheris, M.; Paparistodemou, E.; Stylianou, E. The prevalence and use of emerging technologies in STEAM education: A systematic review of the literature. *J. Comput. Assist. Learn.* **2023**, *39*, 4. [CrossRef]
27. Zhan, Z.; Hu, Q.; Liu, X.; Wang, S. STEAM Education and the Innovative Pedagogies in the Intelligence Era. *Appl. Sci.* **2023**, *13*, 5381. [CrossRef]
28. Jakubowski, R.; Piotrowski, M. W poszukiwaniu uwarunkowań trwałego wprowadzenia STEM/STEAM w polskich szkołach. *Eduk. Elem. W Teor. I Prakt.* **2019**, *14*, 25–37.
29. Tarkowski, A.; Majdecka, E.; Penza-Gabler, Z.; Sienkiewicz, M.; Stunża, G.D. *Analiza Strategii i Działań Mających na celu Rozwój Kompetencji Cyfrowych w Państwach Unii Europejskiej*; Fundacja Centrum Cyfrowe na zlecenie Centrum Projektów Polska Cyfrowa: Warsaw, Poland, 2018.
30. European Commission. *COUNCIL RECOMMENDATION on the 2020 National Reform Programme of Poland and delivering a Council opinion on the 2020 Convergence Programme of Poland*; European Commission: Luxembourg, 2020.
31. Setting New Directions for the Development of Polish Education 2021–2027. Available online: https://efc.edu.pl/en/news/setting-new-directions-for-the-development-of-polish-education-2021-2027 (accessed on 2 February 2024).
32. Dziennik Ustaw Rzeczypospolitej Polskiej. *Rozporządzenie Ministra Edukacji I Nauki z dnia 14 Września 2023 r. Zmieniające Rozporządzenie w Sprawie Standardu Kształcenia Przygotowującego do Wykonywania Zawodu Nauczyciela*; Dziennik Ustaw Rzeczypospolitej Polskiej: Warsaw, Poland, 2023.
33. Eurydice. Ongoing Reforms and Policy Developments. National Reforms in School Education. Available online: https://eurydice.eacea.ec.europa.eu/national-education-systems/poland/national-reforms-school-education (accessed on 2 February 2024).
34. Panskyi, T.; Korzeniewska, E.; Serwach, M.; Grudzien, K. New realities for Polish primary school informatics education affected by COVID-19. *Educ. Inf. Technol.* **2022**, *27*, 5005–5032. [CrossRef] [PubMed]

35. Surma, B.; Rosati, N.; Menon, S.; Fuertes, M.T.; Farren, M.; Maguire, F. Kitchen Lab for Kids. Program kształtowania umiejętności STEM w przedszkolu. *Eduk. Elem. W Teor. I Prakt.* **2019**, *14*, 61–70.
36. Zdybel, D.; Pulak, I.; Crotty, Y.; Fuertes, M.T.; Cinque, M. Rozwijanie umiejętności STEM w przedszkolu. Możliwości i wyzwania z perspektywy przyszłych nauczycieli. *Eduk. Elem. W Teor. I Prakt.* **2019**, *14*, 71–94. [CrossRef]
37. Plebańska, M.; Szyller, A. *STEAM-Owe Przedszkole*; Difin: Warsaw, Poland, 2021.
38. Bojarska-Sokołowska, A. Wykorzystanie STEAM-owego projektu w kształtowaniu wybranych pojęć geometrycznych u przedszkolaków. *Issues Early Educ.* **2021**, *1*, 52. [CrossRef]
39. Fundacja Rozwoju Systemu Edukacji. *Kreatywność w Systemie Edukacji*; FRSE: Warsaw, Poland, 2019.
40. Minchberg, M. Autorska Interdyscyplinarna Metoda Edukacji przez Sztukę jako odpowiedź na nowe trendy w pedagogice europejskiej i problemy na polu polskiej edukacji kulturalnej. *Ogrody Nauk I Szt.* **2018**, *8*, 192–204. [CrossRef]
41. Plebańska, M. Stan cyfryzacji polskich szkół na podstawie badania "Polska szkoła w dobie cyfryzacji. Diagnoza 2017" w kontekście potrzeby wdrożenia nauczania w modelu STEAM. *Interdyscyplinarne Kontekst-Pedagog. Spec.* **2018**, *23*, 59–75. [CrossRef]
42. Łukasik, J.M.; Jagielska, K.; Mróz, A.; Koperna, P. Creative competence of young people in the perspective of sustainable development. *Ann. Univ. Mariae Curie-Skłodowska. Sect. J. Paedagog.-Psychol.* **2021**, *34*, 2.
43. Tomczyk, Ł.; Wnek-Gozdek, J.; Mroz, A.; Wojewodzic, K. ICT, digital literacy, digital inclusion and media education in Poland. In *ICT for Learning and Inclusion in Latin America and Europe*; Tomczyk, Ł., Oyelere, S.S., Eds.; Pedagogical University of Cracow: Cracow, Poland, 2019.
44. Panskyi, T.; Korzeniewska, E. Bridging the Informatics Gap between School and University with the InfoSukces Contest. *Inform. Educ.* **2022**, *20*, 3. [CrossRef]
45. Panskyi, T.; Rowińska, Z. A Holistic Digital Game-Based Learning Approach to Out-of-School Primary Programming Education. *Inform. Educ.* **2022**, *20*, 2. [CrossRef]
46. Hilai, M. *Project-Based Learning (PBL) as a Promising Challenge for Prospective Mathematics Teachers in Math in Elementary School Education*; Wydział Studiów Edukacyjnych: Poznan, Poland, 2020.
47. Guenaga, M.; Mentxaka, I.; Garaizar, P.; Eguiluz, A.; Villagrasa, S.; Navarro, I. Make world, a collaborative platform to develop computational thinking and steam. In *Learning and Collaboration Technologies. Technology in Education*; Zaphiris, P., Ioannou, A., Eds.; Springer: Cham, Switzerland, 2017; p. 10296.
48. Dudel, B.; Naruszewicz, A. *Zbiór Koncepcji Zajęć "STEAM" Dla Klas I-III Szkoły Podstawowej*; Wydział Nauk o Edukacji Uniwersytetu w Białymstoku: Bialystok, Poland, 2022.
49. Syslo, M.M. The new computer science curriculum in Poland-challenges and solutions. In Proceedings of the International Conference on Informatics in Secondary Schools, Tallinn, Estonia, 16–18 November 2020.
50. Łukaszczyk, Z.; Grebski, M. Comparative Analysis of the Curriculum at Science and Technology (STEM) High Schools in Poland and The United States. *Organ. I Zarządzanie Kwart. Nauk.* **2020**, *2*, 75–85.
51. Kisiel, P. Projektowanie modeli trójwymiarowych w szkole średniej z użyciem oprogramowania open source Blender. *Dydakt. Inform.* **2020**, *15*, 120–129. [CrossRef]
52. Walat, W. Kompetencje wizualne w kształceniu informatycznym. *Eduk.–Tech.–Inform.* **2016**, *3*, 17.
53. Mróz, A.; Ocetkiewicz, I. Creativity for Sustainability: How Do Polish Teachers Develop Students' Creativity Competence? Analysis of Research Results. *Sustainability* **2021**, *13*, 571. [CrossRef]
54. Jagielska, K.; Łukasik, J.M.; Mróz, A.; Duda, A.K.; Koperna, P.; Sobieszczańska, K. The New Model of Teacher Education in Poland the Directions of Changes in the Context of the Existing Research and Teaching Practices. In Proceedings of the ICERI2019, Seville, Spain, 11–13 November 2019.
55. Łukasik, J.M. Zaniedbane obszary w procesie kształcenia do zawodu nauczyciela. *Labor Educ.* **2017**, *5*, 155–165. [CrossRef]
56. Łukasik, J.M. The value of work in the perspective of prospective and active teachers. *Labor Educ.* **2022**, *10*, 79–94.
57. Tomczyk, Ł.; Fedeli, L.; Włoch, A.; Limone, P.; Frania, M.; Guarini, P.; Szyszka, M.; Mascia, M.L.; Falkowska, J. Digital Competences of Pre-service Teachers in Italy and Poland. *Technol. Knowl. Learn.* **2023**, *28*, 651–681. [CrossRef]
58. Tomczyk, Ł.; Fedeli, L. Introduction—On the Need for Research on the Digital Literacy of Current and Future Teachers. In *Digital Literacy for Teachers*; Lecture Notes in Educational Technology; Tomczyk, Ł., Fedeli, L., Eds.; Springer: Singapore, 2022.
59. Tomczyk, Ł. Research Trends in Media Pedagogy: Between the Paradigm of Risk and the Paradigm of Opportunity. *Int. J. Cogn. Res. Sci. Eng. Educ.* **2021**, *9*, 3.
60. Tomczyk, Ł. Declared and Real Level of Digital Skills of Future Teaching Staff. *Educ. Sci.* **2021**, *11*, 619. [CrossRef]
61. Tomczyk, Ł.; Fedeli, L. Digital literacy among teachers—Mapping theoretical frameworks: TPACK, DigCompEdu, UNESCO, NETS-T, DigiLit Leicester. In Proceedings of the 38th International Business Information Management Association (IBIMA), Seville, Spain, 23–24 November 2021.
62. Juńczyk, T. Metodologiczne wyzwania w ewaluacji międzynarodowych działań edukacyjnych na przykładzie projektu, EDU-ARCTIC–Engaging students in STEM education through Arctic research". *Edukacja* **2023**, *2*, 165.
63. Waligóra, A.; Górski, M. Interdisciplinary PBL approach in the education process on the example of Silesian University of Technology in Poland. In Proceedings of the 16th annual International Technology, Education and Development Conference, Valencia, Spain, 7–9 March 2022.
64. Balyk, N.; Shmyger, G.; Vasylenko, Y. Influence of University Innovative Educational Environment on the Development of Digital STEM Competences. *Int. J. Res. E-Learn.* **2018**, *4*, 2. [CrossRef]

65. Grzesiak, E. Priorytetyzacja STEM w edukacji. Perspektywa aksjologiczno-teleologiczna. *Stud. Z Teor. Wych.* **2019**, *4*, 29.
66. Musiał, E. Nowe Trendy W Edukacji–Koncepcja "Głębokiego Uczenia Się". *Zesz. Nauk. Wyższej Szkoły Humanit. Pedagog.* **2018**, *16*, 55–64.
67. Rostek, I. Narracje w edukacji STEM. *Eduk. Elem. W Teor. I Prakt.* **2020**, *14*, 39–48.
68. Mińkowska, E. Gry komputerowe w edukacji STEAM–możliwości i przeszkody. *Eduk. Elem. W Teor. I Prakt.* **2021**, *16*, 69–78. [CrossRef]
69. Basogain, X.; Gurba, K.; Hug, T.; Morze, N.; Noskova, T.; Smyrnova-Trybulska, E. STEM and STEAM in contemporary education: Challenges, contemporary trends and transformation: A discussion paper. In *Innovative Educational Technologies, Tools and Methods for E-Learning*; Smyrnova-Trybulska, E., Ed.; University of Silesia in Katowice: Katowice, Poland, 2020; Volume 12.
70. Hanson, S.L.; Krywult-Albańska, M. Gender and access to STEM education and occupations in a cross-national context with a focus on Poland. *Int. J. Sci. Educ.* **2020**, *42*, 6. [CrossRef]
71. Zembski, S.; Ulewicz, R. Usefulness Of Problem Based Learning In Preparing Engineers For Industry 4.0: Literature Review. *Qual. Prod. Improv.-QPI* **2020**, *2*, 1.
72. Tomczyk, Ł.; Mróz, A.; Potyrała, K.; Wnęk-Gozdek, J. Digital inclusion from the perspective of teachers of older adults-expectations, experiences, challenges and supporting measures. *Gerontol. Geriatr. Educ.* **2022**, *43*, 1. [CrossRef] [PubMed]
73. Fernández-Morante, C.; Fernández-de-la-Iglesia, J.-d.-C.; Cebreiro, B.; Latorre-Ruiz, E. ATS-STEM: Global Teaching Methodology to Improve Competences of Secondary Education Students. *Sustainability* **2022**, *14*, 6986. [CrossRef]
74. Jiang, D.; Dahl, B.; Du, X. A Systematic Review of Engineering Students in Intercultural Teamwork: Characteristics, Challenges, and Coping Strategies. *Educ. Sci.* **2023**, *13*, 540. [CrossRef]
75. Asghar, A.; Huang, Y.-S.; Elliott, K.; Skelling, Y. Exploring Secondary Students' Alternative Conceptions about Engineering Design Technology. *Educ. Sci.* **2019**, *9*, 45. [CrossRef]
76. Jantassova, D.; Churchill, D.; Shebalina, O.; Akhmetova, D. Capacity Building for Engineering Training and Technology via STEAM Education. *Educ. Sci.* **2022**, *12*, 737. [CrossRef]
77. Sukackė, V.; Guerra, A.O.P.d.C.; Ellinger, D.; Carlos, V.; Petronienė, S.; Gaižiūnienė, L.; Blanch, S.; Marbà-Tallada, A.; Brose, A. Towards Active Evidence-Based Learning in Engineering Education: A Systematic Literature Review of PBL, PjBL, and CBL. *Sustainability* **2022**, *14*, 13955. [CrossRef]
78. Kozanitis, A.; Nenciovici, L. Effect of active learning versus traditional lecturing on the learning achievement of college students in humanities and social sciences: A meta-analysis. *High Educ.* **2023**, *86*, 1377–1394. [CrossRef]
79. Wang, Y. A Comparative Study on the Effectiveness of Traditional and Modern Teaching Methods. In Proceedings of the 5th International Conference on Humanities Education and Social Sciences (ICHESS 2022), Chongqing, China, 14–16 October 2022.
80. Chan, D.W.M.; Lam, E.W.M.; Adabre, M.A. Assessing the Effect of Pedagogical Transition on Classroom Design for Tertiary Education: Perspectives of Teachers and Students. *Sustainability* **2023**, *15*, 9177. [CrossRef]
81. Panskyi, T.; Korzeniewska, E. Statistical and clustering validation analysis of primary students' learning outcomes and self-awareness of information and technical online security problems at a post-pandemic time. *Educ. Inf. Technol.* **2023**, *28*, 6423–6451. [CrossRef] [PubMed]
82. Keller, J.M. Development and use of the ARCS model of instructional design. *J. Instr. Dev.* **1987**, *10*, 2–10. [CrossRef]
83. Kumar, A.; Saudagar, A.K.J.; Alkhathami, M.; Alsamani, B.; Khan, M.B.; Hasanat, M.H.A.; Ahmed, Z.H.; Kumar, A.; Srinivasan, B. Gamified Learning and Assessment Using ARCS with Next-Generation AIoMT Integrated 3D Animation and Virtual Reality Simulation. *Electronics* **2023**, *12*, 835. [CrossRef]
84. Hsieh, Y.-Z.; Lin, S.-S.; Luo, Y.-C.; Jeng, Y.-L.; Tan, S.-W.; Chen, C.-R.; Chiang, P.-Y. ARCS-Assisted Teaching Robots Based on Anticipatory Computing and Emotional Big Data for Improving Sustainable Learning Efficiency and Motivation. *Sustainability* **2020**, *12*, 5605. [CrossRef]
85. Chang, Y.-S. Applying the ARCS Motivation Theory for the Assessment of AR Digital Media Design Learning Effectiveness. *Sustainability* **2021**, *13*, 12296. [CrossRef]
86. Chang, Y.-S.; Chen, C.-N.; Liao, C.-L. Enhancing English-Learning Performance through a Simulation Classroom for EFL Students Using Augmented Reality—A Junior High School Case Study. *Appl. Sci.* **2020**, *10*, 7854. [CrossRef]
87. González-Zamar, M.-D.; Jiménez, L.O.; Ayala, A.S. Design and Validation of a Questionnaire on Influence of the University Classroom on Motivation and Sociability. *Educ. Sci.* **2021**, *11*, 183. [CrossRef]
88. Miller, A.L. A self-report measure of cognitive processes associated with creativity. *Creat. Res. J.* **2014**, *26*, 203–218. [CrossRef]
89. Wang, C.; Zhang, X.; Pan, Y. Enhancing Sustainable Arts Education: Comparative Analysis of Creative Process Measurement Techniques. *Sustainability* **2023**, *15*, 9078. [CrossRef]
90. Childs, P.; Han, J.; Chen, L.; Jiang, P.; Wang, P.; Park, D.; Yin, Y.; Dieckmann, E.; Vilanova, I. The Creativity Diamond—A Framework to Aid Creativity. *J. Intell.* **2022**, *10*, 73. [CrossRef]
91. Conradty, C.; Bogner, F.X. From STEM to STEAM: How to Monitor Creativity. *Creat. Res. J.* **2018**, *30*, 3. [CrossRef]
92. Roth, T.; Conradty, C.; Bogner, F.X. Testing Creativity and Personality to Explore Creative Potentials in the Science Classroom. *Res. Sci. Educ.* **2022**, *52*, 1293–1312. [CrossRef]
93. Gabora, L. Creativity. Oxford Research Encyclopedia of Psychology. Available online: https://oxfordre.com/psychology/view/10.1093/acrefore/9780190236557.001.0001/acrefore-9780190236557-e-608 (accessed on 4 February 2024).

94. Bai, J.; Ng, S. Tests for Skewness, Kurtosis, and normality for times series data. *J. Bus. Econ. Stat. Am. Stat. Assoc.* **2005**, *23*, 49–60. [CrossRef]
95. Ma, K.W.; Mak, C.M.; Wong, H.M. Development of a subjective scale for sound quality assessments in building acoustics. *J. Build. Eng.* **2020**, *29*, 101177. [CrossRef]
96. Koo, T.K.; Li, M.Y. A Guideline of Selecting and Reporting Intraclass Correlation Coefficients for Reliability Research. *J. Chiropr. Med.* **2016**, *15*, 155–163. [CrossRef] [PubMed]
97. Gordon, A.D. Cluster Validation. In Data Science, Classification, and Related Methods. In Proceedings of the Fifth Conference of the International Federation of Classification Societies (IFCS-96), Kobe, Japan, 27–30 March 1996.
98. Jain, A.; Dubes, R. *Algorithms for Clustering Data*; Prentice Hall: Englewood Cliffs, NJ, USA, 1988.
99. Chakravarthy, S.V.; Ghosh, J. Scale-based clustering using the radial basis function network. *IEEE Trans. Neural Netw.* **1996**, *7*, 5. [CrossRef] [PubMed]
100. Mojarad, M.; Parvin, H.; Nejatian, S.; Rezaie, V. Consensus function based on clusters clustering and iterative fusion of Base Clusters. *Int. J. Uncertain. Fuzziness Knowl.-Based Syst.* **2019**, *27*, 1. [CrossRef]
101. Kryszczuk, K.; Hurley, P. Estimation of the Number of Clusters Using Multiple Clustering Validity Indices. In *Multiple Classifier Systems. MCS 2010*; Lecture Notes in Computer Science; El Gayar, N., Kittler, J., Roli, F., Eds.; Springer: Berlin/Heidelberger, Germany, 2010; Volume 5997.
102. Yera, A.; Arbelaitz, O.; Jodra, J.; Gurrutxaga, I.; Pérez, J.; Muguerza, J. Analysis of several decision fusion strategies for clustering validation. Strategy definition, experiments and validation. *Pattern Recognit. Lett.* **2017**, *85*, 42–48. [CrossRef]
103. Tan, P.N.; Steinbach, M.; Kumar, V. *Introduction to Data Mining*; Pearson Addison Wesley: Boston, MA, USA, 2006.
104. Panskyi, T.; Mosorov, V. A Step Towards The Majority-Based Clustering Validation Decision Fusion Method. *Inform. Autom. Pomiary W Gospod. I Ochr. Środowiska* **2021**, *11*, 2. [CrossRef]
105. Charrad, M.; Ghazzali, N.; Boiteau, V.; Niknafs, A. NbClust: An R Package for Determining the Relevant Number of Clusters in a Data Set. *J. Stat. Softw.* **2014**, *61*, 6. [CrossRef]
106. Kumar, D.; Bezdek, J.C. Visual Approaches for Exploratory Data Analysis: A Survey of the Visual Assessment of Clustering Tendency (VAT) Family of Algorithms. *IEEE Syst. Man Cybern. Mag.* **2020**, *6*, 2. [CrossRef]
107. Wang, L.; Nguyen, U.T.; Bezdek, J.C.; Leckie, C.A.; Ramamohanarao, K. iVAT and aVAT: Enhanced visual analysis for cluster tendency assessment. In Proceedings of the Pacific-Asia Conference on Knowledge Discovery and Data Mining, Hyderabad, India, 21–24 June 2010; Springer: Berlin/Heidelberg, Germany, 2010.
108. Khalil, R.Y.; Tairab, H.; Qablan, A.; Alarabi, K.; Mansour, Y. STEM-Based Curriculum and Creative Thinking in High School Students. *Educ. Sci.* **2023**, *13*, 1195. [CrossRef]
109. Borg Preca, C.; Baldacchino, L.; Briguglio, M.; Mangion, M. Are STEM Students Creative Thinkers? *J. Intell.* **2023**, *11*, 106. [CrossRef]
110. Baričević, M.; Luić, L. From Active Learning to Innovative Thinking: The Influence of Learning the Design Thinking Process among Students. *Educ. Sci.* **2023**, *13*, 455. [CrossRef]
111. Hartikainen, S.; Rintala, H.; Pylväs, L.; Nokelainen, P. The Concept of Active Learning and the Measurement of Learning Outcomes: A Review of Research in Engineering Higher Education. *Educ. Sci.* **2019**, *9*, 276. [CrossRef]
112. Peng, L.; Deng, Y.; Jin, S. The Evaluation of Active Learning Classrooms: Impact of Spatial Factors on Students' Learning Experience and Learning Engagement. *Sustainability* **2022**, *14*, 4839. [CrossRef]
113. Song, B.; He, B.; Wang, Z.; Lin, R.; Yang, J.; Zhou, R.; Cai, Y. Research on Open Practice Teaching of Off-Campus Art Appreciation Based on ICT. *Sustainability* **2022**, *14*, 4274. [CrossRef]
114. Aguilera, D.; Ortiz-Revilla, J. STEM vs. STEAM Education and Student Creativity: A Systematic Literature Review. *Educ. Sci.* **2021**, *11*, 331. [CrossRef]
115. Altan, E.B.; Tan, S. Concepts of creativity in design based learning in STEM education. *Int. J. Technol. Des. Educ.* **2021**, *31*, 503–529. [CrossRef]
116. Kapliński, O. Architecture: Integration of Art and Engineering. *Buildings* **2022**, *12*, 1609. [CrossRef]
117. Qadir, J.; Al-Fuqaha, A. A Student Primer on How to Thrive in Engineering Education during and beyond COVID-19. *Educ. Sci.* **2020**, *10*, 236. [CrossRef]
118. Sanz-Camarero, R.; Ortiz-Revilla, J.; Greca, I.M. The Impact of Integrated STEAM Education on Arts Education: A Systematic Review. *Educ. Sci.* **2023**, *13*, 1139. [CrossRef]
119. Pirrie, A. Where Science Ends, Art Begins? Critical Perspectives on the Development of STEAM in the New Climatic Regime. In *Why Science and Art Creativities Matter*; Brill: Leiden, The Netherlands, 2019.
120. Kim, J.H.; Nguyen, N.T.; Campbell, R.C.; Yoo, S.; Taraban, R.; Reible, D.D. Developing reflective engineers through an arts-incorporated graduate course: A curriculum inquiry. *Think. Ski. Creat.* **2021**, *42*, 100909. [CrossRef]

applied sciences

MDPI

Article

Identification of Students with Similar Performances in Micro-Learning Programming Courses with Automatically Evaluated Student Assignments

Valerii Popovych and Martin Drlik *

Faculty of Natural Sciences and Informatics, Constantine the Philosopher University in Nitra, Tr. A. Hlinku 1, 9494 01 Nitra, Slovakia; valerii.popovych@ukf.sk
* Correspondence: mdrlik@ukf.sk

Featured Application: The student performance data stored in the micro-learning platform Priscilla, specially developed to improve IT students' knowledge, is analyzed using learning analytics methods to identify meaningful groups of students with similar performances and uncover possible changes in the number of identified groups and their compositions in consecutive term periods and based on different types of activities.

Abstract: The identification of heterogeneous and homogeneous groups of students using clustering analysis in learning analytics is still rare. The paper describes a study in which the students' performance data stored in the micro-learning platform Priscilla are analyzed using learning analytics methods. This study aims to identify the groups of students with similar performances in micro-learning courses focused on learning programming and uncover possible changes in the number and composition of the identified groups of students. The CRISP-DM methodology was used to navigate through the complexity of the knowledge discovery process. Six different datasets representing different types of graded activities or term periods were prepared and analyzed for that purpose. The clustering analysis using the K-Means method found two clusters in all cases. Subsequently, performance metrics, the internal composition, and transfers of the students between clusters identified in different datasets were analyzed. As a result, this study confirms that analyzing student performance data from a micro-learning platform using learning analytics methods can reveal distinct groups of students with different academic performances, and these groups change over time. These findings align with teachers' assumptions that the micro-learning platform with automated evaluation of programming assignments highlights how the students perceive the role of learning tools during learning programming in different term periods. Simultaneously, this study acknowledges that clustering, as an exploratory method, provides a solid basis for further research and can identify distinct groups of students with similar characteristics.

Keywords: learning analytics; clustering; student performance; user behavioral patterns; micro-learning

1. Introduction

Learning analytics (LA) represents a research discipline that gathers, measures, and analyses available educational data about stakeholders for understanding and optimizing the learning process and the whole environment where the learning process occurs [1]. Higher educational institutions (HEIs) worldwide have already understood the benefits of LA, which applies modern statistical and machine learning techniques on different levels of the learning process, targets the institutional level, and tries to support the decision-making processes of the HEIs. In this context, LA research is primarily driven by the data available in different virtual learning environments (VLE) to improve student success and predict early dropout, student retention, and the learning experience.

In this context, the Erasmus Plus project FITPED aimed to design and implement a sustainable educational model, which supports the development of highly specialized skills and competencies of future IT experts in programming and software development. The research outcomes of the project were summarized in [2]. Its successor project, FITPED-AI, continues in this effort and aims to prepare IT students for a career in artificial intelligence, especially in machine and deep learning.

Both projects continually develop a modern micro-learning platform named Priscilla. This platform implements a novel educational model, which utilizes the benefits of micro-learning units, interactive activities, automated evaluation of student assignments, and gamification elements. As a result, this educational model applies selected elements of work-based, collaborative, and problem-based learning to improve students' readiness for lifelong learning. The proposed and implemented model has already shown its potential to decrease the number of students with learning problems and increase the level of knowledge obtained from computer science courses [2]. Therefore, the micro-learning courses provided on the platform can be considered a valuable prerequisite for increasing graduates' highly specialized knowledge and skills.

The Priscilla platform has become an integral part of teaching and learning programming at the universities participating in the projects for three academic years. More than 1800 individual student accounts have been created on the platform. In addition, students have actively participated in most of the 26 micro-learning courses created. The platform provides more than 4330 micro-learning units, 2200 automatically evaluated programs, 180 course topics based on the work-based and problem-based learning strategy, and 560 solved student projects from practice within the educational model.

The positive impact of the educational model implemented in the micro-learning platform Priscilla has been confirmed by surveys and conventional statistical and educational technology research methods. Together with the conceptual architecture of the platform, these are described on the project webpage (www.fitped.eu).

Even though the Priscilla stakeholders positively evaluated the attractivity, interactivity, and usefulness of the platform, the student's behavior in the micro-learning courses enhanced with automated programming assignment evaluation requires further research to confirm that the proposed educational settings have a positive impact on the number of students, who will learn better programming skills and achieve higher academic performances.

The learning analytics researchers previously confirmed the positive impact of the LA methods on the understating of learners, learning outcomes, and the learning process in general. Considering the number of students, micro-learning courses, and assignments mentioned above and the rising amount of stakeholders' data stored in the platform, it is natural to expect that using suitable learning analytics methods can uncover hidden patterns in students' learning behavior or identify students with similar learning patterns. Their understanding can further improve the platform and educational model behind it. It can also influence subsequent platform and educational content development. Although these methods can be applied at the course, topic, or platform level, the most promising results can be expected at the course level because of the direct relationship between the applied educational model, course structure, and students.

Concurrently, it is worth noting that the primary motivation to develop the platform was a belief that providing the learning content in the form of micro-learning units suitably combined with questions and automatically evaluated programming assignments can positively influence students' self-confidence and engagement in learning programming. This expectation was amplified by the teachers' previous negative experiences with teaching programming in two separate systems for providing educational online content and automated programming code evaluation.

Therefore, this paper's main aim is to analyze students' academic performance data using unsupervised machine learning methods to identify the groups of students with similar performances and behaviors in a Java programming micro-learning course and uncover possible changes in the number of identified groups in consecutive periods of the term and based on different types of micro-learning units. The following research questions can be formulated to addresses this issue:

Q1. How do the students' performances in micro-learning units with quizzes and programming assignments in introductory micro-learning courses affect the number of identified meaningful clusters?

Q2. How do identified groups of students with different academic performances change in composition over three periods of term?

According to Shafiq et al. [3], who published a taxonomy of machine learning approaches used in learning analytics in different learning environments, the application of ensemble and unsupervised learning clustering techniques for the identification of the heterogeneous and homogeneous groups of students is still generally lacking. In addition, available studies deal predominantly with data obtained from the LMS environments or surveys. Clustering analysis of the students' partial achievements in learning programming in a tailor-made environment is rare. Therefore, the application of clustering analysis with the evaluation of students' data obtained from a micro-learning environment and comparing changes in cluster distribution found in different term periods can be considered partially novel. This paper contributes to the discussion of whether delivering educational content in the form of micro-learning units with automated programming assignments can identify meaningful groups of students with different academic performances and whether these groups change over time. Consequently, these findings could initialize the adjustment of the educational model and initialize changes in course structure and content sequencing. Simultaneously, the presented results can evoke new ideas, which can be considered in the future development of the platform Priscilla in the context of the planned implementation of the learning analytics module and dashboard. These findings could decrease the dropout rate and achieve better academic performance.

The paper is structured as follows. First, the Related Works section summarizes some interesting findings of similar studies. Consequently, all phases of the CRISP-DM methodology are described to explain the complexity of the problem of the practical adoption of machine learning methods in the case of data obtained from the micro-learning platform. The selected clustering methods and other evaluation and visualization techniques are then applied to different datasets to identify distinct groups of students with similar performances. This section also analyzes and visualizes the changes in the composition of individual groups of students to understand the differences and behavioral patterns of the found groups over time and based on types of activities. Afterwards, this paper summarizes the main findings, the limitations of this study, its implications for future changes in the micro-learning platform and educational model, and recommendations for further research directions.

2. Related Works

This section summarizes the results of other related scientific papers from the educational data mining (EDM) and learning analytics domains, which apply different clustering methods to students' performance data from their interactions with a virtual learning environment. The researchers in both domains systematically analyze specific patterns of students' interactions with VLE, their performances, and the risk of early dropout or achievements [4]. Aldowah et al. [5] also confirmed these trends, especially in higher education. They stated that specific EDM and LA methods could offer the best means of solving specific learning problems. Similarly, Gashevic et al. [1] provided a concise review of the current state of the learning analytics research field and its typical tasks, technologies, and challenges. As a result, they stated that many learning analytics and educational data mining methods, including clustering, have been successfully applied to the student's activity and performance data stored in VLEs during the last decade.

As stated by Shafiq et al. [3], publications that dealt with unsupervised machine learning clustering techniques for the identification of the heterogeneous and homogeneous groups of students as a primary research topic are still rare. Clustering analysis was more frequently applied as a part of the data-understanding phase [6] or data-modeling phase, together with classification or prediction [7,8]. Several papers tried to find the best clustering methods and algorithms using artificial, open-source, or real datasets [9,10].

Regarding the selection of the best clustering method in LA research, K-Means was the most often selected [10]. The reason is that this unsupervised machine learning method is considered simple, flexible, and time-effective for the cluster analysis of educational data, which usually have linear or monotonic relationships. Other, more robust methods like DBSCAN or OPTICS, often applied in industry and engineering, were not frequently applied in the LA domain [11]. Table 1 summarizes the main characteristics of other studies focused on clustering analysis in the context of learning analytics and topics related to the study presented in this paper. As can be seen, a low number of identified clusters is typical for the education domain because of the issues related to the correct interpretation of higher values. Simultaneously, it is surprising that several studies did not apply any internal or external validation metrics, which can cause problems with the generalization of the findings.

Even fewer articles dealt with clustering data about the students' performances in learning programming courses. Avella et al. [12] confirmed the potential of clustering methods to understand students' performances better and identify behavioral patterns. However, they did not deal with programming courses or data from the micro-learning environment. Beena et al. [13] used K-Means clustering to group students with similar performances without collecting their previous academic records or demographic information. They identified 7.7% as slow learners, 31% as average learners and the rest as advanced learners. Simultaneously, they found a correlation between these groups and academic success. Nafuri et al. [14] classified low-income students based on the clustering analysis of their academic performance. They applied three unsupervised models, including K-Means. As a result, they identified five clusters of students and proposed actions that could increase their success. Gonzalez-Nucamendi et al. [15] studied relationships among the self-regulation learning and affective strategies and multiple intelligence dimensions of students using clustering analysis. As a result, they identified features that positively or negatively impacted academic performance. These publications confirm the suitability of clustering analysis and selection of the K-Means method for the identification of groups with similar academic achievements. On the other hand, they did not focus on data from programming learning courses or the micro-learning environment.

The following papers used clustering analysis in close connection with programming. Gupta et al. [16] proposed a rubric based on the learning outcomes like syntactical, logical, conceptual, and advanced JAVA skills. K-Means clustering was applied to the results to classify the students according to their learning preferences and abilities. Bey et al. [17] introduced a three-phase process and a case study in which unsupervised clustering techniques were used to automatically identify learners' programming behavior. They continued their research [18] and analyzed students' programming habits using K-Means clustering. They found six students' profiles, which differed in behavioral trajectories and performance. However, they did not combine data from VLE with the automated programming code evaluation tool.

Similarly, López-Pernas et al. [19] combined learning analytics methods, including clustering and sequence analysis, to examine students' strategies when learning programming. They combined data from VLE and automated assessment tools. They found that the students have special preferences for specific learning resources when learning programming and often use slides that support search and copy-and-paste actions. Kovari et al. [20] applied K-Means clustering to study self-confidence in programming, problem-solving, and engagement in software development courses. They identified two groups of students based on their self-confidence in programming.

Table 1. A summary of related scientific publications that dealt with clustering analysis in the learning analytics domain.

Paper	Summary	Dataset	Clustering Method	Number of Clusters	Validation
Liu [6]	They introduced a study that applies unsupervised learning to identify how students' engagement and demographic features in online learning activities can affect their learning achievement.	University dataset	K-prototypes	7	N/A
López et al. [7]	They analyzed the interaction of university students in VLE, comparing clustering and classification methods. They stated that both groups of methods could contribute to a better comprehension of the learning process.	Moodle	Hierarchical method and K-Means method	4	N/A
Fida et al. [8]	They focused on generating a smarter dataset through a reduced number of features without compromising the number of records. Their approach combined the strengths of classification and clustering for better prediction results.	Students' Academic Performance Dataset	PAM (Partition Around Medoids), EM (Expectation Maximization) and K- Means	3	N/A
Palani et al. [9]	They applied clustering using three different techniques to identify students with a low level of engagement in learning. As a result, they identified clusters that can be further analyzed.	OULAD	Gaussian Mixture, Hierarchical and K-prototype	3	Dunn's Index, Silhouette score, and Davies–Bouldin
Hooshyar et al. [10]	They provided a detailed review of the clustering algorithms in an educational context. They proposed an evaluation approach that compared several clustering methods using multiple internal and external performance measures for nine real-world educational datasets of different sizes.	Moodle	K-Means, K-Medoids, Expectation–maximization (EM), Agglomerative, Spectral Clustering, DBSCAN, OPTICS	4	Dunn's Index, Silhouette Index, Davies–Bouldin Index, Adjusted Mutual Information, Completeness Index, Jaccard Index, Purity Index, Rand Index, Adjusted Rand Index, Fowlkes–Mallows Index, F-measure, Rogers–Tanimoto, Normalized Mutual Information, Mutual Information, Variation in Information, Geometric Accuracy, Overlapping Normalized Mutual Information
Navarro et al. [11]	They compared different clustering algorithms using an educational dataset and recommended the ones with the highest performances.	University Dataset of Students' Profiles	K-Means, Clara, PAM, FANNY, hierarchical, AGNES, DIANA	3	Silhouette coefficient, Dunn's index

This analysis of the related works confirms that there is a general agreement that clustering methods can contribute to a better understanding of the students' behavior and achievements in VLE, but further research is needed. Moreover, despite the popularity of learning programming platforms, including micro-learning platforms, the number of research studies that apply clustering is limited.

3. Methodology

This study used clustering analysis to evaluate students' data from a micro-learning environment and looked for changes in cluster distribution found in different term periods. The study followed individual steps of a well-known knowledge discovery process: the CRISP-DM methodology. This methodology guides the researchers through all phases of the knowledge discovery process. Previous research papers confirmed its suitability for learning analytics projects [21,22]. All phases of the CRISP-DM methodology, such as data understanding, preprocessing, modeling, and evaluation, are described in detail to eliminate the risk of the unintentional underestimation of a particular research phase. At the same time, this methodology can explain the complexity of the issues, as well as the practical use of machine learning methods in learning analytics.

Figure 1 visualizes the necessary actions realized in the individual steps of the CRISP-DM methodology. The output datasets on the right side will be introduced and analyzed in detail in the next sections.

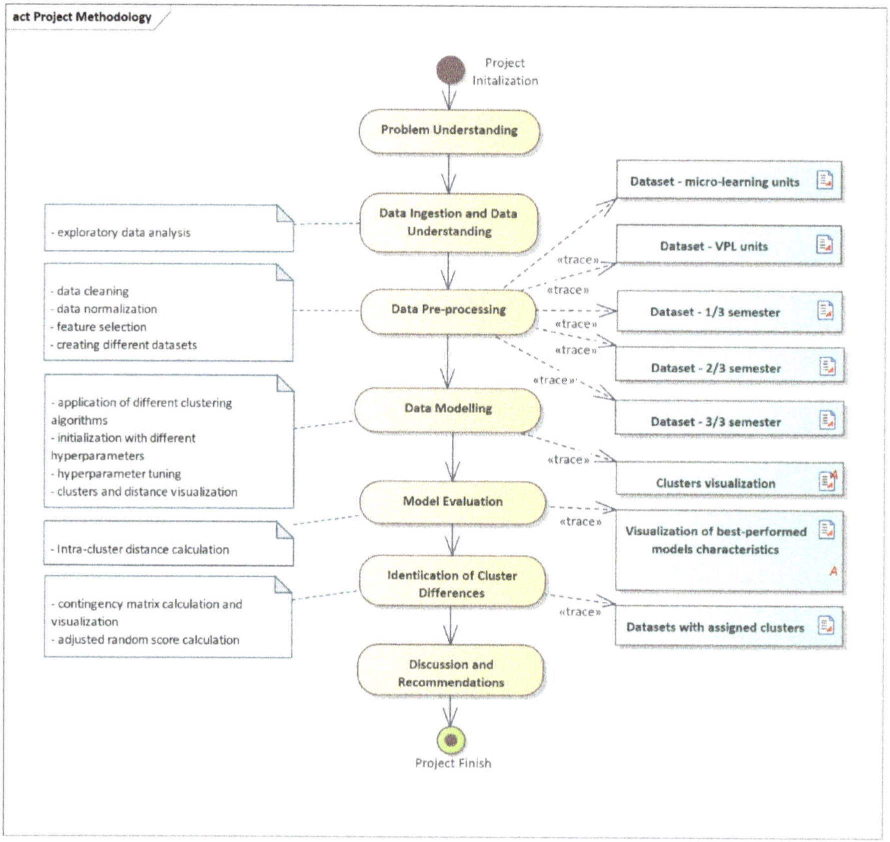

Figure 1. Description of the actions realized in the individual steps of the CRISP-DM methodology and output datasets.

3.1. Problem Understanding

As mentioned in the introduction section, this paper aims to determine whether delivering educational content in micro-learning units with automated programming assignments can identify groups of students with different academic performances and whether these groups change over time. The students' performance data stored in the micro-learning platform Priscilla were analyzed using selected learning analytics methods. Among other things, the results of this study can help to explore the potential and suitability of the collected data for the future implementation of a learning analytics module or dashboard. This module should engage students in achieving their learning goals, support teachers in applying suitable and early forms of intervention, and fulfil the expectations of the micro-learning platform Priscilla community of stakeholders.

Nowadays, the platform stores data about the students' activities in the form of xAPI logs. However, their potential for clustering without massive preprocessing is questionable. On the other hand, the platform provides several reports for teachers and managers through its user interface. These reports provide detailed information about each student's progress, partial and final grades, number of attempts, and experience points as gamification module outputs. Nevertheless, these data resources have not yet been systematically analyzed with machine learning methods.

A programming course, Introduction to Java, was selected for this study. Almost 120 students had started to learn programming in this course. During the term, they gradually went through individual topics suitably ordered to develop the students' skills in Java programming in line with the object-oriented paradigm. Each topic contained micro-learning units with quizzes, which tested students' correct understanding of the topic and provided immediate feedback to them. Afterwards, each student had to solve a set of programming assignments with automated feedback and programming code evaluation to verify whether they correctly understood the given programming concept. Each student could repeat the quizzes and automatically evaluate the programming code without limit, use hints, and receive grades for successfully passing the activities.

This educational strategy can be considered one of the most essential features of the micro-learning platform. It aligns with the expectations of the current generation of students who prefer combining short textual or graphical content with interactive activities. The micro-learning platform replaced programming courses created in LMS Moodle, which used the Virtual Programming Lab (VPL) plugin for the automated evaluation of programming assignments. The main motivation to leave LMS Moodle resulted from the students' dissatisfaction with the complexity of the programming courses enhanced with the VPL plugin. The reason was that many activities and resources made the course confusing. The micro-learning platform tried to eliminate this problem. The automated evaluation of programming assignments based on VPL was improved and successfully implemented in the micro-learning platform during the realization of the FITPED project.

The primary motivation for clustering analysis can be explained using heatmap visualization (Figure 2). This figure shows a normalized heatmap of each student's performance in individual micro-learning units and programming code assignments in individual topics. Topics were ordered chronologically by how they happened in the term. The students with the lowest performance were ordered at the bottom. The enrolled students with zero activity were removed for better clarity. Each student's performance was normalized because each topic contained a different number of graded quizzes and programming assignments.

Unsurprisingly, the students' activity levels and performances declined during the term. However, this distribution simultaneously indicates that the students' activity levels seem to decline after two control tests (Methods and Static class topics). There seem to be differences between grades achieved in micro-learning units with quizzes and programming assignments, which are parts of the thematic topics. These findings should be investigated in more detail to understand the students' behavior better and improve the course structure and its content. Therefore, the clustering analysis was chosen as the first to uncover whether the students with similar performance create clusters and whether these clusters change over time.

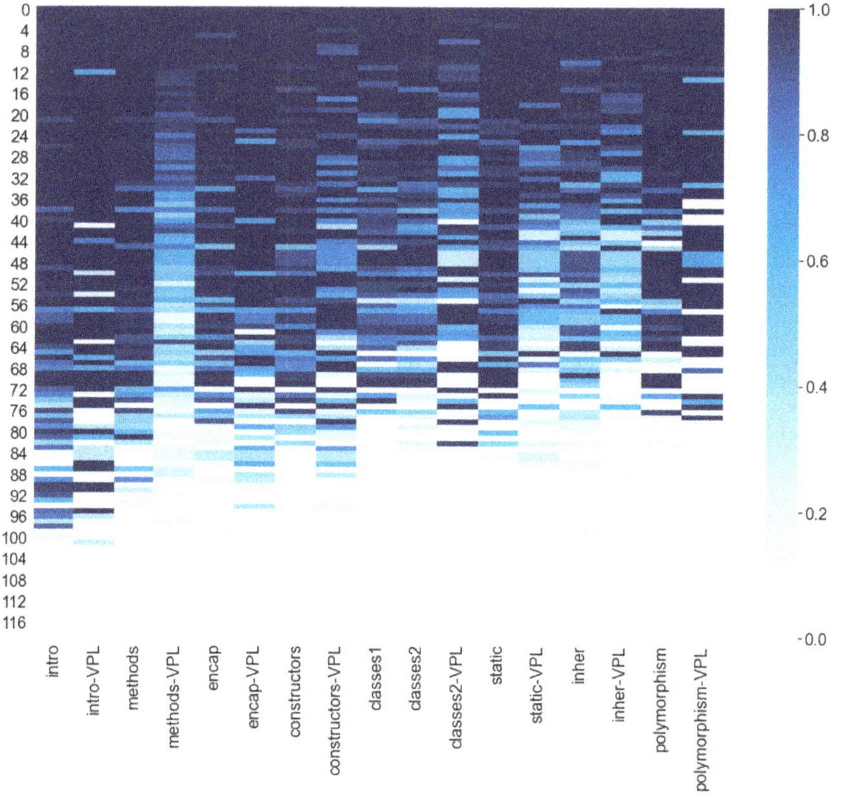

Figure 2. Normalized heatmap of each student's performance in individual micro-learning units and programming code assignments. The students with the lowest performances are ordered at the bottom.

3.2. Data Understanding

The dataset was downloaded in tabular form from the Priscilla platform. The data were anonymized in the first step because this knowledge discovery task did not expect any intervention in which student identification occurs. As a result, it was not possible to reversely identify individual student personal data from the analyzed dataset or identified clusters. The dataset contained detailed information about each student enrolled in the micro-learning course on programming in Java. The course contained more than 400 micro-learning units divided into thematic topics. Each topic provided several categories of units, like an explanation of the theory, quizzes, and programming assignments with automated programming code evaluation. The students could repeatedly use these units and train in practical programming skills. The students received the following two types of grades:

- Grades, which characterize the level of comprehension of the learning content belonging to the individual topics of the course. The platform collected data about the number of attempts in quizzes, received scores, and spent time, which were summarized in these grades.
- Grades from automatically evaluated programming assignments using a modified VPL. The platform collected data about the number of attempts, received scores, use of hints, and spent time, which were reflected in the calculation of these grades.

The dataset contained aggregated data about 119 unique students and 17 attributes, which characterized their performances in graded automated programming assignments and micro-learning units with quizzes. Table 2 shows an exploratory data analysis of the input dataset. Seventeen students enrolled on the course but did not participate in any activity. Therefore, they were removed from the dataset before this analysis to minimize the bias. Each table column represents a learning topic containing micro-learning units with quizzes or programming assignments, which correspond to a particular programming learning construct. These topics are shown in the same order in which they took place in the term. One topic took approximately two weeks.

Table 2. Basic exploratory data analysis of the input dataset.

	Count	Mean	Std	Min	25%	50%	75%	Max
intro	119.0	94.509804	30.101066	0.0	97.25	110.0	110.00	110.0
intro-VPL	119.0	46.196078	23.239621	0.0	40.00	60.0	60.00	60.0
methods	119.0	123.421569	59.649544	0.0	102.50	159.0	160.00	160.0
methods-VPL	119.0	348.843137	224.051875	0.0	107.50	346.0	584.50	630.0
encap	119.0	108.745098	58.633901	0.0	71.75	146.5	150.00	150.0
encap-VPL	119.0	71.598039	38.642115	0.0	40.00	100.0	100.00	100.0
constructors	119.0	79.117647	44.749533	0.0	51.00	105.5	110.00	110.0
constructors-VPL	119.0	112.637255	63.217179	0.0	60.00	139.0	165.00	165.0
classes1	119.0	52.401961	34.236007	0.0	0.00	73.0	80.00	80.0
classes2	119.0	109.450980	71.353482	0.0	32.50	150.0	169.75	170.0
classes2-VPL	119.0	54.166667	41.345462	0.0	0.00	65.0	95.00	95.0
static	119.0	57.362745	33.316303	0.0	40.00	77.0	80.00	80.0
static-VPL	119.0	147.950980	109.225027	0.0	38.00	170.0	270.00	270.0
inher	119.0	116.568627	75.730569	0.0	40.00	161.5	186.50	190.0
inher-VPL	119.0	139.941176	110.768944	0.0	27.00	158.5	270.00	270.0
polymorphism	119.0	45.107843	31.401238	0.0	0.00	69.0	70.00	70.0
polymorphism-VPL	119.0	26.647059	21.434518	0.0	0.00	45.0	45.00	45.0

After initial data exploration and discussion with the course developers and teachers about the exact meanings of individual attributes, the final dataset contained 17 attributes. Nine attributes represented the grades received from micro-learning units with quizzes, and eight attributes characterized the students' performances in programming assignments (with VPL suffix). The first group of attributes was consequently visualized using boxplot to graphically depict the distribution of the numerical data through their quartiles (Figure 3). Moreover, correlation matrix was calculated to ensure that the initial requirements of the planned clustering methods were fulfilled (Figure 4). A similar approach was applied to the second group of attributes (Figures 5 and 6).

Figure 3 confirms our expectations that the students repeatedly answered the questions in the quiz until they reached almost maximal grades. Even though they could ask for a hint, they used this option rarely. As a result, most boxplots had very high values of the mean. Values below average belong to the students who did not finish all quizzes or other graded activities in micro-learning units.

Spearman's rank correlation coefficient is a nonparametric measure of rank correlation. It assesses how well the relationship between two variables can be described using a monotonic function. Spearman's correlation assesses monotonic relationships, whether linear or not. Even though Figure 4 shows positive correlations between individual topics, these attributes were not removed from the dataset for several reasons. Firstly, this correlation was expected because students were motivated to complete all the quizzes as part of their learning process. Secondly, planned clustering methods are not sensitive to the correlation between input features. Finally, removing individual features can significantly impact the final distribution and number of clusters. However, these correlations should be considered if principal component analysis (PCA) is applied.

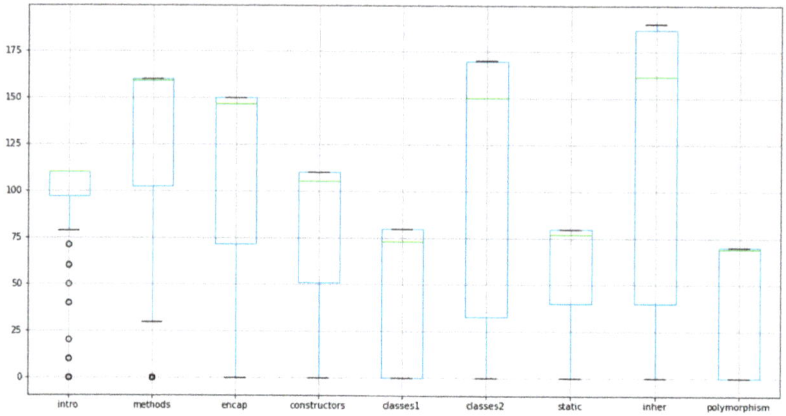

Figure 3. Boxplot of the features related to each student's performance in modules with micro-learning units and quizzes.

Figure 4. Correlation matrix between the attributes, which evaluate the level of comprehension of the micro-learning unit in individual chapters of the course.

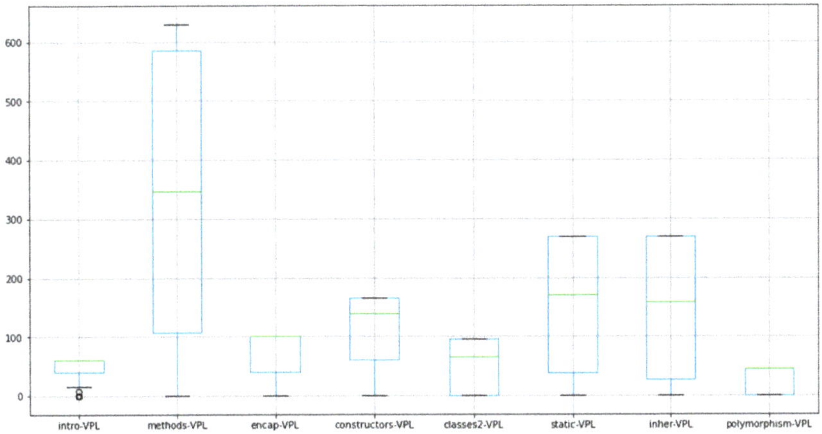

Figure 5. Boxplot of the features related to each student's performance in automatically evaluated programming assignments.

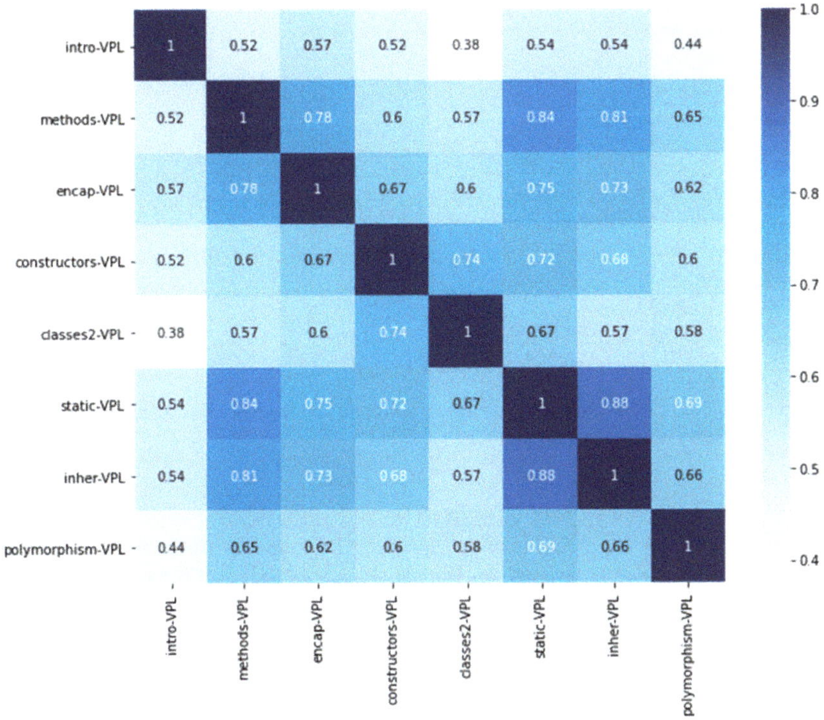

Figure 6. Correlation matrix between the attributes, which belong to programming assignments, automatically evaluated by the platform.

Figure 5 shows the box plots representing grades received from programming assignments. The students could receive different grades depending on the number of unique tasks, number of attempts, and used hints. The lower position of the means in boxplots is worth mentioning, as it differs in comparison to the previous groups of attributes. The higher difficulty of the programming tasks probably caused this difference.

The impact of a wider distribution of the students' achievements in programming assignments is also evident in the correlation matrix (Figure 6). Spearman's rank correlation coefficients were lower. Therefore, it was not necessary to remove any attributes for the same reasons as those previously considered.

3.3. Data Preprocessing

Each machine learning project requires thoughtful analysis of the original dataset and the application of inevitable data preprocessing techniques to prepare an input dataset, which respects the specificity of the planned machine learning methods. In this study, most preprocessing tasks were already realized during the initial data collection and ingestion phase from different database tables. As was mentioned, attributes represented aggregated grades from different topics. Each row in the dataset represented the academic performance of one student. Therefore, grades on individual activities were not scarce and did not contain null values.

Consequently, an open-source low-code machine learning library named PyCaret for Python was used in this case study to automate the machine learning workflow. The preprocessing phase is the right point at which such low-code machine learning libraries can be helpful. The library automates many repeating preprocessing steps and provides many options to correctly set up the parameters of the implemented preprocessing methods [23]. During its application on the input dataset, several features were retyped. Simultaneously, the dataset was normalized. As a result, a final dataset with 102 rows and 17 numeric features was created and prepared for applying selected clustering algorithms.

After cleaning and normalizing the dataset, the following datasets were created in line with the planned main aim of this study:

- Dataset with all features, which represent students' performances in the programming course;
- Dataset MC, which contains only features representing students' performances in micro-learning units and quizzes;
- Dataset VPL, which contains only features representing students' performances in automatically evaluated programming code assignments;
- Dataset T1 has all the features from the first period of the term, including micro-learning units, quizzes, and assignments;
- Dataset T2 with all features from the second period of the term, including micro-learning units, quizzes, and assignments;
- Dataset T3 with all features from the third period of the term, including micro-learning units, quizzes, and assignments.

3.4. Modeling and Model Evaluation

Clustering analysis is an exploratory data mining activity that aims to group a set of objects so that the objects in the same group (a cluster) are more similar than those in other groups. The different clustering methods include density-based, distribution-based, grid-based, connectivity-based, and partitioning clustering. Each type of clustering method has its own strengths and limitations, and the choice of method depends on the study's specific data analysis needs. The most frequently used unsupervised machine learning methods in learning analytics are K-Means, Agglomerative Clustering, DBCAN, and K-Modes.

Involving the low-code machine learning library Pycaret simplified the modeling phase of the CRISP-DM methodology. It automates the repeating application of clustering methods, including K-Means, Affinity Propagation, Mean Shift Clustering, Spectral Clustering, Agglomerative Clustering, Density-Based Spatial Clustering, OPTICS Clustering, Birch Clustering, and K-Modes. Calling individual APIs includes hyperparameter tuning of the selected clustering method and repeating initialization. As a result, the library visualized several characteristics of the found clusters, as well as summarized their performance metrics, considering the unique features of the selected method.

The models were evaluated using the most common performance metrics, such as the Silhouette Coefficient, Davies–Bouldin Index, and Calinski–Harabasz Index. All three provide intra-cluster characteristics of the clusters. In other words, the score is based on the cluster itself and not on external knowledge such as labels. The Calinski–Harabasz index compares the variance between clusters to the variance within each cluster. The Davies–Bouldin index is a validation metric often used to evaluate the optimal number of clusters. It is defined as a ratio between the cluster scatter and the cluster's separation, and a lower value will mean that the clustering is better.

Whereas the Elbow method assumes that the distortion/inertia decreases linearly, the appropriate number of clusters can be found visually, locating the kink or elbow in the plot. However, the Elbow method, in selecting the number of clusters, may work correctly because the error function is monotonically decreasing for all initially selected centroids.

3.4.1. Clustering Analysis Using All Features

The first analyzed dataset contained features from both groups of grades. Table 3 summarizes the results of the application of several clustering methods. Even though the Elbow method suggested five initial clusters for the K-Means method (Figure 7), Table 3 shows that the K-Means model with $k = 2$ was the best after hyperparameter tuning. Other methods, which do not require an initial estimation of the number of clusters, suggested more clusters. However, the intra-clustering characteristics were significantly worse.

Table 3. The application of different clustering algorithms and performance metrics for the case of the automatic programming code evaluation features and features related to each student's performance in micro-learning units with quizzes were considered together.

Algorithm	Clusters	Silhouette	Calinski–Harabasz	Davies–Bouldin
K-Means	2	0.5700	161.6648	0.6984
K-Means	3	0.4313	114.7803	1.2559
K-Means	4	0.3747	93.1811	1.3109
K-Means	5	0.3284	81.0733	1.4830
K-Means	6	0.3465	76.1701	1.2466
DBCAN	3	0.4583	94.3754	1.4969
OPTICS	4	−0.0615	17.0579	1.3876
Affinity Propagation	10	0.3662	57.6724	1.2593
Agglomerative Clustering	4	0.3688	85.9097	1.2241
Meanshift	8	0.3891	41.8296	0.7929
BIRCH	4	0.4499	83.2682	1.0465

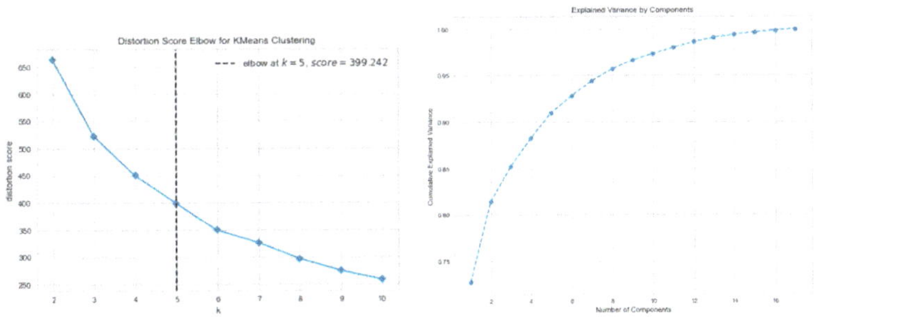

Elbow method　　　　Principal component analysis curve

Figure 7. Visualization of the Elbow method with the suggested number of clusters and principal component analysis curve, which denotes the number of components in relationship with the cumulative explained variance.

Similar results were obtained when principal component analysis (PCA) was applied before clustering. PCA tries to find the best possible subspace, which explains most of the variance in the data. In other words, PCA can avoid the curse of dimensionality. PCA suggested that five components explained more than 90% of the variance (Figure 7).

These findings also confirmed inter-cluster distance maps for different *k* values (Figure 8). Inter-cluster distance maps display the embedding of the cluster centers in a two-dimensional space with the distance to the other centers preserved. The clusters are sized according to the number of instances that belong to each center. Their size expresses the relative importance of clusters. It is worth noting that the clusters have comparable sizes in the case of six clusters, except one. The overlapping of the clusters means the case was wrongly assigned to the correct cluster. It also confirmed that not all students' performance attributes could be included in the clustering to find groups of students with similar behaviors.

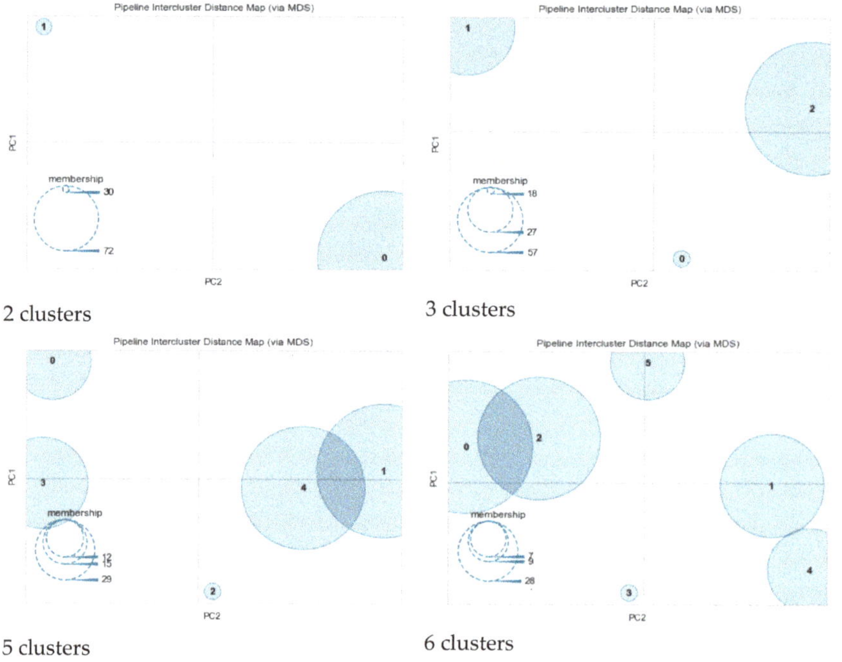

Figure 8. Visualization of inter-cluster distance maps for different *k* values for the clustering dataset with all features.

All metrics confirmed that the distribution of the cases in clusters was not optimal and could potentially be improved using subsets of original features. Therefore, the following sections will analyze other datasets created in the preprocessing phase.

3.4.2. Clustering Analysis of Micro-Learning Units and Quizzes

This dataset contained nine features, representing the grades from micro-learning units with quizzes. Each feature summarized the grades from one topic. Table 4 summarizes the results of K-Means clustering with performance metrics. Again, although the Elbow method suggested five clusters, the method performed the best in the case of two clusters. The value of intra-cluster characteristics continually declined for a higher number of clusters.

Table 4. The K-Means clustering performance metric values for different k values for features representing grades from the micro-learning units with quizzes in individual course topics.

Clusters	Silhouette	Calinski–Harabasz	Davies–Bouldin
2	0.7363	383.3815	0.4624
3	0.6896	286.8699	0.6816
4	0.5769	274.3522	0.8355
5	0.5605	262.4060	0.9686
6	0.5601	241.8890	0.9954

The higher value of the Silhouette coefficient indicates that the data points are well clustered, with clear separation between clusters and tight cohesion within each cluster. The comparison of the Silhouette plots of K-Means clustering for different k values showed uneven data distribution between clusters (Figure 9). Simultaneously, the negative value of the Silhouette coefficient confirmed the problems with correct data distribution and the possible misclassification of data to the correct clusters.

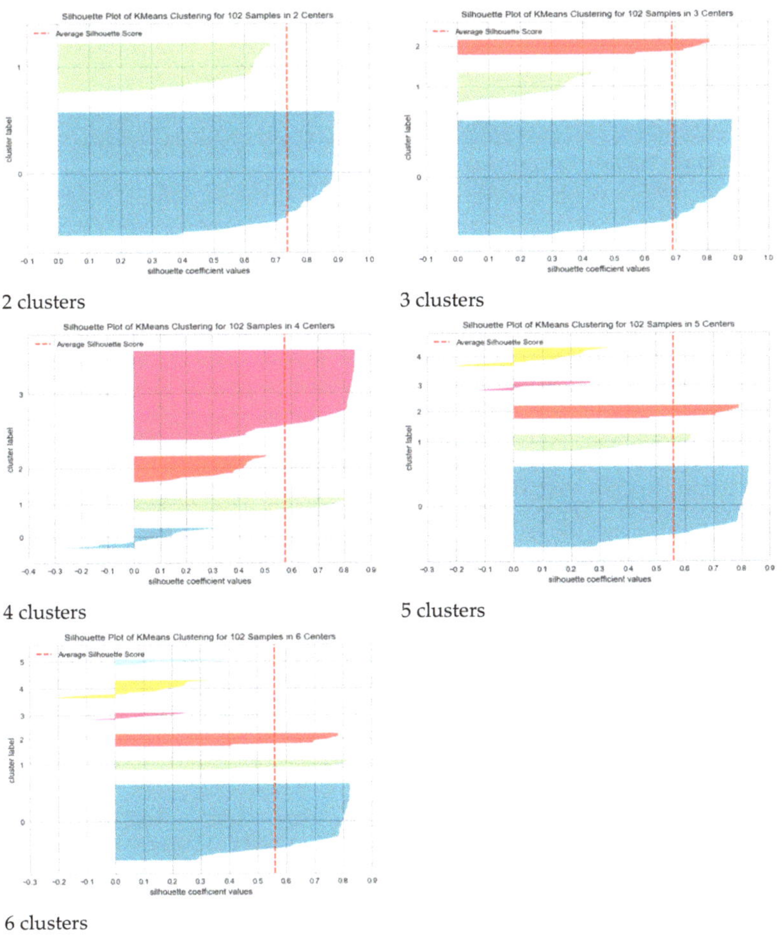

Figure 9. Silhouette plots of K-Means clustering for different values of k for features, which represent the grades from micro-learning units with quizzes.

3.4.3. Clustering Analysis of Automatically Evaluated Programming Code Assignments

The dataset analyzed in this section represents grades from automatically evaluated programming code assignments in individual topics. The dataset contained eight features. The Elbow method curve suggested setting the number of clusters to five. The results of K-Means clustering with performance metrics for different values of k for these features confirmed again that the students' performance results could create two clusters (Table 5). However, the highest Silhouette coefficient value was lower than in the dataset with micro-learning units.

Table 5. The K-Means clustering performance metric values for different k values for features, which represent grades from the programming assignments, automatically evaluated using the micro-learning platform.

Clusters	Silhouette	Calinski–Harabasz	Davies–Bouldin
2	0.5482	162.3892	0.7030
3	0.4275	121.7160	1.1015
4	0.3960	100.6032	1.0784
5	0.4148	91.6369	1.1800
6	0.4228	81.6195	1.1520

The comparison of Silhouette plots of K-Means clustering for different values of k showed the dominancy of one cluster (Figure 10) and uneven distribution of data between other small clusters.

The suggestion to divide the students' grades into two clusters aligns with the previous results. The reasoning for the existence of two clusters was also confirmed in teachers' feedback. They repeatedly noticed that involving a micro-learning platform did not cause a normal distribution of the grades between students. However, it highlighted the differences between students with a positive attitude toward learning tools and students who do not prefer intensive use of tools in learning programming.

More subjectively, students could be divided into more than two clusters, as suggested in Figure 11. A visualization of inter-cluster distance maps for different k values indicates that the dataset contains more groups of students with similar performances. This statement aligns with the survey results among the teachers who implemented the course. They observed a similar number of student groups, as identified using clustering methods. For example, the students who made continual efforts during the semester represented the first group. The second group of students focused mainly on automated programming assignments. The third group is characterized by losing interest as the assignments become harder. Finally, the last group or cluster represents students with minimal effort or self-motivation. However, these observations and subjective conclusions should be further analyzed and confirmed using other learning analytics methods or process mining. The reason is that inter-cluster distance maps use PCA with two components to map multidimensional space to 2D space. This means the clusters with visible intra-cluster distance can look different in higher-dimensional space.

The comparison of how students moved between clusters identified using micro-learning units with quizzes and programming assignments could add an interesting contribution to this discourse about students' behavior. The Rand Index computes a similarity measure between two clustering results by considering all pairs of samples and counting pairs assigned in the same or different clusters. The Adjusted Rand Index represents its normalized value. It has a value close to 0.0 for random labeling independently of the number of clusters and samples and exactly 1.0 when two clustering results are identical. The order of the samples must be the same in both clustering results. The adjusted Rand Index was 0.574, indicating that several students moved between clusters. A contingency matrix can visualize this movement, which shows the relationship between labels. As a result, eight students moved from the first cluster identified in micro-learning units to the second cluster identified in programming assignments. On the other hand, four students moved from the second cluster to the first one. Other students stayed in the same cluster (Figure 12).

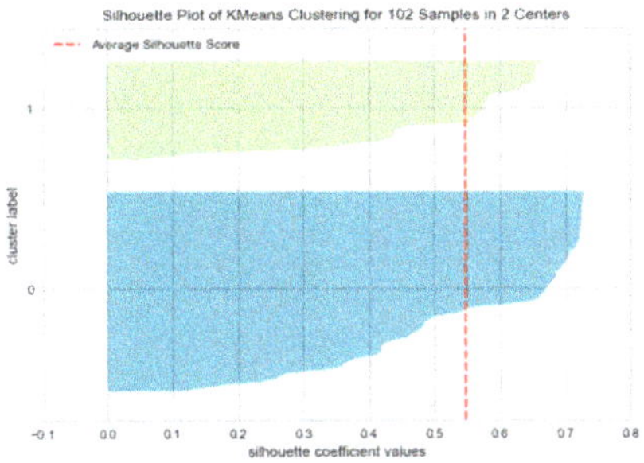

2 clusters

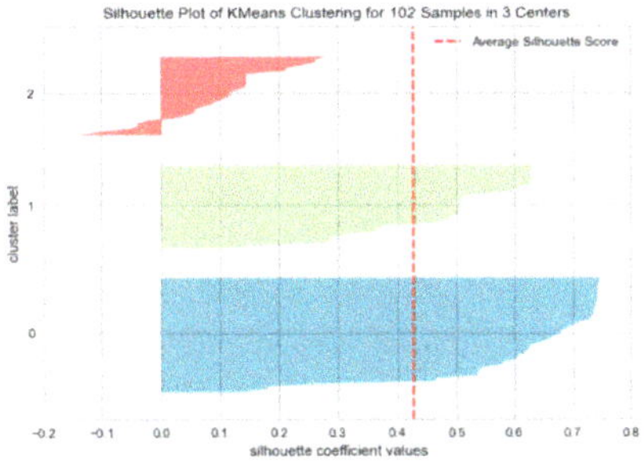

3 clusters

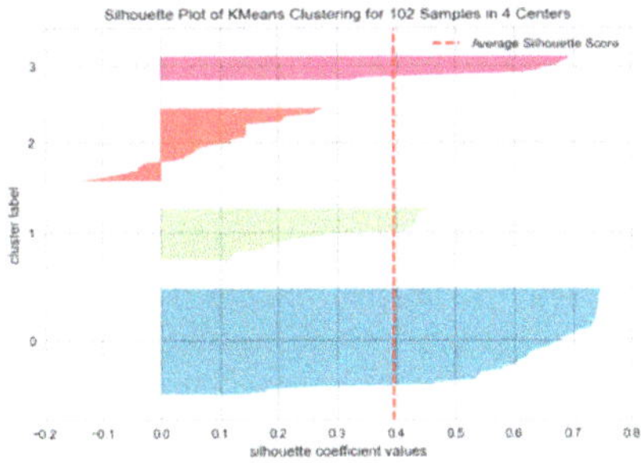

4 clusters

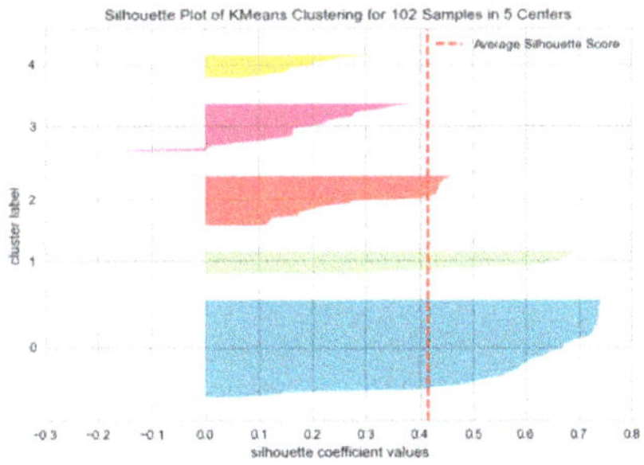

5 clusters

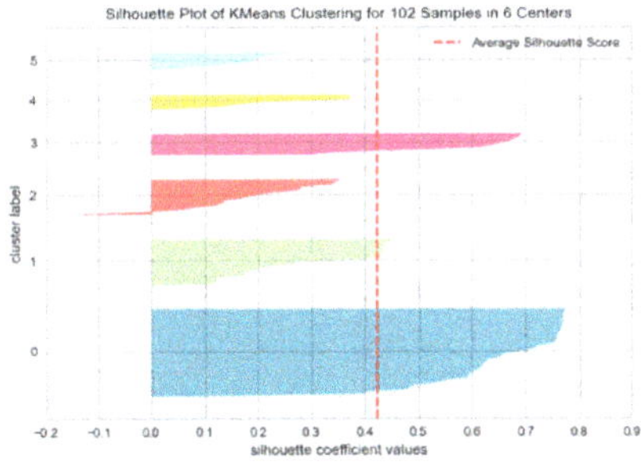

6 clusters

Figure 10. Silhouette plots of K-Means clustering for different *k* values for features representing the grades from automated programming assignments.

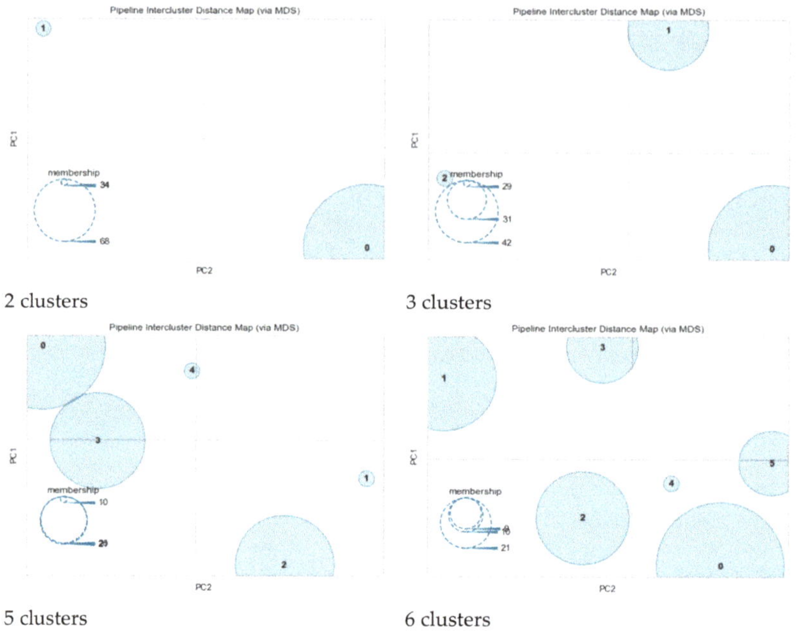

Figure 11. Visualization of inter-cluster distance maps for different k values for features representing grades from automatically evaluated programming assignments.

3.4.4. Clustering Analysis of Students' Academic Performance over Term

As was mentioned in the problem understanding phase, the students' activity levels and performances declined during the term, mainly after two control tests (Methods and Static class topics) (Figure 12). Therefore, the clustering analysis was applied to the following three datasets, which represent the students' grades obtained in different term periods:

- Dataset T1, with all features from the first term period, including micro-learning units, quizzes, and assignments;
- Dataset T2, with all features from the second term period, including micro-learning units, quizzes, and assignments;
- Dataset T3, with all features from the third term period, including micro-learning units, quizzes, and assignments.

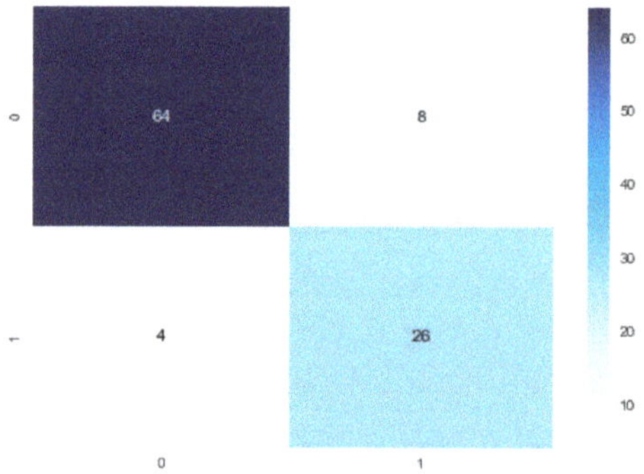

Figure 12. Contingency matrix describing the relationship between labels assigned to the data belonging to the clusters found for features that represent grades from micro-learning units with quizzes (0) and features that represent grades from programming assignments (1).

Tables 6–8 summarize the results of applying the K-Means method for different values of k on datasets from different parts of the term. Similarly, with the models from previous sections, the best performance metrics belong to $k = 2$. These findings confirmed the teachers' observation that students created two groups based on their attitudes toward using the micro-learning platform. They mainly considered the automated evaluation of programming assignments as a part of their learning process.

Table 6. The K-Means clustering performance metric values for different k values for features from the first term period.

Clusters	Silhouette	Calinski–Harabasz	Davies–Bouldin
2	0.5947	132.5591	0.7669
3	0.4654	109.8245	0.8889
4	0.4917	118.7943	0.8273
5	0.5219	128.2155	0.8026
6	0.5316	133.9519	0.7479

Table 7. The K-Means clustering performance metric values for different k values for features from the second term period.

Clusters	Silhouette	Calinski–Harabasz	Davies–Bouldin
2	0.6297	204.7539	0.5791
3	0.5643	167.2016	0.7829
4	0.5701	144.1638	0.9049
5	0.4677	133.5979	1.0457
6	0.4565	123.8955	1.0332

Table 8. The K-Means clustering performance metric values for different k values for features from the third term period.

Clusters	Silhouette	Calinski–Harabasz	Davies–Bouldin
2	0.5956	182.6024	0.6036
3	0.5677	150.9572	0.8597
4	0.6057	167.6004	0.7765
5	0.5775	177.7359	0.7942
6	0.5948	187.3231	0.7725

Moreover, considering the subjectively identified four groups of students with similar performances, the results shown in the tables could motivate us to examine not only the results of clustering with $k = 2$ but also those of clustering with $k = 4$ in more detail. Figure 13 shows the final distribution of data between identified clusters. The most significant changes are visible between the third period and the two previous ones in the case of two identified clusters. Similar movement between clusters is also visible in the case of four identified clusters, namely between cluster 2 and cluster 0 in the last observed period of the term.

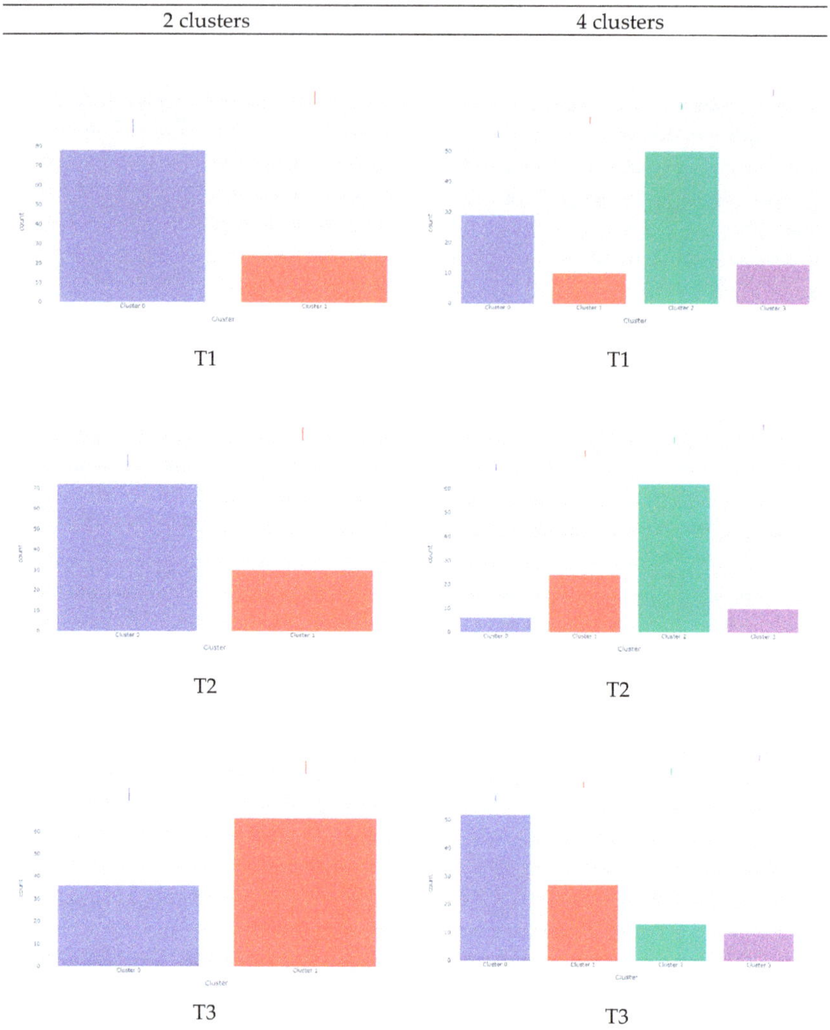

Figure 13. The data distribution between identified clusters for $k = 2$ and $k = 4$.

Table 9 shows the values of the Adjusted Rand Index, which expresses the similarity between the clustering results with $k = 2$. The most significant movement between clusters happened between the first and the third period.

Table 9. The Adjusted Rand Index values express the similarity between clustering results with $k = 2$ for different term periods.

T1 vs. T2	T2 vs. T3	T1 vs. T3
0.7660	0.7043	0.5113

Figure 14 shows that the composition of clusters between the first and second periods did not change significantly. Only six students moved to another cluster based on their results. However, the composition changed significantly between the second and third periods and the first and third periods.

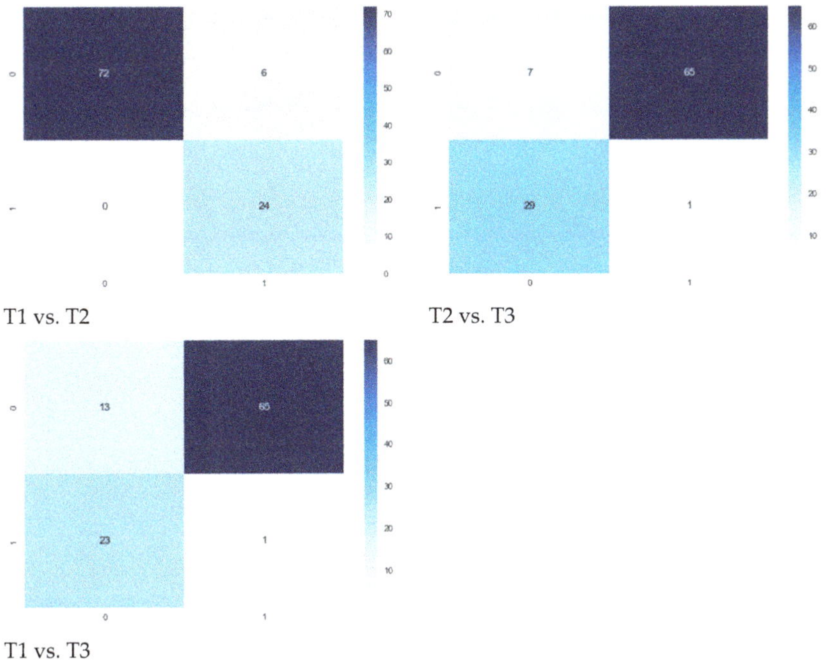

Figure 14. Contingency matrix describing the relationship between labels assigned to the data belonging to the clusters found for features, which represent grades from different term periods. The results are for the K-Means model with $k = 2$.

An analysis of similarities between the clustering results with $k = 4$ confirmed the transfer of the students between clusters after the second period (Table 10). The most significant changes in the composition of the clusters happened between the first and the third periods.

Table 10. The values of the Adjusted Rand Index, which expresses the similarity between the clustering results with $k = 4$ for different term periods.

T1 vs. T2	T2 vs. T3	T1 vs. T3
0.4258	0.6094	0.3720

Figure 15 shows changes in the compositions of clusters in absolute numbers. Again, the most significant change happened towards the third period. The composition of the clusters changed between the second and third periods and between the first and third periods. While most students did not change clusters based on their performances between the first two periods (47 students), they transferred to other clusters during the third period. These findings and possible reasons are analyzed in the Discussion section in more detail.

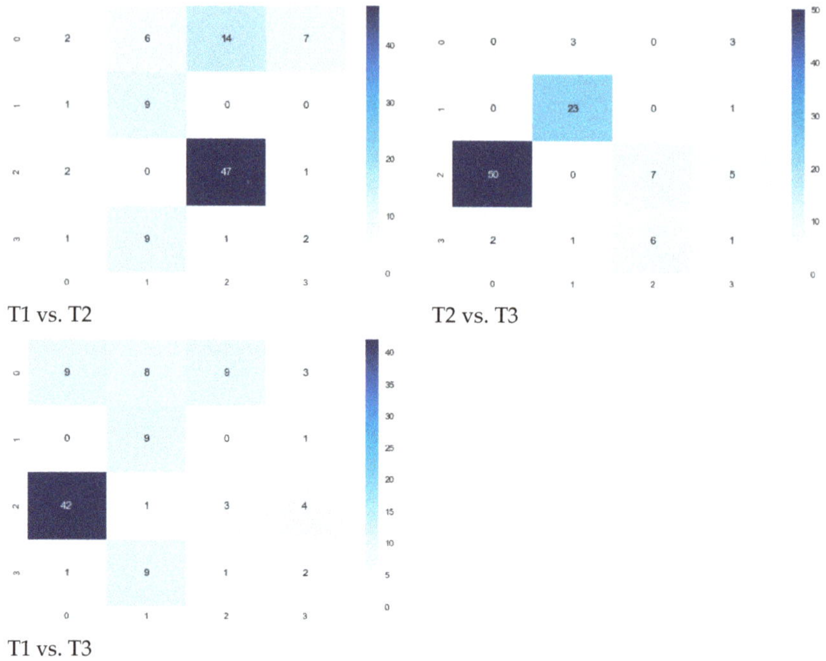

Figure 15. Contingency matrix describing the relationship between labels assigned to the data belonging to the clusters found for features, which represent grades from different term periods. The results are for the K-Means model with $k = 4$.

4. Discussion

This paper's main objective was to analyze students' academic performance data using unsupervised machine learning methods and find potentially useful patterns regarding students' performances in learning programming. The first research question (Q1) tried to find how the students' performances in micro-learning units with quizzes and programming assignments affect the number of identified meaningful clusters. Therefore, several datasets were prepared and analyzed for that purpose. The clustering analysis found two clusters in all cases. This finding confirmed teachers' assumptions and feedback that the micro-learning platform with automated evaluation of programming assignments highlighted the students' attitudes towards using learning tools during learning programming. While the activity of the first group of students on the platform led to higher grades, the activity of the students who did not prefer the intensive support provided by the platform led to lower grades. This observation is comparable with each student's performance before the platform was deployed at the university. It is not surprising and confirms the conclusions of other research publications, which dealt with the role of virtual learning platforms with automated programming code evaluation in the learning process, especially in learning programming. The positive effects of platforms like Priscilla on the learning process will manifest only when the students accept that these tools were developed to help them with the most challenging part of learning programming—associating theoretical knowledge with practical programming skills. While the developers and platforms seem to be better prepared, the teachers and, in particular, students should continually work on improving their learning and teaching methods in line with the main principles of self-regulation learning.

Furthermore, the cluster analysis indicates the existence of more than two clusters of students. Although the performance metrics did not strongly support this statement, further analysis of each student's characteristics and behavior could uncover potentially useful knowledge. For example, using association or sequence rule analysis or process mining methods could confirm the correctness of the teachers' observation that the students naturally group into four clusters based on their performances.

As mentioned earlier, the aim of this study was to identify the groups of students with similar performances in the Java programming micro-learning course and uncover possible changes in the number of identified groups in consecutive periods of the term and based on different types of micro-learning units. However, the methodology and findings can also be helpful outside this course. This course represents a whole family of similar courses developed on the micro-learning platform, which have the same structure and combine micro-learning units with quizzes and programming assignments for different programming languages. Moreover, they contain similar structures of thematic topics, which allow the students to practice and improve their programming skills. Therefore, the analysis of students' performance data from other courses could be extrapolated from these findings.

The second research question (Q2) tried to examine how identified groups of students with different academic performances change in composition over three periods of term. This study confirmed the natural assumption that delivering educational content in micro-learning units with automated programming assignments leads to creating groups of students with different academic performances. The compositions of these groups change based on activity type and over time. The transfer of students between the clusters identified in cluster analyses of datasets with different types of features indicates (Figure 12) that eight students decreased their performances and fell into another cluster. On the other hand, four students increased their performances in programming assignments. The transfer of the students between the found clusters in different periods of the term confirmed these findings and provided more interesting opportunities for further analysis. In the case of two clusters found, the composition of the clusters significantly changed before the third period of the term (Figure 14). Unfortunately, the analysis of input data showed that this movement was not in favor of higher activity of students in activities, particularly programming assignments.

On the contrary, each student's motivation to solve programming assignments declined during the term and after the midterm tests. This conclusion aligns with the teachers' feedback, as they mentioned that they had a problem keeping students interested after they became familiar with the course's difficulty and teaching style and achieved high enough grades to pass the course with minimal effort. These students' behaviors can also be observed from contingency matrices, which represent changes in the cluster composition in the case of the four clusters identified in different term periods (Figure 15). However, these conclusions were based on the analysis of the input data. Further detailed analysis of input data supplemented by interviews with students or surveys is needed before these findings can be generalized.

This study confirmed that data about the students' performance in the micro-learning platform have research potential in connection with learning analytics methods. Applying clustering analysis to student data obtained from a micro-learning environment and comparing changes in cluster distribution found in different periods of the term can be considered valuable. Clustering, as an exploratory method, provides a solid basis for further research, along with other supervised or semi-supervised methods.

On the other hand, the presented research has several limitations that should be considered. This study confirmed the conclusions of other research publications that the volume and veracity of data impact the research findings and their possible generalization. The micro-learning platform Priscilla can be considered a newly developed system that provides valuable reports for teachers and managers. Even though the implemented approach to storing data about students' activity and performance is based on recommendations and standards for educational research, it does not fully support learning analytics research and does not simplify the machine learning process. As a result, any attempt to implement a learning analytics module still requires a precise evaluation of the structure and semantics of stored data and additional preprocessing or computing.

The datasets used in this study did not require massive cleaning because they were directly received from the structured reports available on the platform. They contained aggregated information about the students' grades and other behavioral characteristics in the course. The number of rows corresponds to the number of unique students because of data aggregation. This limited input dataset size can have a negative impact on the results. However, the size of the dataset used in this study was comparable with the dataset often used in the learning analytics domain.

The collinearity or multicollinearity of the input attributes could potentially play an essential role in clustering analysis. Even though Figures 4 and 5 showed positive correlations between individual attributes, these attributes were not removed for several reasons. Firstly, this correlation was expected because students were motivated to complete all the quizzes as part of their learning process. Secondly, the K-Means clustering method was not sensitive to the correlation between input features. Finally, removing individual features could significantly impact the final distribution and number of clusters. Nevertheless, these correlation matrices could initialize a discussion about which correlations are natural or valuable from the learning process point of view.

The CRISP-DM methodology reliably guided the knowledge discovery process in this study. However, using the machine learning library Pycaret in the data preprocessing, modeling, and evaluation phases had several specifics, including missing detailed information about the implemented machine learning algorithms and their hyperparameters. Nevertheless, it speeded up the overall machine learning process after precise familiarization with the hyperparameters of this automated learning machine library. It helped to evaluate the potential of the examined datasets. Simultaneously, the application of this library allowed us to quickly examine and visualize the performance metrics of the most frequently used clustering methods.

As was mentioned in Section 3.4.1, the K-Means clustering method was finally selected as the best unsupervised machine learning method. This method has several limitations, which must be thoroughly considered. K- Means gives more weight to the bigger clusters. It assumes spherical shapes of clusters, with a radius equal to the distance between the centroid and the furthest data point. It does not work well when clusters are in different shapes, such as elliptical clusters. Moreover, if clusters overlap, K-Means does not have an intrinsic measure of uncertainty for the cases belonging to the overlapping regions. Therefore, other popular clustering methods were analyzed. However, they did not provide more precise results or have better performance metrics values.

The evaluation phase of the methodology uncovered several limitations of the proposed approach used for identifying students with similar performances. The selected approach for clustering provided limited added value to the more conservative statistical or visualization methods. From the results point of view, the selected features only partially contributed to answering the question about their potential to identify the groups of students with similar performances, effort made, or behaviors. The quality of the found clusters was evaluated using intra-cluster performance metrics. Although these metrics could be better, the found clusters can be considered useful. Higher values of these metrics could be achieved using additional features that characterize the students' performances in the course. This finding is in line with the results of similar research papers, which recommended combining different characteristics of the students, as well as their achievements and progress. Following the discussion about the relationship between the identified clusters and the heatmap in Figure 2, it can be concluded that the presented approach objectivated the subjective observations of the teachers about each student's performance in the course.

5. Conclusions

This study showed how the selected clustering techniques could be applied to identify groups of students with similar behaviors in the micro-learning environment. Available similar studies deal predominantly with data obtained from LMS or surveys. The study showed no difference between the application of clustering techniques on data from micro-learning, e-learning, or MOOC learning platforms. However, clustering analyses of the students' partial achievements in learning programming in a tailor-made environment are still rare. Therefore, the clustering analyses of students' performance data obtained from a micro-learning environment and comparison of changes in cluster composition found in different periods of the term can be considered novel enough. This study contributes to the discussion about the benefits of using a micro-learning platform for learning programming based on delivering educational content in the form of micro-learning units with automated programming assignments.

Consequently, these findings could initialize future educational model adjustments and changes in course structure and content sequencing. Simultaneously, the presented results can evoke new ideas, which can be considered in the future development of the platform Priscilla in the context of the planned implementation of the learning analytics module. These findings could decrease the dropout rate and achieve better academic performance.

Author Contributions: Conceptualization, V.P. and M.D.; methodology, V.P. and M.D.; software, V.P.; validation, V.P. and M.D.; formal analysis, V.P. and M.D.; investigation, V.P. and M.D.; resources, V.P.; data curation, V.P. and M.D.; writing—original draft preparation, V.P.; writing—review and editing, V.P. and M.D.; visualization, V.P. and M.D.; supervision, M.D.; project administration, M.D.; funding acquisition, M.D. All authors have read and agreed to the published version of the manuscript.

Funding: This work was supported in part by the Science Grant Agency of the Ministry of Education of the Slovak Republic under Grant VEGA 1/0490/22 and in part by the European Commission through the ERASMUS+ Program 2021 under Grant 2021-1-SK01-KA220-HED-000032095.

Institutional Review Board Statement: This study was conducted in accordance with the Declaration of Helsinki and approved by the Ethics Committee of Constantine the Philosopher University in Nitra (protocol code UKF-2023/1136-2:191013 and date of approval 2 October 2023) for studies involving humans.

Informed Consent Statement: Informed consent was obtained from all subjects involved in this study.

Data Availability Statement: The data presented in this study will be openly available in anonymized form at the author's GitHub repository: https://github.com/martin-ukf/mdpi-clustering-data.

Conflicts of Interest: The authors declare no conflicts of interest.

References

1. Gašević, D.; Merceron, A. *The Handbook of Learning Analytics*; Society for Learning Analytics Research: Edinburgh, UK, 2022; ISBN 9780995240834.
2. Smyrnova-Trybulska, E.; Kommers, P.; Drlik, M.; Skalka, J. (Eds.) *Microlearning: New Approaches to A More Effective Higher Education*; Springer: Cham, Switzerland, 2022. [CrossRef]
3. Shafiq, D.A.; Marjani, M.; Habeeb, R.A.A.; Asirvatham, D. Student Retention Using Educational Data Mining and Predictive Analytics: A Systematic Literature Review. *IEEE Access* **2022**, *10*, 72480–72503. [CrossRef]
4. Arizmendi, C.J.; Bernacki, M.L.; Raković, M.; Plumley, R.D.; Urban, C.J.; Panter, A.T.; Greene, J.A.; Gates, K.M. Predicting student outcomes using digital logs of learning behaviors: Review, current standards, and suggestions for future work. *Behav. Res. Methods* **2022**, *55*, 3026–3054. [CrossRef]
5. Aldowah, H.; Al-Samarraie, H.; Fauzy, W.M. Educational Data Mining and Learning Analytics for 21st century higher education: A Review and Synthesis. *Telemat. Inform.* **2019**, *37*, 13–49. [CrossRef]
6. Liu, S.; D'Aquin, M. Unsupervised learning for understanding student achievement in a distance learning setting. In Proceedings of the BT—2017 IEEE Global Engineering Education Conference, EDUCON 2017, Athens, Greece, 25–28 April 2017.
7. López, S.L.S.; Redondo, R.P.D.; Vilas, A.F. Discovering knowledge from student interactions: Clustering vs. classification. In Proceedings of the 5th International Conference on Technological Ecosystems for Enhancing Multiculturality (TEEM 2017), Cádiz, Spain, 18–20 October 2017; Association for Computing Machinery: New York, NY, USA, 2017. Article 40. pp. 1–8. [CrossRef]

8. Fida, S.; Fida, S.; Masood, N.; Masood, N.; Tariq, N.; Tariq, N.; Qayyum, F.; Qayyum, F. A Novel Hybrid Ensemble Clustering Technique for Student Performance Prediction. *JUCS—J. Univers. Comput. Sci.* **2022**, *28*, 777–798. [CrossRef]
9. Palani, K.; Stynes, P.; Pathak, P. Clustering Techniques to Identify Low-Engagement Student Levels. In Proceedings of the 13th International Conference on Computer Supported Education (CSEDU 2021), Virtual Event, 23–25 April 2021.
10. Hooshyar, D.; Yang, Y.; Pedaste, M.; Huang, Y.-M. Clustering Algorithms in an Educational Context: An Automatic Comparative Approach. *IEEE Access* **2020**, *8*, 146994–147014. [CrossRef]
11. Navarro, M.; Moreno-Ger, P. Comparison of Clustering Algorithms for Learning Analytics with Educational Datasets. *Int. J. Interact. Multimed. Artif. Intell.* **2018**, *5*, 9–16. [CrossRef]
12. Nunn, S.; Avella, J.T.; Kanai, T.; Kebritchi, M. Learning Analytics Methods, Benefits, and Challenges in Higher Education: A Systematic Literature Review. *Online Learn.* **2016**, *20*, 13–29. [CrossRef]
13. Joseph, B.; Abraham, S. Identifying slow learners in an e-learning environment using k-means clustering approach. *Knowl. Manag. E-Learn. Int. J.* **2023**, *15*, 539–553. [CrossRef]
14. Nafuri, A.F.M.; Sani, N.S.; Zainudin, N.F.A.; Rahman, A.H.A.; Aliff, M. Clustering Analysis for Classifying Student Academic Performance in Higher Education. *Appl. Sci.* **2022**, *12*, 9467. [CrossRef]
15. Gonzalez-Nucamendi, A.; Noguez, J.; Neri, L.; Robledo-Rella, V.; García-Castelán, R.M.G.; Escobar-Castillejos, D. Learning Analytics to Determine Profile Dimensions of Students Associated with Their Academic Performance. *Appl. Sci.* **2022**, *12*, 10560. [CrossRef]
16. Priyanka, P.G.; Mehrotra, D. Objective Assessment in Java Programming Language Using Rubrics. *J. Inf. Technol. Educ. Innov. Pr.* **2022**, *21*, 155–173. [CrossRef]
17. Bey, A.; Champagnat, R. Analyzing Student Programming Paths using Clustering and Process Mining. In Proceedings of the 14th International Conference on Computer Supported Education (CSEDU 2022), Virtual Event, 22–24 April 2022; Volume 2, pp. 76–84. [CrossRef]
18. Bey, A.; Perez-Sanagustin, M.; Broisin, J. Unsupervised Automatic Detection of Learners' Programming Behavior. In *Transforming Learning with Meaningful Technologies*; EC-TEL 2019, Lecture Notes in Computer Science; Springer: Berlin/Heidelberg, Germany, 2019; Volume 11722, pp. 69–82. [CrossRef]
19. López-Pernas, S.; Saqr, M.; Viberg, O. Putting It All Together: Combining Learning Analytics Methods and Data Sources to Understand Students' Approaches to Learning Programming. *Sustainability* **2021**, *13*, 4825. [CrossRef]
20. Kovari, A.; Katona, J. Effect of software development course on programming self-efficacy. *Educ. Inf. Technol.* **2023**, *28*, 10937–10963. [CrossRef]
21. Calderon-Valenzuela, J.; Payihuanca-Mamani, K.; Bedregal-Alpaca, N. Educational Data Mining to Identify the Patterns of Use made by the University Professors of the Moodle Platform. *Int. J. Adv. Comput. Sci. Appl.* **2022**, *13*, 321–328. [CrossRef]
22. Kabathova, J.; Drlik, M. Towards Predicting Student's Dropout in University Courses Using Different Machine Learning Techniques. *Appl. Sci.* **2021**, *11*, 3130. [CrossRef]
23. Ali, M. An Open Source, Low-Code Machine Learning Library in Python 2020. Available online: https://www.pycaret.org (accessed on 15 January 2024).

applied sciences

MDPI

Article

Tayseer: A Novel AI-Powered Arabic Chatbot Framework for Technical and Vocational Student Helpdesk Services and Enhancing Student Interactions

Abeer Alabbas [1,*] and Khalid Alomar [2]

1 Faculty of Computer and Information Technology, Technical and Vocational Training Corporation, Najran 66253, Saudi Arabia

2 Faculty of Computing and Information Technology, King Abdulaziz University, Jeddah 21589, Saudi Arabia; khalomar@kau.edu.sa

* Correspondence: aalabbas3@tvtc.gov.sa

Abstract: The rise of conversational agents (CAs) like chatbots in education has increased the demand for advisory services. However, student–college admission interactions remain manual and burdensome for staff. Leveraging CAs could streamline the admission process, providing efficient advisory support. Moreover, limited research has explored the role of Arabic chatbots in education. This study introduces Tayseer, an Arabic AI-powered web chatbot that enables instant access to college information and communication between students and colleges. This study aims to improve the abilities of chatbots by integrating features into one model, including responding with audiovisuals, various interaction modes (menu, text, or both), and collecting survey responses. Tayseer uses deep learning models within the RASA framework, incorporating a customized Arabic natural language processing pipeline for intent classification, entity extraction, and response retrieval. Tayseer was deployed at the Technical College for Girls in Najran (TCGN). Over 200 students used Tayseer during the first semester, demonstrating its efficiency in streamlining the advisory process. It identified over 50 question types from inputs with a 90% precision in intent and entity predictions. A comprehensive evaluation illuminated Tayseer's proficiency as well as areas requiring improvement. This study developed an advanced CA to enhance student experiences and satisfaction while establishing best practices for education chatbot interfaces by outlining steps to build an AI-powered chatbot from scratch using techniques adaptable to any language.

Keywords: Arabic chatbot; RASA; transformer model; DIET; vocational education

1. Introduction

Computer programs utilizing NLP have emerged as pivotal tools for simulating human-like conversations in the evolving landscape of technology, chatbots, and virtual assistants [1]. These AI-driven entities have found widespread application in automating customer support and services [1]. The integration of artificial intelligence (AI) across various computing and technological domains [2] notably includes the field of NLP [3]. This branch of AI facilitates human–computer interaction and communication through natural language, enabling chatbots to converse with humans or other bots via text or speech [4].

The utility of chatbots extends across diverse sectors, including e-commerce [5] and education [6–10], with notable applications in customer service [5,11,12] and hospital patient counseling [13–15]. Institutions increasingly rely on corporate websites and social media for student services in the educational domain by utilizing web-based response systems [4]. However, these methods are resource-intensive and time-consuming, leading to delays in response times [5]. This issue has been further compounded by the shift towards online education during the pandemic, intensifying the demand for efficient online assistance and round-the-clock service availability [16]. The growing reliance on intelligent agents, such as chatbots, in daily life makes their integration into social media and instant messaging ecosystems increasingly commonplace and rational [17].

Conversational agents have garnered considerable interest in the education sector because of their capacity to fulfill several educational functions, such as acting as instructors, coaches, and learning companions [18]. These agents possess the ability to analyze and react to human language. These have been incorporated into educational computer systems to facilitate dynamic learning interactions with students [19,20]. Implementing AI-powered chatbots in educational settings can significantly reduce administrative burden and enhance the student experience, potentially boosting student retention rates and satisfaction [6].

This study proposes the Tayseer architecture to explore this field by utilizing advanced technological innovations such as NLP, dialogue systems, and sentiment analysis, all finely tuned for the Arabic language. This study introduces a dual-mode conversational architecture that synergizes NLP capabilities with different interaction methods (write-based, click-based, or both), specifically catering to technical students' needs. Employing the RASA framework, the system adeptly handles open-ended dialogues, whereas the menu option paradigm offers structured guidance for various tasks. This dual approach aims to provide flexible self-service channels that accommodate user preferences, regardless of whether they favor structured or unstructured interactions.

Developing effective chatbots requires a solid understanding of deep learning (DL), a subset of ML within AI [17]. DL, which is often associated with deep artificial neural networks, involves sophisticated algorithms that model data through nonlinear function transformations across multiple layers [21]. Deep learning chatbots are primarily categorized as retrieval-based and generative bots. Retrieval-based bots leverage classification models to discern user intent and retrieve suitable responses from a database, offering greater flexibility than rule-based systems. In contrast, generative bots generate responses based on both current and previous user interactions, displaying more human-like characteristics but often struggling with grammatical accuracy and consistency [22–24].

Focusing on the RASA framework, a key component in chatbot development comprises natural language understanding (NLU) and dialogue management elements. RASA NLU enables customizable NLP through various prebuilt components, including the Dual Intent and Entity Transformer (DIET) [7]. DIET, a transformer-based model, efficiently categorizes intents and extracts entities, which is compatible with pre-trained embedding models such as BERT, GloVe, and ConveRT [8]. However, there is a notable research gap in the application of the RASA framework to the Arabic language, which is characterized by complex syntax and rich morphology, presenting unique challenges for NLP [25–27].

Tayseer was developed using the DIET architecture with attention-based mechanisms to address these challenges, as shown in Figure 1. The integration of pre-trained models within DIET enables the chatbot to effectively handle intent and entity recognition tasks [28–30].

Moreover, the system was designed to securely store individual academic backgrounds and dialogue histories in a structured SQL database, facilitating personalized and context-aware responses. An integrated feedback loop leveraging sentiment analysis of student conversational experiences was employed to incrementally improve the system's effectiveness over time.

This applied research seeks to contribute to the field by integrating data-driven machine-learning methods with custom knowledge representation within a modular Arabic chatbot framework. The framework was specifically designed to address admission inquiries and frequently asked questions (FAQs) in the context of higher education for technical students. The findings of this study are intended to inform the ongoing development of effective and user-friendly conversational interfaces for the Arabic-speaking student population in line with the identified key features and design principles.

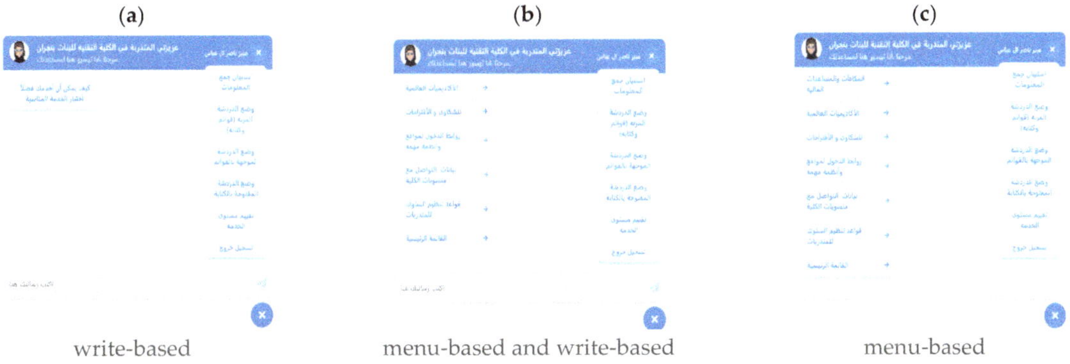

Figure 1. Interactions of different interface methods: (**a**) write-based; (**b**) menu-based and write-based; (**c**) menu-based.

The rest of the paper is structured as follows: Section 2 offers a comprehensive overview of pertinent prior research, providing a background for our study. Section 3 presents the methodology and the proposed model. Section 4 presents and discusses the results and highlights the key findings. Finally, Section 5 concludes the paper and offers recommendations for future research.

2. Related Work

2.1. Chatbot Overview

ELIZA [31] and PARRY [32] were among the first chatbots. These systems do not employ data-driven learning but rely on a combination of rules and patterns; they work exceptionally well when the scope of the discussion is defined, i.e., when the conversation is focused on a specific topic or task [4]. However, learning-based alternatives have been developed to circumvent some of the drawbacks of rule-based (handwritten rules) systems. These technologies enable chatbots to learn from massive volumes of available human-to-human conversations, for example, chat platforms, Twitter, or movie dialogues.

Learning-based approaches can be broadly characterized as retrieval-based or generative-based approaches. Retrieval-based models collect candidate answers from a prebuilt index, rank the candidates, and select the answer from the top-ranked options [22,33]. In contrast, generation-based approaches typically generate answers via deep learning (DL) techniques [17]. The basic premise of generative-based approaches is to create a new sentence word by word in response to a user's query [34].

Ritter et al. [35] employed phrase-based statistical machine translation (SMT) to convert a user query into a system response. They showed that the SMT method was superior to retrieval-based models in generating responses to a Twitter dataset [35,36]. Subsequently, it became apparent that the problem of generating responses differed from that of machine translation [37]. While words or phrases in the source and target sentences usually match well, a user's speech in a coherent response may not have any words or phrases in common with a coherent response [38].

It was later shown that another machine translation technique, sequence-to-sequence (seq2seq) learning, performed better at generating responses [38]. The seq2seq architecture includes an encoder model that compresses user input (query) and represents it as a vector [24]. A decoder model decodes the vector (representing the encoded input) and generates sentences word by word [24]. Typically, the encoder and decoder models are based on RNNs. Attention-based methods have been developed to allow the encoder to concentrate on certain portions of the encoded input, which is crucial for forecasting specific segments of the output during the decoding phase [39]. Bidirectional RNN (BRNN) models have also been used to improve the encoding of the context of the output sentence.

A new, simple network design based on the attention mechanism was later introduced, resulting in significantly improved response quality while requiring significantly less training time [24]. This type of design is called the transformer architecture [40]. The transformer model establishes global relationships between the input and output using only an attention mechanism [24]. This prevents recursion and allows for greater parallelization; however, the transformer model retains the encoder–decoder architecture. The development of transformer-based language models was a turning point in NLP [41]. These models, such as OpenAI GPT3, GPT4 [42], Bert [43], and T5 [44], have achieved state-of-the-art performance in a variety of NLP tasks [41].

2.2. RASA Framework for AI Chatbots

RASA has gained significant popularity as a platform for building AI chatbots across various disciplines [45]. RASA, an open-source framework based on machine learning, has proven its ability to reliably detect user intentions, extract entities, and effectively handle contextual nuances [45,46]. Researchers have examined the implementation of RASA in diverse domains, including academic research, tourism, mental health, and customer service [45,46].

The RASA framework has become a prominent tool in Arabic language processing. In addition, incorporating both morphological and syntactic disambiguation in a unified framework has shown highly favorable outcomes for languages such as Arabic [47]. An integrated strategy is essential for efficiently handling Arabic text and comprehending user input within the context of a chatbot. In addition, incorporating both morphological and syntactic disambiguation in a unified framework has shown highly favorable outcomes for languages such as Arabic [47]. An integrated method is essential for efficiently processing Arabic text and comprehending user inputs within the context of a chatbot.

To address these issues, it is essential to utilize the available tools and technology specifically designed for Arabic language processing. The MADA tool, also known as morphological analysis and disambiguation for Arabic, is utilized to perform part-of-speech tagging, stemming, and lemmatization of Arabic text [48]. Moreover, advancements in Arabic conversational AI, such as chatbots, have broadened the range of text-based conversational AI systems [49].

Multiple studies have focused on the research and implementation of RASA chatbots in Arabic and other languages. Alruily [26] evaluated the performance of the ArRASA framework, which is a deep learning-based Arabic NLU chatbot framework developed using the RASA framework. This research demonstrates the potential of the RASA framework to produce accurate and effective chatbots in Arabic language settings. A survey by Al-Hagbani and Khan [50] emphasized the difficulties associated with creating Arabic chatbots, mainly due to the intricacy of the language. Notwithstanding these difficulties, RASA has been effectively implemented in several fields, including university [51,52] and college inquiry chatbots [53]. The combination of RASA NLU and neural network techniques has also been investigated to create an advanced chatbot system [25,54].

The creation of an Arabic RASA chatbot necessitates profound comprehension of the distinctive linguistic attributes of the Arabic language and the employment of specialized tools and technology specifically designed for processing the Arabic language. Utilizing open-source frameworks such as RASA and including morphological and syntactic disambiguation greatly enhances the creation of a proficient Arabic RASA chatbot [26,53].

Despite these studies, there is still a lack of research that specifically addresses the gaps in using the RASA framework for AI chatbots in Arabic. The existing literature highlights the scarcity of Arabic chatbots and the need for more research on Arabic chatbot development, linguistic complexities, and content availability [55,56].

2.3. Chatbots for Arabic Settings

Arabic is a complex language; therefore, creating Arabic chatbots has proven to be difficult for researchers [57]. To date, only a few researchers have attempted to develop

Arabic chatbots. According to a survey [55], Arabic chatbots are still in their infancy; all of the current work is rule-based or retrieval-based, and the lack of accessible datasets impedes their growth [25].

However, some studies have been able to overcome Arabic language limitations by using translation tools. For example, Mozannar et al. [56] developed a question-answering method, and Biltawi et al. [58] developed an Arabic language model. The authors of [57] developed MidoBot, a deep learning-based generative chatbot in Arabic that uses a sequence-to-sequence model to generate new responses from a dataset. The performance of these developments and that of [59] demonstrates the potential of neural models for understanding the Arabic language.

In addition, Eljundi et al. [60] highlighted the potential of neural models for Arabic language comprehension. This prompted an investigation of neural solutions for open tasks in Arabic, such as empathic response production. Recent advances in transformer-based models have shown that language-specific BERT-based models are incredibly efficient in understanding language, provided they are pre-trained on an extensive corpus [59]. These models have established new benchmarks and have achieved state-of-the-art results for most NLP tasks [41].

Recent research has focused entirely on data-driven end-to-end systems that use neural generative models to generate the appropriate responses [41]. The most popular choices are seq2seq and transformer, and pre-trained models [33,37] and GPT3 [42] are two examples of encoder–decoder models. To explore these models for Arabic, large corpora of conversations with verified and cleaned annotations are needed [44] by creating contextualized word embeddings, transformers, and pre-trained models that can compensate for data deficiencies [33].

Their low computational cost makes them suitable for small datasets that Arabic can handle. Recent contextualized embeddings can be taught on unannotated materials, thereby improving various Arabic NLP tasks [41]. Multilingual BERT [61], AraT5 [62], AraGPT2 [63], ARBERT & MARBERT [64], AraELECTRA [41], and AraBERT [59] are contextualized embedding models that support Arabic.

Controllability is another aspect that can be used to construct sound Arabic conversational AI systems [33]. Hybrid models are an excellent alternative to this, as they integrate the capabilities of retrieval and generative models, where returned answers can be matched with generated answers to help retrieval models find a better answer [37].

2.4. Arabic Chatbots in University Settings

Chatbots are becoming increasingly prevalent across various industries, including product/service domains, healthcare, medicine, and education. Two key factors driving chatbot popularity are advancements in artificial intelligence (AI) and the growth of mobile messaging apps [65]. Recent research reviews indicate a burgeoning focus on exploring chatbot applications for educational purposes [33].

Based on an analysis of 80 studies on educational chatbots, the primary uses identified were administrative services and teaching assistants for delivering information and enhancing student learning [66]. The highlighted advantages of educational chatbots include 24/7 availability, adaptability to learner needs, and individualized support.

Jooka [67] is a bilingual chatbot for education that aims to improve the university admission process in both Arabic and English. The authors of [68] developed a social chatbot called Nabiha. Nabiha acts as an academic advisor who can support conversations with students of the Information Technology Department, engage with them, and respond to their queries about course options or other academic matters at King Saud University using the Saudi Arabic dialect. Nabiha was created on the Pandorabots platform by combining a pattern-matching approach with Artificial Intelligence Markup Language (AIML).

LANA -I [69] is an Arabic chatbot designed to help young people with autism spectrum disorder (ASD) in their educational endeavors. The system is rule-based and performs its functions of pattern matching and text similarity. Jordanian universities have proposed [2] the development of an intelligent Arabic chatbot system to overcome the outbreak of the pandemic COVID-19 and its global consequences. This technology mainly promotes spoken Arabic through the Jordanian dialect among students, especially at Jordan's Al-Zaytoonah Private University. Regrettably, it was not stated whether the students had tried this method.

Labeeb [70] is another chatbot that helps students answer questions about education or academic rules. This chatbot uses the Wikipedia API to retrieve the first paragraph of each requested query in the form of XML documents. However, the authors did not elaborate on the NLP models they used. SeerahBot is an Arabic chatbot specializing in the biography of Prophet Muhammad [8]. It utilizes a retrieval technique incorporating 200 questions and answers, employing machine learning to identify semantic question similarity, thereby optimizing the match between inputs and intentions. However, the study did not address the performance outcomes of this approach. Alazzam's [71] work will primarily concentrate on developing a chatbot tailored to comply with educational regulations in the UAE.

In the context of university admission helpdesks and frequently asked questions (FAQs), chatbots can be used to answer frequently asked questions and provide information about various university-related topics, such as courses, admissions, facilities, and impending events [72]. Powered by artificial intelligence, these avatars can comprehend and communicate using natural language processing techniques [72]. Using machine learning algorithms and accessing vast amounts of data, they can perpetually learn and evolve [72].

Al-Madi et al. [73] proposed an intelligent Arabic chatbot system to alleviate the burden of admission departments in a separate study. The study demonstrated that the chatbot system could provide efficient and gratifying responses to the majority of user queries, indicating its potential to streamline the admission process and enhance the user experience [27].

While there is a growing body of research on AI chatbots for university admission helpdesks and FAQ inquiries, there are still gaps in our comprehension of their efficacy, particularly in Arabic-speaking environments. Future research should concentrate on designing and evaluating chatbot systems that can effectively address the requirements of diverse student populations and enhance the admission process. Guidelines and ethical considerations regarding the use of chatbots in academic settings should be investigated further.

It is important to note that not all chatbots are powered by AI. Rule-based chatbots, for instance, adhere to pre-programmed logic and have limited intelligence [72]. Frequently, they are used to respond to inquiries regarding delivery and transportation [72]. To date, Arabic chatbots in education have been rule-based, which means that there is a need to develop intelligent conversational agents (CAs) capable of meeting academic and practitioner expectations. Moreover, further research is needed to enhance conversational quality using advanced natural language processing techniques [33]. Conversely, chatbots powered by AI can comprehend and communicate in human language using natural language processing techniques [74]. They provide sophisticated features and respond to various inquiries.

However, there are still gaps in the literature regarding the use of AI chatbots in Arabic-speaking settings for admission helpdesks and FAQ inquiries [71,73]. Fuad and Al-Yahya [30] conducted an assessment of recent developments in Arabic conversational AI and identified the need for further research on building human-like Arabic conversational AI systems. Similarly, [52] examined the advancements and challenges of intelligent Arabic chatbots and emphasized the need to surmount the linguistic complexities of Arabic.

Although chatbots have been studied for their potential educational benefits, there is a lack of solutions that cater to the specific requirements of technical and vocational students speaking Arabic. Our research aims to help fill some of these gaps by leveraging state-of-the-art NLP to enrich conversational abilities. Using previous works as a springboard, our framework introduces novel features that emphasize personalization and emotional intelligence.

3. Method

Our study presents the methodology behind the development of "Tayseer", an Arabic AI chatbot designed for educational support, including addressing inquiries on entrance admissions and frequently asked questions. This chatbot features both menu-based and write-based (RASA-based) interfaces to cater to diverse user preferences; it provides a structured navigation system through predefined options and a dynamic conversational interface for natural language interactions, as shown in Figure 1.

Menu-based interaction simplifies user navigation with predefined options, making it ideal for those preferring direct methods or finding natural language processing complex. This facilitates switching between RASA-based and menu-based interactions to suit user preferences, enhancing comfort and usability.

The write-based framework employs RASA, split into RASA NLU for understanding user inputs and RASA Core for dialogue flow management, chosen for its flexibility, natural language handling, and Arabic language support. Tayseer, designed to address queries about entrance and admission, among other frequent topics, incorporates multimodal interactions, feedback mechanisms, and sentiment analysis for comprehensive user engagement. It also supports diverse content types, such as images, links, and text, ensuring a versatile communication platform.

3.1. Menu-Based Interaction

The menu-based chatbot uses JavaScript and follows a hierarchical menu structure for user interaction.

A. Data Collection:

A comprehensive Arabic dataset encompassing intents, entities, and examples is currently being sought for menu-based and write-based modes. This initiative aims to equip the chatbot with the capability to accurately interpret inputs in Arabic. The desired dataset should cover a broad spectrum of college-related subjects, including but not limited to admissions, classes, scheduling, professors, amenities, and events. This dataset is derived from multiple authoritative sources, ensuring a rich and comprehensive compilation.

- Utilization of regulatory frameworks and procedural manuals to create a list of questions and answers for students by closely following the rules and steps found in manuals for technical and vocational training.
- Consultation with domain experts and faculty members to gather suggestions and recommendations.
- A pre-survey aimed at identifying the information priorities of prospective students was conducted among the current student population. The survey sought to understand the specific pieces of information that students consider imperative to know before and after college admission. A notable question within the survey was, "What are the crucial pieces of information you endeavor to obtain before and after being admitted to the college?" The survey garnered responses from 248 students, and a chart accompanying the results delineated the most common inquiries along with their respective percentages, as shown in Figure 2.

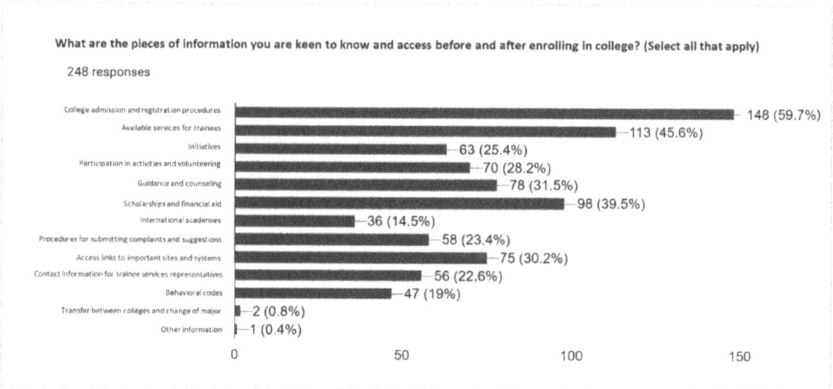

Figure 2. The most common inquiries.

B. Menu Structure:

The menu structure of the Taysser is represented as an object called "**menu**" in the code. Each key in the "**menu**" object represents a "**menu**" item, and the corresponding value is an object containing the "**text**" to display and the menu array representing the submenu items. For example, the top-level menu items include "القبول والتسجيل بالكلية" (College Admission and Registration), "خدمات المتدربات" (Trainee Services), "المبادرات" (Initiatives), and so on. Each menu item has its own submenu, allowing for hierarchical navigation.

C. Menu Rendering:

The "**menu**" options are rendered dynamically based on the "**menu**" object. The code iterates over the keys of the menu object and generates the corresponding UI elements to display the menu items. For instance, the code may generate buttons or list items for each menu item, with the appropriate text and event handlers to handle user interactions. The specific UI components and rendering process would depend on the chosen frontend framework or library.

D. User Interaction:

Tayseer handles user interactions by capturing user clicks or selections on the menu items. When a user selects a menu item, the chatbot navigates to the corresponding submenu or performs the associated action. For example, if the user selects "بالكلية القبول والتسجيل" (College Admission and Registration), the chatbot would display the submenu items related to college admission and registration, such as "تخصصات الكلية" (College Specializations), "شروط القبول في الكلية" (College Admission Requirements), and so on.

E. Menu Text and Links:

The text property of each menu item can contain plain text, URLs, or a combination of both. The chatbot renders the appropriate content based on the value of the text property. For example, let us consider the menu item "التقويم التدريبي" (Training Calendar). The text property of this menu item is set to "https://i.postimg.cc/MKdcZjnJ/1444.jpg" (accessed on 15 March 2023), representing an image URL. When the user selects this menu item, the chatbot will display the image associated with the training calendar.

On the other hand, the menu item "دليل المتدرب" (Trainee's Guide) has a text value of "https://tvtc.gov.sa/ar/MediaCenter/Elan/Documents/Trainee's-Guide.pdf" (accessed on 15 March 2023). This URL points to a document or a web page containing the trainee's guide. When the user selects this menu item, the chatbot provides a link or redirects the user to the specified URL, allowing them to access the trainee's guide.

In some cases, the text property contains plain text without any links. For instance, the menu item "معلومات أساسية للمستجدات" (Basic Information for New Students) has a text value that consists of several paragraphs of plain text. When this menu item is selected, the chatbot will display the text content directly to the user. Figure 3 illustrates an example of how the chatbot may render different types of menu text and links.

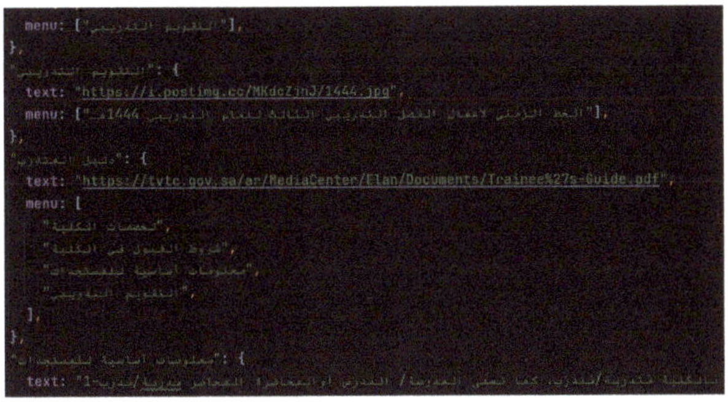

Figure 3. An example of how the chatbot renders different types of menu text and links.

In the example shown in Figure 3, the chatbot would handle each menu item differently based on the text property. For the "التقويم التدريبي" (training calendar) item, it would display the associated image. For the "دليل المتدرب" (trainee's guide) item, it would provide a link or redirect the user to the trainee's guide. And for the "معلومات أساسية للمستجدات" (basic information for new developments) item, it would display the plain text content directly. By leveraging the flexibility of the text property, Tayseer can accommodate various types of content, such as images, links, and plain text, providing a rich and informative user experience.

F. Error Handling and Fallback Mechanisms:

When the user selects an invalid or non-existent menu option, the chatbot provides appropriate error messages or fallback responses to guide the user back to valid menu options.

The chatbot's menu structure is represented as an object, and the menu options are dynamically rendered based on this structure. User interactions are handled by capturing user selections, navigating to the corresponding submenus, or performing associated actions. The menu-based chatbot offers a user-friendly and intuitive interface for users to interact with the system.

G. Evaluation and User Feedback:

To assess the effectiveness and usability of the Arabic AI chatbot, a comprehensive evaluation and user feedback process was integrated into Tayseer's front end using JavaScript, as shown in Figure 4. The evaluation consisted of a feedback questionnaire.

Figure 4. User feedback process.

3.2. Write-Based (RASA-Based)

The choice of RASA for the conversational framework was motivated by its open-source nature, flexibility, and comprehensive support for Arabic. The architecture of RASA, comprising RASA NLU for natural language understanding and RASA Core for managing conversation flows, is particularly adept at handling the complexities of Arabic, including dialectal variations and script nuances. This framework facilitated the development of a chatbot capable of engaging in meaningful Arabic conversations and addressing the specific needs of technical and vocational college inquiries. The key components of the RASA framework are as follows:

- RASA NLU: processes and interprets user inputs in Arabic, accommodating linguistic complexities.
- RASA Core: guides conversation flow using machine learning to provide contextually relevant responses.
- Domain: defines the chatbot's scope, including intents, entities, and actions tailored to college-related queries.
- Rules and Stories: establish conversational guidelines and example paths to train chatbots for diverse interactions.
- Actions: both predefined text messages and custom actions allow for dynamic responses and functionalities such as API calls or database queries.

A. Data Collection

The data collection process for the write-based approach followed the same methodology described in the "Data Collection" sub-section for the menu-based approach. The same sources, including existing college documentation, frequently asked questions (FAQs), and expert knowledge from college staff and administrators, were utilized to gather a diverse and comprehensive dataset. The collected data were preprocessed and prepared for training the write-based chatbot model, ensuring consistency and compatibility with the write-based approach.

B. Data Preprocessing:

- The collected data underwent preprocessing to prepare it for training the chatbot model.
- Text normalization techniques were applied to standardize the text's format, including converting characters to lowercase and removing irrelevant characters or information.
- Tokenization was performed to split the text into individual words or tokens, enabling accurate processing and analysis.
- Arabic-specific challenges, such as discretization and morphological variations, were handled using specialized libraries and techniques to ensure proper processing of the Arabic language.

C. Choosing the RASA NLU Pipeline

The fundamental role of RASA NLU is to interpret the data provided by the user to the chatbot, which includes identifying necessary intents and entities. Within the RASA Open Source, a series of components sequentially process incoming messages. This ordered execution of components, known as the processing pipeline, is established in the config.yml file [75]. The supervised embedding pipeline allows for training in any global language, as it is designed to learn from the ground up. This type of pipeline is similar to that of the Tayseer pipeline, as shown in Figure 5.

```yaml
recipe: default.v1
language: ar
pipeline:
  - name: WhitespaceTokenizer
  - name: RegexFeaturizer
  - name: LexicalSyntacticFeaturizer
  - name: CountVectorsFeaturizer
  - name: CountVectorsFeaturizer
    analyzer: char_wb
    min_ngram: 1
    max_ngram: 4
  - name: DIETClassifier
    epochs: 100
  - name: EntitySynonymMapper
  - name: ResponseSelector
    epochs: 100
policies:
  - name: RulePolicy
    core_fallback_threshold: 0.7
    core_fallback_action_name: action_empathy_response
    enable_fallback_prediction: True
assistant_id: 20230402-172228-some-winning
```

Figure 5. Tayseer supervised embedding pipeline.

The open-source RASA platform provides users with a powerful toolkit for building AI-driven agents that can chat, fine-tuned with a natural language processing pipeline. Figure 6 illustrates the agent training process executed in RASA.

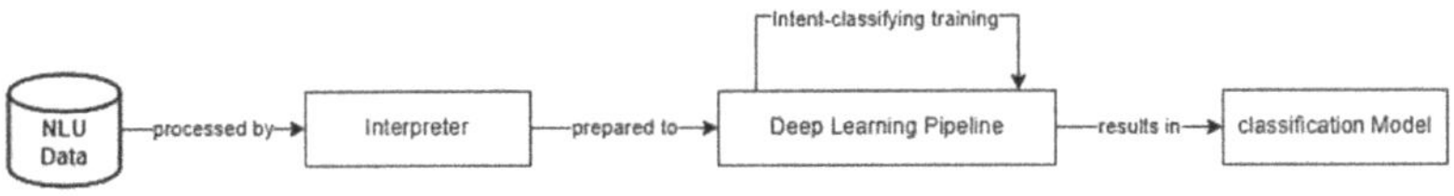

Figure 6. Illustration of the agent training process executed in RASA.

Within the scope of the Arabic language, RASA supports comprehensibility and conversational participation through a set of customizable components specified as follows. First, the preprocessing phase uses the **WhitespaceTokenizer** for simple text segmentation. This is complemented by the **RegexFeaturizer,** which extracts features based on the patterns. The **LexicalSyntacticFeaturizer** performs deep syntactic parsing to ease navigation through the complexities of Arabic syntax. It uses two **CountVectorsFeaturizer** components: the first to produce token-level analysis using the bag of words and the second to analyze the Arabic language's linguistic subtleties using character models ranging from 1- to 4-g. Intent and entity recognition are handled quite well by the **DIETClassifier,** which leverages transformer technology to produce iteratively updated context-sensitive predictions, as shown in Figure 7. Subsequent levels include the **EntitySynonymMapper**, which aligns terms of entities for overall consistency, and the **ResponseSelector**, which is adept at sourcing suitable replies to common queries.

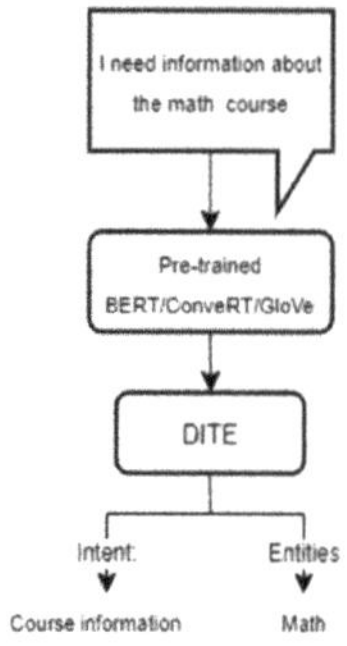

Figure 7. Overview of the DIET classifier architecture.

In instances where predictions may falter with low confidence, a **RulePolicy** steps in to maintain conversational coherence, and there is a unique assistant ID for every part of the pipeline to make it possible to trace any part. As indicated in Table 1, the RASA configuration is designed conscientiously, especially for Arabic conversational AI, with each unit serving a separate function. Designed with educational dialogues to be precisely tuned, the system is engineered to offer an unbeatable, knowledgeable virtual aid for technical and vocational students. In general, RASA's flexible pipeline combines state-of-the-art NLP methodologies that are sensitive to producing next-generation conversational AI abilities in Arabic.

D. Intent and Entity Recognition:

- The intents and entities relevant to the chatbot's domain were carefully defined based on the collected data and the chatbot's desired functionality.
- Intents represent the underlying purpose or goal of a user's query, while entities refer to specific pieces of information within the query.
- Examples of intents defined for the Arabic AI chatbot include "user_ask_about _admission", "user_ask_about_email", and "user_ask_about_major_change".
- Each intent was associated with a set of example user expressions that capture different ways users might express their queries so the chatbot could respond with greater specificity, as shown in Figure 8.

Table 1. Summary of the responsibility of each component in the chosen configuration.

Component	Functionality	
Tokenizer	WhitespaceTokenizer	Splits text into tokens using spaces for clear whitespace languages.
Featurizer	RegexFeaturizer	Uses regex to identify patterns and shapes in text for comprehension.
	LexicalSyntacticFeaturizer	Captures lexical and syntactic text features, like POS tags.
	CountVectorsFeaturizer	Converts text to numerical data using a bag-of-words model.
Classifier	DIETClassifier	Predicts intents and extracts entities simultaneously.
Mapper	EntitySynonymMapper	Maps synonyms to entities to recognize varied expressions as the same.
Selector	ResponseSelector	Picks appropriate responses from predefined options and trains for 100 epochs.

```yaml
- intent: bot_challenge
  examples: |
    - هل انت روبوت ؟
    - هل انت إنسان ؟
    - هل أتحدث إلى روبوت ؟
    - هل أتحدث إلى إنسان ؟

- intent: user_ask_about_admission
  examples: |
    - ماهي شروط القبول بدبلوم الكليات التقنية صباحي ؟
    - كيف أستطيع التسجيل في الكلية ؟
    - كيف اسجل في الكلية ؟
    - كيف ممكن أن الالتحاق بالكلية ؟
    - ماهي آلية القبول والتسجيل ؟

- intent: user_ask_about_email
  examples: |
    - ماهو ايميل المتدربه ؟
    - ماهو البريد الإلكتروني للمتدربة ؟
    - ماهو ايميل الطالبه ؟
    - ماهو البريد الإلكتروني للطالبه ؟
    - ماهو بريد المتدربه
```

Figure 8. A set of example user expressions.

- The intent is specified in the "nlu.yml" and "domain.yml" files. The "nlu.yml" file will encompass the intent data that will be utilized for subsequent training purposes. Table 2 displays the various contexts and quantities of intent within each. For this research, we created 75 Arabic intents. Figure 9 illustrates the distribution of Tayseer's intents by context.

Table 2. Description of intents.

Context	Description
Greeting	Respond to various forms of greetings from users.
College admission and registration	Handle queries related to the college admissions and registration process.
Trainee services	Address queries or requests related to services provided to trainees.
Initiatives	Responding to user inquiries about initiatives for trainee career guidance and college goal achievement.
Activity participation and volunteering	Responding to user inquiries or interests in extracurricular activities and volunteering opportunities.
Guidance and counseling	Address user queries related to advice, support, or information on personal, educational, or career matters.
Rewards and financial aid	Respond to queries about financial support.
Global academies	Inquiries on Adobe, Cisco, and Microsoft academies' programs and courses.
Complaints and suggestions	For handling user feedback and suggestions.
Access key site/system links	Assist users in finding and navigating to essential website or system links.
Faculty staff contact information	Assist users in obtaining faculty members' contact details.
Trainee conduct regulations	Information on trainee behavior rules.

- To manage diverse inquiries, a strategy is established in the "domain.yml" file under "utter", specifying conversational responses and their intents. The templates in this file address intent-specific queries. Tayseer encompasses more than 40 functions, with Table 3 providing samples from its diverse "utter" (user expressions or inputs).
- The training data was annotated with the corresponding intents and entities using the RASA NLU framework, which provides a structured format for training the chatbot's natural language understanding component.

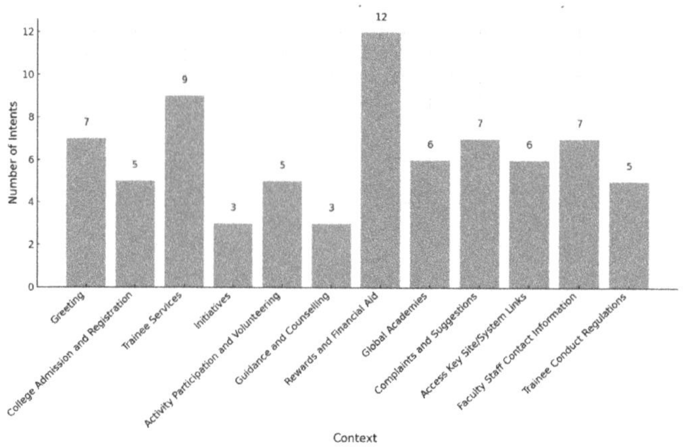

Figure 9. The distribution of chatbot intents by context.

Table 3. List of examples of utter.

#	Ultter Name
1	utter_greet
2	utter_goodbye
3	utter_bot_challenge
4	utter_user_ask_about_admission
5	utter_user_ask_about_major_change
6	utter_user_ask_about_transfer_college
7	utter_user_ask_about_handicapped
8	utter_user_ask_about_admission_comparison
9	utter_user_wants_to_check_marks
10	utter_user_ask_about_re-enrollment
11	utter_user_ask_about_conditions_changing_specialization
12	utter_user_ask_about_studying_in_TVTC
13	utter_user_ask_about_rewarding_for_Student_withdrawn_deferred
14	utter_user_ask_about_instances_of_nonpayment_rewarding
15	utter_user_ask_about_training_record
16	utter_user_college_disciplines
17	utter_user_Training_plans_for_specializations
18	utter_user_Training_Guide
19	utter_user_Trainee's_Guide
20	utter_user_Training_calendar
21	utter_user_Complaints_Suggestions
22	utter_user_activities_volunteering
23	utter_user_uniform
24	utter_user_functional_coordination

E. Dialogue Management:

- The dialogue management component of the chatbot was implemented using the RASA framework.
- The conversation flow was defined by mapping intents to appropriate responses and actions using rules and stories.
- Rules specify the triggering intents and the corresponding actions or responses to be executed.
- Stories represent more extended conversation flows and allow for more complex interactions, enabling the chatbot to handle multi-turn dialogues.

- The dialogue management component enables the chatbot to maintain context, handle user input, and generate appropriate responses based on the identified intents and entities, as shown in Figure 10.

```
rules:
 - rule: when user ask about admission
   steps:
     - intent: user_ask_about_admission
     - action: utter_user_ask_about_admission
```

Figure 10. Intent-based response generation.

F. Response Generation:
- Generating accurate and coherent responses in Arabic was a crucial aspect of the chatbot's development.
- For each intent, predefined response templates were created to provide relevant information to users.
- These templates were designed to include specific details based on the identified intent, such as college admission requirements, email formats, and transfer procedures.
- The response templates were written in Arabic, considering proper grammar, gender agreement, and idiomatic expressions to ensure natural and fluent responses.
- The response generation process involved retrieving the appropriate template based on the user's intent and filling in any necessary dynamic information.

G. Integration and Deployment:
- Chatbot development: In this study, we explored the implementation of a web-based chatbot, "Tayseer", designed to enhance educational support through 24/7 accessibility, self-service convenience, interactive learning, and privacy. The chatbot was developed using the RASA open-source platform, enabling connections with popular messaging platforms and allowing for website integration.
- Front-end development: The front-end user interface of "Tayseer" was built using Next.js, a robust framework that ensures a highly responsive and user-friendly experience. Next.js was chosen for its ability to create dynamic and interactive user interfaces, which is crucial for engaging users and providing a seamless chatbot experience.
- Back-end development: The back end of "Tayseer" was developed using the Nest.js framework. Nest.js was selected for its robustness and ease of integration with the front end. It facilitates the management of critical functions such as user authentication and feedback processing, ensuring smooth communication between the chatbot system's front-end and back-end components.
- Deployment environment setup: To deploy "Tayseer", we utilized Microsoft Azure as the cloud platform. Azure provides a scalable and reliable production environment for hosting the chatbot. The deployment process involved setting up the necessary virtual machine resources; configuring Nginx as a high-performance proxy server; and deploying the RASA NLU, RASA Core, and custom action components.
- Front-end deployment: The web-based front-end interface, built with Next.js, was deployed separately on Vercel. Vercel is a cloud platform that specializes in hosting and deploying front-end applications. By deploying the front end on Vercel, we ensure fast and efficient delivery of the user interface to end users.
- Back-end deployment: The Nest.js back-end code was uploaded to a DigitalOcean Droplet. DigitalOcean is a cloud infrastructure provider that offers scalable and

secure hosting solutions. Hosting the back-end on a DigitalOcean Droplet ensures reliable and efficient processing of user requests and interactions with the chatbot.

- System integration: The final deployment step involved tying the front-end and back-end systems together. This integration ensures the user interface (UI) seamlessly communicates with the back-end components, enabling a unified and effective chatbot platform. Figure 11 depicts the Tayseer system architecture, illustrating the integration of various components.
- User authentication: To enhance the user experience, a user authentication process utilizing an SQL database was developed. User-specific information was stored and retrieved to enable tailored responses based on individual user profiles. Custom actions were developed for efficient data retrieval from the database, facilitating personalized interactions. Figure 12 illustrates the user registration process and the flow of user authentication.

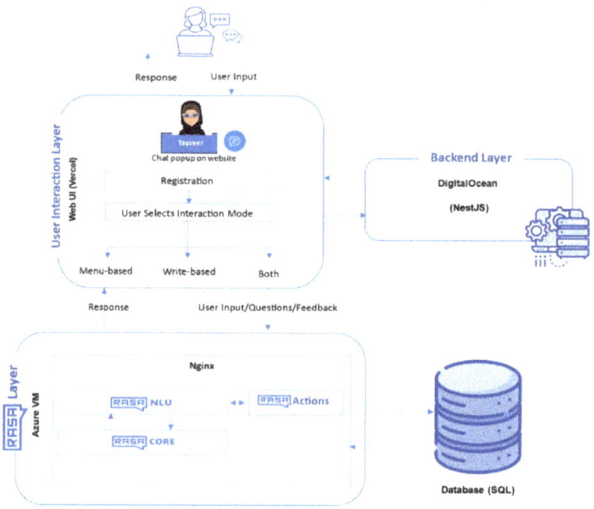

Figure 11. Tayseer system architecture.

Figure 12. Registering a new user profile.

4. Experiment

4.1. Training and Validation

In this work, we experimented with a dataset with 75 intents, inquiring about main issues such as college admission and registration, trainee services, initiatives, guidance and counseling, rewards and financial aid, faculty staff contact information, and trainee conduct regulations. The dataset consisted of approximately 265 examples.

If you want to ask about the conditions for changing specialization at the Technical College for Girls in Najran, you can send the message "ماهي شروط تغيير التخصص" ("What are the conditions for changing specialization?"). RASA NLU will then extract all the necessary information and identify the user's intent as "user_ask_about_conditions_changing_specialization". This enables Tayseer to provide an accurate and relevant response to the user's query.

We created a global 80% train and 20% test split from the nlu.yml file to ensure robustness and reliability. This split was performed to evaluate the model's performance on unseen data and assess its generalization ability. Additionally, we applied k-fold cross-validation to the training data to obtain a more comprehensive assessment of the model's performance.

The training data were divided into five equal-sized groups for k-fold cross-validation. In each iteration, one-fold was used as the testing set, while the remaining folds were used for training. This process was repeated until each of the five folds had been used for testing. Although computationally expensive, this cross-validation strategy provides an adequate evaluation of the model's performance, especially when the dataset is not excessively large.

To handle the scenario where the conversation can start with any intent, we added each story to the stories.yml file with a single intent mentioned in the domain.yml file. This ensured that Tayseer could respond appropriately, regardless of the user's initial intent. The pipeline and policies for the chatbot were defined in the config.yml file.

4.2. User Testing

A user testing phase assessed the chatbot's usability and effectiveness in a real-world scenario. College students and staff members were invited to interact with the chatbot and provide feedback. The objectives of user testing were to evaluate the chatbot's ability to understand user queries, provide relevant and accurate information, and enhance user satisfaction.

Users were given specific tasks to perform, such as asking about admission requirements, requesting information about email formats, or inquiring about major change procedures. Feedback was collected through surveys and interviews, capturing users' experiences, opinions, and suggestions for improvement.

4.3. Sentiment Analysis

After conducting user testing, sentiment analysis was performed on the collected user feedback to gain insights into users' sentiments and opinions regarding their interaction with the chatbot. The sentiment analysis process involved the following steps:

A. Data collection: The user feedback data, including survey responses and interview transcripts, were collected and preprocessed for analysis. The relevant feedback questions and responses were extracted and organized in a structured format.

B. Data preprocessing: The feedback data underwent preprocessing to prepare for sentiment analysis. This involved cleaning the text, removing any irrelevant characters or information, and handling Arabic-specific challenges such as discretization and morphological variations.

C. Sentiment lexicon: A sentiment lexicon specific to Arabic was utilized to assign sentiment scores to words and phrases. The lexicon contained a list of words along with their associated sentiment scores, ranging from positive to negative.

D. Sentiment scoring: Each feedback response was tokenized, and the sentiment score for each word was calculated using the sentiment lexicon. The overall sentiment score for each response was determined by aggregating the sentiment scores of individual words.

E. Sentiment classification: Based on the sentiment scores, each feedback response was classified as positive, negative, or neutral. Responses with scores above a certain threshold were considered positive, those below a threshold were considered harmful, and those within a specific range were considered neutral.
F. Analysis and visualization: The sentiment analysis results were analyzed to identify patterns, trends, and insights regarding users' sentiments towards the chatbot. The distribution of positive, negative, and neutral sentiments was visualized using charts and graphs to provide a clear overview of user opinions.

By conducting sentiment analysis on the user feedback, valuable insights were gained into the users' perceptions and experiences with the chatbot. The sentiment analysis results complemented the user testing findings, providing a comprehensive understanding of user satisfaction and areas for improvement.

5. Evaluation and Discussion

We employed a comprehensive evaluation methodology to ensure Tayseer's efficacy and efficiency. This methodology was designed to assess the chatbot's performance across multiple dimensions, ensuring that it understands user inputs accurately, responds promptly, and can scale to handle many queries. The conversational agent was evaluated on several key performance metrics to assess its natural language understanding capabilities and scalability.

5.1. Accuracy

To calculate the accuracy of our chatbot, we evaluated the two main components of every chatbot using multiple evaluation metrics:

- NLU model evaluation: The chatbot's natural language understanding (NLU) component was rigorously tested using a held-out test set. This test set comprised user inputs that the model had not seen previously during training, simulating real-world interactions. The model's predictions on this test set were then compared with the actual intents and entities to determine their accuracy.
- Dialogue model evaluation: The dialogue management component was evaluated using test stories. These stories represent potential user–bot interactions, and the chatbot's responses in these stories were compared to the expected actions to determine their accuracy in managing dialogues.

To calculate the accuracy of the NLU and dialogue models, we used multiple evaluation metrics: confusion matrix, precision, recall, F1 score, and support. By leveraging these metrics, we were able to gain a holistic understanding of our chatbot's performance, pinpointing areas of strength and potential improvement. Figure 13 shows the results of the intent confusion matrix. This table shows the performance of the classification model. It compares actual and predicted classifications. Our chatbot shows how often each action or intent was correctly predicted and how often an incorrect action or intent was predicted.

The histogram in Figure 14 shows the distribution of confidence scores for correct and incorrect intent predictions made using the conversational AI model. The vertical axis displays the confidence probability, ranging from 0 to 1. The horizontal axis represents the number of predictions within each confidence interval. Conversely, a notable number of incorrect predictions (red bar) align with a confidence level just below the midpoint of 0.5, indicating a higher likelihood of error as the model's confidence diminishes. This distribution illustrates the model's ability to align prediction confidence with accuracy, a desirable characteristic indicating effective model calibration. Figure 15 shows the results of the intent error.

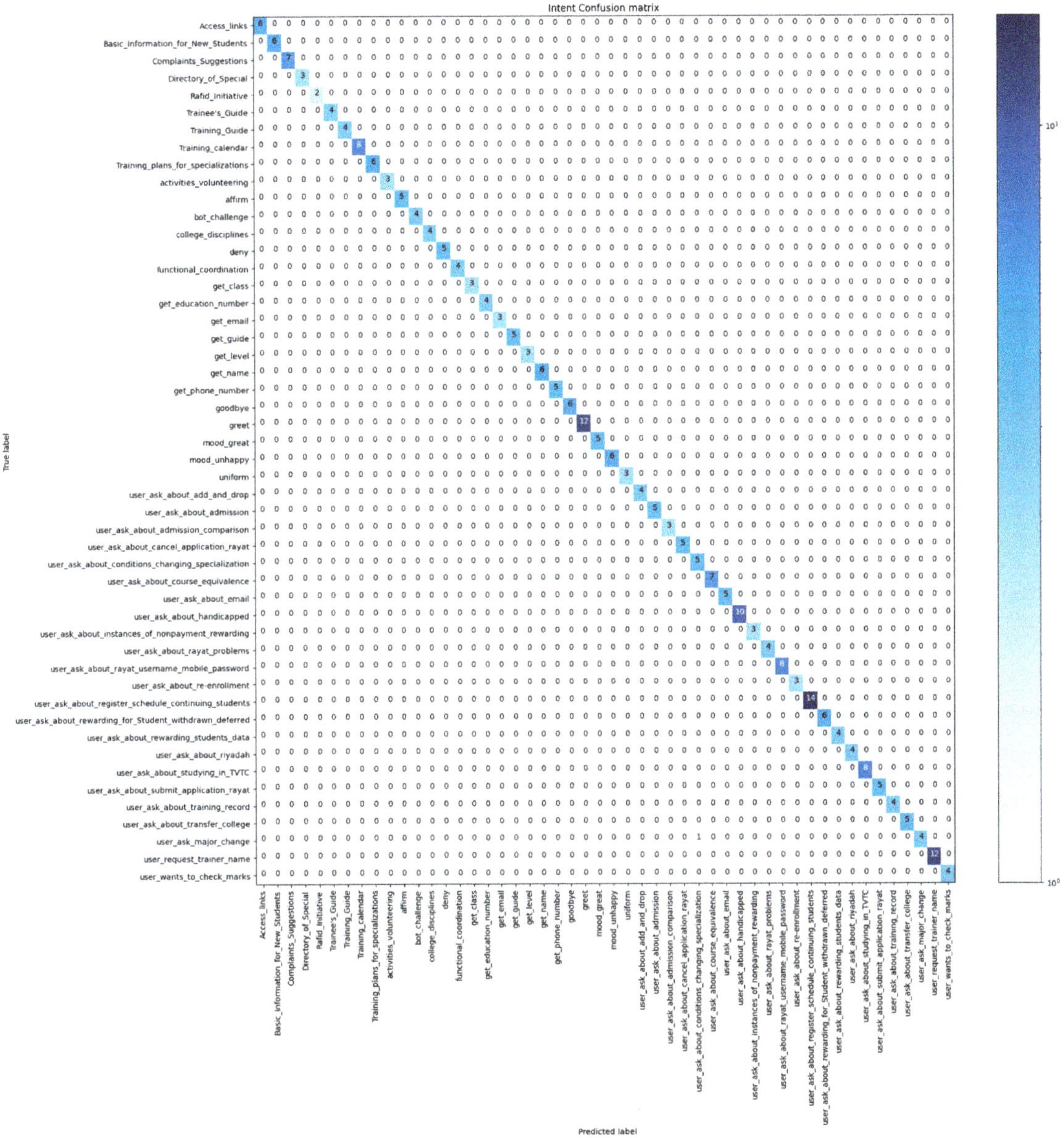

Figure 13. Intent confusion matrix.

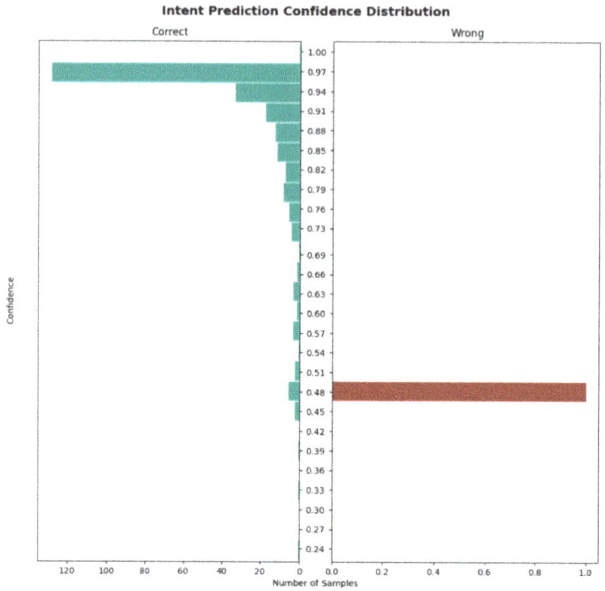

Figure 14. Intent prediction confidence distribution.

```
{
  "text": "ماهي شروط تغيير التخصص ؟",
  "intent": "user_ask_major_change",
  "intent_prediction": {
    "name": "user_ask_about_conditions_changing_specialization",
    "confidence": 0.48637133836746216
  }
```

Figure 15. Intent_Error.json.

The sentence "ماهي شروط تغيير التخصص" is meant to go on the user_ask_major_change intent but is predicted to go on the user_ask_about_conditions_changing_specialization intent. This will affect the confidence level of the intent prediction, resulting in responses that do not align with the inquirer's intended question. Table 4 shows the test result value of F1, recall, support, and precision score of the Tayseer intent.

The F1 score measures the test accuracy by combining precision and recall metrics, with maximum and minimum values of 1.0 and 0, respectively. Testing of the developed chatbot model revealed near-perfect F1 scores of 1.0 for most intent classes. However, two intents, user_ask_major_change and user_ask_about_conditions_changing_specialization, exhibited suboptimal F1 scores below 1.0. As illustrated in Table 4, the reduced F1 scores for these minority classes stem from sporadic errors in the model predictions. The confusion between these intents with similar semantic meanings contributed to diminished precision and recall. While most intents were classified with high accuracy, optimization of the model could potentially improve identification of the more easily confused intents, thereby increasing their F1 scores towards the optimal value of 1.0. Targeted augmentation of the training data for those specific intents may help enhance the precision and recall of the model for those classes. Overall, the F1 metric highlights strengths in broad classification accuracy while revealing specific areas for improvement in minority intents that could benefit from focused refinement.

Table 4. Test results.

Intent	Result			
	Precision	**Recall**	**F1 Score**	**Support**
Access_links	1.0	1.0	1.0	6
greet	1.0	1.0	1.0	12
user_ask_about_rewarding_for_Student_withdrawn_deferred	1.0	1.0	1.0	6
user_ask_about_re-enrollment	1.0	1.0	1.0	3
user_ask_about_admission	1.0	1.0	1.0	5
user_ask_about_rewarding_students_data	1.0	1.0	1.0	4
user_ask_major_change	1.0	0.8	0.888888888888889	5
user_ask_about_rayat_problems	1.0	1.0	1.0	4
mood_great	1.0	1.0	1.0	5
user_ask_about_admission_comparison	1.0	1.0	1.0	3
uniform	1.0	1.0	1.0	3
activities_volunteering	1.0	1.0	1.0	3
Complaints_Suggestions	1.0	1.0	1.0	7
bot_challenge	1.0	1.0	1.0	4
get_guide	1.0	1.0	1.0	5
user_ask_about_riyadah	1.0	1.0	1.0	4
mood_unhappy	1.0	1.0	1.0	6
user_ask_about_submit_application_rayat	1.0	1.0	1.0	5
get_phone_number	1.0	1.0	1.0	5
user_ask_about_rayat_username_mobile_password	1.0	1.0	1.0	8
get_name	1.0	1.0	1.0	6
deny	1.0	1.0	1.0	5
user_ask_about_email	1.0	1.0	1.0	5
user_ask_about_transfer_college	1.0	1.0	1.0	5
get_level"	1.0	1.0	1.0	3
functional_coordination	1.0	1.0	1.0	4
Training_plans_for_specializations	1.0	1.0	1.0	6
goodbye	1.0	1.0	1.0	6
college_disciplines	1.0	1.0	1.0	4
Rafid_Initiative	1.0	1.0	1.0	2
Training_Guide	1.0	1.0	1.0	4
user_ask_about_instances_of_nonpayment_rewarding	1.0	1.0	1.0	3
user_ask_about_cancel_application_rayat	1.0	1.0	1.0	5
get_education_number	1.0	1.0	1.0	4
user_ask_about_training_record	1.0	1.0	1.0	4
get_email	1.0	1.0	1.0	3
Trainee's_Guide	1.0	1.0	1.0	4
get_class	1.0	1.0	1.0	3
user_wants_to_check_marks	1.0	1.0	1.0	4
Directory_of_Special	1.0	1.0	1.0	3
user_ask_about_conditions_changing_specialization	0.8333333333333334	1.0	0.9090909090909091	5
Training_calendar	1.0	1.0	1.0	8
affirm	1.0	1.0	1.0	5
user_ask_about_handicapped	1.0	1.0	1.0	10
user_request_trainer_name	1.0	1.0	1.0	12
Basic_Information_for_New_Students	1.0	1.0	1.0	6
user_ask_about_register_schedule_continuing_students	1.0	1.0	1.0	14
user_ask_about_course_equivalence	1.0	1.0	1.0	7
user_ask_about_add_and_drop	1.0	1.0	1.0	4
user_ask_about_studying_in_TVTC	1.0	1.0	1.0	8

5.2. Response Time

Chatbot responsiveness is crucial for user satisfaction. To evaluate this, we logged the timestamp when a user sent a message and the bot responded. The difference between these timestamps provided the response time, allowing us to assess the chatbot's speed in generating and delivering responses. The response times for intents averaged under 1 s (967 ms), providing reasonably fast interactions, as shown in Figure 16. Any responses above 1 s, such as for user_ask_about_course_equivalence, present optimization opportunities.

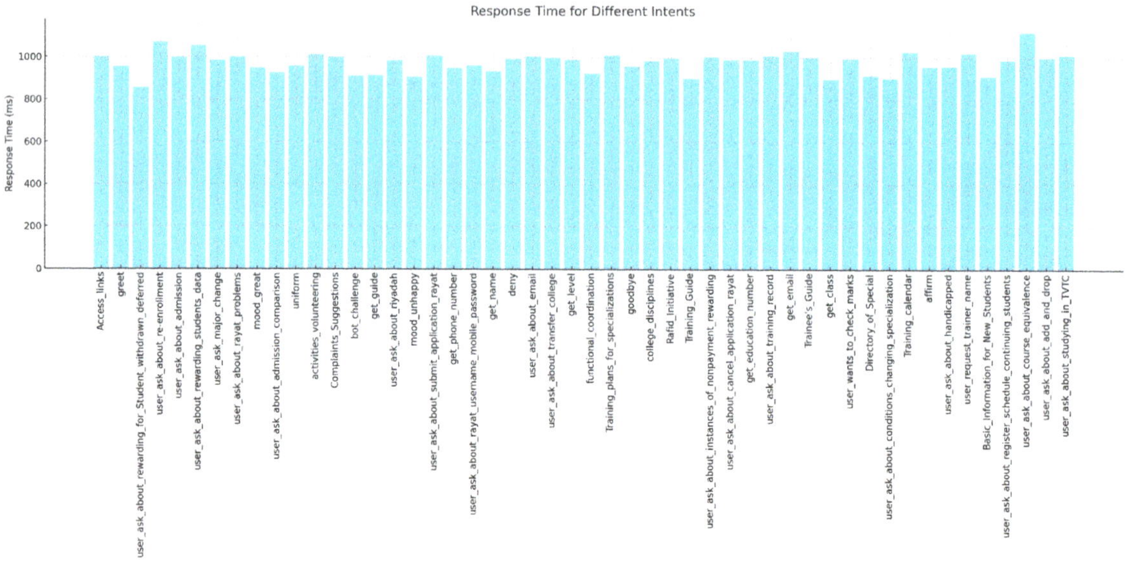

Figure 16. Response time for different intents.

5.3. Scalability

Scalability is a vital metric in chatbot solutions. It determines the chatbot's ability to handle many simultaneous user queries without performance degradation or increased response time. Scalability tests involved subjecting the chatbot to a high volume of user queries in a short timeframe and observing its performance. Testing up to six concurrent users showed minor increases in the average response times from 981 to 1023 ms, as shown in Figure 17. This linear scalability indicates the potential for maintaining performance with higher request volumes.

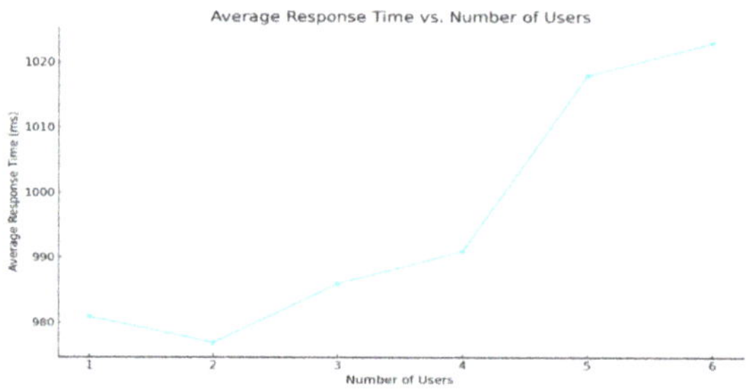

Figure 17. Average response time vs. number of users.

5.4. User Feedback and Sentiment Analysis

A. Collecting User Feedback

Our system integrates a feedback mechanism post-interaction to ascertain the chatbot's effectiveness and enhance user experience. At the conclusion of their chat sessions, users were invited to rate their experiences or offer comments and suggestions. This feedback was then cataloged within an SQL database for subsequent analysis and enhancement.

B. Sentiment Analysis Procedure

We employed sentiment analysis to extract insights into user satisfaction and discern general sentiments from the feedback. This involved utilizing a sentiment analysis model to classify feedback sentiment as positive, negative, or neutral, aiding in identifying areas for improvement and directing future chatbot enhancements.

C. Data Preprocessing Steps

Data preprocessing involved several critical steps, initiated by importing feedback data into a panda DataFrame for the initial analysis. Subsequently, entries marked as "None" in specific columns were excluded to ensure the integrity of the data. The process then focused on refining the dataset for sentiment analysis by filtering out columns, specifically isolating those related to the "rate", and textual responses ranging from "answer1" to "answer19".

D. Sentiment Analysis Model Development

Developing the sentiment analysis model involved creating a lexicon-based approach that utilized a compilation of positive and negative terms in both Arabic and English, with each term assigned a specific sentiment score. This model operated through a series of steps. Initially, feedback texts underwent tokenization, breaking down the content into individual terms for detailed analysis. Subsequently, each term received a sentiment score based on the lexicon, with the collective scores of the terms used to ascertain the overall sentiment of the feedback. Finally, aggregate scores were categorized into three sentiment classes.

- Positive (score > 0)
- Neutral (score = 0)
- Negative (score < 0)

E. Sentiment Analysis of User Feedback Results

The analysis of the sentiment results was structured to provide a comprehensive overview of the feedback sentiments, as shown in Figure 13. We began by calculating the frequency of positive, negative, and neutral sentiments for each item under review. Following this, an overall sentiment distribution analysis was carried out to understand the cumulative sentiment trends across all the received feedback. Specific to the nature of the questions, the analysis included distribution analyses for responses in the "rate" column, categorizing them into classifications such as "Good" and "Normal". The percentages of "Yes/No" responses were computed for the binary questions. In addition, for the open-ended questions, prevalent responses were identified to gauge common themes or opinions. The findings are summarized in Figures 18 and 19, showing the sentiment analysis results and overall sentiment distribution across the feedback dataset.

The evaluation of 210 users' feedback following their interaction with the chatbot prototype revealed favorable outcomes. Among the 1460 sentiment expressions identified in all the questions, 1206 were positive, 184 were negative, and 70 were neutral. These findings suggest that most participants had a favorable encounter when engaging with the chatbot.

The specific characteristic emphasized was the chatbot's ability to provide clear and easily comprehensible responses, which received a favorable sentiment score of 68.

- Prompt and efficient response times to inquiries were recorded (69 good attitudes);
- Seamless and effortless engagement in interactive chats with the chatbot resulted in a favorable sentiment score of 70.

Areas requiring enhancement encompassed the following: encountering specific challenges in engaging with the chatbot resulted in 48 instances of negative sentiment responses. Moreover, the sentiment on the extent of comprehensiveness in the FAQ subjects and alternatives varied, with only 36 instances of positive sentiment compared to 32 instances of negative sentiment. Overall, the sentiment and reaction towards the customized academic support chatbot concept and existing prototype were pretty favorable, with particular areas for enhancing the user experience.

Figure 18. Sentiment analysis results.

Overall Sentiment Distribution:

```
# Calculate the overall sentiment distribution across all questions
overall_sentiment_distribution = df_sentiment_results.sum()
overall_sentiment_distribution
```

```
negative     184
neutral       78
positive    1206
dtype: int64
```

Figure 19. Overall sentiment distribution.

To summarize the evaluation metrics and results, conversational AI underwent an evaluation to assess its proficiency in comprehending natural language by employing a sense test.

- The confusion matrix for intent classification and the distribution of prediction confidence demonstrated robust overall performance, albeit with scope for enhancement in differentiating between particular comparable intentions.
- The F1 score for the majority of intentions was high, approaching 1.0, indicating a robust level of precision and recall. However, two semantically similar intentions exhibited lower F1 scores, suggesting occasional confusion errors. Additional training data augmentation can enhance the optimization of these less common intentions.
- The average response time for user interactions was 967 ms, which is considered reasonably fast. However, certain specific intents, such as course equivalency, had longer response times, indicating areas that could be optimized.
- Scalability testing involved up to six concurrent users and demonstrated satisfactory performance, with only slight increments in response times from 981 to 1023 ms. This suggests an ability to manage large quantities without a decrease in performance.
- Positive user sentiment expressions were predominant, accounting for 83% of interactions. The main advantages were the respondents' clear and comprehensible answers as well as seamless interaction.

- Areas for improvement included difficulties faced with the chatbot and limitations in the range of FAQ topics addressed.
- The positive user response underscores the need for more chatbot improvements to assist students with typical academic inquiries and procedures.

Through the application of these metrics, a comprehensive evaluation of the chatbot's performance was achieved, illuminating its proficiency and areas that necessitate enhancement. This rigorous analysis not only underscores the chatbot's current capabilities but also guides future refinements to augment its effectiveness in user interaction.

6. Conclusions

To summarize, this study has highlighted the vast potential of AI-driven conversational agents in efficiently managing academic student inquiries through natural language processing of the Arabic language. With DL techniques made by the RASA framework and implemented on the web, over 75 data intents in the dataset were handled with care by Tayseer and used in multiple ways to be processed into an effective pipeline and a strategy for data pre-processing.

A thorough assessment suggests that the prototype chatbot created offers precise predictions of user intentions for most scenarios, achieving F1 scores close to 1.0 and delivering rapid responses in less than a second. Scalability tests demonstrate the capacity to sustain performance even when faced with increased levels of requests. Nevertheless, there are still opportunities for improvement, specifically for minority intentions that occasionally have misunderstandings caused by semantic resemblance. We can increase precision and recall by focusing on improving the training data for those particular intentions. There is also the possibility of broadening the chatbot's knowledge base to encompass additional academic question categories that interest pupils and enhancing the dialog system to be more context-aware.

Significantly, user sentiment analysis indicated primarily favorable perceptions of utilizing the chatbot for acquiring university admission and registration aid. Students expressed positive feedback regarding the chatbot's clear and user-friendly responses. However, certain difficulties were faced while dealing with increasingly complex questions, which emphasizes the necessity for continuous improvements in a particular area.

Chatbots with AI are meaningful tools for highly individualized and accessible academic support, leaving no doubt about discovering new ways to achieve a more elevated level of experience through effective management of internal functions. Future work will focus on refining chatbots using user feedback to better adapt to changing educational needs. Open-ended input from students could provide specific suggestions that could be used to determine new question subjects. Moreover, the knowledge base should be expanded, and advanced language models should be integrated. By implementing these enhancements, chatbots have significant potential to function as an exceptionally valuable self-service tool, effectively reducing the time consumption for both students and college administrators. This study highlights the importance of user-centric approaches in creating engaging chatbot experiences that align with human preferences.

Author Contributions: The presented idea was planned by A.A., who also wrote the manuscript with the help of K.A., who supervised the final revisions and contributed to preparing the original draft. All authors have read and agreed to the published version of the manuscript.

Funding: This research received no external funding.

Institutional Review Board Statement: This research project has received approval from the Institutional Review Board (IRB) at the Technical and Vocational Training Corporation of Saudi Arabia, thereby guaranteeing that the study follows ethical guidelines and conforms to the necessary requirements for conducting research with participants.

Informed Consent Statement: Not applicable.

Data Availability Statement: The data presented in this study are available on request from the corresponding author. The data are not publicly available due to privacy.

Conflicts of Interest: The authors declare no conflict of interest.

References

1. Shawar, B.A.; Atwell, E. Chatbots: Are they really useful? *LDV-Forum Z. Comput. Sprachtechnol.* **2007**, *22*, 29–49.
2. Al-Madi, N.A.; Maria, K.A.; Al-Madi, M.A.; Alia, M.A.; Maria, E.A. An Intelligent Arabic Chatbot System Proposed Framework. In Proceedings of the 2021 International Conference on Information Technology, ICIT 2021, Amman, Jordan, 14–15 July 2021; pp. 592–597. [CrossRef]
3. Russell, S.; Nilsson, P.; Nilsson, N.J. Artificial Intelligence: A Modern Approach. *Artif. Intell.* **1996**, *82*, 369–380.
4. Adamopoulou, E.; Moussiades, L. Chatbots: History, technology, and applications. *Mach. Learn. Appl.* **2020**, *2*, 100006. [CrossRef]
5. Skjuve, M.; Brandzaeg, P.B. Measuring user experience in chatbots: An approach to interpersonal communication competence. In *Internet Science*; Lecture Notes in Computer Science (including subseries Lecture Notes in Artificial Intelligence and Lecture Notes in Bioinformatics); Springer: Cham, Switzerland, 2019; pp. 113–120. [CrossRef]
6. Crisp, G.; Baker, V.L.; Griffin, K.A.; Lunsford, L.G.; Pifer, M.J. Mentoring Undergraduate Students. *ASHE High. Educ. Rep.* **2017**, *43*, 7–103. [CrossRef]
7. Wardhana, A.K.; Ferdiana, R.; Hidayah, I. Empathetic Chatbot Enhancement and Development: A Literature Review. In Proceedings of the AIMS 2021, International Conference on Artificial Intelligence and Mechatronics Systems, Jakarta, Indonesia, 28–30 April 2021; pp. 1–6. [CrossRef]
8. Yassin, S.M.; Khan, M.Z. SeerahBot: An Arabic Chatbot About Prophet's Biography SeerahBot: An Arabic Chatbot About Prophet's Biography. *Int. J. Innov. Res. Comput. Sci. Technol.* **2021**, *9*, 89–97. [CrossRef]
9. Moshuber, L. Grätzelbot: Using a Chatbot-Based Scavenger Hunt to Support Students' Transition to University. Master's Thesis, University of Vienna, Vienna, Austria, 2020.
10. Dibitonto, M.; Leszczynska, K.; Tazzi, F.; Medaglia, C.M. Chatbot in a campus environment: Design of LiSa, a virtual assistant to help students in their university life. In *Human-Computer Interaction: Interaction Technologies*; Lecture Notes in Computer Science (including subseries Lecture Notes in Artificial Intelligence and Lecture Notes in Bioinformatics); Springer International Publishing: Berlin/Heidelberg, Germany, 2018; pp. 103–116. [CrossRef]
11. Agarwal, A.; Maiya, S.; Aggarwal, S. Evaluating Empathetic Chatbots in Customer Service Settings. *arXiv* **2021**. [CrossRef]
12. Xu, A.; Liu, Z.; Guo, Y.; Sinha, V.; Akkiraju, R. A new chatbot for customer service on social media. In Proceedings of the Conference on Human Factors in Computing Systems, Denver, CO, USA, 6–11 May 2017; pp. 3506–3510. [CrossRef]
13. Fadhil, A.; AbuRa'Ed, A. Ollobot—Towards a text-based Arabic health conversational agent: Evaluation and results. In Proceedings of the International Conference Recent Advances in Natural Language Processing, RANLP, Varna, Bulgaria, 2–8 September 2019; pp. 295–303. [CrossRef]
14. Høiland, C.G.; Følstad, A.; Karahasanovic, A. Hi, Can I Help? Exploring How to Design a Mental Health Chatbot for Youths. *Hum. Technol.* **2020**, *16*, 139–169. [CrossRef]
15. Vaidyam, A.N.; Wisniewski, H.; Halamka, J.D.; Kashavan, M.S.; Torous, J.B.; Torous, J. Chatbots and Conversational Agents in Mental Health: A Review of the Psychiatric Landscape. *Can. J. Psychiatry* **2019**, *64*, 456–464. [CrossRef]
16. Von Wolff, R.M.; Hobert, S.; Schumann, M. How May I Help You?—State of the Art and Open Research Questions for Chatbots at the Digital Workplace. In Proceedings of the 52nd Hawaii International Conference on System Sciences, Maui, HI, USA, 8–11 January 2019; Volume 6, pp. 95–104.
17. Agarwal, R.; Wadhwa, M. Review of State-of-the-Art Design Techniques for Chatbots. *SN Comput. Sci.* **2020**, *1*, 246. [CrossRef]
18. Kuhail, M.A.; Alturki, N.; Alramlawi, S.; Alhejori, K. Interacting with Educational Chatbots: A Systematic Review. *Educ. Inf. Technol.* **2022**, *28*, 973–1018. [CrossRef]
19. Pérez-Marín, D. A Review of the Practical Applications of Pedagogic Conversational Agents to Be Used in School and University Classrooms. *Digital* **2021**, *1*, 18–33. [CrossRef]
20. Briel, A. Toward an Eclectic and Malleable Multiagent Educational Assistant. *Comput. Appl. Eng. Educ.* **2021**, *30*, 163–173. [CrossRef]
21. Gupta, M.; Verma, S.K.; Jain, P. Detailed Study of Deep Learning Models for Natural Language Processing. In Proceedings of the IEEE 2020 2nd International Conference on Advances in Computing, Communication Control and Networking, ICACCCN, Greater Noida, India, 18–19 December 2020; pp. 249–253. [CrossRef]
22. Luo, B.; Lau, R.Y.K.; Li, C.; Si, Y. A critical review of state-of-the-art chatbot designs and applications. *Wiley Interdiscip. Rev. Data Min. Knowl. Discov.* **2022**, *12*, e1434. [CrossRef]
23. Palasundram, K.; Sharef, N.M.; Nasharuddin, N.A.; Kasmiran, K.A.; Azman, A. Sequence to sequence model performance for education chatbot. *Int. J. Emerg. Technol. Learn.* **2019**, *14*, 56–68. [CrossRef]
24. Sojasingarayar, A. Seq2Seq AI Chatbot with Attention Mechanism. *arXiv* **2020**, arXiv:2006.02767.
25. Alkhurayyif, Y.; Sait, A.R.W. A comprehensive survey of techniques for developing an Arabic question answering system. *PeerJ Comput. Sci.* **2023**, *9*, e1413. [CrossRef] [PubMed]

26. Alruily, M. ArRASA: Channel Optimization for Deep Learning-Based Arabic NLU Chatbot Framework. *Electronics* **2022**, *11*, 3745. [CrossRef]
27. Ahmed, A.; Ali, N.; Alzubaidi, M.; Zaghouani, W.; Abd-alrazaq, A.; Househ, M. Arabic chatbot technologies: A scoping review. *Comput. Methods Programs Biomed. Update* **2022**, *2*, 100057. [CrossRef]
28. Vo, V.M.N.; Ngo, V.S. An Integrated DIET-BO Model for Intent Classification and Entity Extraction. *Int. J. Intell. Syst. Appl. Eng.* **2023**, *11*, 666–673.
29. Arevalillo-Herraez, M.; Arnau-Gonzalez, P.; Ramzan, N. On Adapting the DIET Architecture and the Rasa Conversational Toolkit for the Sentiment Analysis Task. *IEEE Access* **2022**, *10*, 107477–107487. [CrossRef]
30. Fuad, A.; Al-Yahya, M. Recent Developments in Arabic Conversational AI: A Literature Review. *IEEE Access* **2022**, *10*, 23842–23859. [CrossRef]
31. Weizenbaum, J. ELIZA—A Computer Program for the Study of Natural Language Communication Between Man and Machine. *Commun. ACM* **1983**, *26*, 23–28. [CrossRef]
32. Colby, K.M. Modeling a paranoid mind. *Behav. Brain Sci.* **1981**, *4*, 515–534. [CrossRef]
33. Rashkin, H.; Smith, E.M.; Li, M.; Boureau, Y.L. Towards empathetic open-domain conversation models: A new benchmark and dataset. In Proceedings of the ACL 2019—57th Annual Meeting of the Association for Computational Linguistics, Florence, Italy, 28 July–2 August 2019; pp. 5370–5381. [CrossRef]
34. Hussain, S.; Sianaki, O.A.; Ababneh, N. *A Survey on Conversational Agents/Chatbots Classification and Design Techniques*; Springer International Publishing: Berlin/Heidelberg, Germany, 2019; Volume 927. [CrossRef]
35. Ritter, A.; Cherry, C.; Dolan, W.B. Data-driven response generation in social media. In Proceedings of the EMNLP 2011—Conference on Empirical Methods in Natural Language Processing, Edinburgh, UK, 27–31 July 2011; pp. 583–593.
36. Unsupervised Modeling of Twitter Conversations—ACL Anthology. Available online: https://aclanthology.org/N10-1020/ (accessed on 21 February 2022).
37. Huang, M.; Zhu, X.; Gao, J. Challenges in Building Intelligent Open-domain Dialog Systems. *ACM Trans. Inf. Syst.* **2020**, *38*, 21. [CrossRef]
38. Sutskever, I.; Vinyals, O.; Le, Q.V. Sequence to sequence learning with neural networks. *Adv. Neural Inf. Process. Syst.* **2014**, *4*, 3104–3112.
39. Smith, J. Empathetic Dialogue Generation Using Generation-Based Models. Diploma Thesis, Dept. Signals, Control, and Robotics, School of Electrical and Computer Eng., National Technical Univ. of Athens, Athens, Greece, 2020.
40. Vaswani, A.; Shazeer, N.; Parmar, N.; Uszkoreit, J.; Jones, L.; Gomez, A.; Kaiser, L.; Polosukhin, I. Attention is all you need. *Adv. Neural Inf. Process. Syst.* **2017**, *30*, 5998–6008.
41. Farha, I.A.; Magdy, W. Benchmarking Transformer-based Language Models for Arabic Sentiment and Sarcasm Detection. In Proceedings of the Arabic Natural Language Processing Workshop, Virtual, 19 April 2021; pp. 21–31.
42. Bubeck, S.; Chandrasekaran, V.; Eldan, R.; Gehrke, K.; Horvitz, E.; Kamar, E.; Lee, P.; Lee, Y.T.; Lundberg, S.; Nori, H.; et al. Sparks of Artificial General Intelligence: Early experiments with GPT-4. *arXiv* **2023**, arXiv:2303.12712.
43. Devlin, K.; Chang, M.W.; Lee, K.; Toutanova, K. Bert: Pre-training of deep bidirectional transformers for language understanding. *arXiv* **2018**, arXiv:1810.04805.
44. Raffel, C.; Shazeer, N.; Roberts, A.; Lee, K.; Narang, S.; Matena, M.; Zhou, Y.; Li, W.; Liu, P. Exploring the limits of transfer learning with a unified text-to-text transformer. *J. Mach. Learn. Res.* **2020**, *21*, 1–67.
45. Escall, F.; Mello, D.A.A.; Arantes, D.S.; Rothenberg, C.E. Chat-IBN-RASA: Building an Intent Translator for Packet-Optical Networks based on RASA. In Proceedings of the IEEE 9th International Conference on Network Softwarization (NetSoft), Madrid, Spain, 19–23 June 2023; pp. 534–538. [CrossRef]
46. Bocklisch, T.; Faulkner, J.; Pawlowski, N.; Nichol, A. Rasa: Open Source Language Understanding and Dialogue Management. *arxiv* **2017**. [CrossRef]
47. Mishra, D.S.; Agarwal, A.; Swathi, B.; Akshay, K.C. Natural Language Query Formalization to SPARQL for Querying Knowledge Bases Using Rasa. *Prog. Artif. Intell.* **2022**, *11*, 193–206. [CrossRef]
48. Tsarfaty, R. Integrated Morphological and Syntactic Disambiguation for Modern Hebrew. In Proceedings of the COLING/ACL 2006 Student Research Workshop, Sydney, Australia, 20 July 2006. [CrossRef]
49. Almasoud, A.M.; Al-Khalifa, H.S. SemSignWriting: A Proposed Semantic System for Arabic Text-to-Sign Writing Translation. *J. Softw. Eng. Appl.* **2012**, *5*, 604–612. [CrossRef]
50. AL-Hagbani, E.S.; Khan, M.B. Support of Existing Chatbot Development Framework for Arabic Language: A Brief Survey. In *Advances in Intelligent Systems and Computing*; Springer International Publishing: Berlin/Heidelberg, Germany, 2018; pp. 26–35.
51. Alqahtani, Q.; Alrwais, O. Building a Machine Learning Powered Chatbot for KSU Blackboard Users. *Int. J. Adv. Comput. Sci. Appl.* **2023**, *14*, 781–789. [CrossRef]
52. Abdulkader, Z.; Al-Irhayim, Y. A Review of Arabic Intelligent Chatbots: Developments and Challenges. *Al-Rafidain Eng. J.* **2022**, *27*, 178–189. [CrossRef]
53. Aloqayli, A.; Abdelhafez, H. Intelligent Chatbot for Admission in Higher Education. *Int. J. Inf. Educ. Technol.* **2023**, *13*, 1348–1357. [CrossRef]
54. Jiao, A. An Intelligent Chatbot System Based on Entity Extraction Using RASA NLU and Neural Network. *J. Phys. Conf. Ser.* **2020**, *1487*, 012014. [CrossRef]

55. Alhumoud, S.; Al Wazrah, A.; Aldamegh, W. Arabic Chatbots: A Survey. *Int. J. Adv. Comput. Sci. Appl.* **2018**, *9*. [CrossRef]
56. Mozannar, H.; El Hajal, K.; Maamary, E.; Hajj, H. Neural Arabic question answering. In Proceedings of the Fourth Arabic Natural Language Processing Workshop, Florence, Italy, 1 August 2019; pp. 108–118. [CrossRef]
57. Boussakssou, M.; Ezzikouri, H.; Erritali, M. Chatbot in Arabic language using seq to seq model. *Multimed. Tools Appl.* **2022**, *81*, 2859–2871. [CrossRef]
58. Biltawi, M.; Awajan, A.; Tedmori, S. Towards Building an Open-Domain Corpus for Arabic Reading Comprehension Towards Building an Open-Domain Corpus for Arabic Reading Comprehension. In Proceedings of the 35th IBIMA Conference, Seville, Spain, 1–2 April 2020.
59. Antoun, W.; Baly, F.; Hajj, H. AraBERT: Transformer-based Model for Arabic Language Understanding. *arXiv* **2021**, arXiv:2003.00104.
60. Eljundi, O.; Antoun, W.; El, E. Hulmona: The Universal Language Model in Arabic. In Proceedings of the Fourth Arabic Natural Language Processing Workshop, Florence, Italy, 1 August 2019; Eljundi, O., Antoun, W., El, E., Hajj, H., El-Hajj, W., Shaban, K., Eds.; Association for Computational Linguistics: Kerrville, TX, USA, 2019; pp. 68–77.
61. Kushwaha, A.K.; Kar, A.K. MarkBot—A Language Model-Driven Chatbot for Interactive Marketing in Post-Modern World. *Inf. Syst. Front.* **2021**, 1–18. [CrossRef]
62. Nagoudi, E.M.B.; Elmadany, A.; Abdul-Mageed, M. AraT5: Text-to-Text Transformers for Arabic Language Understanding and Generation. *arXiv* **2021**, arXiv:2109.12068.
63. Antoun, W.; Baly, F.; Hajj, H. AraGPT2: Pre-Trained Transformer for Arabic Language Generation. In Proceedings of the Sixth Arabic Natural Language Processing Workshop, Kiev, Ukraine, 19 April 2021; pp. 196–207.
64. Abdul-Mageed, M.; Elmadany, A.; Nagoudi, E.M.B. ARBERT&MARBERT: Deep bidirectional transformers for Arabic. *arXiv* **2020**, arXiv:2101.01785.
65. Brandtzaeg, P.; Følstad, A. Chatbots: Changing user needs and motivations. *Interactions* **2018**, *25*, 38–43. [CrossRef]
66. Pérez, J.Q.; Daradoumis, T.; Puig, J.M.M. Rediscovering the use of chatbots in education: A systematic literature review. *Comput. Appl. Eng. Educ.* **2020**, *28*, 1549–1565. [CrossRef]
67. El Hefny, W.; Mansy, Y.; Abdallah, M.; Abdennadher, S. Jooka: A Bilingual Chatbot for University Admission. In *Advances in Intelligent Systems and Computing*; Springer: Cham, Switzerland, 2021; pp. 671–681. [CrossRef]
68. Al-Ghadhban, D.; Al-Twairesh, N. Nabiha: An Arabic Dialect Chatbot. *Int. J. Adv. Comput. Sci. Appl.* **2020**, *11*. [CrossRef]
69. Aljameel, S.; O'Shea, J.; Crockett, K.; Latham, A.; Kaleem, M. LANA-I: An Arabic Conversational Intelligent Tutoring System for Children with ASD. In *Advances in Intelligent Systems and Computing*; Springer: Berlin/Heidelberg, Germany, 2019; pp. 498–516. [CrossRef]
70. Almurtadha, Y. LABEEB: Intelligent Conversational Agent Approach to Enhance Course Teaching and Allied Learning Outcomes Attainment. *J. Appl. Comput. Sci. Math.* **2019**, *13*, 9. [CrossRef]
71. Alazzam, B.A.; Alkhatib, M.; Shaalan, K. Arabic Educational Neural Network Chatbot. *Inf. Sci. Lett.* **2023**, *12*, 2579–2589.
72. Tuning Your NLU Model. Available online: https://rasa.com/docs/rasa/tuning-your-model/ (accessed on 4 February 2024).
73. Al-madi, N.A.; Maria, K.A.A.; Al-madi, M.A.; Maria, E.A.A. An Intelligent Arabic Chatbot System to Reduce Admissions Department Burdens. *Authorea* **2023**. [CrossRef]
74. Castillo, D.; Canhoto, A.I.; Said, E. The Dark Side of AI-powered Service Interactions: Exploring the Process of Co-Destruction from the Customer Perspective. *Serv. Ind. J.* **2020**, *41*, 900–925. [CrossRef]
75. Nguyen, T.T.; Le, A.D.; Hoang, H.T.; Nguyen, T. NEU-chatbot: Chatbot for admission of National Economics University. *Comput. Educ. Artif. Intell.* **2021**, *2*, 100036. [CrossRef]

applied sciences

Article

Predicting Student Performance in Online Learning: A Multidimensional Time-Series Data Analysis Approach

Zhaoyu Shou [1,*], Mingquan Xie [2], Jianwen Mo [2] and Huibing Zhang [3]

1 Guangxi Wireless Broadband Communication and Signal Processing Key Laboratory, Guilin University of Electronic Technology, Guilin 541004, China
2 School of Information and Communication, Guilin University of Electronic Technology, Guilin 541004, China; 21022303168@mails.guet.edu.cn (M.X.); jwmo@guet.edu.cn (J.M.)
3 School of Computer and Information Security, Guilin University of Electronic Technology, Guilin 541004, China; zhanghuibing@guet.edu.cn
* Correspondence: guilinshou@guet.edu.cn

Abstract: As an emerging teaching method, online learning is becoming increasingly popular among learners. However, one of the major drawbacks of this learning style is the lack of effective communication and feedback, which can lead to a higher risk of students failing or dropping out. In response to this challenge, this paper proposes a student performance prediction model based on multidimensional time-series data analysis by considering multidimensional data such as students' learning behaviors, assessment scores, and demographic information, which is able to extract the characteristics of students' learning behaviors and capture the connection between multiple characteristics to better explore the impact of multiple factors on students' performance. The model proposed in this paper helps teachers to individualize education for students at different levels of proficiency and identifies at-risk students as early as possible to help teachers intervene in a timely manner. In experiments on the Open University Learning Analytics Dataset (OULAD), the model achieved 74% accuracy and 73% F1 scores in a four-category prediction task and was able to achieve 99.08% accuracy and 99.08% F1 scores in an early risk prediction task. Compared with the benchmark model, both the multi-classification prediction ability and the early prediction ability, the model in this paper has a better performance.

Keywords: online learning; multidimensional time-series data; student performance prediction; multi-classification prediction; individualized education; early prediction

1. Introduction

With the continuous improvement of online learning platforms, a proliferation of schools has opened up online course teaching, providing students with more flexible learning methods. However, there is a significant problem with the online learning process: due to the lack of face-to-face interaction between instructors and students, it is difficult for instructors to accurately grasp the level of understanding and mastery of course content, which may lead to students' poor performance on exams. Therefore, it is particularly important to capture the hidden information related to student performance from massive educational data collected by online learning platforms. Using this information, students at risk of failing and dropping out of school can be identified and an intervention can be carried out as early as possible to effectively improve their learning status and enhance learning effectiveness [1]. Data mining techniques aid researchers in discovering and understanding latent critical information within massive educational datasets, enabling the identification and prediction of students' performance trends [2]. Hughes et al. [3] mined high/active engagement, attendance, and multiple interactions as key factors for students to achieve higher marks on a Massive Open Online Learning

(MOOC) platform by using data mining techniques. Such information extraction is not confined solely to the field of data mining but also extends into the domain of learning analytics. With the rise in learning analytics, there has been a growing emphasis on investigating student behaviors and aggregating various methodologies to enhance the comprehension of student behaviors, consequently enabling the prediction of students' academic performance [4]. As a category of data reflecting student behaviors, clickstream data illustrate the pathways students take while navigating one or multiple learning websites on an online learning platform [5]. Casalino et al. [6] used data from students' interactions with the virtual learning environment to analyze learning through a neuro-fuzzy system, showing that students who regularly visit the course homepage and actively participate in the teaching and learning activities through evaluation tests are more likely to pass the exams. In addition to clickstream data, students' demographic information and assessment scores have also been verified to be closely correlated with students' academic performance [7,8]. Considering more dimensions of information beyond students' learning behaviors, modeling and analyzing students' academic performance as a whole can more accurately predict students' academic performance.

The early prediction of student performance allows for the proactive identification of at-risk students, enabling instructors to devise additional courses, and assignments, or implement other measures to enhance students' performance before the conclusion of the course. Currently, most research only applies to predicting student performance at the end of the course, which does not allow for timely intervention. Therefore, it is particularly important to predict student performance as early as possible before the end of the course [9]. Furthermore, most studies have predicted student performance (pass, fail) on a dichotomous basis, which would result in students at an excellent level losing their upward mobility and failing to maximize benefits from the course [10]. To promote the sustainable development of online education and provide a basis for teachers to intervene with at-risk students promptly, this paper synthesizes multidimensional data, such as students' learning behaviors, assessment scores, and demographic information, and proposes a prediction model for student performance based on the analysis of multidimensional time-series data (MTAPSP), aiming at efficiently and accurately predicting students' performances in an online learning platform. The main contributions of this study can be summarized in the following three points:

1. A student performance prediction model for multi-dimensional time-series data analysis is proposed, which uses the multi-head self-attention (MHSA) mechanism to better integrate features such as time-series behaviors, assessment scores, and demographic information, avoiding the limitations of single-dimensional analysis while enhancing the model's ability to predict students' performance in multi-classification.
2. In this paper, we use temporal behavioral features extracted from students' chronological behavior and assessment scores using multilayer LSTM and further enhance the model's nonlinear mapping ability to these features through ANN, which improves the model's early prediction ability to identify at-risk students as early as possible and assists teachers in making timely interventions.
3. A large number of experiments have been conducted on the publicly available dataset OULAD, and the experimental results demonstrate that the model proposed in this paper outperforms the benchmark model in terms of both multi-classification prediction and early prediction ability.

The paper is organized as follows: Section 2 provides an overview of related research work. Section 3 presents the dataset used for experiments. Section 4 provides a detailed description of the multi-dimensional time-series analysis-based student performance prediction model (MTAPSP). Section 5 describes the experimental process and analyzes the results to evaluate the performance of the proposed algorithm. Finally, Section 6 summarizes the paper and proposes future research.

2. Related Work

Student performance prediction research can be categorized into regression and classification problems, where the former predicts a score value, and the latter predicts a category [11]. Earlier studies in student performance prediction focused on traditional machine learning methods, which used logistic regression [12], support vector machines [13], and decision trees [14] to establish models for predicting student performance. In a study [15], an automatic method for observing and predicting student grades was proposed. This method utilized a genetic algorithm to capture the 30 best attributes from students' historical learning data and trained a K-NN regression model and a decision tree using these features and labels to predict students' performance score and categories. PEK et al. [16] used naive Bayes, random forest, decision tree, AdaBoost classifier, logistic regression, and KNN algorithms as basic learners, support vector machine (SVM) as meta-learner, and created a stacking method to develop a hybrid ensemble model. By analyzing the data, they discovered important features that influence student learning outcomes and successfully helped teachers identify students at risk. Jawad et al. [17] and Bujang et al. [18] combined the SMOTE technique with machine learning techniques to improve the impact of data imbalance on the model and enhance the accuracy of student performance prediction. Hung et al. [19] proposed a method for predicting student performance based on time-series clustering. The method aggregates learning behavior data such as the frequency of accessing course materials, frequency of reading forums, number of discussions, and number of replies posted to identify at-risk students and predicts student performance more accurately than traditional frequency aggregation methods. Traditional machine learning methods ignore time information in the original data, which cannot effectively capture the impact of time features on student performance. However, deep neural networks can effectively address the problem of time information missing [20]. Therefore, some researchers have begun to utilize deep learning techniques to predict student performance. For example, He et al. [21] used a GRU network to extract time-series features from clickstream data and assessment score data and combined them with demographic features to predict student performance. Liu et al. [22] proposed a hybrid deep learning model that can extract time-behavioral and overall behavioral information from learning behavior data to more accurately predict high-risk students. Qu et al. [23] constructed a student performance prediction framework with an attention mechanism, in which an LSTM neural network was used to reflect students' learning processes, and a DSP-based adapter was used to enhance the importance of key information and improve the accuracy of student performance prediction. Kusumawardani et al. [24] proposed a transformer-based method for predicting student performance by converting the learning behavior data of students into a sequential feature vector. In some studies [25–27], authors used convolutional neural networks (CNN) to extract high-dimensional time information from the time series of student activities to better extract spatio-temporal features and utilized LSTM to capture the sequence information of student learning dynamics to more accurately predict student performance. Chen et al. [28] proposed a performance prediction model based on the differences in patterns of student behavioral features for students with significant performance changes, which uses a multi-head attention mechanism to automatically select more important higher-order behavioral combinations of features, maintain higher temporal accuracy, and can predict student performance more accurately. Li et al. [29] considered the potential relationships between students and used multiple graphs with different topologies to reflect the relationships between students and proposed a student performance prediction model based on a multi-topological graph neural network (MTGNN). Yang et al. [30] utilized clickstream data and assessment scores as input data to train a time series neural network to capture the unique features of each student's learning pattern. The results showed that the correlation between behavior features and scores was not very high, and it needed to consider the overall effect and nonlinear predictive factors.

In summary, most existing performance prediction models have not taken into account age group, residence location, and other demographic information, and the influence of assessment scores on student performance has not been fully considered in many of them. Additionally, many of these models have limited early prediction abilities. Most existing research has primarily focused on binary classification of student performance, with insufficient granularity in terms of dividing student performance. This paper considers additional information beyond a student's learning behavior to achieve the early prediction of student performance and multi-classification prediction of performance, providing effective support for individualized teaching.

3. Dataset

3.1. Introduction to the Dataset

The validation of the MTAPSP student performance prediction model using the Open University Learning Analytics Dataset (OULAD) provided by the Open University UK. OULAD comprises detailed learning behavior log data of 32,593 students in seven courses during the 2013–2014 academic year, including demographic data and clickstream data of student interactions in the VLE, which provided effective data support for analyzing the relationship between student behavior and student performance [31]. In the OULAD dataset, the seven courses are subdivided into 22 instructional modules covering a wide range of areas in science, social sciences, technology, engineering, and mathematics. Each module is named according to the year and month in which it is offered, e.g., a module offered in February 2013 is named 2013B, and a module offered in October 2013 is named 2013J, reflecting the fact that each module is offered multiple times during the year. Course information is provided in Table 1.

Table 1. Course information table.

Module	Domain	Presentations	Students
AAA	Social Sciences	2	748
BBB	Social Sciences	4	7909
CCC	STEM	2	4434
DDD	STEM	4	6272
EEE	STEM	3	2934
FFF	STEM	4	7762
GGG	Social Sciences	3	2534
Total	7	22	32,593

Each course has resources in the VLE, and there are a total of 20 activities for students in the VLE, and information on their activity types is shown in Table 2. Students are evaluated on the Teacher-Marked Assessment (TMA), the Computer-Marked Assessment (CMA), and the Final Exam (Exam), with the final performance categorized into four outcomes: "Distinction", "Pass", "Fail", and "Withdrawn". In addition to this, student information and student registration information is called demographic information, and the information is shown in Table 3.

Table 2. Type of learning behavior statistics.

Type of Learning Behavior	Description
Folder	Open folder
Forumng	Clicks on the discussion forum
Oucollaborate	Clicks on the online video discussions
Oucontent	Clicks on the contents of the assignment
Ouwiki	Query with Wikipedia
Ouelluminate	Participate in simulation course seminars
Quiz	Clicks on the course quiz
Questionnaire	Participate in simulation course seminars
Dataplus	Supplementary data
Dualpane	Access double window
Homepage	Clicks on the course homepage

Table 2. *Cont.*

Type of Learning Behavior	Description
Htmlactivity	Web activity
Page	Clicks on the information related to course
Resource	Clicks on the course homepage
RepeatActivity	Repetitive activity
Glossary	Access the glossary
Url	Clicks on the links to audio/video contents
Subpage	Clicks on the other sites enabled in the course
Sharedsubpage	Shared pagination
Externalquiz	Complete extracurricular quizzes

Table 3. Introduction to student demographics.

Type of Student Demographics	Description
Code_module	Module identification code
Code_presentation	Presentation identification code
Gender	Student's gender
Region	The geographic region, where the student lived while taking the module presentation
Highest Education	The highest student education level on entry to the module presentation
Imd_band	Band of statistical data
Age_band	Band of student's age
Num_of_prev_attempts	The number of times the student has attempted this module
Studied_credits	The total number of credits for the modules the student is currently studying
Disability	Indicates whether the student has declared a disability

3.2. Data Processing

In this paper, the data were preprocessed while using the OULAD dataset for model validation. First, samples with incorrect date fields were removed; then all course lengths were standardized to 270 days. To verify the early prediction ability of the model, this paper divided the data by course time length into five segments of 20%, 40%, 60%, 80%, and 100% lengths and summarized the clickstream data and weekly average click volume for each segment as representative variables for the student's virtual learning environment (VLE) interaction. In addition, to reduce the dimensionality and complexity of the model, integer encoding was used instead of the classification data in the demographic data. Finally, the clickstream data, assessment scores, and coded demographic data were concatenated to form the training data for the model.

4. MTAPSP Model

4.1. MTAPSP Model Structure

The student performance prediction model proposed in this paper consists of two parts: a multilayer LSTM network with a multi-head self-attention mechanism and an artificial neural network (ANN). First, multi-layer LSTM was used to extract time-series learning behavior features from the learning behavior time series data. Then, a multi-head self-attention mechanism was used to evaluate the importance of time-behavior features, assessment score features, and demographic features, and extract key feature information. Finally, ANN was used to integrate temporal behavioral features, assessment score features, and demographic data features to improve the performance of the model. The MTAPSP model structure is shown in Figure 1.

4.2. Multilayer LSTM Based on a Multi-Head Self-Attention Mechanism

This module aims to extract temporal behavioral features from the time-series data of learning behaviors through a three-layer LSTM. It utilizes a multi-head self-attention mechanism to capture the importance of temporal behavioral features, assessment score features, and demographic data features—three distinct dimensions—thereby further extracting more salient features.

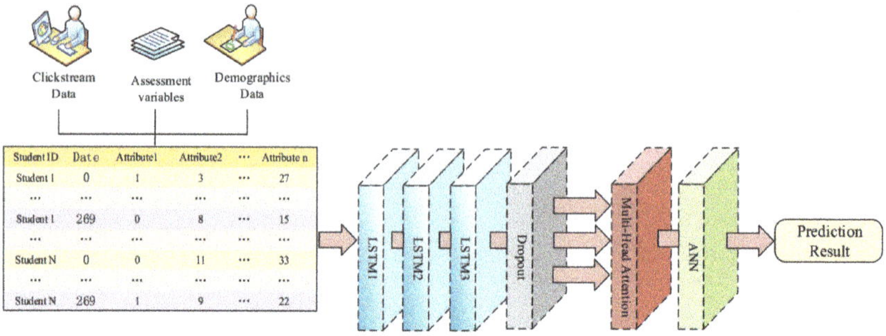

Figure 1. MTAPSP model structure.

4.2.1. Multilayer LSTM

Traditional recurrent neural networks (RNN) suffer from the problem of gradient vanishing for long sequence data like OULAD, in contrast, LSTM can efficiently convey and express the information of long-time sequences and solve the problem of long-term dependency [4]. As shown in Figure 2, compared to single-layer LSTM, a three-layer LSTM can enhance the extraction of long-distance memory and contextual information, better capture the long-distance dependencies in the sequence of students' learning time behaviors, and improve the model's generalization ability and robustness.

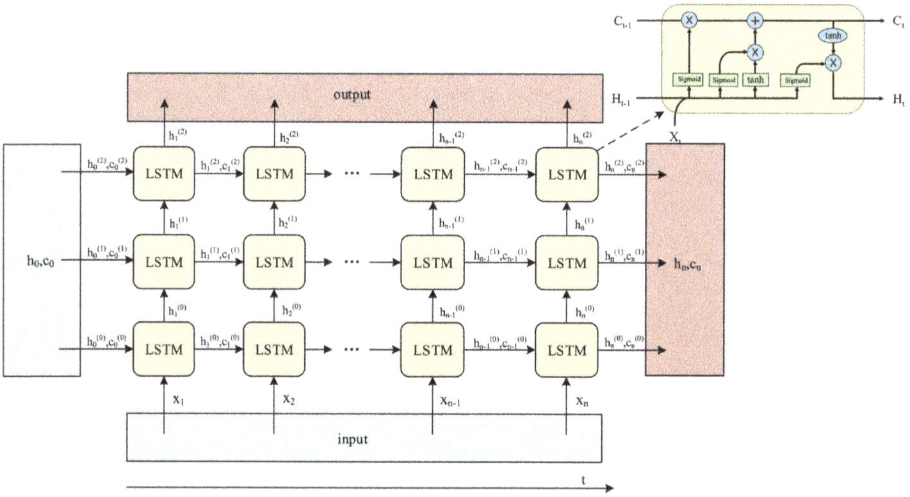

Figure 2. Three-layer LSTM network structure.

The input data $X_t \in R^{t \times m}$ is a two-dimensional tensor, where t represents the number of course days for a student, and m denotes the number of features in the input data. LSTM uses a gating mechanism to control the flow of information by feeding data X_t into the LSTM network, which sequentially goes through the computational process of forgetting gates, input gates, and output gates in LSTM. In the forget gate, LSTM integrates new input data and the output from the previous time step to determine the omission of irrelevant information. The computational formula is expressed as Equation (1).

$$f_t = Sigmoid(W_f[h_{t-1}, x_t] + b_f) \quad (1)$$

In the input gate, LSTM filters and updates the information by multiplying the current input data with the hidden state of the previous moment, which is calculated as shown in Equations (2)–(4).

$$i_t = Sigmoid(W_i[h_{t-1}, x_t] + b_i) \quad (2)$$

$$\widetilde{C}_t = \tanh(W_c[h_{t-1}, x_t] + b_c) \quad (3)$$

$$C_t = f_t * C_{t-1} + i_t * \widetilde{C}_t \quad (4)$$

In the output gate, LSTM extracts relevant information from the vector obtained by integrating the current input data with the previous time step's output value, determining the information to be outputted. The computational formulas as shown in Equations (5) and (6).

$$o_t = Sigmoid(W_o[h_{t-1}, x_t] + b_o) \quad (5)$$

$$h_t = o_t * \tanh(C_t) \quad (6)$$

The above equations, where b_f, b_i, b_c, b_o, W_f, W_i, W_c, W_o are the biases and weights that correspond to the three gates, and denote the activation function.

4.2.2. MHSA

The self-attention mechanism enables the model to automatically learn and focus on the most relevant information within the input data [32]. A multi-head self-attention mechanism that can map a sequence into multiple specific spaces separately perform multiple attention computations, and finally stitch the results together to obtain richer contextual information.

As shown in Figure 3, in MTAPSP, when the multi-head self-attention mechanism receives the output sequence from LSTM, the sequence is linearly transformed into vector matrices Q (Query), K (Key), and V (Value), facilitating multiple attention computations. The single self-attention is calculated as shown in Equation (7).

$$Attention(Q, K, V) = softmax(\frac{QK^T}{\sqrt{d_k}})V \quad (7)$$
$$d_k = \frac{d_{model}}{h}$$

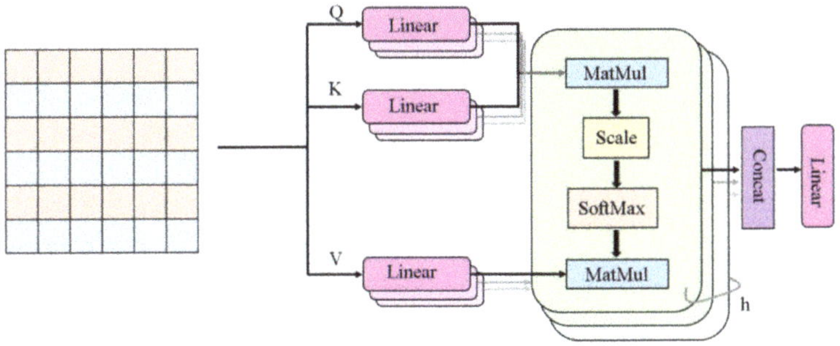

Figure 3. Structure of the multi-head self-attention mechanism.

In the equation, d_k represents the dimension of the vectors in a single attention computation, $softmax$ denotes the weight normalization function, and the product of the dimension d_k and the number of heads h in a single attention computation equals the model's dimension d_{model}. The multi-head self-attention mechanism stitches together the results of each set of single self-attention calculations to perform a linear transformation and finally output the final result, which is calculated by the following formula:

$$\begin{aligned} MultiHead(Q,K,V) &= Concat(head_1,\ldots,head_h)W^o \\ head_i &= Attention(QW_i^Q, KW_i^K, VW_i^V) \end{aligned} \tag{8}$$

where W^O, W_i^Q, W_i^K, $W_i^V \in R^{d_{model} \times d_k}$ is the weight matrix.

4.3. ANN

ANN learning is performed through multiple iterations, which prioritizes previous biases iteratively adjusting weights to classify potential outcomes [33]. Using ANN to integrate student learning behavior features, assessment score features, and demographic features, as shown in Figure 4, the algorithm can nonlinearly map the tensor resulting from computations through the multi-head attention mechanism. This process extracts correlated information between features and outputs the predicted student performance.

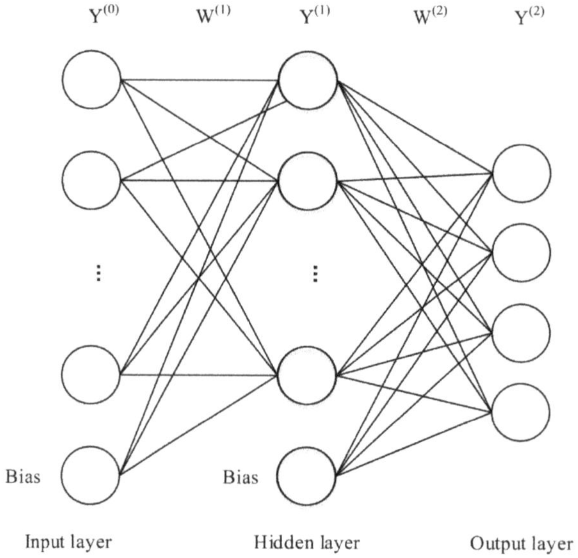

Figure 4. ANN structure.

ANN consists of an input layer, hidden layers, and an output layer. The output expression of the hidden layer is shown in Equation (9) and the output expression of the output layer is shown in Equation (10).

$$Y^{(1)} = \delta[W^{(1)}Y^{(0)} + b_1] \tag{9}$$

$$Y^{(2)} = \delta[W^{(2)}Y^{(1)} + b_2] \tag{10}$$

where δ is the activation function, $W^{(1)} \in R^{m \times r}$, $W^{(2)} \in R^{r \times n}$, m are the dimensions of the output vectors of the multi-head attention mechanism, r is the dimension of the output vectors of the hidden layer, n is the dimension of the output vectors of the output layer, and b_1, b_2 are the bias vectors.

The algorithmic flow for multi-classification prediction using the MTAPSP model is shown in Algorithm 1 below.

Algorithm 1 MTAPSP Algorithm

Input: N: Number of students; T: Number of course days; U: Student clickstream data matrix, which has a shape $N \times T \times 20$; C: Student assessment variables matrix, which has shape $N \times T \times 3$; D: Demographics data matrix, $N \times T \times 10$; K: Max epoch
Output: Y: Predicted Results of Student Performance

Splice U, C, D to get matrix X;
Initialization learning rate, hyper-parameter randomly and LSTM parameter;
for epoch $\leftarrow$ 1 to K do
    $Y_1 \leftarrow$ LSTM1($X[n]$);
    $Y_2 \leftarrow$ LSTM2($Y_1[n]$);
    $Y_2 \leftarrow$ LSTM3($Y_2[n]$);
    $Q_i, K_i, V_i \leftarrow Y_3[n]$; $//i \in [0,3]$, i is the number of heads of the multi-head self-attention mechanism
    Atti $\leftarrow$ softmax($\frac{Q_i K_i^T}{\sqrt{d_k}}$)$V_i$; // d_k is the dimension of X
    $Y_{Att} \leftarrow$ concat Atti; $i \in [0,3]$;
    $Y \leftarrow$ ANN(Y_{Att})
end
Return Y;

5. Experimental

5.1. Evaluation Indicators

To evaluate the performance of the MTAPSP prediction model proposed in this paper, accuracy, precision, recall, and F1-score, which are commonly used in student performance prediction models, are used in this paper as evaluation metrics [34]. The formula for accuracy is shown in Equation (11):

$$Accuracy = \frac{TP + TN}{TP + TN + FP + FN} \tag{11}$$

where TP represents the number of correctly predicted positive classifications, FP represents the number of incorrectly predicted positive classifications, TN represents the number of correctly predicted negative classifications, and FN is the number of incorrectly predicted negative classifications. The precision rate represents the model's prediction accuracy in positive sample results, which is calculated as shown in Equation (12):

$$Precision = \frac{TP}{TP + FP} \tag{12}$$

Recall is the probability that the model correctly predicts a positive sample among positive samples and is calculated as shown in Equation (13):

$$Recall = \frac{TP}{TP + FN} \tag{13}$$

F1-scoreis the reconciled mean of precision and recall, which is calculated as shown in Equation (14):

$$F1\text{-}Score = 2 \times \frac{Precision \times Recall}{Precision + Recall} \tag{14}$$

5.2. Experimental Environment and Hyper-Parameter Settings

The experimental environment of this study is shown in Table 4. The dataset is randomly split into 80% training data and 20% testing data for experiments. The initial learning rate of the MTAPSP model is set to 0.001, the number of neurons in the LSTM hidden layer is set to 32, the number of heads in the multi-head attention mechanism is set to 4, the number of neurons in the hidden layer of the ANN is set to 64, the batch size of the model is set to 50, the cross-entropy loss function is used, the optimizer is set to Adam, and the number of iterations of the model is set to 100.

Table 4. Experimental environment.

Experimental Environment	Environment Configuration
Operating systems	Windows 10
CPU	Intel core i5-6700HQ
Video Cards	NVIDIA GTX 960
RAM	16 GB
DISK	1T SSD
Programming Languages	Python 3.6
Framework	Scikit-learn 0.24.2 + Pytorch 1.10.2

5.3. Experimental Results and Analysis

This section provides a detailed description of the experimental results of the MTAPSP model and the benchmark model for two scenarios: firstly, the multi-classification prediction performance. In the experiments, the "Distinction", "Fail", "Pass", and "Withdrawn" categories of samples are selected for verification. Second, the performance of the model in terms of binary prediction performance is explored, specifically validating the effectiveness of the MTAPSP model in predicting students who may be at risk in the first few weeks of school and its potential benefits. In addition, ablation experiments aimed at exploring the role and impact of MHSA and ANNs in the MTAPSP model.

5.3.1. Multi-Classification Prediction Performance

To validate the advantages of the MTAPSP model for multi-classification prediction, four benchmark models, random forest (RF), naive Bayes (NB), deep feedforward neural network (DFFNN) [34], and sequential engagement-based prediction network for academic performance (SEPN) [26], were used for comparative experiments, using the students' clickstream data, their assessment scores, and the data of the students' demographic characteristics, and each of these models' performance is shown in Table 5.

Table 5. Comparison of model multi-classification prediction performance.

Precision	RF	NB	DFFNN	SEPN	MTAPSP
Distinction	0.66	0.24	0.66	0.40	0.58
Fail	0.59	0.34	0.55	0.55	0.62
Pass	0.78	0.62	0.74	0.71	0.79
Withdrawn	0.73	0.72	0.76	0.74	0.78
Macro avg.	0.69	0.48	0.68	0.60	0.69
Weighted avg.	0.71	0.55	0.70	0.66	0.73
Recall	**RF**	**NB**	**DFFNN**	**SEPN**	**MTAPSP**
Distinction	0.54	0.57	0.47	0.07	0.40
Fail	0.43	0.37	0.36	0.29	0.51
Pass	0.89	0.454	0.90	0.96	0.89
Withdrawn	0.77	0.65	0.80	0.83	0.83
Macro avg.	0.66	0.51	0.63	0.54	0.66
Weighted avg.	0.72	0.50	0.71	0.70	0.74
F1-score	**RF**	**NB**	**DFFNN**	**SEPN**	**MTAPSP**
Distinction	0.59	0.34	0.55	0.12	0.47
Fail	0.50	0.36	0.44	0.38	0.56
Pass	0.83	0.52	0.81	0.82	0.84
Withdrawn	0.75	0.68	0.78	0.79	0.80
Macro avg.	0.67	0.47	0.64	0.53	0.67
Weighted avg.	0.71	0.52	0.70	0.65	0.73
Accuracy	0.72	0.50	0.71	0.70	0.74

Table 5 shows that the MTAPSP model obtained the best performance in most cases. Although the RF model performs best in differentiating between "Distinction" performance categories, given the imbalance of the data, we must consider evaluation metrics based on F1 scores. The MTAPSP model has slightly higher F1 scores than the RF model for the categories "Failed", "Passed", and "Withdrawn", indicating that the MTAPSP model has better predictive power in the face of data imbalance. Compared to the deep learning models DFFNN and SEPN, the MTAPSP model achieved superior results in all metrics, which confirms that it is more capable of capturing temporal features of learning behaviors and integrating multidimensional features to more accurately predict student performance.

The multi-classification prediction experiments clearly demonstrate the excellent performance of the MTAPSP model in this area. Teachers can predict student performance by utilizing the MTAPSP model and are able to gain insight into the academic level of their students and develop targeted instructional programs accordingly. For instance, students who are predicted to have "Distinction" results are recommended higher-level courses to deepen their learning and improve their abilities; students who are predicted to withdraw from school are provided with the necessary support and assistance by finding out whether they have any difficulties in their academic life. The application of the MTAPSP model not only provides an effective tool for predicting student performance but also provides a powerful means of guidance for educators, enabling them to counsel students in a more comprehensive and individualized manner, and promoting the improvement of teaching quality.

5.3.2. Early Risk Prediction

To validate the performance of the MTAPSP model in terms of binary prediction performance, three validation methods were used in this study based on relying only on students' clickstream data and assessment score information:

1. The WF-PD task, which categorizes "Distinction" and "Pass" as one category and "Fail" and "Withdrawn" as another. The purpose of the WF-PD task is to identify whether a student has passed the course.
2. The F-PD task, which identifies students at risk of "Fail" by placing "Fail" in a separate category and "Distinction" and "Pass" in the same category, excluding dropouts. "The F-PD task identifies students at risk of "Fail".
3. The W-PD task, with "Withdrawn" in a separate category and "Distinction" and "Pass" combined in one category, excluding "Fail" students. "Fail" students were excluded, to distinguish between students at risk of dropping out and those not at risk.

To demonstrate the early risk prediction capability of the MTAPSP model, this paper divided the dataset by percentage of course length and created two models to predict student performance: (1) the "daily" model using student activity data as input daily; (2) the "weekly" model using students' weekly activity data as input. A comparison was made with the latest student performance prediction model, TEnc [24], and the results of the comparison are shown in Table 6.

Table 6. Comparison of prediction performance for model binary classification.

		Stages and Performance (%)									
	Models	20%		40%		60%		80%		100%	
		Acc	F1	Acc	F1	Acc	F1	Acc	F1	Acc	F1
TEnc	WF-PD (daily)	75.41	71.50	82.83	80.91	88.23	86.77	90.78	89.93	93.42	93.01
MTAPSP	WF-PD (daily)	91.79	91.80	92.50	92.50	92.70	92.70	93.51	93.51	94.63	94.63
TEnc	F-PD (daily)	77.80	52.50	82.95	67.70	87.04	75.92	89.29	81.04	91.85	86.22
MTAPSP	F-PD (daily)	91.38	91.18	91.40	91.20	91.53	91.32	91.87	91.78	92.71	92.55
TEnc	W-PD (daily)	83.17	68.03	90.49	83.84	94.46	90.77	96.78	94.80	98.02	96.88
MTAPSP	W-PD (daily)	95.73	95.71	96.08	96.07	97.28	97.27	98.08	98.08	99.08	99.08
TEnc	WF-PD (weekly)	76.76	73.37	83.93	81.87	88.96	87.62	91.81	91.10	93.57	93.21
MTAPSP	WF-PD (weekly)	91.02	91.03	91.99	91.99	92.59	92.59	93.17	93.18	94.32	94.33
TEnc	F-PD (weekly)	78.53	55.68	83.73	69.20	87.85	77.41	90.43	83.09	91.75	86.27
MTAPSP	F-PD (weekly)	91.00	90.77	91.17	91.00	91.15	91.01	91.95	91.83	92.71	92.54
TEnc	W-PD (weekly)	83.79	70.63	91.01	84.79	95.51	92.61	97.22	95.58	98.14	97.10
MTAPSP	W-PD (weekly)	94.46	94.43	96.10	96.09	96.85	96.84	97.94	97.94	98.92	98.92

An analysis of Table 6 reveals the following:

1. The proposed method in this paper achieved optimal performance across various course durations, demonstrating the superior effectiveness of MTAPSP in binary predicting student performance.
2. Compared with the TEnc model, the three-layer LSTM network owned by MTAPSP performs better in dealing with long sequence data, indicating that the three-layer LSTM is more adept at capturing the patterns of long sequence data.
3. Within the first 20% of the course start time, MTAPSP embodies a high prediction performance, which indicates that the ANN in the MTAPSP model can more accurately map the relationship between the assessment scores characteristics and the students' final performance and map the potential influence of the assessment scores characteristics on the students' final performance.
4. Numerically, MTAPSP's daily model performs better than the weekly model, with performance improvements ranging from 0.1% to 1%, which suggests that the aggregation of student activity data changes important feature values, resulting in MTAPSP's inability to learn better features from this aggregated data.

To further demonstrate the early prediction ability of MTAPSP, the prediction performance of the MTAPSP model was compared with that of the TEnc model over three months after the start of the course, and the results are shown in Figures 5 and 6.

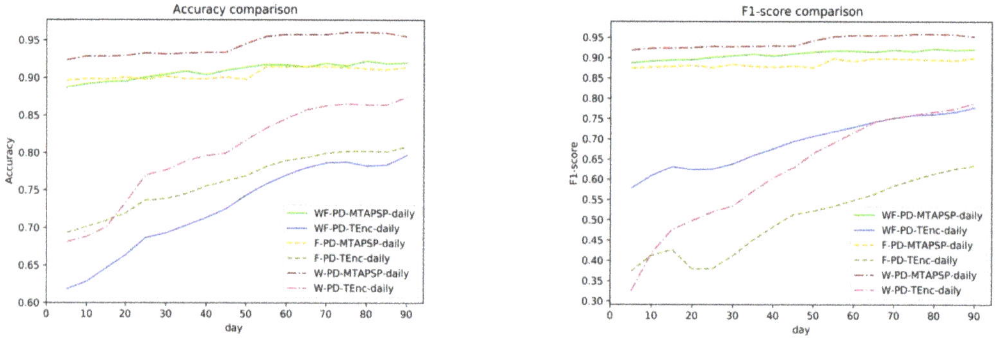

(a) (b)

Figure 5. (**a**) Comparison of the accuracy of "daily" models based on the WF-PD task, based on the F-PD task, and based on the W-PD task three months after the start of the course; (**b**) comparison of the F1-scores of "daily" models based on the WF-PD task, based on the F-PD task, and based on the W-PD task three months after the start of the course.

As can be seen from Figures 5 and 6, in the WF-PD, F-PD, and W-PD tasks, at the beginning of the course, the daily and weekly models of MTAPSP achieve more than 85% in terms of accuracy and recall in predicting students' performance, and gradually increase the indexes over time, outperforming the TEnc model.

In the early risk prediction experiment, the MTAPSP model demonstrated excellent early prediction capabilities, enabling instructors to predict final student performance early in the course. Once the model predicts that a student is likely to end up with a "fail" performance, the instructor can develop an intervention plan to support these students to adjust their learning before the end of the course to ensure that they can successfully pass the exam. With the predictive results of the MTAPSP model, teachers can identify potential student problems ahead of time and help them overcome their difficulties through effective measures, thus improving overall student performance and learning experience. This early prediction and intervention approach helps to maximize the optimization of the student learning process and improve teaching and learning outcomes.

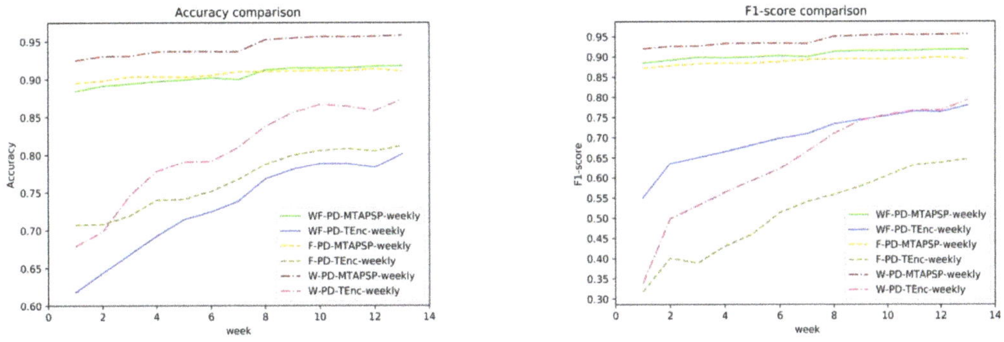

(**a**) (**b**)

Figure 6. (**a**) Comparison of the accuracy of "weekly" models based on the WF-PD task, based on the F-PD task, and based on the W-PD task three months after the start of the course; (**b**) comparison of the F1-scores of "weekly" models based on the WF-PD task, based on the F-PD task, and based on the W-PD task three months after the start of the course.

5.3.3. Ablation Experiments

To verify the effect of different modules on MTAPSP, ablation experiments were performed on MTAPSP models. Where MTAPSP-M-A denotes the MTAPSP model after removing MHSA and ANN, MTAPSP-A denotes the MTAPSP model after removing ANN, and the MTAPSP-M model denotes the MTAPSP model after removing MHSA. The experimental results are shown in Table 7.

Table 7. Results of ablation experiments.

Model	Precision	Recall	F1-Score	Accuracy
MTAPSP-M-A	0.643	0.641	0.626	0.703
MTAPSP-A	0.650	0.588	0.594	0.719
MTAPSP-M	0.697	0.576	0.572	0.723
MTAPSP	0.692	0.656	0.668	0.744

Figure 7a–d shows the confusion matrix of the predictive model at the time of the ablation experiment, with the diagonal line being the number of correctly predicted categories. Combined with Table 7, it can be seen that compared to MTAPSP-M-A, MTAPSP-A's accuracy and precision are increased, but its recall and F1 score are decreased, a phenomenon that stems from the problem of imbalance in the distribution of the sample categories in the dataset. As shown in Figure 7b, MTAPSP-A is inferior to MTAPSP-M-A in identifying the categories "Distinction" and "Withdrawn", while the number of correct identifications is significantly higher than that of MTAPSP-M-A in the categories "Fail" and "Pass", and "Pass" is significantly higher than in MTAPSP-M-A. Although the overall accuracy increased, the overall recall and F1 score decreased due to the decrease in the performance of the metrics in the other categories. The significant difference in precision and accuracy of the MTAPSP-M model compared to the MTAPSP-M-A model is because the ANN can fit arbitrary nonlinear functions, which allows it to learn the assessment scores and demographic features well and improve the model performance. Compared with MTAPSP-M-A, the overall number of correctly predicted categories of MTAPSP is substantially increased and the overall performance metrics are significantly improved. This is because MHSA can focus on different dimensions of features such as student learning behavior, assessment scores, and demographic information, and, when paired with ANN, can better integrate student learning behavior features, assessment scores, and demographic features to further improve model performance.

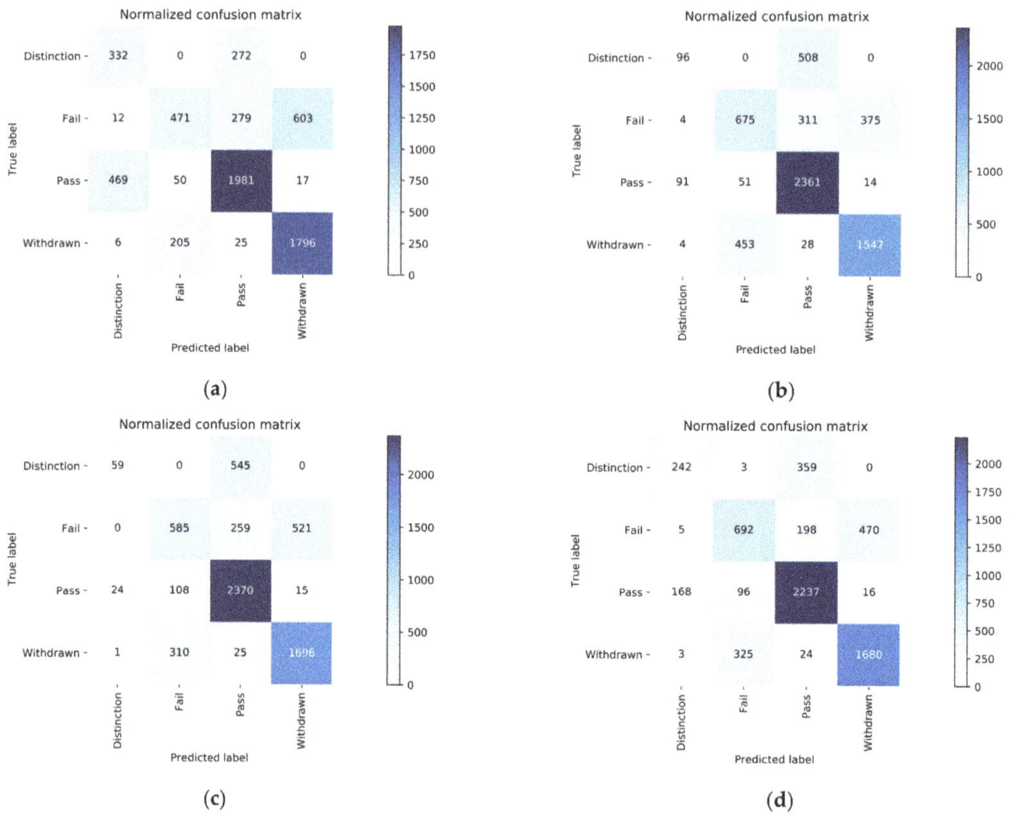

Figure 7. (**a**) Normalized confusion matrix for MTAPSP-M-A model predictions; (**b**) normalized confusion matrix for MTAPSP-A model predictions; (**c**) normalized confusion matrix for MTAPSP-M model predictions; (**d**) normalized confusion matrix for MTAPSP model predictions.

6. Conclusions

This paper proposes a student performance prediction model (MTAPSP) based on multidimensional time-series data analysis. The model incorporates multidimensional data such as students' learning behaviors, assessment scores, and demographic information (e.g., age group, place of residence) to achieve the multi-classification prediction of students' performance, and the multi-classification prediction performance of the MTAPSP model is verified by comparison with the baseline model. In addition, this paper tests the model's binary early prediction ability by dividing the dataset by course duration, and the experimental results show that it exhibits excellent performance in multi-classification prediction and binary early prediction. In future research, there is a need to address the imbalance in the data sample and to explore the impact of student–teacher interaction and student-to-student interaction data on student academic performance.

Author Contributions: Conceptualization, Z.S. and M.X.; methodology, M.X.; software, M.X.; validation, M.X., J.M. and H.Z.; formal analysis, H.Z.; investigation, J.M.; resources, Z.S.; data curation, M.X.; writing—original draft preparation, M.X.; writing—review and editing, Z.S.; visualization, M.X.; supervision, Z.S.; project administration, Z.S.; funding acquisition, Z.S. All authors have read and agreed to the published version of the manuscript.

Funding: This work was supported by the National Natural Science Foundation of China (62177012, 61967005, 62267003) and was supported by the Project of Guangxi Wireless Broadband Communication and Signal Processing Key Laboratory (GXKL06240107).

Data Availability Statement: The raw data supporting the conclusions of this article will be made available by the authors on request.

Conflicts of Interest: The authors declare no conflicts of interest.

References

1. Alhothali, A.; Albsisi, M.; Assalahi, H.; Aldosemani, T. Predicting student outcomes in online courses using machine learning techniques: A review. *Sustainability* **2022**, *14*, 6199. [CrossRef]
2. Tao, T.; Sun, C.; Wu, Z.; Yang, J.; Wang, J. Deep Neural Network-Based Prediction and Early Warning of Student Grades and Recommendations for Similar Learning Approaches. *Appl. Sci.* **2022**, *12*, 7733. [CrossRef]
3. Hughes, G.; Dobbins, C. The utilization of data analysis techniques in predicting student performance in massive open online courses (MOOCs). *Res. Pract. Technol. Enhanc. Learn.* **2015**, *10*, 10. [CrossRef]
4. Aljohani, N.R.; Fayoumi, A.; Hassan, S.U. Predicting at-risk students using clickstream data in the virtual learning environment. *Sustainability* **2019**, *11*, 7238. [CrossRef]
5. Liu, Y.; Fan, S.; Xu, S.; Sajjanhar, A.; Yeom, S.; Wei, Y. Predicting Student performance using clickstream data and machine learning. *Educ. Sci.* **2022**, *13*, 17. [CrossRef]
6. Casalino, G.; Castellano, G.; Zaza, G. Explainable Fuzzy Models for Learning Analytics. In *International Conference on Intelligent Systems Design and Applications*; Springer: Cham, Switzerland, 2022; pp. 394–403.
7. Arashpour, M.; Golafshani, E.M.; Parthiban, R.; Lamborn, J.; Kashani, A.; Li, H.; Farzanehfar, P. Predicting individual learning performance using machine-learning hybridized with the teaching-learning-based optimization. *Comput. Appl. Eng. Educ.* **2023**, *31*, 83–99. [CrossRef]
8. Hussain, S.; Khan, M.Q. Student-performulator: Predicting students' academic performance at secondary and intermediate level using machine learning. *Ann. Data Sci.* **2023**, *10*, 637–655. [CrossRef]
9. Conijn, R.; Snijders, C.; Kleingeld, A.; Matzat, U. Predicting student performance from LMS data: A comparison of 17 blended courses using Moodle LMS. *IEEE Trans. Learn. Technol.* **2016**, *10*, 17–29. [CrossRef]
10. Meier, Y.; Xu, J.; Atan, O.; Van der Schaar, M. Predicting grades. *IEEE Trans. Signal Process.* **2015**, *64*, 959–972. [CrossRef]
11. Ahmad, M.S.; Asad, A.H.; Mohammed, A. A Machine Learning Based Approach for Student Performance Evaluation in Educational Data Mining. In Proceedings of the 2021 International Mobile, Intelligent, and Ubiquitous Computing Conference (MIUCC), Cairo, Egypt, 26–27 May 2021; IEEE: Piscataway, NJ, USA, 2021; pp. 187–192.
12. Zhang, W.; Huang, X.; Wang, S.; Shu, J.; Liu, H.; Chen, H. Student performance prediction via online learning behavior analytics. In Proceedings of the 2017 International Symposium on Educational Technology (ISET), Hong Kong, China, 27–29 June 2017; IEEE: Piscataway, NJ, USA, 2017; pp. 153–157.
13. Al-Shehri, H.; Al-Qarni, A.; Al-Saati, L.; Batoaq, A.; Badukhen, H.; Alrashed, S.; Alhiyafi, J.; Olatunji, S.O. Student performance prediction using support vector machine and k-nearest neighbor. In Proceedings of the 2017 IEEE 30th Canadian Conference on Electrical and Computer Engineering (CCECE), Windsor, ON, Canada, 30 April–3 May 2017; IEEE: Piscataway, NJ, USA, 2017; pp. 1–4.
14. Wang, C.; Wei, X.; Yang, A.; Zhang, H. Construction and Analysis of Discrete System Dynamic Modeling of Physical Education Teaching Mode Based on Decision Tree Algorithm. *Comput. Intell. Neurosci.* **2022**, *2022*, 2745146. [CrossRef] [PubMed]
15. Rao, G.M.; Kumar, P.K.K. Students Performance Prediction in Online Courses Using Machine Learning Algorithms. *United Int. J. Res. Technol* **2021**, *2*, 74–79.
16. Pek, R.Z.; Özyer, S.T.; Elhage, T.; Özyer, T.; Alhajj, R. The role of machine learning in identifying students at-risk and minimizing failure. *IEEE Access* **2022**, *11*, 1224–1243. [CrossRef]
17. Jawad, K.; Shah, M.A.; Tahir, M. Students' Academic Performance and Engagement Prediction in a Virtual Learning Environment Using Random Forest with Data Balancing. *Sustainability* **2022**, *14*, 14795. [CrossRef]
18. Bujang, S.D.A.; Selamat, A.; Ibrahim, R.; Krejcar, O.; Herrera-Viedma, E.; Fujita, H.; Ghani, N.A.M. Multiclass prediction model for student grade prediction using machine learning. *IEEE Access* **2021**, *9*, 95608–95621. [CrossRef]
19. Hung, J.L.; Wang, M.C.; Wang, S.; Abdelrasoul, M.; Li, Y.; He, W. Identifying at-risk students for early interventions—A time-series clustering approach. *IEEE Trans. Emerg. Top. Comput.* **2015**, *5*, 45–55. [CrossRef]
20. Yu, C.C.; Wu, Y. Early warning system for online stem learning—A slimmer approach using recurrent neural networks. *Sustainability* **2021**, *13*, 12461. [CrossRef]
21. He, Y.; Chen, R.; Li, X.; Hao, C.; Liu, S.; Zhang, G.; Jiang, B. Online at-risk student identification using RNN-GRU joint neural networks. *Information* **2020**, *11*, 474. [CrossRef]
22. Liu, T.; Wang, C.; Chang, L.; Gu, T. Predicting High-Risk Students Using Learning Behavior. *Mathematics* **2022**, *10*, 2483. [CrossRef]
23. Qu, S.; Li, K.; Wu, B.; Zhang, S.; Wang, Y. Predicting student achievement based on temporal learning behavior in MOOCs. *Appl. Sci.* **2019**, *9*, 5539. [CrossRef]

24. Kusumawardani, S.S.; Alfarozi, S.A.I. Transformer Encoder Model for Sequential Prediction of Student Performance Based on Their Log Activities. *IEEE Access* **2023**, *11*, 18960–18971. [CrossRef]
25. Chen, H.-C.; Prasetyo, E.; Tseng, S.-S.; Putra, K.T.; Prayitno; Kusumawardani, S.S.; Weng, C.-E. Week-Wise Student Performance Early Prediction in Virtual Learning Environment Using a Deep Explainable Artificial Intelligence. *Appl. Sci.* **2022**, *12*, 1885. [CrossRef]
26. Song, X.; Li, J.; Sun, S.; Yin, H.; Dawson, P.; Doss, R.R.M. SEPN: A sequential engagement based academic performance prediction model. *IEEE Intell. Syst.* **2020**, *36*, 46–53. [CrossRef]
27. Aljaloud, A.S.; Uliyan, D.M.; Alkhalil, A.; Elrhman, M.A.; Alogali, A.F.M.; Altameemi, Y.M.; Altamimi, M.; Kwan, P. A deep learning model to predict Student learning outcomes in LMS using CNN and LSTM. *IEEE Access* **2022**, *10*, 85255–85265. [CrossRef]
28. Chen, Y.; Wei, G.; Liu, J.; Chen, Y.; Zheng, Q.; Tian, F.; Zhu, H.; Wang, Q.; Wu, Y. A prediction model of student performance based on self-attention mechanism. *Knowl. Inf. Syst.* **2023**, *65*, 733–758. [CrossRef]
29. Li, M.; Wang, X.; Wang, Y.; Chen, Y.; Chen, Y. Study-GNN: A novel pipeline for student performance prediction based on multi-topology graph neural networks. *Sustainability* **2022**, *14*, 7965. [CrossRef]
30. Yang, T.-Y.; Brinton, C.G.; Joe-Wong, C.; Chiang, M. Behavior-based grade prediction for MOOCs via time series neural networks. *IEEE J. Sel. Top. Signal Process.* **2017**, *11*, 716–728. [CrossRef]
31. Kuzilek, J.; Hlosta, M.; Zdrahal, Z. Open university learning analytics dataset. *Sci. Data* **2017**, *4*, 170171. [CrossRef] [PubMed]
32. Vaswani, A.; Shazeer, N.; Parmar, N.; Uszkoreit, J.; Jones, L.; Gomez, N. Attention is all you need. In Proceedings of the Advances in Neural Information Processing Systems 30: Annual Conference on Neural Information Processing Systems 2017, Long Beach, CA, USA, 4–9 December 2017.
33. Alsalman, Y.S.; Halemah NK, A.; AlNagi, E.S.; Salameh, W. Using decision tree and artificial neural network to predict students academic performance. In Proceedings of the 2019 10th International Conference on Information and Communication Systems (ICICS), Irbid, Jordan, 11–13 June 2019; IEEE: Piscataway, NJ, USA, 2019; pp. 104–109.
34. Adnan, M.; Habib, A.; Ashraf, J.; Mussadiq, S.; Raza, A.A.; Abid, M.; Bashir, M.; Khan, S.U. Predicting at-risk students at different percentages of course length for early intervention using machine learning models. *IEEE Access* **2021**, *9*, 7519–7539. [CrossRef]

applied sciences

MDPI

Article

Development and Evaluation of an Image Processing-Based Kinesthetic Learning System

Deniz Yıldız [1], Uğur Fidan [2], Mehmet Yıldız [3], Büşra Er [4], Gürbüz Ocak [5], Fatih Güngör [6], İjlal Ocak [7] and Zeki Akyildiz [3,*]

1 Graduate School of Social Sciences, Afyon Kocatepe University, 03204 Afyonkarahisar, Türkiye; dnz-dikkaya@hotmail.com
2 Department of Biomedical Engineering, Faculty of Engineering, Afyon Kocatepe University, 03204 Afyonkarahisar, Türkiye; ufidan@aku.edu.tr
3 Department of Coaching, Faculty of Sports Sciences, Afyon Kocatepe University, 03204 Afyonkarahisar, Türkiye; mehmetyildiz@aku.edu.tr
4 Department of Electrical and Electronics Engineering, Graduate School of Natural Sciences, Afyon Kocatepe University, 03204 Afyonkarahisar, Türkiye; busraerr0@gmail.com
5 Department of Curriculum and Instruction, Faculty of Education, Afyon Kocatepe University, 03204 Afyonkarahisar, Türkiye; gocak@aku.edu.tr
6 Department of Foreign Languages Education, Faculty of Education, Afyon Kocatepe University, 03204 Afyonkarahisar, Türkiye; fgungor@aku.edu.tr
7 Department of Mathematics and Science Education, Faculty of Education, Afyon Kocatepe University, 03204 Afyonkarahisar, Türkiye; iocak@aku.edu.tr
* Correspondence: zekiakyldz@hotmail.com

Abstract: This study aims to develop an interactive language learning game and explore its efficacy for English language learners. A computer-generated playground was projected onto a large classroom floor (4 × 3 m) with a wide-angle projection device. A Kinect depth camera determined the spatial positions of the playground and the positions of the students' heads, feet, and bodies. Then, we evaluated the system's effect on English education through pre- and post-tests. While there was no significant difference between the groups in terms of achievement in the pre-tests, the experimental group exhibited significantly greater improvement in the post-tests (F: 14.815, $p < 0.001$, η^2p: 0.086). Also, both groups demonstrated significant learning gains in post-tests compared to pre-tests (F: 98.214, $p < 0.001$, η^2p: 0.383), and the group x time interaction of the experimental group increased more in percentage (32.32% vs. 17.54%) compared to the control group (F: 9.166, $p < 0.003$, η^2p: 0.055). Qualitative data from student views indicated enhanced learning pace, vocabulary acquisition, enjoyment of the learning process, and increased focus. These findings suggest that a kinesthetic learning environment can significantly benefit English language learning in children.

Keywords: visual processing; kinesthetic; movement; learning; projection device

1. Introduction

Recently, many researchers in psychology and education have focused on individual differences in learning processes and suggested activities that take these differences into account [1–4]. The theory of multiple intelligences can help explain individual differences in abilities and skills [5], especially in language learning contexts where learners come together from different linguistic and cultural backgrounds [5]. Individuals may perceive their environment visually, auditorily, or kinesthetically (bodily), and their learning styles may change according to how they perceive their environment [6]. Fleming and Mills [6] proposed four learning styles, including the kinesthetic style, and the following studies [6] showed how kinesthetic learning improves students' learning outcomes.

Traditional instruction across a range of subjects, including foreign language teaching, typically relies on static visual materials and occasionally employs audiovisual content.

This learning paradigm commonly assigns students to remain seated for extended periods within classroom environments. However, research shows that prolonged sitting (20–30 min) leads to a significant redistribution of blood flow, with roughly 80% accumulating in the lower body [7]. This seating design reduces the brain's oxygen amount, induces drowsiness, and decreases focus [1]. In order to eliminate this disadvantage, many studies recommend incorporating physical activity into the learning environment [8,9]. In parallel with technological advances, recent studies on neurophysiology have supported the idea that movement can be an effective cognitive teaching model by improving memory and increasing student motivation and interest in the lesson [10,11]. Similarly, Davis argued that stimulation of the vestibular (inner ear) and cerebellar (motor activity) systems by movement activities can lead to significant gains in attention and reading [12].

The region of the brain primarily linked to motor control is the cerebellum. Interestingly, the segment responsible for processing movement is also the one involved in learning [12]. Contrary to expectations, visual data indicate the generation of new brain cells during physical activity. Pre-physical activity examination of the hippocampal region shows heightened blood flow following such activity. Activities involving play contribute to increased blood volume, providing evidence of the formation of new cells in this area. Magnetic resonance imaging (MRI) scans visually demonstrate the tangible growth of new neuronal connections following physical activity [12]. One of the effects of movement on the brain is amines (weak derivatives of ammonia). Amines are one of the brain's primary fuels in the attention system [13]. Physical movement increases the amine ratio [14]. The research on amines has revealed that a simple walk supports an increase in amines, leading to increased attention levels [13].

Moreover, Jensen [14] reported that physical movements trigger the release of stress hormones such as epinephrine and norepinephrine (adrenaline), increasing students' arousal levels and making them ready to learn. Another new theory [6] proposed that the brain is like a muscle, growing when used and weakening with inactivity. Similar to cognitive activities, physical and playful activities contribute to brain growth. Although physical movement does not make people more intelligent, it helps people focus better and optimizes the brain for learning [6]. These studies, conducted for different purposes and populations, show a potential relationship between physical activity and learning. The students who learn academic content with movement are better motivated by discriminating important information from unnecessary information, depending on the nature of the activity [6]. In psychology, this state of "arousal" can be achieved by incorporating physical activities into classroom activities, because movement increases sensory awareness, motivation, and attention, leading to effective learning [15]. Kinesthetic learning intends to strengthen students physically, emotionally, and academically. Through an active learning environment, students are more willing to cooperate because of the fun instruction, leading to less misbehavior [16]. Moreover, educators have begun incorporating kinesthetic learning activities in their classrooms because of the increased pressure for high test scores and reductions in physical education programs [17].

In the literature, movements used for educational purposes at school are divided into learning-based and non-learning-based movements. Learning-based movements aim to trigger learning and improve its quality through movement [18–23]. These movements are primarily used in mathematics, foreign language, and science courses. For example, Caterino and Polak [24] divided the students in a class into control and experimental groups. The experimental group did stretching and physical movements in the first 15 min of the lesson and then participated in the lesson again. Achievement tests were applied to both groups before and after the applications. At the end of the study, the experimental group achieved significantly more than the control group. The researchers also stated that students participating in physical activity can dramatically increase their mental concentration [24]. In a study by Hillman et al. [25], the control group was tested after 20 min of sitting, while the experimental group was tested after 20 min of running on a treadmill. The achievement test scores and the attention measurement results of the participants showed that physical activity increased attention and academic achievement more positively. The results of an American Academy of Pediatrics study of first and second graders suggest that exercise and voluntary movement may be associated with higher academic test scores. In a pre-arranged movement-based learning laboratory, students participated in activities such as saying the names of colors on colored steps while climbing stairs and moving on scooters over shapes on the floor to draw and name the shapes. The study reported that students' academic achievement tests increased significantly [8]. Studies on hyperactive students show that using therapy balls in lessons increases students' focus and thus contributes to academic achievement [26,27]. In their study, Vazou et al. [11] showed that physical movement practices in the classroom environment motivated students by entertaining them much more, increasing their academic achievement. In a study conducted by Flippin [16], the impact of employing kinesthetic equipment on students' task behaviors was explored. Using exercise balls and standing desks in the classroom revealed a noticeable increase in task behaviors. These findings suggest that incorporating more movement into the classroom environment can prove advantageous for teachers, without requiring extensive planning and instruction time.

Another investigation by Beserra [28] delved into the acceptability and feasibility of delivering physically active academic lessons through dance. Thirty-seven teachers developed a series of dance routines based on mathematical functions. They asserted that synchronizing the mind and body enhances motor, coordination, memory, reading, speech, language, and mathematical skills. Additionally, the study noted a growing interest among teachers in integrating physically active academic lessons into their regular school routine.

Interactive motion-based games have recently attracted more attention as an alternative to computer-based educational games. One of the most popular applications in this field is the XBOX game system developed by Microsoft. The essential feature of the Kinect camera set for interaction in this system is its ability to detect the movements of the human body and transfer them to a computer environment in a non-contact way. In this system, a person enters the game in three dimensions and interactively participates. This system is used not only for gaming purposes but also for education and training activities. Zhang [29] reported that Kinect allows students to communicate more with games through their movements. As a learning tool, XBOX Kinect potentially increases student motivation and creates a pleasant classroom environment [29]. Numerous studies [30,31] have documented the potential benefits of Kinect-based interactive systems for language learning, demonstrating efficacy when compared to traditional methods. Through this technology, a person can create movement in the system by using only their bodily limbs from the point where they are located. This leads to movement limitation and does not reflect real movements. Similarly, while touchscreen smart boards offer multisensory interaction to teachers and students, their limited screen size and static positioning present constraints on movement diversity and range, potentially hindering kinesthetic learning. Therefore, it is important to create computer game-based programs embedded within classroom settings. Such programs can leverage the inherently engaging nature of games while promoting active participation and natural, embodied language use, potentially offering a more holistic and effective approach to foreign language education. Recently, the development of projection technology has helped us obtain wide-angle images with a high resolution from short distances. Also, the fact that depth cameras can determine people's limbs and spatial positions motivates the idea that a large area can be used as a touch screen. Based on this idea, we developed a kinesthetic learning system in which we created a training and game area on the floor (4 $\times$ 3 m) with a high-resolution projection device mounted on the ceiling of a classroom to perceive the limb movements and spatial positions (x, y, and z) of students in a game area with the depth camera, thus creating a large touch area and allowing us to develop game software containing the words and dialogues of the "Fun With Science", "Prepositions", and "Jobs" units. With this system, students can interact with the learning content through movement within the designated area. Based on our review of the literature, this study is the first in its field and will make valuable contributions to the literature. Therefore, this study aims to investigate the effect of the developed system on English language learning achievement and tries to answer the following research questions:

- How does teaching English with the newly developed system affect students' word learning levels compared to the classical method?
- What are the students' attitudes and opinions toward this system?

2. Materials and Methods

2.1. Experimental Design

This study was carried out in two phases. The first phase was the development of the system, and the second phase was the evaluation process of the developed system. The learning environment designed in this planning process was created based on [32] "Generic Model" instructional design model. When the stages of this model, called Analysis, Design, Development, Implementation, and Evaluation (ADDIE) were examined, this model completely fit with the purpose of this study. In the analysis step, the learning environment (technological equipment in the school, existing equipment, and classroom layout) was analyzed, and pre-tests were administered to determine students' prior knowledge of the vocabulary in the relevant units of the curriculum. In the design stage, we designed the learning environment, selected the appropriate visuals and data to be used in the system, coded software, reviewed the measurement tools, and finalized them. In the development step, the final data entries for the system were provided, and a piloting application was tried with a few students. This step helped to bring the system to its last version. In the implementation phase, the image processing-based kinesthetic learning system was put into use, the active participation of the learners was ensured, and appropriate prompts and reinforcements were given through the system. In this process, data on the effectiveness of the process were collected through student opinions and video recordings. Finally, in the evaluation step, post-tests were applied to determine the system's effectiveness in the experimental group and the lesson taught in the control group. Following implementation, we collected students' opinions about their motivation, focus, and attitude toward the course.

2.2. First Phase

Development of the Image Processing-Based Kinesthetic Learning System

The first phase was to develop the hardware and software contents of the image processing-based interactive kinesthetic learning system.

Physical Layout of the Projection Device

A projection device Xbox 360 (HD29He; Optoma, Fremont, CA, USA) with 4000 lumens was used in this study. The physical layout in Figure 1 was transferred to the school environment. The lens of the projection device mounted at a height of 3.8 m from the entrance wall of the activity area, centered on the activity area, was oriented so that the part where the lens of the projection device was located would see the floor, and a reflection area of approximately 15 m^2 was created on the floor. While determining the projection area, the field of view that the Kinect sensor could detect during movement was taken into account. The projection device settings in Table 1 were used for image enhancement of the projected image. In order to transfer the data received from the Kinect device to the projection device accurately and to minimize the deviations in the vectorial positions of the limbs, the lens of the projection device hanging on the wall and the camera in Kinect One were aligned. In this way, the positions of the body parts in the game were transferred accurately. The distance from the Kinect sensor to the projection mirror was never below the sensor's field of view. For this sensor, which can detect objects at a minimum distance of 1 m, the distance between the projection and the Kinect sensor was set to 1.4 m.

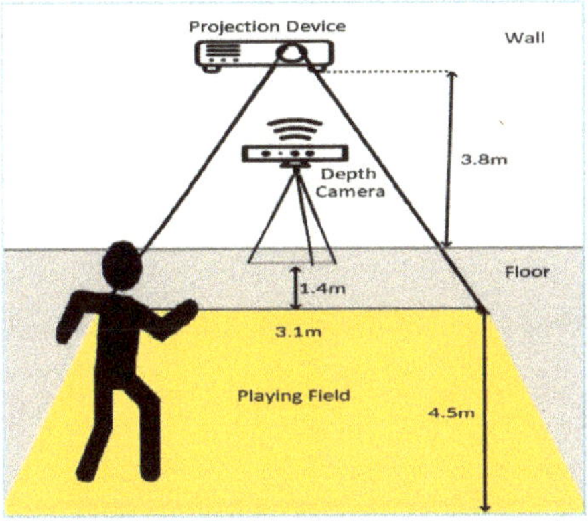

Figure 1. Physical layout of the system (Yellow; Playing Field, Brown; Floor, White; Wall).

Table 1. Technical details of the projection device.

Menu	Sub-Menu	Rate	Preference
Screen	Aspect ratio	4:3 16:9 LBX Natural Auto	16:9
	Edge masks	0~10	1
	Keystone correction	−40~40	−17
Audio	Silent	Open Closed	Open
	Audio	0~10	10
Settings	Projection	Front Rear Ceiling Rear top	Rear top

Optimization of the Data in the System

In the design of the projection-based kinesthetic English language learning module, the data received from the Xbox Kinect One sensor were converted into coordinate data through the Kinect for the Windows SDK v1.8 program. This plugin made the data meaningful, and the limb positions on the x, y, and z axes were accurately transferred to the computer environment. The x data from the Kinect sensor define the position of the selected limb on the horizontal axis, the data on the y-axis depict the position of the selected limb on the vertical axis, and the data on the z-axis define the distance between the selected limb and the sensor. The communication between the program plugin and the data received in the Unity game engine used in the realization of the games was provided by the Kinect for the Windows Unity Pro 2.0 library.

Figure 2 shows the skeletal system structure that can be obtained from the Kinect sensor. In this study, after creating the test setup, we predicted that the data from joints 0, 3, 15, and 19 in Figure 2 could carry the appropriate characteristics for the system requirements. However, after the experiments, the data from limbs 0 and 3 caused incorrect responses. These limbs may have remained in front or behind when the person bended or jumped. As a result, the best coordinates were obtained from the right and left foot joints numbered 15 and 19, respectively. In this way, the margin of error between the data received from the sensor and the position of the person in the game was reduced to ±8 cm.

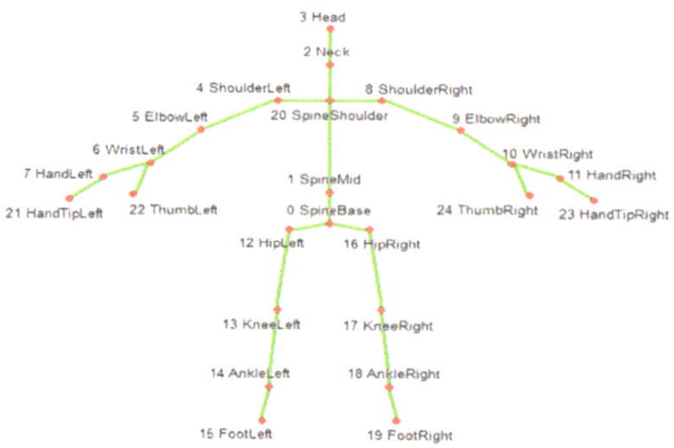

Figure 2. Positions of the limbs in the skeletal system from the Kinect One sensor.

Projection-Based Kinesthetic Learning System

The software interface (Figure 3) was designed to be quite simple and plain in order to be user-friendly. When the software is started, the entire curriculum and the activity content of the curriculum are listed. As explained in Table 2, the program includes a total of 13 different activities belonging to 3 different topics. High-frequency and low-decibel background music (jazz and/or classical) was added to the games to contribute to people's attention and focus. Figures 4–6 depict the interface of the activity-based games belonging to professions, prepositions, and science sections. Table 2 shows the vocabulary and sentence groups determined as the outcomes of these sections.

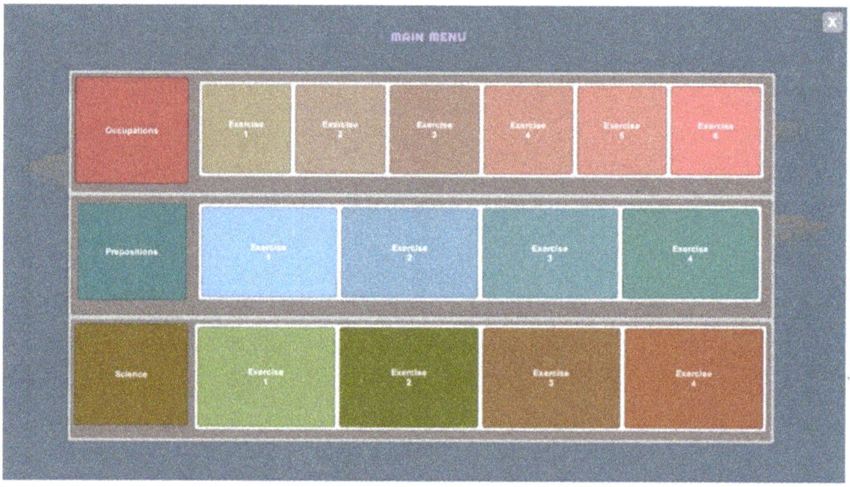

Figure 3. Main menu designed for the application.

Unity engine 2021.3.21f1 LTS version was used to create the activities in the design of the projection-based kinesthetic English language learning module. Game scenarios were created in line with predetermined game scenarios and activity objectives. The elements that provide the game dynamics in the programming of the games are the right and left foot coordinate data received from the Kinect sensor and subjected to a certain pre-processing. The developed activities can be used in 64-bit and 32-bit Windows operating systems.

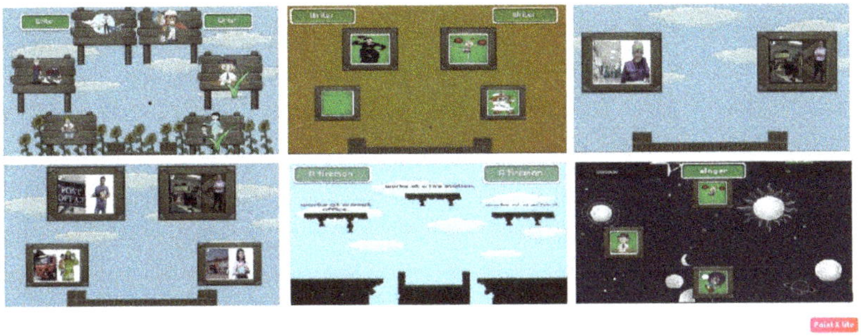

Figure 4. Professions section game designs.

Table 2. Topics, target vocabulary items, and target sentence groups used in the English lesson.

Topic	Target Vocabulary Items	Target Sentence Group
Professions	Doctor, Farmer, Writer, Painter, Teacher, Businessman, Singer, Actor, Actress, Pilot, Policeman, Nurse, Fireman, Vet, Waiter, Chef	A nurse works at a hospital.
		A chef works at a restaurant.
		A teacher works at a school.
		A fireman works at a fire station.
		A policeman works at a police station.
		A postman works at a post office.
Prepositions	In, On, Behind, Under, In front of, Near	It is under the television.
		It is in front of the sofa.
		It is between the three chicks.
		It is behind the sofa.
		They are on the bookshelf.
		It is on the sofa.
		It is near the sofa.
		It is on the wall.
Science	Break, Cup, Experiment, Flower, Fold, Freeze, Glass, Goggles, Jar, Melt, Mix, Paper, Pebbles, Peel, Plant, Pour water, Put, Science, Toothbrush, Scissors	Fold the paper.
		Melt the ice.
		Do an experiment.
		Plant a flower.
		Pour some water.
		Break an egg.
		Peel an orange.
		Freeze the water.

Figure 7 shows the physical environment where the system was implemented in the pilot school. The system allows one user to six users to perform the activities simultaneously. Thus, the level of entertainment was increased by creating competition among students. Also, there are different levels in the games, and the activity levels can be adjusted according to the level of the student, from easy to difficult. Similarly, the duration and difficulty level of the games can be adjusted by the teacher according to the level of the students.

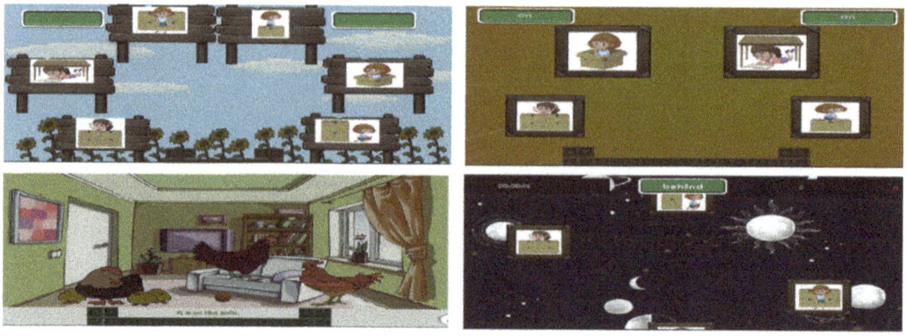

Figure 5. Prepositions section game designs.

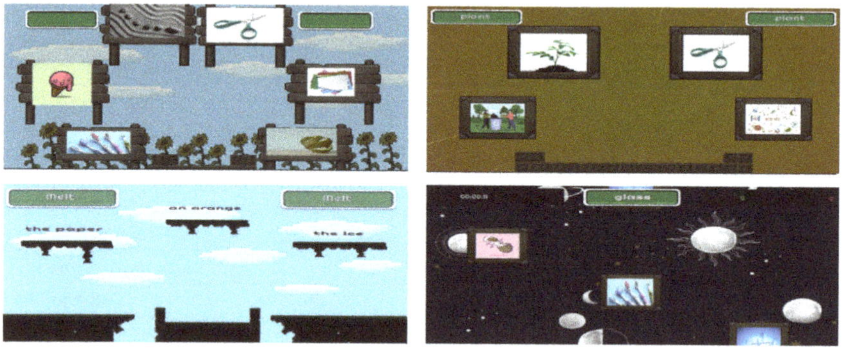

Figure 6. Science section game designs.

Figure 7. Application in the classroom.

How the System Works

Since the position of the image is clear when any image to be shown in the software is reflected on the floor, the system is designed to perform a new operation when the position of the student's feet is in the same position as the image. Figure 7 shows that the spelling of an English word was projected on the floor, and an audio stimulus was given through the projection software. Then, the correct image of the given word and the wrong images were projected at various points of the projection area. As soon as the student moved to the correct image and put his/her foot on it, the depth camera detected the spatial position of his/her feet, and a new operation was performed. If the student selected the wrong image, a red cross appeared next to the warning buzzer. The visuals were animated (the visuals moved around in the playground) in some of the applications in the system software, and the students were asked to react according to these animated visuals. For example, when the spelling of an English word and the audio stimulus were given, one correct image and

3–4 incorrect images started to flow in the upper part of the projection projected on the floor in the game-based exercises. The student was asked to select the correct image by moving quickly without leaving the projection area or stepping on the wrong images. The applications such as games, exercises, tests, group work, and so on were developed based on fixed and moving images.

2.3. Second Phase

Evaluation of the System

In implementing and evaluating the system, we designed the research process following an explanatory sequential design, which started with a quantitative stage and continued with a qualitative stage to explain the quantitative results. Qualitative data were analyzed to help explain the quantitative results [9]. In the quantitative stage of the study, we calculated statistical significance, confidence intervals, and effect sizes and presented the main findings. The research was finalized by continuing with the qualitative stage to explain the quantitative results. In this process, a semi-structured interview method was applied to obtain the students' opinions about the system's effectiveness. Thus, this two-stage evaluation process built on itself helped us understand the effect of the image processing-based kinesthetic learning system more reliably.

The first stage of the evaluation was conducted on the experimental and control groups formed in accordance with the quasi-experimental design, one of the quantitative research designs. The effects of the developed image processing-based kinesthetic learning system on students' foreign language learning levels were determined with the experimental group's applications. The experimental and control groups, whose application conditions were equalized, were administered the pre-test developed by the researchers regarding the "Fun With Science", "Prepositions", and "Jobs" units before the procedure. Then, the experimental group students were taught language through the image processing system, while the control group was taught in accordance with the curriculum. After three weeks of implementation, a post-test was administered to both groups, and student scores were compared in terms of significant differences. After analyzing the quantitative data in the first stage of the evaluation, the second stage, the qualitative stage, was started. In the qualitative stage, the data on the implementation, usability, and perception of the image processing-based kinesthetic learning system were collected and evaluated through focus group interviews with the experimental group students.

2.4. Participants

The study group of the research consisted of 81 students in two different classes studying in the fourth grade of primary school (experimental group n: 41 and control group n: 40). The socioeconomic level of the students was mainly at the middle level and consisted of parents who were closely interested in student achievement.

Since the experimental group was underage, a "Family Consent Form" was prepared for them to be allowed to participate in the study. In addition, the ethics committee approval report numbered 86,082 was obtained from the Afyon Kocatepe University's ethics committee at the Graduate School of Social Sciences.

2.5. Application Procedure

This study was conducted for five weeks with the participation of fourth-grade primary school students. A total of three applications were conducted for three weeks after the pre-test, lasting one class hour (40 min) per week. Within the scope of the fourth-grade English curriculum, a total of eighteen unknown words in two units in the textbooks were identified. The words were taught with the developed system in the experimental group during one class hour of the three-hour weekly English lesson for three weeks and then reinforced with different activities. The selected words were taught and reinforced in the control group with the traditional method (slide show). Before the study, both groups' prior knowledge (pre-test) about the words to be used was determined. The same test was administered to the groups again as a post-test after the activities were completed. While planning the activities, we endeavored to design different activities for the control and experimental groups to teach the same vocabulary for the same periods of time.

2.6. Quantitative and Qualitative Research

This study used a semi-structured interview form, video recordings, and the researchers' diary as qualitative data collection tools. The Vocabulary Knowledge Scale was used as a quantitative data collection tool. This scale was developed by Paribakht and Wesche (1993) [33].

2.7. Data Analysis

We used descriptive analysis in this study. It is an analysis technique in which the conceptual structure of the research is clearly determined in advance. The primary purpose of descriptive analysis is to quantify qualitative data, analyze descriptive possibilities with frequencies, and transfer the findings to the reader in an organized and interpreted manner. Finally, the findings were enriched and interpreted with relevant texts according to their importance in answering the research questions, and the findings were supported with direct quotations. In order to increase the reliability of the determined themes and codes, more than one researcher independently marked the interview forms and the interview coding key by marking the appropriate option for the interviewed students' opinions on the relevant interview coding key. To determine the consistency of the markings by the researchers on the interview coding key, we checked the answers to each question one by one. We marked them as "Consensus" or "Disagreement", and the calculation was made using Miles and Huberman's reliability formula P (Percentage of Consensus) = Na (Consensus)/Na (Consensus) + Nd (Disagreement) $\times$ 100 to ensure validity and reliability(reference).

The SPSS 18.0 (Statistical Package for the Social Sciences) package program was used in the statistical analysis of the quantitative data. Descriptive statistics such as the percentage, frequency, and arithmetic mean were used to analyze the quantitative data collected at different stages of the study. The repeated measures ANOVA test was applied to determine the differences according to group and time interaction. A value of $p < 0.05$ was taken as a significance value. A partial η^2 was used to determine the effect size. For the partial η^2 values, 0.01 was considered a low effect size, 0.06, a medium effect size, and 0.14, a large effect size. Pre-test and post-test proportional (%) differences in the groups were calculated by the following formula: ((post-test − pre-test)/pre-test $\times$ 100).

3. Results

Figure 8 shows a significant difference between the repeated measurements of the experimental and control groups in terms of Group: F: 14.815, $p < 0.001$, η^2p: 0.086; Time: F: 98.214, $p < 0.001$, η^2p: 0.383; GroupxTime: F: 9.166, $p < 0.003$, η^2p: 0.055. Although the Bonferroni (post hoc) test showed no significant difference between the experimental and control groups in the pre-tests (<0.10), a significant difference was found in favor of the experimental group in the post-tests (68.317 $\pm$ 8.033 vs. 59.450 > 8.670). Furthermore, there was a significant increase in the post-tests compared to the pre-tests ($p < 0.001$) in both groups ((experimental group pre-test: 51.634 $\pm$ 6.647 vs. post-test: 68.317 $\pm$ 8.033), (control group pre-test: 50.575 $\pm$ 9.279 vs. post-test: 59.450 $\pm$ 8.670)); however, the groupxtime interaction results revealed that the experimental group showed a higher increase in percentage (32.32% vs. 17.54%) compared to the control group.

How Did Students React to the Image Processing-Based Kinesthetic Learning System?

In this section, we analyzed the answers the students gave to the questions in the questionnaire form and present them in line with the study's aims. The forms collected from the students were coded as S1, S2, S3,. . .S38, and the statements revealing the views of the students were quoted in accordance with these codes. At this stage, similar and different parts of the data were noted, and the coding process was carried out. The questions in the interview form formed the categories of the research. The categories identified are the following: "the contribution of learning English through the kinesthetic learning system to students", "students' views on English vocabulary learning activities through the kinesthetic learning system", "the use of this application in other courses", and "students' views on the differences in kinesthetic learning system application from other educational tools". The data under these categories were analyzed descriptively.

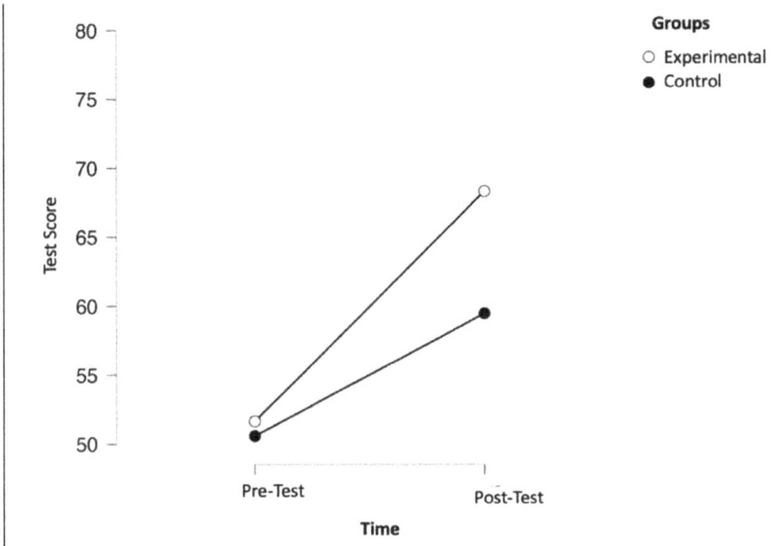

Group: F:14.815; $p < 0.001$; η^2_p: 0.086
Time: F:98.214; $p < 0.001$; η^2_p: 0.383
Group x Time: F:9.166; $p < 0.003$; η^2_p: 0.055
Change percentage: Experimental group: 32.32%; Control group: 17.54%

Figure 8. Change values of experimental and control groups according to group time.

The students were asked what they gained from learning English through the kinesthetic learning system, and their views on this issue were determined. Table 3 shows students' views on learning English through the kinesthetic learning system.

Table 3. Contributions of the kinesthetic learning system to students' English learning.

Category	F	%
Faster learning	14	28
Contributed to my learning	12	24
Learning vocabulary items permanently	10	20
Fun	6	12
Did not contribute to my learning	4	8
Learning by moving	4	8
Total	50	100

Table 4 shows that the students learned faster (28%), learned unknown words permanently (20%), found it fun (12%), and learned by moving (8%). While some students reported the system's contribution to their learning (24%), some students (8%) mentioned that the system did not contribute to their learning.

Table 4. Students' views on the entertainment level of the system.

Category	F	%
Fun	29	52.72
Learning by using the whole body	7	12.72
Learning by playing	7	12.72
Not fun	4	7.27
Learning unknown vocabulary items	3	5.45
Increasing motivation	3	5.45
Relaxing learning environment	2	3.63
Total	55	100

The following include examples of views of students of the system:

S2: It contributed to me because we play with our whole body, so it is more fun, and I can learn more quickly.
S4: It contributed because when we use our whole body, the words stay in mind.
S6: It contributed to me because I both learned and had fun.
S14: It contributed because the words became more permanent in my mind.
S26: It contributed to me. I learn very well when I use visuals and my body.

The students were asked whether learning English vocabulary through the kinesthetic learning system was fun or not, and the students' views on this issue were determined. The students' views are summarized in Table 4.

Table 4 shows that students found the kinesthetic learning system fun (52.72%), learned by using their whole body (12.72%), learned by playing games (12.72%), did not find the activities fun (7.27%), learned unfamiliar words (5.45%), increased their motivation (5.45%), and found the learning environment relaxing (3.63%).

The following include examples of views of students of the entertainment level of the system:

S3: It was a lot of fun because when I am moving, I get excited to learn, and I get excited when I am moving, and that's what makes it fun.
S6: I liked it very much because I am moving and it's a lot of fun.
S11: It was fun because, let's face it, a textbook lesson is a bit boring.
S13: Yes, because you are moving and learning English at the same time.
S17: It was fun because we are learning by playing games.
S23: I had fun because I was moving inside while doing it.
S29: I didn't think it was fun because we have weekly lessons in the books.
S33: It was fun because it made me understand the words better.
Ö34: It was fun because the pictures looked like a game to me. Or rather, it was fun because it was a game.
Ö36: It was a lot of fun because having fun and learning with friends is very nice.

Students were asked what they thought about the use of the kinesthetic learning system in other courses. Table 5 summarizes their views on the use of this application in other courses.

Table 6 reveals that students wanted the kinesthetic learning system to be used in mathematics (34.84%), Turkish (16.66%), science (13.63%), physical education (7.57%), social studies (7.57%), music (4.54%), arts (6.05%), and all courses (4.54%). The percentage of students who did not want it to be used in other courses was 4.54%.

Table 5. Views on the use of the kinesthetic learning system in other courses.

Category	f	%
Mathematics	23	34.84
Turkish	11	16.66
Science	9	13.63
Physical education	5	7.57
Social Studies	5	7.57
Music	3	4.54
Arts	4	6.05
All courses	3	4.54
I do not want to use it in other courses	3	4.54
Total	66	100

Table 6. Student views about the differences that distinguish the kinesthetic learning system application from other educational tools.

Category	f	%
Mobilizing the whole body	22	41.50
Learning by having fun with games	15	28.30
Ensuring attention and focus in the lesson	14	26.41
Not being boring	2	3.77
Total	53	100

The following include examples of students' views on the use of the kinesthetic learning system in other courses:

S2: I think mathematics is suitable for it. For example, we can mark the answer 2 + 2 = with our body.
S7: It can facilitate my learning in Turkish, mathematics, and science courses.
S11: In fact, I would like it to be used in all other courses except for only one drawing course (mathematics, Turkish, science, physical education, and social studies).
S18: I would like it in Turkish class because it would be more fun.
S24: I'd like to, especially if it's in math.
S27: I would like it to be prepared, especially for experiments in science class.
S32: Yes, I would like to, and I would like to do it in every course.
S33: I wouldn't want to because that's how I understand other subjects better.

Students were asked about the differences distinguishing the kinesthetic learning system application from other educational tools. Table 6 summarizes student views about the differences distinguishing the kinesthetic learning system application from other educational tools.

Table 6 shows that the activities mobilized students' whole bodies (41.50%), they learned by having fun with games (28.30%) and increased their attention and focus (26.41%), and the activities were not boring (3.77%).

The following include examples of students' views on the differences discriminating the kinesthetic learning system from other educational tools:

S2: It's different because we don't just use our hands; we do it with our whole body. I think that's the difference.
S4: We use our whole body, and the most significant difference is that it is very fun like a game.

S7: It is more fun and not boring. I have fun and learn even while waiting for my friends to do an activity.
S16: It is more fun than other English classes and it was very nice to mark with my foot without having to write.
S28: It made learning English words more fun, but I would have gotten bored if it was a blackboard or a textbook.
S33: I think the biggest difference is that it attracts our attention. For example, when we get bored of sitting, we cannot pay attention to the lesson. In this method, we did the lesson by jumping while standing so we could pay attention to the lesson with this method.
S35: Because in the other lessons, the mind and intelligence are used. In this one, our whole body is actively involved in the lesson.
S38: It's on the ground, and it's playful, so it stands out from other vehicles. I think it's a lot of fun.

4. Discussion

This study implemented an immersive and interactive English language learning environment projected onto a large classroom floor. To ensure accurate interaction between students and the projected content, we tackled two key challenges: pixel distortion and shadow interference. Initially, a pixel distortion approach was considered but abandoned due to limitations in its scalability and accuracy. Instead, we opted for determining participants' real-time three-dimensional coordinates (X, Y, and Z) utilizing a depth camera. The Kinect One camera, the most widely used camera on the market, was chosen to identify user positions and bodily movements, overcoming the limitations of traditional pixel-based interactions. Additionally, the camera's high frame rate (30 fps) ensured smooth and responsive gameplay within the projected area. This approach effectively mitigated shadow interference, a potential issue with projected floor displays. By directly tracking individual body positions rather than relying on pixel analysis, the system disregarded shadows cast by users, preventing inaccurate inputs and ensuring reliable interaction with the learning content.

While the existing literature on interactive learning games is limited, several studies offer relevant comparisons. Si et al. (2013) [32] employed a Kinect sensor-based system for Chinese language education, transitioning away from traditional keyboard–mouse and touch screens. The virtual environments created in the study only worked with a computer interface however, and they remained confined to the computer screen, limiting opportunities for physical activity. In the sports field, Düking et al. (2016) [34] used a system to improve and measure the agility values of athletes by applying contact plates to certain parts of the floor. In this system, the athlete sees which contact plates to go to on the television screen in front of him/her. However, our system employed projected images instead of contact plates, allowing for easier manipulation and flexibility in pattern and object design. This eliminates limitations on the number and complexity of elements within the learning environment. By incorporating physical movement and interaction with projected images, our system encourages active participation and potentially enhances student engagement compared to screen-based solutions. In addition, our system facilitates collaborative learning through its multi-user functionality, enabling group activities and fostering social interaction within the learning process.

Following the system development and pre-testing phase, the application phase was implemented. Upon completion, post-test achievement assessments were administered to both the experimental and control groups, and qualitative data were collected through student interviews. Analysis of pre-test scores revealed no significant differences between the groups. However, post-test results demonstrated a marked improvement in the experimental group compared to the control group ($p < 0.001$), indicating a positive impact of the kinesthetic learning system on learning outcomes. This finding was further corroborated by a significant group-by-time interaction effect (F: 9.166, $p < 0.003$), suggesting that the intervention specifically influenced the trajectory of learning over time. Qualitative data gleaned from student interviews offered valuable insights into the students' experiences with the system. Participants reported faster learning, enhanced vocabulary retention, increased enjoyment of learning activities, improved focus and attention during lessons, and a positive perception of the system's contribution to their learning. Notably, students perceived the system as facilitating learning through both game-based elements and physical movement. Additionally, they expressed a desire for similar systems to be implemented in other subjects, including mathematics, Turkish, and science. Similar to this study, studies in which the Kinect system was used for teaching purposes concluded that it was much more effective than classical teaching methods with a significant difference, especially in English language teaching [30,31]. Shakroum, Wong, and Fung (2018) [35] found that a gesture-based learning system with a Kinect sensor (GBLS) positively affected students' intrinsic motivation. Hsu (2011) reported that XBOX Kinect, as a learning tool, can potentially increase student motivation and create a pleasant classroom environment. Zhang (2012) [29] reported that Kinect allows students to communicate more with games through their own movements. In almost all of the studies in the literature where the Kinect sensor was used for educational purposes, the participants were involved in games with a television screen on the opposite side in 3D. In addition, they realized data flow with their hands without moving too much from their position. This leads to movement limitation and cannot reflect real movements. Therefore, both the motivation and focus of individuals and kinesthetic learning will remain limited. However, in this study, the participants moved in a 4×3 m area and provided the data flow with their feet. The qualitative data and video recordings show that the students' focus and enjoyment levels were very high. These levels were directly reflected in the achievement test results.

Numerous studies in the existing literature indirectly support the findings of this research, demonstrating that kinesthetic learning contributes to increased active engagement in English as a second language (ESL) classes [36,37]. Additionally, multiple investigations have found that kinesthetic learning fosters active participation among ESL students in various activities [38]. A study by Brady et al. [39] reported that engaging in physically active classroom lessons could enhance students' comprehension of the content, aiding in long-term retention. Beyond academic achievement, integrating movement may be beneficial for students facing challenges in grasping academic content. The observed enhancement in academic performance and positive student attitudes align with the existing literature, suggesting that the kinesthetic approach influences the brain, thereby improving learning and cognitive functions. The positive impact of physical activity on memory, learning, anxiety, stress, depression, and attention deficit is evident, all of which significantly affect students' academic performance [40–42]. Research indicates a connection between the brain region associated with "body learning" and cognitive processes, highlighting the natural stimulant effect of exercise, play, and activity on circulatory and neurovascular systems [14]. Ratey [43] further reported that physical activity enhances learning by optimizing mindsets for improved alertness, attention, and motivation, promoting the connection of nerve cells in response to new information, and fostering the development of stem cells into nerve cells for efficient internal information transfer.

However, a notable limitation of this system is its dependency on sufficient classroom space for the projection onto the floor. Its applicability is constrained in classrooms with limited space and areas featuring a multicolored background. Another drawback is the necessity for a high-resolution projection device and depth camera. Additionally, crowded classrooms may impede student rotation and complicate classroom management.

5. Conclusions

This study investigated the efficacy of a novel "Projection-Based Kinesthetic English Language Learning Module" in promoting second language learning. The experimental group engaging with the interactive learning environment demonstrated significantly greater improvement in post-test scores compared to the control group, indicating enhanced learning outcomes. Furthermore, group-by-time interaction analysis revealed a statistically significant interaction effect, suggesting the intervention's specific impact on learning trajectory. Qualitative data from student interviews corroborated the quantitative findings, highlighting self-reported faster learning, vocabulary acquisition, increased enjoyment, and improved focus during lessons. These findings collectively support the potential of kinesthetic, immersive learning environments in facilitating English language acquisition.

While acknowledging this study's limitations, we recommend utilizing this module as an active learning tool to complement traditional second language instruction. Additionally, student feedback points toward the potential application of similar software in diverse educational domains, including mathematics, science, and preschool education. Further research could explore the efficacy of such systems in promoting cognitive development in elderly populations and enhancing motor skills in sports training. In conclusion, this study offers valuable insights into the benefits of kinesthetic interaction for language learning, paving the way for further exploration of immersive technologies in educational settings.

Author Contributions: Conceptualization, D.Y., U.F., M.Y., B.E., G.O., F.G., İ.O. and Z.A.; methodology, D.Y., U.F., M.Y., B.E., G.O., F.G., İ.O. and Z.A.; formal analysis, D.Y., U.F., M.Y., B.E., G.O., F.G., İ.O. and Z.A.; investigation, D.Y., U.F., M.Y., B.E, G.O., F.G., İ.O. and Z.A.; writing—original draft preparation, D.Y., U.F., M.Y., B.E., G.O., F.G., İ.O. and Z.A.; writing—review and editing, D.Y., U.F., M.Y., B.E., G.O., F.G., İ.O. and Z.A.; visualization, D.Y., U.F., M.Y., B.E., G.O., F.G., İ.O. and Z.A. All authors have read and agreed to the published version of the manuscript.

Funding: This research was funded by the Scientific and Technological Research Council of Türkiye (TÜBİTAK) 1002 Short Term R&D Funding Program.

Institutional Review Board Statement: This study was conducted in accordance with the Declaration of Helsinki and approved by the institutional ethics committee of Afyon Kocatepe University's Graduate School of Social Sciences (86082). Date of approval from the ethics committee: 11 March 2022.

Informed Consent Statement: Informed consent was obtained from all subjects involved in this study.

Data Availability Statement: Data are available from the authors and can be provided upon request.

Acknowledgments: We thank the participants of the study.

Conflicts of Interest: The authors declare no conflicts of interest.

References

1. Richard, L.; Allington, R.G. *Classroom Influences on Individual Differences*; Routledge: London, UK, 2015.
2. Barbot, B.; Lubart, T.I.; Besançon, M. "Peaks, Slumps, and Bumps": Individual Differences in the Development of Creativity in Children and Adolescents. *New Dir. Child Adolesc. Dev.* **2016**, *2016*, 33–45. [CrossRef]
3. Costley, T.; Gkonou, C.; Myles, F.; Roehr-Brackin, K.; Tellier, A. Multilingual and Monolingual Children in the Primary-Level Language Classroom: Individual Differences and Perceptions of Foreign Language Learning. *Lang. Learn. J.* **2020**, *48*, 643–655. [CrossRef]
4. Kim, Y.-S.; Petscher, Y.; Foorman, B. The Unique Relation of Silent Reading Fluency to End-of-Year Reading Comprehension: Understanding Individual Differences at the Student, Classroom, School, and District Levels. *Read. Writ.* **2015**, *28*, 131–150. [CrossRef]
5. Emmiyati, N.; Rasyid, M.A.; Rahman, M.A.; Arsyad, A.; Dirawan, G.D. Multiple Intelligences Profiles of Junior Secondary School Students in Indonesia. *Int. Educ. Stud.* **2014**, *7*, 103. [CrossRef]
6. Fleming, N.D.; Mills, C. Not Another Inventory, Rather a Catalyst for Reflection. *Improv. Acad.* **1992**, *11*, 137–155. [CrossRef]
7. Corbin, B. *Unleashing the Potential of the Teenage Brain: Ten Powerful Ideas*; Corwin Press: Thousand Oaks, CA, USA, 2008.
8. Hillman, C.H.; Erickson, K.I.; Kramer, A.F. Be Smart, Exercise Your Heart: Exercise Effects on Brain and Cognition. *Nat. Rev. Neurosci.* **2008**, *9*, 58–65. [CrossRef] [PubMed]
9. Tomporowski, P.D.; Davis, C.L.; Miller, P.H.; Naglieri, J.A. Exercise and Children's Intelligence, Cognition, and Academic Achievement. *Educ. Psychol. Rev.* **2008**, *20*, 111–131. [CrossRef] [PubMed]
10. Robinson, L.E.; Wadsworth, D.D. Stepping toward Physical Activity Requirements: Integrating Pedometers into Early Childhood Settings. *Early Child Educ. J.* **2010**, *38*, 95–102. [CrossRef]
11. Vazou, S.; Gavrilou, P.; Mamalaki, E.; Papanastasiou, A.; Sioumala, N. Does Integrating Physical Activity in the Elementary School Classroom Influence Academic Motivation? *Int. J. Sport Exerc. Psychol.* **2012**, *10*, 251–263. [CrossRef]
12. Davis, C.L.; Tomporowski, P.D.; McDowell, J.E.; Austin, B.P.; Miller, P.H.; Yanasak, N.E.; Allison, J.D.; Naglieri, J.A. Exercise Improves Executive Function and Achievement and Alters Brain Activation in Overweight Children: A Randomized, Controlled Trial. *Health Psychol.* **2011**, *30*, 91–98. [CrossRef]

13. Saklofske, D.H.; Blomme, G.C.; Kelly, I.W. The Effects of Exercise and Relaxation on Energetic and Tense Arousal. *Pers. Individ. Dif.* **1992**, *13*, 623–625. [CrossRef]
14. Jensen, E. *Learning With the Body in Mind: The Scientific Basis for Energizers, Movement, Play, Games, and Physical Education*; SAGE Publications: Sauzendoaks, CA, USA; Pennsylvania State University: Philadelphia, PA, USA, 2000.
15. Lambourne, K.; Hansen, D.M.; Szabo, A.N.; Lee, J.; Herrmann, S.D.; Donnelly, J.E. Indirect and Direct Relations between Aerobic Fitness, Physical Activity, and Academic Achievement in Elementary School Students. *Ment. Health Phys. Act.* **2013**, *6*, 165–171. [CrossRef] [PubMed]
16. Flippin, M.; Clapham, E.D.; Tutwiler, M.S. Effects of Using a Variety of Kinesthetic Classroom Equipment on Elementary Students' on-Task Behaviour: A Pilot Study. *Learn Environ. Res.* **2021**, *24*, 137–151. [CrossRef]
17. Webster, C.A.; Zarrett, N.; Cook, B.S.; Egan, C.; Nesbitt, D.; Weaver, R.G. Movement Integration in Elementary Classrooms: Teacher Perceptions and Implications for Program Planning. *Eval. Program Plan.* **2017**, *61*, 134–143. [CrossRef]
18. Holt, E.; Bartee, T.; Heelan, K. Evaluation of a Policy to Integrate Physical Activity Into the School Day. *J. Phys. Act. Health* **2013**, *10*, 480–487. [CrossRef] [PubMed]
19. Reed, J.A.; Einstein, G.; Hahn, E.; Hooker, S.P.; Gross, V.P.; Kravitz, J. Examining the Impact of Integrating Physical Activity on Fluid Intelligence and Academic Performance in an Elementary School Setting: A Preliminary Investigation. *J. Phys. Act. Health* **2010**, *7*, 343–351. [CrossRef]
20. Shoval, E. Using Mindful Movement in Cooperative Learning While Learning about Angles. *Instr. Sci.* **2011**, *39*, 453–466. [CrossRef]
21. Trost, S.G.; Fees, B.; Dzewaltowski, D. Feasibility and Efficacy of a "Move and Learn" Physical Activity Curriculum in Preschool Children. *J. Phys. Act. Health* **2008**, *5*, 88–103. [CrossRef]
22. Peiris, C.L.; O'Donoghue, G.; Rippon, L.; Meyers, D.; Hahne, A.; De Noronha, M.; Lynch, J.; Hanson, L.C. Classroom Movement Breaks Reduce Sedentary Behavior and Increase Concentration, Alertness and Enjoyment during University Classes: A Mixed-Methods Feasibility Study. *Int. J. Environ. Res. Public Health* **2021**, *18*, 5589. [CrossRef]
23. Infantes-Paniagua, Á.; Silva, A.F.; Ramirez-Campillo, R.; Sarmento, H.; González-Fernández, F.T.; González-Víllora, S.; Clemente, F.M. Active School Breaks and Students' Attention: A Systematic Review with Meta-Analysis. *Brain Sci.* **2021**, *11*, 675. [CrossRef]
24. Caterino, M.C.; Polak, E.D. Effects of Two Types of Activity on the Performance of Second-, Third-, and Fourth-Grade Students on a Test of Concentration. *Percept. Mot. Ski.* **1999**, *89*, 245–248. [CrossRef] [PubMed]
25. Hillman, C.H.; Pontifex, M.B.; Raine, L.B.; Castelli, D.M.; Hall, E.E.; Kramer, A.F. The Effect of Acute Treadmill Walking on Cognitive Control and Academic Achievement in Preadolescent Children. *Neuroscience* **2009**, *159*, 1044–1054. [CrossRef] [PubMed]
26. Bagatell, N.; Mirigliani, G.; Patterson, C.; Reyes, Y.; Test, L. Effectiveness of Therapy Ball Chairs on Classroom Participation in Children With Autism Spectrum Disorders. *Am. J. Occup. Ther.* **2010**, *64*, 895–903. [CrossRef] [PubMed]
27. Schilling, D.L.; Schwartz, I.S. Alternative Seating for Young Children with Autism Spectrum Disorder: Effects on Classroom Behavior. *J. Autism Dev. Disord.* **2004**, *34*, 423–432. [CrossRef] [PubMed]
28. Beserra, V.; Nussbaum, M.; Navarrete, M.; Alvares, D. Teaching through Dance: An Opportunity to Introduce Physically Active Academic Lessons. *Teach. Teach. Educ.* **2021**, *106*, 103450. [CrossRef]
29. Zhang, Z. Microsoft Kinect Sensor and Its Effect. *IEEE Multimed.* **2012**, *19*, 4–10. [CrossRef]
30. Homer, B.D.; Kinzer, C.K.; Plass, J.L.; Letourneau, S.M.; Hoffman, D.; Bromley, M.; Hayward, E.O.; Turkay, S.; Kornak, Y. Moved to Learn: The Effects of Interactivity in a Kinect-Based Literacy Game for Beginning Readers. *Comput. Educ.* **2014**, *74*, 37–49. [CrossRef]
31. Aksoy, H. The Impact of Kinect-Based Game On L2 Speakıng Skills And Its Implications on Bodily Kinesthetic Intelligence, Learner Motivation and Learner Autonomy. Master Thesis, Gazi University, Ankara, Turkey, 2015.
32. Si, J.; Mukherjee, A.; Liu, B.; Li, Q.; Li, H.; Deng, X. Exploiting Topic Based Twitter Sentiment for Stock Prediction. In Proceedings of the 51st Annual Meeting of the Association for Computational Linguistics (Volume 2: Short Papers), Sofia, Bulgaria, 4–9 August 2013.
33. Paribakht, T.S.; Wesche, M.B. Reading Comprehension and Second Language Development in a Comprehension-Based ESL Program. *TESL Can. J.* **1993**, *11*, 09–29. [CrossRef]
34. Düking, P.; Born, D.-P.; Sperlich, B. The SpeedCourt: Reliability, Usefulness, and Validity of a New Method to Determine Change-of-Direction Speed. *Int. J. Sports Physiol. Perform.* **2016**, *11*, 130–134. [CrossRef]
35. Shakroum, M.; Wong, K.W.; Fung, C.C. The Influence of Gesture-Based Learning System (GBLS) on Learning Outcomes. *Comput. Educ.* **2018**, *117*, 75–101. [CrossRef]
36. Culp, B.; Oberlton, M.; Porter, K. Developing Kinesthetic Classrooms to Promote Active Learning. *J. Phys. Educ. Recreat. Dance* **2020**, *91*, 10–15. [CrossRef]
37. Abi Nader, P.; Hilberg, E.; Schuna, J.M.; John, D.H.; Gunter, K.B. Association of Teacher-Level Factors With Implementation of Classroom-Based Physical Activity Breaks. *J. Sch. Health* **2019**, *89*, 435–443. [CrossRef]
38. Yukselturk, E.; Altıok, S. Zeynep Başer Using Game-Based Learning with Kinect Technology in Foreign Language Education Course. *Educ. Technol. Soc.* **2018**, *21*, 159–173.
39. Brady, M.; Miller, S.C.; Lindt, S.F. *Movement Integration: What Is It and Why Should We Do It*; Springer: Singapore, 2018; pp. 1–17.
40. Edwards, S. Active Learning in the Middle Grades Classroom: Overcoming the Barriers Toimplementation. *Middle Grades Res. J.* **2015**, *10*, 58–65.

41. Dugger, R.; Rafferty, A.; Hunt, E.; Beets, M.; Webster, C.; Chen, B.; Rehling, J.; Weaver, R.G. Elementary Classroom Teachers' Self-Reported Use of Movement Integration Products and Perceived Facilitators and Barriers Related to Product Use. *Children* **2020**, *7*, 143. [CrossRef] [PubMed]
42. Benes, S.; Finn, K.E.; Sullivan, E.C.; Yan, Z. Teachers' Perceptions of Using Movement in the Classroom. *Phys. Educ.* **2016**, *73*, 110. [CrossRef]
43. Ratey, J.J.; Hagerman, E. *The Revolutionary New Science of Exercise and the Brain*; Little, Brown Spark: New York, NY, USA; Boston, MA, USA; London, UK, 2008.

applied sciences

MDPI

Article

Enhanced Student Admission Procedures at Universities Using Data Mining and Machine Learning Techniques

Basem Assiri [1,*], Mohammed Bashraheel [2] and Ala Alsuri [2]

1 Computer Science Department, College of Computer Science and Information Technology, Jazan University, Jazan 82817, Saudi Arabia
2 Department of Information Technology and Security, College of Computer Science and Information Technology, Jazan University, Jazan 82817, Saudi Arabia; msbashraheel@jazanu.edu.sa (M.B.); ayusef@jazanu.edu.sa (A.A.)
* Correspondence: babumussmar@jazanu.edu.sa

Abstract: The progress of technology has played a crucial role in enhancing various fields such as education. Universities in Saudi Arabia offer free education to students and follow specific admission policies. These policies usually focus on features and scores such as the high school grade point average, general aptitude test, and achievement test. The main issue with current admission policies is that they do not fit with all majors, which results in high rates of failure, dropouts, and transfer. Another issue is that all mentioned features and scores are cumulatively calculated, which obscures some details. Therefore, this study aims to explore admission criteria used in Saudi Arabian universities and the factors that influence students' choice of major. First, using data mining techniques, the research analyzes the relationships and similarities between the university's grade point average and the other student admission features. The study proposes a new Jaccard model that includes modified Jaccard and approximated modified Jaccard techniques to match the specifications of students' data records. It also uses data distribution analysis and correlation coefficient analysis to understand the relationships between admission features and student performance. The investigation shows that relationships vary from one major to another. Such variations emphasize the weakness of the generalization of the current procedures since they are not applicable to all majors. Additionally, the analysis highlights the importance of hidden details such as high school course grades. Second, this study employs machine learning models to incorporate additional features, such as high school course grades, to find suitable majors for students. The K-nearest neighbor, decision tree, and support vector machine algorithms were used to classify students into appropriate majors. This process significantly improves the enrolment of students in majors that align with their skills and interests. The results of the experimental simulation indicate that the K-nearest neighbor algorithm achieves the highest accuracy rate of 100%, while the decision tree algorithm's accuracy rate is 81% and the support vector machine algorithm's accuracy rate is 75%. This encourages the idea of using machine learning models to find a suitable major for applicants.

Keywords: students; university admission; major selection; data mining analysis; machine learning models

1. Introduction

This paper is an extension of a work originally presented at the 7th International Conference on Data Science and Machine Learning Applications (CDMA) [1]. In today's world, technology plays a crucial role in the development of various fields, including medicine, education, industry, economy, and securities. Consequently, technological advancement has become essential as it strengthens the quality, facilities, and overall improvement of these fields. Education is one of the most significant fields, so developed nations typically place emphasis on their educational systems. This contributes to a capable, efficient, and thriving society. Countries provide different kinds of educational systems and teaching methods suiting all fields of science. Guiding people correctly to suitable educational systems and fields will improve people's achievements, leading to success and satisfaction [2,3].

Studying at university is a significant stage for students, as it enables them to secure employment opportunities. Every year, a large number of high school graduates pursue job opportunities or enroll in universities. Students who enroll in universities must avoid favoring prestigious majors since they may not always align with their interests and capabilities. According to research, several factors can affect a student's decision when selecting a major at a university. These may include their background, economic status, time and financial constraints, psychological factors, gender, skills, job market conditions, and the impact of family and peers [4,5]. The factors mentioned above can have an impact on student choices and the processes of decision-making.

Free higher education is provided to students in public universities in Saudi Arabia, and earning a degree enhances the chances of acquiring appropriate employment opportunities with a decent income. As a result, a considerable number of high school graduates seek admission to universities. As an illustration, the Admission and Registration Deanship at Jazan University reported an annual enrollment of approximately 13,000 students from 2017 to 2020. However, unfortunately, there are many cases where the academic major that students favor may not be the optimal selection, leading to potential failure or the need to transfer to a different major after several years, where the failure rates reach 28–33% [6–8]. As a result, universities permit students to choose their preferred areas of study while assessing their aptitude through evaluations such as the high-school grade point average (GPAH), general aptitude test (GAT), and achievement test (AT). The GAT is a widely used test around the world [9,10] that assesses students' cognitive abilities, e.g., their comprehension and analytical skills. In addition, it evaluates students' performance in mathematics and Arabic. The AT evaluates students' success in mathematics, physics, chemistry, and biology during their high school education. In fact, it is not mandatory for every candidate, e.g., those who seek admission to Arabic Language or Islamic Studies programs, to take the test [9,10].

Following this, a student's weighted score (WS) is calculated by combining their GPAH (50%) and GAT score (50%). Furthermore, a student's qualifying score (QS) is determined by combining their GPAH (30%), GAT score (30%), and AT score (40%). These percentages may be modified in certain situations, such as during the COVID-19 pandemic.

The primary concern with these scores is that they conceal certain crucial details that influence a student's choice of major at university. For instance, some students may have the same WS, but among them, some may have an average GPAH and a high GAT score, while others may have a high GPAH and an average GAT score. Additionally, individuals with an average GPAH and a high AT score might not be suitable for an English language major, but they could still be admitted to this major based on their overall score. Moreover, this study attempts to mitigate the influence of human factors by utilizing data mining and machine learning methods.

Furthermore, machine learning is applied in various fields, including education, healthcare services, marketing, and finance [11]. Machine learning models are heavily dependent on data and statistics. As a subset of artificial intelligence, machine learning can learn from data, recognize patterns, and make decisions with minimal human involvement. Indeed, there are two categories of machine learning models: supervised and unsupervised learning. Supervised learning involves providing the model with training examples to teach the algorithms how to generate the correct output. In unsupervised learning, algorithms can learn and discover hidden patterns on their own [12].

This research examines the admission standards for universities in Saudi Arabia, focusing on the impact of concealed information (which underlies students' GPAH, WS, and QS scores) on their selection of majors. First, a questionnaire is presented to experts to comprehend the ideal process of selecting a student's major. Second, we gather actual data about Jazan University students to examine the current state of the process of selecting a major. Here, we use data mining techniques to analyze the collected data and establish connections between student performance in the university, indicated by GPAU, and admission features such as GPAH, GAT, AT, WS, and QS. Actually, four types of relationship analysis are employed: modified Jaccard analysis, approximated modified Jaccard analysis, data distribution analysis, and correlation coefficient analysis. The research modifies Jaccard index [13] to suit the characteristics of the students' data records. Indeed, Jaccard analysis measures the similarity between two sets of data. The objective is to find the percentage of similarity between the elements in the sets. Our data sets include numerous attributes, including GPAU, GPAH, GAT, AT, WS, and QS. Each attribute is considered as a set, and the degree of similarity is calculated. However, since our data set includes students' records, we initially propose the notion of modified Jaccard analysis, which measures the similarity between the elements of two sets and of the same record. For instance, it examines the similarity between GPAU and GPAH for the same student. Next, we use approximated modified Jaccard analysis to measure similarity in a more lenient manner, treating values that are near each other as similar. Moving forward, we conduct an analysis of the data distribution, followed by using correlation coefficient analysis. Accordingly, the investigation shows that current admission procedure has a generalization issue, where the relationships tend to differ significantly from one major to another.

Thirdly, this article suggests incorporating additional features, like the grades earned in high school courses, into supervised machine learning models to accurately find suitable majors. The paper specifically applies K-nearest neighbor (KNN), decision tree (DT), and support vector machine (SVM) algorithms to classify students into their respective major categories [14–16]. The experimental simulation revealed that KNN provides the most accurate results, with a 100% success rate. In comparison, the DT has an accuracy rate of 81% and the SVM has an accuracy rate of 75%. Employing this process has the potential to increase enrolment rates for students in majors that are best suited to their individual strengths and abilities. This enables students to graduate on time with a wealth of knowledge, practical experience, and overall satisfaction. Additionally, this technique can substantially reduce the frequency of students transferring, withdrawing, or failing courses.

1.1. K-Nearest Neighbor

The initial application of the KNN model was in statistical settings in the early 1970s [14]. The model operates by identifying a collection of K instances that are the closest in distance to a given point. The optimal K-value is used to cluster the given data. The variable K represents the number of neighboring data points that are taken into account during the classification process.

1.2. Decision Tree

A decision tree is a type of supervised learning method that is commonly used for both classification and regression [15]. The DT algorithm finds the value of an unknown data point by analyzing its attributes. Each attribute is evaluated as true or false using an if statement. In this study, the DT model is utilized specifically for classification purposes.

1.3. Support Vector Machine

SVM is a commonly used supervised learning algorithm that is applicable for various tasks such as classification, regression, and outlier detection. In SVM, the objective is to construct a hyperplane, which is basically a line that separates data points into different classes, and this hyperplane is used to predict the classes of new data points. The SVM algorithm is able to handle data that have multiple features represented in high-dimensional space, as mentioned in reference [16].

2. Related Work

Many researchers have examined the challenge of selecting a suitable university major. Some studies have specifically emphasized the difficulty of making this decision due to external factors such as familial or cultural pressure, as referenced in studies [4,5]. Social and community pressures, including the influence of friends, instructors, and family members, can significantly impact a student's choice of major, particularly in the fields of science, technology, engineering, and mathematics. Research has indicated that mothers, even those who have not attended college themselves, have the most significant influence in guiding their sons' and daughters' decisions regarding their choice of major [6]. According to research, gender plays a role in selecting a major. Malgwi et al. performed a survey and discovered that females tend to consider their abilities when selecting a major, whereas males are more inclined to choose majors that have better job prospects [11]. Many factors influence this issue, including economic factors, familial considerations, and personal preferences, as well as gender. Montmarquette et al. found that male students tend to be more careful in their choice of major compared to their female counterparts [12]. Overall, these studies demonstrate that the process of choosing a major is complex and multifaceted.

In addition, universities have various plans in place to help students select an appropriate major. While the specific criteria may vary among institutions, many universities make major selection a part of the admissions process [3,4]. Conversely, some universities adopt a general curriculum in the first year, allowing students to explore various fields before deciding on a major. However, some students may still struggle to make a decision after the first year, and as a result, their GPA may suffer if they take courses in diverse areas like mathematics or physics [4]. Furthermore, a different study indicates that certain universities utilize various methods to assist students in choosing their majors. One approach involves selecting majors based on enrollment essays, which are evaluated using automated computer systems to analyze their content, logic, and psychological information [17]. Moreover, the availability of jobs and the demand for certain majors rely heavily on countries' policies and marketplaces. While various countries employ different education strategies, a student's decision to select a suitable major is crucial, as it helps a country advance and enables students to succeed in their careers [18–20]. Thus, enrolling in a major without considering its suitability can result in failure, which can have detrimental effects on students, families, educational systems, and even countries. Therefore, various studies provide guidance to students on selecting a suitable major [21,22].

Many countries utilize pre-college examinations to assess students' academic abilities and accomplishments. In certain countries, assessments like the American College Test (ACT) and Scholastic Assessment Test (SAT) are employed to authenticate the accuracy of student GPAH. According to research, it is not sufficient to solely rely on either tests or GPAH. However, for a more precise evaluation of students' abilities and academic achievements, we can consider both pre-college examinations and GPAH together [23–26].

In the Kingdom of Saudi Arabia, universities provide free education, and to maintain the large number of applicants, they have special admission criteria. They allow applicants to rank their preferable majors. Then, enrollment is based on GPAH and two pre-university tests, which are the GAT and AT. A study was conducted on students at Damam University, and it was found that there is a strong relation between students' GPAU and their GAT, AT, and GPAH scores. The study shows that the influence of the GAT and AT is important for students pursuing medical and science majors but has less significance for students pursuing humanities majors [9]. The influence of the GAT and AT on students' performance at King Saud University was also studied. It was found that GPAH affects the college GPAU more than the GAT and AT [10].

Furthermore, many works use artificial intelligence techniques such as data mining, machine learning, and deep leaning models to analyze the issue of choosing majors at university and to guide students. A proposed study will focus on a comparison between first-year students and other students. It will analyze some factors such as a student's general ability and personality characteristics to help students select suitable majors. It will use artificial intelligence and an expert system rather than traditional methods [5,27]. Another work categorized student performance and learning aspects using an intelligent system, which contained adaptive neuro fuzzy and learning vector quantization network methods, to help institutions classify students based on their ability to progress [18]. Another work used game theory for admission polices [28]. Xu et al. considered the influence of pedagogical process of students on their background, academic performance, and subject choices. This research applied a machine learning model to make predictions related to an acquired program relying on students' progress during the college period [29]. Some others used parallelism to enhance machine learning models and their distribution processes [30,31]. Ahmad et al. used data mining to predict students' ability to progress while studying at university. Their research applied decision tree, naïve Bayes, and rule-based models, and they found that the rule-based model was the best in terms of prediction accuracy, which reached 71% [32].

The purpose of this work is to examine admission procedures for universities. It measures the relation between the main features of current students' admission procedures, such as GPAU and other scores such as GPAH, GAT, AT, WS, and QS. A Jaccard analysis is used to measure the similarities among datasets [33–35]. Some works modify or relax the Jaccard similarity index to fit it to different kinds of problems. This also helps in finding the similarity and relations between structured and unstructured data [35–37].

Additionally, this study employs data mining techniques to compile and analyze the provided data. Finally, this paper presents machine learning models that help to find the most fitting majors for university students during the admission process. These models suggest a suitable major based on GPAH, individual scores of high school courses, and some other qualitative and quantitative standardized tests.

3. Methodology

The process of this investigation involved six stages, namely, collecting data, preparing data, data mining, creating a relevant dataset, developing machine learning models, and making assumptions. In this research, data collection and analysis took two paths to investigate the major selection process, which assessed how it should be and how it is currently. First, a questionnaire was given to domain experts to investigate the major selection process, which assisted in determining the correct process and procedure. Second, we gathered real-word information about students and used data mining to examine the present state of procedures related to the selection of majors.

3.1. Data Collection

This section looks into the viewpoints of professionals and experts in 19 different majors at Jazan University, as outlined in Table 1. This study was carried out using a survey that is available in both Arabic and English and is presented below:

- Goal: The objective of the survey was to enable specialists to identify the key requirements for admission in every major. Specialists in the field ensured data accuracy and the comprehension of all aspects of the issue. This facilitated the creation of a precise dataset and provided proper guidance to machine learning models.
- Participants: The respondents, who were the primary stakeholders, consisted of 61 male and female faculty members from various majors at Jazan University, as demonstrated in the first and second columns of Table 1.
- Design: Our proposal involved an internet-based survey that was administered to 61 faculty members from 19 different majors at Jazan University, as depicted in the first column of Table 1. The survey was conducted between April and June of 2020 and comprised seven straightforward questions to identify the primary prerequisites that students must meet to be admitted into majors. The main inquiry sought faculty members' input about the significant high school subjects that concern each university major (based on their respective fields). The second question concerned the minimum scores that students must achieve in these subjects (determined by the first question).

The third question concerned the minimum GAT and AT scores that students must obtain to be eligible for these majors. The fourth question inquired about the pre-admission exams that are mandatory for specific college majors. The answers show that pre-admission exams are not required for about 95% of the university's majors. The fifth question focused on students' success factors. Lastly, the survey concluded the input with an open question.

- Results: Table 1 provides a summary of the questionnaire results. The first column indicates the colleges and majors, while the second column displays the number of faculty members participating from various majors. The third column outlines the significant courses associated with each major. This information was used to create our dataset (which provided training examples for machine learning models) to help students choose an appropriate major at university. The fourth column of the table displays the minimum scores required for each major, including GPAH, GAT, and AT. However, it should be noted that the responses to this question were not entirely accurate, indicating that the faculty members may not have a complete understanding of the structure of the GAT and AT. For instance, a Mathematics faculty member indicated that a GAT score of 60 and an AT score of 60 are sufficient, even though this falls below the acceptance threshold at Jazan University. The last column in the table outlines the essential skills and qualities that students need to possess to excel in each major. The findings of the questionnaire suggest that each major has specific courses that are of utmost importance, while other courses may not be as necessary.
- Practical aspects: A comprehensive dataset needed to be designed, including specific training examples based on the survey's results and further analyzed data. After the initial questionnaire results and real-world data examples were combined, the accuracy of the training process within the machine learning models could be improved. By analyzing the overlap between majors, 60 training examples were used and categorized into four labels or classes, as shown in Tables 2 and 3.

Table 1. The questionnaire results.

Colleges and Majors		Participants	Important Courses	GPAH, GAT, and AT	Characteristics and Skills
Majors of Medicine		5	Biology and Chemistry	90, 90, 90	- Analysis - Memorizing - Discipline - Being up to date - Time management - Patience - Hard working
Majors of Dentistry		3	Biology and Chemistry	90, 90, 85	
Majors of Public Health and Tropical Medicine		4	Biology	90, 90, 85	
Majors of Nursing		3	Biology	90, 85, 85	
Majors of Applied Medical Science		3	Biology	90, 85, 85	
Majors of Pharmacy		5	Chemistry and Biology	87.5, 85, 85	
Majors of Engineering		3	Math, Physics, Chemistry, and English Lang	85, 80, 85	- Critical thinking - Analysis - Discipline - Being up to date - Time management - Hard working - Creativity
Majors of Computer Science and IT		4	Math and English Lang	80, 85, 80	
Majors of Design and Architecture		3	Art, History, Geography, and Science	80, 80, 80	
Science	Math	8	Math and English Lang	80, 80, 80	- Analysis - Memorizing - Discipline - Time management - Hard working
	Physics		Math, Physics, and English Lang		
	Chemistry		Math, Chemistry, and English Lang		
	Biology		Biology		
Majors of Business Administration		3	Math and English Lang	80, 75, 70	
Majors of Education		3	No Preference	75, 75	
Majors of Sharia and Law		3	Arabic Lang and Religious Studies	75, 75	
Arts and Humanities	English	11	English Lang	75, 75	- Analysis - Memorizing - Discipline - Time management - Hard working - Creativity
	Arabic		Arabic Lang		
	Others		Arabic Lang, History, and Geography		

Table 2. Dataset sample that includes features and labels (the first row includes the number of rows and columns as well as the classes).

# of Examples	**# of Columns**	**Classes**																	
60	20	1	2	3	4														
Quran	Tafseer	Hadith	Toheed	Fegh	Arabic	Math	Physics	Chemistry	Biology	Geology	English	Computer Science	Sociology	Research Skills	Applied Skills	Physical Education	GAT	AT	Class
80	80	80	80	80	80	75	75	75	75	75	80	80	75	80	75	80	75	75	4
95	99	98	99	99	100	96	98	100	100	99	97	96	99	94	96	100	91	87	1
80	85	85	85	85	85	85	85	85	85	85	85	85	85	85	85	85	80	80	3

Table 3. The four classes in the dataset.

#	**Labels (Classes)**	**Colleges and Majors**
1	Medicine and Healthcare	Majors of Medicine, Dentistry, Public Health and Tropical Medicine, Nursing, Pharmacy, and Applied Medical Science
2	Engineering and Computation	Majors of Engineering, Computer Science and IT, Design and Architecture
3	Science and Management	Majors of Science and Business Administration
4	Theoretical Science	Majors of Education, Sharia (and Law), and Arts (and Humanities)

3.2. Real Data Preparation

To support the survey, student information was also required to analyze and measure the correlation between GPAU and various other factors such as GPAH, GAT, AT, WS, and QS. This investigation helped in dataset (training examples and labels) preparation. This process was conducted as illustrated below:

- Data specification: This involved identifying the necessary data for this study, which included high school records, GPAU, GPAH, GAT, AT, WS, and QS. The data covered both male and female students across all undergraduate majors at Jazan University.
- Requesting data: This involved seeking permission from the Jazan University administration to access the requested data, ensuring that this study used real-world data.
- Data cleaning: Once the data were obtained, data cleaning was conducted to address any missing values and to eliminate noisy data. This study considered data from 2018, 2019, and 2020.
- Data ordering: This re-arranged the data based on the domain experts' answers (in the survey). We divided students' data into two categories, scientific majors and theoretical (literature) majors, since they have different properties. For example, students pursuing scientific majors use QS, while theoretical major students use WS.

3.3. Data Mining Process

After receiving the students' information from Jazan University, we investigated the patterns and relationships among the data as follows:

(1) The correlation between the GPAU and GPAH for students pursuing all majors.
(2) The correlation between the GPAU and GAT for students pursuing all majors.

(3) The correlation between the GPAU and WS for students pursuing theoretical majors.
(4) The correlation between the GPAU and AT for students pursuing scientific majors.
(5) The correlation between the GPAU and QS for students pursuing scientific majors.

In this part, we applied data mining to analyze the given data and to find the relations between student performance at university, represented by GPAU, and other scores such as GPAH, GAT, AT, WS, and QS. We used four kinds of relation analysis, which were modified Jaccard analysis, approximated modified Jaccard analysis, data distribution analysis, and correlation coefficient analysis. As mentioned earlier, this study proposed modified Jaccard and approximated modified Jaccard analyses to suit the structure of our data. Our data consisted of students' records in different majors, and each record had many attributes such as GPAU, GPAH, GAT, AT, WS, and QS.

This investigation transformed the data and information into knowledge, providing deep insights into how these individual scores influence students' major selections and how these scores are reflected in students' success in their majors. This helped to diagnose the presence and the size of the problem and to adjust the measurements while devising a solution, which will be explained in detail later.

3.4. Dataset Preparation

We followed the guidance of the questionnaire and correlation results to prepare our dataset in a way that was suitable for use with a machine learning model. This involved building training examples in which all data were numerically represented and divided into features and labels.

Our dataset was made up of 20 columns and 60 rows. Of these columns, 19 represented features, including scores of students in various high school courses such as Quran, Tafseer, Hadith, Toheed, Feqh, Arabic, Math, Physics, Chemistry, Biology, Geography, English, Computer Science, Research Skills, Applied Skills, and Physical Education. Additionally, GAT and AT scores were also included among the features. The last column represented the label, which was the suitable class, as illustrated in Table 2. The dataset contained 60 rows. The first row provided a summary of the numbers of rows, columns, and classes. The second row displayed the features, while the remaining rows presented training examples. As previously mentioned, there are four classes represented in Table 3.

3.5. Machine Learning Procedures

To assist students in selecting a suitable major, this paper suggests using supervised machine learning models. These models aim to enhance the precision of finding students' university majors. This study used KNN, DT, and SVM and compares the accuracy of these three classifiers in relation to our problem. This study analyzed GPAH features by evaluating individual scores in all high school courses, as well as GAT and AT scores, to anticipate the correct major. Additionally, certain courses (features) that were significant for each major (as demonstrated in Table 1) were assigned greater importance for each major.

3.6. Assumption

This study assumes that using individual scores of high school courses (which are hidden behind GPAH) has a notable influence on finding the appropriate major for university applicants.

4. Data Mining Analysis and Experimental Simulation

The analysis and the experimental simulation used two methods, which were data mining and machine learning processes. The first method used data mining to determine the correlation among the dataset's main features, focusing on the correlations between GPAU and other scores such as GPAH, GAT, AT, WS, and QS. We emphasize the positive and negative influences of some features and illustrate the importance of conducting more investigations into current university student acceptance procedures.

The second approach utilized machine learning models to demonstrate the recommended process for admitting university students. The experimental simulation investigated the consideration of additional features such as high school course grades. The proposed admission procedures were analyzed using three machine learning models.

4.1. Data Mining (Correlation Analysis)

This section involves analyzing the connections and similarities among the principal features of students' admission procedures. The analysis explored the relationships between students' GPAU and GPAH, GAT, AT, WS, and QS. To facilitate this investigation, all GPAU scores were converted to a 100-point scale. Then, we conducted modified Jaccard, approximated modified Jaccard, data distribution, and correlation coefficient analyses. With the use of Microsoft Visual C++17 software, the code was written in C++ to find the mentioned relations.

4.1.1. Modified Jaccard Analysis (MJA)

As mentioned before, Jaccard analysis measures the similarity between two sets of data. It determines the similarity percentage of the elements of the sets. A and B are two sets, and Jaccard analysis measures the similarity between A and B using the following formula [13]:

$$J(A,B) = |A \cap B| / |A \cup B|$$

Since our datasets consisted of many features, such GPAU, GPAH, GAT, AT, WS, and QS, we considered each feature as a set, and we measured the similarity. We measured the similarities between GPAU from one side and GPAH, GAT, AT, WS, and QS from the other side. However, our dataset consisted of students' records, so we proposed the concept of MJA, which measures the similarities between the elements of two sets and the same records. For example, it checks the similarity of GPAU and GPAH for the same student. Figure 1 shows the difference between Jaccard analysis and the proposed MJA. For four students, GPAU = {99, 91, 86, 66} and GPAH = {91, 86, 85, 80}. It is clear that the first elements of the two sets belong to one student (one record), the second elements of the two sets belong to another student (one record), and so on.

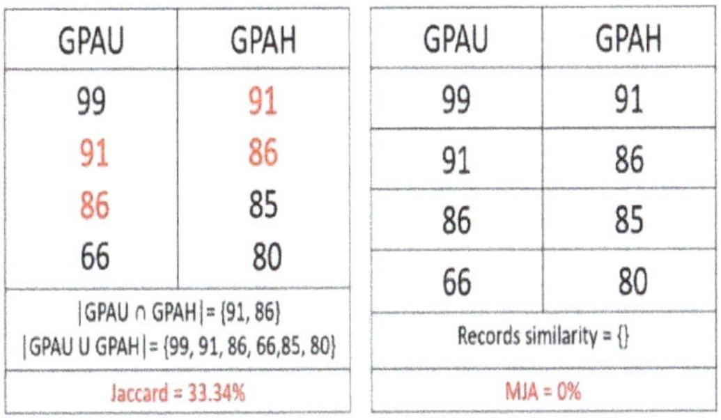

Figure 1. Illustrates a comparison between Jaccard analysis and MJA.

Therefore, the result of the traditional Jaccard analysis was approximately 33.34%, as it found two similarities {91, 86} out of six elements {99, 91, 86, 66, 85, 80}. It compared the elements of the two sets regardless of their records. The result of the MJA was 0%, as it checked the similarity of the first elements in the two sets, then the second elements, and so on. In other words, it considered the similarity only if the student's GPAU matched his/her GPAH. Then, we separated the investigation results into two groups, which were scientific colleges, as shown in Table 4, and theoretical colleges, as shown in Table 5. Clearly, Table 4 shows the number of students and the MJA calculation between the GPAU and other scores (which are GPAH, GAT, AT, and QS), and Table 5 also shows the MJA calculation between the GPAU and GPAH, GAT, and WS. The general results of the MJA for scientific colleges in Table 4 show very low similarities, as they do not exceed 9%, and the most similar percentages lay between 3% and 5%. In addition, the MJA results differ from one college to another. The MJA results between GPAU and QS are the highest for 45% of scientific majors such as dentistry, public health and tropical medicine, pharmacy, design and architecture, and business administration. The table also shows that for 27% of majors, such as medicine, nursing, and applied medical sciences, GPAU to GPAH is the best indicator, while GPAU to AT is the best indicator for engineering, computer science and information technology, and science colleges and GPAU to GAT is the best indicator for some other colleges. However, there are some colleges that have the same MJA results for two or more indicators. For instance, the nursing college has the same result for GPAU to GPAH and GPAU to GAT. Moreover, most of the findings also apply to the MJA results for theoretical colleges that appear in Table 5.

Table 4. Illustrates MJA between GPAU and GPAH, GAT, AT, and QS for scientific colleges.

College	No. of Students	GPAU—GPAH	GPAU—GAT	GPAU—AT	GPAU—QS
Medicine	458	0.09	0.04	0.03	0.04
Dentistry	185	0.02	0.04	0.04	0.05
Public Health Tropical Medi	620	0.02	0.02	0.02	0.03
Nursing	521	0.03	0.03	0.02	0.02
Applied Medical Sciences	1363	0.03	0.02	0.02	0.02
Pharmacy	454	0.04	0.03	0.03	0.05
Engineering	945	0.01	0.02	0.03	0.02
Computer Science and IT	1515	0.01	0.03	0.04	0.03
Design and Architecture	409	0.02	0.02	0.03	0.05
Science	2056	0.01	0.02	0.03	0.02
Business Administration	2637	0.03	0.03	0.02	0.03

Table 5. Illustrates MJA between GPAU and GPAH, GAT, and WS for theoretical colleges.

College	No. of Students	GPAU—GPAH	GPAU—GAT	GPAU—WS
Education	603	0.05	0.01	0.03
Sharia and Law	1216	0.04	0.02	0.02
Arts and Humanities	2957	0.01	0.03	0.02

To summarize, the MJA does not show strong relations between student performance at university (GPAU) and the other features. Thus, this study proposes further investigating feature relations using scores behind these features.

4.1.2. Approximated Modified Jaccard Analysis (AMJA)

The low MJA results encouraged us to relax this analysis and instead propose AMJA. Since MJA considers a similarity only if the value of the element in set A matches the exact value of the corresponding element in set B, this paper proposes AMJA, which measures the approximated similarities in a range (+10/−10) of values. We relaxed the similarity measure to (+10/−10) under the guidance of the university's grading system since each grade represents about 10 points. For example, grade A lies between 90 and 100, grade B lies between 80 and 89, and so on. This relaxation allowed us to find more relations among our features.

AMJA was applied to the example in Figure 1, in which GPAU = {99, 91, 86, 66} and GPAH = {91, 86, 85, 80}. The AMJA resulted in 75% approximated similarity because the difference between the first elements of the two sets was less than 10. This was also true for the second and the third elements. However, it did not count the fourth elements {66–80}, as the difference exceeded 10 points. As usual, we separated the AMJA results into two groups, which were scientific colleges, as shown in Table 6, and theoretical colleges, as shown in Table 7.

Table 6 shows the AMJA calculations between the GPAU and the other scores (GPAH, GAT, AT, and QS), and Table 7 shows the MJA calculations between the GPAU and GPAH, GAT, and WS. The numbers of students are the same as those in Tables 4 and 5.

Table 6. Illustrates AMJA between GPAU and GPAH, GAT, AT, and QS for scientific colleges.

College	GPAU—GPAH	GPAU—GAT	GPAU—AT	GPAU—QS
Medicine	38.56	35.73	38.56	40.52
Dentistry	30.65	31.18	44.09	47.31
Public Health and Tropical Medi	12.56	21.09	25.44	28.34
Nursing	24.14	23.37	29.17	27.97
Applied Medical Sciences	22.36	22.07	27.27	29.18
Pharmacy	31.37	23.75	31.81	40.74
Engineering	12.37	27.17	28.75	28.96
Computer Science and IT	11.15	26.25	28.96	28.96
Design and Architecture	19.51	17.07	25.85	42.2
Science	6.17	24.99	29.7	28.44
Business Administration	19.98	20.02	18.99	27.29

Table 7. Illustrates AMJA between GPAU and GPAH, GAT, and WS for theoretical colleges.

College	GPAU—GPAH	GPAU—GAT	GPAU—WS
Education	50.67	3.31	9.77
Sharia and Law	25.72	2.71	8.46
Arts and Humanities	12.95	1.69	6.22

Although the AMJA is very relaxed, the relations between GPAU and other scores are not very tight. The results of the AMJA show that the relation between GPAU and QS is the best indicator, even if it does not suit some majors such as nursing and science. Also, we cannot ignore the huge percentages of non-similarities that range between 52 and 72% for all colleges.

For theoretical colleges, the AMJA of GPAU and GPAH is the best indicator, even if non-similarities lay between 49 and 87% for all colleges. The relation is very low for GPAU and WS, which is the main measure in the current acceptance procedures at Saudi universities.

4.1.3. Non-Similarity Distribution Using Approximated Modified Jaccard Analysis (DAMJA)

Before proceeding, we investigated the non-similarity percentages to see the data distribution of these non-similar values. In Tables 8 and 9, non-similar values are classified as high or low. If the difference between a student's GPAU and the corresponding score (GPAH, GAT, AT, WS, or QS) was greater than +10 points, it was classified as high, and if the difference between a student's GPAU and the corresponding score was smaller than −10 points, it was classified as low; otherwise, it was classified as an approximated similarity (the approximated similarities already appear in Tables 6 and 7). The numbers of students are the same as those in Tables 4 and 5.

Table 8 shows a classification of similar and non-similar AMJA results. For the major of medicine, first, the approximated similarity of GPAU and GPAH was 38.56%. No student had +10 GPAU in comparison with GPAH, and for 61.4% of students the GPAU was more than 10 points less than the GPAH. Second, for GPAU and GAT, the approximated similarity was 35.73%, where 29.19% was high and 35.1% was low. Third, for GPAU and AT the approximated similarity was 38.56%, where 18.95% was high and 44.5% was low. Fourth, for GPAU and QS the approximated similarity was 40.52%, where 11.98% was high and 47.49% was low. According to this investigation, a student who is classified as high performs better at university, which is a good indicator for major selection, even if they are not classified as an approximated similarity. This means that even if GPAU and QS have the highest approximated similarity, GPAU and AT are better since the total number of both the approximated similarity and high classification is the highest. This is also applicable for the majors of dentistry, public health, engineering, computer science and information technology, and business administration. In addition, the approximated similarity of GPAU and GAT is the best indicator for the majors of nursing, applied medical sciences, pharmacy, design and architecture, and science.

Table 8. Illustrates AMJA and non-similar distribution that lays between GPAU and GPAH, GAT, AT, and QS for scientific colleges.

College	GPAU—GPAH			GPAU—GAT			GPAU—AT			GPAU—QS		
	Compatible	High	Low	Compatible	High	Low	Compatible	High	Low	Compatible	High	Low
Medicine	38.56	0.00	61.41	35.73	29.16	35.16	38.56	18.95	42.58	40.52	11.98	47.49
Dentistry	30.65	0.54	68.83	31.18	46.22	22.65	44.09	26.88	29.00	47.31	18.82	33.87
Public Health and Tropical Medicine	12.56	0.48	87.00	21.09	42.84	36.16	25.44	39.77	34.83	28.34	21.74	49.92
Nursing	24.14	0.00	75.91	23.37	54.25	22.44	29.17	42.15	28.76	27.97	33.14	38.89
Applied Medical Sciences	22.36	0.22	77.45	22.07	48.56	29.42	27.27	40.84	31.92	29.18	27.86	42.96
Pharmacy	31.37	0.22	68.43	23.75	57.32	19.00	31.81	45.75	22.41	40.74	32.03	27.23
Engineering	12.37	0.11	87.52	27.17	29.81	43.00	28.75	34.46	36.88	28.96	15.75	55.29
Computer	11.15	0.99	87.96	26.25	46.80	26.90	28.96	44.53	26.57	28.96	25	46.04
Design and Architecture	19.51	0.49	80.00	17.07	75.17	7.80	25.85	64.15	10.00	42.2	41.22	16.59
Science	6.17	0.78	93.11	24.99	45.19	29.91	29.7	40.25	30.00	28.44	21.29	50.27
Business Administration	19.98	2.19	77.80	20.02	61.45	18.66	18.99	69.47	11.53	27.29	44.84	27.87

On the other hand, for the majors of, Sharia and law, and arts and humanities, the approximated similarity of GPAU and GPAH is the highest, but by considering the advantages of those students who have high performances at university, the approximated similarity of GPAU and GAT becomes the best indicator for education majors and Sharia and law majors.

In short, such variations demonstrate the weakness and inconsistency of current admission procedures and features.

Table 9. Illustrates AMJA and non-similar distribution that lays between GPAU and GPAH, GAT, AT, and QS for theoretical colleges.

College	GPAU—GPAH			GPAU—GAT			GPAU—WS		
	Compatible	High	Low	Compatible	High	Low	Compatible	High	Low
Education	50.67	44.87	4.47	3.31	96.69	0.00	9.77	9.23	0.00
Sharia and Law	25.72	72.14	2.14	2.71	97.12	0.16	8.46	91.45	0.08
Arts and Humanities	12.95	85.16	1.89	1.69	98.11	0.20	6.22	93.71	0.07

4.1.4. Correlation Analysis

In this part, we analyze the relations among the main features of student acceptance procedures using the correlation coefficient. The correlation coefficient is used to measure statistical values, which range between 1 and −1, and to create a relationship between them using the following formula [38]:

$$P(X,Y) = E\,(X - \mu x)(Y - \mu y)/\sigma X \cdot \sigma Y$$

The Pearson product–moment correlation coefficient (ρ) is a measure of the correlation between two variables, X and Y. Here, E is the expectation; μx and μy are the mean values of X and Y, respectively; σX is the standard deviation of X; and σY is the standard deviation of Y. This formula considers the means and the expectations of X and Y.

We separated the investigation results into two groups, which are scientific and theoretical colleges. Tables 10 and 11 show the correlation coefficient calculation between GPAU and the other scores (GPAH, GAT, and WS).

Table 10. Illustrates the correlation between GPAU and GPAH, GAT, AT, and QS for scientific colleges.

College	GPAU—GPAH	GPAU—GAT	GPAU—AT	GPAU—QS
Medicine	0.33	0.16	0.20	0.26
Dentistry	0.28	−0.06	0.21	0.16
Nursing	0.45	0.13	0.23	0.31
Pharmacy	0.40	0.10	0.20	0.30
Applied Medical Sciences	0.42	0.21	0.41	0.45
Public Health	0.38	0.28	0.39	0.46
CSIT	0.50	0.47	0.52	0.61
Design	0.53	0.33	0.44	0.53
Engineering	0.50	0.43	0.50	0.57
Science	0.52	0.36	0.53	0.58
Business	0.36	0.21	0.43	0.41

Table 11. Illustrates the correlation between GPAU and GPAH, GAT, AT, and QS for theoretical colleges.

College	GPAU—GPAH	GPAU—GAT	GPAU—WS
Education	0.34	0.01	0.26
Sharia	0.38	0.31	0.43

The correlation coefficient between GPAU and GPAH is the highest for the medicine, dentistry, nursing, and pharmacy majors, as well as some theoretical majors such as education and Sharia. Meanwhile, for majors such as applied medical sciences, public health, computer science and information technology, design, engineering, and science the correlation coefficient between GPAU and QS is the best indicator, while for the art major the correlation coefficient between GPAU and WS is the best indicator. In addition, for the business major the correlation coefficient between GPAU and AT is the best indicator.

In short, the correlation between GPAU and GPAH is suitable for some colleges, and considering other scores has a negative impact on the correlation. The qualifying score is an appropriate indicator for other scientific colleges, while the AT is a suitable indicator for the business college. These results show that the scores currently considered for students' acceptance are not suitable for many majors. Thus, this study proposes further investigating the detailed scores behind GPAH, AT, and QS.

4.2. Machine Learning Models

This study focused on the admission procedures of universities, considering additional features such as high school course grades. This research employed three machine learning models to analyze the suggested acceptance procedures. This study was conducted using Windows 10, and Python was the programming language. It utilized a 2.90 GHz Intel Core (TM) i7 CPU and 4 GB of RAM for processing. Each test was conducted five times, and the average results are reported.

This section of the study aimed to enhance the university admission procedures by testing the dataset and features. This research evaluated the dataset using three machine learning models, which were KNN, DT, and SVM. The parameters for these models were set to their default values. The dataset used for this study comprised 19 features, which included high school courses, GAT, and AT. The classifier generated four classes that corresponded to specific domains, as listed in Table 3.

During the KNN processing, an accuracy rate of 100% was achieved when K = 1. However, increasing the value of K resulted in a reduction in accuracy. For instance, when

K = 3, the accuracy of the KNN model was 91%, as demonstrated in Figure 2. In Figure 2, the data points represent examples from the dataset, while the background areas indicate the four classes, with class 1 represented by white, class 2 represented by yellow, class 3 represented by pink, and class 4 represented by gray. The figure demonstrates that the majority of the points were classified accurately in the intended classes.

Figure 3 shows the result of the DT, where the accuracy rate reached 81%. It uses the same class backgrounds as Figure 2. Obviously, some points were classified on the borders between the classes, and one of the points was misclassified. Furthermore, when the SVM classifier was applied to the dataset, an accuracy rate of 75% was achieved, as illustrated in Figure 4, which also uses the same class backgrounds as Figure 2. The figure reveals that the SVM struggled with classification due to the narrow margins between classes. Additionally, a point was misclassified, and some points were located on the borders between the classes. Based on the results obtained by running the three classifiers, it was proven that the KNN machine learning model provided the highest accuracy rate and was the most suitable classifier for the proposed method.

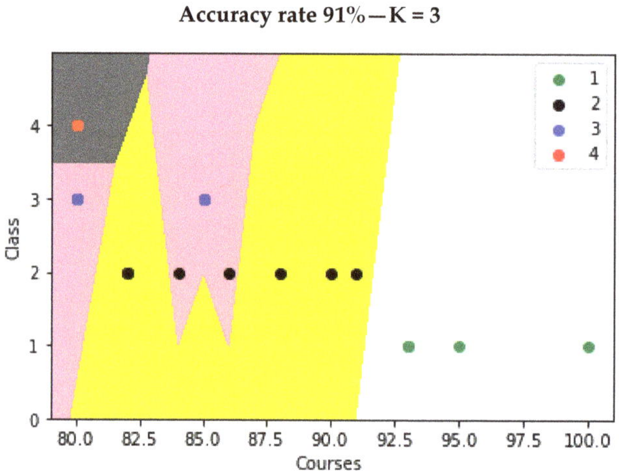

Figure 2. Illustrates KNN with an accuracy rate of 91% with K = 3.

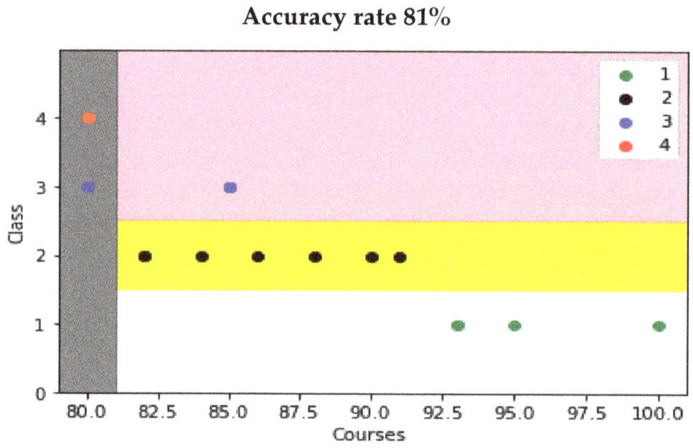

Figure 3. Illustrates DT with an accuracy rate of 81%.

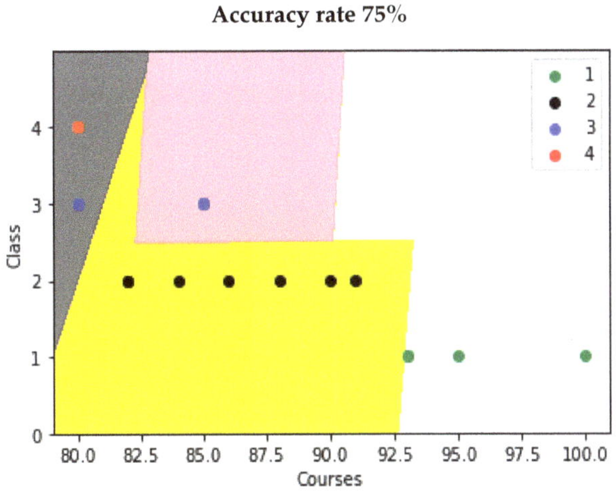

Figure 4. Illustrates SVM with an accuracy rate of 75%.

5. Discussion and Recommendation

According to the findings of this research, the current admission policies at universities are not accurate. These polices rely on some composed numerical scores, which are WS and QS. Such composed scores are calculated according to scores such as GPAH, GAT, and AT. These scores are calculated as a projection of a set of data, so they hide important information. Indeed, GPAH represents the scores of about 17 courses. GAT is a number that represents students' aptitude in mathematics and Arabic. AT is another number representing mathematics, physics, chemistry, and biology. However, the medical major gives more importance to English, chemistry, and biology. The scores of the three courses are hidden behind GPAH and AT.

The main issue with accumulated scores is that they do not accurately represent the values they are based on. For example, there are two stores (Store 1 and Store 2). The average sales figure for Store 1 for the last two years (y1 and y2) is 100,000, and the average sales figure for Store 2 for the last two years is also 100,000. However, the sales for Store 1 in the first year were Store1.y1 = 20,000 and the sales for Store 1 in the second year were Store1.y2 = 180,000, while the sales for Store 2 were Store2.y1 = 99,000 and Store2.y2 = 10,100. Obviously, the sales of Store 1 increased incredibly last year, while the sales of Store 2 were stable for the last two years. In this way, the average does not show some important details behind the value. The same issue is applicable to student GPAH and other scores, as shown in Figure 5. Figure 5 shows the course scores and GPAH for two students (GPAH = 85 for both students, represented by red line). Figure 5 highlights the distances between GPAH and individual course scores. It is clear that the GPAH for student 2 is more representative (represented by the green line), while the course scores for student 1 (represented by the blue line) have more distance from the GPAH. In fact, standardized measures can be used, such as variance and standard deviation. Clearly, the variance and standard deviation for student 2 are smaller than those for student 1, which means that the GPAH for student 2 is more representative of the course scores.

Another example demonstrating the issues of WS and QS is shown in Table 12. Table 12 illustrates that two students applied to medical college. Student x has GPAH = 95, GAT = 91, and AT = 89, while student y has GPAH = 94, GAT = 89, and AT = 89. Using the current enrollment policy, student x has a better chance to be accepted as a result of having higher scores for GPAH and GAT. In Table 6, we can find the important courses for each major. Obviously, student x has higher GPAH and GAT scores because of their high scores in mathematics, physics, Arabic, and others. However, considering the hidden (detailed) scores, student y is better in chemistry and biology, which are more important when joining medical college. In addition, student y has a lower GAT score because of their performance in mathematics, while y's English score is better than x's. English scores are more important than mathematics scores for students who apply to medical college. Indeed, the same concept is applicable to engineering, computer science and information technology, and business colleges.

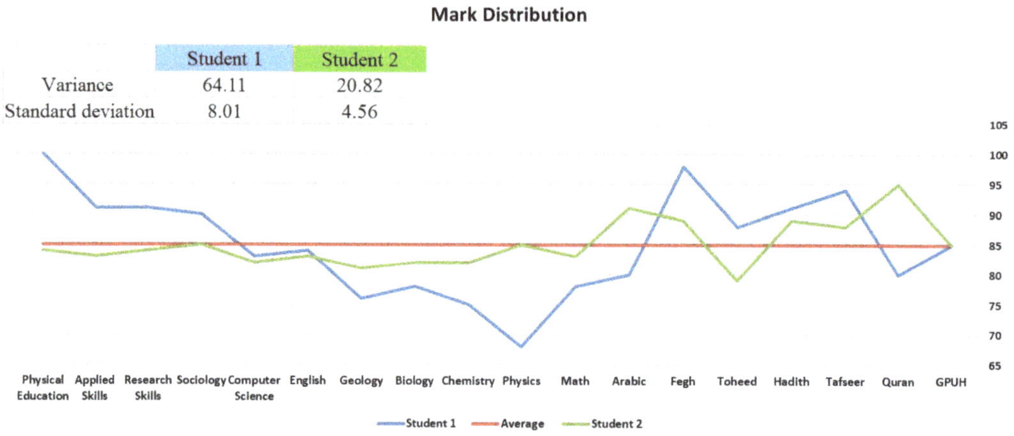

Figure 5. Illustrates distributions of course scores for two students and compares them to the GPAH, which is 85 for both students. It also calculates the variance and standard deviation.

On the other hand, the majors of education, Sharia, and arts and humanities do not consider AT. GPAH and GAT affect the enrollment in these three majors negatively. GPAH has the highest correlation with GPAU for education and Sharia majors, while WS has the highest correlation for arts and humanities majors (as shown previously in Table 5). However, GPAH and QS scores give more credit to mathematics, physics, chemistry, and biology, which are not relevant to theoretical majors. In addition, an English major requires good performance in English courses, while the scores of mathematics, physics, chemistry, and biology are not important. Moreover, AT tests students' aptitude in mathematics and Arabic, which are not important for an English major.

In practice, the proposed machine learning model can be deployed in the form of an application programming interface, where a student's scores can be entered as inputs and the output is a suitable class (major). In addition, the current admission exams and scores should be changed to show more hidden details. Therefore, we end this paper with the following recommendations:

1. To enhance the accuracy of the classification, the GAT score should be divided into two separate scores: GAT-Math and GAT-Language.
2. For students applying to an English major, a specific English test should be incorporated or the high school English course score should be used for evaluation.
3. The AT score should take into account the proficiency in the English language.
4. To provide a more detailed evaluation, the AT score should be divided into five separate scores: AT-Math, AT-Physics, AT-Chemistry, AT-Biology, and AT-English.
5. The WS and QS should be replaced by a relative course-based model, as depicted in Table 1, rather than using them for evaluation purposes.

Table 12. Example of scores of two students showing the detail scores.

Student x			**Student y**		
Mathematics		99	Mathematics		90
Physics		100	Physics		91
Chemistry		90	Chemistry		100
Biology		91	Biology		100
English		92	English		97
Arabic		100	Arabic		94
GPAH		95	**GPAH**		94
GAT = 91	Math	91	**GAT = 89**	Math	87
	Arabic	91		Arabic	93
AT = 89	Math	92	**AT = 89**	Math	88
	Physics	90		Physics	86
	Chemist	86		Chemist	90
	Biology	88		Biology	92

6. Conclusions

In conclusion, this research enhances the process of student admission at universities. This research investigated the current acceptance procedures using GPAH, QS, and WS. This paper highlighted the issue of free education, a large number of applicants, and high failure rates. We first considered experts' points of view using a survey. Then, we used data mining analysis to investigate the relations between student performance and the admission features. Four data mining techniques were applied to real-world data. We created suitable methods such as modified Jaccard and approximated modified Jaccard analyses that fit our problem. The results show that the current admission procedures are not adjusted for all majors, as the relations vary from one major to another. Next, we determined the importance of considering the individual grades of high school courses and linked them to each major. Finally, we suggested using successful students' records as training examples for each major and we applied machine learning models. In the future, this model can be the basis of a smart application to guide students and universities in the admission process.

Author Contributions: Conceptualization, B.A., M.B. and A.A.; methodology, B.A., M.B. and A.A.; software, B.A., M.B. and A.A.; validation, B.A., M.B. and A.A.; formal analysis, B.A., M.B. and A.A.; investigation, B.A., M.B. and A.A.; resources, B.A., M.B. and A.A.; data curation, B.A.; writing—original draft preparation, M.B. and A.A.; writing—review and editing, B.A.; visualization, B.A., M.B. and A.A.; supervision, B.A.; project administration, B.A.; funding acquisition, B.A. All authors have read and agreed to the published version of the manuscript.

Funding: The authors extend the appreciation to the Deputyship for Research Innovation, Ministry of Education in Saudi Arabia for funding this research work and covering the APC through project number ISP-2024.

Data Availability Statement: Restrictions apply to the availability of these data. Data was obtained from Jazan University with the permission of Jazan University HABO-10-Z-001, with Reference No: REC-43/05/094.

Conflicts of Interest: The authors declare no conflict of interest.

References

1. Assiri, B.; Bashraheel, M.; Alsuri, A. Improve the Accuracy of Students Admission at Universities Using Machine Learning Techniques. In Proceedings of the 7th International Conference on Data Science and Machine Learning Applications (CDMA), Riyadh, Saudi Arabia, 1–3 March 2022; pp. 127–132.
2. Kumar, M.; Singh, A.J.; Handa, D. Literature survey on student's performance prediction in education using data mining techniques. *Int. J. Educ. Manag. Eng.* **2017**, *7*, 40–49. [CrossRef]
3. Kumar, M.; Singh, A.J.; Handa, D. Literature survey on educational dropout prediction. *Int. J. Educ. Manag. Eng.* **2017**, *7*, 8. [CrossRef]
4. Miller, G.A. Undergraduates' Decision-making Processes in College Major Selection. Ph.D. Thesis, Hofstra University, Hempstead, NY, USA, 2018.
5. Albakri, B.; Abuhamdeiyeh, S.; Mousa, A. Rule-Based Expert System to Lead Freshmen Students in Choosing a Suitable College Major. In Proceedings of the 10th IADIS International Conference on Information Systems, Budapest, Hungary, 10–12 April 2017.
6. Whitehead, A. Examining influence of family, friends, and educators on first-year college student selection STEM major selection. *J. Mason Grad. Res.* **2018**, *5*, 58–84.
7. Casinillo, L. Factors affecting the failure rate in mathematics: The case of Visayas State University (VSU). *Rev. Socio-Econ. Res. Dev. Stud.* **2019**, *3*, 1–18.
8. Bennedsen, J.; Caspersen, M.E. Failure rates in introductory programming. *AcM SIGcSE Bull.* **2007**, *39*, 32–36. [CrossRef]
9. Hassan, S.M.; Al-Razgan, M. Pre-university exams effect on students GPA: A case study in IT department. *Procedia Comput. Sci.* **2016**, *82*, 127–131. [CrossRef]
10. Alghamdi, A.; Hamdan, K.; Al-Hattami, A.A. The Accuracy of Predicting University Students' Academic Success. *J. Saudi Educ. Psychol. Assoc.* **2016**, *186*, 1–10.
11. Malgwi, C.A.; Howe, M.A.; Burnaby, P.A. Influences on students' choice of college major. *J. Educ. Bus.* **2005**, *80*, 275–282. [CrossRef]
12. Montmarquette, C.; Cannings, K.; Mahseredjian, S. How do young people choose college majors? *Econ. Educ. Rev.* **2002**, *21*, 543–556. [CrossRef]
13. Niwattanakul, S.; Singthongchai, J.; Naenudorn, E.; Wanapu, S. Using of Jaccard coefficient for keywords similarity. In Proceedings of the International Multiconference of Engineers and Computer Scientists, Hong Kong, China, 13–15 March 2013; Volume 1.
14. Thanh Noi, P.; Kappas, M. Comparison of random forest, k-nearest neighbor, and support vector machine classifiers for land cover classification using Sentinel-2 imagery. *Sensors* **2018**, *18*, 18. [CrossRef]
15. Decision Trees. Scikit. (n.d.). Available online: https://scikit-learn.org/stable/modules/tree.html (accessed on 3 November 2021).
16. Noble, W.S. What is a support vector machine? *Nat. Biotechnol.* **2006**, *24*, 1565–1567. [CrossRef]
17. Boyd, R.L.; Pennebaker, J.W. Building a Personalized College Major Selection Web Page. *PsyArXiv* **2018**. [CrossRef]
18. Iraji, M.S.; Aboutalebi, M.; Seyedaghaee, N.R.; Tosinia, A. Students classification with adaptive neuro fuzzy. *Int. J. Mod. Educ. Comput. Sci.* **2012**, *4*, 42. [CrossRef]
19. Adu, E.O.; Oshati, T. Psychological Variables as Correlate of Students' Academic Achievement in Secondary School Economics in Oyo State Nigeria. *J. Psychol.* **2014**, *5*, 125–132. [CrossRef]
20. Kiaghadi, M.; Hoseinpour, P. University admission process: A prescriptive analytics approach. *Artif. Intell. Rev.* **2023**, *56*, 233–256. [CrossRef]
21. Abu Saa, A.; Al-Emran, M.; Shaalan, K. Factors affecting students' performance in higher education: A systematic review of predictive data mining techniques. *Technol. Knowl. Learn.* **2019**, *24*, 567–598. [CrossRef]
22. Ribera, A.K.; Miller, A.L.; Dumford, A.D. Sense of peer belonging and institutional acceptance in the first year: The role of high-impact practices. *J. Coll. Stud. Dev.* **2017**, *58*, 545–563. [CrossRef]
23. Kutscher, M.; Nath, S.; Urzúa, S. Centralized admission systems and school segregation: Evidence from a national reform. *J. Public Econ.* **2023**, *221*, 104863. [CrossRef]
24. Noble, J.; Sawyer, R. *Predicting Grades in Specific College Freshman Courses from ACT Test Scores and Self-Reported High School Grades*; ACT Research Report Series; American College Testing Program: Iowa City, IA, USA, 1999.
25. Noble, J.; Sawyer, R. *Predicting College Grades from ACT Assessment Scores and High School Course Work and Grade Information*; American College Testing Program: Iowa City, IA, USA, 1991.
26. Betts, J.R.; Morell, D. The determinants of undergraduate grade point average: The relative importance of family background, high school resources, and peer group effects. *J. Hum. Resour.* **1999**, *34*, 268–293. [CrossRef]
27. Sadiq, F.A.; Mitlif, R.J.; Abbas, J. A computational mechanism for making admission decisions in the centralized admission system. In *AIP Conference Proceedings*; AIP Publishing: Melville, NY, USA, 2023; Volume 2414.
28. Lee, F.X.; Suen, W. Gaming a Selective Admissions System. *Int. Econ. Rev.* **2023**, *64*, 413–443. [CrossRef]
29. Xu, J.; Moon, K.H.; Van Der Schaar, M. A machine learning approach for tracking and predicting student performance in degree programs. *IEEE J. Sel. Top. Signal Process.* **2017**, *11*, 742–753. [CrossRef]
30. Hossain, M.A.; Assiri, B. Emotion specific human face authentication based on infrared thermal image. In Proceedings of the 2020 2nd International Conference on Computer and Information Sciences (ICCIS), Sakaka, Saudi Arabia, 13–15 October 2020.

31. Numan, M.; Subhan, F.; Khan, W.Z.; Assiri, B.; Armi, N. Well-organized bully leader election algorithm for distributed system. In Proceedings of the 2018 International Conference on Radar, Antenna, Microwave, Electronics, and Telecommunications (ICRAMET), Serpong, Indonesia, 1–2 November 2018; pp. 5–10.
32. Fadhilah, A.; Ismail, N.H.; Abdul Aziz, A. The prediction of students' academic performance using classification data mining techniques. *Appl. Math. Sci.* **2015**, *9*, 6415–6426.
33. Jaccard, P. Nouvelles recherches sur la distribution florale. *Bull. Soc. Vaud. Sci. Nat.* **1908**, *44*, 223–270.
34. Jaccard, P. The distribution of the flora of the alpine zone. *New Phytol.* **1918**, *11*, 37–50. [CrossRef]
35. Albatineh, A.N.; Niewiadomska-Bugaj, M.; Mihalko, D.P. On similarity indices and correction for chance agreement. *J. Classif.* **2016**, *23*, 301–313. [CrossRef]
36. Albatineh, A.N.; Niewiadomska-Bugaj, M. Correcting Jaccard and other similarity indices for chance agreement in cluster analysis. *Adv. Data Anal. Classif.* **2011**, *5*, 179–200. [CrossRef]
37. Eelbode, T.; Bertels, J.; Berman, M.; Vandermeulen, D.; Maes, F.; Bisschops, R.; Blaschko, M.B. Optimization for medical image segmentation: Theory and practice when evaluating with dice score or jaccard index. *IEEE Trans. Med. Imaging* **2020**, *39*, 3679–3690. [CrossRef]
38. Benesty, J.; Chen, J.; Huang, Y.; Cohen, I. Pearson correlation coefficient. In *Noise Reduction in Speech Processing*; Springer: Berlin/Heidelberg, Germany, 2009; pp. 1–4.

applied sciences

MDPI

Article

Enhanced Chinese Domain Named Entity Recognition: An Approach with Lexicon Boundary and Frequency Weight Features

Yan Guo [1], Shixiang Feng [1], Fujiang Liu [2,*], Weihua Lin [3], Hongchen Liu [1], Xianbin Wang [4], Junshun Su [5] and Qiankai Gao [6]

1 School of Computer Science, China University of Geosciences, Wuhan 430078, China; guoyan49@cug.edu.cn (Y.G.); fengsx@cug.edu.cn (S.F.); codernewhoc@cug.edu.cn (H.L.)
2 School of Geography and Information Engineering, China University of Geosciences, Wuhan 430078, China
3 Hubei Key Laboratory of Regional Ecology and Environmental Change, School of Geography and Information Engineering, China University of Geosciences, Wuhan 430078, China; linweihua@cug.edu.cn
4 Piesat Information Technology Co., Ltd., Beijing 100195, China; wangxianbin@piesat.cn
5 Xi'ning Natural Resources Comprehensive Survey Center, China Geological Survey, Xi'ning 810016, China; sujunshun@163.com
6 Kunming Natural Resources Comprehensive Survey Center, China Geological Survey, Kunming 650111, China; gaoqiankai@126.com
* Correspondence: liufujiang@cug.edu.cn

Abstract: Named entity recognition (NER) plays a crucial role in information extraction but faces challenges in the Chinese context. Especially in Chinese paleontology popular science, NER encounters difficulties, such as low recognition performance for long and nested entities, as well as the complexity of handling mixed Chinese–English texts. This study aims to enhance the performance of NER in this domain. We propose an approach based on the multi-head self-attention mechanism for integrating Chinese lexicon-level features; by integrating Chinese lexicon boundary and domain term frequency weight features, this method enhances the model's perception of entity boundaries, relative positions, and types. To address training prediction inconsistency, we introduce a novel data augmentation method, generating enhanced data based on the difference set between all and sample entity types. Experiments on four Chinese datasets, namely Resume, Youku, SubDuIE, and our PPOST, show that our approach outperforms baselines, achieving F1-score improvements of 0.03%, 0.16%, 1.27%, and 2.28%, respectively. This research confirms the effectiveness of integrating Chinese lexicon boundary and domain term frequency weight features in NER. Our work provides valuable insights for improving the applicability and performance of NER in other Chinese domain scenarios.

Keywords: Chinese named entity recognition; dual pointer network; lexicon enhancement; Chinese paleontology popular science

1. Introduction

Over the past few years, with the development of the Internet and information technology, various domains have accumulated a large amount of textual data, containing valuable information and knowledge [1]. Effectively utilizing this textual data can have a significant impact on various domains. Information extraction refers to automatically extracting key information from massive data, named entity recognition (NER), as a subtask, and has become a research hotspot in various domains [2].

In the domain of education, popular science (also known as popsci) is a scientific explanation targeted at the general audience [3]. Popular science education can cultivate the public's interest in science and promote the popularization and dissemination of science and technology in society, thereby fostering social development. Faced with the growing data in the domain of popular science education, how to quickly and accurately identify keywords and reveal deep semantic relationships is an urgent issue for intelligent systems in popular science education. Many discoveries and research results in the domain of paleontology contribute to understanding the evolution of the Earth, as well as the origin and evolution of life and humanity [4], and one of the advantages of paleontology is popularization. In order to promote the latest research results to the public and advance the development of popular science education, the efficient extraction of key information from texts in the domain of paleontology popular science using deep learning-based information extraction technology has gained widespread attention from domain experts. One of the primary tasks in the domain of Chinese paleontology popular science NER is to identify different types of entities, such as a person, geological time, strata, and fossil, etc., from unstructured textual data. NER involves preprocessing, feature extraction, and classification to identify entities from unstructured textual data. By recognizing entities, valuable information can be extracted from textual data, laying the foundation for subsequent tasks [5], such as relationship extraction, sentiment analysis [6,7], and knowledge graph construction.

The NER methods can be summarized from rule-based methods to deep learning-based methods. In the early stage, rule-based methods utilized predefined and inductive assumptions and rules to identify entity names in the text and classify them [8,9]. However, with the increase in annotated data, rule-based methods need constant updates to discover more entities. Traditional machine learning methods, mainly based on supervised learning, involve training classification or sequence labeling models on annotated datasets to learn features of positive or negative instances, or type features. For example, based on a support vector machine (SVM) [10], hidden Markov model (HMM) [11,12], and conditional random field (CRF) [13], etc., those methods no longer require the manual construction of rules, but they still require feature selection. In recent years, with the development of deep learning, various neural networks have been used to address NER problems, such as convolutional neural networks (CNNs) [14], recurrent neural networks (RNNs), such as long short-term memory (LSTM) [15] and bidirectional LSTM (BiLSTM) [16], graph neural networks (GNNs) [17], and attention mechanisms [18], etc. Neural network methods typically use vectors to learn the syntax and context information of language. Pretrained models, represented by BERT (bidirectional encoder representation from transformers) [19], have a significant advantage in extracting contextual information features from text due to training on large-scale datasets. This notably improves various metrics for natural language processing (NLP) tasks. In the field of NER, significant achievements have been made by methods based on fine-tuning pretrained models, like BERT [19,20], BERT-CRF [21,22], and BERT-BiLSTM-CRF [23]. Pretrained models can learn contextual features of text on the basis of training on large-sample data, enabling fine-tuning in situations with limited annotated data and subsequently learning contextual features for downstream tasks.

The models mentioned above are based on sequence labeling to accomplish NER, which means performing a multi-classification task for each token. However, in the domain of Chinese paleontology popular science, as shown in Figure 1, the entity "古脊椎动物与古人类研究所" (Institute of Vertebrate Paleontology and Paleoanthropology, Chinese Academy of Sciences) consists of 12 Chinese characters; the span "寒武纪地层" (Cambrian Strata) includes the Geological Time entity, the Cambrian, and the Strata entity, Cambrian Strata; the span "古生物学家 Grant Zazula" (the Paleontologist named Grant Zazula) contains both Chinese and English characters. Due to the existence of the mentioned phenomena and problems, directly applying sequence labeling models will significantly degrade the model's performance, and these problems directly hinder the further development of NER in this domain. To address this issue, the pointer labeling task is proposed. Pointer labeling is a framework for NER. It sets start and end pointers to record the starting and ending positions of each entity, and simultaneously annotates the entity type. This method excels in recognizing long entities and nested entities. However, using only the pointer labeling for NER results in the loss of Chinese lexical boundary information, causing the model to erroneously output any two boundaries as entities. Recently, the integration of lexicon-level features into Chinese NER has attracted widespread attention. Integrating lexicon-level features involves incorporating lexicons as external information, helping to determine the span and type of entities. At the same time, in our domain, we found that the distribution of domain-specific lexicon also affects the performance of NER. Currently, the utilization of distributional features of a domain-specific lexicon is still in its early stages.

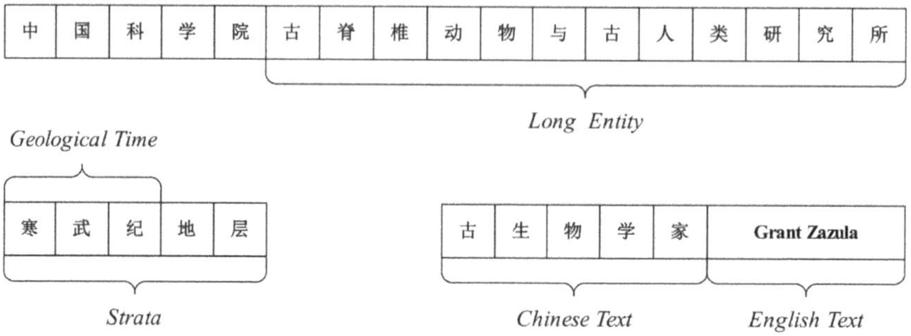

Figure 1. An example containing a long entity, nested entities, and a mixed text of Chinese and English.

To address the above-mentioned issues, in this paper, we propose an enhanced NER method based on lexicon boundary and frequency weight features, named the BERT-Lexicon-PointerNetwork (BERT-LPN) model. The model is proposed for extracting Chinese paleontology popular science entities and it consists of an encoder based on BERT, a lexicon feature fusion layer and a fully connected classifier layer from the bottom up: the encoder is a Chinese character-level model based on BERT, and it maps Chinese characters to a low-dimensional, highly dense real-vector space to extract latent semantic information from Chinese entity elements; the lexicon feature fusion layer takes lexicon boundary and weight information from Chinese word segmentation as input to capture the lexicon-level features of the input; finally the fully connected classifier layer takes BERT-encoded embeddings along with lexicon-level features as input to generate start position and end position labels for entities. The experimental results indicate that our proposed model outperforms previous models, and our approach achieves state-of-the-art (SOTA) performance on the constructed dataset.

The main contributions of this research are summarized as follows:

- Based on the characteristics of texts in the domain of Chinese paleontology popular science, we propose a NER model enhanced with Chinese lexicon boundary and frequency weight features. This model is designed for identifying and extracting entities from unstructured textual data.
- Based on the structural characteristics of our model, we propose a data augmentation method utilizing all entity types and sample entity types to alleviate the inconsistency between training and prediction tasks.

To comprehensively evaluate the performance of our model, we established a new dataset, namely the PPOST dataset, specifically designed for NER in the Chinese paleontology popular science domain. The dataset primarily consists of data from authoritative institutions in the domain of paleontology popular science in China. Experimental results on both public datasets and the PPOST dataset validate the effectiveness of our approach. In addition, we also analyze the possibility of applying our method to other domains, providing a reference for enhancing NER performance in other domains.

The remaining sections of this paper are organized as follows. Section 2 introduces NER work based on deep learning that is closely related to our work. Section 3 presents our proposed model and core algorithms. Section 4 provides a detailed description of the datasets used in our experiments and analyzes the experimental results. Finally, we conclude the article.

2. Related Work

In this section, we introduce works closely related to our proposed model, including NER models that integrate lexicon-level features, methods for addressing domain data scarcity, and NER models based on pointer labeling, as follows.

2.1. Lexicon-Based Chinese NER

Chinese, unlike languages, such as English, that use spaces as separators, encounters the issue of word segmentation, where Chinese takes words as the basic semantic units. To better adapt to Chinese NER, recent studies have found that introducing Chinese lexicon-level features can enhance the performance of NER models. For example, Zhang et al. [24] proposed a lattice LSTM model for Chinese NER. This model is an extension of character-based NER models, incorporating words as input and additional gates to control information flow. The model explicitly utilizes word and word order information. However, due to the RNN structure adopted by the lattice LSTM model, it cannot capture long-distance dependencies, and the introduction of lexical information is lossy. To address these limitations, Li et al. [25] proposed the FLAT model for Chinese NER. This model is based on a transformer [26] to solve the lexical loss problem in lattice LSTM and uses relative position encoding to adapt a transformer to NER, enabling efficient GPU parallel computation. These methods, combined with pre-trained language models, explore Chinese NER and achieve SOTA performance on several Chinese benchmark datasets. However, Guo et al. [27] and Liu et al. [28] believe that existing lexicon-based models only conform to lexicon features through shallow and randomly initialized encoding layers without integrating them into the underlying layers of pre-trained language models to explore deep lexicon knowledge. To address this, they proposed methods to deeply integrate external lexicon features into the pre-trained language model BERT, achieving more effective fusion of entity boundaries and lexical information.

All of these studies have explored the integration of Chinese lexicon-level features, but they did not address the practical issues faced by more detailed domains. Moreover, these methods rely on the quality of lexicons and have certain requirements for computational resources. In contrast, this paper focuses on extracting entities from the Chinese paleontology popular science domain, integrating Chinese lexicon-level features in a simple and effective manner while enhancing relative positional information. It provides a reference for addressing the practical issues faced by this domain.

2.2. Prompt and Pointer Network

Typically, there is not much annotated data for downstream tasks, and there is often a gap between pre-trained BERT models on large corpora and fine-tuning on downstream tasks. To address the gap, some researchers have proposed modeling downstream tasks in the form of pre-training tasks [29]. Taking BERT as an example, some tasks can be modeled as masked language modeling (MLM) or next sentence prediction (NSP), and this approach is particularly effective in scenarios with small datasets [30]. For instance, Cui et al. [31] and others proposed the TemplateNER method, which completes NER using prompt learning. This method employs a template-based approach to solve the problem of few-shot NER: a template is pre-defined, for example, "<candidate_span>is a <entity_type> entity". With the help of a suitable prompt, this method reduces the difference between pre-training and fine-tuning, allowing the model to achieve good results with a small number of samples. It significantly outperforms traditional sequence labeling methods and distance-based few-shot NER methods in cross-domain and few-shot scenarios. Subsequently, methods based on the prompt mechanism have been widely used in NER [32,33], such as modeling the task as machine reading comprehension and question answering (MRC-QA) format, fitting the NSP task for NER [34].

The NER model based on the dual pointer network aims to address both the challenges of recognizing long entities and nested entities. Lu et al. [35] proposed a unified extraction framework, where the NER module is based on a pre-trained model, employing a multi-layer dual pointer network. Input samples are constructed based on the template "[SPOT]<entity_type>[text]", predicting the start and end positions of possible entities in the samples, and NER is completed through the decoding algorithm. However, real-world datasets often exhibit a long-tail distribution problem, with a few types representing the majority of the samples. The multi-layer dual pointer network requires setting up a classifier for each entity type, potentially leading to some classifiers not being adequately trained, thus impacting the overall performance of NER model.

To mitigate this issue, some researchers proposed a model based on a single-layer dual pointer network and the prompt mechanism. Gong et al. [36] introduced a model based on a single-layer dual pointer network, constructing the input in the format "<entity_type>[SEP]<text>" to accomplish NER. However, Su et al. [37] argue that the conventional design of the dual-pointer network, when performing NER or MRC, typically employs two modules to separately identify the start and end positions of entities. This can lead to inconsistencies during training and prediction. To address this issue, they proposed global pointer model which uses a globally normalized approach for NER. In non-nested scenarios, it achieves results comparable to CRF, and in nested scenarios, it performs well. Moreover, it allows for fully parallelized training, significantly accelerating the training process. This method addresses the inconsistency between training and prediction tasks from a model perspective.

All these studies have explored NER methods based on pre-trained models and proposed unique insights and solutions for various issues. However, most of them optimized from the perspective of the model without considering the scarcity and differences in datasets across various domains in real-world applications. In contrast, this paper introduces a data augmentation method that enhances NER performance by increasing training data, alleviating the inconsistency between training and prediction tasks. This approach provides a reference for improving NER performance in domain-specific applications.

3. Methods

In this section, we provide a detailed description of the proposed model architecture, and the overall structure of our model is illustrated in Figure 2. Firstly, we introduce some preprocessing operations specific to this domain to acquire the necessary experimental foundation. Secondly, based on the WordPiece [38] algorithm, we use BERT to obtain character-level embeddings and acquire lexicon-level features through Chinese word segmentation; lexicon-level features fusion are accomplished using a multi-head self-attention mechanism. Finally, the embeddings based on BERT and the lexicon-level features serve as inputs to the dual pointer network, which is applied to complete the NER through decoding in the domain of Chinese paleontology popular science.

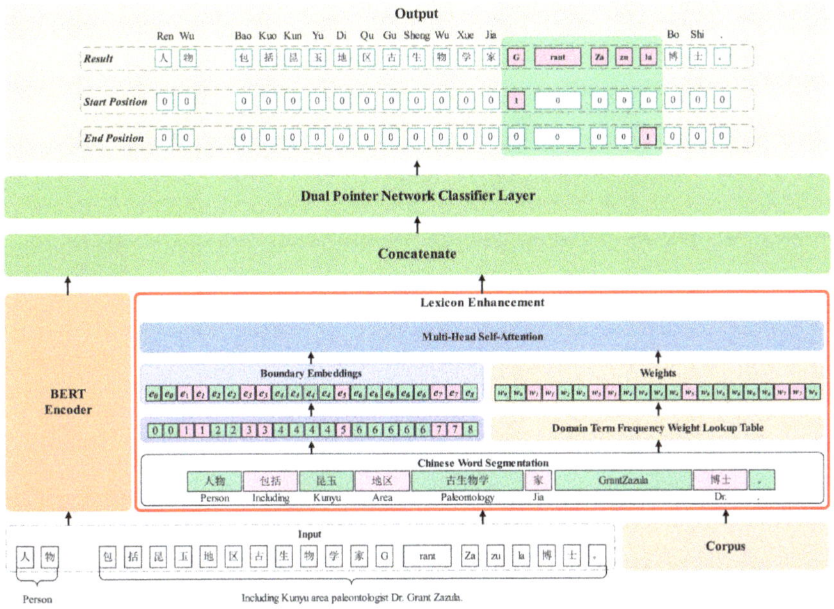

Figure 2. BERT-LPN: The input is processed through BERT to obtain character-level embedding vectors and the Chinese lexicon-level feature module to acquire lexicon-level embedding vectors (red box). After concatenation, a dual pointer network composed of two fully connected networks is used to accomplish the NER.

3.1. Preprocessing

To obtain domain term frequency weight features, we base our calculations on the collected corpus in the domain of Chinese paleontology popular science. TF-IDF [39] is a statistical method that consists of term frequency (TF) and inverse document frequency (IDF) and evaluates the importance of a term to a document set or one of the documents in a corpus, the TF-IDF scores of domain terms are computed as domain term frequency weights in our approach. Specifically, for the raw text corpus we collected, we employ the Chinese word segmentation tool pkuseg [40] for segmentation. After removing stop words and segmenting the Chinese text, the TF-IDF scores of each term are calculated as the weight feature for domain term frequency. Since there are N documents in total, term t has N TF-IDF scores, and we integrate these scores using the averaging strategy because this approach preserves more features, thereby better assessing the weight of a term in the corpus and obtaining the final feature for domain term frequency. The method for calculating domain term frequency weights is as follows:

$$W_{t,d} = tf_{t,d} \cdot log(\frac{N}{1 + df_t}), \tag{1}$$

$$Weight_t = \frac{\sum_{i=0}^{N} W_{t,d}^{i}}{N}, \tag{2}$$

where $tf_{t,d}$ represents the term frequency of term t in document d, log is the natural logarithm, df_t represents the number of documents out of N that contain term t, $W_{t,d}$ is the TF-IDF score of term t in document d, and $Weight_t$ represents the average weighted domain term frequency weight of term t. The final domain term frequency weights are shown in Table 1.

Table 1. Domain term frequency weight lookup table (DTFW-lookup Table).

Domain Term	Weight
恐龙 (Dinosaur)	$W_{恐龙} = 1.36 \times 10^{-3}$
侏罗纪 (Jurassic)	$W_{侏罗纪} = 3.67 \times 10^{-4}$
中国科学院 (Chinese Academy of Sciences)	$W_{中国科学院} = 7.38 \times 10^{-5}$
…	…
$Term_i$	$W_i = x$

Usually, the NER based on a dual pointer network and the MRC-QA mechanism faces the problem of inconsistency between training and prediction tasks. That is, during training, fitting is carried out with the correct entity type as a prompt, while in prediction tasks, it is often not possible to specify the correct prompt for prediction, meaning the prediction task also includes an entity classification task. To alleviate this inconsistency and enhance the model's generalization ability, we propose a data augmentation method based on the set difference between all entity types and sample entity types.

As shown in Figure 3, our dataset includes 13 types of entities, denoted as the set $AllTypes$ = {Strata, FossilClassification (FossilClass), Subject, Event, Position, Time, Feature, Person, Organization, GeologicalTime (GeoTime), Address, Fossil, BiologicalClassification (BioClass)}, while the current set of entity types included in the samples is denoted as set $Gold_{sample}$. Through set difference operations on the sets, the entity types not included in the current samples are generated. In this way, after calculating the difference set for all samples, the labels of the enhanced data are set to an empty set to avoid introducing noisy data, as follows :

$$Diff_{sample} = \left\{type \in AllTypes : type \notin Gold_{sample}\right\}, \tag{3}$$

$$DA^{item}_{sample} = \left\{Text_{sample}, EntityType = Diff^{item}_{sample}, Label = \varnothing\right\}, \tag{4}$$

where $Diff^{item}_{sample}$ is an element in the collection $Diff_{sample}$, and DA^{item}_{sample} is a piece of data generated when the entity type is $Diff^{item}_{sample}$.

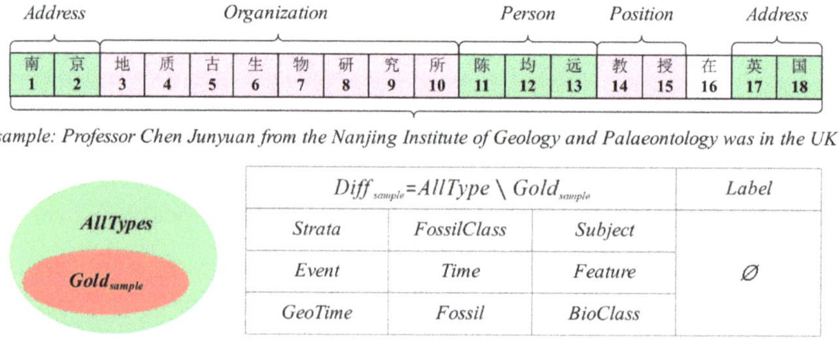

Figure 3. Data augmentation: Calculate the set difference between the set $AllTypes$ and the set $Gold_{sample}$ to obtain augmented data.

Then, based on the given predefined data augmentation ratio, a certain number of augmented data are generated through a random sampling algorithm and merged into the training dataset for model training. By introducing augmented data, the model's perception of entity types not present in the sample data is strengthened, effectively alleviating the inconsistency between the training and prediction tasks.

3.2. BERT Encoder

To effectively capture textual information from text, the encoder is constructed based on a pretrained language model, BERT. Given an input (i.e., a sentence or a word) $S = \left[s_1, s_2, \ldots, s_{|S|}\right]$, the encoder tokenizes it using a predefined vocabulary and encodes s_i into a vector representation $\overline{s_i}$. For detailed steps, please refer to the literature [19]. The encoder embeds the text S into vector representations $\Gamma^S = \mathbb{R}^{len \times dim}$, where len represents the maximum acceptable input length for the model, dim represents the dimension of BERT token embedding vectors, and in the case of BERT-base dim is set to 768. Generally, Γ^S_j can be considered as the embedding of the jth token, and it is worth noting that due to the usage of the WordPiece tokenization algorithm, variations may occur in our methods. Unlike other Chinese NER models, we use the WordPiece tokenization algorithm to represent Chinese characters as independent tokens and English as their subword representations, and the method can simplify the model input and reduce the complexity of model. Additionally, based on BERT, we fine-tune the model in conjunction with the MRC-QA task. The detailed steps are as follows:

$$input = \text{“} < entity\_type >, < text > \text{”}, \tag{5}$$

$$tokenized = WordPiece(input) = \{t_1, t_1, \ldots, t_n\}, \tag{6}$$

$$\chi_i = e(t_i). \tag{7}$$

In the above, using the provided *text* sequence and its corresponding entity type *entity_type*, we construct the *input* and obtain *tokenized* representations for the input using the WordPiece algorithm. The variable χ_i represents the embedding vector by BERT for the given *tokenized* representation.

3.3. Lexicon Encoder

To better harness Chinese lexicon-level information, we have introduced lexicon boundary and frequency weight features. Given the input, it is segmented into individual lexicons or characters after applying a segmentation tool, as follows:

$$SegId_{input} = \{(0, a), (a+1, b), \ldots, (x, y), \ldots, 0 \ldots, 0\}, \tag{8}$$

where (x, y) represents the segment composed of the characters from the xth to the yth position in the input sequence, in other words, in the ordered set of $SegId_{input}$, each element represents a Chinese lexicon or character. Let us define a function $f(i, j)$—element i repeats j times results in an ordered set, as follow:

$$f(i, j) = \{i \ repeats \ j \ times\}, \tag{9}$$

giving $(x, y)_{id} \in SegId_{input}$, where id is the index of $(x, y)_{id}$ in the ordered set $SegId_{input}$, and we substitute $f(i, j)$ to obtain $f(id, y-x)$, where $y - x$ is the length of the current lexicon or character segment. In summary, we obtain the lexicon boundary features of the input sequence through Algorithm 1, and its time complexity is $O(n)$, where n is the length of the input sequence. At the same time, if n does not reach the maximum allowed length of the model, padding with zeros is applied after the last element of *boundary* until it reaches its maximum. On the other hand, as shown in Figure 4, when there are three entities which belong to the same type (BioClass) in a given input (lexicons 0, 2, and 4), without considering relative positional information, the model may arbitrarily combine their boundaries in the output. For example, the phrase “恐龙位于爬行动物和鸟类” (Dinosaurs are between reptiles and birds, with characters between the start of lexicon 0 and the end of lexicon 4) could be incorrectly combined into a single BioClass entity. With the inclusion of boundary information, the model becomes sensitive to the length and span of entities.

Dinosaur	are between	reptiles	and	birds	crossroads between them		
恐龙$^{\mathbf{BioClass}}$	位于	爬行动物$^{\mathbf{BioClass}}$	和	鸟类$^{\mathbf{BioClass}}$	之间	的	十字路口
0	**1**	**2**	**3**	**4**	**5**	**6**	**7**

Figure 4. Boundary features incorporate simultaneously lexical boundary information and relative positional information: the method introduces relative positional information from another perspective, making it sensitive to the length and span of entities, reducing the likelihood of the model predicting the combination of any two entity boundaries as the target.

Algorithm 1 Lexicon Boundary Generation

```
Input: Ordered Set of SegId_input
Output: Ordered Set of Lexicon Boundary
boundary = {}
for each (x, y)_id ∈ SegId_input do
    item = {}
    for each index ∈ [0, y − x] do
        item = item ∪ {i}
    end for
    boundary = boundary ∪ item
end for
return boundary
```

Given the $boundary_i$, we look up the boundary index embedding e_i from the specialized embedding matrix. This matrix contains fixed-size embeddings for each boundary index 0, 1, 2, ..., and these embeddings are learned through backpropagation.

Similarly, $(x, y)_{weight} \in SegId_{input}$, where $weight$ represents the term frequency weight of the xth to y^{th} subsequence of the input, and this weight value is obtained through the DTFW-lookup Table in Section 3.1. Substitute $f(i, j)$ to obtain $f(weight, y - x)$, where $y - x$ is the length of the current segment. This step will result in the lexicon or character frequency weight representation w_i for the input. Specifically, when the lexicon or character do not exist in the DTFW-lookup Table, we use zero as the default value, as follows:

$$w_i = \begin{cases} DTFW - Lookup\ Table_{term}, & if\ term \in DTFW - Lookup\ Table \\ 0, & else. \end{cases} \quad (10)$$

Ultimately, for the given $boundary_i$, the lexicon-level feature representation at position i is as follows (whereas $\circ$ denotes concatenation):

$$\theta_i = e_i \circ w_i. \quad (11)$$

Simultaneously, in order to capture features of different dimensions for boundaries and weights, we employ a multi-head self-attention mechanism for feature fusion. The workflow of the above algorithm is illustrated in Figure 5.

In this mechanism, the multi-head self-attention mechanism calculates attention weights and outputs through the query (Q), key (K), and value (V) inputs. Specifically, for the inputs Q, K, and V, the calculation formula for the output vectors is as follows:

$$MultiHead(Q, K, V) = head_1 \circ head_2 \circ \ldots \circ head_h \times W^o, \quad (12)$$

where $W^O \in \mathbb{R}^{hd_v \times d_{model}}$, and $head_i$ is calculated as follows:

$$head_i = Attention\left(QW_i^Q, KW_i^K, VW_i^V\right), \tag{13}$$

$$where\ Attention(Q, K, V) = softmax(\frac{QK^T}{\sqrt{d_k}})V, \tag{14}$$

where $W_i^Q \in \mathbb{R}^{d_{model} \times d_k}$, $W_i^K \in \mathbb{R}^{d_{model} \times d_k}$, $W_i^V \in \mathbb{R}^{d_{model} \times d_v}$ (h represents the number of heads, and d_{model} is the dimension of the input vector. Generally, $d_k = d_v = d_{model}$/h).

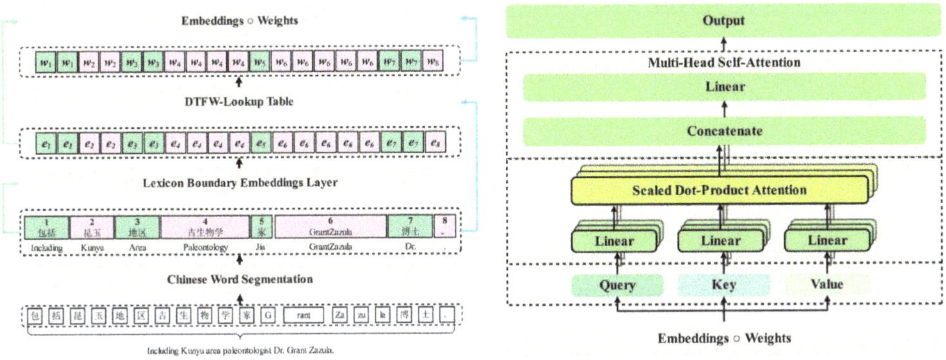

(a) (b)

Figure 5. (**a**) Method for enhancing Chinese lexicon-level features: after undergoing Chinese word segmentation processing, we obtain the lexicon-level representation. We encode and embed its boundaries while simultaneously looking up the DTFW-lookup Table to acquire domain term frequency weights, and concatenate embeddings and weights; (**b**) after concatenating them, we use the multi-head self-attention mechanism for feature fusion, with the final output serving as the Chinese lexicon-level features vector.

3.4. Classifier

To address the challenges posed by long entities and nested entities, we apply a dual pointer fully connected network classifier. The classifier consists of two pointers: one indicating the start position of an entity and the other indicating the end position. In addition, a sigmoid function is applied to map the fully connected network output into the [0, 1] range. Finally, based on a specified *threshold*, the output is discretized into {0, 1} for representation, where "1" indicates that the position is the start position or end position of the entity, and "0" indicates the opposite. The specific classification process is illustrated in Figure 6. Given the embedding vector representation of the input, *context*, is determined as follows:

$$context_i = \chi_i \circ \theta_i. \tag{15}$$

We utilize the *context* vector as the input for the classifier and decode based on predefined *threshold*, as follows:

$$output = classifier(context), \tag{16}$$

$$position_i^s, position_i^e = \begin{cases} 1, & if\ position_i > threshold \\ 0, & else, \end{cases} \tag{17}$$

where $position_i^s$ denotes the start position classifier's output at position i, and $position_i^e$ denotes the end position classifier's output at position i.

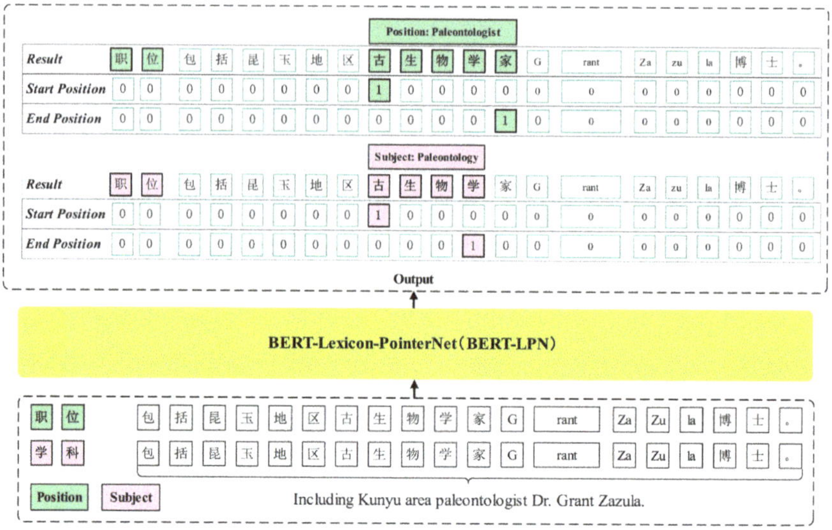

Figure 6. For nested-type entities, our model processes multiple pairs of "<entity_type>,<text>" inputs, and the classifier ultimately outputs the start and end position labels for the entities.

3.5. Loss Function

The loss function is applied to gauge the performance of the model, specifically the disparity between the model's predictions and the actual targets. This discrepancy is typically referred to as the loss value, and the objective of the loss function is to minimize this value. When the model's predictions perfectly align with the targets, the loss value is zero, indicating optimal model performance. We apply the binary cross-entropy loss (BCELoss) function to compute the disparity between the model's outputs and the actual labels.

Additionally, the AdamW [41] optimizer, which is a variant of the Adam [42] optimizer, introduces a weight decay term to control the regularization of model parameters. Weight decay helps prevent overfitting by encouraging the model to use simpler parameter settings, thereby enhancing generalization capabilities. It is important to note that we do not explicitly introduce a regularization term in the loss function. In our loss function, we calculate the difference between the model's outputs and the true labels using binary cross-entropy, and we apply the AdamW optimizer with weight decay to enhance the model's generalization without explicitly introducing a regularization term in the loss function, as follows:

$$\mathcal{L}oss = \frac{1}{2}(\mathcal{L}^s + \mathcal{L}^e), \tag{18}$$

where $\mathcal{L}^s$ denotes the start position classifier's loss and $\mathcal{L}^e$ denotes the end position classifier's loss. A sentence consists of N tokens, $\mathcal{L}$ is calculated as follows:

$$\mathcal{L}(Y,P) = -\frac{1}{N}\sum_{i=1}^{N}[\mathcal{Z}(y_i, p_i) + \mathcal{Z}(1-y_i, 1-p_i)], \tag{19}$$

$$\text{where } \mathcal{Z}(y,p) = y * log(p), \tag{20}$$

where log is the natural logarithm.

4. Experiments

4.1. Preparation

4.1.1. Dataset

We evaluated our proposed method on the following datasets: Resume [24], Youku [43], DuIE [44], and our own PPOST dataset.

The Resume dataset is a Chinese NER dataset focused on resumes. Youku provides an open NER dataset in the entertainment domain. DuIE is a large-scale manually annotated dataset, and all sentences in this dataset are extracted from Baidu Baike and Baidu News Search. The text in this dataset covers various domains found in real-world applications, including news, entertainment, and user-generated content. Due to the large size of the dataset, we randomly sampled a portion of the data, referred to as SubDuIE, for evaluation. Table 2 shows the details of these datasets.

Table 2. Statistics on the four dataset for the experiments.

Dataset	Resume	Youku	SubDuIE	PPOST [1]
Train	3821	8001	19,521	13,124
Eval	463	1000	1950	1396
Test	477	1001	1948	1395
Types	8	9	13	13
Avg [2]	33	18	66	83

[1] The PPOST dataset includes 9% of data generated using the data augmentation method; [2] "Avg" represents the average number of characters in the samples of this dataset.

The PPOST dataset consists of manually labeled entity data in the domain of Chinese paleontology popular science. The label distribution of the PPOST dataset is illustrated in the Figure 7. It can be seen that even after data augmentation processing, the phenomenon of a long-tail distribution still exists in practical production scenarios.

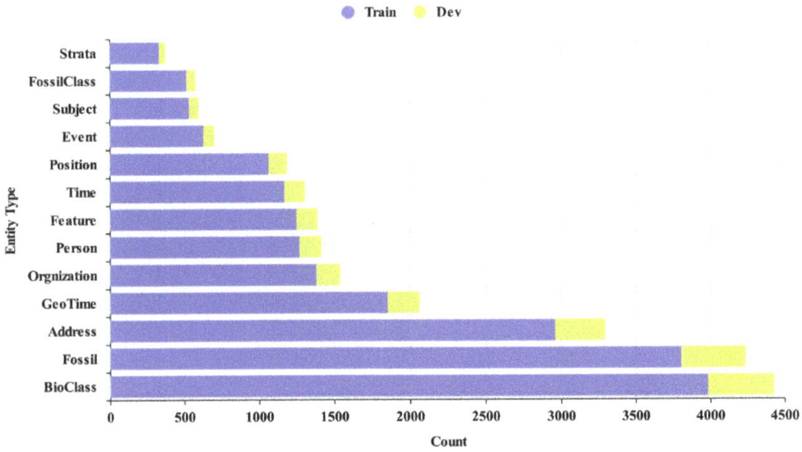

Figure 7. The label distribution of the PPOST dataset.

4.1.2. Evaluation Metrics

To evaluate the performance on Chinese NER, following most of the baselines, we use precision (P), recall (R) and F1-score (F1) as the metrics, they are computed on the entity-level number of true positives (TP), false positives (FP) and false negatives (FN):

$$Precision = \frac{TP}{TP + FP} * 100\%, \tag{21}$$

$$Recall = \frac{TP}{TP + FN} * 100\%, \tag{22}$$

$$F_1 = 2 * \frac{Precision \times Recall}{Precision + Recall} * 100\%, \tag{23}$$

where F_1 is obtained by directly averaging the F1 scores of all types, that is, Macro-F1 [45].

4.1.3. Baseline Methods

Aiming to measure and analyze the proposed method, we compare its performance on the four datasets mentioned above with mainstream models in recent years:

- Methods without lexicon features: BiLSTM-CRF [16], BERT [19], BERT-CRF [23], BERT-BiLSTM-CRF [46], BERT-GlobalPointer [37] and BERT-Pointer [36]. These models are based on deep learning, such as BiLSTM or BERT, using sequence labeling or pointer labeling frameworks. In particular, these models do not introduce Chinese lexicon-level features.
- Methods with lexicon features: lattice LSTM [24] and lexicon-enhanced BERT (LEBERT) [28]. Lattice LSTM can be seen as an extension of the character-based NER model, which adds words as input and additional gates to control information flow; LEBERT directly integrates external lexicon knowledge into the BERT layer through the lexicon adapter layer.

4.1.4. Training

In all experiments, we used the same BERT pre-trained model. Due to differences between datasets, we adjusted batch size and max_seq_len separately for each dataset, and all experiments were conducted on a single NVIDIA (Santa Clara, CA, USA) GeForce RTX 3070Ti GPU. For specific details and settings of other hyperparameters, please refer to Tables A1–A3 in Appendix A.

4.2. *Results and Analysis*

4.2.1. Comparative with Other Models

To validate the effectiveness of the proposed method in the Chinese NER, we applied an early stopping mechanism and calculated the average results over five runs of experiments. Tables 3 and 4 present the experimental results for the models mentioned in Section 4.1.3.

Table 3. Precision, recall, and F1-score statistics on the Resume and Youku datasets.

Model	Resume			Youku		
	P(%)	R(%)	F1(%)	P(%)	R(%)	F1(%)
BiLSTM-CRF	93.70	93.30	93.50 [47]	80.31	79.22	79.76
BERT	94.20	95.80	95.00 [19]	85.06	76.75	80.69
BERT-CRF	-	-	**96.87** [48]	83.00	81.70	82.40 [43]
BERT-BiLSTM-CRF	-	-	95.59 [49]	89.56	80.48	84.78
Lattice LSTM	94.81	94.11	94.46 [24]	84.43	81.28	82.82
LEBERT	-	-	96.08 [28]	91.93	82.10	86.74
BERT-GlobalPointer	96.94	93.71	95.30	92.76	82.02	87.06
BERT-Pointer	97.15	94.27	95.69 [27]	95.06	83.47	**88.89**
BERT-LPN	98.12	95.70	**96.90**	95.74	83.23	**89.05**

Significant bold values are the current SOTA in performance achieved, and the bold in the last row of these tables denotes the new achievements we obtained in our model.

Experimental results on two shorter-text datasets, Resume and Youku, and two longer-text datasets, SubDuIE and PPOST, show that our model improves F1-score by 0.03%, 0.16%, 1.27%, and 2.28%, respectively, compared to previous models. This demonstrates that our feature fusion strategy effectively utilizes Chinese lexicon-level information, enhances Chinese NER performance. However, the advantages of our method are not fully reflected in the Resume and Youku datasets, where the sample texts are shorter, and entities are simpler. Moreover, the DTFW-lookup Table for the original corpus of these datasets was not calculated. The more significant improvement on longer-text datasets may be attributed to the use of the WordPiece algorithm, simplifying the model input and integrating Chinese lexicon-level features, enhancing the model's generalization performance and feature capture capability. Specifically, on the PPOST dataset, compared to sequence labeling models, the performance of pointer labeling models has significantly improved. This experimental result indicates that pointer labeling models can effectively improve the recognition performance of long and nested entities in the domain of Chinese paleontology popular science.

Table 4. Precision, recall, and F1-score statistics on the SubDuIE and PPOST datasets.

Model	SubDuIE			PPOST		
	P(%)	R(%)	F1(%)	P(%)	R(%)	F1(%)
BiLSTM-CRF	72.37	58.71	64.83	61.14	47.37	53.38
BERT	79.48	55.49	65.35	66.79	46.88	55.09
BERT-CRF	78.60	57.66	66.52	68.82	46.40	55.43
BERT-BiLSTM-CRF	76.65	56.71	65.19	67.78	46.22	54.96
Lattice LSTM	79.00	62.68	69.90	71.07	50.38	58.96
LEBERT	83.31	61.58	70.81	79.48	53.13	63.69
BERT-GlobalPointer	83.20	61.84	70.95	80.40	67.98	73.67
BERT-Pointer	83.67	61.81	**71.10**	81.50	71.60	**76.23**
BERT-LPN	84.47	63.30	**72.37**	83.50	74.07	**78.51**

4.2.2. Comparative with Various Entity Types

To comprehensively evaluate our model's recognition performance on various entities in our domain, we collected the performance of our model in identifying different entities on the PPOST dataset, as shown in Figure 8. The experimental results indicate that the recognition accuracy of various entities is mainly positively correlated with the number of labels in the dataset. This may be due to the scarcity of samples for certain types, leading to the model's inability to learn richer features of these type entities. Specifically, since we define the Feature type mostly for the attributes of things, such as the volume of fossils or the appearance of ancient organisms, and with diverse characteristics, the recognition performance for this part of entities is relatively lower. Therefore, it is more appropriate to consider conducting attribute extraction tasks for these entities.

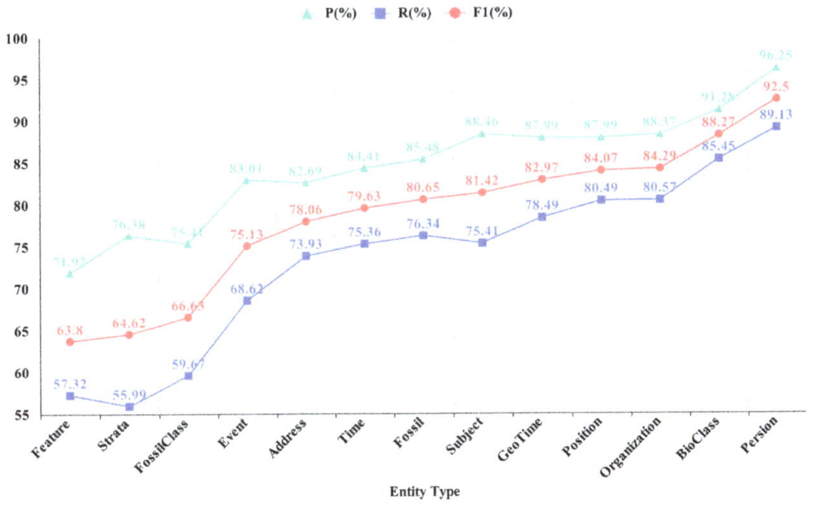

Figure 8. Precision, recall, and F1-score of 13 types statistics on the PPOST.

To evaluate the recognition performance of our model for long entities and nested entities, we sampled 324 instances of long entities with an entity length exceeding 8 characters and an average length of 10 characters and 213 instances of nested entities from the PPOST dataset. The relevant experimental results are shown in Table 5. The results indicate that compared to models based on CRF, models based on pointer labeling can effectively enhance the recognition performance of long and nested entities. Moreover, our BERT-LPN model achieves the best performance among the models mentioned.

Table 5. Precision, recall, and F1-score statistics of long entities and nested entities on the PPOST.

Model	Long Entities			Nested Entities		
	P(%)	R(%)	F1(%)	P(%)	R(%)	F1(%)
BiLSTM-CRF	43.36	39.76	41.48	47.68	40.01	43.50
BERT	46.90	41.47	44.02	45.19	37.31	40.87
BERT-CRF	48.43	41.29	44.58	42.17	35.31	38.44
BERT-BiLSTM-CRF	49.00	40.39	44.28	43.16	37.99	40.41
Lattice LSTM	51.49	45.62	48.38	44.11	41.01	42.50
LEBERT	52.04	45.84	48.74	45.80	42.61	44.15
BERT-GlobalPointer	57.42	53.73	55.51	65.80	56.76	60.95
BERT-Pointer	59.00	54.71	**56.77**	67.67	58.30	**62.64**
BERT-LPN	60.39	55.19	**57.67**	68.32	59.26	**63.47**

4.2.3. Comparative with Complexity of Various Models

Compared to neural networks, CRF models themselves have a relatively high time performance overhead [50], with an overall time complexity of $O(m * n^2)$, where m represents the length of the observed sequence, and n denotes the number of hidden states. To assess the computational requirements of our model, we only recorded the trainable parameters and training speed of the pointer labeling models, as shown in Table 6. The experimental results indicate that the additional time performance overhead of our method is negligible. This is because, through preprocessing, we obtained the DTFW-lookup Table, eliminating the need for repetitive computation of lexicon frequency weight during training, requiring only table lookup operations. Additionally, compared to the BERT-GlobalPointer model, we used only two fully connected layers for positional marking, reducing the parameters of fully connected layers and improving the training speed by 15%. In contrast to the BERT-Pointer model, we need to calculate the boundaries of lexicon and look up the DTFW-Lookup Table to obtain lexicon frequency weights. Moreover, the boundary embedding layer is trainable, resulting in an additional 1.7% performance overhead.

Table 6. Parameters and training speed statistics of the GlobalPointer, LPN, and Pointer models based on BERT on the PPOST dataset.

Model	Parameters	Speed
BERT-GlobalPointer	29.00 M	1× *
BERT-LPN	28.97 M	1.15× *
BERT-Pointer	28.94 M	1.17× *

* 1.15× and 1.17× represent 15% and 17% faster than 1×, respectively.

4.2.4. Comparative with Various Strategies

In order to investigate the contribution of each strategy to our approach, we conducted a series of ablation experiments on the SubDuIE and PPOST datasets, and the results are shown in Figure 9.

The ablation studies were designed as follows:

- Strategy 1: Strategy of adding only domain term frequency weight features without additional methods for feature fusion;

- Strategy 2: Strategy of adding only Chinese lexicon boundary features without additional methods for feature fusion;
- Strategy 3: Strategy of simultaneously adding Chinese lexicon boundary and domain term frequency weight features without additional methods for feature fusion;
- Strategy 4: Strategy of simultaneously adding Chinese lexicon boundary and domain term frequency weight features, using the BiLSTM for feature fusion.

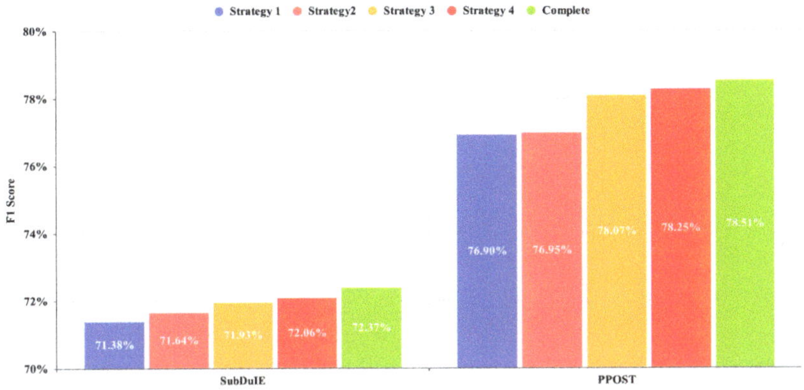

Figure 9. Performance comparison of different strategies on the SubDuIE and PPOST datasets in terms of F1-score. The Complete represents our proposed BERT-LPN model, simultaneously adding Chinese lexicon boundary and domain term frequency weight features and using the multi-head self-attention for feature fusion.

We remove any strategy that will lead to a decrease in entity recognition accuracy. Specifically, compared to the original pointer network, Strategy 1, Strategy 2, and Strategy 3 have all shown improvements on our experimental SubDuIE and PPOST datasets. Their F1-scores increased by 0.28%, 0.54%, 0.67% and 0.67%, 0.72%, 1.84%, respectively. This indicates that lexicon boundary features and domain term frequency weight features can effectively integrate Chinese lexicon-level features, improving the performance of the NER model. Compared to Strategy 3, Strategy 4 and the Complete model showed F1-score improvements of 0.13% and 0.44% and 0.18% and 0.44%, respectively, on the above two datasets. This suggests that the feature fusion of Chinese lexicon-level feature vectors can more effectively integrate contextual information features, enhancing the performance of the NER model. Compared to Strategy 4, the Complete model showed F1-score improvements of 0.31% and 0.26% on the above two datasets, respectively, indicating that the multi-head self-attention mechanism better handles dependencies between information.

4.2.5. Comparative with Various Data Augmentation Ratios

To verify the effectiveness of the data augmentation method, we conducted an additional set of experiments on the BERT-LPN model. The experiments were performed on the PPOST dataset, and the results are shown in Figure 10. The experimental results indicate that, on the PPOST dataset, our proposed data augmentation method has a certain performance improvement within the enhancement ratio range of (0, 0.1], where the ratio value is the ratio of the generated data to the original total data. This improvement may be attributed to the model further learning the ability to identify entities that do not exist in the input within this enhancement ratio range. However, when the ratio is more than 0.1, the model's performance sharply declines, and the performance is even worse than that of the original dataset. This may be due to an excess of augmented data in the training set, causing the model to preferentially predict the absence of entities in the input.

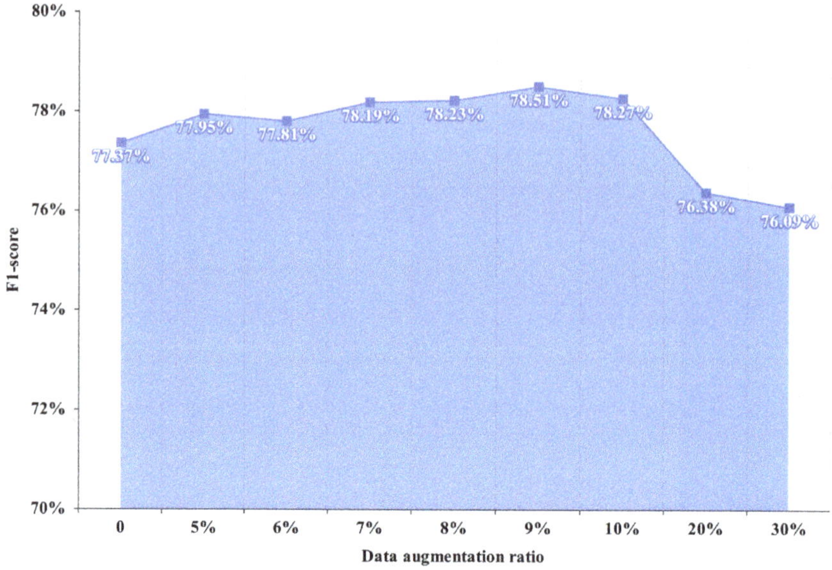

Figure 10. F1-score of the BERT-LPN at different data augmentation ratios on the PPOST dataset.

4.3. Extraction Results and Error Analysis

In this study, a set of experiments were conducted to validate the extracted NER results and analyze errors. Some examples of experimental results are shown in Table 7. From the experimental data with IDs 1 and 2, it can be observed that our model can effectively recognize nested entities, such as "寒武纪地层" (Cambrian strata), and long entities, like "古脊椎动物与古人类研究所" (Institute of Vertebrate Paleontology and Paleoanthropology). From the experimental data with IDs 3 and 4, it is evident that when the input data contains multiple entities of the same type (Address: "湖南" (Hunan), "江西" (Jiangxi), "浙江" (Zhejiang), "安徽" (Anhui)), our model, with the introduction of relative positional information, can effectively distinguish the boundaries of these four entities, reducing the likelihood of erroneously combining arbitrary start and end positions to form incorrect entities.

Additionally, we summarized and analyzed the model's recognition errors, identifying the following main issues:

1. We cannot accurately identify some continuous entity characters separated by the symbol "—". For example, experimental data with ID 5 indicates that in the entity "中—上奥陶统" (Middle—Upper Ordovician), "中—" (Middle—) is not considered part of the entity. This may be due to a limited number of samples for such entities or the negative impact of the "—" symbol on the recognition of contextual features;
2. We cannot accurately identify entities with uncommon terms as boundaries. For instance, experimental data with ID 6 shows that in the entity "獬豸盘角鹿" (Discokeryx xiezhi), the model fails to recognize "獬豸" (xiezhi) as part of the entity. This could be because "獬豸" (xiezhi) is an uncommon term, and its domain term frequency weight is relatively low, preventing the model from fully learning the features of uncommon lexicons.

Table 7. Illustrative examples of recognition results.

ID	Sentence	Ground Truth	Prediction
1	"寒武纪地层记录" (Record of Cambrian strata)	寒武纪 (Cambrian), 寒武纪地层 (Strata).	寒武纪 (Cambrian) √, 寒武纪地层 (Strata) √.
2	"中国科学院古脊椎动物与古人类研究所" (Institute of Vertebrate Paleontology and Paleoanthropology, Chinese Academy of Sciences)	中国科学院 (Chinese Academy of Sciences), 古脊椎动物与古人类研究所 (Institute of Vertebrate Paleontology and Paleoanthropology).	中国科学院 (Chinese Academy of Sciences) √, 古脊椎动物与古人类研究所 (Institute of Vertebrate Paleontology and Paleoanthropology)√.
3	"主要分布于我国湖南, 江西、浙江、安徽等地" (Mainly distributed in Hunan, Jiangxi, Zhejiang, Anhui and other places in China)	湖南 (Hunan), 江西 (Jiangxi), 浙江 (Zhejiang), 安徽 (Anhui).	湖南 (Hunan) √, 江西 (Jiangxi) √, 浙江 (Zhejiang) √, 安徽 (Anhui) √.
4	"恐龙位于爬行动物和鸟类之间的十字路口" (Dinosaurs were at the crossroads between reptiles and birds)	恐龙 (Dinosaurs), 爬行动物 (Reptiles), 鸟类 (Birds).	恐龙 (Dinosaurs) √, 爬行动物 (Reptiles) √, 鸟类 (Birds) √.
5	"我国华南地区中—上奥陶统黑色页岩" (Middle-Upper Ordovician black shale in South China)	华南地区 (South China), 中—上奥陶统 (Middle-Upper Ordovician).	华南地区 (South China) √, 上奥陶统 (Upper Ordovician)×.
6	"獬豸盘角鹿属于长颈鹿科" (Discokeryx xiezhi belongs to the family Giraffeidae)	獬豸盘角鹿 (Discokeryx xiezhi), 长颈鹿科 (Giraffeidae).	盘角鹿 (Discokeryx) ×, 长颈鹿科 (Giraffeidae) √.

"×" represents entity type or boundaries recognition errors. "√" represents complete and accurate recognition of entity type and boundaries.

5. Conclusions and Future Work

In the domain of Chinese paleontology popular science, NER is a fundamental step for extracting information and knowledge from a massive amount of popular science articles or books. In this study, we researched and compared mainstream Chinese deep learning-based NER methods to address the challenges in our domain. We constructed a corpus for the Chinese paleontology popular science domain and its annotated NER dataset, PPOST. We proposed a domain NER model, BERT-LPN, based on the BERT pre-trained language model. The model was applied to four different datasets, and its performance was evaluated in terms of precision, recall, and F1-score. The experimental results show that the proposed BERT-LPN method significantly outperforms baseline models and other deep learning models in both general domain and Chinese paleontology popular science texts.

The contributions of this research can be viewed from two perspectives. Methodologically, this study introduces a deep learning model named BERT-LPN. Experimental results demonstrate that, compared to other deep learning models, BERT-LPN performs better in extracting entities, providing multiple advantages for Chinese NER, including the following:

1. Domain-specific information: In the task of Chinese NER within a specific domain, lexicon-level features can encompass domain-specific lexicons, thereby enhancing the model's performance within that domain;
2. Improved semantic understanding: Lexicon-level features can assist the model in better comprehending the semantic information within the text, capturing the meanings of lexicons and their contextual relationships. This helps the model in more accurately distinguishing between different entity types in longer texts.

From an application perspective, this study developed a Chinese paleontology popular science NER dataset, promoting the development of NER, popular science knowledge graphs, and popular science education in the Chinese paleontology domain. Moreover, in other NER tasks in different domains, researchers typically have easy access to original corpora, and our method relies only on the target domain's corpus, greatly increasing the possibility of generalizing our model to other domains. Future research will focus on the following three aspects:

- Corpus construction is an ongoing process, and we will further optimize the Chinese paleontology popular science corpus and PPOST dataset. Specifically, for the Feature entity type, it will be split into more specific attributes to further improve NER performance;
- Studying the performance of our model in other Chinese domains to promote the development of Chinese NER;
- Studying how to handle ambiguous entity boundaries and uncommon terms situations based on our approach.

Author Contributions: Conceptualization, Y.G. and F.L.; methodology, S.F. and H.L.; software, Y.G. and S.F.; validation, X.W., J.S. and Q.G.; formal analysis, Y.G. and S.F.; investigation, S.F. and H.L.; resources, W.L.; data curation, S.F. and H.L.; writing—original draft preparation, Y.G. and S.F.; writing—review and editing, Y.G. and F.L.; visualization, S.F.; supervision, W.L. and X.W.; project administration, W.L. and X.W.; funding acquisition, W.L. and X.W. All authors have read and agreed to the published version of the manuscript.

Funding: This research was funded by International Research Center of Big Data for Sustainable Development Goals, grant number CBAS2022GSP05, State Key Laboratory of Remote Sensing Science, grant number 6142A01210404, Hubei Key Laboratory of Intelligent Geo-Information Processing, grant number KLIGIP-2022-B03, and Metallogenic patterns and mineralization predictions for the Daping gold deposit in Yuanyang County, Yunnan Province, grant number 2022026821.

Institutional Review Board Statement: Not applicable.

Informed Consent Statement: Not applicable.

Data Availability Statement: This manuscript utilizes publicly available datasets, which can be accessed at the following links: Resume dataset (https://tianchi.aliyun.com/dataset/144345 (accessed on 17 July 2023)), Youku dataset (https://github.com/allanj/ner_incomplete_annotation (accessed on 10 August 2023)), and DuIE dataset (https://www.luge.ai/#/luge/dataDetail?id=5 (accessed on 15 September 2023)). Restrictions apply to the availability of the PPOST data, and the data are available from the authors.

Acknowledgments: We gratefully acknowledge the support of the School of Computer Science, China University of Geosciences, Wuhan, and the School of Geography and Information Engineering, China University of Geosciences, Wuhan. We also thank the Piesat Information Technology Co., Ltd., and the China Geological Survey.

Conflicts of Interest: Author Xianbin Wang was employed by the company Piesat Information Technology Co., Ltd. The remaining authors declare that the research was conducted in the absence of any commercial or financial relationships that could be construed as a potential conflict of interest.

Appendix A

There are some details about the experimental process using systems, such as BERT, embedding layer, and multi-head self-attention, that require fine-tuning of the hyperparameters, and some details about this process are given in the following Tables A1 and A2. Some hardware and software environments for the experiments are given in the following Table A3.

Table A1. Some common hyperparameter settings about the experiments.

Hyperparameter	Value
Data Augmentation Ratio	0.09
Epoch	20
BERT Model	bert-base-chinese
BERT LR	1×10^{-5}
BERT Dropout Rate	0.35
Linear LR	3×10^{-4}
Optimizer	AdamW
Boundary Embedding Dim	16
Multi-head self-attention head	8
Threshold	0.6

Table A2. Some hyperparameter settings about the experiments on the four datasets.

Dataset	max_seq_len	Batch Size
Resume	64	64
Youku	64	64
SubDuIE	256	32
PPOST	256	32

Table A3. Some hardware and software environments about the experiments.

Environment	Value
Processor	12th Gen Intel(R) Core(TM) i5-12600KF
RAM	32.0 GB
GPU	NVIDIA GeForce RTX 3070Ti GPU 8GB
Python Version	3.8.16
PyTorch Version	2.0.0

References

1. Tao, D.; Yang, P.; Feng, H. Utilization of text mining as a big data analysis tool for food science and nutrition. *Compr. Rev. Food Sci. Food Saf.* **2020**, *19*, 875–894. [CrossRef] [PubMed]
2. Singh, S. Natural language processing for information extraction. *arXiv* **2018**, arXiv:1807.02383.
3. Contributors, W. Popular Science—Wikipedia, the Free Encyclopedia. 2023. Available online: https://en.wikipedia.org/wiki/Popular_science (accessed on 1 July 2023).
4. Zhai, X. Research on Tourism Promotion of Shandong Zhucheng Dinosaur National Paleontologic Geopark. In Proceedings of the 2015 International Conference on Education, Management and Computing Technology, Tianjin, China, 13–14 June 2015; Atlantis Press: Amsterdam, The Netherlands, 2015.
5. Mansouri, A.; Affendey, L.S.; Mamat, A. Named entity recognition approaches. *Int. J. Comput. Sci. Netw. Secur.* **2008**, *8*, 339–344.
6. Ye, J.; Zhou, J.; Tian, J.; Wang, R.; Zhou, J.; Gui, T.; Zhang, Q.; Huang, X. Sentiment-aware multimodal pre-training for multimodal sentiment analysis. *Knowl. Based Syst.* **2022**, *258*, 110021. [CrossRef]
7. Chennafi, M.E.; Bedlaoui, H.; Dahou, A.; Al-Qaness, M.A.A. Arabic Aspect-Based Sentiment Classification Using Seq2Seq Dialect Normalization and Transformers. *Knowledge* **2022**, *2*, 388–401. [CrossRef]
8. Saha, S.K.; Chatterji, S.; Dandapat, S.; Sarkar, S.; Mitra, P. A hybrid approach for named entity recognition in indian languages. In Proceedings of the IJCNLP-08 Workshop on NER for South and South East Asian languages, Hyderabad, India, 12 January 2008.
9. Tanabe, L.; Xie, N.; Thom, L.H.; Matten, W.; Wilbur, W.J. GENETAG: A tagged corpus for gene/protein named entity recognition. *BMC Bioinform.* **2005**, *6*, S3. [CrossRef] [PubMed]
10. Ju, Z.; Wang, J.; Zhu, F. Named entity recognition from biomedical text using SVM. In Proceedings of the 2011 5th International Conference on Bioinformatics and Biomedical Engineering, Wuhan, China, 10–12 May 2011.
11. Morwal, S.; Jahan, N.; Chopra, D. Named entity recognition using hidden Markov model (HMM). *Int. J. Nat. Lang. Comput. (IJNLC)* **2012**, *1*. [CrossRef]
12. Zhou, G.; Su, J. Named entity recognition using an HMM-based chunk tagger. In Proceedings of the 40th Annual Meeting of the Association for Computational Linguistics, Philadelphia, PA, USA, 6–12 July 2002.
13. Konkol, M.; Konopík, M. CRF-Based Czech Named Entity Recognizer and Consolidation of Czech NER Research. In *International Conference on Text, Speech and Dialogue*; Springer: Berlin/Heidelberg, Germany, 2013; pp. 153–160.

14. Khan, A.; Sohail, A.; Zahoora, U.; Qureshi, A.S. A survey of the recent architectures of deep convolutional neural networks. *Artif. Intell. Rev.* **2020**, *53*, 5455–5516. [CrossRef]
15. Staudemeyer, R.C.; Morris, E.R. Understanding LSTM--a tutorial into long short-term memory recurrent neural networks. *arXiv* **2019**, arXiv:1909.09586.
16. Huang, Z.; Xu, W.; Yu, K. Bidirectional LSTM-CRF models for sequence tagging. *arXiv* **2015**, arXiv:1508.01991.
17. Wu, Z.; Pan, S.; Chen, F.; Long, G.; Zhang, C.; Philip, S.Y. A comprehensive survey on graph neural networks. *IEEE Trans-Actions Neural Netw. Learn. Syst.* **2020**, *32*, 4–24. [CrossRef] [PubMed]
18. Qiu, X.; Sun, T.; Xu, Y.; Shao, Y.; Dai, N.; Huang, X. Pre-trained models for natural language processing: A survey. *Sci. China Technol. Sci.* **2020**, *63*, 1872–1897. [CrossRef]
19. Devlin, J.; Chang, M.-W.; Lee, K.; Toutanova, K. Bert: Pre-training of deep bidirectional transformers for language understanding. *arXiv* **2018**, arXiv:1810.04805.
20. Darji, H.; Mitrović, J.; Granitzer, M. German BERT Model for Legal Named Entity Recognition. *arXiv* **2023**, arXiv:2303.05388.
21. Souza, F.; Nogueira, R.; Lotufo, R. Portuguese named entity recognition using BERT-CRF. *arXiv* **2019**, arXiv:1909.10649.
22. Song, Z.; Xu, W.; Liu, Z.; Chen, L.; Su, H. A BERT-Based Named Entity Recognition Method of Warm Disease in Traditional Chinese Medicine. In Proceedings of the 2023 IEEE 18th Conference on Industrial Electronics and Applications (ICIEA), Ningbo, China, 18–22 August 2023; pp. 1226–1231.
23. Dai, Z.; Wang, X.; Ni, P.; Li, Y.; Li, G.; Bai, X. Named Entity Recognition Using BERT BiLSTM CRF for Chinese Electronic Health Records. In Proceedings of the 2019 12th International Congress on Image and Signal Processing, BioMedical Engineering and Informatics (CISP-BMEI), Suzhou, China, 19–21 October 2019; pp. 1–5.
24. Zhang, Y.; Yang, J. Chinese NER Using Lattice LSTM. *arXiv* **2018**, arXiv:1805.02023.
25. Li, X.; Yan, H.; Qiu, X.; Huang, X.-J. FLAT: Chinese NER Using Flat-Lattice Transformer. *arXiv* **2020**, arXiv:2004.11795.
26. Vaswani, A.; Shazeer, N.; Parmar, N.; Uszkoreit, J.; Jones, L.; Gomez, A.N.; Kaiser, Ł.; Polosukhin, I. Attention is all you need. In *Advances in Neural Information Processing Systems 30 (NIPS 2017)*; Curran Associates Inc.: Long Beach, CA, USA, 2017.
27. Guo, Q.; Guo, Y. Lexicon enhanced Chinese named entity recognition with pointer network. *Neural Comput. Appl.* **2022**, *34*, 14535–14555. [CrossRef]
28. Liu, W.; Fu, X.; Zhang, Y.; Xiao, W. Lexicon Enhanced Chinese Sequence Labeling Using BERT Adapter. *arXiv* **2021**, arXiv:2105.07148.
29. Sun, Y.; Zheng, Y.; Hao, C.; Qiu, H. NSP-BERT: A Prompt-based Few-Shot Learner Through an Original Pre-training Task--Next Sentence Prediction. *arXiv* **2021**, arXiv:2109.03564.
30. Liu, P.; Yuan, W.; Fu, J.; Jiang, Z.; Hayashi, H.; Neubig, G. Pre-train, Prompt, and Predict: A Systematic Survey of Prompting Methods in Natural Language Processing. *ACM Comput. Surv.* **2023**, *55*, 1–35. [CrossRef]
31. Cui, L.; Wu, Y.; Liu, J.; Yang, S.; Zhang, Y. Template-Based Named Entity Recognition Using BART. *arXiv* **2021**, arXiv:2106.01760.
32. Huang, Y.; He, K.; Wang, Y.; Zhang, X.; Gong, T.; Mao, R.; Li, C. Copner: Contrastive learning with prompt guiding for few-shot named entity recognition. In Proceedings of the 29th International Conference on Computational Linguistics, Gyeongju, Republic of Korea, 12–17 October 2022.
33. Ding, N.; Chen, Y.; Han, X.; Xu, G.; Wang, X.; Xie, P.; Zheng, H.; Liu, Z.; Li, J.; Kim, H.-G. Prompt-learning for Fine-grained Entity Typing. *arXiv* **2021**, arXiv:2108.10604.
34. Li, X.; Feng, J.; Meng, Y.; Han, Q.; Wu, F.; Li, J. A Unified MRC Framework for Named Entity Recognition. *arXiv* **2019**, arXiv:1910.11476.
35. Lu, Y.; Liu, Q.; Dai, D.; Xiao, X.; Lin, H.; Han, X.; Sun, L.; Wu, H. Unified Structure Generation for Universal Information Extraction. *arXiv* **2022**, arXiv:2203.12277.
36. Gong, O. Chinese Information Extraction Using Pointer Network, in GitHub Repository. 2022. Available online: https://github.com/taishan1994/PointerNet_Chinese_Information_Extraction (accessed on 10 December 2023).
37. Su, J.; Murtadha, A.; Pan, S.; Hou, J.; Sun, J.; Huang, W.; Wen, B.; Liu, Y. Global Pointer: Novel Efficient Span-based Approach for Named Entity Recognition. *arXiv* **2022**, arXiv:2208.03054.
38. Song, X.; Salcianu, A.; Song, Y.; Dopson, D.; Zhou, D. Fast WordPiece Tokenization. *arXiv* **2020**, arXiv:2012.15524.
39. Rajaraman, A.; Ullman, J.D. *Mining of Massive Datasets*; Cambridge University Press: Cambridge, UK, 2011.
40. Luo, R.; Xu, J.; Zhang, Y.; Zhang, Z.; Ren, X.; Sun, X. Pkuseg: A toolkit for multi-domain chinese word segmentation. *arXiv* **2019**, arXiv:1906.11455.
41. Loshchilov, I.; Hutter, F. Decoupled weight decay regularization. *arXiv* **2017**, arXiv:1711.05101.
42. Kingma, D.P.; Ba, J. Adam: A method for stochastic optimization. *arXiv* **2014**, arXiv:1412.6980.
43. Jie, Z.; Xie, P.; Lu, W.; Ding, R.; Li, L. Better modeling of incomplete annotations for named entity recognition. In Proceedings of the 2019 Conference of the North American Chapter of the Association for Computational Linguistics: Human Language Technologies, Minneapolis, MN, USA, 2–7 June 2019; Long and Short Papers. Volume 1.
44. Li, S.; He, W.; Shi, Y.; Jiang, W.; Liang, H.; Jiang, Y.; Zhang, Y.; Lyu, Y.; Zhu, Y. Duie: A large-scale chinese dataset for in-formation extraction. In Proceedings of the Natural Language Processing and Chinese Computing: 8th CCF International Conference, NLPCC 2019, Dunhuang, China, 9–14 October 2019. Proceedings, Part II 8.

45. Lewis, D.D.; Schapire, R.E.; Callan, J.P.; Papka, R. Training algorithms for linear text classifiers. In Proceedings of the 19th Annual International ACM SIGIR Conference on Research and Development in Information Retrieval, Zurich, Switzerland, 18–22 August 1996.
46. Tang, X.; Huang, Y.; Xia, M.; Long, C. A Multi-Task BERT-BiLSTM-AM-CRF Strategy for Chinese Named Entity Recognition. *Neural Process. Lett.* **2023**, *55*, 1209–1229. [CrossRef]
47. Lample, G.; Ballesteros, M.; Subramanian, S.; Kawakami, K.; Dyer, C. Neural Architectures for Named Entity Recognition. *arXiv* **2016**, arXiv:1603.01360.
48. Wang, X.; Jiang, Y.; Bach, N.; Wang, T.; Huang, Z.; Huang, F.; Tu, K. Improving Named Entity Recognition by External Context Retrieving and Cooperative Learning. *arXiv* **2021**, arXiv:2105.03654.
49. Zhao, J.; Cui, M.; Gao, X.; Yan, S.; Ni, Q. Chinese Named Entity Recognition Based on BERT and Lexicon Enhancement. In Proceedings of the RICAI 2022: 2022 4th International Conference on Robotics, Intelligent Control and Artificial Intelligence, Dongguan China, 16–18 December 2022.
50. Strubell, E.; Verga, P.; Belanger, D.; McCallum, A. Fast and Accurate Entity Recognition with Iterated Dilated Convolutions. *arXiv* **2017**, arXiv:1702.02098.

applied sciences

MDPI

Article

Using LSTM to Identify Help Needs in Primary School Scratch Students

Luis Eduardo Imbernón Cuadrado *, Ángeles Manjarrés Riesco and Félix de la Paz López

Department of Artificial Intelligence, Universidad Nacional de Educación a Distancia, 28040 Madrid, Spain; amanja@dia.uned.es (Á.M.R.); delapaz@dia.uned.es (F.d.l.P.L.)
* Correspondence: limbernon3@alumno.uned.es

Featured Application: This work can be used to build a model capable of identifying when a student needs help while performing block-based programming exercises, and providing personalized pedagogical help, or an automatic system that can guide the student when he/she is stuck.

Abstract: In the last few years, there has been increasing interest in the use of block-based programming languages as well as in the ethical aspects of Artificial Intelligence (AI) in primary school education. In this article, we present our research on the automatic identification of the need for assistance among primary school children performing Scratch exercises. For data collection, user experiences have been designed to take into account ethical aspects, including gender bias. Finally, a first-in-class distance calculation method for block-based programming languages has been used in a Long Short-Term Memory (LSTM) model, with the aim of identifying when a primary school student needs help while he/she carries out Scratch exercises. This model has been trained twice: the first time taking into account the gender of the students, and the second time excluding it. The accuracy of the model that includes gender is 99.2%, while that of the model that excludes gender is 91.1%. We conclude that taking into account gender in training this model can lead to overfitting, due to the under-representation of girls among the students participating in the experiences, making the model less able to identify when a student needs help. We also conclude that avoiding gender bias is a major challenge in research on educational systems for learning computational thinking skills, and that it necessarily involves effective and motivating gender-sensitive instructional design.

Keywords: distance calculation; block-based programming language; LSTM model; teaching with Scratch; ethics in AI

1. Introduction

In the last decade, online learning environments have gained attention, even more so since the start of the lockdowns in 2020 caused by the coronavirus (SARS-CoV2 or COVID-19) pandemic [1]. Additionally, Computer Science (CS) is experiencing exponential growth, and Scratch, a graphical block-based programming language, is a widely-used approach to improving students' Computational Thinking (CT) skills in primary school [2,3].

In an educational environment, and even more so in primary education, feedback is an important factor, as it allows students to evaluate their progress in learning. Identifying student's mental blocks and providing personalized feedback can be difficult when teachers and learners are separated by space and time, such as in online learning settings [4]. As concluded in [5], automatic feedback provided by the software can improve students' performance in programming activities.

How far a student is from finding the solution to an exercise is a key factor in knowing if he/she needs help. For this purpose, there are methods that calculate the distance between the student's workspace and the given solution of the exercise at hand. However, though different existing distance calculation approaches take into account the name and the position of the program blocks, they ignore other attributes like the block family, or the value of the block inputs. For this reason, in [6], we proposed a novel method for calculating distances in block-based programming languages.

One of the most important elements of the learning process is knowing when to ask for help. When a student has exhausted his/her capacity for understanding and still cannot find a solution for the problem he/she is facing, he/she has to ask for help from a more competent person. However, some students avoid asking for help because they think they are incompetent and not smart enough [7]. This can prevent the student from progressing, causing him/her to lose motivation and interest in the subject. Therefore, identifying the moment when a student needs help to complete the task he/she is carrying out is a crucial aspect of effective learning.

In Ref. [6], the defined distance between the student's workspace and the solution varies depending on the difficulty of the exercise and the student's knowledge of Scratch. So each student will need help at different points in time. In Ref. [8] it is concluded that investigating the time and circumstances in which a learner requests help in an intelligent tutor system allows us to understand the type of help they need, and therefore provide high-quality hints. However, we did not find in the literature any model for the identification of help needs in block-based programming languages. This leads us to propose a help identification mechanism based on a time-series model and on user interactions in block-based programming languages. In the last few years, time-series models have seen increasing use in forecasting [9] and are widely implemented in education for identifying key elements at a certain moment, for example to analyse the concentration of students during English courses [10], or for predicting school absenteeism of autistic students [11].

Until recently it was not common to integrate ethical considerations into the development processes of artificial intelligence applications. However, nowadays experts agree on the importance of embedding ethical requirements into the design of artificial intelligence systems [12]. One of the key ethical concerns in education is discrimination based on gender and towards minority groups. An educational system that has a bias can cause student frustration and lead to loss of interest in the subject, which can in turn lead to academic failure. In recent years, researchers have focused their efforts on improving the ethical frameworks of artificial intelligence, and implementing them in different areas, including in education.

The work presented here was carried out in the context of the project Affective Robot Tutor Integrated Environment (ARTIE), in which we are developing an environment for configuring an affective robot tutor, to be used to teach block-based programming in primary school education. This environment must be able to identify when a student needs help, what their emotional state is, and how the robot should act according to each situation and emotional state [13,14]. Therefore, the robot is "affective" because it deals with or arouses emotions. The objective of this work is to create an artificial intelligence model capable of identifying the need for help of a primary school student, while performing Scratch exercises. This objective involves contributions in different research areas, in particular in the field of intelligent educational systems for teaching block-based programming and, in general, in the field of intelligent educational systems that provide automatic feedback, since the results of this work can be generalised to apply to other disciplines. The application of LSTM architectures in this domain, as well as the function proposed in [6] to evaluate the distance between the student's workspace and a given block-based programming exercise, are innovative.

The rest of the paper is structured as follows. In Section 2, we present the research context and the technological choices, we explain the user experience carried out to evaluate our proposal, we detail the distance calculation for block-based programming languages that we have used, and we explain the construction of the help identification model. In Section 3 we report on the results of our research. In Section 4 we include a discussion of our work and outline some future research directions suggested by our experience. Finally, in Section 5 we draw conclusions from the results and discussion, and propose future work to expand on aspects of this research.

2. Materials and Methods

All the code developed for the different services to carry out the user experiences and for the implementation of the model presented in this work is available in open source at the following GitHub link: https://github.com/orgs/ARTIEROCKS/repositories (accessed on 27 November 2023).

2.1. Research Context and Technological Choices

In this section, we present the current state of the research areas of interest for this paper. Such areas include block-based programming interfaces in primary school education, distance calculation methods, time-series based models, and ethics in education.

2.1.1. Block-Based Programming Interfaces in Primary School

Learning programming is beneficial for primary school children, providing them with tools to improve problem-solving, as observed in [15] where students solve arithmetic and geometric sequences and series using block-based programming codes instead of using algebra. It also has other benefits such as improvement in efficiency and learning, and an increase of enthusiasm, motivation, a sense of fun and CT [16,17].

It has been observed in previous studies that, in order to increase student's achievements and abilities, programming should start at earlier stages than high school [18]. There is not much research involving primary school learners, which has been considered a critical stage to develop an interest in this subject. Block-based programming interfaces are useful in reducing the complexity of programming, with respect to text-based programming, and they are also effective at creating interest in programming at primary school [19].

Our experiment is based on Scratch, which is a free web-based programming tool focused on young people that allows for the creation of games, interactive stories, and animations [3]. This tool has been used in many CT research projects in which primary school children were involved, such as [2,3,20].

2.1.2. Distance Calculation Methods in Block-Based Programming Languages

In order to identify how "far" or "close" a block-based program is from the solution to a programming problem, we have to measure in some way the distance between the two. Block-based programming programs can be transformed into Abstract Syntax Trees (ASTs).

There are several methods that can be used to measure the "closeness" between two different ASTs, such as pg-gram implemented in [21], Tree Edit Distance (TED) applied in [22], or Tree Inherited Distance (TID) as defined in [23].

Though these methods measure the distance between two ASTs, they only consider two dimensions: the position and the label of the node. In block-based programming, as we will further explain, there are other dimensions that should be part of the distance calculation.

2.1.3. Time-Series Based Models

Time series data are samples taken over time from a real process. Many time series are temporal dependent and non-stationary [24], and constitute a discipline of data modeling where past values of a given variable are analyzed to predict the future values of this variable [9].

One of the most popular methods for processing time series is Recurrent Neural Networks (RNNs) [25]. This method is widely used for Time Series Classification (TSC), for example in forecasting [26,27], as RNNs are able to preserve information between neurons over the time line [28].

There are different types of architecture in RNNs, but the most common and currently used are LSTM and Gated Recurrent Unit (GRU) [25].

LSTM is a solution of the problem of vanishing gradient and gradient explosion in the training of RNNs. This architecture has two gates: an "input gate" and a "forgetting gate". The "input gate" is used to add information to the neurons, and the "forgetting gate" decides which information is important and should be stored and which information to forget.

GRU is more general than LSTM and simplifies both the mathematical model and the parameter mechanism [25].

When there is a limited amount of historical sequence data, problems such as gradient explosion or gradient disappearance may occur when processing s long-term sequence of samples. This fact causes instability in the model, making it unable to converge to the optimal result. LSTM, with a strong feedback mechanism solves the problem of the long-term dependency in the Recurrent Neural Network (RNN) [10].

In education, LSTM has been implemented in different domains. For example, in [10,11], LSTM has been implemented together with a Multilayer Perceptron (MLP) and a Convolutional Neural Network (CNN) to increase their accuracy. In [28] a new framework based on the LSTM algorithm was proposed for performing personalized and adaptive learning, giving better results than the Deep Knowledge Tracing (DKT) method. And in [29], a method based on LSTM helps programming students by providing the next word following an incomplete program.

When a student is carrying out an exercise, it is crucial to identify the moment when the student will need help, therefore time is a dimension to take into account. Because we have a limited dataset, that may cause the gradient vanish or gradient explosion problem, we will implement a LSTM architecture.

RNNs have different configurations, as is described in [30]:

- One-to-one (Figure 1): One input and one output are suitable for classification tasks, but not for time series models;
- One-to-many (Figure 2): This model converts an input into a sequence;
- Many-to-many (Figure 3): This model is a sequence to sequence generator;
- Many-to-one (Figure 4): This model is suitable for a prediction or classification from a sequence.

In this study, the input is a sequence of data and the output is a boolean that indicates if a student needs help or not, therefore we will implement a many-to-one configuration.

Figure 1. LSTM one-to-one.

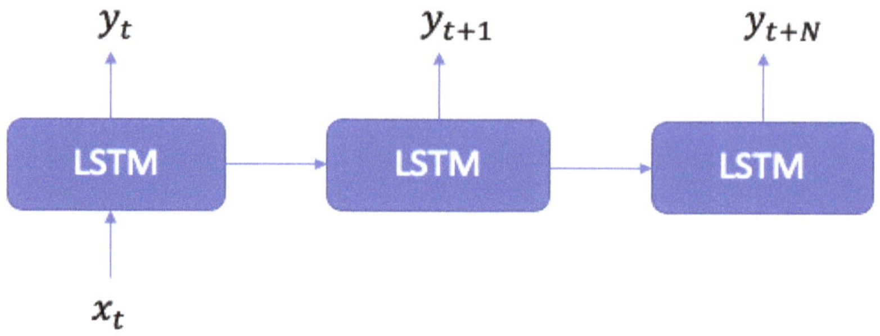

Figure 2. LSTM one-to-many.

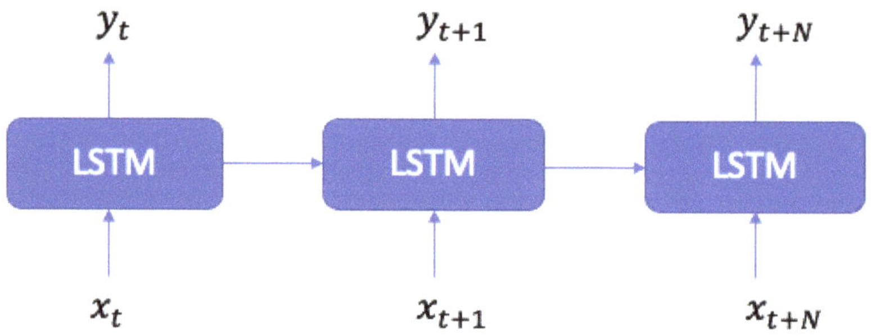

Figure 3. LSTM many-to-many.

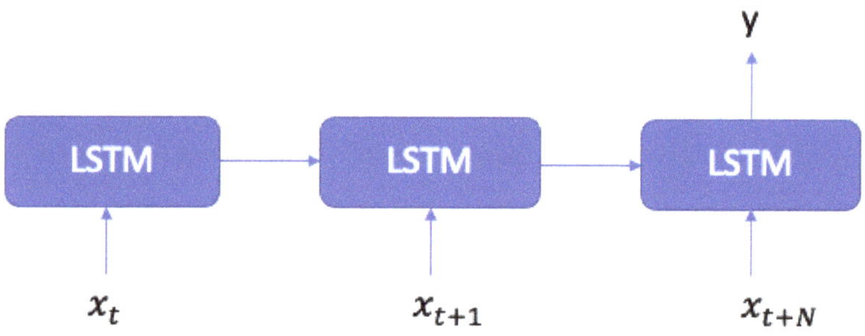

Figure 4. LSTM one-to-many.

2.1.4. Ethics in Education

Greater use of technology in education has led to an increase in the data available for training AI models. These datasets sometimes contain sensitive information. The lack of attention to ethical aspects during the design of these models has produced errors, such as bias in machine learning models, that could have a negative impact on people, industry, or society. Ref. [31] highlights a disproportionate lack of representation of black women in commercial datasets used for facial recognition. Hence, all the models fully trained with these datasets also have the same bias. Today, the biggest challenges that AI is facing in education are its huge implementation costs, ethics, transparency, and security [32].

Relatedly, we expect technology to treat students fairly, so that it does not favor some students over others. This will only be possible if an ethical design in AI is correctly implemented. We need to know how to differentiate between doing ethical things and doing things ethically. Ethical intentions, like good intentions, do not always lead to ethical designs [12].

Within the ethical aspects of education, one of the most analyzed topics is related to bias, usually aligned to gender discrimination and discrimination against minority groups. At present, there are many stereotyped educational environments, that cause negative effects on individuals belonging to minority groups. These negative effects could produce a decrease in performance of students of these groups while they are interacting with such environments.

Educational technologies are often developed with a strong unconscious gender bias, focused on the male gender. In this sense, gender must be a very important factor when designing educational technologies, in order to make them more inclusive and egalitarian [33].

Whereas AI has been integrated into education for more than a decade, the concern around ethics in this area is quite recent. Despite this, today, a large number of researchers dealing with AI in education recognize the importance and value of ethics [12].

According to [34], educational AI is divided into three large groups, within each of which there are ethical aspects that must be considered:

- Tools focused on student assistance: they help students to understand the subject being studied and increase their motivation. An incorrect categorization of students by groups can lead to presenting students with tasks of inappropriate difficulty;
- Tools focused on teacher assistance: they help teachers to focus their time and effort on identifying which students need more help. An incorrect diagnosis about which students need more help can lead the teacher to focus on those who, in fact, require less attention;
- Tools focused on educational administration: they relate to problems found in the classroom with the didactic material. In these tools, errors usually come from an incorrect identification of the student profile or from an incorrect evaluation of the didactic material.

According to [12], the AI education community needs to acknowledge the value of developing an ethical framework and practical guides. Currently, many courses include the study of ethical frameworks for the promotion of ethical AI systems in education. This is the case in [32] in primary school education, or [31] in high school education, which implemented Developed AI Literacy (DAILy), incorporating ethics into the teaching of AI, with the aim of raising awareness among students about the importance of ethical aspects within this discipline. According to the general consensus, there are five fundamental principles for ethics in AI [35,36]:

- Beneficence: technology must be beneficial to humanity;
- No maleficence: technology must be prevented from causing harm, particularly focused on privacy violations;
- Autonomy: technology should empower people's autonomy, allowing them to choose what they delegate to technology and what not;
- Justice: technology must be fair, avoiding any kind of discrimination;
- Explicability: technology must be transparent, understandable, and interpretable by people, so that they understand the logic that the AI has followed, and are thus able to identify and correct errors.

Of the five points cited [34] highlights autonomy and explicability as elements that must play a central role. The integration of ethics in the teaching of AI is a fundamental element to increase the awareness of the population about the impact of AI in industry and society [31].

Regarding ethical aspects in the field of education, our aim is to achieve an AI model capable of identifying when a student needs help, avoiding gender discrimination as much as possible. To achieve this goal, we trained two models and compared the results: the first model was trained with the gender attribute, and the second model was trained without the gender attribute.

With regard to the ethical aspects of our research, we aim to:

- Use an ethical approach to the user experiences conducted to collect data for our research, taking into account the standards of the Universidad Nacional de Educación a Distancia (UNED) Research Ethics Committee and the Guide for research ethics committee members of the Council of Europe (https://www.coe.int/en/web/bioethics/guide-for-research-ethics-committees-members, accessed on 27 November 2023).
- Respect the five standard fundamental ethical principles to guide the development of AI applications, with a focus on developing affective robot tutors that support autonomous learning in a transparent way by acting as a support and not as a substitute for the human teacher (avoiding dehumanisation of teaching), and that do not act in a discriminatory way and take into account the idiosyncrasies of each student.

As far as the work presented in this article is concerned, our aim is to achieve an AI model capable of identifying when a student needs help, avoiding gender discrimination as much as possible. As a first step towards achieving this goal, to identify any gender bias, we trained two models and compared the results: the first with the gender attribute, and the second without.

It is not our intention in this paper to address in depth the elimination of gender bias in intelligent educational systems for learning computational skills, but to take a first step in that direction by highlighting the existence of such bias and the influence of considering the gender attribute as a training parameter in the models. We recognise the difficulty of addressing this multifaceted issue.

There is a wealth of academic literature, as well as various reports from international organisations, on the gender digital divide, which is of great concern at this historical moment of the digital revolution [37–43]. Although progress has been made in reducing the gender gap in education, the fact remains that women's participation in ICT is not growing in proportion to women's education in general. This is particularly true in the case of AI, whose disruptive potential makes women's participation essential. Occupational segregation by gender is increasing, so that in Spain, for example, only 0.5% of AI professionals are women.

Among the causes of this gap, there is agreement in the literature on the importance of the educational factor. Ref. [37] points to girls' lack of self-esteem/self-confidence/self-sufficiency, ignorance of their own needs/interests, imposed codes of conduct/stereotypes, fewer learning opportunities for girls inside and outside educational settings, and lack of recognition that they are capable of the same achievements but have different ways of learning. Many studies agree on the need to promote gender equality through education, culture, and communication. Educational institutions need to identify gender barriers in the STEM ecosystem and promote digital literacy for women at all levels of education with a proactive and personalised gender approach.

2.2. User Experience with Scratch

The purpose of the user experience is to collect data to assist in the development of an intelligent model for identifying, through the use of machine learning techniques and the block-based programming language distance calculation, when a primary school student has need of pedagogical assistance.

Since our ultimate goal is to help elementary students in a real learning environment while performing Scratch exercises, the data collection must be done in a real learning environment. To this end, we reached a collaboration agreement with two learning centers, and chose 10 Scratch exercises, designed and validated by primary school technology teachers, of varying degrees of difficulty for students to perform, with the goal of learning the basics of programming and computational thinking. The realization of the user experiences took a total of three months, and a total of four teachers specialized in teaching technology to primary school students participated in them.

Since not all students have the same knowledge of Scratch, they will need help at different points of the process, so the model must be customized according to the knowledge level. To evaluate this, we also designed three level assessments, also of varying degrees of difficulty, that must be completed prior to executing the aforementioned exercises. These assessments measure the student's prior knowledge of Scratch.

The students who participated in these user experiences were very aware that during the experiment they should ask for help only when they needed it. In order for students to be able to ask for help, and for this request to be recorded in the database, modifications were made to the Scratch project to enable a button for students to ask for help and to display hints about the exercise they were currently working on. In order to clarify the procedure of the sessions, an explanatory session was held prior to the user experiences to explain the purpose of the research and the importance of requesting help when it was really needed. In addition, the teachers supervised the development of these experiences and also validated the students' requests for help.

As described in Section 2.1, the application of AI ethical principles must play a central role in education. In this study, in addition to building a model capable of identifying when a student needs help, we relied on ethical methods.

In user experiences, two of the most critical ethical elements are biases (discrimination by gender and by minority groups), and transparency. To avoid any bias due to gender or minority status, participation in user experiences was offered to any elementary school student in the collaborating centers, without any kind of discrimination based on gender, race, or social group. Despite this, we are aware that given the low number of participants in the user experiences, there is no equity in the representation of different races, social groups, or minority groups. In this study we have just focused on gender, trying to ensure that both genders are represented as equally as possible. In addition, to provide maximum transparency, parents or legal guardians and students who wanted to participate in the experiences received an informed consent form that included a description of the benefits of the study and its possible benefits and risks for the student, a detailed description of the study itself and of the data protection policy, and finally a channel through which to ask questions. Parents or legal guardians can request at any time the withdrawal of a student from the experience, as well as the desire to keep or delete the data already collected. Likewise, a certificate was provided by the Research Ethics Committee of the UNED.

In the case of this research, the Research Ethics Committee of the UNED required the declaration of a "data treatment activity", which in our case referred to the automated treatment of academic data of individuals under the age of 14. This declaration implies the commitment to keep the data strictly for the time necessary to fulfill the purpose for which it was collected and to determine any responsibilities that may arise from this purpose and from the processing of the data.

Another stipulation of the Ethics Committee is that the user experience does not cause any harm or discomfort to the participants. In this research, the user experiences were carried out during school hours, in the case of the collaborating school Los Peñascales (https://www.colegiolospenascales.com, accessed on 27 November 2023), and in the case of the company Rockbotic (https://rockbotic.com, accessed on 27 November 2023), during already scheduled after-school courses. Furthermore, they could be carried out both in person and remotely, since the tool was accessible via the web. The competences exercised during the user experiences were those of the training program that the students were following, and the duration of each session varied between 25 min and 1 h, coinciding with the duration of the usual classes.

In addition, at the beginning of the experience at each center, the students received an introductory talk about the research project in which they would be participating. The talk explained the details of the scientific method, the models of artificial intelligence, and the potential of this project within artificial intelligence and education. At the end of the talk, all the students' questions were answered. The opportunity to participate in a research project in the field of artificial intelligence contributed to the motivation of participating students and their confidence in intelligent technologies, as well as to the development of their computational thinking skills, which could be very useful for their professional future.

The Scratch tool was adapted to facilitate the recording of student profiles and the actions performed during the completion of the exercises through keyboard and mouse interaction. It was designed to be as easy to use as possible and was accessed through a login with a username, password, and student number. The student numbers were assigned by the teacher. In addition, at the end of the sessions, the codes were randomly renamed by an automatic system to ensure that there was no record of the origin of the data.

The Research Ethics Committee also supervised the data-storage procedures. All data collected was stored exclusively in a remote database (never on the local computers used by the students in the schools or academies) in order to be able to retrieve and analyze it throughout the analysis phase. The server hosting the database also hosted the web platform used by the teachers and students. Once the research is completed, the data will be deleted from the server and stored on a hard drive in the researchers' custody.

The committee also requires a commitment that any publications resulting from the research will avoid any explicit reference to student data that could uniquely identify students through cross-referencing. Therefore, these future publications will present only anonymized, aggregated data and will never include images of students, unless their legal guardians have given their explicit consent to do so. The data collected will only be used for the purposes of this research and will not be shared with third parties. Finally, the data will only be kept for the legal period and for the time necessary to fulfill the purposes of the study. Finally, on the recommendation of the Committee, the researchers commit themselves to communicate the results of the research in which they participate to the collaborators.

A summary of the user experience process is described below:

- We first sent the informed consent to the parents or legal guardians of the students. Before the students participated, both parents or legal guardians had enough time to consider participation;
- Then, the students whose parents had authorized their participation started by filling out a questionnaire.
- After that, they started the experience by carrying out the assessments. During them, they could not request any kind of help to evaluate the real prior knowledge of Scratch, but they could exit from the assessment at any time. Once the students had completed or exited from all the assessments, the knowledge level was set;
- Finally, the students could start performing the Scratch exercises in their preferred order. In those exercises they could ask for help and also switch to another exercise at any time.

From the first interaction with Scratch up to the end of the exercises, all the interactions were recorded in a database with the aim of analyzing them later. All this information was made anonymous by previously giving a random number to each participant. The stored information of this experiment is hereunder listed:

- Student data:
  - Student number: the student number was randomly assigned to the student;
  - Gender: provided by the student in the opening questionnaire and stored in numerical format (1 for boy and 2 for girl);
  - Mother tongue: provided by the student in the opening questionnaire. It is a boolean field that indicates if the student's mother tongue is Spanish or not. Given that the exercises and the user experiences were carried out in Spanish, we think that the variable "language" may affect whether a student understands the statements of the exercises and the teacher's explanations. Therefore, this variable could influence the number of help requests made by the students;
  - Age: provided by the student in the opening questionnaire;
  - Competence: set by the aforementioned starting assessments.

- Exercise data:
  - Name: exercise name;
  - Description: exercise description;
  - Skills: each exercise may help to improve students' CT skills to different degrees. These skills and their degrees (which are numerical values from 0 to 10) were set by experts;
  - Evaluation: indicates if the exercise is an evaluation level exercise or not.
- Solution distance data: family distance, element distance, position distance, input distance, total distance (explained in Section 2.3).
- Interaction data:
  - Date time: date and time of the interaction;
  - Request help: indicates if the student requested help or not;
  - Seconds help open: when a student requested help, this parameter indicates how many seconds the help popup was open before the student closed it;
  - Last login: date and time of the student's last login.
- Workspace data:
  - Elements: block elements that the student currently has in their workspace. This information is then used to calculate the distance between the student workspace and the solution.

2.3. *Distance Calculation for Block-Based Programming Languages*

In programming learning activities there may be multiple solutions for a single task. The learning progress can be measured by calculating the distance between the student's workspace and the best candidate solution, i.e., the solution that is closest to the student's workspace according to the distance measure. Below we present a normalized version of the distance function presented in [6].

The elementary blocks of a block-based programming language are divided into families (Figure 5) that group a common functionality, as we can see from example in Figure 6. Elementary blocks can be combined into higher-level blocks. These blocks can also have an input, and each input can contain one or many fields. When designing a program in a block-based programming language, blocks can be combined into a single block for a sequential execution, as shown in Figure 7, or in different groups for a parallel execution, as illustrated in Figure 8.

Figure 5. Block families in Scratch.

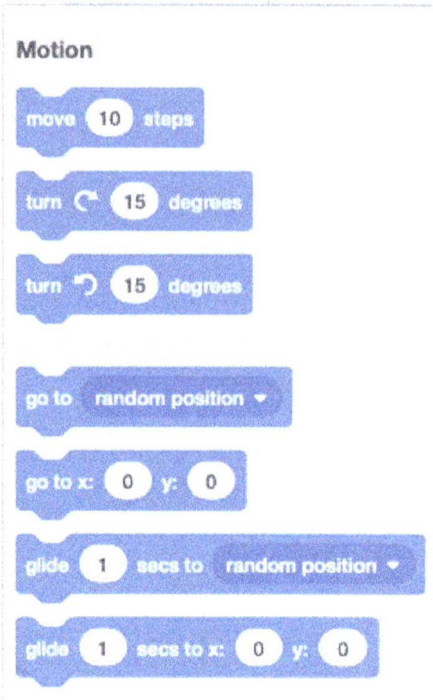

Figure 6. Movement family blocks in Scratch.

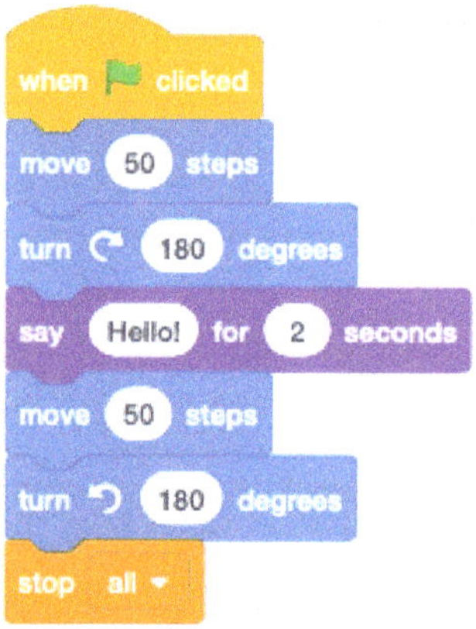

Figure 7. Single element block in Scratch.

Figure 8. Multiple blocks of elements in Scratch.

In summary, the dimensions that influence the distance in a block-based programming language are:

1. Block family;
2. Block;
3. Block position;
4. Block inputs values.

2.3.1. Block Family Distance Calculation

Let F_w be the block family set of the student's workspace, and F_s the block family set of the exercise solution. The difference between both family groups is fixed as the symmetric difference between the set F_w and the set F_s.

We consider the correct family set as the block families in the student's workspace, that are also present in the exercise solution; and the incorrect family set as the block families in the student's workspace that are not present in the exercise solution. The distance calculation is carried out as follows:

Let: $F_i = \{family \mid family \in F_w \triangle F_s\}$, the incorrect family set.

If x is a set of elements, and $n(x)$ is the function returning the number of elements in x, the distance between the families of the solution and the families of the student's workspace is defined as:

$$d_f = n(F_i) \tag{1}$$

2.3.2. Block Distance Calculation

As mentioned previously, blocks are grouped by families, which means that blocks are elements of these families. We consider the correct block set as the block elements in the student's workspace, that are also present in the exercise solution; and the incorrect block set as the block elements in the student's workspace that are not present in the exercise solution.

Let:

B_w, the set of blocks in the student's workspace.

B_s, the set of blocks in the solution.

$B_i = \{block \mid block \in B_w \triangle B_s\}$, the incorrect block set.

The distance calculation between the solution block set and the student's workspace block set is defined by:

$$d_b = n(B_i) \tag{2}$$

2.3.3. Block Position Distance Calculation

As noted above, blocks can be organized sequentially, establishing an order for their execution. We must check that the correct blocks of the student's workspace are executed in the proper order. To do this, we compare the position of each block in the student's workspace with the position of the same block in the solution.

Let $p_s(block)$, $p_w(block)$, be the positions of the blocks in the solution and in the workspace, respectively, where we assume that the position is a natural number higher than zero.

The distance calculation for the correct block sets is carried out by subtracting from the position of each block in the student's workspace, the position of that same block in the solution. We use the term correct block position for the position of a block in the student's workspace that matches the position of the same block in the exercise solution; and the term incorrect block position for the position of a block in the student's workspace that does not coincide with the position of the same block in the exercise solution. Therefore, the positional distance of the correct block sets is calculated as follows:

$$d_{pc} = \sum_{block \in B_s \cap B_w} |p_s(block) - p_w(block)| \tag{3}$$

The positional distance of the incorrect block sets B_i is also taken into account in a similar way. To do this, the value of each position of the blocks belonging to B_i is added, as follows:

$$d_{pi} = \sum_{block \in B_s - B_w} p_s(block) + \sum_{block \in B_w - B_s} p_w(block) \quad (4)$$

The final positional distance will be the sum of the correct block position distance and the incorrect block positional distance:

$$d_p = d_{pc} + d_{pi} \quad (5)$$

2.3.4. Block Input Values Distance Calculation

As mentioned in Section 2.3, blocks can have alphanumerical and numerical inputs. The calculation of the input distance between the input values is performed in two steps: In a first step, we compare the input values of the correct blocks in the student's workspace and the solution, and in a second step, we add the input values for the incorrect blocks.

Because there can be both alphanumerical and numerical inputs, we defined two different ways to calculate the distance depending on the input type.

Let $i_s(block)$, $i_w(block)$ be the inputs of a block in the solution and in the workspace, respectively. The calculation of the input-value distance between the student's workspace and the solution in the correct set of blocks is done in two different ways. In the case of alphanumerical inputs, for different values we simply consider that the distance is 1. In the case of numerical inputs, we calculate the numerical difference between the input value of the student's workspace and that of the solution.

$$d_{ic} = \sum_{block \in B_s \cap B_w} \begin{cases} \frac{|i_s(block) - i_w(block)|}{|i_s(block)| + |i_w(block)|}, \text{ if the input is numerical} \\ 1, \text{ if the input is alphanumerical, and } i_w(block) \neq I_s(block) \end{cases} \quad (6)$$

As in Equation (6), the calculation of the input value distance in the incorrect block set is also done in two different ways as follows:

$$d_{ii_s} = \sum_{block \in B_s - B_w} \begin{cases} i_s(block), \text{ if the input is numerical} \\ 1, \text{ if the input is alphanumerical} \end{cases} \quad (7)$$

$$d_{ii_w} = \sum_{block \in B_w - B_s} \begin{cases} i_w(block), \text{ if the input is numerical} \\ 1, \text{ if the input is alphanumerical} \end{cases} \quad (8)$$

The final input value distance is calculated as the sum of the input value distance in the correct block set (d_{ic}), and the input-value distance in the incorrect block set ($d_{ii_s} + d_{ii_w}$):

$$d_i = d_{ic} + d_{ii_s} + d_{ii_w} \quad (9)$$

2.3.5. Total Distance

As outlined above, there is an order of relevance in the parameters when determining the distance between the workspace and the solution.

So, the total distance is calculated as the weighted sum of Equations (1), (2), (5) and (9).

The total distance between the student's workspace and the solution is defined as follows:

$$d_t = d_f + \left(\frac{1}{2} \cdot d_b\right) + \left(\frac{1}{4} \cdot d_p\right) + \left(\frac{1}{8} \cdot d_i\right) \quad (10)$$

2.4. *Help Identification Model Development*

Once the user experiences defined in Section 2.2 were completed, we used the collected data to develop a model that can determine when learners need help. This was achieved through feature selection and analysis of the data, and development and training of the model, as shown in Figure 9.

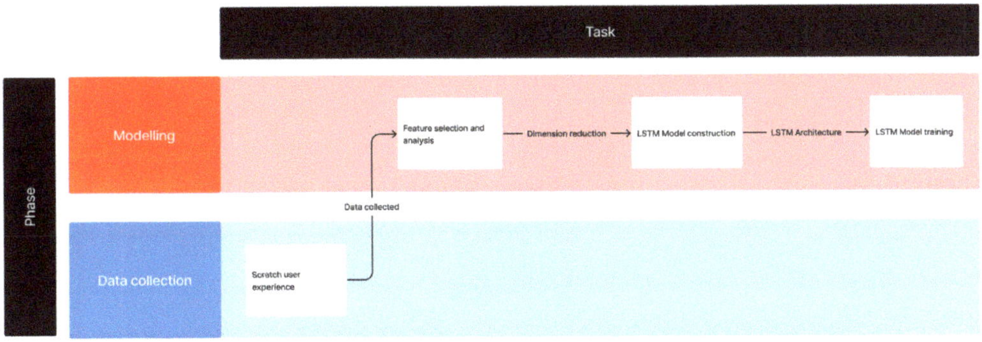

Figure 9. LSTM model building flow.

2.4.1. Feature Selection and Analysis

To the list of features described in Section 2.2, we added the calculated feature total_seconds. This feature measures the time between the first interaction that occurs during the completion of an exercise, and the current interaction. The purpose of this feature is to know how much time the student has spent on the current exercise.

Feature selection is performed in two stages: the first stage consists of eliminating those features that do not improve the precision of the model, and the second stage consists of removing those characteristics that can pose an ethical risk. The first stage of feature selection was performed by combining an initial manual selection, based on our own knowledge of the features, with an automatic feature selection. In the manual selection, we discarded the following features: exercise_is_evaluation, exercise_name, exercise_description and last_login. In the second stage, student_gender was manually removed, because it can cause the system to take into account the student's gender to give more or less support, leading to a bias. At this stage, the effect on the accuracy of the model of removing the gender was analysed. For the automatic feature selection, we implemented a filter method based on Pearson correlation. This method filters out all features that are not relevant by using a correlation matrix. In our study, we filtered all the features with a correlation above or equal to 0.85. The correlation matrix heatmap result is shown in Figure 10.

Figure 10, shows that gender is slightly correlated with the attributes exercise_skill_logical_thinking, exercise_skill_flow_control, exercise_skill_user_interactivity and exercise_skill_information_representation. It also shows that the mother tongue attribute has almost no correlation with any attribute, and that there is also a correlation between the student competencies and the different skills. Once features were filtered, we analyzed them by calculating the feature impact, using the SHapley Additive exPlanations (SHAP) framework [44]. The results show which features are driving model decisions, and they were obtained as follows:

1. Takes a sample of records from the training data;
2. Computes SHAP values for each record in the sample, generating the local importance of each feature in each record;
3. Computes global importance by taking the average of abs(SHAP values) for each feature in the sample;
4. Normalizes the results.

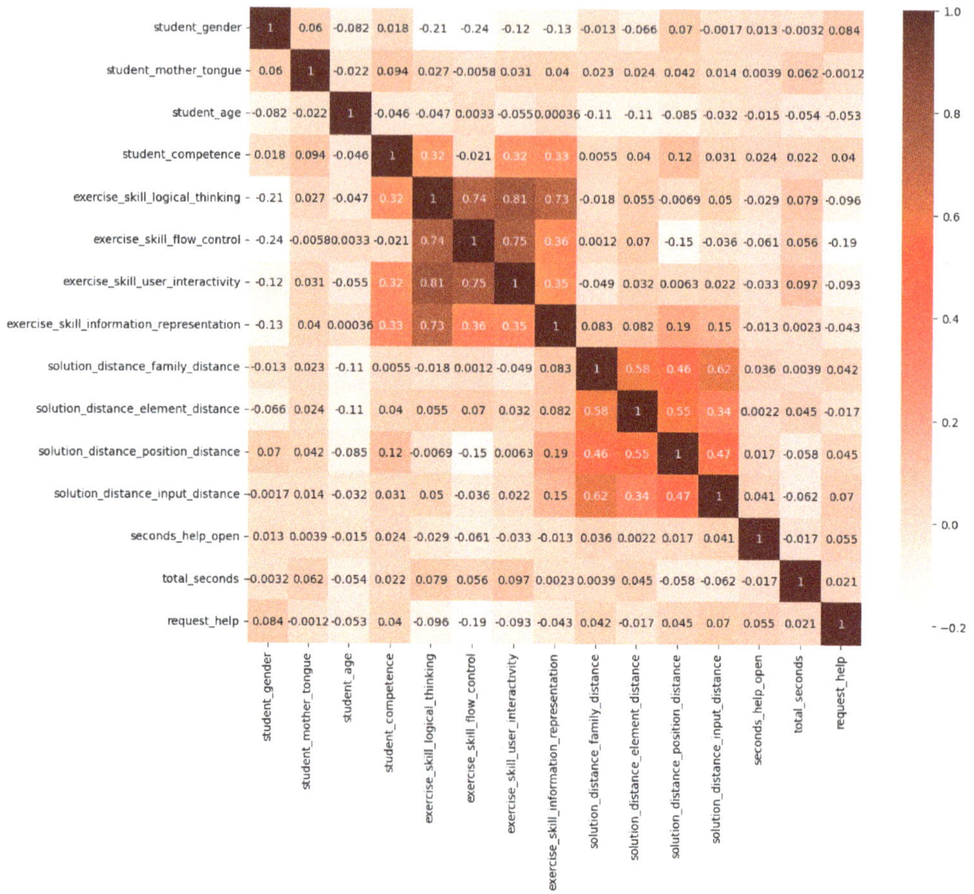

Figure 10. Correlation Matrix Heatmap.

Figure 11 shows the feature impact results. Notice that total_seconds, the different skills, and the student_competence have a big impact on the model, which means that these variables are important when determining if a student needs help. Additionally, the gender and student_mother_tongue attributes have a very low impact.

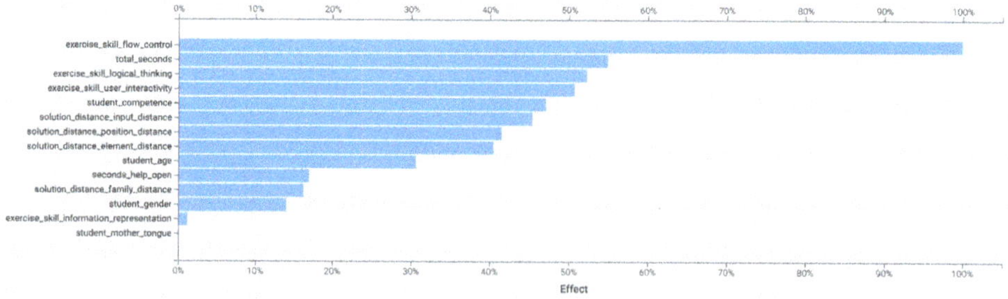

Figure 11. Feature impact.

2.4.2. Model Construction

For the construction and training of the model, the dataset in the correlation matrix shown in Figure 10 was used. Since we are looking for the model to predict whether the student needs help or not, we use the variable "request_help" as the target variable, and we look for the output of the model to be a boolean value.

Given that all the student actions are recorded over time and reflect when a student needs help, the model implementation is based on a RNN, making use of LSTM. Since we have many inputs (interactions in Scratch over time), and just one output (whether the student needs help or not), we applied a many-to-one architecture, as shown in Figure 4. In order to identify the architecture model that best fits the problem and the current data, we conducted some experiments with different parameters. In these experiments, we combined the number of LSTM layers (one or two LSTM layers), and none, or a few dropout layers (with a value of 0.5). Taking into account that the model must identify if a student needs help or not in a particular moment, we were faced with a binary problem, hence, we implemented a binary cross entropy loss function with an adam optimizer in each experiment. The different model architectures we tested are shown in Figures 12 and 13.

The following parameters were used for the training of the different architectures:

- LSTM units (in each LSTM layer): 256;
- Activation function of the LSTM layers: sigmoid;
- Return sequences: true;
- Dropout layer value (in each dropout layer if applicable): 0.5;
- Dense layer units: 1;
- Dense layer activation function: sigmoid;
- Dense layer dropout function: binary_crossentropy;
- Dense layer optimizer: adam;
- Dense layer metrics: binary_accuracy;
- Training batch size: 1;
- Number of training epochs: 50;
- Percentage of training data: 70%;
- Percentage of validation data: 30%.

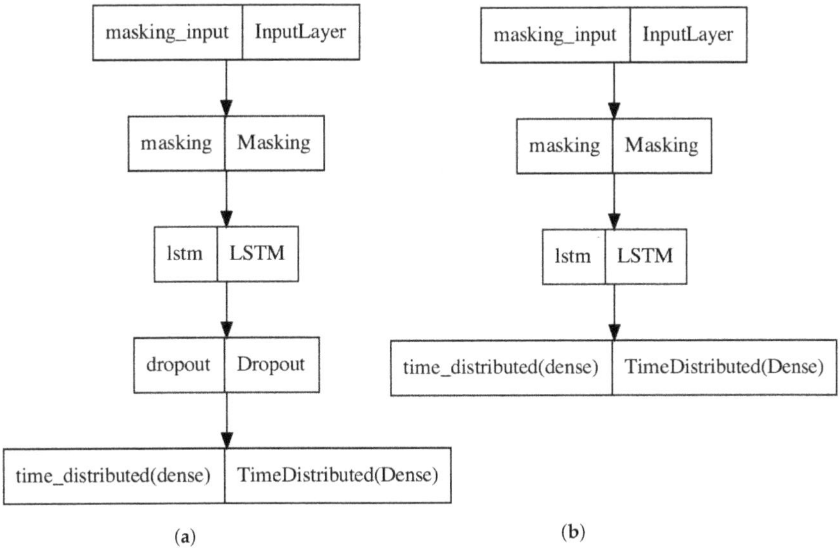

Figure 12. Architectures with 1 LSTM layer. (**a**) 1 LSTM layer with dropout layer; (**b**) 1 LSTM layer with no dropout layers.

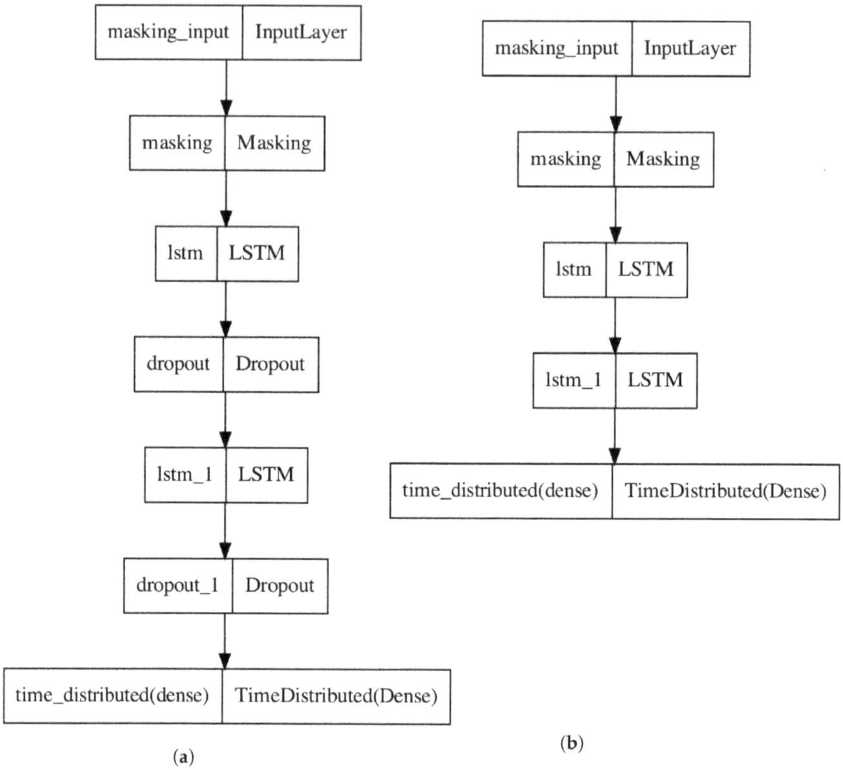

Figure 13. Architectures with 2 LSTMs layers. (**a**) 2 LSTMs layer with dropouts layer; (**b**) 2 LSTMs layer with no dropout layers.

During the user experiences described in Section 2.2, we collected 11,790 interactions, taken from 82 students, who generated 695 help requests. In Table 1 we present the data collected.

Table 1. Data collected.

Interactions	Help Requests	Students	Average Age	Girls	Boys
11,790	695	82	13.3	30	52

Because each user performs many interactions in a single exercise and day, grouping these interactions by user, date, and exercise, we finally obtained 316 time series. Therefore, we have an average of 2.19 help requests per time series, and 37.31 interactions per time series. Finally, for the training of our model we used 70% of the time series, and 30% for validation, as shown in Table 2.

Table 2. Dataset distribution.

Phase	Time Series	Help Requests	Students	Average Age	Girls	Boys
Training	221	450	52	13.5	16	36
Validation	95	245	30	13	14	16

Finally, in order to analyze whether gender is a bias in the models, it was decided to train the four architectures shown in Figures 12 and 13 twice. In the first training run the dataset includes the gender of the students, while in the second run it does not.

3. Results

After executing the different training runs (including and excluding gender) of the four architectures shown in Figures 12 and 13, the results were divided into two different tables. Table 3 shows the results of the training model including the gender. In the majority of the architectures we exceeded 99% accuracy. Table 4 shows the results of the training model without the gender feature. It can be observed that in all the architectures we exceeded 91% accuracy.

Table 3. Model training results with gender feature.

LSTM Layers	Dropout Layers	Loss	Binary Accuracy	Val Loss	Val Binary Accuracy
1	False	0.017185	0.94459	0.0071762	0.99144
1	True	0.019589	0.94277	0.0077519	0.99249
2	False	0.017146	0.94621	0.0078189	0.98975
2	True	0.020001	0.94273	0.0085079	0.99266

Table 4. Model training results without gender feature.

LSTM Layers	Dropout Layers	Loss	Binary Accuracy	Val Loss	Val Binary Accuracy
1	False	0.015799	0.9574	0.04275	0.91122
1	True	0.013381	0.95972	0.047054	0.91095
2	False	0.015337	0.95727	0.045792	0.91122
2	True	0.013264	0.94273	0.045034	0.91116

4. Discussion

In this research, we used the distance calculation presented in [6], and we implemented a first prototype model for predicting when a primary school student will need help while performing block-based programming exercises.

One important motivation for this model is to ensure that the students will be properly helped when they are stuck. To achieve this goal, the model must be able to correctly identify the moment when a student has a mental block while he/she is carrying out Scratch exercises.

During the user experiences, despite the fact that we sought to have an equal number of boys and girls, and to have a significant number of students whose mother tongue is different from Spanish, the reality is that there was an imbalance in the gender of the participants, as well as a large majority of students whose mother tongue is Spanish. User experiences were designed taking into account ethical aspects such as transparency and avoiding any kind of discrimination. In this first approach, we focused on trying to have a significant number of boys and girls, and studying the influence of gender, being mindful of its relevance to learning in the field of computer science, as studies like [45] show. In this sense, we intend to conduct larger experiments in the future. Even with such data, we consider that this first approach, which includes the data collection with the distance-calculation attribute and the predictive help model, is relevant, because there is no mention in the literature of repositories containing user interactions with block-based programming software, and solution distance metrics may help to build the suggested model.

Looking at the calculated impact of the different features in Figure 11, it can be seen that the most important features are the student's skills, the total time, and the distance calculations. This means that the distance calculation introduced in Section 2.3.2 is a relevant feature that helps to improve the accuracy, reaching more than 40% impact. Our predictive model trained with the data collected from user experiences in a real learning environment obtained 99.2% accuracy if we consider the gender attribute in training, and 91.1% accuracy if we exclude the gender attribute in training. Despite the impact of gender being relatively low, as shown in Figure 11, it can be seen in Tables 3 and 4 that this attribute influences the accuracy of the model by around 8%. All this seems to indicate that gender may be a relevant attribute. Also, in Figure 10, gender is inversely correlated with the different skills, which means that, in general, in our experiment girls had a lower score in these skills. The small amount of data collected, and the high precision of the model when gender is present, makes us suspect that the model is overfitting, leading to it not properly identifying when a student requires help, due to gender, and this fact could cause frustration and loss of interest in the subject. In addition, we believe that in our experiment variables correlated with gender, such as learning style, additional skills, etc., were lacking.

5. Conclusions and Future Work

The model that has been presented has 99.2% accuracy taking gender into account and 91.1% accuracy without taking gender into account. This could indicate that the gender present in the dataset may be introducing an overfitting into the model.

By including the gender variable, given the lower representation of the female gender in the student body participating in our experiments, the trained models will not generalise adequately and their predictions will be biased by the specific characteristics of the participating girls, who had low computational thinking skills in the experiments conducted. By removing the gender variable, predictions of competence in completing the exercises would correlate with computational skills without gender bias.

We can conclude from the above that there are hidden variables, determinants of low computational competence, that it would be interesting to make explicit in our models; variables that, in the current context, define women's idiosyncrasies but do not imply a lower real capacity of women for computational thinking. These variables would be related, for example, and as indicated in [37], to having been taught computational skills through ineffective and demotivating instructional designs.

From what can be observed in this first approximation, students' skills, total time, and distance calculations could be relevant in determining whether a primary school student needs help while performing block-based programming exercises.

Since the distance calculations shown are common to all block-based programming languages, and the results are not specific to Scratch, nor to elementary education, this same study can be replicated with any other block-based programming tool and for any educational level.

Therefore, this work could serve as a starting point to investigate models to identify when a primary school student (even at other levels) gets stuck during block-based programming exercises. This may allow for more personalized interventions when they are really needed, through tutor interventions, or help systems, improving the students' learning experience and increasing their motivation.

In future work, we intend to perform new user experiences to collect more data with more participants; we will also strive to have an equal number of boys and girls, and to explore feature selection methods, as well as RNN architectures to improve the accuracy of the model. In addition, for these models, more performance metrics, such as precision, recall, and f1 score, as well as statistical techniques to measure the loss of accuracy resulting from shuffling, will be taken into account. We will also refine and validate the model in a real learning environment, and a more in-depth analysis of the impact of the different attributes in determining whether a student needs help will be conducted. Finally, this work will also support the ARTIE project [13], introduced in Section 1, so that the system will be able to automatically identify when a primary school student needs help during block-based programming exercises, and send the appropriate pedagogical intervention to the affective tutor robot.

Author Contributions: Conceptualization, L.E.I.C.; Methodology, L.E.I.C.; Software, L.E.I.C.; Validation, L.E.I.C., Á.M.R. and F.d.l.P.L.; Formal analysis, L.E.I.C.; Investigation, L.E.I.C.; Resources, Á.M.R. and F.d.l.P.L.; Writing—review & editing, Á.M.R. and F.d.l.P.L.; Visualization, L.E.I.C.; Supervision, Á.M.R. and F.d.l.P.L. All authors have read and agreed to the published version of the manuscript.

Funding: This research was funded by MCIN/AEI/10.13039/501100011033 and, as appropriate, by "ERDF A way for making Europe", by the "European Union" or by the "European Union Next Generation EU/PRTR" grant number PID2020-115220RB-C22.

Institutional Review Board Statement: All subjects gave their informed consent for inclusion before they participated in the study. The study was conducted in accordance with the Declaration of Helsinki and the protocol was approved by the Ethics Committee of UNED (10/02/2022 and Project identification code: 1-ETSIINF-2021).

Informed Consent Statement: Informed consent was obtained from all subjects involved in the study.

Data Availability Statement: On request: The data presented in this study are available on request from the corresponding author. The data are not publicly available due to privacy restrictions.

Acknowledgments: The authors also wish to thank the staff of Rockbotic and Los Peñascales primary school of Las Rozas (Madrid) for their collaboration, in particular David Moreno from Rockbotic and Ignacio Martín from Los Peñascales, the regular teachers of the students who took part in the experiments, as well as the children who collaborated in this research during the sessions. Finally, we want to express our gratitude to Hipoo, the company that provided us with the tools to train the model.

Conflicts of Interest: The authors declare no conflict of interest.

Abbreviations

The following abbreviations are used in this manuscript:

ARTIE	Affective Robot Tutor Integrated Environment
ASTs	Abstract Syntax Trees
AI	Artificial Intelligence
CS	Computer Science
CT	Computational Thinking
CNN	Convolutional Neural Network
DAILy	Developing AI Literacy
DKT	Deep Knowledge Tracing
GRU	Gated Recurrent Unit
LSTM	Long Short-Term Memory
MLP	Multilayer Perceptron
RNN	Recurrent Neural Network
SHAP	SHapley Additive exPlanations
TED	Tree Edit Distance
TID	Tree Inherited Distance
TSC	Time Series Classification
UNED	Universidad Nacional de Educación a Distancia

References

1. Uddin, I.; Imran, A.S.; Muhammad, K.; Fayyaz, N.; Sajjad, M. A Systematic Mapping Review on MOOC Recommender Systems. *IEEE Access* **2021**, *9*, 118379–118405. [CrossRef]
2. Jiang, B.; Li, Z. Effect of Scratch on computational thinking skills of Chinese primary school students. *J. Comput. Educ.* **2021**, *8*, 505–525. [CrossRef]
3. Fagerlund, J.; Häkkinen, P.; Vesisenaho, M.; Viiri, J. Computational thinking in programming with Scratch in primary schools: A systematic review. *Comput. Appl. Eng. Educ.* **2021**, *29*, 12–28. [CrossRef]
4. Howard, N.R. How Did I Do?: Giving learners effective and affective feedback. *Educ. Technol. Res. Dev.* **2021**, *69*, 123–126. [CrossRef]
5. Cavalcanti, A.P.; Barbosa, A.; Carvalho, R.; Freitas, F.; Tsai, Y.-S.; Gašević, D.; Mello, R.F. Supporting Teachers Through Social and Emotional Learning. *Comput. Educ. Artif. Intell.* **2021**, *2*, 100027. [CrossRef]
6. Cuadrado, L.-E.I.; Riesco, A.M.; de la Paz López, F. A first-in-class block-based programming language distance calculation. In *Lecture Notes in Computer Science*; Including Subseries Lecture Notes in Artificial Intelligence and Lecture Notes in Bioinformatics; Springer: Berlin/Heidelberg, Germany, 2019; Volume 13259, pp. 423–432. [CrossRef]
7. Kamil, K.; Sevimli, M.; Aydin, E. An investigation of primary school students' self regulatory learning skills. In Proceedings of the 7th International Conference on Education and Education on Social Sciences, Taipei, Taiwan, 15–17 June 2020.[CrossRef]

8. Wiggins, J.B.; Fahid, F.M.; Emerson, A.; Hinckle, M.; Smith, A.; Boyer, K.E.; Mott, B.; Wiebe, E.; Lester, J. Exploring Novice Programmers' Hint Requests in an Intelligent Block-Based Coding Environment. In Proceedings of the 52nd ACM Technical Symposium on Computer Science Education (SIGCSE 2021), Virtual Conference, 13–20 March 2021; pp. 52–58. [CrossRef]
9. Alsharef, A.; Aggarwal, K.; Kumar, M.; Mishra, A. Review of ML and AutoML Solutions to Forecast Time-Series Data. *Arch. Comput. Methods Eng.* **2022**, *29*, 5297–5311. [CrossRef]
10. He, K.; Gao, K. Analysis of Concentration in English Education Learning Based on CNN Model. *Sci. Program.* **2022**, *2022*, 1489832. [CrossRef]
11. Jarbou, M.; Won, D.; Gillis-Mattson, J.; Romanczyk, R. Deep learning-based school attendance prediction for autistic students. *Sci. Rep.* **2022**, *12*, 1431. [CrossRef]
12. Holmes, W.; Porayska-Pomsta, K.; Holstein, K.; Sutherland, E.; Baker, T.; Buckingham, S.; Santos, S.O.C.; Rodrigo, M.T.; Cukurova, M.; Bittencourt, I.I.; et al. Ethics of AI in Education: Towards a Community-Wide Framework. *Int. J. Artif. Intell. Educ.* **2022**, *32*, 504–526. [CrossRef]
13. Cuadrado, L.-E.I.; Riesco, A.M.; de la Paz López, F. ARTIE: An Integrated Environment for the Development of Affective Robot Tutors. *Front. Comput. Neurosci.* **2016**, *10*, 77. [CrossRef]
14. Cuadrado, L.-E.I.; Riesco, A.M.; de la Paz López, F. FER in primary school children for affective robot tutors. In *From Bioinspired Systems and Biomedical Applications to Machine Learning*; Springer: Berlin/Heidelberg, Germany, 2019; pp. 461–471. [CrossRef]
15. Ng, O.L.; Cui, Z. Examining primary students' mathematical problem-solving in a programming context: Towards computationally enhanced mathematics education. *Zdm -Math. Educ.* **2021**, *53*, 847–860. [CrossRef]
16. López, J.M.S.; Otero, R.B.; García-Cervigón, S.D.L. Introducing robotics and block programming in elementary education. *Rev. Iberoam. Educ. Distancia* **2020**, *24*, 95. [CrossRef]
17. Jen-I, C.; Mengping, T. Meta-analysis of children's learning outcomes in block-based programming courses. In *HCI International 2020—Late Breaking Posters*; Springer: Berlin/Heidelberg, Germany, 2020; pp. 259–266. _33. [CrossRef]
18. Demirkiran, M.C.; Hocanin, F.T. An investigation on primary school students' dispositions towards programming with game-based learning. *Educ. Inf. Technol.* **2021**, *26*, 3871–3892. [CrossRef]
19. Wang, J. Use Hopscotch to Develop Positive Attitudes Toward Programming For Elementary School Students. *Int. J. Comput. Sci. Educ. Sch.* **2021**, *5*, 48–58. [CrossRef]
20. Jo, Y.; Chun, S.J.; Ryoo, J. Tactile scratch electronic block system: Expanding opportunities for younger children to learn programming. *Int. J. Inf. Educ. Technol.* **2021**, *11*, 319–323. [CrossRef]
21. Obermüller, F.; Heuer, U.; Fraser, G. Guiding Next-Step Hint Generation Using Automated Tests. *Annu. Conf. Innov. Technol. Comput. Sci. Educ.* **2021**, *1293*, 220–226. [CrossRef]
22. Fahid, M.F.; Tian, X.; Emerson, A.; Wiggins, J.B.; Bounajim, D.; Smith, A.; Wiebe, E.; Mott, B.; Boyer, K.E.; Lester, J. Progression trajectory-based student modeling for novice block-based programming. In Proceedings of the 29th ACM Conference on User Modeling, Adaptation and Personalization (UMAP 2021), Utrecht Netherlands, 21–25 June 2021; pp. 189–200. [CrossRef]
23. Mlinaric, D.; Milasinovic, B.; Mornar, V. Tree Inheritance Distance. *IEEE Access* **2020**, *8*, 52489–52504. [CrossRef]
24. Yu, W.; Yong, I.; Mechefske, K.C. *Analysis of Different RNN Autoencoder Variants for Time Series Classification and Machine Prognostics*; Academic Press: Cambridge, MA, USA, 2021; Volume 149. [CrossRef]
25. Pudikov, A.; Brovko, A. Comparison of LSTM and GRU Recurrent Neural Network Architectures. In *Recent Research in Control Engineering and Decision Making*; Springer: Cham, Switzerland, 2021; Volume 337, pp. 114–124. [CrossRef]
26. Affonso, F.; Rodrigues, T.M.; Pinto, D.A.L. Financial Times Series Forecasting of Clustered Stocks. *Mob. Netw. Appl.* **2021**, *26*, 256–265. [CrossRef]
27. Bandara, K.; Hewamalage, H.; Hao, Y.; Kang, L.Y.; Bergmeir, C. Improving the accuracy of global forecasting models using time series data augmentation. *Pattern Recognit.* **2021**, *120*, 108148. [CrossRef]
28. Yongjun, M.; Wei, L. Design and Implementation of Learning System Based on T-LSTM. In Proceedings of the Advances in Web-Based Learning—ICWL 2021, Macau, China, 13–14 November 2021; pp. 148–153. [CrossRef]
29. Terada, K.; Watanobe, Y. Code completion for programming education based on deep learning. *Int. J. Comput. Intell. Stud.* **2021**, *10*, 78–98. [CrossRef]
30. Smagulova, K.; James, A.P. A survey on LSTM memristive neural network architectures and applications. *Eur. Phys. J. Spec. Top.* **2019**, *228*, 2313–2324. [CrossRef]
31. Zhang, H.; Lee, I.; Ali, S.; DiPaola, D.; Cheng, Y.; Breazeal, C. Integrating Ethics and Career Futures with Technical Learning to Promote AI Literacy for Middle School Students: An Exploratory Study. *Int. J. Artif. Intell. Educ.* **2022**, *33*, 290–324. [CrossRef]
32. Kazi, K.; Solutions, K.K.; Solapur, M.; Devi, S.; Sreedhar, B.; Arulprakash, P.; Kazi, K. A Path Towards Child-Centric Artificial Intelligence based Education. *Int. J. Early Child. Spec. Educ. (INT-JECS)* **2022**, *14*, 9915–9922. [CrossRef]
33. Santos, J.; Bittencourt, I.; Reis, M.; Chalco, G.; Isotani, S. Two billion registered students affected by stereotyped educational environments: An analysis of gender-based color bias. *Humanit. Soc. Sci. Commun.* **2022**, *9*, 249. [CrossRef]
34. Du Boulay, B. Artificial intelligence in education and ethics. In *Handbook of Open, Distance and Digital Education*; Springer: Berlin/Heidelberg, Germany, 2022; pp. 1–16. [CrossRef]
35. Floridi, L.; Cowls, J. A Unified Framework of Five Principles for AI in Society. *Harv. Data Sci. Rev.* **2019**, *1*, 1. [CrossRef]
36. Schiff, D. Education for AI, not AI for Education: The Role of Education and Ethics in National AI Policy Strategies. *Int. J. Artif. Intell. Educ.* **2022**, *32*, 527–563. [CrossRef]

37. Samuel, Y.; George, J.; Samuel, J. *Beyond Stem, How Can Women Engage Big Data, Analytics, Robotics and Artificial Intelligence?—An Exploratory Analysis of Confidence and Educational Factors in the Emerging Technology Waves Influencing the Role of, and Impact Upon, Women*; Leibniz Center for Informatics: Wadern, Germany, 2018.
38. Hajibabaei, A.; Schiffauerova, A.; Ebadi, A. Women, artificial intelligence, and key positions in collaboration networks: Towards a more equal scientific ecosystem. *arXiv* **2022**, arXiv:2205.12339.
39. West, M.; Kraut, R.; Ei Chew, H. I'd Blush if I Could: Closing Gender Divides in Digital Skills Through Education. 2019. Available online: https://en.unesco.org/Id-blush-if-I-could (accessed on 27 November 2023).
40. UNESCO. AI and Gender Equality: Key Findings of UNESCO'S Global Dialogue. Available online: https://unesdoc.unesco.org/ark:/48223/pf0000387610.locale=es (accessed on 27 November 2023).
41. ONTSI. Brecha Digital de Género. Available online: https://www.ontsi.es/es/publicaciones/brecha-digital-de-genero-2022 (accessed on 27 November 2023).
42. UGT. Informe Agosto 2022. Available online: https://www.fesmcugt.org/wp-content/uploads/2022/09/Resumen-Estadistico-242-AGO22.pdf (accessed on 27 November 2023).
43. WEF. Global Gender Gap Report 2021. Available online: https://www.weforum.org/publications/global-gender-gap-report-2021/ (accessed on 27 November 2023).
44. Lundberg, S.M.; Allen, P.G.; Lee, S.-I. A Unified Approach to Interpreting Model Predictions. 2017. Available online: https://github.com/slundberg/shap (accessed on 27 November 2023).
45. Kordaki, M.; Berdousis, I. Identifying Barriers for Women Participation in Computer Science. *Int. J. Educ. Sci.* **2020**, *2*, 5–20. [CrossRef]

applied sciences

MDPI

Article

A Fusion Framework for Confusion Analysis in Learning Based on EEG Signals

Chenlong Zhang, Jian He, Yu Liang *, Zaitian Wang and Xiaoyang Xie

Faculty of Information, Beijing University of Technology, Beijing 100124, China;
zhangchenlong@emails.bjut.edu.cn (C.Z.); jianhee@bjut.edu.cn (J.H.); wangzaitian@emails.bjut.edu.cn (Z.W.); xiexiaoyang@bjut.edu.cn (X.X.)
* Correspondence: yuliang@bjut.edu.cn

Abstract: Human–computer interaction (HCI) plays a significant role in modern education, and emotion recognition is essential in the field of HCI. The potential of emotion recognition in education remains to be explored. Confusion is the primary cognitive emotion during learning and significantly affects student engagement. Recent studies show that electroencephalogram (EEG) signals, obtained through electrodes placed on the scalp, are valuable for studying brain activity and identifying emotions. In this paper, we propose a fusion framework for confusion analysis in learning based on EEG signals, combining feature extraction and temporal self-attention. This framework capitalizes on the strengths of traditional feature extraction and deep-learning techniques, integrating local time-frequency features and global representation capabilities. We acquire localized time-frequency features by partitioning EEG samples into time slices and extracting Power Spectral Density (PSD) features. We introduce the Transformer architecture to capture the comprehensive EEG characteristics and utilize a multi-head self-attention mechanism to extract the global dependencies among the time slices. Subsequently, we employ a classification module based on a fully connected layer to classify confusion emotions accurately. To assess the effectiveness of our method in the educational cognitive domain, we conduct thorough experiments on a public dataset CAL, designed for confusion analysis during the learning process. In both subject-dependent and subject-independent experiments, our method attained an accuracy/F1 score of 90.94%/0.94 and 66.08%/0.65 for the binary classification task and an accuracy/F1 score of 87.59%/0.87 and 41.28%/0.41 for the four-class classification task. It demonstrated superior performance and stronger generalization capabilities than traditional machine learning classifiers and end-to-end methods. The evidence demonstrates that our proposed framework is effective and feasible in recognizing cognitive emotions.

Keywords: human–computer interaction; electroencephalographic; emotion recognition; confusion analysis; self-attention

1. Introduction

In modern education, human–computer interaction (HCI) plays a crucial role, with emotion recognition being particularly significant in the field of HCI. By accurately identifying and understanding students' emotional states, educational systems can better respond to their needs and provide personalized support. Emotion recognition technology can assist educators in determining whether students are experiencing confusion, frustration, or focus during the learning process, enabling timely adoption of appropriate teaching strategies and supportive measures [1–3]. Therefore, the importance of emotion recognition in HCI and education is self-evident. It optimizes the teaching process, enhances learning outcomes, and provides students with more personalized support and guidance. Confusion is more common than other emotions in the learning process [4–6]. Although confusion is an unpleasant emotion, addressing confusion during controllable periods has been shown to be beneficial for learning [7–9], as it promotes active student engagement in learning activities. However, research on learning confusion is still in its early stages and requires further exploration.

Electroencephalography (EEG) is considered a physiological indicator of the aggregated electrical activity of neurons in the human brain's cortex. EEG is employed to record such activities and, compared to non-physiological indicators like facial expressions and gestures, offers a relatively objective assessment of emotions, making it a reliable tool for emotion recognition [10].

Traditionally, the classification of EEG signals relies on manual feature extractors and machine learning classifiers [11], such as Naive Bayes, SVM, and Random Forest. Although deep-learning architectures are a more recent introduction, they have consistently improved performance [12]. Convolutional Neural Networks (CNNs) and Long Short-Term Memory Networks (LSTMs) are the primary architectures employed [13]. However, employing CNNs for feature extraction primarily focuses on local aspects, hindering temporal information perception. Although LSTM-based approaches exhibit commendable performance, they also struggle with global temporal representation. Various attempts with end-to-end hybrid networks [14] have been made. However, these endeavors have resulted in models with excessively intricate architectures, leading to sluggish convergence rates or even failures to converge. Furthermore, end-to-end methodologies lack the advantages of conventional feature extraction methods in representing EEG signals. The Transformer [15] has showcased its formidable capabilities in natural language processing (NLP), owing to its significant advantage in comprehending global semantics. However, its application in EEG systems is still an area that requires further exploration.

In light of the limitations of end-to-end network models, as well as the disadvantages of CNN and LSTM, and considering the advantages of traditional feature extraction methods and the Transformer network structure, this paper proposes a fusion framework combining feature extraction and self-attention mechanism. Precisely, The EEG signals are first sliced, and frequency-domain features are extracted. These features are then tokenized into temporal tokens. Subsequently, the self-attention mechanism of the Transformer encoder layer is employed to capture temporal correlations. Finally, the extracted features are integrated using a fully connected layer to derive classification results, subsequently subjected to confusion analysis. We summarize the contributions of this paper as follows:

1. We present a fusion framework that integrates the strengths of traditional feature extraction and deep learning to analyze confusion during the learning process. This framework enables targeted guidance by assessing the cognitive level of students.
2. By harnessing the robust capabilities of the multi-head self-attention mechanism, we capture global contextual representations of long EEG segments, which proves beneficial for predicting confusion emotions.
3. In both subject-dependent and subject-independent experiments, we compare our framework with traditional machine learning classifiers and end-to-end methods. The experimental results demonstrate the superiority of our framework.

The rest of this paper is organized as follows: Section 2 presents the related work about the topic. Section 3 describes the methods. Section 4 presents the experiments and discusses the results. Section 5 describes the conclusion and future research directions.

2. Related Work

Confusion in learning refers to feeling perplexed or uncertain while absorbing knowledge or solving problems. Given its shared attributes with emotions, it is a nascent study area, primarily exploring confusion's classification as an emotion or affective state. Confusion is deemed a cognitive emotion, indicating a state of cognitive imbalance [9,16]. Individuals are encouraged to introspect and deliberate upon the material to redress this imbalance and facilitate progress, enabling a more profound comprehension. Consequently, when confused, individuals tend to activate profound cognitive processes to pursue enhanced learning outcomes. The investigation into confusion within the learning context remains in its preliminary stages.

Using EEG to recognize human emotions during various activities, including learning, is an area currently being explored. Recent research has focused on using electroencephalography to study cognitive states and emotions for educational purposes. These studies focus on attention or engagement [17,18], cognitive load, and some basic emotions such as happiness and fear. For example, researchers [19] used an EEG-based brain-computer interface (BCI) to record EEG in the FP1 region to track changes in attention. By utilizing visual and auditory cues, such as rhythmic hand raising, adaptive proxy robots can help students shift their attention when their attention falls below a preset threshold. The research results indicate that this BCI can improve learning performance.

Most traditional EEG-based classification methods rely on two steps: feature extraction and classification, and emotion classification is no exception. Many researchers have focused on exploring effective EEG features for classification, and the advancement of machine learning methods and technologies has significantly contributed to the development of these traditional methods. There have been attempts using the Common Spatial Pattern (CSP) algorithm [20], such as the FBCSP algorithm [21], which filters signals through filter banks, computes CSP energy features for each signal output through time filters, and then selects and classifies these features. Despite enhancements to the original CSP method, these techniques solely focus on analyzing the CSP energy dimension, disregarding the incorporation of temporal contextual information. Kaneshiro et al. [11] proposed Principal Component Analysis (PCA), extracting feature vectors of specific sizes from minimally preprocessed EEG signals, followed by training a classifier based on Linear Discriminant Analysis (LDA). Karimi-Rouzbahani et al. [22] explored the discriminative power of many statistical and mathematical features, and their experiments on three datasets showed that multi-valued features like wavelet coefficients and the theta frequency band performed better. Zheng et al. [23] investigated the pivotal frequency bands and channels of multi-channel EEG data in emotion recognition. Jensen & Tesche [24] and Bashivan et al. [25] demonstrated through experiments that cortical oscillatory activity associated with memory operations primarily exists in the theta (4–7 Hz), alpha (8–13 Hz), and beta (13–30 Hz) frequency bands. The studies above utilize traditional machine learning classifiers to explore critical frequency bands and channels; nevertheless, traditional machine learning classifiers do not demonstrate any performance advantages. In addition, separately optimizing feature extraction and classifier could potentially result in suboptimal global optimization.

Compared to traditional methods, end-to-end deep networks eliminate the need for manual feature extraction. For most EEG applications, it has been observed that shallow models yield good results, while deep models might lead to performance degradation [12,13]. Especially for classification based on CNNs, despite the shallow architectures of CNNs with few parameters, they have been widely utilized: DeepConvNet [12], EEGNet [26], ResNet [27], and other variants [28]. However, due to the limitations imposed by kernel size, CNNs can learn features with local receptive fields. However, they cannot capture the crucial long-term dependencies for time series analysis. Furthermore, Recurrent neural networks(RNNs) and long short-term memory(LSTM) are introduced to capture the temporal features of EEG classification [29,30]. However, these models cannot be trained in parallel, and the dependencies calculated by hidden states quickly vanish after a few time steps, making it challenging to capture global temporal dependencies. Moreover, end-to-end methods insist on utilizing deep networks to learn from raw signals, often overlooking the advantages of manual feature extraction, and complex networks can lead to difficulties in model convergence.

3. Methods

Transformer [15] is an emerging neural network architecture that originated in machine translation tasks. In recent years, it has gained remarkable prominence in natural language processing. However, its application to emotion recognition based on EEG data remains an area requiring further research. In this paper, we combine feature extraction with

Transformer for EEG classification. Drawing on the idea of Transformer, we first extract local temporal and frequency features and then adopt the self-attention mechanism to capture global temporal features.

The overview of the proposed framework is depicted in Figure 1. It comprises three modules: preprocessing, multi-head self-attention, and a fully connected classifier. In the preprocessing module, the noise and artifacts of EEG signals are filtered out. Then, the temporal and frequency-domain information, encapsulating crucial local features, is extracted. Next, the multi-head self-attention module extracts long-term features by learning the global correlations between different temporal positions. Finally, utilizing the features extracted in the previous steps, which encapsulate spatial, frequency, and temporal information, a classifier composed of fully connected layers is adopted to output the classification results.

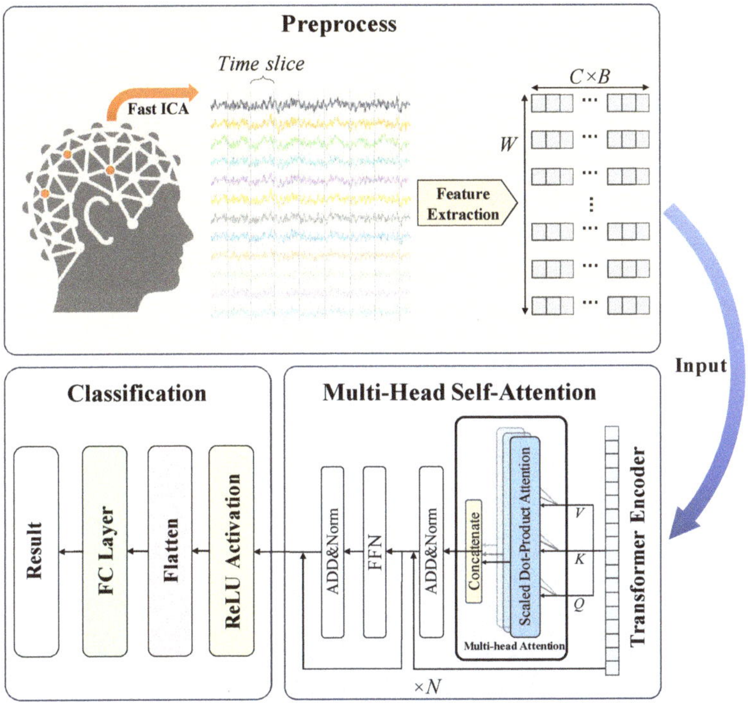

Figure 1. Overview of the proposed framework.

3.1. Preprocessing

Since the majority of EEG signals are concentrated within the range of 1 Hz to 50 Hz, a bandpass filter with a range of 1 Hz to 50 Hz was selected. This filtering procedure serves a dual purpose, eliminating low-frequency baseline drift, electrocardiographic (ECG) interference, and other high-frequency noise while effectively removing the most prominent power line interference (typically 50 Hz in China). Furthermore, EEG signals overlap with the electrooculogram (EOG) and electromyogram (EMG) signals in the frequency band. Therefore, relying solely on a single bandpass filter is insufficient to eliminate interference from EOG and EMG. This study adopted the fast, independent component analysis (fast ICA) to eliminate artifacts from EOG and EMG.

EEG comprises multiple time series corresponding to different spatial positions on the cerebral cortex where different electrodes are located on the collection device. Like audio signals, frequency-domain features are the most salient features. Thus, the spectrogram of the signals is typically employed for analysis. In frequency-domain analysis methods,

Power Spectral Density (PSD) analysis is a typically adopted method, and most previous studies have used this method to investigate epilepsy and hypnosis [31,32]. This method extracts frequency features that effectively detect cognitive and motor tasks [33]. Moreover, the PSD method consistently exhibits the highest robustness and effectiveness in extracting distinctive spectral patterns to differentiate motor imagery EEG signals accurately [34].

In this paper, Welch's method is used to extract the power spectrum features of EEG signals. The data sequence is applied to data windowing, producing modified periodograms [35]. For the EEG signal $x(n)$ of a certain channel, first divide it into L segments, with each segment having M sampling points, then the i-th small segment $x_i(n)$ can be denoted as:

$$x_i(n) = x(m + iM), 0 \leqslant m < M, 0 \leqslant i < L \tag{1}$$

take iD to be the point of start of the i th sequence. Then, L of length $2M$ represents data segments that are formed. The resulting output periodograms give:

$$\overset{\approx(i)}{P_{xx}}(f) = \frac{1}{MU}\left|\sum_{n=0}^{M-1} x_i(n)w(n)e^{-j2\pi fn}\right|^2 \tag{2}$$

Here, in the window function, U gives the normalization factor of the power and is chosen such that:

$$U = \frac{1}{M}\sum_{n=0}^{M-1} w^2(n) \tag{3}$$

where $w(n)$ is the window function. The average of these modified periodograms gives Welch's power spectrum as follows:

$$P_{xx}^{W} = \frac{1}{L}\sum_{i=0}^{L-1} \overset{\approx(i)}{P_{xx}}(f) \tag{4}$$

EEG signals predominantly reside within the 1 Hz to 50 Hz range, categorized into five frequency bands, depicted in Table 1. In this paper, the above method is applied to extract the PSD features of all channels and five frequency bands of EEG signals. Instead of manually selecting several of the five frequency bands for PSD feature extraction, the key features are extracted with the powerful ability of the transformer deep-learning network architecture. We divided each sample into 0.25 s slices. The sum of values within the five frequency bands is computed for each time slice and used as a separate measurement for each channel. This approach ensures that the extracted features encompass both frequency-domain and temporal information. PSD features for each time slice are extracted separately. This results in a sample with dimensions of $[W, B \times C]$, where W is the number of time slices, B is the number of frequency bands, and C is the number of channels. Figure 1 illustrates the above process.

3.2. Multi-Head Self-Attention

Due to the continuous nature of neural activity, the context-dependent representation between different time segments would contribute to EEG classification. This module uses self-attention to learn global temporal information of EEG features. The multi-head self-attention mechanism enables the model to attend to information from different representation subspaces from various channels. Each self-attention head, $h \in [1, 2, \ldots, H]$, with H being the total number of heads, relies on Q_c (queries), K_c (keys), V_c (values) vectors for token assessment. Within the features of each time segment, the frequency band features of different channels are concatenated sequentially. Given this, H is set to the number of channels, so The time slice representations are projected to latent representations of $Q_c, K_c, V_c \in \mathbb{R}^{1\times B}$, where $c \in [1, 2, \ldots, C]$. The "queries-keys" pair aims to map the key slices to the query slices based on their in-between representational similarity, calculated as their scaled dot-product, followed by SoftMax operation [15]. The resultant matrix is again multiplied with V_c to calculate the representational context as the aggregation of the self-attentional interactions. For a set of C slices and a single self-attention head h, the representational context is calculated as

$$Head(Q_c, K_c, V_c)_h = Softmax(\frac{Q_c \cdot K_c^{Transpose}}{\sqrt{B}}) \cdot V_c \tag{5}$$

Multiple projections of Q_c, K_c, and V_c calculate the respective self-attention heads, and their outputs become concatenated to form the aggregate of multiple heads, as given by

$$\begin{aligned} &MultiHeadAttention(Q_c^{[1,2,\ldots,H]}, K_c^{[1,2,\ldots,H]}, V_c^{[1,2,\ldots,H]}) \\ &= Concat(Head_1, Head_2, \ldots, Head_H) \end{aligned} \tag{6}$$

In this module, N multi-head self-attention layers are employed. Finally, the output vector of this module is flattened, and a fully connected layer is utilized as the classifier to obtain EEG classification results.

Table 1. Five frequency bands of EEG signals.

Bands	Frequencies	States	Examples	
delta	1–4 Hz	Sleep and dreaming	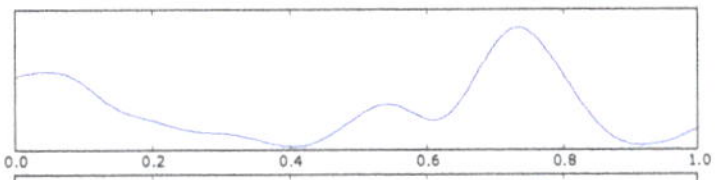	*slow*
theta	4–8 Hz	Deep relaxation or meditative states	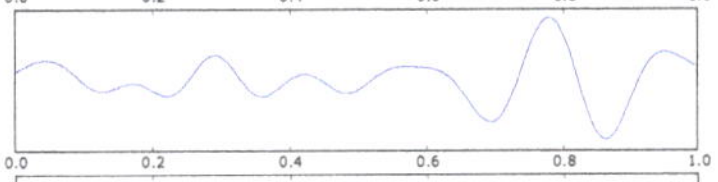	
alpha	8–14 Hz	Resting or relaxed	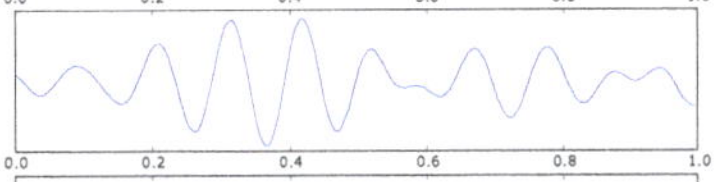	
beta	14–31 Hz	Alert, active mind	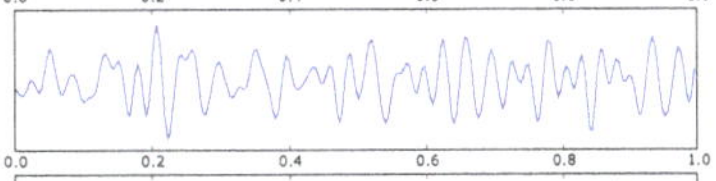	
gamma	31–50 Hz	Intense focus, problem solving	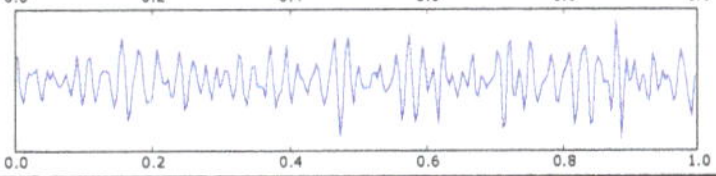	*fast*

4. Experiments and Discussions

4.1. Dataset

We utilize a publicly available dataset called CAL, designed explicitly for confusion analysis in learning. The CAL dataset is first used in [1]. It focuses on cognitive emotions during the learning process, including four categories (confused, non-confused, guess, and think-right). Raven's Progressive Matrices [36] is employed as confusion stimuli to design the experiment. A total of 25 subjects participated in this experiment. Subjects watch ten scene pictures, each of which lasts 10 s. Next, they view and perform 48 tests, each lasting a maximum of 15 s. There are 23 subjects' data obtained because the unexpected equipment problem caused a failed collection for two persons. The participants evaluate their level of confusion for each test item at the end of the trials. OpenBCI is employed as the EEG collector, which has eight channels (Fp1, Fp2, C3, C4, T5, T6, O1, O2) and a 250 Hz sampling rate, depicted in Figure 2. Table 2 summarizes the relevant information of the CAL dataset. Each trial is segmented with a non-overlapped three-second time window. Each segment is regarded as one data sample during the model training.

Table 2. Summary of information on the CAL dataset.

Emotion Categories	Emotion Stimuli	#Subjects	#Channels	Sampling Rate
confused, non-confused, think-right, guess	tests	23 male/female: 12/11	8	250 Hz

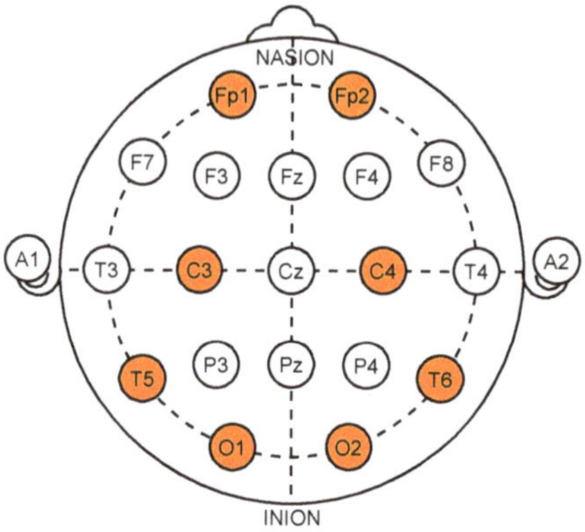

Figure 2. The eight channel locations (colored orange) of the CAL dataset, using the international 10–20 system.

4.2. Experiment Settings

We follow the settings outlined in [1] and conduct experiments involving binary classification (confused and non-confused) and four classification (confused, non-confused, guess, and think-right). In addition, we consider the following two scenarios to validate the proposed methods.

(1) subject-dependent: the data are trained across multiple subjects in the subject-dependent experiments. Specifically, 70% of the EEG data from all experiments for each participant are allocated as the training set, while the remaining 30% serve as the testing set.

(2) subject-independent: In the subject-independent model, the experiment emphasizes the differences between different subjects to test the method's generalization ability. Specifically, EEG data are divided into a cross-subject validation set with a split of 70%/30%, where the data from 16 subjects is used for training, and the data from the remaining seven subjects is used for testing.

To evaluate the classification performance of various methods, we consider the outcomes of conventional machine learning classifiers (Naive Bayes, SVM, and Random Forest) based on PSD features as presented in [1], as well as end-to-end methods (LSTM [30], ResNet [37], and EEGNet [26]). Furthermore, we conduct experiments using our approach without extracting PSD features to explore the benefits of feature extraction.

When extracting the samples of each category, we set a sliding window of 4 s to segment the data according to the setting in [1,38,39] to increase the sample size, i.e., the experimental samples of 4 s of EEG data. At the same time, to solve the data imbalance problem, we set overlapping parts of different lengths: 0.25 s for confused, 0.75 s for non-confused, 0.5 s for think-right, and 0.75 s for guess.

In this paper, the MNE [40] library in Python is adopted for data preprocessing operations. Our method is implemented with the PyTorch framework in Python 3.8 with an NVIDIA Geforce 3090 GPU. Using the same hyperparameter settings, we fix random seeds to repeat the experiment for different methods. We train the model using Adam optimizer with a learning rate of $1 \times e^{-4}$. The Adam optimizer combines the benefits of Momentum and RMSprop and adaptively adjusts the learning rate. $1 \times e^{-4}$ is a common starting learning rate. During training, batch size and dropout rates are set to 32 and 0.5.

The batch size is 32, which is a relatively small value to improve the model's generalization performance. A dropout rate of 0.5 is chosen, a common regularization technique that can reduce overfitting.

We set the number of multi-head self-attention layers N to 6, the number of heads H to the number of channels with a value of 14, and the dimension of the feed Forward layer to 2048. The number of parameters of the model in this configuration is 1M.

4.3. Analysis of Results

Tables 3 and 4 present the experimental results of different methods in subject-dependent and subject-independent experiments on the CAL dataset.

Table 3. Subject-dependent experiment of different methods on CAL dataset (Acc/F1 score). "w/" for "with" and "w/o" for "without". The best results are highlighted in **bold**.

Methods	Binary Classification	Four Classification
Naive Bayes w/ PSD	57.30/0.52	69.72/0.69
SVM w/ PSD	67.43/0.67	48.10/0.33
Random Forest w/ PSD	69.72/0.69	37.72/0.26
EEGNet	72.02/0.71	49.81/0.43
LSTM	73.45/0.73	53.29/0.49
ResNet	80.61/0.80	73.10/0.73
Our method w/o PSD	85.53/0.85	85.99/0.86
Our method	**90.49/0.90**	**87.59/0.87**

Table 4. Subject-independent experiment of different methods on CAL dataset (Acc/F1 score). "w/" for "with" and "w/o" for "without". The best results are highlighted in **bold**.

Methods	Binary Classification	Four Classification
Naive Bayes w/ PSD	60.59/0.55	40.83/0.27
SVM w/ PSD	55.71/0.53	38.05/0.31
Random Forest w/ PSD	55.98/0.52	35.94/0.25
EEGNet	64.46/0.61	39.14/0.20
LSTM	61.25/0.59	40.53/0.24
ResNet	57.95/0.55	40.05/0.23
Our method w/o PSD	65.85/0.63	40.88/0.39
Our method	**66.08/0.65**	**41.28/0.41**

The binary classification results of the subject-dependent experiments demonstrate that our approach significantly improves accuracy by 33.19%, 23.06%, and 20.77%, respectively, compared to conventional machine learning methods (Naive Bayes, SVM, and Random Forest). Additionally, we can observe that end-to-end deep-learning methods based on CNN, ResNet, and EEGNet perform well (with accuracies of 80.61% and 72.02%, respectively), indicating strong feature extraction capabilities of CNN-based methods. However, due to the limited receptive field of CNN, it struggles to capture global features. In contrast, the Transformer architecture based on the self-attention mechanism excels at capturing global information. Experimental results further confirm the advantages of the Transformer architecture: our method, based on Transformer, achieves an average accuracy increase of 18.47% and 9.88% compared to the CNN-based EEGNet and ResNet. Furthermore, we can observe that the end-to-end LSTM network performs impressively in the binary classification tasks of the subject-dependent experiments, achieving an accuracy of 73.45%. However, when faced with ultra-long time series data such as physiological signals, the end-to-end LSTM still loses information during training, leading to the inability to capture global features. In contrast, our Transformer-based approach can capture global temporal context information, resulting in a 17.04% improvement in accuracy compared to the end-to-end LSTM method.

In the subject-dependent experiments for the four-classification task, our method also exhibits strong performance with an accuracy of 87.59%. Compared to the best-performing traditional machine learning method, Naive Bayes, our method achieves an improvement of 17.87% in accuracy. Additionally, compared to the top-performing end-to-end method, ResNet, our method shows an accuracy improvement of 14.49%.

Moreover, in subject-independent experiments, our approach achieves the highest accuracy of 68.08% and 41.74% in binary and four-class classifications, respectively. Compared to other methods, our approach achieves an accuracy improvement of 1.62% in binary classification and 0.45% in four-class classification.

The F1 score, which is based on precision and recall, is also an important evaluation metric [41]. We present the F1 score of the binary and four-class classification tasks in Figures 3 and 4. Our approach achieved F1 scores of 0.90, 0.87, 0.65, and 0.41 in the experiments. Compared to the best-performing methods, our approach demonstrates improvements of 0.10, 0.14, 0.04, and 0.10, respectively.

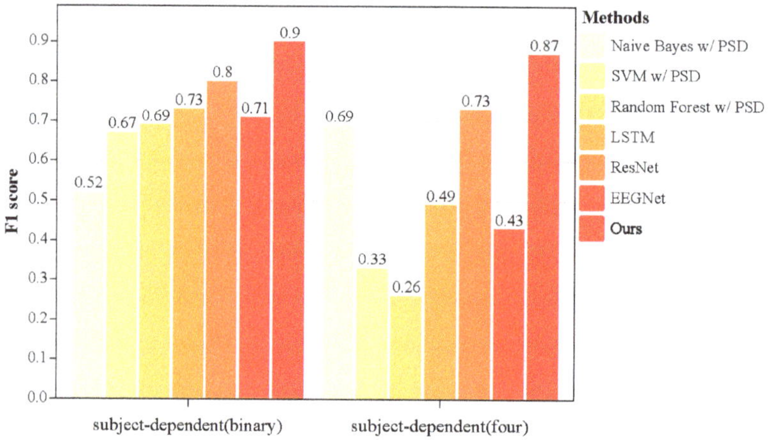

Figure 3. F1 score of different methods for the subject-dependent experiments.

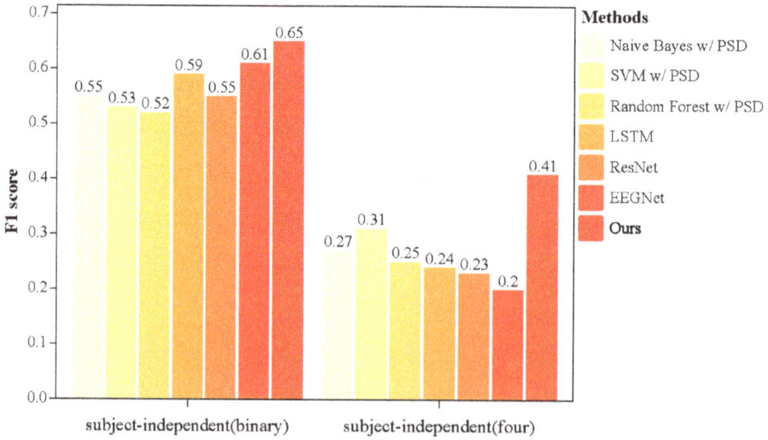

Figure 4. F1 score of different methods for the subject-independent experiments.

To evaluate the contribution of the feature extraction, we compare the performance of directly inputting the raw signal into the Transformer encoding layer without extracting

PSD features. Comparing the last two rows of Tables 3 and 4, it is found that there is a decrease in performance. These ablation study results indicate PSD's effectiveness in representing EEG. The results also demonstrate the effectiveness of the feature extraction part in our proposed framework.

We also provide the accuracy and loss during the model training process. In Figure 5, the accuracy and loss curves are plotted to visualize the model's performance during training. The accuracy curve shows how well the model can correctly classify the data, while the loss curve indicates the error the model makes during training. We can see that both the accuracy and loss are improving as the number of epochs increases. It suggests that the model is learning and becoming more accurate over time. The convergence point, reached at around 2000 epochs, indicates that the model has achieved a stable performance and that further training may not lead to significant improvements. Overall, the provided accuracy and loss curves demonstrate that the model performs well and converges satisfactorily after approximately 2000 training epochs.

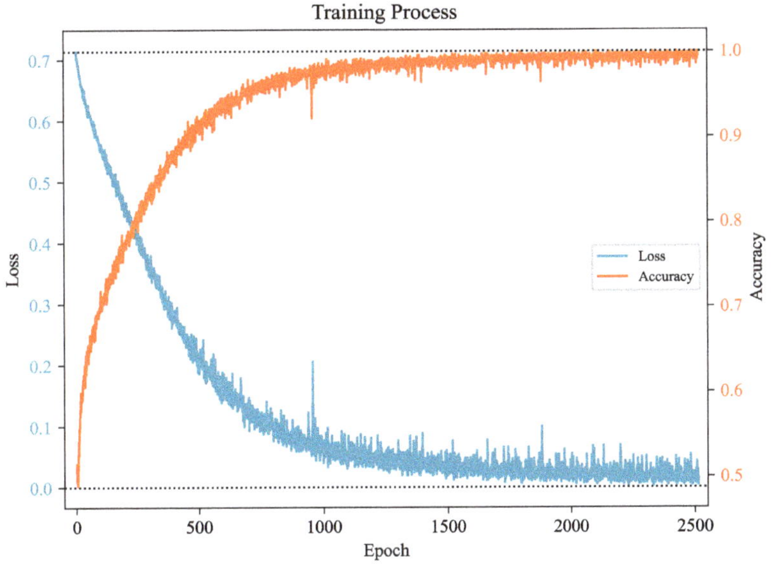

Figure 5. Accuracy and loss during the model training process.

The training time of the deep-learning model is also an important parameter. We compared the convergence time of all methods, as shown in Figure 6. Our study shows that our method has a significant advantage over the end-to-end deep-learning method in terms of training time. Our method offers a more efficient model that can save computing resources and time.

In addition to the faster training time, our method demonstrates better performance. The combination of faster training time and improved performance makes our method compelling for cognitive emotion identification. By reducing the computational burden and achieving better results, our method provides a practical and effective solution for this task. These advantages can have significant implications for real-world applications where efficiency and accuracy are crucial.

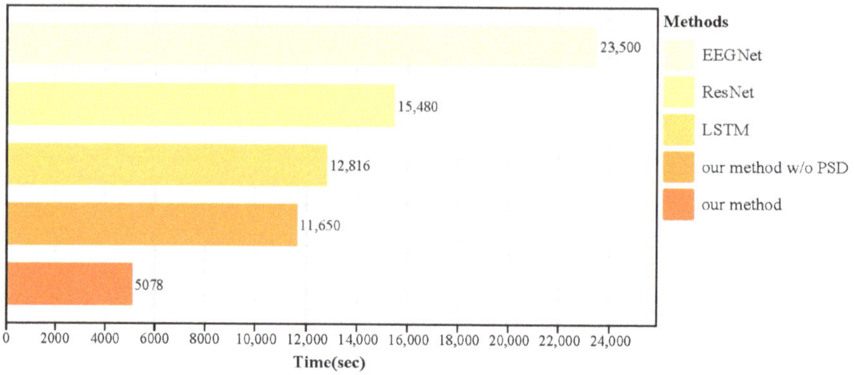

Figure 6. Training time for convergence of different methods.

4.4. Analysis of Confusion Matrix

We provide the confusion matrices of our approach for binary and four-class classification tasks for the subject-dependent and subject-independent experiments, as shown in Figures 7 and 8, respectively.

Our method performs well in identifying cognitive emotions during the subject-dependent experiment in binary and four-class setups. In the binary task, the model exhibits excellent discrimination between confused and non-confused emotions. In the four-class task, the recognition effect of confused emotions is slightly worse than that of the other three types of emotions. A total of 13% of the confused sample identified as think-right, suggesting that similar EEG patterns may be generated when subjects are confused or think-right. However, the identification performance is relatively poor in the subject-independent experiment due to the variations among different subjects. It is evident that in the case of binary classification, the model demonstrates good recognition of confused emotions among different subjects, but it struggles with identifying non-confused emotions. The same pattern can be observed in the four-class scenario, where the recognition performance is relatively better for confused emotions. This suggests that there is a certain degree of similarity in the EEG of different individuals when they are confused during learning, while there is a significant difference in the EEG of emotions other than confusion. These findings illustrate that addressing the differences among subjects poses a significant challenge.

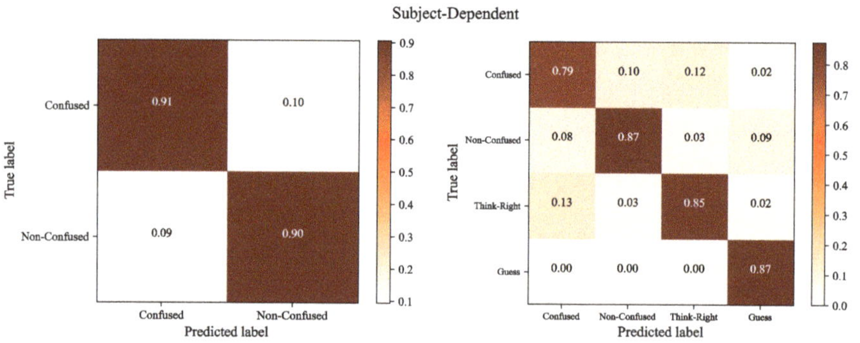

Figure 7. Confusion matrix of our method for the subject-dependent experiments.

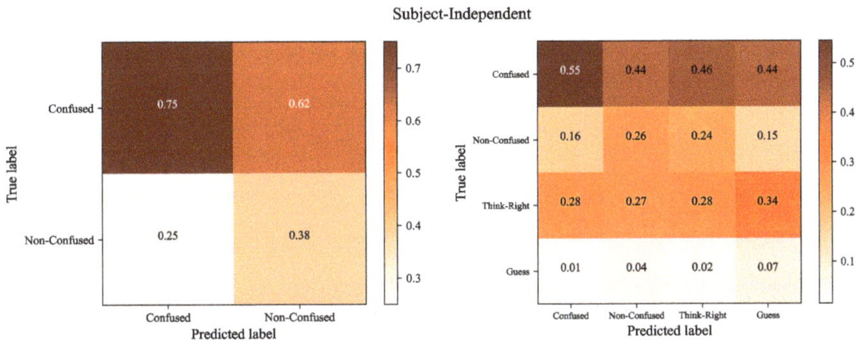

Figure 8. Confusion matrix of our method for the subject-independent experiments.

5. Conclusions and Future Work

In this work, we propose a fusion framework to analyze the presence of confusion in students during the learning process, aiming to expand the research on emotion recognition in education. This framework combines feature extraction and deep learning for analyzing confusion during learning. We first extract time-domain features with temporal information from EEG signals using a time-slicing approach and then utilize a multi-head self-attention mechanism to capture global-level temporal context representations. Extensive experiments conducted on the public dataset CAL demonstrate that our approach outperforms state-of-the-art methods in terms of performance and generalization ability. Furthermore, the effectiveness of our EEG representation approach can be extended to other physiological signal representations.

In future work, for the preprocessing module, we plan to extract multiple types of features to complement PSD features. For the subject-independent experiment, further research can focus on developing techniques that can effectively deal with the differences between different subjects to improve the universality and robustness of the model in identifying different individual cognitive emotions. It may involve collecting more diverse and representative data from more subjects. In addition, our work has certain limitations, including the small dataset size, the lack of additional physiological features, and any other potential constraints that may have influenced the findings. We plan to design experiments to collect multi-modal data (including EEG, ECG, facial expression, eye movement data, etc.) to build our dataset and explore emotion recognition in education by fusing multi-modal data.

Author Contributions: Conceptualization, C.Z. and Y.L.; methodology, C.Z.; visualization, Z.W.; validation, X.X.; formal analysis, X.X.; data curation, Z.W.; writing, C.Z. and Y.L.; supervision, J.H. and Y.L.; funding acquisition, Y.L. All authors have read and agreed to the published version of the manuscript.

Funding: This work is supported in part by the Beijing Postdoctoral Research Foundation (No. 2023-22-97) and the State Key Laboratory of Software Development Environment (SKLSDE-2022KF-10).

Institutional Review Board Statement: The study was approved by the administrators of the public dataset used in the article.

Informed Consent Statement: Informed consent was obtained from all subjects involved in the study.

Data Availability Statement: The data presented in this study are openly available in CAL, MNE at https://iopscience.iop.org/article/10.1088/1741-2552/acbfe0 and https://www.frontiersin.org/articles/10.3389/fnins.2013.00267/full [1,40].

Acknowledgments: We thank the Beijing Engineering Research Center for IoT Software and Systems for providing the experimental environment and equipment.

Conflicts of Interest: The authors declare no conflict of interest.

References

1. Xu, T.; Wang, J.; Zhang, G.; Zhang, L.; Zhou, Y. Confused or not: Decoding brain activity and recognizing confusion in reasoning learning using EEG. *J. Neural Eng.* **2023**, *20*, 026018. [CrossRef]
2. Peng, T.; Liang, Y.; Wu, W.; Ren, J.; Pengrui, Z.; Pu, Y. CLGT: A graph transformer for student performance prediction in collaborative learning. In Proceedings of the AAAI Conference on Artificial Intelligence, Washington, DC, USA, 7–14 February 2023; Volume 37, pp. 15947–15954.
3. Liang, Y.; Peng, T.; Pu, Y.; Wu, W. HELP-DKT: An interpretable cognitive model of how students learn programming based on deep knowledge tracing. *Sci. Rep.* **2022**, *12*, 4012. [CrossRef]
4. Baker, R.S.; D'Mello, S.K.; Rodrigo, M.M.T.; Graesser, A.C. Better to be frustrated than bored: The incidence, persistence, and impact of learners' cognitive–affective states during interactions with three different computer-based learning environments. *Int. J. Hum.-Comput. Stud.* **2010**, *68*, 223–241. [CrossRef]
5. Han, Z.M.; Huang, C.Q.; Yu, J.H.; Tsai, C.C. Identifying patterns of epistemic emotions with respect to interactions in massive online open courses using deep learning and social network analysis. *Comput. Hum. Behav.* **2021**, *122*, 106843. [CrossRef]
6. Lehman, B.; Matthews, M.; D'Mello, S.; Person, N. What are you feeling? Investigating student affective states during expert human tutoring sessions. In Proceedings of the International Conference on Intelligent Tutoring Systems, Montreal, QC, Canada, 23–27 June 2008; Springer: Berlin/Heidelberg, Germany, 2008; pp. 50–59.
7. Lehman, B.; D'Mello, S.; Graesser, A. Confusion and complex learning during interactions with computer learning environments. *Internet High. Educ.* **2012**, *15*, 184–194. [CrossRef]
8. D'Mello, S.; Lehman, B.; Pekrun, R.; Graesser, A. Confusion can be beneficial for learning. *Learn. Instr.* **2014**, *29*, 153–170. [CrossRef]
9. Vogl, E.; Pekrun, R.; Murayama, K.; Loderer, K.; Schubert, S. Surprise, curiosity, and confusion promote knowledge exploration: Evidence for robust effects of epistemic emotions. *Front. Psychol.* **2019**, *10*, 2474. [CrossRef] [PubMed]
10. Gunes, H.; Piccardi, M. Bi-modal emotion recognition from expressive face and body gestures. *J. Netw. Comput. Appl.* **2007**, *30*, 1334–1345. [CrossRef]
11. Kaneshiro, B.; Perreau Guimaraes, M.; Kim, H.S.; Norcia, A.M.; Suppes, P. A representational similarity analysis of the dynamics of object processing using single-trial EEG classification. *PLoS ONE* **2015**, *10*, e0135697. [CrossRef]
12. Schirrmeister, R.T.; Springenberg, J.T.; Fiederer, L.D.J.; Glasstetter, M.; Eggensperger, K.; Tangermann, M.; Hutter, F.; Burgard, W.; Ball, T. Deep learning with convolutional neural networks for EEG decoding and visualization. *Hum. Brain Mapp.* **2017**, *38*, 5391–5420. [CrossRef] [PubMed]
13. Roy, Y.; Banville, H.; Albuquerque, I.; Gramfort, A.; Falk, T.H.; Faubert, J. Deep learning-based electroencephalography analysis: A systematic review. *J. Neural Eng.* **2019**, *16*, 051001. [CrossRef] [PubMed]
14. Yang, Y.; Wu, Q.; Qiu, M.; Wang, Y.; Chen, X. Emotion recognition from multi-channel EEG through parallel convolutional recurrent neural network. In Proceedings of the IEEE 2018 International Joint Conference on Neural Networks (IJCNN), Rio de Janeiro, Brazil, 8–13 July 2018; pp. 1–7.
15. Vaswani, A.; Shazeer, N.; Parmar, N.; Uszkoreit, J.; Jones, L.; Gomez, A.N.; Kaiser, Ł.; Polosukhin, I. Attention is all you need. *Adv. Neural Inf. Process. Syst.* **2017**, *30* , 5998–6008.
16. Xu, T.; Zhou, Y.; Wang, Z.; Peng, Y. Learning emotions EEG-based recognition and brain activity: A survey study on BCI for intelligent tutoring system. *Procedia Comput. Sci.* **2018**, *130*, 376–382. [CrossRef]
17. Huang, J.; Yu, C.; Wang, Y.; Zhao, Y.; Liu, S.; Mo, C.; Liu, J.; Zhang, L.; Shi, Y. FOCUS: Enhancing children's engagement in reading by using contextual BCI training sessions. In Proceedings of the SIGCHI Conference on Human Factors in Computing Systems, Toronto, ON, Canada, 26 April–1 May 2014; pp. 1905–1908.
18. Xu, T.; Wang, X.; Wang, J.; Zhou, Y. From textbook to teacher: An adaptive intelligent tutoring system based on BCI. In Proceedings of the 2021 43rd Annual International Conference of the IEEE Engineering in Medicine & Biology Society (EMBC), Guadalajara, Mexico , 1–5 November 2021; pp. 7621–7624.
19. Xu, J.; Zhong, B. Review on portable EEG technology in educational research. *Comput. Hum. Behav.* **2018**, *81*, 340–349. [CrossRef]
20. Ramoser, H.; Muller-Gerking, J.; Pfurtscheller, G. Optimal spatial filtering of single trial EEG during imagined hand movement. *IEEE Trans. Rehabil. Eng.* **2000**, *8*, 441–446. [CrossRef] [PubMed]
21. Ang, K.K.; Chin, Z.Y.; Wang, C.; Guan, C.; Zhang, H. Filter bank common spatial pattern algorithm on BCI competition IV datasets 2a and 2b. *Front. Neurosci.* **2012**, *6*, 39. [CrossRef] [PubMed]
22. Karimi-Rouzbahani, H.; Shahmohammadi, M.; Vahab, E.; Setayeshi, S.; Carlson, T. Temporal codes provide additional category-related information in object category decoding: A systematic comparison of informative EEG features. *bioRxiv* **2020**. . [CrossRef]
23. Zheng, W.L.; Lu, B.L. Investigating critical frequency bands and channels for EEG-based emotion recognition with deep neural networks. *IEEE Trans. Auton. Ment. Dev.* **2015**, *7*, 162–175. [CrossRef]
24. Jensen, O.; Tesche, C.D. Frontal theta activity in humans increases with memory load in a working memory task. *Eur. J. Neurosci.* **2002**, *15*, 1395–1399. [CrossRef]
25. Bashivan, P.; Bidelman, G.M.; Yeasin, M. Spectrotemporal dynamics of the EEG during working memory encoding and maintenance predicts individual behavioral capacity. *Eur. J. Neurosci.* **2014**, *40*, 3774–3784. [CrossRef]
26. Lawhern, V.J.; Solon, A.J.; Waytowich, N.R.; Gordon, S.M.; Hung, C.P.; Lance, B.J. EEGNet: A compact convolutional neural network for EEG-based brain–computer interfaces. *J. Neural Eng.* **2018**, *15*, 056013. [CrossRef]

27. Tian, T.; Wang, L.; Luo, M.; Sun, Y.; Liu, X. ResNet-50 based technique for EEG image characterization due to varying environmental stimuli. *Comput. Methods Programs Biomed.* **2022**, *225*, 107092. [CrossRef]
28. Kalafatovich, J.; Lee, M.; Lee, S.W. Decoding visual recognition of objects from eeg signals based on attention-driven convolutional neural network. In Proceedings of the 2020 IEEE International Conference on Systems, Man, and Cybernetics (SMC), Toronto, ON, Canada, 11–14 October 2020; pp. 2985–2990.
29. Chowdary, M.K.; Anitha, J.; Hemanth, D.J. Emotion recognition from EEG signals using recurrent neural networks. *Electronics* **2022**, *11*, 2387. [CrossRef]
30. Lu, P. Human emotion recognition based on multi-channel EEG signals using LSTM neural network. In Proceedings of the IEEE 2022 Prognostics and Health Management Conference (PHM-2022 London), London, UK, 27–29 May 2022; pp. 303–308.
31. Fraiwan, L.; Lweesy, K.; Khasawneh, N.; Wenz, H.; Dickhaus, H. Automated sleep stage identification system based on time–frequency analysis of a single EEG channel and random forest classifier. *Comput. Methods Programs Biomed.* **2012**, *108*, 10–19. [CrossRef] [PubMed]
32. Deivanayagi, S.; Manivannan, M.; Fernandez, P. Spectral analysis of EEG signals during hypnosis. *Int. J. Syst. Cybern. Informatics* **2007**, *4*, 75–80.
33. Brodu, N.; Lotte, F.; Lécuyer, A. Exploring two novel features for EEG-based brain–computer interfaces: Multifractal cumulants and predictive complexity. *Neurocomputing* **2012**, *79*, 87–94. [CrossRef]
34. Duan, L.; Zhong, H.; Miao, J.; Yang, Z.; Ma, W.; Zhang, X. A voting optimized strategy based on ELM for improving classification of motor imagery BCI data. *Cogn. Comput.* **2014**, *6*, 477–483. [CrossRef]
35. Faust, O.; Acharya, R.; Allen, A.R.; Lin, C. Analysis of EEG signals during epileptic and alcoholic states using AR modeling techniques. *Irbm* **2008**, *29*, 44–52. [CrossRef]
36. Raven, J. The Raven's progressive matrices: Change and stability over culture and time. *Cogn. Psychol.* **2000**, *41*, 1–48. [CrossRef]
37. Wang, P.; Guo, C.; Xie, S.; Qiao, X.; Mao, L.; Fu, X. EEG emotion recognition based on knowledge distillation optimized residual networks. In Proceedings of the 2022 IEEE 6th Advanced Information Technology, Electronic and Automation Control Conference (IAEAC), Beijing, China, 3–5 October 2022; pp. 574–581.
38. Zheng, W.L.; Liu, W.; Lu, Y.; Lu, B.L.; Cichocki, A. Emotionmeter: A multimodal framework for recognizing human emotions. *IEEE Trans. Cybern.* **2018**, *49*, 1110–1122. [CrossRef]
39. Liu, W.; Qiu, J.L.; Zheng, W.L.; Lu, B.L. Comparing recognition performance and robustness of multimodal deep learning models for multimodal emotion recognition. *IEEE Trans. Cogn. Dev. Syst.* **2021**, *14*, 715–729. [CrossRef]
40. Gramfort, A.; Luessi, M.; Larson, E.; Engemann, D.A.; Strohmeier, D.; Brodbeck, C.; Goj, R.; Jas, M.; Brooks, T.; Parkkonen, L.; et al. MEG and EEG data analysis with MNE-Python. *Front. Neurosci.* **2013**, *7*, 267. [CrossRef] [PubMed]
41. Powers, D. Evaluation: From Precision, Recall and F-Measure to ROC, Informedness, Markedness & Correlation. *J. Mach. Learn. Technol.* **2011**, *2*, 37–63.

applied sciences

MDPI

Article

A Parallel Open-World Object Detection Framework with Uncertainty Mitigation for Campus Monitoring

Jian Dong [1,2,†], Zhange Zhang [1,†], Siqi He [3], Yu Liang [4], Yuqing Ma [1,5,*], Jiaqi Yu [6], Ruiyan Zhang [6] and Binbin Li [2]

1 State Key Lab of Software Development Environment, Beihang University, Beijing 100191, China; dongjian@cesi.cn (J.D.); zhangezhang@buaa.edu.cn (Z.Z.)
2 China Electronics Standardization Institute, Beijing 100007, China; libb@cesi.cn
3 School of Computer Science, Peking University, Beijing 100871, China; hesq47@foxmail.com
4 Faculty of Information Technology, Beijing University of Technology, Beijing 100124, China; yuliang@bjut.edu.cn
5 Institute of Artificial Intelligence, Beihang University, Beijing 100191, China
6 Beijing Institute of Control and Electronic Technology, Beijing 100038, China; yujq2004@163.com (J.Y.); zhangruiyan65@foxmail.com (R.Z.)
* Correspondence: mayuqing@buaa.edu.cn
† These authors contributed equally to this work.

Abstract: The recent advancements in artificial intelligence have brought about significant changes in education. In the context of intelligent campus development, target detection technology plays a pivotal role in applications such as campus environment monitoring and the facilitation of classroom behavior surveillance. However, traditional object detection methods face challenges in open and dynamic campus scenarios where unexpected objects and behaviors arise. Open-World Object Detection (OWOD) addresses this issue by enabling detectors to gradually learn and recognize unknown objects. Nevertheless, existing OWOD methods introduce two major uncertainties that limit the detection performance: the unknown discovery uncertainty from the manual generation of pseudo-labels for unknown objects and the known discrimination uncertainty from perturbations that unknown training introduces to the known class features. In this paper, we introduce a Parallel OWOD Framework with Uncertainty Mitigation to alleviate the unknown discovery uncertainty and the known discrimination uncertainty within the OWOD task. To address the unknown discovery uncertainty, we propose an objectness-driven discovery module to focus on capturing the generalized objectness shared among various known classes, driving the framework to discover more potential objects that are distinct from the background, including unknown objects. To mitigate the discrimination uncertainty, we decouple the learning processes for known and unknown classes through a parallel structure to reduce the mutual influence at the feature level and design a collaborative open-world classifier to achieve high-performance collaborative detection of both known and unknown classes. Our framework provides educators with a powerful tool for effective campus monitoring and classroom management. Experimental results on standard benchmarks demonstrate the framework's superior performance compared to state-of-the-art methods, showcasing its transformative potential in intelligent educational environments.

Keywords: intelligent campus monitoring; uncertainty mitigation; objectness-driven object discovery; parallel network; open-world object detection; artificial intelligence

1. Introduction

The continuous advancement of deep learning [1–4] technology has enabled the widespread application of artificial intelligence (AI) in multiple domains [5–7], with the field of education as a notably impacted area. Object detection [8–12] stands out as one of the key technologies driving this transformation in education, which plays a significant role in shaping intelligent campus and can be harnessed for various purposes, including campus monitoring and class behavior detection. Zhou et al. [8] design an object detection algorithm to detect the raising hand action in real classroom scenarios. Zhao et al. [9] propose CNPH-Net with a multiscale feature extraction module to enhance the capability of small object detection in classroom scenarios. Xu et al. [13] applied object detection and gaze-tracking technology to analyze the students' attention in the classroom. These methods are all based on traditional closed-set object detection techniques, meaning that the models can only detect objects that are already present in the training set. However, due to the openness and diversity of objects in real campus and classroom scenarios, instances of objects and behaviors not included in the training datasets may arise. Traditional object detection methods are ill-equipped to detect such unexpected objects, making it challenging to meet the demands of real-world campus monitoring.

To address the challenges of detection in open environments, such as campuses and classrooms, a new object detection task, Open-World Object Detection (OWOD) [14], is proposed to focus on the detection of objects in real-world open environments. This problem requires detection models not only to detect known objects but also to discover and incrementally learn about previously unseen unknown objects during the training phase. In the classroom and similar education scenarios, the OWOD models are required to discover and locate unknown targets and behaviors, and subsequently engage in incremental learning, which helps to facilitate intelligent monitoring of the teaching process.

Previous OWOD works mainly realized unknown class detection by designing complex unknown proposal generation mechanisms. The ORE method [14] incorporates an unknown object-aware Region Proposal Network (RPN), selecting top-k region proposals sorted by the objectness of the RPN from background as unknown objects, thereby endowing the model with the ability to detect unknown objects. Zhao et al. [15] further introduces an auxiliary proposal advisor that combines an unknown-aware RPN with the selective search method to generate more accurate pseudo-labels for unknown categories. Building on the recent success of closed-set detector D-DETR [16], OW-DETR [17] proposes an unknown pseudo-labeling method that selects query boxes with high attention scores which not match with the ground truth of known class.

However, in these OWOD works, two major uncertainties that limit the detection performance are commonly observed: the unknown discovery uncertainty and the known discrimination uncertainty. The unknown discovery uncertainty indicates the uncertain bias from the manual generation of pseudo-labels for unknown objects. In the previous OWOD works, the unknown regions were largely derived from high-score region proposals generated by RPN, which may potentially cover some background areas that introduce great uncertain bias. The known discrimination uncertainty arises from the uncertain perturbations that the learning of unknown classes introduces to the features of known class discrimination. The differentiation among known classes primarily relies on discriminative features, whereas the learning of unknown classes guides the model to extract more generic features among objects, which are only distinguished from the background. Therefore, while granting the model the ability to detect unknown classes, previous methods often resulted in an uncertain perturbation in the discrimination performance of known classes. These two kinds of uncertainty have impeded further improvements in detection performance in open-world scenarios, making it challenging to apply them to real-world environments like campus.

In this paper, we introduce a Parallel OWOD Framework to mitigate the unknown discovery uncertainty and the known discrimination uncertainty within the OWOD task. Specifically, the framework employs a parallel architecture to decouple the process of learning unknown classes from the detection of known classes, thereby preventing the influence of uncertain features learned through unknown guidance on the known classes. Through the parallel architecture, our framework successfully mitigates the known discrimination uncertainty during open-world training. To alleviate the unknown discovery uncertainty, we have designed and added an objectness-driven discovery module to replace the uncertain unknown object selection, driving the model to focus on the objectness of various objects derived from the deterministic supervision of known classes. Therefore, the Objectness-Driven Discovery Module can learn objectness scores from stable supervision of known classes. For all objects, including both known and unknown classes, their objectness scores are expected to be significantly higher than those of the background regions. Consequently, the objectness-driven discovery module is capable of identifying potential objects with objectness more certainly from the background, including unknown objects. Furthermore, the output from the objectness-driven discovery module and a close-world known object detector is subsequently reconciled via a collaborative open-world classifier, which eliminates redundant unknown detection instances and facilitates collaborative learning of known and unknown classes. As shown in Figure 1, this framework may provide a powerful tool for educators and administrators to effectively monitor campus and manage classroom behaviors.

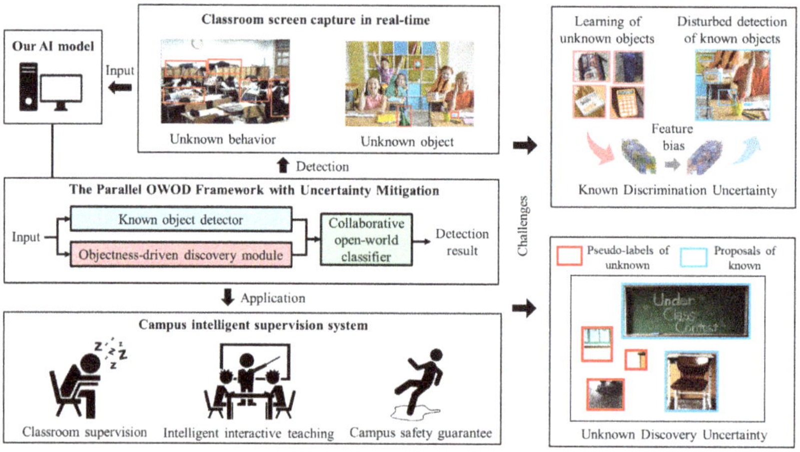

Figure 1. Introduction of our parallel OWOD framework. Our framework will provide a powerful tool for educators and administrators to effectively campus monitoring. The framework effectually mitigates the unknown discovery uncertainty and the known discrimination uncertainty in the OWOD task.

We summarize our contributions as follows:

(1) We propose a Parallel OWOD Framework comprised of two different detectors to mitigate the unknown discovery uncertainty and the known discrimination uncertainty within OWOD, which can be used for effective management and monitoring of campus to ensure a conducive and productive learning atmosphere.

(2) An objectness-driven discovery module is trained to guide the model to focus on the objectness of various objects derived from deterministic supervision of known classes, which could help alleviate the unknown discovery uncertainty.

(3) A known object detector and a collaborative open-world classifier are designed to accomplish the collaborative learning of known and unknown classes, helping mitigate the known discrimination uncertainty.

(4) Experiments on common benchmarks demonstrate that our framework shows superior results compared to state-of-the-art methods.

The upcoming sections of this article will delve into our parallel open-world object detection framework designed for campus monitoring, with a specific focus on uncertainty mitigation. Section 2 offers a review of existing methods for object detection in campus scenarios, open-set object detection, and open-world object detection. Following that, Section 3 details our proposed parallel open-world object detection framework, including its components and training process. In Section 4, we present the experimental evaluation of the proposed framework on standard benchmarks, as well as an ablation study, a parameter analysis, and a visualization of the results. Finally, Section 5 summarizes our findings, outlines limitations, and explores potential avenues for future research.

2. Related Work

The continual progress in deep learning technologies has extensively applied AI across diverse domains, such as autonomous driving [18–20], medicine [21–23], and intelligent education. In the field of intelligent education, artificial intelligence plays a pivotal role in predicting students' performance [24,25], comprehending how students learn [26], detecting their behaviors [27], and fulfilling a series of other educational requirements. Notably, the development of object detection technology has elevated intelligent campus monitoring to a notable research and application domain within the field of intelligent education. In order to achieve intelligent campus monitoring, previous efforts were primarily based on the application of traditional closed-set object detection techniques in educational settings such as classrooms. On the other hand, to handle the challenge in the real-world object detection that an object detector may encounter several unknown objects which not appear in the training process, previous approaches have explored open-set and open-world settings. In this section, we discuss related work on object detection in campus scenarios, as well as work related to open-set and open-world target detection.

2.1. Object Detection in Campus Scenarios

With the advancement of deep learning, object classification and detection techniques have found application in various domains [28,29]. Existing object detection approaches in campus scenarios often involve the direct application of traditional closed-set detection models to the classroom environment. Ref. [9] proposes a single-stage object detector called CBPH-Net and designs a feature extraction module to capture more channel information and relevant features to enhance the multiscale recognition capability in classrooms. Ref. [8] proposes an automatic hand-raiser recognition algorithm to show who raises their hands in real classroom scenarios, which is of great importance for further analyzing the learning states of individuals. Ref. [13] proposes a fusion model based on gaze tracking and object detection to intelligently analyze the students' attention in the classroom from the first-person perspective and promote teachers' precise teaching and students' personalized learning. Although these methods successfully apply object detection to campus scenarios, they lack the capability to detect and learn from unknown targets, rendering them inadequate for meeting the detection requirements within the open environments of campus and classrooms.

2.2. Open-Set Object Detection

In the open-set setting, the incomplete information extracted from the training set causes the model couldn't classify the unknown categories that are not encountered during the training phase. Based on several assumptions, a number of previous works [30–35] attempted to step into open-set tasks. A metric learning framework [36] and a model with OpenMax classifier [37] have been proposed to address open-set classification. Some works also exploited self-supervised learning [38] and unsupervised learning [39] to handle open-set classification tasks. Open-set object detection protocol was first proposed in [40], followed by works [41–43] that improved detection performance by measuring object uncertainties. OpenDet [44] designed a manual unknown-discover strategy based on feature density to enhance the identification of unknown proposals.

2.3. Open-World Object Detection

Compared with open-set tasks, open-world tasks are more demanding since these tasks require not only identifying unknown classes, but also learning incrementally based on newly obtained category data. Open-world tasks go beyond open-set by requiring incremental learning based on newly obtained category data. Bendale et al. [45] proposed the first open-world classification model, and Xu et al. [46] introduced a meta-learning method for identifying unknown classes. Previous works [36,47,48] explored more complex scenarios, e.g., long-tail distribution [49], few-shot learning [50] and zero-shot learning [51] on the open-world classification, respectively.

ORE, the first OWOD method proposed by Joseph et al. [14] designed a specific RPN, selecting top-k region proposals sorted by the objectness of the RPN from the background to discover unknown objects. SA proposed by Yang et al. [52] defined semantic centroids in feature space to exploit embedded semantic topology. Wu et al. [53] designed a two-stage detector distinguishing diverse classes on the basis of clustering and similarity to handle the raised Unknown-Classified OWOD problem. Zhao et al. [15] proposed a model coupling an auxiliary proposal advisor and a class-specific expelling classifier to improve the detection performance of unknown classes. Inspired by the powerful presentation ability of transformer [54], Gupta et al. [17] first extended transformer to OWOD problem and proposed a transformer-based open-world detector OW-DETR, which combines an attention-driven pseudo-labeling scheme for selecting unknown query boxes and an objectness branch to effectively separate foreground objects and background.

Previous methods used complex strategies to generate pseudo-labels for unknown classes but introduced excessive uncertainties, harming the learning of unknown objects and influencing the known classification as well. In contrast, our method only explores the unknown information from the known instances through a reasonable decoupling framework, which improves the detection performance of the unknown objects while maintaining that of the known objects.

3. Method

In this section, we introduce our parallel OWOD framework which is designed to mitigate the detection uncertainties of known and unknown objects in open environments such as campuses and classrooms, with the primary objective of facilitating campus monitoring. We begin by analyzing the task of open-world object detection in Section 3.1. Subsequently, we provide a comprehensive overview of the proposed framework's structure in Section 3.2. Finally, Section 3.3 outlines the training process of our framework.

3.1. Problem Analysis

In the setting of open-world object detection, the whole training process is separated into T tasks. At any task $t \in \{1, ..., T\}$, we consider the set of known classes as $\mathcal{K}^t$ and the set of unknown classes as $\mathcal{U}^t$, where $\mathcal{K}^t \cap \mathcal{U}^t = \varnothing$. During the training phase at task t, the annotations of known classes $\mathcal{K}^t$ are accessible to ensure the model has the capacity to classify the known instances into correct classes accurately, i.e., an instance $\mathbf{x}_i^{(\mathcal{K}^t)}$ of class $c \in \mathcal{K}^t$ is annotated as $\mathbf{y}_i^{(\mathcal{K}^t)} = [l_i^{(\mathcal{K}^t)}, b_i^{(\mathcal{K}^t)}]$, where $l_i^{(\mathcal{K}^t)} = c \in \mathcal{K}^t$ denotes corresponding label and $b_i^{(\mathcal{K}^t)} = [x, y, w, h]$ denotes the coordinates of corresponding bounding box. In contrast, the unknown classes $\mathcal{U}^t$ appear during the training phase without annotations but are required to be classified as an *unknown* denoted by label $l_i^{(\mathcal{K}^t)} = c \in \mathcal{K}^t$ when inferring.

For incremental learning, part of the unknown classes $\bar{\mathcal{U}}^t \in \mathcal{U}^t$ are labeled and update the known classes as $\mathcal{K}^{t+1} = \mathcal{K}^t \cup \bar{\mathcal{U}}^t$ and unknown classes as $\mathcal{U}^{t+1} = \mathcal{U}^t \backslash \bar{\mathcal{U}}^t$ at task $t + 1$. The model adaptively updates itself with new knowledge and identifies current known classes $\mathcal{K}^{t+1}$ accurately while classifying current unknown classes $\mathcal{U}^{t+1}$. This cycle continues over the life of the object detector.

3.2. Parallel OWOD Framework with Uncertainty Mitigation

Our parallel OWOD framework is based on the state-of-the-art object detection model D-DETR [16]. As illustrated in Figure 2, our framework comprises an objectness-driven discovery module, a known object detector, and a collaborative open-world classifier. The framework utilizes a parallel architecture to decouple the learning processes to address the challenges of uncertainties within OWOD. The objectness-driven discovery module converts all known objects into the same class and guides the model to focus on the objectness derived from the supervision of different objects, thereby alleviating the unknown discovery uncertainty. The objectness-driven discovery module effectively aids in the discovery of more potential objects distinct from the background including unknown objects. The known object detector retains the fundamental approach of D-DETR to ensure precise detection performance for known classes. Detection of known and unknown classes is carried out separately by different sub-modules within the parallel architecture, which ensures that the learning of known and unknown classes does not interfere with each other, effectively alleviating the known discrimination uncertainty.

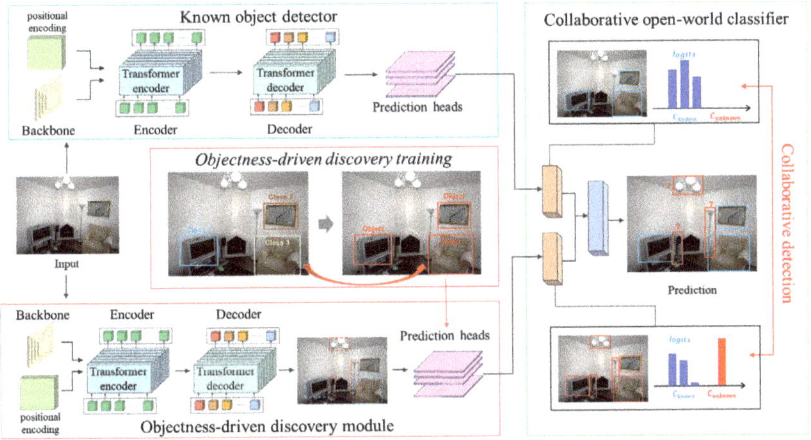

Figure 2. Architecture of our parallel OWOD framework. The parallel OWOD framework comprises an objectness-driven discovery module, a known object detector, and a collaborative open-world classifier.

During the inference phase, the image is separately processed by both the objectness-driven discovery module and the known object detector. The outputs are then integrated through the collaborative open-world classifier, facilitating cooperative learning for both known and unknown classes. The framework ultimately leads to strong detection performance for both known and unknown classes.

3.2.1. Objectness-Driven Discovery Module

To address the challenge posed by the unknown discovery uncertainty introduced from the strategy of unknown pseudo-labeling, we introduced an objectness-driven discovery method to learn the unknown objects certainly from the deterministic supervision of known classes.

Since both unknown and known classes inherently share the common characteristic of being "generalized objects", we can learn objectness from known classes to guide the learning of unknown classes. We simply convert all known object classes to a generalized class and only distinguish them from background regions without objects, extracting generalized features from all known classes while attenuating discriminative features to enhance the model's potential for object discovery. Through this approach, the detector can not only detect known objects but also discover other unknown objects with universities that are similar to known ones. Compared to previous methods for unknown object discovery, our objectness-driven discovery module extracts stable objectness from the supervised information of known classes, rather than manually selecting potential regions for unknown objects from the background. Therefore, our framework effectively reduces the uncertainties and enhances the detection performance of unknown objects.

Specifically, during the training phase of the detector, the instances of all known classes except the background region are converted to a universal class "generalized object". Given the i-th instance of known classes $\mathbf{x}_i^{(\mathcal{K}^t)}$ with its annotations $\mathbf{y}_i^{(\mathcal{K}^t)}$, the supervision label used for classification regularization could be denoted as $\widetilde{\mathbf{y}}_i$:

$$\widetilde{\mathbf{y}}_i = \mathbf{y}_i^{(\mathcal{G}^t)} = [l^{(\mathcal{G}^t)}, b_i^{(\mathcal{K}^t)}] \tag{1}$$

where $\mathbf{y}_i^{(\mathcal{G}^t)}$ denotes the generalized annotation of the instance after convert, and $l^{(\mathcal{G}^t)} = o$ denotes the universal label "generalized object" of every known instance. The classification regularization for the objectness-driven discovery module can be presented as follows:

$$\begin{aligned} \widetilde{\ell}^{cls}_{\text{universal}} &= -\sum_i \widetilde{\mathbf{y}}_i \log \mathbf{x}_i^{(\mathcal{K}^t)} \\ &= -\sum_i \mathbf{y}_i^{(\mathcal{G}^t)} \log \mathbf{x}_i^{(\mathcal{K}^t)} \end{aligned} \tag{2}$$

where $\widetilde{\ell}^{cls}_{\text{universal}}$ is the loss of classification regularization calculated by the universal label for the objectness-driven discovery module.

Through the objectness-driven discovery module, our framework is capable of identifying generalized objects possessing the attribute of "being an object" in the input image, including numerous unknown objects that have never been encountered in the training phase.

3.2.2. Known Object Detector

To mitigate the known discrimination uncertainty, a known object detector built upon the D-DETR [16] is employed to operate in parallel with the objectness-driven discovery module to accurately detect known classes. In order to preserve the precise detection performance of known category objects in the framework. The Known Object Detector is designed to learn discriminative features that are as relevant as possible to the categories, enabling it to distinguish between different known categories effectively.

Specifically, during the training phase of the known object detector, we treat the known proposals as their origin label. The supervision label used for classification regularization of known proposal i could be denoted as $\mathbf{y}_i$:

$$\mathbf{y}_i = \mathbf{y}_i^{(\mathcal{K}^t)} \tag{3}$$

The classification regularization for the known object detector can be presented as follows:

$$\begin{aligned} \ell^{cls}_{\text{known}} &= -\sum_i \mathbf{y}_i \log \mathbf{x}_i^{(\mathcal{K}^t)} \\ &= -\sum_i \mathbf{y}_i^{(\mathcal{K}_i^t)} \log \mathbf{x}_i^{(\mathcal{K}^t)} \end{aligned} \tag{4}$$

where $\ell^{cls}_{\text{known}}$ is the loss of classification regularization for the known object detector, and $\mathbf{y}^{(\mathcal{K}^t)_i}$ denotes original label of proposal i.

At each task, the known object detector is independently trained. During testing, the bounding boxes and logits produced by the Known Object Detector are input into the Collaborative open-world Classifier to facilitate the fusion of results with the detection of unknown categories.

3.2.3. Collaborative Open-World Classifier

To accomplish the collaborative learning of known classes and unknown classes, we introduce a collaborative open-world classifier to integrate the output of the two detectors of our framework and eliminate the impact of learning unknown classes on the detection performance of known classes.

Considering that the generalized objects logits predicted by the objectness-driven discovery module may overlap with the output of the known object detector with high probability, the collaborative open-world classifier first filters the generalized objects logits by calculating an Intersection over Union(IoU) matrix $\mathcal{P} = \{p_{mn}\}$ between the generalized object boxes of the objectness-driven discovery module $\mathcal{B}^G = \{b_m^G\}$ and the known object boxes $\mathcal{B}^K = \{b_n^K\}$. The calculation of matrix $\mathcal{P}$ could be formulated as follows:

$$p_{ij} = IoU(b_m^G, b_n^K) \tag{5}$$

where b_m^G denotes each generalized object box in $\mathcal{B}^U$ and b_n^K denotes each known object box in $\mathcal{B}^K$. Subsequently, any generalized bounding box is labeled as a known class if it satisfies that its IOU with any predicted bounding box of a known class exceeds a threshold P_t. Conversely, the remaining generalized bounding boxes are categorized as unknown class targets. The filter process of the generalized object box could be presented as follows:

$$\mathcal{B}^U = \left\{b_m^G | \forall n, p_{mn} \leq P_t\right\} \tag{6}$$

The final output $\mathcal{B}^{out}$ of the collaborative open-world classifier is formed by merging the filtered unknown class predicted boxes with the original known class predicted boxes:

$$\mathcal{B}^{out} = \mathcal{B}^U \cup \mathcal{B}^K \tag{7}$$

The objective of the collaborative open-world classifier is to achieve collaborative detection of both known and unknown class targets by extracting unknown class targets with the attribute of "object" from the generalized objects outputted by the objectness-driven discovery module, excluding objects belonging to the known class.

3.3. Training Process

In accordance with the OWOD setup, the training of our framework consists of two stages: the initial training stage of the task 1 and the incremental learning stage. During the initial training of the task 1, the known object detector and the objectness-driven discovery module are trained separately. The training loss for the known object detector ℓ_{known} is the same as that of D-DETR [16] and can be represented as:

$$\ell_{\text{known}} = \ell_{\text{known}}^{cls} + \ell_{\text{known}}^{reg} \tag{8}$$

where $\ell_{\text{known}}^{reg}$ represents the loss for bounding box regression of the known object detector. The objectness-driven discovery module is trained using labels that have undergone the objectness-driven discovery training, and its loss $\widetilde{\ell}_{\text{universal}}$ can be represented as:

$$\widetilde{\ell}_{\text{universal}} = \widetilde{\ell}_{\text{universal}}^{cls} + \ell_{\text{universal}}^{reg} \tag{9}$$

where $\ell_{\text{universal}}^{reg}$ represents the loss for bounding box regression calculated by the universal label of the objectness-driven discovery module.

In this way, our objectness-driven discovery module learns to detect generalized objects regardless of their class. The loss curve during the initial training of the task 1 of the known object detector and the objectness-driven discovery module is shown in Figure 3a.

During the Incremental Learning stage, for training efficiency, only the known object detector is actively trained, and the parameters of the objectness-driven discovery module are frozen. The training process of the known object detector for incremental learning is similar to the task 1, and it employs the data replay method used in OW-DETR [17] to overcome catastrophic forgetting. This strategy allows the model to retain knowledge of previously learned categories while adapting to new data and categories introduced during the incremental learning phase. The loss curve during incremental learning from task 2 to task 4 of the known object detector is shown in Figure 3b.

After training the two detectors separately, we use the collaborative open classifier to perform post-processing to obtain the final predicted output of the model. The collaborative open-world classifier does not partake in training but rather solely contributes to the inference process of the model.

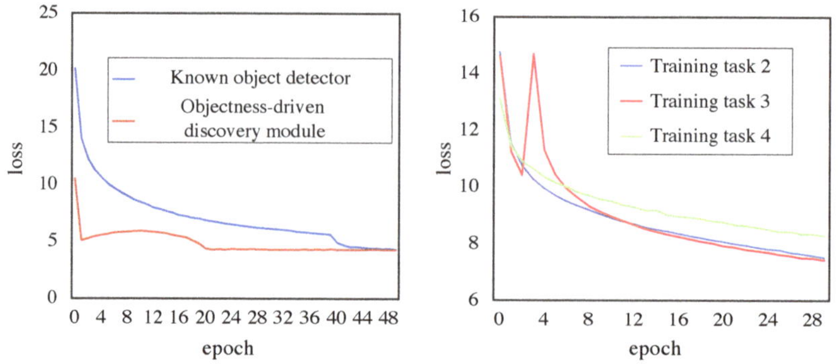

Figure 3. The loss curve during the training. (**a**) The loss curve of the known object detector and the objectness-driven discovery module during the initial training stage of the task 1. (**b**) The loss curve of the known object detector during the incremental learning stage of the task 1, task 2 and task 4.

4. Results

In this section, we perform comprehensive experiments and detailed analyses to demonstrate the effectiveness of the proposed method for open-world object detection.

4.1. Experiment Settings

Datasets. According to the typical setup of OWOD, all classes from the training set are grouped in T incremental tasks. Following [14], we set T as 4, and adopt the Pascal-VOC [55] and MS-COCO [56] dataset. When learning the task 1, we treat classes and data from Pascal-VOC as the training set, and the remaining 60 classes of MS-COCO are considered unknown. For subsequent tasks, the class division strategy is exactly the same as that in ORE. For evaluation, we use the Pascal-VOC test set and MS-COCO validation set.

Implementation details. We implement our method based on closed world detection model D-DETR [16]. During the training of the known object detector and the objectness-driven discovery module, the SGD optimizer is used and the batch size is set to 8. The IOU threshold P_t in the collaborative open-world classifier is set to 0.4. During both the initial training and the incremental learning stage, the initial learning rate of the known object detector and the objectness-driven discovery module is set to 2×10^{-4}. We train the known object detector for 50 epochs in the initial training stage of the task 1, and train 50 epochs to learn new known classes during the incremental learning. Additionally, we trained for 50 epochs in each stage for data replay to address catastrophic forgetting, following a strategy similar to OW-DETR [17]. About the objectness-driven discovery module, we train it for 80 epochs only in the initial training stage.

Evaluation metrics. To conduct a fair and comprehensive evaluation of the proposed method, we first employ widely recognized metrics in object detection, such as mean Average Precision (mAP) and Recall, as evaluation benchmarks for known and unknown classes. Additionally, we utilize other common-used open-set evaluation criteria, such as Wilderness Impact (WI) [40], Unknown Detection Recall (UDR) and Unknown Detection Precision (UDP) [15].

4.2. State-of-the-Art Comparison

Table 1 shows the comparison of our method respectively with the state-of-the-art methods according to traditional detection metrics, such as mAP and Recall. Note that after completing incremental learning of the task 4, all our classes have become known, and there are no longer any unknown targets in the test set. As a result, metrics related to unknown targets, such as U-mAP, UDR, UDP WI, etc., are not included. From Table 1, our method outperforms other state-of-the-art OWOD methods in most cases. Specifically, the unknown mAP of our method (4.84) is almost 7 times that for the model ORE (0.71) with the better known mAP performance, while the unknown class recall of our method reaches 17.40 and is higher than OW-DETR by 9.75, which is the best method in the existing works. These phenomena clearly demonstrate the effectiveness of our parallel decoupling OWOD framework. As novel known classes are added in the subsequent tasks, the recognition ability of unknown classes decreases, which is consistent with the performance of ORE. However, our method still outperforms the comparison methods by a large margin in U-mAP and unknown Recall metrics. And our known mAP performance on the newly-annotated and previous known classes is maintained. That is to say, our model is capable of promoting the collaborative learning of both known and unknown classes with fewer uncertainties.

Table 1. State-of-the-art comparison for OWOD according to traditional detection metrics. "K-" indicates the known classes, and "U-" represents the unknown classes. Our model achieves superior performance in terms of traditional evaluation metrics in most cases. Note that U-mAP and U-Recall are not calculated because all classes are known in the task 4. The bold represents the best performance on the metric.

Task IDs (→)	Task 1			Task 2			Task 3			Task 4
	K-mAP (↑)	U-mAP (↑)	U-Recall (↑)	K-mAP (↑)	U-mAP (↑)	U-Recall (↑)	K-mAP (↑)	U-mAP (↑)	U-Recall (↑)	K-mAP (↑)
Faster RCNN [57]	56.94	0	0	41.56	0	0	32.41	0	0	27.03
ORE [14]	56.49	0.71	5.72	39.64	0.14	2.66	20.17	0.12	3.34	25.95
SA [52]	55.56	0.20	1.93	39.02	0.03	0.79	31.54	0.003	0.12	26.42
D-DETR [16]	59.75	0	0	46.08	0	0	38.28	0	0	30.60
OW-DETR [17]	58.78	0.07	7.65	44.11	0.04	5.83	35.96	0.03	5.97	27.94
Ours	**59.84**	**4.84**	**17.40**	**46.38**	**1.60**	**13.53**	**38.37**	**1.45**	**14.30**	**32.63**

Table 2 further shows the comparison of our method respectively with the state-of-the-art methods according to common-used open-set evaluation metrics, such as WI, UDR, and UDP. From Table 2, it is evident that our method achieves similar WI scores to state-of-the-art methods while obtaining higher UDR and UDP scores. Specifically, our framework approaches or outperforms existing methods comprehensively in UDR at every stage, indicating that our framework helps discover more unknown objects. Moreover, our framework exhibits a significant improvement in UDP, suggesting that our framework has a more accurate classification capability for the discovered unknown objects.

Table 2. State-of-the-art comparison for OWOD according to open-set metrics. Our model achieves superior performance in terms of traditional evaluation metrics in most cases. Note that these metrics are only computed in the first three tasks because all classes are known in the task 4. The bold represents the best performance on the metric.

Task IDs (→)	Task 1			Task 2			Task 3		
	WI-0.8 (↓)	UDR (↑)	UDP (↑)	WI-0.8 (↓)	UDR (↑)	UDP (↑)	WI-0.8 (↓)	UDR (↑)	UDP (↑)
Faster RCNN [57]	0.0645	17.58	0	0.0273	16.32	0	0.0164	**24.69**	0
ORE [14]	**0.0528**	18.58	31.28	0.0315	17.30	15.37	0.0209	23.67	14.95
SA [52]	0.0563	8.51	22.73	**0.0181**	5.74	13.83	**0.0136**	9.12	1.30
D-DETR [16]	0.0600	**20.74**	0	0.0245	14.41	0	0.0187	34.48	0
OW-DETR [17]	0.0599	18.31	41.77	0.0319	16.24	35.88	0.0220	21.53	27.72
Ours	0.0553	20.00	**86.97**	0.0251	**19.35**	**69.89**	0.0179	22.59	**63.30**

Overall, our method has significantly improved its performance in unknown class detection, indicating that the objectness-driven discovery module we added has successfully suppressed the unknown discovery uncertainty. At the same time, the performance in known class detection has been more effectively maintained, indicating that our parallel decoupled architecture effectively suppresses the known discrimination uncertainty.

4.3. Ablation Study

As shown in Table 3, we investigate different components in our proposed framework. The table respectively lists the performance of our framework without the known object detector, our model without the objectness-driven discovery module, and the full model. The findings from the table indicate that the framework without the known object detector possesses only the capability to detect unknown targets, lacking the ability to discriminate known classes. The framework without the objectness-driven discovery module fails to detect unknown objects. In contrast, our framework possesses the combined abilities of known class discrimination and unknown class detection, achieving state-of-the-art performance. In conclusion, our full model achieves the best performance, and each module contributes to the proposed model.

Table 3. Ablation study of our model. "w/o KO detector" indicates our model without the known object detector. "w/o CTG detector" indicates our model without the objectness-driven discovery module. Our full model achieves the best performance, and each module contributes to the proposed model.

	WI-0.8	K-mAP	U-mAP	U-Recall
w/o KO detector	0	0	5.74	17.79
w/o CTG detector	0.0553	59.84	0	0
Ours	0.0553	59.84	4.84	17.40

4.4. Parameter Analysis

As shown in Figure 4, we delve into a comprehensive analysis of our model's behavior under varying IOU threshold P_t in the collaborative open-world classifier. The performance of our framework in terms of unknown class mAP and Recall for different P_t values is shown in Figure 4a, while UDP and UDR metrics are illustrated in Figure 4b. From the figures, it can be observed that when P_t is set to 0 or 0.2, the framework exhibits lower UDR, UDP, and U-Recall values. As the value of P_t increases, U-Recall and UDR increase accordingly. This phenomenon can be attributed to the larger number of retained unknown-class candidate boxes with high Intersection over Union (IOU) scores. However, it also implies a higher occurrence of known-class detections being misclassified as unknown-class, resulting in fluctuations in UmAP and UDP with the increasing value of P_t. In general, when P_t takes other values, UDR, UDP, and U-Recall exhibit relatively stable performance, indicating that our method can detect many unknown objects with small overlaps with known objects, and these detections are less affected by changes in P_t. When P_t is set to 0.4, the framework achieves its highest U-mAP value of 4.83. Therefore, our method effectively achieves the detection of unknown categories with good stability and fewer uncertainties.

4.5. Visualization

Figure 5 displays a visualization comparison between our model and other state-of-the-art models. It is evident that our model demonstrates the capability to simultaneously detect both known and unknown objects with a higher degree of accuracy in bounding box localization and classification labeling. In addition, ORE exhibits an inclination to over-generate bounding boxes with the goal of localizing more unseen objects. Conversely, SA tends to under-generate bounding boxes to minimize misclassification. While OW-DETR is capable of detecting some unknown classes, its precision in detection is relatively poor, often resulting in misalignment of bounding box positions or misclassifying certain portions of known classes as unknown classes. In contrast to these methods, our model excels in generating appropriate bounding boxes and providing precise predictions.

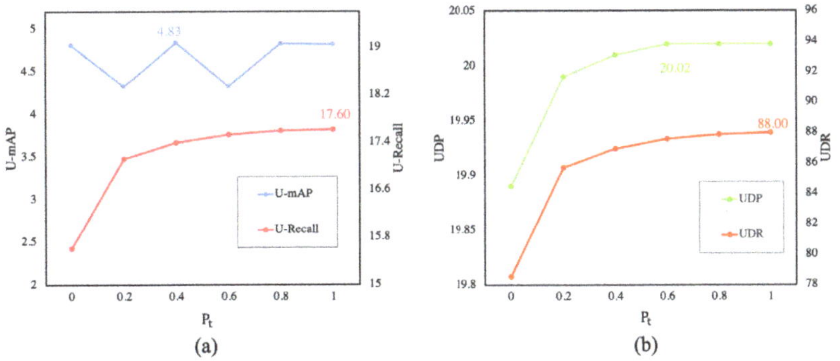

Figure 4. Parameter analysis of our model. P_t indicates the IOU threshold we use in our model. (**a**) The curve of U-mAP and U-Recall with P_t growth. (**b**) The curve of UDP and UDR with P_t growth. The larger the P_t is, the better U-Recall and UDR performance the framework achieves. Our model can consistently detect a rich set of unknown objects with minimal overlap with the known classes.

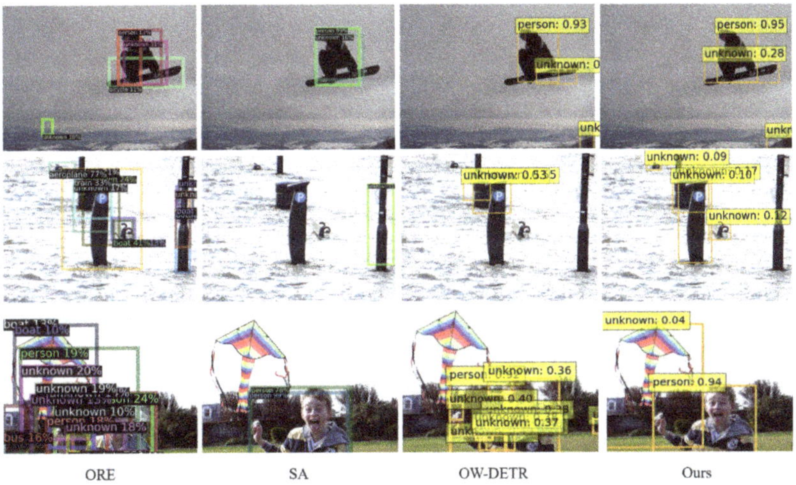

Figure 5. Visualization comparison of ORE, SA, OW-DETR and ours. Our model demonstrates the capability to simultaneously detect both known and unknown objects with a higher degree of accuracy in bounding box localization and classification labeling.

Figure 6 illustrate the visualization of the proposed methodology in the task 1 and the task 4, respectively. In the context of the task 1, the model demonstrates its capability to detect known classes and identify unfamiliar instances. In the task 4, our model exhibits its ability to progressively acquire the semantic class of all instances.

Figure 7 shows some failure cases of our model. Our model easily detects a combination of multiple objects or a part of a known object as a single object without appropriate guidance. In addition, there must be many missing-detection situations according to 20% UDR performance. In the future, we aspire to enhance the detection performance of unknown classes and address instances of false positives or misses through approaches like self-supervised learning.

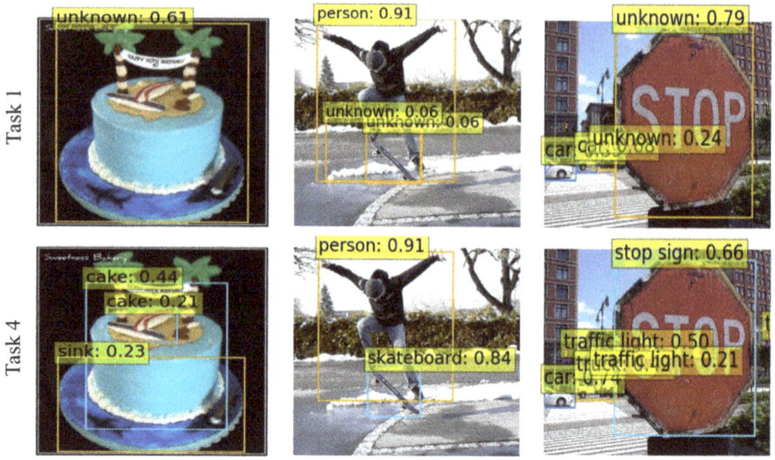

Figure 6. Visualization of our model in the task 1 and the task 4. After the initial training, our model recognizes the cake, the skateboard, and the stop sign as unknown. After the incremental learning, they can be labeled correctly.

(a) (b) (c)

Figure 7. Failure cases of our model. (**a**) The model mistakenly detected the combination of a person and a surfboard as a bird. (**b**) Even though people and unknown objects (surfboards) have been detected, their combination is still mistakenly detected as a boat. (**c**) Part of this unknown object (upper right part) is mistakenly detected as a separate unknown object.

5. Conclusions and Future Work

In conclusion, our parallel OWOD framework presented effectively addresses the challenges of uncertainties in open campus scenarios, enabling the detection of known and unknown objects. The objectness-driven discovery module alleviates the unknown discovery uncertainty by driving the model to learn the objectness derived from the deterministic supervision of known classes. The known object detector and the collaborative open-world classifier mitigate the known discrimination uncertainty and accomplish collaborative detection. This framework has the potential to transform campus monitoring and classroom behavior detection in intelligent educational settings.

It should be pointed out that there are still limitations of our parallel OWOD framework. While our framework has superior performance compared to state-of-the-art methods and notably enhances the detection of unknown classes, it still demonstrates apparent shortcomings in unknown object detection compared to known classes, with one of the obvious shortcomings lying in a noticeable decrease in precision. To address this issue, a substantial improvement in the model's unsupervised learning capabilities by exploring more effective methods to enhance the model's ability to detect unknown classes through learning is essential.

In addition, in the context of future prospects, several promising research directions emerge for the refinement and extension of the parallel OWOD framework. Subsequent investigations may center on augmenting the framework's adaptability across a spectrum of diverse campus scenarios, expanding its capability to detect a broader array of unknown objects, and minimizing the incidence of false positives. Moreover, the exploration of the framework's applicability in domains beyond education, such as healthcare, surveillance, and autonomous systems, represents an intriguing avenue for future exploration. The development of more efficient training strategies and the exploration of advanced neural network architectures may also make valuable contributions to the ongoing advancement of OWOD technology.

Author Contributions: Conceptualization, J.D., Z.Z. and Y.L.; methodology, J.D., Z.Z. and Y.M.; software, J.D. and Z.Z.; validation, J.D., S.H., R.Z. and Y.L.; formal analysis, J.D., S.H. and J.Y.; investigation, S.H., Y.L. and J.Y.; resources, Y.M., J.Y. and B.L.; data curation, Z.Z., S.H. and R.Z.; writing—original draft preparation, J.D., Z.Z. and S.H.; writing—review and editing, Y.M., R.Z. and B.L.; visualization, Z.Z., S.H. and R.Z.; supervision, J.Y., Y.M. and B.L.; project administration, J.D., Y.M. and B.L.; funding acquisition, Y.M. All authors have read and agreed to the published version of the manuscript.

Funding: This work was supported by the National Natural Science Foundation of China under Grant 62206010, 62022009, and the National Key R&D Program of China (2021ZD0110603).

Institutional Review Board Statement: Not applicable.

Informed Consent Statement: Not applicable.

Data Availability Statement: Publicly available datasets were analyzed in this study. This data can be found here: http://host.robots.ox.ac.uk/pascal/VOC/ (accessed on 24 October 2023) and https://cocodataset.org/#home (accessed on 24 October 2023).

Acknowledgments: We thank the State Key Lab of Software Development Environment and the China Electronics Standardization Institute for providing the experimental environment and equipment.

Conflicts of Interest: The authors declare no conflict of interest.

References

1. He, K.; Zhang, X.; Ren, S.; Sun, J. Deep residual learning for image recognition. In Proceedings of the IEEE Conference on Computer Vision and Pattern Recognition, Las Vegas, NV, USA, 26 June–1 July 2016; pp. 770–778.
2. Arcucci, R.; Zhu, J.; Hu, S.; Guo, Y.K. Deep data assimilation: Integrating deep learning with data assimilation. *Appl. Sci.* **2021**, *11*, 1114. [CrossRef]
3. Li, F.; He, F.; Wang, F.; Zhang, D.; Xia, Y.; Li, X. A novel simplified convolutional neural network classification algorithm of motor imagery EEG signals based on deep learning. *Appl. Sci.* **2020**, *10*, 1605. [CrossRef]
4. Shieh, C.S.; Lin, W.W.; Nguyen, T.T.; Chen, C.H.; Horng, M.F.; Miu, D. Detection of unknown ddos attacks with deep learning and gaussian mixture model. *Appl. Sci.* **2021**, *11*, 5213. [CrossRef]
5. Chiu, M.T.; Xu, X.; Wei, Y.; Huang, Z.; Schwing, A.G.; Brunner, R.; Khachatrian, H.; Karapetyan, H.; Dozier, I.; Rose, G.; et al. Agriculture-vision: A large aerial image database for agricultural pattern analysis. In Proceedings of the IEEE/CVF Conference on Computer Vision and Pattern Recognition, Seattle, WA, USA, 13–19 June 2020; pp. 2828–2838.
6. Ni, J.; Chen, Y.; Chen, Y.; Zhu, J.; Ali, D.; Cao, W. A survey on theories and applications for self-driving cars based on deep learning methods. *Appl. Sci.* **2020**, *10*, 2749. [CrossRef]
7. Rehman, A.; Iqbal, M.A.; Xing, H.; Ahmed, I. COVID-19 detection empowered with machine learning and deep learning techniques: A systematic review. *Appl. Sci.* **2021**, *11*, 3414. [CrossRef]
8. Zhou, H.; Jiang, F.; Shen, R. Who are raising their hands? Hand-raiser seeking based on object detection and pose estimation. In Proceedings of the Asian Conference on Machine Learning, PMLR, Beijing, China, 14–16 November 2018; pp. 470–485.
9. Zhao, J.; Zhu, H. CBPH-Net: A Small Object Detector for Behavior Recognition in Classroom Scenarios. *IEEE Trans. Instrum. Meas.* **2023**, *72*, 2521112. [CrossRef]
10. Liu, H.; Yu, Y.; Liu, S.; Wang, W. A Military Object Detection Model of UAV Reconnaissance Image and Feature Visualization. *Appl. Sci.* **2022**, *12*, 12236. [CrossRef]
11. Park, Y.; Shin, Y. Applying Object Detection and Embedding Techniques to One-Shot Class-Incremental Multi-Label Image Classification. *Appl. Sci.* **2023**, *13*, 10468. [CrossRef]
12. Miao, B.; Chen, Z.; Liu, H.; Zhang, A. A target re-identification method based on shot boundary object detection for single object tracking. *Appl. Sci.* **2023**, *13*, 6422. [CrossRef]
13. Xu, H.; Zhang, J.; Sun, H.; Qi, M.; Kong, J. Analyzing students' attention by gaze tracking and object detection in classroom teaching. *Data Technol. Appl.* **2023**, *57*, 643–667. [CrossRef]

14. Joseph, K.; Khan, S.; Khan, F.S.; Balasubramanian, V.N. Towards open world object detection. In Proceedings of the IEEE/CVF Conference on Computer Vision and Pattern Recognition, Nashville, TN, USA, 20–25 June 2021; pp. 5830–5840.
15. Zhao, X.; Liu, X.; Shen, Y.; Ma, Y.; Qiao, Y.; Wang, D. Revisiting open world object detection. *arXiv* **2022**, arXiv:2201.00471.
16. Zhu, X.; Su, W.; Lu, L.; Li, B.; Wang, X.; Dai, J.F.D.D. Deformable transformers for end-to-end object detection. In Proceedings of the 9th International Conference on Learning Representations, Virtual Event, Austria, 3–7 May 2021; OpenReview.net.
17. Gupta, A.; Narayan, S.; Joseph, K.; Khan, S.; Khan, F.S.; Shah, M. OW-DETR: Open-world detection transformer. In Proceedings of the IEEE/CVF Conference on Computer Vision and Pattern Recognition, New Orleans, LA, USA, 18–24 June 2022; pp. 9235–9244.
18. Chen, C.; Seff, A.; Kornhauser, A.; Xiao, J. Deepdriving: Learning affordance for direct perception in autonomous driving. In Proceedings of the IEEE International Conference on Computer Vision, Santiago, Chile, 7–13 December 2015; pp. 2722–2730.
19. Guo, S.; Wang, S.; Yang, Z.; Wang, L.; Zhang, H.; Guo, P.; Gao, Y.; Guo, J. A Review of Deep Learning-Based Visual Multi-Object Tracking Algorithms for Autonomous Driving. *Appl. Sci.* **2022**, *12*, 10741. [CrossRef]
20. Lee, Y.; Park, S. A deep learning-based perception algorithm using 3D lidar for autonomous driving: Simultaneous segmentation and detection network (ssadnet). *Appl. Sci.* **2020**, *10*, 4486. [CrossRef]
21. Smistad, E.; Falch, T.L.; Bozorgi, M.; Elster, A.C.; Lindseth, F. Medical image segmentation on GPUs—A comprehensive review. *Med. Image Anal.* **2015**, *20*, 1–18. [CrossRef] [PubMed]
22. Nagi, A.T.; Awan, M.J.; Mohammed, M.A.; Mahmoud, A.; Majumdar, A.; Thinnukool, O. Performance analysis for COVID-19 diagnosis using custom and state-of-the-art deep learning models. *Appl. Sci.* **2022**, *12*, 6364. [CrossRef]
23. Qureshi, S.A.; Raza, S.E.A.; Hussain, L.; Malibari, A.A.; Nour, M.K.; Rehman, A.U.; Al-Wesabi, F.N.; Hilal, A.M. Intelligent ultra-light deep learning model for multi-class brain tumor detection. *Appl. Sci.* **2022**, *12*, 3715. [CrossRef]
24. Peng, T.; Liang, Y.; Wu, W.; Ren, J.; Pengrui, Z.; Pu, Y. CLGT: A graph transformer for student performance prediction in collaborative learning. In Proceedings of the AAAI Conference on Artificial Intelligence, Washington, DC, USA, 7–14 February 2023; Volume 37, pp. 15947–15954.
25. Tsiakmaki, M.; Kostopoulos, G.; Kotsiantis, S.; Ragos, O. Transfer learning from deep neural networks for predicting student performance. *Appl. Sci.* **2020**, *10*, 2145. [CrossRef]
26. Liang, Y.; Peng, T.; Pu, Y.; Wu, W. HELP-DKT: An interpretable cognitive model of how students learn programming based on deep knowledge tracing. *Sci. Rep.* **2022**, *12*, 4012. [CrossRef]
27. Si, J.; Lin, J.; Jiang, F.; Shen, R. Hand-raising gesture detection in real classrooms using improved R-FCN. *Neurocomputing* **2019**, *359*, 69–76. [CrossRef]
28. Ma, Y.; Liu, X.; Bai, S.; Wang, L.; Liu, A.; Tao, D.; Hancock, E.R. Regionwise generative adversarial image inpainting for large missing areas. *IEEE Trans. Cybern.* **2022**, *53*, 5226–5239. [CrossRef]
29. Hu, S.; Ma, Y.; Liu, X.; Wei, Y.; Bai, S. Stratified rule-aware network for abstract visual reasoning. In Proceedings of the AAAI Conference on Artificial Intelligence, Virtual, 2–9 February 2021; Volume 35, pp. 1567–1574.
30. Li, F.; Wechsler, H. Open set face recognition using transduction. *IEEE Trans. Pattern Anal. Mach. Intell.* **2005**, *27*, 1686–1697. [PubMed]
31. Heflin, B.; Scheirer, W.; Boult, T.E. Detecting and classifying scars, marks, and tattoos found in the wild. In Proceedings of the 2012 IEEE Fifth International Conference on Biometrics: Theory, Applications and Systems (BTAS), Arlington, VA, USA, 23–27 September 2012; IEEE: Piscataway, NJ, USA, 2012; pp. 31–38.
32. Pritsos, D.A.; Stamatatos, E. Open-set classification for automated genre identification. In Proceedings of the European Conference on Information Retrieval, Moscow, Russia, 24–27 March 2013; Springer: Berlin/Heidelberg, Germany, 2013; pp. 207–217.
33. Scherreik, M.D.; Rigling, B.D. Open set recognition for automatic target classification with rejection. *IEEE Trans. Aerosp. Electron. Syst.* **2016**, *52*, 632–642. [CrossRef]
34. Fei, G.; Liu, B. Breaking the closed world assumption in text classification. In Proceedings of the 2016 Conference of the North American Chapter of the Association for Computational Linguistics: Human Language Technologies, San Diego, CA, USA, 12–17 June 2016; pp. 506–514.
35. Vareto, R.; Silva, S.; Costa, F.; Schwartz, W.R. Towards open-set face recognition using hashing functions. In Proceedings of the 2017 IEEE International Joint Conference on Biometrics (IJCB), Denver, CO, USA, 1–4 October 2017; IEEE: Piscataway, NJ, USA, 2017; pp. 634–641.
36. Liu, Z.; Miao, Z.; Zhan, X.; Wang, J.; Gong, B.; Yu, S.X. Large-scale long-tailed recognition in an open world. In Proceedings of the IEEE/CVF Conference on Computer Vision and Pattern Recognition, Long Beach, CA, USA, 15–20 June 2019; pp. 2537–2546.
37. Bendale, A.; Boult, T.E. Towards open set deep networks. In Proceedings of the IEEE Conference on Computer Vision and Pattern Recognition, Las Vegas, NV, USA, 26 June–1 July 2016; pp. 1563–1572.
38. Perera, P.; Morariu, V.I.; Jain, R.; Manjunatha, V.; Wigington, C.; Ordonez, V.; Patel, V.M. Generative-discriminative feature representations for open-set recognition. In Proceedings of the IEEE/CVF Conference on Computer Vision and Pattern Recognition, Seattle, WA, USA, 13–19 June 2020; pp. 11814–11823.
39. Yoshihashi, R.; Shao, W.; Kawakami, R.; You, S.; Iida, M.; Naemura, T. Classification-reconstruction learning for open-set recognition. In Proceedings of the IEEE/CVF Conference on Computer Vision and Pattern Recognition, Long Beach, CA, USA, 15–20 June 2019; pp. 4016–4025.
40. Dhamija, A.; Gunther, M.; Ventura, J.; Boult, T. The overlooked elephant of object detection: Open set. In Proceedings of the IEEE/CVF Winter Conference on Applications of Computer Vision, Snowmass Village, CO, USA, 1–5 March 2020; pp. 1021–1030.

41. Miller, D.; Nicholson, L.; Dayoub, F.; Sünderhauf, N. Dropout sampling for robust object detection in open-set conditions. In Proceedings of the 2018 IEEE International Conference on Robotics and Automation (ICRA), Brisbane, Australia, 21–25 May 2018; IEEE: Piscataway, NJ, USA, 2018; pp. 3243–3249.
42. Miller, D.; Dayoub, F.; Milford, M.; Sünderhauf, N. Evaluating merging strategies for sampling-based uncertainty techniques in object detection. In Proceedings of the 2019 International Conference on Robotics and Automation (ICRA), Montreal, QC, Canada, 20–24 May 2019; IEEE: Piscataway, NJ, USA, 2019; pp. 2348–2354.
43. Hall, D.; Dayoub, F.; Skinner, J.; Zhang, H.; Miller, D.; Corke, P.; Carneiro, G.; Angelova, A.; Sünderhauf, N. Probabilistic object detection: Definition and evaluation. In Proceedings of the IEEE/CVF Winter Conference on Applications of Computer Vision, Snowmass Village, CO, USA, 1–5 March 2020; pp. 1031–1040.
44. Han, J.; Ren, Y.; Ding, J.; Pan, X.; Yan, K.; Xia, G.S. Expanding Low-Density Latent Regions for Open-Set Object Detection. In Proceedings of the IEEE/CVF Conference on Computer Vision and Pattern Recognition, New Orleans, LA, USA, 18–24 June 2022; pp. 9591–9600.
45. Bendale, A.; Boult, T. Towards open world recognition. In Proceedings of the IEEE Conference on Computer Vision and Pattern Recognition, Boston, MA, USA, 7–12 June 2015; pp. 1893–1902.
46. Xu, H.; Liu, B.; Shu, L.; Yu, P. Open-world learning and application to product classification. In Proceedings of the The World Wide Web Conference, San Francisco, CA, USA, 13–17 May 2019; pp. 3413–3419.
47. Willes, J.; Harrison, J.; Harakeh, A.; Finn, C.; Pavone, M.; Waslander, S. Bayesian embeddings for few-shot open world recognition. *IEEE Trans. Pattern Anal. Mach. Intell.* **2022**, 1–16.. [CrossRef] [PubMed]
48. Mancini, M.; Naeem, M.F.; Xian, Y.; Akata, Z. Open world compositional zero-shot learning. In Proceedings of the IEEE/CVF Conference on Computer Vision and Pattern Recognition, Nashville, TN, USA, 20–25 June 2021; pp. 5222–5230.
49. Zhang, S.; Li, Z.; Yan, S.; He, X.; Sun, J. Distribution alignment: A unified framework for long-tail visual recognition. In Proceedings of the IEEE/CVF Conference on Computer Vision and Pattern Recognition, Nashville, TN, USA, 20–25 June 2021; pp. 2361–2370.
50. Wang, Y.; Yao, Q.; Kwok, J.T.; Ni, L.M. Generalizing from a few examples: A survey on few-shot learning. *ACM Comput. Surv.* **2020**, *53*, 1–34. [CrossRef]
51. Xian, Y.; Schiele, B.; Akata, Z. Zero-shot learning-the good, the bad and the ugly. In Proceedings of the IEEE Conference on Computer Vision and Pattern Recognition, Honolulu, HI, USA, 21–26 July 2017; pp. 4582–4591.
52. Yang, S.; Sun, P.; Jiang, Y.; Xia, X.; Zhang, R.; Yuan, Z.; Wang, C.; Luo, P.; Xu, M. Objects in semantic topology. *arXiv* **2021**, arXiv:2110.02687.
53. Wu, Z.; Lu, Y.; Chen, X.; Wu, Z.; Kang, L.; Yu, J. UC-OWOD: Unknown-Classified Open World Object Detection. *arXiv* **2022**, arXiv:2207.11455.
54. Vaswani, A.; Shazeer, N.; Parmar, N.; Uszkoreit, J.; Jones, L.; Gomez, A.N.; Kaiser, Ł.; Polosukhin, I. Attention is all you need. *Adv. Neural Inf. Process. Syst.* **2017**, *30*, 5999–6009.
55. Everingham, M.; Van Gool, L.; Williams, C.K.; Winn, J.; Zisserman, A. The pascal visual object classes (voc) challenge. *Int. J. Comput. Vis.* **2010**, *88*, 303–338. [CrossRef]
56. Lin, T.Y.; Maire, M.; Belongie, S.; Hays, J.; Perona, P.; Ramanan, D.; Dollár, P.; Zitnick, C.L. Microsoft coco: Common objects in context. In Proceedings of the European Conference on Computer Vision, Zurich, Switzerland, 6–12 September 2014; Springer: Berlin/Heidelberg, Germany, 2014; pp. 740–755.
57. Ren, S.; He, K.; Girshick, R.; Sun, J. Faster r-cnn: Towards real-time object detection with region proposal networks. *Adv. Neural Inf. Process. Syst.* **2015**, *28*, 91–99. [CrossRef]

applied sciences

MDPI

Article

Developing an Intelligent Recommendation System for Non-Information and Communications Technology Major University Students

TaeYoung Kim [1] and JongBeom Lim [2,*]

[1] College of Education, Hankuk University of Foreign Studies, 107 Imun-ro, Seoul 02450, Republic of Korea; ktyoung66@hanmail.net
[2] Division of ICT Convergence, Pyeongtaek University, 3825 Seodong-daero, Pyeongtaek-si 17869, Republic of Korea
* Correspondence: jblim@ptu.ac.kr

Abstract: Various services and applications based on information and communications technology (ICT) are converging with cultural aspects of historical implementations. At the same time, developing a convergence course for non-ICT majors is becoming increasingly popular in universities. In this paper, we develop an AI application course for non-ICT major university students toward convergence with recommendation systems and Silk Road studies. Based on our five-year research on the martial arts, dance, and play of seven Silk Road countries, we have created and categorized an accessible database for 177 items in those countries. For our convergence course, we measure the similarity between the items for summary and perform collaborative filtering based on alternating least squares (ALS) matrix factorization so that our prototyped intelligent recommendation engine can predict the items in which a user might be interested. The course is designed to teach non-ICT major university students not only historical aspects of the Silk Road but also implementation aspects of recommendation systems with web services.

Keywords: AI application course; convergence course; recommendation system; Silk Road

1. Introduction

Convergence education refers to an interdisciplinary approach that combines diverse fields of study, such as technology, the arts, the sciences, and the humanities [1,2]. It aims to foster a well-rounded understanding of complex issues by integrating different perspectives, knowledge, and skills. This approach encourages students to think critically, solve problems creatively, and make connections between seemingly unrelated subjects.

In a world where technology is rapidly changing and becoming more integrated into various fields, a basic understanding of ICT is essential [3,4]. In this regard, convergence education is important for non-ICT major students because it allows them to develop a broader set of skills that are relevant in today's technology-driven world. Convergence education provides students with an interdisciplinary approach that combines different fields of study, such as computer science, engineering, design, and business [5,6]. This allows students to develop a deeper understanding of how technology can be used to solve real-world problems and create new opportunities.

Developing a convergence course for non-ICT major students contributes to the interdisciplinary field by fostering a diverse workforce, addressing skills gaps, broadening perspectives, and future-proofing careers in an increasingly interconnected world [7]. Specifically, non-ICT major students may not have a strong background in technology, but by exposing them to convergence education, they can acquire a basic understanding of ICT-related concepts. This helps create a diverse workforce that possesses a broader range of skills, fostering innovation and adaptability within different industries.

At the same time, as artificial intelligence is growing to a great extent, recommendation systems have also received much attention in various areas, such as e-commerce, media, banking, telecom, and more [8,9]. The recommendation systems market is anticipated to reach 54 billion dollars by 2030, growing at a compound annual growth rate of 37% from 2022 [10,11]. While artificial intelligence and recommendation systems do not likely fall into the same research area, they share something in common in some dimensions in that both leverage predictive analytics based on machine learning techniques in a user-centric approach [12–15].

There are three major approaches to recommendation systems: (1) content-based filtering [16,17], (2) collaborative filtering [18,19], and (3) hybrid approaches [20–22]. Literally, content-based filtering methods use items' contents, such as summary, description, and profile information. The content-based methods are suitable for applications that use well-known data sets, including identifiers, descriptions, GPS locations, or metadata [23]. Occasionally, content-based methods utilize data from users. For example, a movie recommendation system utilizes the reviews of moviegoers. A recommender engine then provides suggestions and recommends options to those who have similar tastes based on some type of similarity measurement.

The basic assumptions of the collaborative filtering methods are that a particular user group is likely to use similar items in the future [24,25]. Therefore, it elaborates on making well-classified user groups based on complex criteria. Occasionally, collaborative methods use other user profile information that contains a past rating history. Hybrid recommendation approaches combine two or more filtering methods. In this paper, we consider hybrid recommendation systems that use both content-based filtering and collaborative filtering methods. The benefits of using hybrid recommendation systems are apparent. They provide better user experiences while achieving business goals. The target for our recommendation system is the martial arts, dance, and play of seven of the Silk Road countries (Korea, Japan, China, Mongolia, Kazakhstan, Uzbekistan, and Iran).

In this paper, we develop a convergence course for students not majoring in ICT toward convergence with intelligent recommendation systems and Silk Road studies. We expect that our convergence course can help them develop skills that are highly valued in the job market, such as problem-solving, critical thinking, creativity, and collaboration.

The Silk Road played an essential role in the embracement and development of cultures, in addition to being the main route of trade. Based on the accumulated resources, we provide a convergence course for the martial arts, dances, and plays of the Silk Road countries. The expected outcome is to highlight the importance of the cultural heritage (martial arts, dances, and plays) of the Silk Road. It is important to pay attention to and raise awareness about the intangible heritage of the Silk Road, as the potential gains are significant for countries regardless of their level of international integration. In this regard, understanding the martial arts, dances, and plays of the Silk Road can be a vital tool to dissolve the barriers between countries through physical and emotional communication.

The long-term outlook for developing the convergence course of martial arts, dances, and plays on the Silk Road is promising. We can establish a firm foothold in martial arts, dance, and play on the Silk Road, which can act as a stepping stone into the world of integration. This implies that our research will support the transmission of traditional cultures from generation to generation and lay the foundation to produce a brave new world in harmony with modern cultures.

The rest of the paper is organized as follows: Section 2 discusses our research motivation and related work. Section 3 provides implementation details for our intelligent hybrid recommendation system. Next, we present the proposed convergence course for non-ICT major university students in Section 4. Finally, Section 5 gives the conclusions and future research directions.

2. Research Motivation and Related Work

In this section, we provide the research motivation to develop a hybrid recommendation system for the martial arts, dance, and play of seven Silk Road countries for our convergence course. Then, we review previous work related to our study.

2.1. Research Motivation

The research motivation for developing a convergence course for non-ICT major students that involves developing a hybrid recommendation system for the martial arts, dance, and play of Silk Road countries could be to promote cultural exchange and understanding among countries along the Silk Road while also fostering technological innovation. By developing a hybrid recommendation system, students can learn both the cultural elements and the technical skills necessary for designing and implementing such a system. Additionally, the course would encourage students to engage in interdisciplinary learning, which could lead to creative problem-solving and innovation.

After five years of research and development, we would build datasets and databases for the martial arts, dance, and play of seven Silk Road countries. In addition to the database, we are also planning to make our database useful. To this end, we are making a convergence course (a mixture of cultures and information and communications technology) for non-ICT major university students. As a part of the course development, we built an intelligent hybrid recommendation system for the martial arts, dance, and play of seven Silk Road countries. Since the convergence course is for non-ICT major university students, we make it understandable and straightforward rather than complex and complicated. In the next subsection, we review the existing methods for recommendation systems.

2.2. Content-Based Filtering

The content-based filtering method often uses users' profile information in the system. There are two general methods to construct users' profile information. One is the preference model, and the other involves users' data history [26,27]. Technically, the profile information can be converted to a discrete score or number rating. The well-known representation for user profile information is the term frequency-inverse document frequency (tf-idf) or vector space representation [28,29]. The tf-idf method converts user profile information to a vector space based on item features.

In this paper, we use the tf-idf method to develop content-based filtering. While there are more sophisticated methods, such as support vector machines [30,31], Bayesian classifiers [32,33], and deep learning [8,34,35], we leave those for future work since these methods often require a deeper level of technical expertise and are more appropriate for advanced students who have already acquired a foundational understanding of the subject matter. Additionally, introducing these methods too early may overwhelm students and discourage them from continuing with the course. By leaving these methods as future research work, students can build on their existing knowledge and skills and gradually work toward more advanced topics as they become more confident and proficient.

The important thing when implementing the content-based filtering method is whether the recommendation system can learn and use the profile information in an effective manner. To solve the issue, we build summaries for 177 items related to martial arts, dance, and play in seven Silk Road countries and use them for item similarities. The details and descriptions of the tf-idf and the similarity measure are provided in Section 3.2.

2.3. Collaborative Filtering

The collaborative filtering method provides recommended items for a specific page (web page or application panel). The collaborative filtering method uses rating history information from all the users in the recommendation system. Explicit or implicit data collection methods can be used to achieve this [36,37]. The explicit data collection method allows users to respond to the score or rating for items (e.g., movie review and rating system).

On the other hand, the implicit data collection method uses various data, including item view history, online cart or purchase information, and likes and dislikes in social network services. There is a key issue in implementing the implicit data collection method: privacy. The operator of the recommendation system must obtain the users' agreement and specify the usage in the Terms and Conditions. To avoid this issue, we use the explicit data collection method in our recommendation system.

The basic method to implement collaborative filtering is matrix factorization [38–40]. Mathematically, matrix factorization is able to represent information about users and items in a latent feature space or embedding space, which is similar to dimensionality reduction methods such as principal component analysis (PCA). This method is called latent-factor collaborative filtering. One potential problem with matrix factorization is selecting the proper number of columns in factor matrices. To deal with the issue, the authors of [41] proposed an automatic hyper-parameter learning method based on the Bayesian model.

The authors of [42] introduced matrix factorization techniques for recommender systems, aiming to capture latent factors in user–item interactions and provide accurate predictions for personalized recommendations. The authors of [43] proposed neural collaborative filtering, a hybrid recommendation approach that combines matrix factorization and deep learning techniques to capture both user–item interactions and item–item correlations for enhanced recommendation performance. Although these state-of-the-art studies involve sophisticated techniques and methods, we focus on basic and straightforward methods for non-ICT major students.

2.4. Hybrid Recommendation Systems

Hybrid recommendation systems combine content-based and collaborative filtering approaches to leverage the strengths of both methods to provide more accurate, diverse, and adaptable recommendations while addressing the limitations of each individual approach. Hybrid recommendation systems offer certain advantages: (1) more accurate recommendations, (2) overcome data sparsity, (3) more diverse recommendations, (4) mitigate the cold-start problem, and (5) flexibility and adaptability [44,45].

In this paper, we implement a hybrid recommendation system by combining content-based filtering and collaborative filtering methods. There are seven approaches to building a hybrid recommendation system [46,47]: (1) weighted, (2) switching, (3) mixed, (4) feature combination, (5) feature augmentation, (6) cascade, and (7) meta-level. From them, we chose a mixed approach that combines different techniques or sources of information to enhance the recommendation process. Each recommendation technique has strengths and weaknesses. By combining them, a mixed approach can leverage the complementary aspects of different methods, which leads to a synergistic effect and potentially improves the overall quality of the recommendations.

2.5. Silk Road and Convergence Education

The Silk Road played a crucial role in the exchange of cultures, including martial arts, dance, and play, among the countries along its route. Martial arts like Kung Fu, Karate, and Taekwondo have historical roots in the Silk Road region, spreading through trade and cultural interactions [48]. Similarly, various forms of traditional dance and play originated from the diverse cultures along the Silk Road [49,50]. Traditional games and sports were also exchanged, fostering cultural understanding and unity among the Silk Road countries [51,52]. These artistic and recreational practices not only enriched the cultural heritage of the Silk Road but also contributed to the mutual influence and appreciation between nations.

As far as convergence education is concerned, which combines technical and non-technical subjects, it is increasingly important in today's world. It offers students a well-rounded education that integrates different disciplines and helps them develop a broader skill set. By merging technical and non-technical contents, convergence education promotes critical thinking, creativity, problem-solving abilities, and interdisciplinary understanding.

It prepares students to navigate complex real-world challenges that often require a combination of technical expertise and a deep understanding of social, cultural, and ethical implications [53].

Convergence education in this study also cultivates innovation by encouraging students to explore connections between seemingly unrelated fields. In a rapidly evolving global society, convergence education equips students with the versatility and adaptability needed to thrive in various professional domains. It bridges the gap between different areas of knowledge and promotes lifelong and sustainable learning [54].

3. The Proposed Recommendation System

In this section, we provide our hybrid recommendation system for non-ICT major university students. After reviewing the overall architecture and its implementation details, we provide detailed procedures and examples with a working example.

3.1. Overall Architecture and Implementations

The overall architecture for our hybrid recommendation system is depicted in Figure 1. We use both Silk-road datasets and user data, including profiles, view history, and rating information. For content-based filtering, similarity measures (tf-idf and cosine similarity) are used. For collaborative filtering, the k-nearest neighbors (KNN) classifier and ALS factorization are used. In the collaborative filtering procedure, the user similarity is also calculated. As our recommendation system uses a hybrid approach, both methods (content-based filtering and collaborative filtering) are combined with additional online data. That is, users' item rating information and users' view history data are stored in the database, and the stored data are also intelligently used for calculating top-N recommendations.

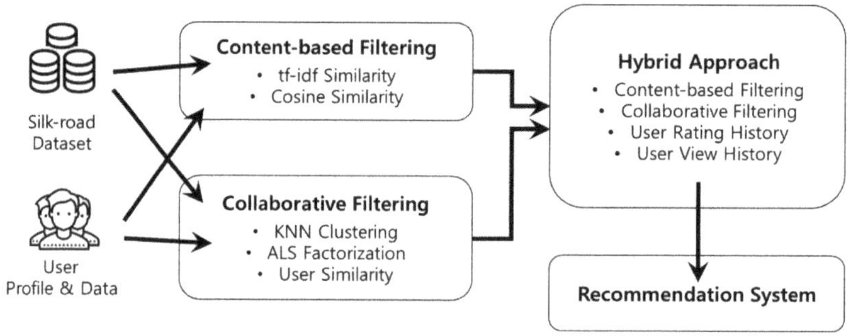

Figure 1. The overall architecture of the recommendation system.

Implementing a hybrid recommendation system requires careful consideration of data preprocessing, feature selection, similarity calculation, and weighting strategies. The high-level algorithmic procedures for implementing a hybrid recommendation system based on content-based filtering and collaborative filtering are described as follows:

1. Data collection
    - Gather user–item interaction data (ratings, reviews, etc.) for collaborative filtering.
    - Collect item features (types, keywords, etc.) for content-based filtering.
2. User profiling
    - Create user profiles based on their historical interactions using collaborative filtering techniques.
    - Analyze user preferences, behaviors, and similarities with other users.
3. Item profiling
    - Extract relevant features from items using content-based filtering techniques.
    - Represent items in a feature space based on their attributes.

4. Hybrid score calculation
    - Calculate a similarity score between the target user and other users using collaborative filtering.
    - Compute a similarity score between the target item and other items using content-based filtering.
5. Weighted combination
    - Assign weights to the similarity scores from collaborative filtering and content-based filtering based on their importance.
    - Combine the scores using a weighted sum or other fusion techniques.
6. Top-N recommendations
    - Rank the items based on the combined scores from the previous step.
    - Select the top-N items with the highest scores as recommendations for the target user.

3.2. Preprocessing

To effectively process content-based and/or collaborative filtering, stop words should be removed. Since the data sets for the recommendation system are written in Korean, libraries such as Natural Language Toolkit (NLTK) [55,56], which provides natural language processing only for English in the Python programming language, cannot be used. Thus, we included the most-used Korean stop words in addition to special characters (e.g., '{', '}', '[', ']', '.', '?', etc.). The number of stop words in Korean is about 700. With the collected stop words ready, we perform the iteration process to remove stop words for the data sets and the tokenizing process.

3.3. Content-Based Filtering

To examine the similarity between items, we use tf-idf, which is defined as Equation (1) [28,29].

$$tf - idf(t, d) = tf(t, d) \times idf(t,d), \tag{1}$$

where t is a term, d is a document, $tf(t, d)$ is the number of occurrences of t in d, and $idf(t, d)$ is the inverse document frequency. Simply put, tf-idf is the product of the term frequency and the inverse document frequency. The idf(t, d) is defined as Equation (2).

$$idf(t,d) = log\frac{n_d}{1 + df(d,t)}, \tag{2}$$

where n_d is the number of documents in the data set, and $df(d, t)$ is the number of documents that contain t.

For example, we assume that there are three simple documents:

1. Kevin likes drinking.
2. Jack likes drinking warm juice.
3. Kevin and Jack like drinking cold water.

The total word frequency can be summarized in Table 1.

Table 1. An example of tf.

	And	Cold	Drinking	Jack	Juice	Kevin	Like	Warm	Water
Document 1	0	0	1	0	0	1	1	0	0
Document 2	0	0	1	1	1	0	1	1	0
Document 3	1	1	1	1	0	1	1	0	1

However, instead of using raw term frequencies, it is typical to normalize data, such as L2-normalization, to get a vector of length 1. The L2-normalization can be calculated as Equation (3) [57].

$$v_{norm} = \frac{v}{||v||_2} = \frac{v}{\sqrt{v_1^2 + v_1^2 + \cdots + v_n^2}} = \frac{v}{\left(\sum_{i=1}^{n} v_i\right)^{1/2}} \tag{3}$$

Next, we measure similarities between items. In the context of tf-idf, an item can be considered a document for our data sets as shown in Table 2. There are numerous similarity measures in the data science field. Among them, we use the cosine similarity measure, which is calculated as Equation (4) [58,59].

$$cosine\ similarity = \cos(\theta) = \frac{A \cdot B}{||A||||B||} = \frac{\sum_{i=1}^{n} A_i B_i}{\sqrt{\sum_{i=1}^{n} A_i^2}\sqrt{\sum_{i=1}^{n} B_i^2}}, \tag{4}$$

where A and B are two separate documents. The result of cosine similarity for our example is listed in Table 3. Note that, unlike the Euclidean similarity measure, the cosine similarity measure is scale-insensitive, and thus, it can be used for general use cases.

Table 2. An example of tf-idf.

	And	Cold	Drinking	Jack	Juice	Kevin	Like	Warm	Water
Document 1	0	0	1	0	0	1.41	1	0	0
Document 2	0	0	1	1.41	2.10	0	1	2.10	0
Document 3	2.10	2.10	1	1.41	0	1.41	1	0	2.10

Table 3. An example of document similarity using $\cos(\theta)$.

	Document 1	Document 2	Document 3
Document 1	1	0.28	0.46
Document 2	0.28	1	0.25
Document 3	0.46	0.25	1

3.4. Collaborative Filtering

In addition to content-based filtering, we also use collaborative filtering. The advantage of using collaborative filtering is that it is able to recommend things based on the patterns of others without requiring personal or item information. Collaborative filtering has become popular because of the Netflix Prize, an open competition for the best collaborative filtering algorithm.

We use two notable algorithms for collaborative filtering. The first algorithm for collaborative filtering is the KNN classifier [60–62]. The KNN algorithm is one example of lazy learning since it memorizes the data sets instead of learning a discriminative function from the data sets. As the name implies, the KNN algorithm is straightforward.

1. Get the system parameters, such as k (the number of neighbors) and a metric for measuring the distance.
2. Locate reference points for data sets in a feature space.
3. Find the k-nearest neighbors based on the distance metric.
4. Calculate the majority and assign a label to a class.

Simply put, the KNN algorithm uses the following formula for the expected similarity to weigh the score based on the k-nearest neighbors in a feature space:

$$Expected(r_{u,i}) = \frac{\sum_v Similarity(u,v) \times r_{v,i}}{\sum_v Similarity(u,v)} \tag{5}$$

In Equation (5), the expected rating ($r_{u,i}$) for user u for item i can be calculated with the average of the neighbors' rating scores for $r_{v,i}$. Note that it can use various similarity measures, such as cosine similarity and the Pearson correlation. One requirement of using the KNN classifier is that the data sets must contain the score or rating information of the items for users. To this end, we allow users to rate items on the user page, as shown in Figure 2.

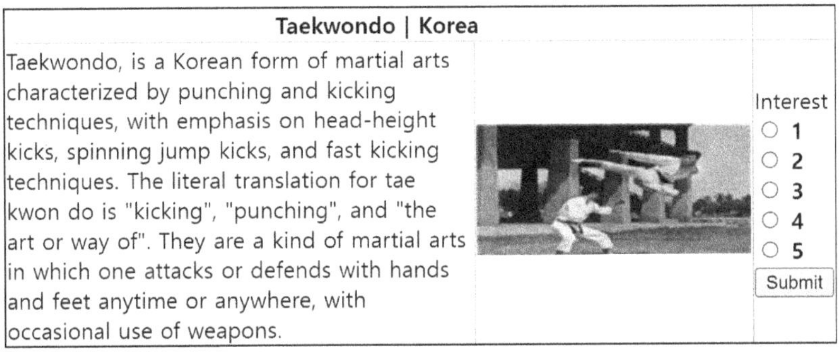

Figure 2. User page example with interest ratings. Users' interest rating scale is 1–5 (1 is the lowest and 5 is the highest value).

The second algorithm for collaborative filtering is the latent factor model [63,64]. The basic concept of latent factor collaborative filtering is to find a latent factor that represents users and items to fill empty slots in the rating matrix. The core principle of latent factor collaborative filtering is matrix factorization, which is also used in the Netflix Prize.

The well-known methods for latent factor collaborative filtering are singular value decomposition (SVD) [65], stochastic gradient descent (SGD) [66,67], and alternating least squares (ALS) [68,69]. The downside of SVD is that it neither captures biases for users and items nor allows missing values. Note that SVD can be defined for complete matrices; thus, zeroing missing values is used instead of removing entire rows. The SGD method performs matrix diagonalization, like eigenvalue decomposition. In fact, the SGD method optimizes two matrices (user latent and item latent) simultaneously.

Meanwhile, the ALS method fixes one of the two matrixes and optimizes the other matrix. Hence, the optimization problem can be converted to a convex type, which makes convergence faster than SGD. Thus, we use the ALS method for our collaborative filtering. The procedures for ALS are as follows:

1. Initialize both the user latent and the item latent.
2. Optimize the user latent while fixing the item latent.
3. Optimize the item latent while fixing the user latent.
4. Iterate steps 2 and 3.

For example, we consider four users and three items (4 × 3 matrix), as shown in Table 4. It initializes the user latent and the item latent. Then it optimizes the user latent while fixing the item latent and vice versa with Equations (6) and (7), respectively.

$$u_i = \left(V^T \times V \times \lambda I\right)^{-1} \times V^T \times R_i \quad (6)$$

$$v_i = \left(U^T \times U \times \lambda I\right)^{-1} \times U^T \times R_j \quad (7)$$

Table 4. An example of rating scores for four users.

	Item 1	Item 2	Item 3
User 1	1	0	5
User 2	0	0	3
User 3	3	2	4
User 4	2	1	5

After 10 iterations of steps 2 and 3, the user latent and the item latent are equivalent in Tables 5 and 6, respectively. The result of ALS can be calculated with the dot product of the item latent and the user latent, as shown in Table 7.

Table 5. The user latent matrix after 10 iterations.

	Item 1	Item 2	Item 3
User 1	1	0	5
User 2	0	0	3
User 3	3	2	4
User 4	2	1	5

Table 6. The item latent matrix after 10 iterations.

	Item 1	Item 2	Item 3
User 1	1	0	5
User 2	0	0	3
User 3	3	2	4
User 4	2	1	5

Table 7. The result of the dot product (item latent · item latent).

	Item 1	Item 2	Item 3
User 1	0.57	0.29	1.02
User 2	0.22	−0.16	0.97
User 3	1.03	0.95	0.99
User 4	1.02	0.94	0.99

3.5. Working Example

In this subsection, we provide a working example of our hybrid recommendation system for martial arts, dance, and play in seven Silk Road countries. For our prototype and convergence course for non-ICT major university students, we built a synthetic user history data set for 177 items. We assume that 2000 users are using the recommendation system for martial arts, dance, and play from seven Silk Road countries with a Poisson distribution where $\lambda = 1$, as shown in Figure 3. The x-axis is the number of unique items, and the y-axis is the number of users.

For KNN clustering, choosing the proper k is essential. To figure out the effect of k in KNN, we calculate the cluster sum of squares (WCSS), and the result is defined as Equation (8).

$$WCSS = \sum_{x \in C} (x - \mu_C)^2, \tag{8}$$

where x is a sample, C is a cluster, and μ_C is the center of the cluster.

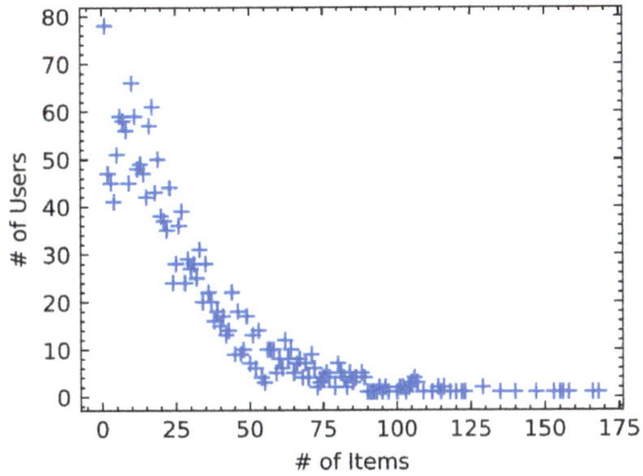

Figure 3. The distribution of users' item view history. Data points are marked as blue cross.

Figure 4 shows WCSS with the number of k increasing. The WCSS is inversely proportional to the number of k. The slope is steepest between $k = 1$ and $k = 2$, and the slopes decrease as the number of k increases. In this case, the proper value of k can be 2 or 3.

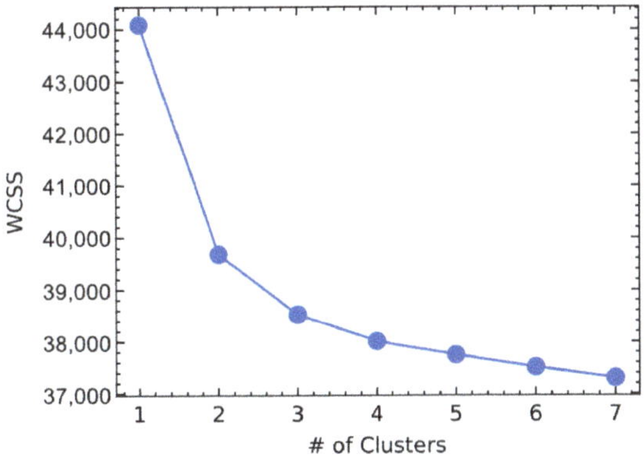

Figure 4. WCSS with the number of clusters. Data points are marked as blue dots.

In addition to WCSS, we use agglomerative and divisive hierarchical clustering to figure out the properties of the synthetic data set. To visualize the hierarchical clustering, we show the dendrogram for sample data. In order to calculate the distance between two clusters, we use Equation (9).

$$d(u,v) = \sqrt{\frac{|v|+|s|}{T}d(v,s)^2 + \frac{|v|+|t|}{T}d(v,t)^2 - \frac{|v|}{T}d(s,t)^2}, \quad (9)$$

where u (the union of s and t), v, s, and t are clusters, and T is $|v| + |s| + |t|$.

Figure 5 shows the dendrogram for hierarchical clustering with the Euclidean distance. Note that we only show a limited number of samples for visibility (100 in Figure 5a and 10 in Figure 5b). Note that the dendrogram does not indicate whether the clustering result is good or bad. Rather, it demonstrates the process of clustering to determine proper clustering.

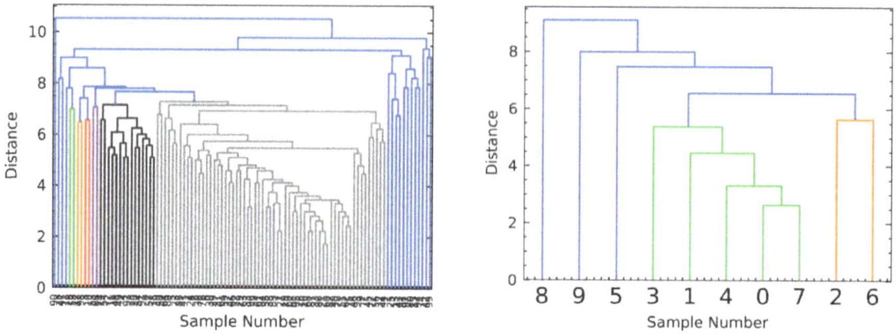

(**a**) # of sample: 100 (**b**) # of sample: 10

Figure 5. Dendrogram for hierarchical clustering. Different colors indicate different subtrees.

Figure 6 shows the prototype example of our hybrid recommendation system for martial arts, dance, and play in seven Silk Road countries. The figure shows the item for Taekwondo (a Korean form of martial arts) with its title, summary, representative figure, interest rating form, and four recommended items (at the bottom of the page). Note that the first two recommended items are from content-based filtering, and the last two items are from collaborative filtering.

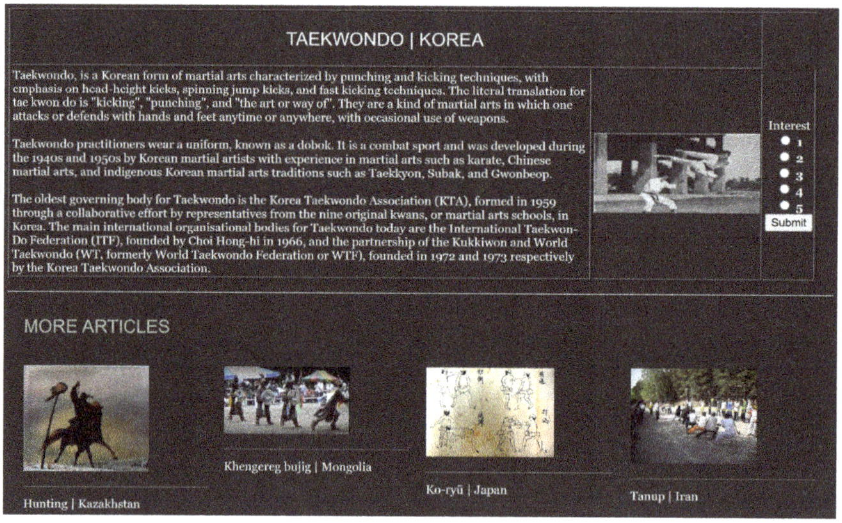

Figure 6. A prototype example of the recommendation system.

4. Curriculum and Discussion

In this section, we discuss the result of the proposed hybrid recommendation system for martial arts, dance, and play in seven Silk Road countries and its usefulness by proposing the convergence course for non-ICT major university students. Table 8 shows the outline of the convergence course for non-ICT university students. As a one-semester course for three credits, the course consists of many topics that should be covered for non-ICT major university students.

Table 8. Outline of the convergence course for non-ICT university students.

Week	Contents	Note
1	Programming environment setup - Installing Anaconda (Python, JupyterLab, etc.) - Construct virtual environments in Anaconda	Including a course introduction Homework Assignment #1 (Programming Environment Setup)
2–4	Python programming basics - Variables, assignment statements, and arithmetic operations - Input and output statements - Control statements	Use IDE, IPython, and Jupyter
5–7	Functions - Built-in functions and defining functions Classes - Defining user-defined classes, controlling access to attributes, and case studies Libraries - NumPy, Matplotlib, Pandas, etc.	Homework Assignment #2
8	Mid-term examination	Practical-based
9–11	Introduction to recommendation systems - Content-based filtering - Collaborative filtering	Homework Assignment #3
12–14	Implementing a hybrid recommendation system - Implementing content-based filtering in Python - Implementing collaborative filtering in Python	Homework Assignment #4
15	Build a website for the recommendation system	Presentation
16	Final examination	Practical-based

Hence, no advanced topics are covered in detail. Rather, it reviews essential topics and programming basics in Python. The students are expected to have studied web programming skills (e.g., HTML, Javascript, and Cascading Style Sheets) because the web programming course is set as a prerequisite. Therefore, the convergence course can be composed of the essential topics to implement the hybrid recommendation system for martial arts, dance, and play in seven Silk Road countries.

While the proposed convergence course for non-ICT university students is based on ICT technologies and programming languages (e.g., Python, HTML, and PHP), the main contents and data sets are about martial arts, dance, and play in seven Silk Road countries. Hence, it promotes learning both the cultural aspects of martial arts, dance, and play in the seven Silk Road countries and the programming concepts needed to develop the hybrid recommendation system.

The core concept of the recommendation system stems from mathematics. However, the students are non-ICT majors; some students may have no mathematical background. With consideration of that, the convergence course does not dive deep into mathematical topics. Rather, it focuses on the concept and guidance of using mathematical libraries in Python. Furthermore, the convergence course utilizes JupyterLab to provide a connection between the offline course and independent studying. With JupyterLab installed in university classrooms and on students' laptops or PCs (Homework Assignment #1), students can learn and study in an efficient way.

Unlike the previous studies [70,71], the learning of historical aspects of martial arts, dance, and play in Silk Road countries can contribute to the interdisciplinary field in different ways. First, martial arts and dance were often used as means of self-defense and physical exercise by people who traveled along the Silk Road. By learning about these practices, we can understand how people adapted to the challenges of their environment, developed resilience, and maintained their physical and mental well-being. This knowledge can help us promote healthy lifestyles and eco-friendly practices.

Second, play and entertainment were essential components of cultural exchange along the Silk Road. People from different cultures and backgrounds would come together to share their traditions, skills, and values through music, games, and performances. By learning about these practices, we can appreciate the diversity of human expression, foster cross-cultural understanding and respect, and promote cultural preservation and inclusivity.

Lastly, the historical aspects of martial arts, dance, and play in Silk Road countries can also inspire innovation and creativity in modern-day practices. By learning about the historical aspects of Silk Road countries, we can understand how different cultures and civilizations interacted with each other and how they adapted to the challenges of their environment. This knowledge can help us promote ethical practices in industry culture.

As far as non-ICT major students are concerned, there are learning challenges faced by non-ICT major students [72–74]. That is, (1) technical terminology (Non-ICT students may find it difficult to understand and grasp the technical jargon and terminology used in ICT subjects.), (2) lack of foundation (Non-ICT students may not have a strong foundation in computer science concepts, programming languages, and other fundamental ICT knowledge, making it harder for them to grasp advanced ICT topics.), and (3) rapid technological changes (ICT is a rapidly evolving field, and keeping up with the latest technologies and trends can be challenging for non-ICT students, especially if they do not have regular exposure to ICT-related subjects.).

Hence, we consider the following teaching principles for non-ICT major university students for the course:

- Fundamentals first: start with foundational concepts and ensure that non-ICT students have a solid understanding of basic ICT principles before moving on to more advanced topics.
- Practical application: emphasize hands-on projects and real-world applications of ICT concepts to help non-ICT students understand the practical relevance of what they are learning.
- Contextualization: relate ICT concepts to other disciplines and fields of study to help non-ICT students see the interdisciplinary nature of ICT and how it can be applied in various domains.

Developing a convergence course that balances technical and cultural/historical aspects can be challenging. Therefore, we suggest guidance on teaching instructions and assessments.

- Define learning objectives for both technical and cultural/historical aspects and communicate them clearly to students.
- Use project-based learning or case studies to contextualize technical skills within cultural and historical contexts.
- Incorporate diverse perspectives and voices in course materials and discussions to broaden students' cultural understanding.
- Provide opportunities for students to reflect on their learning and engage in critical thinking about the intersections between technology and culture/historical contexts.
- Use a variety of assessments, such as coding assignments, research papers, and presentations, to evaluate both technical skills and cultural/historical knowledge.

Regarding the guidance, our proposed curriculum meets the basic requirements of a convergence course and the needs of the era of the ICT convergence industry. In the context of convergence education, which encompasses both technical and non-technical aspects, Bloom's Taxonomy [75] is important. Bloom's Taxonomy provides a framework for categorizing educational objectives and learning outcomes into different levels of cognitive complexity. It can be applied to various fields, including the convergence of technical and non-technical subjects, to promote critical thinking, problem-solving, creativity, and holistic learning experiences. In the context of Bloom's Taxonomy, we suggest and recommend the following structures for our proposed convergence course:

1. **Remembering**
    - Encourage students to create mind maps or concept maps that connect technical and non-technical concepts.
    - Use multimedia resources, such as videos or interactive websites, to reinforce understanding and retention of information.
2. **Understanding**
    - Engage students in discussions that explore the connections between technical concepts and their real-life applications or societal impacts.
    - Assign case studies or project-based tasks that require students to analyze and explain the interplay between technical and non-technical factors.
3. **Applying**
    - Design hands-on activities or simulations that integrate technical skills with problem-solving in real-world contexts.
    - Encourage students to develop innovative solutions by combining technical knowledge with insights from non-technical fields.
4. **Analyzing**
    - Assign tasks where students evaluate the ethical, social, or environmental implications of technical decisions or advancements.
    - Provide opportunities for students to analyze and compare different approaches to solving a problem, considering both technical and non-technical factors.
5. **Evaluating**
    - Encourage students to critically assess the effectiveness and limitations of technical solutions for addressing complex societal challenges.
    - Foster debates or presentations where students evaluate the impact of emerging technologies on various aspects of society.
6. **Creating**
    - Assign open-ended projects that require students to design and develop innovative solutions by integrating technical and non-technical elements.
    - Encourage interdisciplinary collaboration, where students from different backgrounds contribute their expertise to create something new.

5. Conclusions

Developing an intelligent recommendation system for non-ICT major university students is not a trivial task since it requires ICT skills (data science and programming skills). In this paper, we present a convergence course for non-ICT major university students toward convergence with recommendation systems and Silk Road studies. Simultaneously, we try to integrate cultural aspects of martial arts, dance, and play from seven Silk Road countries into the proposed intelligent recommendation system. To achieve the two discrete goals, we built our database for martial arts, dance, and play from seven Silk Road countries, and the database is used in the recommendation system as the main content.

By bridging the gap between non-ICT disciplines and technology, this course has enabled students to acquire valuable skills in both domains. Through the course, students have gained a deeper understanding of the cultural heritage of Silk Road countries, particularly in the realms of martial arts, dance, and play. They have also learned various ICT techniques and tools necessary for developing a hybrid recommendation system. This hands-on experience has not only enhanced their technical proficiency but also fostered their appreciation for the rich diversity of cultural expressions.

When teaching the convergence course for non-ICT major students, there are a few limitations to consider. Since the convergence course aims to cover a wide range of topics within a limited time, it can be difficult to strike a balance between covering topics and providing sufficient depth in each subject area. In addition, students in the course may have varying skill levels and learning abilities. Hence, teaching methods to accommodate this diversity can be a challenge.

In this case, the convergence course can be a co-taught class that includes both technical and non-technical experts, which greatly enhances the effectiveness of learning in convergence education. The reasons for recommending a co-taught class are: (1) the combination of technical and non-technical experts brings a wide range of knowledge and skills to the classroom. Technical experts can provide an in-depth understanding of technology-related concepts, tools, and applications, while non-technical experts can contribute their expertise in other subject areas and pedagogical approaches; this comprehensive expertise allows for a more holistic and well-rounded learning experience. (2) Having both technical and non-technical experts in the classroom enables a seamless integration of different subject areas, fostering interdisciplinary learning and connections between various fields. (3) Students can engage in meaningful teamwork, bringing together diverse perspectives and skills to tackle complex challenges with both technical and non-technical experts since convergence education emphasizes collaborative problem-solving and real-world application of knowledge. (4) In a co-taught class, technical and non-technical experts can provide individualized support to students based on their unique needs; thus, students can receive guidance and assistance in both technical and non-technical aspects of the curriculum, ensuring a more personalized and tailored learning experience.

Author Contributions: Methodology and writing-original draft preparation, T.K.; conceptualization and writing-review, J.L. as the corresponding author. All authors have read and agreed to the published version of the manuscript.

Funding: This work was supported by the ministry of Education of the Republic of Korea and the National Research Foundation of Korea (NRF-2017-S1A5B4055939 and NRF-2021R1F1A1063307) and by Hankuk University of Foreign Studies Research Fund of 2023.

Institutional Review Board Statement: Not applicable.

Informed Consent Statement: Not applicable.

Data Availability Statement: Data available on request. The data are not publicly available due to privacy.

Acknowledgments: We thank the anonymous reviewers for their careful reading and insightful suggestions to help improve the manuscript.

Conflicts of Interest: The authors declare no conflict of interest.

References

1. Herr, D.J.C.; Akbar, B.; Brummet, J.; Flores, S.; Gordon, A.; Gray, B.; Murday, J. Convergence education—An international perspective. *J. Nanoparticle Res.* **2019**, *21*, 229. [CrossRef]
2. Verónica, A.S.; Natalizio, B.; Cibele, V.F.; Pulford, S.; Raquell, M.H. STEAM: Using the Arts to Train Well-Rounded and Creative Scientists. *J. Microbiol. Biol. Educ.* **2018**, *19*, 1–7. [CrossRef]
3. Szymkowiak, A.; Melović, B.; Dabić, M.; Jeganathan, K.; Kundi, G.S. Information technology and Gen Z: The role of teachers, the internet, and technology in the education of young people. *Technol. Soc.* **2021**, *65*, 101565. [CrossRef]
4. Balaji, S.; Nathani, K.; Technology, R.S.I. Applications and Challenges: A Contemporary Survey. *Wirel. Pers. Commun.* **2019**, *108*, 363–388. [CrossRef]
5. Sarooghi, H.; Sunny, S.; Hornsby, J.; Fernhaber, S. Design Thinking and Entrepreneurship Education: Where Are We, and What Are the Possibilities? *J. Small Bus. Manag.* **2019**, *57*, 78–93. [CrossRef]
6. Raj, R.; Sabin, M.; Impagliazzo, J.; Bowers, D.; Daniels, M.; Hermans, F.; Kiesler, N.; Kumar, A.N.; MacKellar, B.; McCauley, R.; et al. Professional Competencies in Computing Education: Pedagogies and Assessment. In Proceedings of the 2021 Working Group Reports on Innovation and Technology in Computer Science Education, Virtual, 26 June–1 July 2021. [CrossRef]
7. Liao, C.H.; Chiang, C.-T.; Chen, I.C.; Parker, K.R. Exploring the relationship between computational thinking and learning satisfaction for non-STEM college students. *Int. J. Educ. Technol. High. Educ.* **2022**, *19*, 43. [CrossRef]
8. Batmaz, Z.; Yurekli, A.; Bilge, A.; Kaleli, C. A review on deep learning for recommender systems: Challenges and remedies. *Artif. Intell. Rev.* **2019**, *52*, 1–37. [CrossRef]
9. Portugal, I.; Alencar, P.; Cowan, D. The use of machine learning algorithms in recommender systems: A systematic review. *Expert Syst. Appl.* **2018**, *97*, 205–227. [CrossRef]

10. Al-Mushayt, O.S.; Gharibi, W.; Armi, N. An E-Commerce Control Unit for Addressing Online Transactions in Developing Countries: Saudi Arabia—Case Study. *IEEE Access* **2022**, *10*, 64283–64291. [CrossRef]
11. Straits Research. Recommendation Engines Market Size Is Projected to Reach USD 54 Billion by 2030, Grow at a CAGR of 37%: Straits Research. Available online: https://www.globenewswire.com/en/news-release/2022/07/07/2476198/0/en/Recommendation-Engines-Market-Size-is-projected-to-reach-USD-54-Billion-by-2030-grow-at-a-CAGR-of-37-Straits-Research.html (accessed on 23 November 2022).
12. Zhang, Q.; Lu, J.; Jin, Y. Artificial intelligence in recommender systems. *Complex Intell. Syst.* **2021**, *7*, 439–457. [CrossRef]
13. Shi, S.; Gong, Y.; Gursoy, D. Antecedents of Trust and Adoption Intention toward Artificially Intelligent Recommendation Systems in Travel Planning: A Heuristic–Systematic Model. *J. Travel Res.* **2020**, *60*, 1714–1734. [CrossRef]
14. Del Carmen Rodríguez-Hernández, M.; Ilarri, S. AI-based mobile context-aware recommender systems from an information management perspective: Progress and directions. *Knowl.-Based Syst.* **2021**, *215*, 106740. [CrossRef]
15. Kim, T.-Y.; Ko, H.; Kim, S.-H.; Kim, H.-D. Modeling of Recommendation System Based on Emotional Information and Collaborative Filtering. *Sensors* **2021**, *21*, 1997. [CrossRef] [PubMed]
16. Javed, U.; Shaukat, K.; Hameed, I.A.; Iqbal, F.; Alam, T.M.; Luo, S. A Review of Content-Based and Context-Based Recommendation Systems. *Int. J. Emerg. Technol. Learn. (IJET)* **2021**, *16*, 274–306. [CrossRef]
17. Navin, K.; Krishnan, M.B.M. Context-Adaptive Content-Based Filtering Recommender System Based on Weighted Implicit Rating Approach. In *Computational Intelligence in Data Mining*; Nayak, J., Behera, H.S., Naik, B., Vimal, S., Pelusi, D., Eds.; Springer Nature: Singapore, 2022; pp. 295–307.
18. Iwendi, C.; Ibeke, E.; Eggoni, H.; Velagala, S.; Srivastava, G. Pointer-Based Item-to-Item Collaborative Filtering Recommendation System Using a Machine Learning Model. *Int. J. Inf. Technol. Decis. Mak.* **2022**, *21*, 463–484. [CrossRef]
19. Ghasemi, N.; Momtazi, S. Neural text similarity of user reviews for improving collaborative filtering recommender systems. *Electron. Commer. Res. Appl.* **2021**, *45*, 101019. [CrossRef]
20. Afoudi, Y.; Lazaar, M.; Al Achhab, M. Hybrid recommendation system combined content-based filtering and collaborative prediction using artificial neural network. *Simul. Model. Pract. Theory* **2021**, *113*, 102375. [CrossRef]
21. Tahmasebi, F.; Meghdadi, M.; Ahmadian, S.; Valiallahi, K. A hybrid recommendation system based on profile expansion technique to alleviate cold start problem. *Multimed. Tools Appl.* **2021**, *80*, 2339–2354. [CrossRef]
22. Sharma, S.; Rana, V.; Malhotra, M. Automatic recommendation system based on hybrid filtering algorithm. *Educ. Inf. Technol.* **2022**, *27*, 1523–1538. [CrossRef]
23. Vani, A.; Kumar, S.A. A Comprehensive Survey Using Novel Content-Based Filtering Technique to Improve Effectiveness on Social Media Datasets for Group Recommendation System Based on External Social Trust Networks. *ECS Trans.* **2022**, *107*, 13381. [CrossRef]
24. Azeroual, O.; Koltay, T. RecSys Pertaining to Research Information with Collaborative Filtering Methods: Characteristics and Challenges. *Publications* **2022**, *10*, 17. [CrossRef]
25. Alnadem, M.W.; LaythKamil, A.; Raad, R.A.; Saad, A. A hybrid approach towards movie recommendation system with collaborative filtering and association rule mining. *Acta Sci. Technol.* **2022**, *44*, 58925. [CrossRef]
26. Tiwari, R.G.; Khullar, V.; Garg, K.D. Web Recommendation by Exploiting User Profile and Collaborative Filtering. In *Recent Innovations in Computing*; Singh, P.K., Singh, Y., Kolekar, M.H., Kar, A.K., Gonçalves, P.J.S., Eds.; Springer: Singapore, 2022; pp. 335–345.
27. Pujahari, A.; Sisodia, D.S. Item feature refinement using matrix factorization and boosted learning based user profile generation for content-based recommender systems. *Expert Syst. Appl.* **2022**, *206*, 117849. [CrossRef]
28. Mee, A.; Homapour, E.; Chiclana, F.; Engel, O. Sentiment analysis using TF–IDF weighting of UK MPs' tweets on Brexit. *Knowl.-Based Syst.* **2021**, *228*, 107238. [CrossRef]
29. Zhang, L. Research on case reasoning method based on TF-IDF. *Int. J. Syst. Assur. Eng. Manag.* **2021**, *12*, 608–615. [CrossRef]
30. Ren, L.; Wang, W. An SVM-based collaborative filtering approach for Top-N web services recommendation. *Future Gener. Comput. Syst.* **2018**, *78*, 531–543. [CrossRef]
31. Dai, T.; Xiao, Y.; Liang, X.; Li, Q.; Li, T. ICS-SVM: A user retweet prediction method for hot topics based on improved SVM. *Digit. Commun. Netw.* **2022**, *8*, 186–193. [CrossRef]
32. Valdiviezo-Diaz, P.; Ortega, F.; Cobos, E.; Lara-Cabrera, R. A Collaborative Filtering Approach Based on Naïve Bayes Classifier. *IEEE Access* **2019**, *7*, 108581–108592. [CrossRef]
33. Yu, J.; Xuan, Z.; Feng, X.; Zou, Q.; Wang, L. A novel collaborative filtering model for LncRNA-disease association prediction based on the Naïve Bayesian classifier. *BMC Bioinform.* **2019**, *20*, 396. [CrossRef]
34. Gündoğan, E.; Kaya, M. A novel hybrid paper recommendation system using deep learning. *Scientometrics* **2022**, *127*, 3837–3855. [CrossRef]
35. Fu, M.; Qu, H.; Yi, Z.; Lu, L.; Liu, Y. A Novel Deep Learning-Based Collaborative Filtering Model for Recommendation System. *IEEE Trans. Cybern.* **2019**, *49*, 1084–1096. [CrossRef] [PubMed]
36. Chen, S.; Peng, Y. Matrix factorization for recommendation with explicit and implicit feedback. *Knowl.-Based Syst.* **2018**, *158*, 109–117. [CrossRef]
37. Aljunid, M.F.; Huchaiah, M.D. IntegrateCF: Integrating explicit and implicit feedback based on deep learning collaborative filtering algorithm. *Expert Syst. Appl.* **2022**, *207*, 117933. [CrossRef]

38. Khaledian, N.; Mardukhi, F. CFMT: A collaborative filtering approach based on the nonnegative matrix factorization technique and trust relationships. *J. Ambient Intell. Humaniz. Comput.* **2022**, *13*, 2667–2683. [CrossRef]
39. Ngaffo, A.N.; Choukair, Z. A deep neural network-based collaborative filtering using a matrix factorization with a twofold regularization. *Neural Comput. Appl.* **2022**, *34*, 6991–7003. [CrossRef]
40. Duan, R.; Jiang, C.; Jain, H.K. Combining review-based collaborative filtering and matrix factorization: A solution to rating's sparsity problem. *Decis. Support Syst.* **2022**, *156*, 113748. [CrossRef]
41. Chen, Y.; Cheng, L.; Wu, Y.-C. Bayesian low-rank matrix completion with dual-graph embedding: Prior analysis and tuning-free inference. *Signal Process.* **2023**, *204*, 108826. [CrossRef]
42. Xu, H. Enhancing Recommender Systems with NLP-based Biased Singular Value Decomposition. In Proceedings of the 2023 3rd International Symposium on Computer Technology and Information Science (ISCTIS), Chengdu, China, 7–9 July 2023; pp. 808–811. [CrossRef]
43. Do, P.M.T.; Nguyen, T.T.S. Semantic-enhanced neural collaborative filtering models in recommender systems. *Knowl.-Based Syst.* **2022**, *257*, 109934. [CrossRef]
44. Troussas, C.; Krouska, A.; Koliarakis, A.; Sgouropoulou, C. Harnessing the Power of User-Centric Artificial Intelligence: Customized Recommendations and Personalization in Hybrid Recommender Systems. *Computers* **2023**, *12*, 109. [CrossRef]
45. Kumari, T.S.; Sagar, K. A Semantic Approach to Solve Scalability, Data Sparsity and Cold-Start Problems in Movie Recommendation Systems. *Int. J. Intell. Syst. Appl. Eng.* **2023**, *11*, 825–837. Available online: https://www.ijisae.org/index.php/IJISAE/article/view/2917 (accessed on 20 September 2023).
46. Nikiforos, M.N.; Malakopoulou, M.; Stylidou, A.; Alvanou, A.G.; Karyotis, V.; Kourouthanassis, P. Enhancing Collaborative Filtering Recommendations for Web-based Learning Platforms with Genetic Algorithms. In Proceedings of the 2020 15th International Workshop on Semantic and Social Media Adaptation and Personalization (SMA), Zakynthos, Greece, 29–30 October 2020; pp. 1–6. [CrossRef]
47. Burke, R. Hybrid Recommender Systems: Survey and Experiments. *User Model. User-Adapt. Interact.* **2002**, *12*, 331–370. [CrossRef]
48. Ye, G. Canton Kung Fu: The Culture of Guangdong Martial Arts. *SAGE Open* **2019**, *9*, 2158244019861459. [CrossRef]
49. Mayer, M. China's Rise as Eurasian Power: The Revival of the Silk Road and Its Consequences. In *Rethinking the Silk Road: China's Belt and Road Initiative and Emerging Eurasian Relations*; Mayer, M., Ed.; Springer: Singapore, 2018; pp. 1–42.
50. Winter, T. The geocultural heritage of the Silk Roads. *Int. J. Herit. Stud.* **2021**, *27*, 700–719. [CrossRef]
51. Dellios, R. Silk Roads of the Twenty-first Century: The Cultural Dimension. *Asia Pac. Policy Stud.* **2017**, *4*, 225–236. [CrossRef]
52. Akiner, S. Silk roads, great games and Central Asia. *Asian Aff.* **2011**, *42*, 391–402. [CrossRef]
53. Kim, S.-W.; Lee, Y. Developing Students' Attitudes toward Convergence and Creative Problem Solving through Multidisciplinary Education in Korea. *Sustainability* **2022**, *14*, 9929. [CrossRef]
54. Deev, M.; Gamidullaeva, L.; Finogeev, A.; Finogeev, A.; Vasin, S. The Convergence Model of Education for Sustainability in the Transition to Digital Economy. *Sustainability* **2021**, *13*, 11441. [CrossRef]
55. Kharade, K.G.; Katkar, S.V.; Patil, N.S.; Sonawane, V.R.; Kharade, S.K.; Pawar, T.S.; Kamat, R.K. Text Summarization of an Article Extracted from Wikipedia Using NLTK Library. In *Advances in Computing and Data Sciences*; Singh, M., Tyagi, V., Gupta, P.K., Flusser, J., Ören, T., Sonawane, V.R., Eds.; Springer International Publishing: Cham, Stwizerland, 2021; pp. 195–207.
56. Kumar, S.; Jailani, N.A.; Singh, A.R.; Panchal, S. Sentiment Analysis on Online Reviews using Machine Learning and NLTK. In Proceedings of the 2022 6th International Conference on Trends in Electronics and Informatics (ICOEI), Tirunelveli, India, 28–30 April 2022; pp. 1183–1189. [CrossRef]
57. Yu, L.; Gao, X.-S. Improve Robustness and Accuracy of Deep Neural Network with L2,∞ Normalization. *J. Syst. Sci. Complex.* **2022**, *36*, 3–28. [CrossRef]
58. Mahmood, T.; Ilyas, M.; Ali, Z.; Gumaei, A. Spherical Fuzzy Sets-Based Cosine Similarity and Information Measures for Pattern Recognition and Medical Diagnosis. *IEEE Access* **2021**, *9*, 25835–25842. [CrossRef]
59. Chen, J.; Guo, Z.; Hu, J. Ring-Regularized Cosine Similarity Learning for Fine-Grained Face Verification. *Pattern Recognit. Lett.* **2021**, *148*, 68–74. [CrossRef]
60. Tang, Y.; Yu, Y. A Personalized Recommendation System for English Teaching Resources Based on Multi-K Nearest Neighbor Regression Algorithm. *Secur. Commun. Netw.* **2022**, *2022*, 7077123. [CrossRef]
61. Hao, Y.; Shi, J. Jointly Recommendation Algorithm of KNN Matrix Factorization with Weights. *J. Electr. Eng. Technol.* **2022**, *17*, 3507–3514. [CrossRef]
62. Airen, S.; Agrawal, J. Movie Recommender System Using K-Nearest Neighbors Variants. *Natl. Acad. Sci. Lett.* **2022**, *45*, 75–82. [CrossRef]
63. Gómez-Pulido, J.A.; Durán-Domínguez, A.; Pajuelo-Holguera, F. Optimizing Latent Factors and Collaborative Filtering for Students' Performance Prediction. *Appl. Sci.* **2020**, *10*, 5601. [CrossRef]
64. Mongia, A.; Jhamb, N.; Chouzenoux, E.; Majumdar, A. Deep latent factor model for collaborative filtering. *Signal Process.* **2020**, *169*, 107366. [CrossRef]
65. Figuera, V.P.; García, B.P. On the Probabilistic Latent Semantic Analysis Generalization as the Singular Value Decomposition Probabilistic Image. *J. Stat. Theory Appl.* **2020**, *19*, 286–296. [CrossRef]
66. Shi, X.; He, Q.; Luo, X.; Bai, Y.; Shang, M. Large-Scale and Scalable Latent Factor Analysis via Distributed Alternative Stochastic Gradient Descent for Recommender Systems. *IEEE Trans. Big Data* **2022**, *8*, 420–431. [CrossRef]

67. Luo, X.; Wang, D.; Zhou, M.; Yuan, H. Latent Factor-Based Recommenders Relying on Extended Stochastic Gradient Descent Algorithms. *IEEE Trans. Syst. Man Cybern. Syst.* **2021**, *51*, 916–926. [CrossRef]
68. Wang, S.; Yan, Q.; Zhang, J.; Wang, J.; Song, L. Coded Alternating Least Squares for Straggler Mitigation in Distributed Recommendations. In Proceedings of the 2021 IEEE International Symposium on Information Theory (ISIT), Melbourne, Australia, 12–20 July 2021; pp. 1058–1063. [CrossRef]
69. Che, R.; Yuan, Y. A Game-Incorporated Alternative Least Squares-Based Approach to Latent Factor Analysis. In Proceedings of the 2021 IEEE 6th International Conference on Big Data Analytics (ICBDA), Xiamen, China, 5–8 March 2021; pp. 249–256. [CrossRef]
70. Syed, J.; Ying, Y.-H. Introduction: Diversity, Convergence and Opportunities Along the Belt and Road. In *China's Belt and Road Initiative in a Global Context: Volume I: A Business and Management Perspective*; Syed, J., Ying, Y.-H., Eds.; Springer International Publishing: Cham, Switzerland, 2019; pp. 1–19.
71. Amarasinghe, P. Revival of the Silk Road legacy. Examining how China uses history to legitimize its ‹‹Belt and Road Initiative››. *Historia Świat* **2023**, *12*, 379–396. [CrossRef]
72. Huang, Y.-H.; Tu, Y.-L.; Wu, H.-J.; Wan, C.-P.; Yeh, C.-H.; Lu, L.-H.; Tsai, I.-C. Applying an Innovative Blended Model to Develop Cross-Domain ICT Talent for University Courses. In Proceedings of the 2019 IEEE Frontiers in Education Conference (FIE), Covington, KY, USA, 16–19 October 2019; pp. 1–5. [CrossRef]
73. Ferdinandy, B.; Guerrero-Higueras, Á.M.; Verderber, É.; Rodríguez-Lera, F.J.; Miklósi, Á. Exploratory study of introducing HPC to non-ICT researchers: Institutional strategy is possibly needed for widespread adaption. *J. Supercomput.* **2021**, *77*, 4317–4331. [CrossRef]
74. Karunakaran, S.; Dhanawardana, R. Integration of ICT in the Teaching-Learning Process: Challenges and Issues Faced by Social Science Teachers. *Eur. J. Educ. Pedagog.* **2023**, *4*, 24–30. [CrossRef]
75. Krathwohl, D.R. A Revision of Bloom's Taxonomy: An Overview. *Theory Into Pract.* **2002**, *41*, 212–218. [CrossRef]

applied sciences

MDPI

Article

Composing Multiple Online Exams: The Bees Algorithm Solution

Manar Hosny [1,*], Rafa Hayel [1,2] and Najwa Altwaijry [1]

1 Department of Computer Science, College of Computer and Information Sciences, King Saud University, Riyadh 12371, Saudi Arabia; rafa.hayel@taiz.edu.ye (R.H.); ntwaijry@ksu.edu.sa (N.A.)
2 Department of Computer Science, Faculty of Applied Sciences, Taiz University, Taiz P.O. Box 6803, Yemen
* Correspondence: mifawzi@ksu.edu.sa; Tel.: +966-118052427

Abstract: Online education has gained increasing importance in recent years due to its flexibility and ability to cater to a diverse range of learners. The COVID-19 pandemic has further emphasized the significance of online education as a means to ensure continuous learning during crisis situations. With the disruption of traditional in-person exams, online examinations have become the new norm for universities worldwide. Among the popular formats for online tests are multiple-choice questions, which are drawn from a large question bank. However, creating online tests often involves meeting specific requirements, such as minimizing the overlap between exams, grouping related questions, and determining the desired difficulty level. The manual selection of questions from a sizable question bank while adhering to numerous constraints can be a laborious task. Additionally, traditional search methods that evaluate all possible solutions are impractical and time-consuming for such a complex problem. Consequently, approximate methods like metaheuristics are commonly employed to achieve satisfactory solutions within a reasonable timeframe. This research proposes the application of the Bees Algorithm (BA), a popular metaheuristic algorithm, to address the problem of generating online exams. The proposed solution entails creating multiple exam forms that align with the desired difficulty level specified by the educator, while considering other identified constraints. Through extensive testing and comparison with four rival methods, the BA demonstrates superior performance in achieving the primary objective of matching the desired difficulty level in most test cases, as required by the educator. Furthermore, the algorithm exhibits robustness, indicated by minimal standard deviation across all experiments, which suggests its ability to generalize, adapt, and be practically applicable in real-world scenarios. However, the algorithm does have limitations related to the number of successful solutions and the achieved overlap percentage. These limitations have also been thoroughly discussed and highlighted in this research.

Keywords: metaheuristics; Bees Algorithm; online exams; swarm intelligence

1. Introduction

Over the past few years, the education system has experienced a significant transformation, primarily attributable to the impact of the COVID-19 pandemic. The shift towards online exams has been motivated by the imperative to prioritize the safety and well-being of both students and staff, while also ensuring the uninterrupted progression of education. Online exams provide students with a secure environment in comparison to traditional modes of examination. In general, from an instructor standpoint, multiple-choice questions (MCQ) can be advantageous with respect to the ease of scoring, perceived objectivity in grading, fast return of scores in large classes, and the capacity to ask more questions [1,2].

The process of selecting exam questions, though, can be time-consuming, particularly when multiple exam models are necessary. Furthermore, when exams are administered to a large number of students at different times, it becomes essential to have a question bank that encompasses a considerable number of questions. Question banks facilitate the inclusion of questions in the exam at different intervals, offering a variety of selection criteria. This approach enhances the diversity of the generated exams. However, a challenge faced in both automatic and manual online exam generation is the question bank's size. To ensure exam variety and minimize repetition, having a question bank larger than the total number of questions required is generally recommended [3,4]. While there is no specific guideline regarding the exact question bank size in relation to the number of questions needed, educational experts and researchers suggest having a substantial number of questions available. A larger question bank offers broader coverage of the curriculum and ensures adequate assessment of all important topics [5,6]. The specific question bank size will depend on factors such as the subject matter, questions' difficulty level, and desired variety in the exams [3,4]. Generating various exam models from a question bank can be accomplished through manual or automated methods. The manual process is typically intricate and time-consuming, while employing an algorithm to automate the task allows for precise adherence to specific criteria instead of random question selection. This approach guarantees that the exam models maintain consistent difficulty levels and meet the required constraints.

The multiple-exam composition problem has recently attracted the attention of many researchers [2]. The problem at hand involves the automated generation of multiple MCQ exam papers or test sheets. These test sheets cover the same subject matter and maintain an identical or nearly identical difficulty level, ensuring a fair assessment process. It is crucial that the difficulty levels of the generated exams align with the desired test level established by the educator. Furthermore, to promote diversity and mitigate the potential for cheating, the overlap between the different tests, in terms of identical questions, should not exceed a specified percentage.

In this study, we propose a method to simultaneously create multiple forms of MCQ exams while adhering to specific criteria, including the difficulty level, avoidance of question repetition, and a predetermined number of questions in each section. Given that traditional search methods that exhaustively examine all feasible solutions are often impractical and time-consuming for larger problem sizes, we introduce a metaheuristic algorithm as a solution, namely, the Bees Algorithm (BA) [7].

In brief, the concept of the BA revolves around a set of initial solutions. The algorithm then focuses on exploiting promising and elite solutions by making modifications to enhance the likelihood of discovering improved solutions. The ultimate objective is to achieve the optimal or near-optimal solution through these iterative refinements [8]. The motivation for employing the BA algorithm lies in its ability to effectively tackle real-world problems that have proven challenging for other classical algorithms [8]. This algorithm, in particular, demonstrates promise in addressing combinatorial optimization problems [7]. Moreover, for the multiple-exam composition problem specifically, we opted for the BA for two primary reasons. Firstly, despite its popularity and simplicity, the BA has not been previously applied to solve this particular problem, unlike other well-known methods such as Simulated Annealing (SA) [9], the Genetic Algorithm (GA) [10], Particle Swarm Optimization (PSO) [2,11,12], and Ant Colony Optimization (ACO) [13]. Secondly, we believe that the BA possesses an inherent advantage over other methods which directly aligns with the multiple-exam composition problem. Specifically, it naturally selects the best and elite solutions as part of its optimization process, thereby readily generating multiple exams (i.e., solutions) in each run. In other words, the BA does not require any additional steps to generate multiple tests simultaneously, as seen in other algorithms, such as the random algorithm [14], SA [9], GA [10], and ACO [13], which involve repeating runs and/or sorting the population to select the best individuals, or incorporating parallelization, as in PSO [2,11,12].

The problem tackled in this research is defined as follows [15]:

Let a question bank $QB = \{Q_1, Q_2, Q_3, \ldots, Q_k\}$ and the generated test $T_i = \{Q_1, Q_2, Q_3, \ldots, Q_n\}$ where $n \leq k$. The difficulty level requirement of the exam set by the educator is $DLR \in [0, 1]$; the difficulty level of each question, $DLx \in [0, 1]$, is also determined by the educator. Moreover, each question belongs to a certain chapter (part). The objective function that determines the quality of the solution (exam) computes the average of the difficulty values of all questions in the exam and then compares it to the required difficulty level *DLR*, as shown in Equation (1) [15].

$$f(Q_{ij} \cdot DL) = \left| \frac{\sum_{j=1}^{n} Q_{ij} \cdot DL}{n} - DLR \right| \tag{1}$$

where $Q_{ij} \cdot DL$ represents the difficulty level of question Q_{ij}. The smaller $f(Q_{ij} \cdot DL)$ becomes, the better the fitness of the solution. In addition, each generated exam must adhere to the following constraints, where C1, C2, and C3 are hard (i.e., mandatory) constraints, while C4 is a soft (i.e., desired but not mandatory) constraint:

C1: Each question in the exam must be unique.

C2: The difficulty level of each question should not be concentrated around the required difficulty level of the exam, ensuring diversity. In other words, selecting questions all of the same difficulty level for the test is not allowed.

C3: The number of questions extracted from each chapter (or section) must match the desired number specified by the educator.

C4: To generate multiple exams while ensuring a high level of uniqueness in each test, it is desirable to restrict the percentage of overlapping questions between tests. The acceptable overlap percentage can be determined using Equation (2). This approach guarantees that the exams exhibit minimal redundancy and maintain a diverse collection of questions. Following [2], the maximum acceptable overlap percentage is 0.3. Assuming we have m generated tests, the overlap percentage P_{test} between the m tests is calculated as follows:

$$P_{test} = \frac{O - U}{m \cdot n} \cdot 100 \tag{2}$$

where

- O is the number of overlapping questions between all tests;
- U is the number of unique questions within O;
- m is the number of tests generated;
- n is the number of questions in each test.

The $O - U$ term accounts for the diversity of the questions involved in the overlapping count. Its purpose is to slightly mitigate the impact of overlap when the questions involved in the overlapping count are different. This approach is used to address situations in which the overlapping portion consists of a small number of questions, indicating a higher similarity between the exams. To illustrate this concept, consider the following example:

Let us consider five exams, each consisting of ten questions. Suppose there are 15 overlapping questions, all of which come from Q1, Q2, and Q3. In this case, the overlap percentage would be calculated as $((15 - 3)/50) \times 100 = (12/50) \times 100 = 24\%$.

On the other hand, suppose we have 15 overlapping questions, but these questions include Q1, Q2, Q3, Q4, and Q5. In this case, the overlap percentage would be $((15 - 5)/50) \times 100 = (10/50) \times 100 = 20\%$. The second scenario is considered better than the first because it demonstrates more diversity in the questions included in the overlapping. If we remove the "U" (unique) component from Equation (2), both scenarios would yield the same overlap percentage, which we believe is not fair.

In this research, the BA is employed to construct a number of MCQ exams from a question bank while adhering to specific constraints. The primary objective of the solution is to attain the desired difficulty level determined by the educator. As a metaheuristic algorithm, the BA offers an approximate solution that can be optimal or near-optimal. As previously highlighted, the BA is well-suited for simultaneously generating multiple exam forms as it interacts with a population of solutions and selects the best and elite solutions during an intensified search process.

This study presents four main contributions. Firstly, it introduces a unique adaptation of the BA, specifically designed to address this problem. Secondly, it enables the generation of multiple online tests simultaneously, meeting the user requirements without resorting to parallelization techniques. Thirdly, it introduces problem-specific neighborhood moves, crafted to adhere to the constraints of this particular problem. Finally, the experimental results demonstrate that our proposed method performs favorably compared to existing approaches, delivering high-quality solutions to the problem at hand.

The remainder of this paper is organized as follows. Section 2 reviews some background information, as well as previous related work. Section 3 details the proposed BA approach for the multiple-exam composition problem. Section 4 describes our experiments and results. Section 5 discusses our findings, and finally Section 6 concludes the study.

2. Background and Related Work

This section provides an overview of the fundamentals of metaheuristics and the BA, while the related work part briefly covers the single-exam generation problem, in which the proposed algorithm generates only one exam paper, and multiple-exam generation, where multiple versions of the exam are generated.

2.1. Metaheuristics

A metaheuristic is a problem-independent algorithmic framework that can solve intricate and complex problems by providing "acceptable" solutions in a reasonable amount of time; it follows a set of principles or strategies for guiding algorithms using heuristic optimization [16]. Metaheuristic algorithms generally outperform simple heuristics. Numerous different classifications exist for metaheuristic algorithms. One approach classifies them as population-based versus single-solution-based. Population-based algorithms search a problem domain by processing a population of individuals (i.e., possible problem solutions) [17]. Examples of this category include Genetic Algorithms (GAs) and Particle Swarm Optimization (PSO). On the other hand, there exist algorithms with a single solution that is traversed over the search space like Hill-Climbing (HC) and Simulated Annealing (SA) [18]. This work focuses on the Bees Algorithm (BA) as a population-based metaheuristic.

2.2. Bees Algorithm (BA)

Bee-inspired algorithms are commonly employed as metaheuristic optimization techniques, leveraging the intelligent behavior of honeybees [19]. These algorithms belong to the category of swarm intelligence optimization methods, drawing inspiration from the cooperative foraging behavior of natural bee colonies [20]. The Bee Algorithm (BA) and its closely related variant, the Artificial Bee Colony (ABC), have demonstrated successful applications in various optimization problems. They have been particularly effective in domains such as vehicle routing and transportation, as exemplified by [21–24], as well as in timetabling and scheduling problems, as demonstrated in [25–27].

In nature, a bee colony is characterized as a collective entity that exhibits a wide-ranging presence, extending over significant distances (often exceeding ten kilometers) in multiple directions simultaneously to explore a vast area in search of food. The success of a colony's foragers relies on their ability to be directed towards productive food sources. In theory, flower patches that can attract a large number of bees are expected to offer abundant nectar and pollen, making them easily accessible with minimal energy expenditure. Conversely, flower patches that provide limited nectar or pollen tend to attract fewer bees [28].

In a bee colony, foraging begins with scout bees searching for suitable flowers. These scouts explore their surroundings in a random manner, moving from one patch to another. When a scout bee discovers a source of nectar or pollen, it returns to the hive and performs a unique behavior known as the waggle dance on the dance floor. This peculiar dance holds significance in colony communication, as it conveys three essential pieces of information about the flower patches: direction, distance, and source quality, as depicted in Figure 1.

Equipped with this information, the colony can effectively guide its bees without the need for external supervision or a predetermined blueprint. Follower bees inside the hive, who have been awaiting the dance, as well as the scout bees, then return to the flower patch. While collecting food from the patch, the bees assess its quality and continue to refine the waggle dance, if necessary, upon their return to the colony. This iterative process ensures that the colony continues its search for the most suitable patches, with bees performing additional waggle dances to recruit more individuals if the patch still contains ample food and requires additional workers [28]. The BA tries to mimic this process by applying the following steps [28]:

- Initialize the population with random solutions.
- Evaluate the fitness of the population.
- While (stopping condition not met):
    - o Select the best sites for neighborhood search.
    - o Select elite sites from the best sites.
    - o Recruit bees around selected sites and evaluating fitness.
    - o Select the fittest bee from each site.
    - o Assign the remaining bees to a random search and assessing their fitness.
- End While.
- Return the best bee as the final solution.

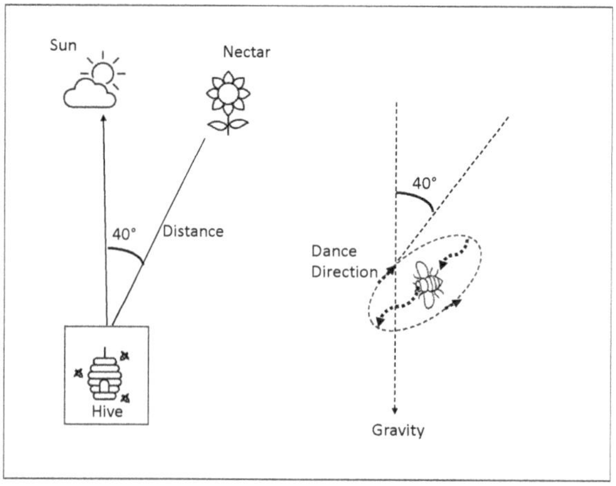

Figure 1. Information that appears in the waggle dance (inspired by [29]).

The BA begins with a population of solutions as its initial starting point. During the first iteration, bees make numerous local moves in order to explore potential solutions to the problem. This process is repeated until a predetermined number of solutions have been discovered. The best solutions, determined by their fitness, are retained. Subsequently, new neighboring solutions are generated around these top-performing solutions. These newly generated solutions are then evaluated to estimate the solutions that should be aimed for. The process continues iteratively until the algorithm satisfies a specified termination condition, ultimately returning a satisfactory solution to the problem [30].

2.3. The Single-Exam Generation Problem

The demand for performing online exams has increased remarkably in recent times. As a result, there has been a growing interest in developing optimal and efficient methods to generate online exams. Several approaches have been employed to address the single-exam generation problem. One straightforward method is the randomization approach utilized in [14], in which a Shuffling Algorithm was employed to detect and prevent the duplication or repetition of randomly generated questions. Additionally, Simulated Annealing (SA) has been employed in [9] to select test items in a web-based environment, aiming to achieve a balanced distribution of information, content coverage, and item exposure (i.e., a measure of the frequency with which a test question is selected), within the generated test. Nevertheless, more sophisticated metaheuristic optimization algorithms are the current trend for solving this problem, as evident by [2,10–13,15,31–34]. Particle Swarm Optimization (PSO) is a widely used optimization algorithm that imitates a flock of birds to locate an optimal solution. In PSO, each bird is referred to as a particle that flies searching for nourishment, which then shares the location of a favorable site. By doing so, the rest of the flock, or the swarm, can profit until they locate the optimal spot or solution. PSO was used in [15] to generate exams based on an objective difficulty level, while satisfying several constraints, such as the required number of questions in the exam and the number of questions used per section. A fitness function for each particle, i.e., exam, was calculated. The swarm keeps moving towards the global best until it reaches an optimal solution or satisfies the termination condition. The results of this approach showed that PSO succeeds in finding the best test to fit the requirements from a question bank with great speed, accuracy, and stability.

The Genetic Algorithm (GA) has also proven to be powerful for generating online exams [10,15,31–33]. GA is an algorithm based on natural selection, in which the best individuals are selected for reproduction. The offspring then inherit characteristics from their parents. At the start of each generation, several individuals are selected and sorted according to their respective fitness levels. If the individual with the best fitness satisfies the requirements, the algorithm will stop. Otherwise, crossover and mutation will take place. The algorithm keeps iterating this process until a satisfactory solution is found.

Studies [31–33] used GAs to generate online tests. The work in [31] attempted to generate tests with many requirement constraints, such as the number of questions, question types, difficulty, answer time, and exposure. The fitness function measures the difficulty level of the test, the class distribution, the total time needed to complete the test and the average exposure to the test paper. This approach yielded reasonable, diverse test papers, with a fast-running time of approximately 10 s.

The researchers in [32] developed two distinct models to generate tests, each considering a problem with a different objective. The first model solves the Specified Length of Assessment Time problem (SLAT), in which the generated exam must be solved in a fixed period of time. The second is the Fixed Number of Test Items (FNTI) problem, wherein the exam must consist of a specific number of questions. The proposed models are the Concept Lower Bound First Genetic (CLFG) approach, which addresses SLAT problems, and the Feasible Item First Genetic (FIFG) approach, used to solve FNTI problems. Both models attempt to make all questions in an exam meet the lower bound of the expected relevance of each concept to be addressed.

Another work that also used GA to generate a test is [33], in which each question in the question bank had the following attributes: a question score, difficulty, teaching requirement, item type, unit of knowledge point, estimated answer time, and item number. The fitness function depends on the deviation between the required values set by the user and the actual values of the following constraints: knowledge points, difficulty coefficient, teaching requirement, and the time required. The algorithm generated tests that met all the requirements and were vigorously tested by altering the size of the question bank and changing the type of questions.

2.4. The Multiple-Exam Generation Problem

When it comes to tests, educators often create multiple versions of each test to prevent cheating among students. However, this introduces a new challenge where the generated tests must maintain similar difficulty levels while covering the same content to ensure fairness. To address this challenge, researchers have explored various metaheuristic algorithms, such as PSO, GA, and ACO, which have shown promising results in generating appropriate versions of tests.

In order to produce multiple unique exams in one run, researchers opted for parallelism in [2], using multi-swarm PSO. In their approach, a thread represents a swarm responsible for forming an exam that satisfies certain constraints, such as uniqueness, the diversity of the difficulty levels between questions, and multi-part questions appearing together. Also, the number of questions extracted from each section is equal to a pre-specified number. Each particle within a swarm thread represents a nominee exam, for which the average difficulty level is the objective function. To ensure the variety and uniqueness of the exam, the overlap percentage between exams should not exceed the overlap threshold set by the educator. Throughout the program, threads are connected via migration, where more robust swarms help the weaker ones. A thorough examination of the results has been conducted, through which the approach in [2] proved to be superior to many others, including Simulated Annealing (SA) [9] and Random Search algorithms [14].

The work in [11] also utilized a single objective PSO approach. The constraints that the resulting exams had to meet were identical to those in [2]. The fitness function calculates the average difficulty level of the exam and subtracts it from the required difficulty level. As the fitness value decreases, the quality of the exam improves. In order to generate multiple exams, the researchers employed parallelism, assigning each thread to a separate swarm. This approach yielded satisfactory solutions within a short timeframe. However, the algorithm faced challenges in terms of generating sufficient variety. Consequently, none of these migration theories were applied in this study.

The study in [12] proposed using Multiswarm Multiobjective Particle Swarm Optimization (MMPSO) to generate multiple tests in one run. The outcome of this approach depends on the tests' overall difficulty and the total testing time. To generate multiple tests, several subswarms were created. SA was employed to create the initial population, thus increasing the diversity of the generated tests. Here the constraints set were also identical to those in [2]. The objective function combines the average difficulty of the exam and the total exam duration. The MMPSO algorithm with SA and migration was tested against several PSO algorithms such as MMPSO, MMPSO with SA, and MMPSO with migration (without SA). The results indicated that MMPSO with migration and SA achieved excellent results with respect to the variety and number of successful attempts.

The study [10] presents an automated generator for multiple test questions using the GA. The questions were categorized in the question bank based on the six levels of Bloom's classification. Previous exam questions were processed and stored in the question bank based on Bloom's classification level. As shown in the results, the decrease in the number of questions impacts each level in Bloom's taxonomy, while a test is of good quality when it covers at least three levels of Bloom's taxonomy.

In [34], a different metaheuristic algorithm was employed. A Tabu-Search-based algorithm called the Bit Map Selection Tabu Approach (BMST) was used to search the solution space effectively. Tabu search prohibits revisiting solutions that were previously encountered. The constraint set was the gross relevance of the chosen questions between the maximum and minimum boundaries of the related concept to be learned. The fitness function stands for the average discrimination of the resulting test. BMST was tested against the feasible-time-first Algorithm (FTF), Random Search, and exhaustive search algorithms. The results showed that near-optimal test sheets can be obtained in much shorter times when using BMST.

Another popular metaheuristic algorithm that can solve such problems is the Ant Colony Optimization (ACO) algorithm. It emulates the attitude of ants while searching for food to find the optimized solution. The main procedure of this algorithm, as described in [13], is based on constructing a graph, where each question from the question bank is represented as a node, and the connecting lines denote the paths formed by the ants to create a test. During the construction process, as the ant builds a solution, the heuristic and pheromone information is influenced by the discriminative nature of the question and signifies the suitability of including the question in the current solution. Once the colony completes the construction of tests, the best solution is chosen. Subsequently, a local search is performed on the best solution to enhance it by exchanging certain questions. Finally, the pheromone values associated with the questions in the best test are updated in the global pheromone memory. The researchers highlight that the results demonstrate the effectiveness of the ACO method in generating tests that meet the requirements and provide satisfactory solutions, which are not impacted by the size of the question bank.

As observed from the review of related work, none of the previous studies have employed the BA to solve the multi-exam generation problem. We firmly believe that the BA is particularly well-suited for addressing this problem due to its capability of handling multiple preferred solutions concurrently, such as the best and elite solutions. This feature facilitates the extraction of multiple tests simultaneously without the need for parallelization, as will be discussed shortly. Furthermore, the BA stands out for its simple implementation and straightforward parameter configuration compared to other types of metaheuristics. It possesses the advantage of effectively handling both discrete and continuous optimization problems, while also adapting its behavior based on the quality of the generated solutions [35]. Therefore, in our research, we aim to explore the feasibility of utilizing the BA as an approach to tackling this problem.

3. Methodology

This section provides a detailed description of the algorithm employed to address the multiple-exam composition problem. Firstly, we will outline the main steps followed by the BA to solve this problem. After specifying the number of desired exams (m), the required difficulty level (DLR), the number of questions in each part, and the overlap threshold, the algorithm proceeds as follows:

1. Generate an initial population of bees ($s > m$), in which each bee represents a potential exam, and s denotes the scout bees. Sort the population based on each bee's fitness, i.e., the exam's difficulty level, calculated using Equation (1).
2. Select the fittest bees (m) for neighborhood search. During neighborhood search, some questions in the selected exam are replaced, resulting in a new exam or neighbor. The number of neighbors generated for the elite (fittest) bees (e) within m is higher than the number of neighbors generated for the remaining bees (m-e). The fittest bee from each of the m patches or neighborhoods is then chosen to transfer to the next generation. Two neighborhood search methods are applied: random for the best and elite bees, and deterministic when the set of m exams fails the overlap condition. In the deterministic method, overlapping questions are replaced with new questions from the same chapters with similar difficulty levels. This preserves the fitness value while reducing the overlap percentage.
3. Test the overlap condition by comparing the m exams and keeping track of overlapping questions. If the overlap condition is satisfied, select the fittest bees from the m patches to transfer to the next generation and discard the remaining (s-m) bees. However, if the ratio of overlapping questions to the remaining questions in the exams exceeds the overlap threshold, perform a new neighborhood search operation on the previously selected m bees until the overlap condition is met. The overlap percentage is calculated by dividing the number of repeated questions by the total number of questions in the m exams, as shown in Equation (2).
4. Generate new (s-m) bees randomly to maintain the original population size.

The stopping condition test memorizes the best DLR and checks whether the overlap percentage has changed. If not, the algorithm returns the desired m exams. The process (steps 2–4) repeats until the termination condition is met. The termination condition can be satisfied when the resulting set of bees includes m exams that fulfill all the conditions specified by the educator, or when the algorithm reaches a maximum pre-defined number of iterations, or when the overlapping percentage remains unchanged for five consecutive iterations. Algorithm 1 presents the pseudocode of the proposed method.

Algorithm 1: BA for Multiple Exam Generation.

Input: number of the desired exams (*m*), required difficulty level, overlap percentage, number of required chapters and the number of questions in each chapter.
Output: multiple tests satisfying the conditions.
Initialize the number of scout bees *s*, number of best sites *e* and patch size.
Initialize a population of *s* scout bees.
Repeat until maximum iterations is reached.
    If iteration ! = 0 **Then**
        If the overlap condition is not satisfied **Then**
            Apply overlap neighborhood move.
        End if
        If stopping condition is satisfied **Then**
            Return the *m* exams generated.
        End if
    End if
    Calculate the objective function of the population.
    Choose *m* best sites for neighborhood search.
    Assign the size of the neighborhood (patch size).
    Apply neighborhood search; search more around the best *e* sites.
    Select the fittest bee from each patch.
    Assign the (*s*-*m*) remaining bees to random search.
    Generate a new population of scout bees.
End Repeat
Return the *m* exams generated.

A crucial aspect of the algorithm is the representation of a problem's solution, which, in our case, corresponds to an exam or test. The exam is structured into parts or chapters, with each part containing multiple questions of varying difficulty levels. The educator specifies the number of questions per part, the overall difficulty level of the test, and the desired overlap percentage between tests. Hence, the solution is represented as a test object comprising parts or chapters, where each part consists of a collection of questions identified via a question ID. Figure 2 provides an illustration of a solution representation for an exam.

As mentioned earlier, our neighborhood move involves replacing one question in the exam to generate a new exam. In the following section, we present an example of a question bank that will be utilized to illustrate and describe the behavior of our algorithm.

Ch1	Q1	Q2	Q3	Q4	Q5	Q6	Q7	Q8	Q9	Q10
DL	0.56	0.93	0.45	0.12	0.98	0.71	0.24	0.25	0.2	0.62
Ch2	Q11	Q12	Q13	Q14	Q15	Q16	Q17	Q18	Q19	Q20
DL	0.37	0.54	0.5	0.36	0.86	0.13	0.9	0.67	0.76	0.28
Ch3	Q21	Q22	Q23	Q24	Q25	Q26	Q27	Q28	Q29	Q30
DL	0.42	0.21	0.33	0.60	0.13	0.59	0.78	0.94	0.81	0.63

The initial step involves initializing a random population. In this example, population size = 3, and there are two exams to be generated (*m* = 2). Each exam consists of five questions, with two questions allocated to chapter 1, two questions to chapter 2, and one question to chapter 3. The desired difficulty level is set to 0.65, and the overlap limit is defined as 0.2. The fitness of each exam is calculated by computing the difference between the average difficulty level of the exam and the required difficulty level specified by the user.

Exam					
Ch1	Q1	Q2	Q3	…	Qn1
Ch2	Q1	Q2	Q3	…	Qn2
Ch3	Q1	Q2	Q3	…	Qn3
Ch4	Q1	Q2	Q3	…	Qn4
:	:	:	:	:	:
Chc	Q1	Q2	Q3	…	Qnc

Figure 2. An example of a solution representation.

Exam 1	Question	Q8	Q9	Q13	Q11	Q24	Fitness
	Chapter	Ch1	Ch1	Ch2	Ch2	Ch3	0.266
	Difficulty	0.25	0.2	0.5	0.37	0.60	

Exam 2	Question	Q1	Q7	Q17	Q14	Q25	Fitness
	Chapter	Ch1	Ch1	Ch2	Ch2	Ch3	0.212
	Difficulty	0.56	0.24	0.9	0.36	0.13	

Exam 3	Question	Q2	Q6	Q12	Q18	Q29	Fitness
	Chapter	Ch1	Ch1	Ch2	Ch2	Ch3	0.016
	Difficulty	0.93	0.71	0.54	0.67	0.32	

Next, a neighborhood search is conducted on the fittest *m* bees (exams), for which *m* equals 2 in this instance. During this step, one question is exchanged within the fittest *m* exams, and replaced by another question from the same chapter. As Exam 3 is the fittest (or elite) exam, we will extract two neighbors from it. For clarity, the exchanged questions are denoted in bold text.

Exam 4 (neighbor 2)	Question	Q1	Q7	Q17	**Q19**	Q25	Fitness
	Chapter	Ch1	Ch1	Ch2	Ch2	Ch3	0.132
	Difficulty	0.56	0.24	0.9	0.76	0.13	

Exam 5 (neighbor 3)	Question	Q2	Q6	Q12	Q18	**Q21**	Fitness
	Chapter	Ch1	Ch1	Ch2	Ch2	Ch3	0.004
	Difficulty	0.93	0.71	0.54	0.67	0.42	

Exam 6 (neighbor 3)	Question	Q2	Q6	**Q11**	Q18	Q29	Fitness
	Chapter	Ch1	Ch1	Ch2	Ch2	Ch3	0.05
	Difficulty	0.93	0.71	0.37	0.67	0.32	

The next step is to select the fittest exams from each neighborhood or patch, specifically Exams 4 and 5. We then evaluate the overlap ratio for each exam. In this case, as there are no overlapping questions between them, the overlap ratio is 0, which does not exceed the defined overlap threshold. Consequently, we need to generate a new random population of (*s*-*m*) bees. Since the *m* bees are Exams 4 and 5, only one new bee is created randomly. Thus, the new set of bees comprises Exams 4, 5, and 7.

Exam 4	Question	Q1	Q7	Q17	Q19	Q25	Fitness
	Chapter	Ch1	Ch1	Ch2	Ch2	Ch3	0.132
	Difficulty	0.56	0.24	0.9	0.76	0.13	

Exam 5	Question	Q2	Q6	Q12	Q18	Q21	Fitness
	Chapter	Ch1	Ch1	Ch2	Ch2	Ch3	0.004
	Difficulty	0.93	0.71	0.54	0.67	0.42	

Exam 7	Question	Q5	Q1	Q20	Q15	Q30	Fitness
	Chapter	Ch1	Ch1	Ch2	Ch2	Ch3	0.126
	Difficulty	0.98	0.56	0.85	0.86	0.63	

Then, we proceed to repeat the neighborhood search operation on Exams 5 and 7, since they are the best exams.

Exam 8 (neighbor 5)	Question	**Q5**	Q6	Q12	Q18	Q21	Fitness
	Chapter	Ch1	Ch1	Ch2	Ch2	Ch3	0.014
	Difficulty	0.98	0.71	0.54	0.67	0.42	

(neighbor 7)	Question	**Q10**	Q1	Q20	Q15	Q30	Fitness
	Chapter	Ch1	Ch1	Ch2	Ch2	Ch3	0
	Difficulty	0.35	0.56	0.85	0.86	0.63	

Exam 9 (neighbor 5)	Question	Q2	Q6	Q12	Q18	**Q29**	Fitness
	Chapter	Ch1	Ch1	Ch2	Ch2	Ch3	0
	Difficulty	0.93	0.71	0.54	0.67	0.40	

The fittest exams from each neighborhood are Exams 9 and 10. Since there are no overlapping questions between them, the overlap percentage is calculated as 0%. At this point, we check the stopping condition. Since we have two distinct exams, both with a fitness value of 0, the algorithm terminates, resulting in two unique exams that have the same difficulty level and the required distribution of questions across chapters.

Furthermore, we provide an example of the overlap neighborhood move, which occurs when the algorithm exceeds the predefined overlap percentage. When the algorithm fails to satisfy the overlap condition, it focuses on diversifying the top m exams. This is accomplished by replacing as many overlapping questions as possible with new questions from the same chapter. The replacement questions are carefully selected to ensure a similar or closely matched difficulty level to the questions they replace. The process of changing the overlapping questions is carried out individually for each question and one exam at a time. In other words, for every overlapping question, an attempt is made sequentially on each overlapping exam to replace the corresponding question with another unique question from the same chapter, with a similar or closely matched difficulty level. If the attempt is successful, there is no need to consider other exams. However, if the attempt fails, the next exam in the sequence is taken into consideration.

The parameters used in this example are as follows: population size = 3, $m = 2$ (number of exams), and five questions per exam. The difficulty level is set to 0.65, and the overlap limit is defined as 0.15. Exams 1 and 2 are identified as the best exams (m best exams). It is noteworthy that both Q9 and Q13 appear twice in the selected best exams. The total number of overlapping questions (O) is determined as 2 + 2 = 4. The variable U represents the number of unique questions within the set of overlapping questions (O). Since all overlapping questions are either Q9 or Q13, $U = 2$. To calculate the overlap percentage, we use the formula in Equation (2), which in this case yields $(4-2)/(5\times2) = 0.2$. Consequently, the overlap percentage exceeds the maximum overlap threshold specified above.

Exam 1	Question	Q2	**Q9**	**Q13**	Q18	Q29	Fitness
	Chapter	Ch1	**Ch1**	**Ch2**	CH2	CH3	0.126
	Difficulty	0.93	**0.2**	**0.5**	0.67	0.32	

Exam 2	Question	Q8	**Q9**	**Q13**	Q11	Q24	Fitness
	Chapter	Ch1	**Ch1**	**Ch2**	Ch2	Ch3	0.266
	Difficulty	0.25	**0.2**	**0.5**	0.37	0.60	

Exam 3	Question	Q1	Q9	Q12	Q14	Q25	Fitness
	Chapter	Ch1	Ch1	Ch2	Ch2	Ch3	0.292
	Difficulty	0.56	0.2	0.54	0.36	0.13	

The overlap neighborhood move is then performed on the best *m* exams, specifically Exams 1 and 2. During this process, the overlapping questions are replaced with new and unique questions from the same chapter. The replacement questions are chosen to have a difficulty level that is either identical to or within a 0.05 margin of the question being replaced. In this case, since Q7 is available and meets the specified criteria (from chapter 1 and within the margins of the difficulty level), it is an appropriate replacement for Q9. Similarly, Q13 is replaced with Q12 for the same reasons. These changes were applied to Exam 1.

Exam 1	Question	Q2	**Q7**	**Q12**	Q18	Q29	Fitness
	Chapter	Ch1	**Ch1**	**Ch2**	Ch2	Ch3	0.11
	Difficulty	0.93	**0.24**	**0.54**	0.67	0.32	

Exam 2	Question	Q8	Q9	Q13	Q11	Q24	Fitness
	Chapter	Ch1	Ch1	Ch2	Ch2	Ch3	0.266
	Difficulty	0.25	0.2	0.5	0.37	0.60	

After the execution of the overlap neighborhood move, the resulting best *m* exams are entirely distinct from each other. As a result, the overlap percentage is now reduced to zero.

4. Experimental Design and Results

To evaluate the performance of the proposed algorithm, we utilized two datasets obtained from references [2,15]. These datasets are outlined in Table 1. The first dataset, the small question bank (small QB) consists of 1000 questions with varying difficulty levels, distributed across 14 chapters. The second dataset, the large question bank (large QB) consists of 12,000 questions with diverse difficulty levels across 12 chapters.

To run the algorithm, the educator needs to provide several parameters, including the number of desired exams (*m*), the difficulty level requirement for each exam (*DLR*), the desired chapters (*Ch1*, *Ch2*, … *Chc*) for each exam, and the number of questions covering each selected chapter (*Qn1*, *Qn2*, … *Qnc*). Additionally, there are other parameters specific to our Bee Algorithm (BA) that need to be specified, such as the number of scout bees (initial population size) (*s*), the number of best solutions from the *m* exams (*e*), the neighborhood (patch) size for both *e* and *m-e*, and the maximum allowed overlap percentage between generated tests. We conducted a parameterized test, exploring different values for these parameters to determine the optimal configuration for our algorithm.

Table 1. Summary of experimental datasets; dataset 1 for the small question bank [2], dataset 2 for the large question bank [15].

Dataset 1 (Small Question Bank)		Dataset 2 (Large Question Bank)	
Level of Difficulty	**Num Questions**	**Level of Difficulty**	**Num Questions**
0.1	77	0.1	1890
0.2	101	0.2	1333
0.3	170	0.3	1357
0.4	142	0.4	1379
0.5	154	0.5	1361
0.6	153	0.6	1369
0.7	131	0.7	1358
0.8	51	0.8	1375
0.9	21	0.9	578
Total	1000	**Total**	12,000

The performance criterion considered was that the quality of each exam, as measured by the fitness function, should have a margin of error no greater than 0.0001 [2]. We also introduced additional measures to handle the trade-off between fitness value and runtime. If the algorithm fails to produce m exams with a fitness value within the range of [0, 0.0001] within the allocated time, the margin of error is increased to 0.15. The resulting exams should not exceed a maximum overlap percentage of 0.3, as assumed in reference [2].

Initially, the algorithm prioritizes the fitness value, which is based on the difficulty of the exam. The algorithm keeps track of its solutions and saves the overall best solutions. Once an acceptable fitness value is achieved, the algorithm shifts its focus towards improving the overlap percentage. If the overlap percentage does not improve over five iterations, the algorithm terminates and outputs its best solution.

We implemented the proposed approach using Python 3.6.12 in the Colab environment, leveraging several libraries that support machine learning algorithms.

To evaluate the effectiveness of the proposed approach, we conducted a comparative analysis with four other algorithms: Parallel Migration PSO (PMPSO) [2], sequential PSO (SPSO) [2], the random algorithm (RA) [14], and the Simulated Annealing (SA) [9] algorithm. The results of these four methods were documented in reference [2].

The performance measurements used for evaluation were the quality of the solution and the standard deviation. Multiple runs of each algorithm were performed to obtain robust results. It is important to note that the population size and difficulty level requirement (*DLR*) were kept consistent across all algorithms in this comparison. However, the size of the question bank was varied to evaluate the algorithms' performance across different search space sizes (i.e., different datasets). The objective function computed the quality of the solution, with a decreasing fitness function indicating an improved solution. The fitness value of each test was measured to assess the quality of the results.

The standard deviation served as a measure of the generated exam's proximity to the desired difficulty level (*DLR*). It acted as an indicator of the algorithm's stability or robustness [2]. To evaluate the stability of the algorithm, we measured the standard deviation across multiple *DLR*s.

Table 2 presents the parameters utilized in the conducted experiments. We performed two distinct experiments, each executed twice: once on the large question bank and once on the small question bank. The first experiment aimed to assess the algorithms' performance when varying the number of generated exams, while the second experiment evaluated their performance when altering the required difficulty level. As the BA is a stochastic algorithm that can produce different results with each run, each experiment was repeated ten times to ensure accurate and reliable outcomes, which aligns with the number of runs conducted in [2].

Table 2. Parameters used in the testing phase.

Parameters	Values
Number of exams (m)	As entered by the user
Number of initial scout bees	$(m) \times 4$
Number of elite bees	Ceil $((m)/4)$
Number of neighbors for (m-e)	5
Number of elite (e) neighbors	8
Overlap percentage limit	0.3
Required difficulty level range	[0.3, 0.7]
Number of exams range	[100, 400]
Number of maximum iterations	250

Table 3 shows the detailed performance of all algorithms when changing the number of exams while keeping the difficulty level the same. The large question bank was used in this experiment. The success rate measures the number of times the algorithm reaches a successful solution, i.e., a solution with a fitness $\in [0, 0.0001]$. Figure 3 shows the average fitness of all algorithms.

Table 3. Algorithms' results when changing the number of exams, using the large question bank.

Algorithm Large QB	Number of Exams	Difficulty Level	Number of Runs	Success Rate	Average Fitness	Std Deviation	Average Overlap
Bees Algorithm (BA)	100	0.5	10	942	3.85×10^{-5}	3.14×10^{-5}	0.36
	200	0.5	10	1905	3.72×10^{-5}	3.08×10^{-5}	0.56
	400	0.5	10	3375	3.82×10^{-5}	3.15×10^{-5}	0.75
Parallel Migration PSO (PMPSO) [2]	100	0.5	10	1000	4.93×10^{-5}	2.92×10^{-5}	0.009
	200	0.5	10	2000	4.74×10^{-5}	2.89×10^{-5}	0.009
	400	0.5	10	4000	4.76×10^{-5}	2.88×10^{-5}	0.009
Sequential PSO (SPSO) [2]	100	0.5	10	1000	4.72×10^{-5}	2.86×10^{-5}	0.009
	200	0.5	10	2000	4.77×10^{-5}	2.88×10^{-5}	0.009
	400	0.5	10	4000	4.78×10^{-5}	2.87×10^{-5}	0.009
Random Algorithm (RA) [14]	100	0.5	10	151	6.41×10^{-4}	6.40×10^{-4}	0.009
	200	0.5	10	263	6.55×10^{-4}	6.39×10^{-4}	0.009
	400	0.5	10	552	6.27×10^{-4}	6.29×10^{-4}	0.009
Simulated Annealing (SA) [9]	100	0.5	10	178	9.07×10^{-4}	9.48×10^{-4}	0.009
	200	0.5	10	373	8.93×10^{-4}	9.29×10^{-4}	0.009
	400	0.5	10	711	8.94×10^{-4}	8.92×10^{-4}	0.009

As evident from the outcomes presented in Table 3 and Figure 3, the BA exhibits superior performance in terms of average fitness values across all dataset sizes compared to all other methods. The PMPSO and the SPSO demonstrate comparable performance, while the RA and SA underperform across all test cases. With regard to the standard deviation, the BA and PSO methods exhibit similar values, with slightly smaller deviations observed for the PSOs. This suggests that all three algorithms demonstrate stability, as the variation between different runs is minimal. Conversely, the RA and SA once again underperform in this aspect.

For the success rate, the BA was able to achieve an approximately 93% success rate on average, compared to the 100% success rate for the PSOs and much worse success rates for the RA and SA.

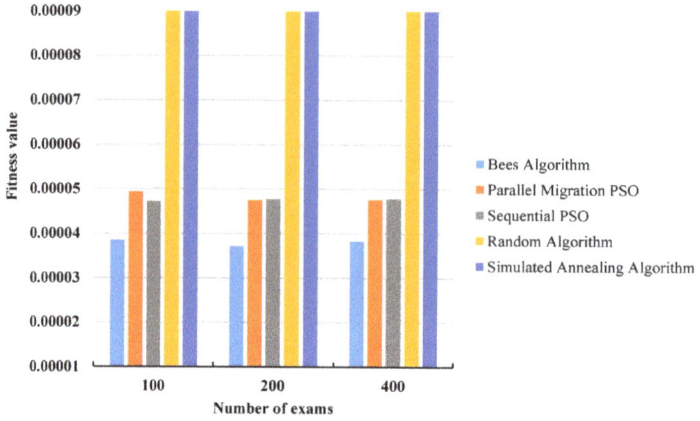

Figure 3. Average fitness values for different numbers of exams using the large question bank.

For the overlap percentage, it is important to highlight that our method for calculating the overlap percentage differs from the approach used by [2]. The reason behind this discrepancy is that their calculation involves dividing the number of overlapping questions by the number of tests, without considering the total number of questions in the tests. We believe that this method is not accurate for calculating the overlap percentage because when calculating a percentage, it is necessary to consider the relationship between a part and the whole. In other words, both the numerator and the denominator should be of the same type. Since exams are distinct entities from questions, we have developed our own formula for determining the overlap percentage, as shown in Equation (2). Therefore, a direct comparison of the overlap percentage between the algorithms cannot be provided. However, since we have established a baseline for comparison, which is the overlap threshold defined by the user, our ultimate objective is to ensure that the overlap percentage does not exceed this threshold, regardless of the calculation method employed. Observing the obtained results, it is evident that the BA was unable to attain the targeted overlap percentage of 30% for the larger test cases. This outcome is somewhat expected as the likelihood of question repetition increases with a larger number of exams.

Table 4 shows the performance of all algorithms when varying the number of exams using the small question bank, while Figure 4 shows a comparison of the algorithms in terms of average fitness values with respect to the number of exams.

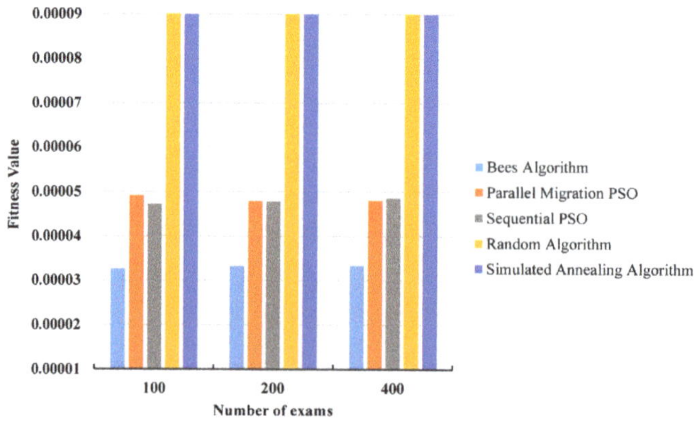

Figure 4. Average fitness values for different numbers of exams using the small question bank.

Table 4. Algorithms' results when changing the number of exams using the small question bank.

Algorithm Small QB	Number of Exams	Difficulty Level	Number of Runs	Successful Solutions	Average Fitness	Std Deviation	Average Overlap
BA	100	0.5	10	949	3.26×10^{-5}	3.08×10^{-5}	0.93
	200	0.5	10	1346	3.32×10^{-5}	2.95×10^{-5}	0.96
	400	0.5	10	3841	3.34×10^{-5}	2.95×10^{-5}	0.98
PMPSO [2]	100	0.5	10	1000	4.91×10^{-5}	2.85×10^{-5}	0.16
	200	0.5	10	2000	4.79×10^{-5}	2.92×10^{-5}	0.16
	400	0.5	10	4000	4.80×10^{-5}	2.89×10^{-5}	0.16
SPSO [2]	100	0.5	10	1000	4.72×10^{-5}	2.84×10^{-5}	0.16
	200	0.5	10	2000	4.78×10^{-5}	2.88×10^{-5}	0.16
	400	0.5	10	4000	4.85×10^{-5}	2.89×10^{-5}	0.16
RA [14]	100	0.5	10	63	1.28×10^{-3}	1.25×10^{-3}	0.16
	200	0.5	10	125	1.32×10^{-3}	1.28×10^{-3}	0.16
	400	0.5	10	291	1.33×10^{-3}	1.32×10^{-3}	0.16
SA [9]	100	0.5	10	94	1.56×10^{-3}	1.53×10^{-3}	0.16
	200	0.5	10	205	1.47×10^{-3}	1.46×10^{-3}	0.16
	400	0.5	10	388	1.55×10^{-3}	1.55×10^{-3}	0.16

Based on the outcomes presented in Table 4 and Figure 4, it is evident once again that the BA generated exams with smaller fitness values compared to all other competing methods. This indicates that the exams produced by the BA were closer to the desired *DLR*. However, it should be noted that the two variants of PSO achieved an optimal success rate, which the BA did not attain, albeit with a small margin. Nevertheless, the BA exhibited a significantly higher success rate compared to the RA and SA. For standard deviation, both the BA and PSO methods demonstrated comparable values, suggesting stability across different runs for the obtained results. In contrast, the RA and SA exhibited much higher deviations, indicating instability in the outcomes across multiple runs. On the other hand, the BA once again fell short of achieving the desired overlap percentage, as calculated using our method.

Figure 5 illustrates the runtime behavior (in seconds) of the Bees Algorithm when varying the number of exams for question banks of different sizes. It can be observed from the figure that as the size of the question bank increases, the runtime also significantly increases, which is expected.

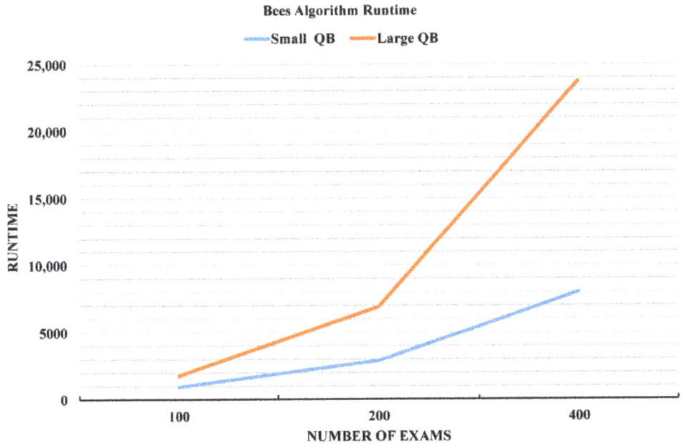

Figure 5. Bees Algorithm's runtime when changing the number of exams.

The subsequent experiment involves modifying the required difficulty level (*DLR*) using the large question bank. The results of all algorithms for this experiment are presented in Table 5, while Figure 6 provides a side-by-side comparison of the algorithms in terms of their average fitness values. From these findings, it can be observed that the BA outperforms the PSO methods in terms of average fitness in three out of five instances (*DLR* = 0.4, 0.5, and 0.6), while having a slightly worse fitness for a *DLR* = 0.7, and much worse fitness for a *DLR* = 0.3. Nevertheless, the BA outperforms the RA and SA consistently in all instances. Also, the BA and both PSO algorithms demonstrate comparable standard deviations in most instances, outperforming the RA and SA.

Table 5. Algorithms' results when changing the *DLR* using the large question bank.

Algorithm Large QB	Number of Exams	Difficulty Level	Number of Runs	Successful Solutions	Average Fitness	Std Deviation	Average Overlap
	100	0.3	10	888	7.31×10^{-5}	2.38×10^{-4}	0.47
	100	0.4	10	905	4.15×10^{-5}	4.82×10^{-5}	0.39
BA	100	0.5	10	922	4.00×10^{-5}	3.36×10^{-5}	0.36
	100	0.6	10	895	4.39×10^{-5}	4.45×10^{-5}	0.39
	100	0.7	10	888	4.88×10^{-5}	7.89×10^{-5}	0.50
	100	0.3	10	1000	4.89×10^{-5}	2.89×10^{-5}	0.015
	100	0.4	10	1000	4.86×10^{-5}	2.89×10^{-5}	0.010
PMPSO [2]	100	0.5	10	1000	4.72×10^{-5}	2.84×10^{-5}	0.009
	100	0.6	10	1000	4.70×10^{-5}	2.89×10^{-5}	0.010
	100	0.7	10	1000	4.73×10^{-5}	2.87×10^{-5}	0.014
	100	0.3	10	1000	4.92×10^{-5}	2.93×10^{-5}	0.015
	100	0.4	10	1000	4.80×10^{-5}	2.91×10^{-5}	0.010
SPSO [2]	100	0.5	10	1000	4.90×10^{-5}	2.88×10^{-5}	0.009
	100	0.6	10	1000	4.95×10^{-5}	2.95×10^{-5}	0.010
	100	0.7	10	1000	4.67×10^{-5}	2.90×10^{-5}	0.014
	100	0.3	10	-	-	-	-
	100	0.4	10	-	-	-	-
RA [14]	100	0.5	10	151	6.39×10^{-4}	6.42×10^{-4}	0.009
	100	0.6	10	-	-	-	-
	100	0.7	10	-	-	-	-
	100	0.3	10	0	1.43×10^{-1}	1.23×10^{-2}	0.009
	100	0.4	10	0	4.30×10^{-2}	1.21×10^{-2}	0.009
(SA) [9]	100	0.5	10	177	9.30×10^{-4}	9.49×10^{-4}	0.009
	100	0.6	10	0	3.95×10^{-2}	1.21×10^{-2}	0.009
	100	0.7	10	0	1.39×10^{-1}	1.22×10^{-2}	0.009

Regarding the success rate, the BA achieved an average success rate of approximately 90% across the various test cases, significantly outperforming the RA and SA, which failed to obtain any successful solutions for the majority of the test cases. The PSO methods, though, had a 100% success rate for all test cases. In terms of the overlap percentage, the BA achieves better results in this experiment compared to the previous one on the same question bank (Table 3), as is discussed in more detail in Section 5.

The performance of the algorithms when varying the required difficulty levels using the small question bank is presented in Table 6. Similar to previous experiments, both PSO methods exhibited higher success rates than the BA in all instances, except for the instance in which *DLR* = 0.3, where both PSO methods significantly underperformed in comparison to the BA.

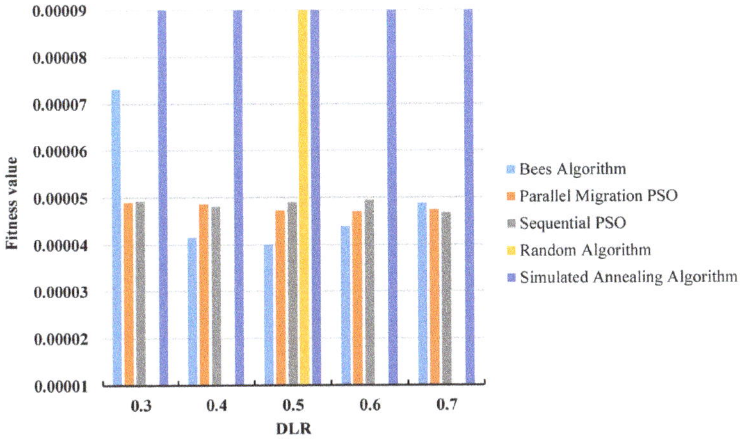

Figure 6. Average fitness values for different *DLRs* using the large question bank.

Table 6. Algorithms' results when changing the *DLR* using the small question bank.

Algorithm Large QB	Number of Exams	Difficulty Level	Number of Runs	Successful Solutions	Average Fitness	Std Deviation	Average Overlap
	100	0.3	10	864	1.87×10^{-4}	7.71×10^{-4}	0.95
	100	0.4	10	854	6.92×10^{-5}	2.05×10^{-4}	0.93
BA	100	0.5	10	874	4.70×10^{-5}	4.14×10^{-5}	0.93
	100	0.6	10	855	5.65×10^{-5}	8.26×10^{-5}	0.93
	100	0.7	10	865	1.88×10^{-4}	8.33×10^{-4}	0.94
	100	0.3	10	41	1.69×10^{-2}	1.09×10^{-2}	0.39
	100	0.4	10	1000	4.70×10^{-5}	2.83×10^{-5}	0.22
PMPSO [2]	100	0.5	10	1000	4.81×10^{-5}	2.93×10^{-5}	0.16
	100	0.6	10	1000	4.75×10^{-5}	2.82×10^{-5}	0.18
	100	0.7	10	998	4.77×10^{-5}	4.85×10^{-5}	0.28
	100	0.3	10	38	1.64×10^{-2}	1.06×10^{-2}	0.39
	100	0.4	10	1000	4.84×10^{-5}	2.87×10^{-5}	0.22
SPSO [2]	100	0.5	10	1000	4.74×10^{-5}	2.86×10^{-5}	0.16
	100	0.6	10	1000	4.94×10^{-5}	2.80×10^{-5}	0.18
	100	0.7	10	998	5.84×10^{-5}	3.27×10^{-4}	0.28
	100	0.3	10	-	-	-	-
	100	0.4	10	-	-	-	-
RA [14]	100	0.5	10	78	1.35×10^{-3}	1.29×10^{-3}	0.16
	100	0.6	10	-	-	-	-
	100	0.7	10	-	-	-	-
	100	0.3	10	0	1.87×10^{-1}	8.28×10^{-3}	0.16
	100	0.4	10	0	8.78×10^{-2}	8.18×10^{-3}	0.16
SA [9]	100	0.5	10	104	1.52×10^{-3}	1.53×10^{-3}	0.16
	100	0.6	10	0	3.52×10^{-2}	8.08×10^{-3}	0.16
	100	0.7	10	0	1.34×10^{-1}	8.03×10^{-3}	0.16

Figures 7 and 8 provide a comparison of the average fitness values for each method across different instances, with Figure 8 separating the outlier instance (*DLR* = 0.3) for clarity. From these figures, it can be observed that the BA demonstrated better performance in terms of average fitness in two instances (*DLR* = 0.3 and 0.5), while it had a worse average fitness than the PSO methods in the remaining three instances. However, it is worth noting that the BA consistently outperformed the RA and SA in all instances. Additionally, the PSO methods exhibited slightly better standard deviations than the BA in most cases. Again in this experiment, the desired threshold for the overlap percentage was still not achieved by the BA.

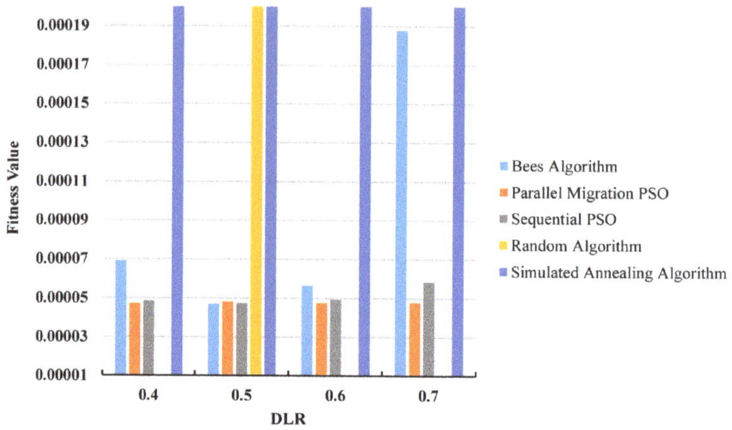

Figure 7. Average fitness values when changing the *DLR* [0.4–0.7] for the small question bank.

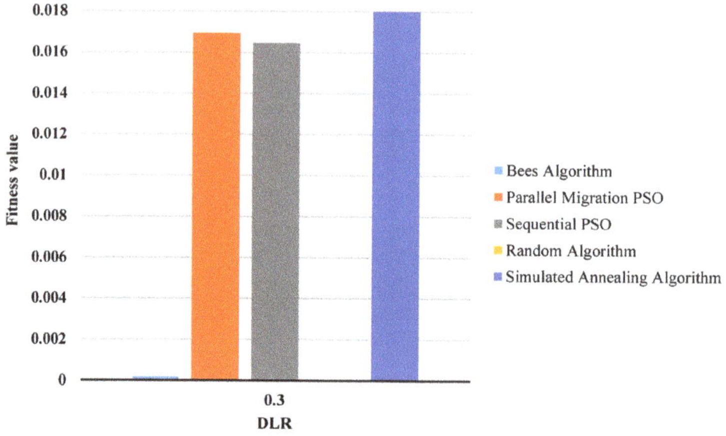

Figure 8. Average fitness for *DLR* = 0.3 for the small question bank.

5. Discussion

In our study, we evaluated the performance of the Bees Algorithm (BA) based on four key criteria: the number of successful solutions, fitness, overlap percentage, and standard deviation. The computational experimentation demonstrated that the BA achieved favorable results in terms of successful solutions, fitness, and standard deviation compared to its rival methods. Specifically, the BA outperformed all competing methods in terms of average fitness in most test cases. However, the PSO methods (both parallel and sequential versions) exhibited a higher success rate, highlighting their advantage over the BA in this particular metric, although the BA outperformed the RA and SA in this metric.

Amongst the different difficulty levels (*DLR*s) considered, the case of *DLR* = 0.5 consistently showed the best results in terms of the fitness function, standard deviation, and successful solutions, as indicated in (Tables 3–6). This can be attributed to the nature of our neighborhood move, which selects replacement questions at random. Therefore, it is relatively easier to balance difficulty level values for *DLR* = 0.5, since there is an approximately equal number of questions with higher and lower difficulty levels than 0.5. On the other hand, in the case of a *DLR* = 0.7, the majority of the questions in the database have a difficulty level lower than 0.7, making it more challenging to generate an exam with that difficulty level.

Furthermore, the standard deviation results were comparable to those of the PMPSO, according to the computational results. Once again, the case of *DLR* = 0.5 exhibited greater a robustness of the algorithm. Additionally, the successful solutions were generally close to the optimal solution, as evident in the results, particularly for *DLR* = 0.5. It is worth noting that even better results could have been achieved if we had set a margin of error less than 15%. Furthermore, the BA did not meet the specified overlap threshold reported in [2]. Despite this, our results can be considered reasonable, especially considering the small question bank, which has a limited number of questions per chapter. For instance, taking the first row of Table 4 as an example, the dataset of the small bank for the first ten chapters comprised 614 unique questions, while the total number of generated questions was 10,000. This resulted in a significant overlap percentage.

Moreover, in line with the results presented in Table 3, there is a direct correlation between the number of exams and the overlap percentage. As expected, as the number of exams increased, the overlap percentage also increased. Additionally, Table 5 revealed that the selected difficulty level (*DLR*) could impact the overlap percentage, as previously explained.

In conclusion, we believe that the BA successfully achieved our research goals by producing near-optimal solutions that satisfied the required difficulty level set by the educator. It displayed robustness with respect to the difficulty level, as evidenced by the small standard deviation.

6. Conclusions

Automatically generated online exams have garnered significant interest among researchers and educators. However, there is a scarcity of resources that can produce multiple versions of the same exam automatically. Educators require such tools for fair assessments and to prevent cheating during exams. This study aims to contribute to the field of education by providing a method to quickly generate unique and fair exams compared to manual solutions. We propose the utilization of the Bees Algorithm (BA) to compose multiple exams from a question bank, with a focus on achieving the desired difficulty level as the primary objective. Each generated exam also adheres to several constraints defined by the educators to ensure an equitable examination process. The contribution of this research lies in presenting a customized version of the BA, specifically designed to address the problem at hand. Our proposed method enables the generation of multiple online tests simultaneously, effectively meeting user requirements without the need for parallelization techniques. Additionally, we introduce innovative problem-specific neighborhood moves that are designed to adhere to the constraints of the problem.

Through extensive testing on benchmark data and comparison with the four rival methods reported in [2], the experimental results indicate that our algorithm outperforms the competing methods in terms of the primary objective, i.e., the difficulty level of the exams, in the majority of test cases. Furthermore, the algorithm exhibits a small standard deviation across all experiments, demonstrating its robustness. However, it has a slightly lower success rate compared to the Particle Swarm Optimization (PSO) methods, although it outperforms the Random and Simulated Annealing (SA) algorithms in this aspect. Furthermore, the BA fell short of attaining the targeted overlap percentage, as assumed in [2].

For future work, we aim to enhance the design of the BA by employing smarter neighborhood moves, particularly when dealing with overlapping questions between exams. Fine-tuning this approach and exploring alternative methods may mitigate the overlap percentage and further improve the algorithm's performance in terms of exam construction. One possible approach is to utilize question selection heuristics instead of randomly selecting replacement questions. An example heuristic is the Least Recently Used (LRU) heuristic, which prioritizes questions that have been used the least number of times in previous exams. By favoring less recently used questions, the LRU heuristic can maximize question diversity and minimize overlap. Additionally, runtime improvements can be achieved through parallelization or other optimization techniques (e.g., adaptive parameter tuning). Moreover, we intend to test the algorithm's scalability by evaluating larger-sized instances.

In conclusion, the BA is a metaheuristic algorithm that proves useful in solving our problem. It stands out for its simplicity, adaptability, and ease of use. We believe that our BA competes well with other metaheuristic algorithms that have been successfully applied to this problem. Our research findings hold promise for improving educational outcomes by reducing the workload for instructors, enhancing student satisfaction, and optimizing resource allocation.

Author Contributions: Conceptualization, M.H.; methodology, M.H.; validation, M.H. and R.H.; formal analysis, M.H.; investigation, M.H. and R.H; resources, M.H. and N.A.; data curation, M.H and R.H.; writing—original draft preparation, R.H.; writing—review and editing, M.H. and N.A.; visualization, M.H.; supervision, M.H.; project administration, M.H; funding acquisition N.A. All authors have read and agreed to the published version of the manuscript.

Funding: This work has been funded by King Saud University, Riyadh, Saudi Arabia, through the Researchers Supporting Project number (RSPD2023R857).

Institutional Review Board Statement: Not applicable.

Informed Consent Statement: Not applicable.

Data Availability Statement: The data presented in this study are available in 10.1109/ACCESS.2021.3057515, reference number 2 and 10.6688/JISE.201811_34(6).0004, reference number 15.

Acknowledgments: This work has been funded by King Saud University, Riyadh, Saudi Arabia, through the Researchers Supporting Project number (RSPD2023R857). The authors acknowledge the contributions of Rafeef Alazzaz, Demah Alawadh, Atheer Mohammad, Beshayer Al-mutawa, Ghadah Alshabib for their role in reviewing the literature, implementing the algorithm, and reporting the results.

Conflicts of Interest: The authors declare no conflict of interest.

References

1. Stanger-Hall, K.F. Multiple-Choice Exams: An Obstacle for Higher-Level Thinking in Introductory Science Classes. *CBE—Life Sci. Educ.* **2012**, *11*, 294–306. [CrossRef] [PubMed]
2. Nguyen, T.; Nguyen, L.T.T.; Bui, T.; Loc, H.D.; Pedrycz, W.; Snasel, V.; Vo, B. Multi-Swarm Optimization for Extracting Multiple-Choice Tests From Question Banks. *IEEE Access* **2021**, *9*, 32131–32148. [CrossRef]
3. Raje, S.; Stitzel, S. Strategies for Effective Assessments While Ensuring Academic Integrity in General Chemistry Courses during COVID-19. *J. Chem. Educ.* **2020**, *97*, 3436–3440. [CrossRef]
4. Beerepoot, M.T.P. Formative and Summative Automated Assessment with Multiple-Choice Question Banks. *J. Chem. Educ.* **2023**, *100*, 2947–2955. [CrossRef]
5. Wen, J.; Lu, W.; Chen, Z. Innovation and Construction of Examination Database of Pharmacology. *Indian J. Pharm. Educ. Res.* **2020**, *54*, 279–283. [CrossRef]
6. Krzic, M.; Brown, S. Question Banks for Effective Online Assessments in Introductory Science Courses. *Nat. Sci. Educ.* **2022**, *51*, e20091. [CrossRef]
7. Yuce, B.; Mastrocinque, E.; Packianather, M.S.; Lambiase, A.; Pham, D.T. The Bees Algorithm and Its Applications. In *Handbook of Research on Artificial Intelligence Techniques and Algorithms*; IGI Global: Pennsylvania, PA, USA, 2015; pp. 122–151, ISBN 978-1-4666-7258-1.

8. Al-Negheimish, S.; Alnuhait, F.; Albrahim, H.; Al-Mogherah, S.; Alrajhi, M.; Hosny, M. An Intelligent Bio-Inspired Algorithm for the Faculty Scheduling Problem. *Int. J. Adv. Comput. Sci. Appl.* **2018**, *9*, 151–159. [CrossRef]
9. Lu, P.; Cong, X.; Zhou, D. The Research on Web-Based Testing Environment Using Simulated Annealing Algorithm. *Sci. World J.* **2014**, *2014*, 167124. [CrossRef]
10. Rahim, T.N.T.A.; Aziz, Z.A.; Rauf, R.H.A.; Shamsudin, N. Automated Exam Question Generator Using Genetic Algorithm. In Proceedings of the 2017 IEEE Conference on e-Learning, e-Management and e-Services (IC3e), Miri, Malaysia, 16–17 November 2017; pp. 12–17.
11. Nguyen, T.; Bui, T.; Vo, B. Multi-Swarm Single-Objective Particle Swarm Optimization to Extract Multiple-Choice Tests. *Vietnam J. Comp. Sci.* **2019**, *6*, 147–161. [CrossRef]
12. Bui, T.; Nguyen, T.; Huynh, H.M.; Vo, B.; Chun-Wei Lin, J.; Hong, T.-P. Multiswarm Multiobjective Particle Swarm Optimization with Simulated Annealing for Extracting Multiple Tests. *Sci. Program.* **2020**, *2020*, e7081653. [CrossRef]
13. Hu, X.-M.; Zhang, J.; Chung, H.S.-H.; Liu, O.; Xiao, J. An Intelligent Testing System Embedded With an Ant-Colony-Optimization-Based Test Composition Method. *IEEE Trans. Syst. Man Cybern. Part C (Appl. Rev.)* **2009**, *39*, 659–669. [CrossRef]
14. Naik, K.; Sule, S.; Jadhav, S.; Pandey, S. Automatic Question Paper Generation System Using Randomization Algorithm. *Int. J. Eng. Technol. Res.* **2014**, *2*, 192–194.
15. Bui, T.; Nguyen, T.; Vo, B.; Pedrycz, W.; Snasel, V. Application of Particle Swarm Optimization to Create Multiple-Choice Tests. *J. Inf. Sci. Eng.* **2018**, *34*, 1405–1423. [CrossRef]
16. Glover, F.; Sörensen, K. Metaheuristics. *Scholarpedia* **2015**, *10*, 6532. [CrossRef]
17. Cheng, S.; Liu, B.; Ting, T.O.; Qin, Q.; Shi, Y.; Huang, K. Survey on Data Science with Population-Based Algorithms. *Big Data Anal.* **2016**, *1*, 3. [CrossRef]
18. Gandomi, A.; Yang, X.-S.; Talatahari, S.; Alavi, A. Metaheuristic Algorithms in Modeling and Optimization. In *Metaheuristic Applications in Structures and Infrastructures*; Elsevier: Amsterdam, The Netherlands, 2013; pp. 1–24. ISBN 978-0-12-398364-0.
19. Ziarati, K.; Akbari, R.; Zeighami, V. On the Performance of Bee Algorithms for Resource-Constrained Project Scheduling Problem. *Appl. Soft Comput.* **2011**, *11*, 3720–3733. [CrossRef]
20. Li, T.; Zhou, C.; Hu, M. An Improved Artificial Bee Colony Algorithm for 3D Protein Structure Prediction. In Proceedings of the 2017 International Conference on Biometrics Engineering and Application, Hong Kong, China, 21–23 April 2017; Association for Computing Machinery: New York, NY, USA, 2017; pp. 7–12.
21. Korkmaz, E.; Akgüngör, A.P. The Forecasting of Air Transport Passenger Demands in Turkey by Using Novel Meta-heuristic Algorithms. *Concurr. Comput.* **2021**, *33*, e6263. [CrossRef]
22. Korkmaz, E.; Akgüngör, A.P. Comparison of Artificial Bee Colony and Flower Pollination Algorithms in Vehicle Delay Models at Signalized Intersections. *Neural Comput. Appl.* **2020**, *32*, 3581–3597. [CrossRef]
23. Korkmaz, E.; Doğan, E.; Akgüngör, A. Estimation of Car Ownership in Turkey Using Artificial Bee Colony Algorithm. In Proceedings of the 3rd International Conference on Traffic and Transport Engineering (ICTTE), Belgrade, Serbia, 24–25 November 2016.
24. Dogan, E.; Korkmaz, E.; Akgungor, A.P. Comparison Of Different Approaches In Traffic Forecasting Models For The D-200 Highway In Turkey. *Sci. J. Silesian Univ. Technol. Ser. Transp.* **2018**, *99*, 25–42. [CrossRef]
25. Li, J.-Q.; Song, M.-X.; Wang, L.; Duan, P.-Y.; Han, Y.-Y.; Sang, H.-Y.; Pan, Q.-K. Hybrid Artificial Bee Colony Algorithm for a Parallel Batching Distributed Flow-Shop Problem With Deteriorating Jobs. *IEEE Trans. Cybern.* **2020**, *50*, 2425–2439. [CrossRef]
26. Alhuwaishel, N.; Hosny, M. A Hybrid Bees/Demon Optimization Algorithm for Solving the University Course Timetabling Problem. In Proceedings of the 3rd NAUN International Conference on Mathematical, Computational and Statistical Sciences, Dubai, United Arab Emirates, 22–24 February 2015.
27. Almaneea, L.I.; Hosny, M.I. A Two Level Hybrid Bees Algorithm for Operating Room Scheduling Problem. In *Intelligent Computing*; Arai, K., Kapoor, S., Bhatia, R., Eds.; Advances in Intelligent Systems and Computing; Springer International Publishing: Cham, Switzerland, 2019; Volume 858, pp. 272–290. ISBN 978-3-030-01173-4.
28. Pham, D.; Ghanbarzadeh, A.; Koç, E.; Otri, S.; Rahim, S.; Zaidi, M. *The Bees Algorithm Technical Note*; Manufacturing Engineering Centre, Cardiff University: Cardiff, UK, 2005; pp. 1–57.
29. Talbi, E.-G. *Metaheuristics: From Design to Implementation*; John Wiley & Sons, Inc.: Hoboken, NJ, USA, 2009; Volume 74, ISBN 978-0-470-27858-1.
30. Teodorović, D.; Dell'Orco, M. Bee Colony Optimization—A Cooperative Learning Approach to Complex Transportation Problems. *Adv. OR AI Methods Transp.* **2005**, *51*, 60–69.
31. Ma, L.; Zhu, X.; Feng, Q. Improvement and Design of Genetic Algorithm in Personalized Test Paper Composition System. In Proceedings of the 4th International Conference on Computer Science and Application Engineering, Sanya, China, 20–22 October 2020; Association for Computing Machinery: New York, NY, USA, 2020; pp. 1–5.
32. Hwang, G.-J.; Lin, B.M.T.; Tseng, H.-H.; Lin, T.-L. On the Development of a Computer-Assisted Testing System with Genetic Test Sheet-Generating Approach. *IEEE Trans. Syst. Man Cybern. Part C (Appl. Rev.)* **2005**, *35*, 590–594. [CrossRef]
33. Shushu, L.; Fengying, W. Strategy and Realization of Auto-Generating Exam Paper Based on Genetic Algorithm. In Proceedings of the 2010 International Conference on Artificial Intelligence and Computational Intelligence, Sanya, China, 23–24 October 2010; Volume 2, pp. 478–482.

34. Hwang, G.-J.; Yin, P.-Y.; Yeh, S.-H. A Tabu Search Approach to Generating Test Sheets for Multiple Assessment Criteria. *IEEE Trans. Educ.* **2006**, *49*, 88–97. [CrossRef]
35. Baronti, L.; Castellani, M.; Pham, D.T. An Analysis of the Search Mechanisms of the Bees Algorithm. *Swarm Evol. Comput.* **2020**, *59*, 100746. [CrossRef]

applied sciences

MDPI

Article

AI Enhancements for Linguistic E-Learning System

Jueting Liu [1], Sicheng Li [2], Chang Ren [2], Yibo Lyu [3], Tingting Xu [1], Zehua Wang [1] and Wei Chen [1,*]

[1] Department of Computer Science and Technology, China University of Mining and Technology, Xuzhou 221000, China; 6476@cumt.edu.cn (J.L.); tingting_xu@cumt.edu.cn (T.X.)
[2] Department of Computer Science and Software Engineering, Auburn University, Auburn, AL 36830, USA; czr0072@auburn.edu (C.R.)
[3] Department of Electrical and Computer Engineering, Auburn University, Auburn, AL 36830, USA; laolvyenq@gmail.com
* Correspondence: chenwdavior@163.com

Abstract: E-learning systems have been considerably developed after the COVID-19 pandemic. In our previous work, we developed a linguistic interactive E-learning system for phonetic transcription learning. In this paper, we propose three artificial-intelligence-based enhancements to this system from different aspects. Compared with the original system, the first enhancement is a disordered speech classification module; this module is driven by the MFCC-CNN model, which aims to distinguish disordered speech and nondisordered speech. The accuracy of the classification is about 83%. The second enhancement is a grapheme-to-phoneme converter. This converter is based on the transformer model and designed for teachers to better generate IPA words from the regular written text. Compared with other G2P models, our transformer-based G2P model provides outstanding PER and WER performance. The last part of this paper focuses on a Tacotron2-based IPA-to-speech synthesis system, this deep learning-based TTS system can help teacher generate high-quality speech sounds from IPA characters which significantly improve the functionality of our original system. All of these three enhancements are related to the phonetic transcription process. and this work not only provides a better experience for the users of this system but also explores the utilization of artificial intelligence technologies in the E-learning field and linguistic field.

Keywords: linguistic E-learning; phonetic transcription; Mel frequency cepstrum coefficient; grapheme-to-phoneme; transformer; speech synthesis

1. Introduction

Phonetic transcription, a process that represents speech sounds using special symbols, plays an important role in the linguistic education field. Generally, International Phonetic Alphabet (IPA) characters are utilized in the process of phonetic transcription [1,2]. The IPA is an alphabet system generated from Latin script that aims to indicate the pronunciation of words; for example, the phonetic format of /Phonetic/ is /fə'nɛtik/.

In a previous work, we developed an interactive E-learning system focused on phonetic transcription and pronunciation for language learners. This system, named APTgt, is an online exam system that provides phonetic transcription exams for IPA language students and automated grading tools for teachers [3]. To improve the intelligence and extensibility of the original system, in this paper, we propose three enhancements for the system based on machine learning and deep learning technology. Figure 1 illustrates the function of our original system and the proposed enhancements. The primary system includes two parts: in the teacher's part, a teacher can create a question by attaching an audio file of word/phrase pronunciation and uploading its corresponding phonetic format as the answer; the student listens to the questions and types the answers on an IPA keyboard. The system then automatically calculates the similarity between the student's answers and the prestored correct answer using the edit distance algorithm and generates the grade [3,4].

Our study mainly focused on the teacher part and aimed at improving the experience of teachers. The prototype of the linguistic system was designed for the Communication Disorder Department. Teachers always require large amounts of time to distinguish disordered speech from nondisordered speech. Moreover, thonetic transcription is a process that generates words in phonetic format (IPA characters) from their written format. For teachers, adding questions in the system also requires them to add the corresponding correct answer in phonetic format. A grapheme-to-phoneme converter assists teachers in finishing this step and significantly improves the user experience. In the last part, the phonetic transcription exams in our system require the speech sound of English words or brief phrases; using the recordings of students or staff may cause privacy issues. A high-quality speech synthesis system perfectly solves this problem. All of the above challenges inspired us to build the following enhancements for our E-learning system.

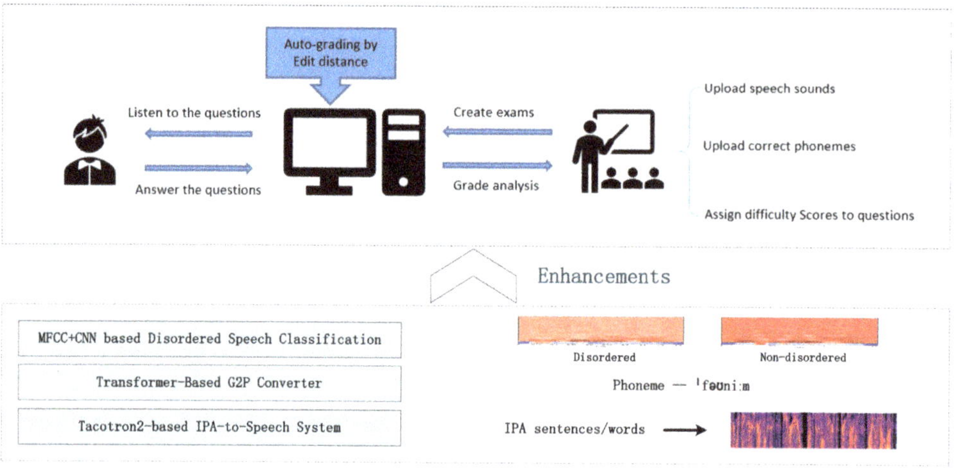

Figure 1. The enhancements to the linguistic E-learning system.

The three enhancements that we propose include the following:

- A MFCC+CNN-based disordered speech classification module.
- A Transformer-based grapheme-to-phoneme (G2P) converter module.
- A Tacotron2-based IPA-to-speech synthesis module.

The disordered speech classification module aims to to distinguish disordered speech from nondisordered speech. As an intelligent linguistic E-learning system, this module helps students understand the difference between correct pronunciation and disordered speech. During this study, we employed the Mel frequency cepstrum coefficient (MFCC) as the feature to represent the speech sound and a convolutional neural network (CNN) as the classification model [5].

The second enhancement module for our E-learning system is a transformer-based grapheme-to-phoneme (G2P) converter. In the phonetic transcription exam, teachers need to upload the pronunciation of English words, phrases, and sentences, where the pronunciation of speech sounds can be represented as phonemes. G2P conversion is a process that converts written words (grapheme) to their pronunciations (phonemes) [6]. With this G2P converter, teachers can easily extract the words in their phonetic format. We used neural machine translation ideas to design this G2P converter. After comparing several models, the transformer model, an encoder–decoder model with self-attention mechanisms, can provide superb performance with low word error rate (WER) and phoneme error rate (PER) [7].

The Tacotron2-based IPA-to-speech synthesis (text-to-speech) system is the last enhancement for our system. In order to help students better understand the pronunciation of IPA symbols, this module was introduced for directly generating high-quality speech audio files from IPA symbols. Furthermore, this module helps teachers easily acquire audio files as a part of the question in the exam system. It took two steps to build the IPA-to-speech system:

- Build a grapheme-to-phoneme system to convert all the English text to IPA format;
- Build the TTS system with the processed data.

2. Related Literature

2.1. Linguistic E-Learning

During the peak of the COVID-19 pandemic, according to data from UNESCO, over 1 billion children were affected and out of the classroom globally. The shift to E-learning or online learning is also significantly increasing with increases in Internet access. For example, Zhejiang University deployed over 5000 courses online in two weeks on the platform DingTalk ZJU [8], and the Imperial College London started offering courses on Coursera starting in 2020.

E-learning is an approach that delivers knowledge or skills remotely and interactively using electrical devices such as smartphones, tablets, and laptops. Compared with traditional classroom learning, E-learning can offer students flexible topics or subjects; students can interact with teachers or professors through email or platforms without the restrictions imposed by physical distance. It cannot entirely replace traditional classroom learning but provides an augmented learning environment. Coursera and Udemy are both successful in E-learning (online learning) platformsonin the market that provide high-quality courses with quizzes and interactive exams [9].

Linguistics is a scientific subject of human language. E-learning can play an important role in linguistics education since the advantages of E-learning benefit linguistics pedagogy. Automated Phonetic Transcription—The Grading Tool (APTgt) is a well-designed interactive web-based E-learning system that focuses on phonetic transcription for students (learners) and teachers (faculty). Phonetic transcription is a process that represents the speech sounds using special characters or symbols [3].

2.2. Speech Disorders Classification

A speech disorder is a condition in which a person experiences problems creating or forming the speech sounds needed to communicate with others. It is a subproblem of speech classification. To solve speech classification problems, both a feature extraction function and a classification algorithm are required. There are two major features in speech classification/recognition subjects: the linear prediction coding (LPC) and the Mel grequency cepstrum coefficient (MFCC) [10]. Classification algorithms include dynamic time warping (DTW) [11], hidden Markov models (HMMs) [12], and deep-learning-based classification.

2.3. Grapheme-to-Phoneme Conversion

In linguistics, a grapheme is the smallest unit of a written language, while a phoneme is the smallest unit of speech sound. Grapheme-to-phoneme (G2P) conversion is a process that converts a spelled-out word to its phonetic format (a sequence of IPA symbols) [13]. G2P plays an essential role in the natural language processing (NLP) field including in text-to-speech (TTS) systems and automated speech recognition (ASR) systems. Generally, the International Phonetic Alphabet (IPA) characters are employed to represent phonemes.

G2P conversion has long been a popular top in the NLP field. Researchers have investigated different approaches for G2P conversion. The phoneme error rate (PER) and word error rate (WER) can be utilized to evaluate the performance of a G2P conversion system. In 2005, the hidden Markov model was employed for G2P conversion by Paul Taylor with a 9.02% PER and 42.69% WER [14]. In 2008, Maximilian Bisani introduced joint-sequence models for G2P conversion. A joint-sequence model is a theoretically stringent probabilistic framework that is applicable to this problem. On different English data sets, joint-sequence models provide better performance than hidden Markov models; for example, the PER on CMUdict was 5.88% and the WER on CMUdict was 24.53% [15]. With the development of neural network technology, deep learning models are playing important roles in the NLP field. G2P conversion, as a text-to-text task, was studied by training different deep learning models. In 2015, the Seq2Seq model was employed for G2P conversion by Kaisheng Yao from Microsoft Research; in this study, the PER on CMUDict was 5.45%, while the WER was 23.55% [16]. In the same year, long short-term memory recurrent neural networks were utilized for the same task, achieving a 9.1% PER and 21.3%, by Kanishka Rao [17]. In 2020, with a start-of-the-art model, the Transformer, the PER was increased to 5.23% and the WER to 22.1% [18].

2.4. Speech Synthesis Systems

Speech synthesis, also known as text-to-speech (TTS), is a process that generates human speech sounds from text. Speech synthesis has been a hot topic since the later part of the 20th century. The early computer-based speech synthesis approaches include articulatory synthesis, formant synthesis, concatenative synthesis, and statistical parametric synthesis [19].

With the development of neural network technology, deep-learning-based end-to-end speech synthesis models have been proposed and have become the main methods used in TTS research. A modern TTS system usually consists of three basic components: a text analysis module, an acoustic model, and a vocoder. As shown in Figure 2, the text analysis module converts a text sequence into linguistic features; the acoustic models generate acoustic features from linguistic features; then, the vocoders synthesize waveforms from acoustic features. Tacotron 1, Tacotron 2, Deep Voice, and Fast Speech are all end-to-end TTS examples [20–23].

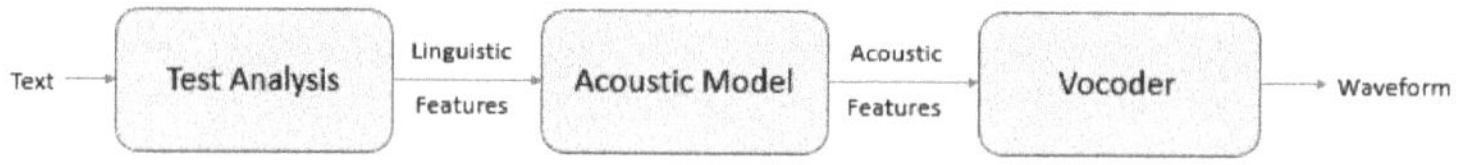

Figure 2. The structure of end-to-end TTS systems.

3. MFCC+CNN-Based Disordered Speech Classification

The first enhancement we developed is a speech classification module. This module has the ability to classify disordered speech and nondisordered speech. The speech classification problem can be divided into two subproblems: feature extraction and classification. In this study, we chose MFCC in image format to represent the features of human speech and the CNN model to perform the classification function.

3.1. Feature Extraction

In sound processing, the Mel frequency cepstrum (MFC) is a representation of the short-term power spectrum of a sound based on a linear cosine transform of a log power spectrum on a nonlinear Mel scale of frequency. Mel frequency cepstral coefficients (MFCCs) are coefficients that collectively make up an MFC. Figure 3 illustrates the steps to generate MFCCs from audio [24].

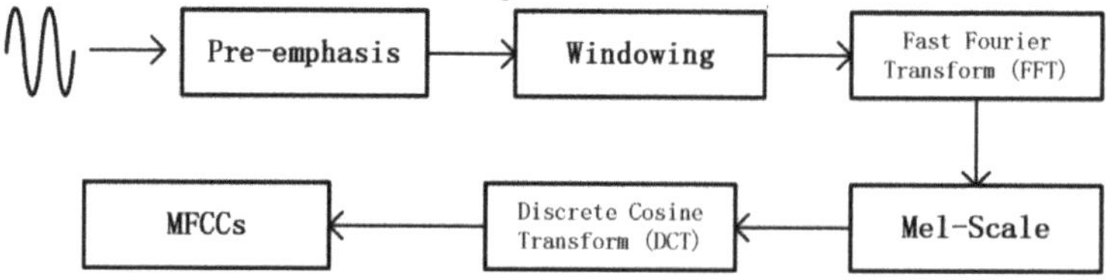

Figure 3. MFCC extraction process.

- Pre-emphasize the audio signal to increase to energy of the signal at a higher frequency.
- Break the sound signal into an overlapping window.
- Take the Fourier transform to transfer the signal from the time domain to the frequency domain.
- Compute the Mel spectrum by passing the Fourier-transformed signal through the Mel filter bank. The transformation from the Hertz scale to the Mel scale is:

$$Mel(f) = 2595 log(1 + \frac{f}{700})$$

- Take the discrete cosine transform of the Mel log signals, and the result of this conversion is MFCCs.

3.2. Data Selection

The Speech Exemplar and Evaluation Database (SEED) dataset was utilized to train our classification model. The SEED contains about 16,000 recorded speech samples, grouped by age (child vs. adult) and speech health status (with or without speech disorder). The children's speech disorders were determined through parent reports and standardized assessments. Speakers in the SEED are between the ages of 2 and 85 years. A significant aspect of SEED is that it provides samples with and without speech disorders [25].

3.3. Implementation and Evaluation

About 1000 samples from the SEED were selected: 80% were used for training and the rest were used for validation. We used the Python Librosa library to process the MFCC values into their image format. Figure 4 shows two MFCC images with the same content but recorded by different recorders (with and without speech disorder).

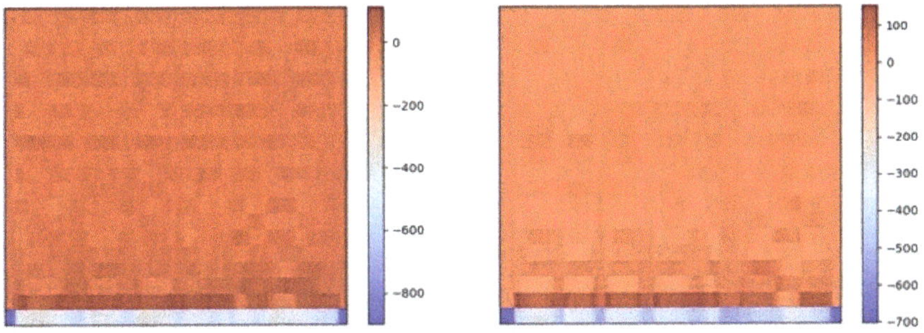

Figure 4. MFCC images of disordered and nondisordered speech.

Thus, this disordered speech classification problem could be transformed into an image classification problem. We then employed a CNN model to build a classification module. Convolutional neural network (ConvNet/CNN) is a deep learning algorithm widely used for image classification and computer vision tasks. Table 1 shows the structure of the CNN model used in our classification module, which took 150 epochs for training. Figure 5 shows the loss and accuracy of our model. The average classification accuracy was about 83%; this disordered speech classification module can neglect the content and the recorder of the speech, which means it is quite efficient and extensive.

Table 1. CNN model utilized in classification module.

Layer	Output Shape	Parameter Number
conv2d	(None, 148, 148, 32)	896
max_pooling2d	(None, 74, 74, 32)	0
conv2d_1	(None, 72, 72, 64)	18,496
max_pooling2d_1	(None, 36, 36, 64)	0
conv2d_2	(None, 34, 34, 128)	73,856
max_pooling2d_2	(None, 17, 17, 128)	0
conv2d_3	(None, 15, 15, 128)	147,584
max_pooling2d_3	(None, 7, 7, 128)	0
flatten	(None, 6272)	0
dense	(None, 512)	3,211,776
dense_1	(None, 1)	513

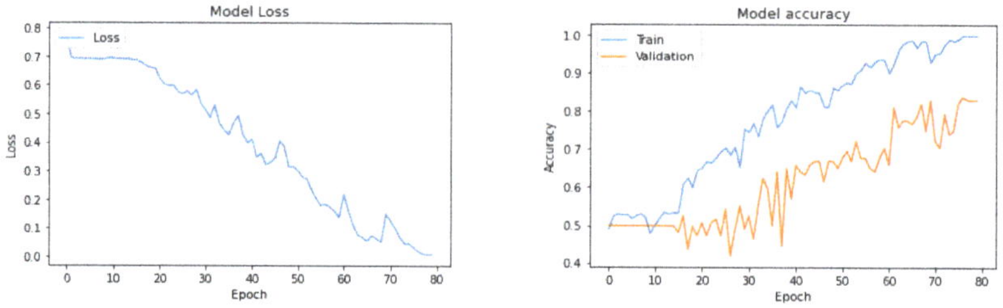

Figure 5. Loss and accuracy of speech classification module.

3.4. Discussion

In our first enhancement, we designed an MFCC+CNN-based disorder speech classification module. It can be regarded as a variant of speech classification but focuses on disordered and nondisordered speech. We developed a new solution in this classification field, where MFCCs ideally represent the features of human speech, and the image of MFCCs is classified by the CNN model. This classification module can help teachers remove disordered speech files from exams and improve the quality of the questions using our E-learning system.

4. Transformer-Based Multilingual G2P Converter

As we discussed above, the core function of our E-learning system is the provision of interactive phonetic transcription exams. The questions in this exam consist of the audio of words/phrases and their corresponding pronunciation (presented in IPA format). The teacher needs to preinput the correct answer to the system to activate the auto-grading module; so, generating IPA characters from written language can be a challenge for teachers. This was our inspiration for building the G2P converter. Grapheme-to-phoneme (G2P) conversion is a process of generating words in their IPA format from written format. G2P converters can be regarded as variants of machine translators [5].

4.1. Data Selection

There are two different kinds of characters used to represent the pronunciation of words and phrases: CMUDict characters and IPA characters. The Carnegie Mellon University Pronouncing Dictionary is an open-source machine-readable pronunciation dictionary for North American English that contains over 134,000 words and their pronunciations; on the other hand, IPA symbols are more widely used in multiple languages. Table 2 illustrates some samples of CUMdict and IPA symbols:

Table 2. CMUDict and IPA symbols.

Written Format	CMUDict	IPA Symbols
eat	IY T	it
confirm	K AH N F ER M	kən'fɜrm
minute	M IH N AH T	'minət
quick	K W IH K	kwik
maker	M EY K ER	'meɪker
relate	R IH L EY T	rɪ'leɪt

In our system, we chose IPA symbols for the representation of pronunciation. Furthermore, to build a multilingual G2P system, we also investigated the French–IPA and Spanish–IPA converters. Table 3 lists all the datasets we employed in this study. The first two English–IPA datasets were used to investigate how the size of the data influences the G2P systems' performance, and the French–IPA and Spanish–IPA datasets were utilized to inspect the feasibility of the multilinguistics.

Table 3. Datasets used for training.

Dataset	Number of Word Pairs	For Validation
English–IPA	125,912	20%
French–IPA	122,986	20%
Spanish–IPA	99,315	20%

4.2. *Transformer-Based G2P Converter*

The Transformer model is an encoder–decoder model with an attention mechanism. Without using any recurrent layers, the self-attention mechanism allows the model to process the input text as a whole rather than word-by-word/character-by-character. This structure makes the Transformer model avoid long dependency issues. The encoder in Transformer is composed of two major elements: the self-attention mechanism (multihead attention) and a feed-forward layer. The decoder includes two multihead attention layers and one feed-forward layer. The encoder maps input sequences/words into attention-based representations; the decoder then takes the continuous representations and generates the output. Figure 6 shows the structure of the attention mechanism in the Transformer.

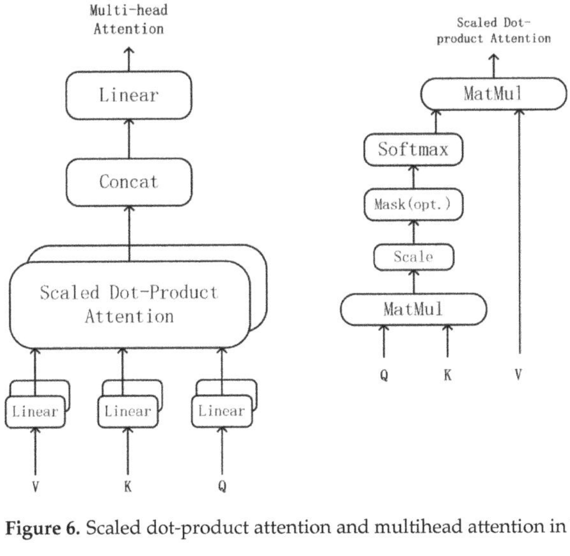

Figure 6. Scaled dot-product attention and multihead attention in Transformer model.

The scaled dot-product attention mechanism means the dot products are scaled down by $\sqrt{d_k}$. **Query Q** represents a vector word, **keys K** are all other words in the sequence, and **value V** illustrates the vector of the word. The attention function can be represented as:

$$Attention(Q, K, V) = softmax(\frac{QK^T}{\sqrt{d_k}})V$$

Multihead attention is a module that runs through an attention mechanism multiple times in parallel, concatenates the results, and produces the result. Each head of the multihead attention extracts the specific representation, which allows the whole model to receive information from different subspaces. For multihead attention:

$$multihead(Q, K, V) = concat(head_1, head_2, \ldots, head_n)W_0$$

$$wherehead_i = Attention(QW_i^Q, KW_i^K, VW_i^V);$$

W_i^Q, W_i^K, and W_i^V are the respective weight matrices calculated from Q, K, and V [26].

4.3. Implementation and Evaluation

Multiple pieces of training were implemented with the Nvidia Tesla P100 graphic card. We employed a six-layer Transformer model and the Adam optimizer in Keras with a learning rate of 0.0001. The phoneme error rate (PER) and word error rate (WER) were utilized for evaluating the performance of our G2P converter. The PER is the distance between two phonetic words calculated by the edit distance divided by the total number of phonemes, while the WER is a standard parameter used for measuring the accuracy in an ASR system. The formulation of WER is:

$$WER = \frac{S + D + I}{N}$$

where S is the number of substitutions, D is the number of deletions, I is the number of insertions, and N refers to the total number of words [7].

Figures 7–9 display the performance of our multilingual G2P converter. For the English G2P converter, it took 220 epochs of training, the PER was about 2.6%, and the WER was 10.7%. For the French and Spanish G2P converters, 190 epochs of training were needed. The PER and WER for the French–IPA converter were 2.1% and 12.3%, while the PER and WER for the Spanish–IPA converter were 1.7% and 12.7%. Compared with the other models discussed in Section 2, our models outperformed the others in converting accuracy. Table 4 provides the comparison among different G2P converters.

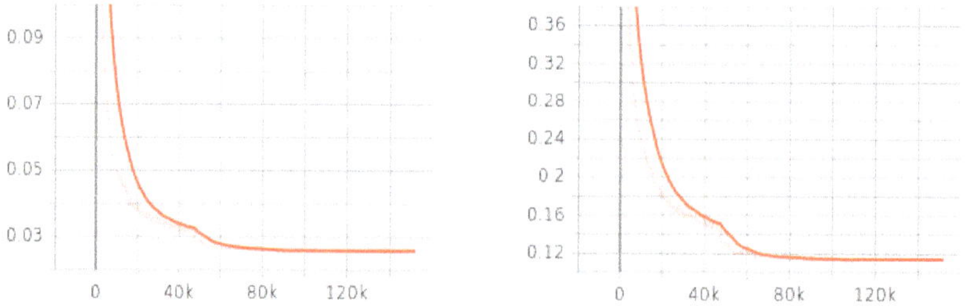

Figure 7. The PER and WER for English–IPA converter.

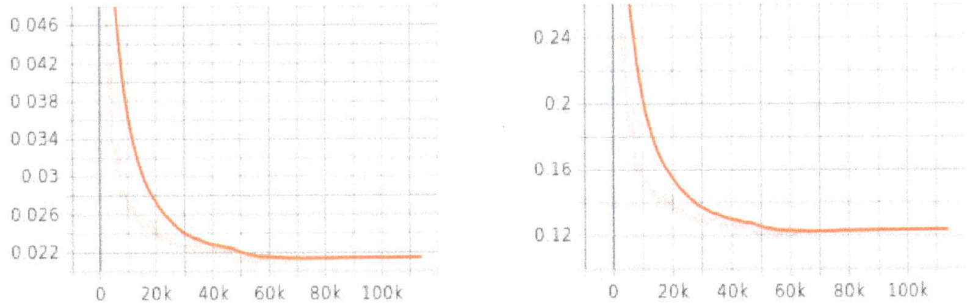

Figure 8. The PER and WER for French–IPA converter.

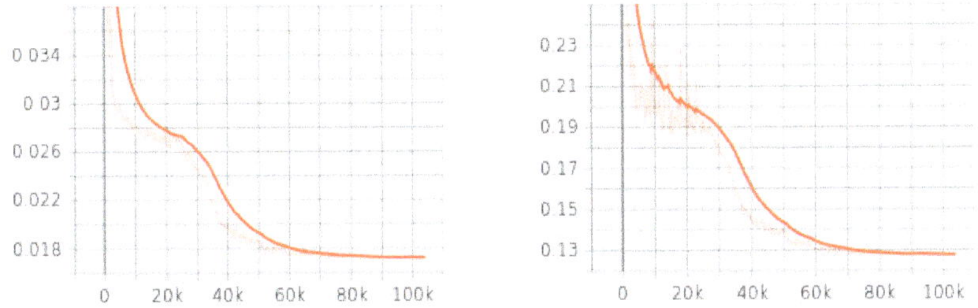

Figure 9. The PER and WER for Spanish–IPA converter.

Table 4. The different models utilized in English G2P conversion.

Models	PER	WER
Hidden Markov Model	9.02%	42.69%
Joint sequence	5.88%	24.53%
Seq2Seq	5.45%	23.55%
LSTM	9.1%	21.3%
Transformer (ours)	**2.6%**	**10.7%**

Table 5 shows the results of our multilingual G2P converter.

Table 5. The results of the multilingual G2P converter.

Language	Written Format	Correct Phonemes	Generated Phonemes
English	displeasure buoyant immortal	dɪspl'ɛʒə b'ɔɪənt ɪm'ɔːtəl	dɪspl'ʒə b'ɔɪənt ɪm'ɔːtəl
Spanish	ababillarais cacofónicos cadañega	aβaβiʎaɾis kako'fonikos kaðaeɣa	aβaβiʎaɾis kako'fonikos kaðaɲeɣa
French	câlineriez damasquiner effrangé	kalinəʁje damaskine efʁɑ̃ʒe	kalinəʁje damaskine efʁɑ̃ʒe

4.4. Discussion

In this part, we describe our second enhancement, a Transformer-based G2P converter. The results showed that the performance of our model is superior to that of other statistical methods and deep learning approaches. Grapheme-to-phoneme conversion is a process that is closely related to phonetic transcription. The multilingual G2P converter is a tremendous contribution to teachers teaching phonetic transcription. With this converter, teachers can easily generate IPA characters from one regular word without browsing a dictionary; moreover, our G2P converter has the ability to be extended to other languages. This feature gives our system the potential to be used as a multilingual E-learning system.

5. Tacotron2-Based IPA-to-Speech System

As we mentioned above, the questions in phonetic transcription exams consist of speech audio and IPA symbols. From the teacher's view, searching and acquiring appropriate speech audio with high quality along with their text are challenging. Additionally, the text of the audio must be converted to IPA format. From this perspective, we designed a text-to-speech (TTS) system that can directly generate speech sounds from words/phrases/sentences in IPA format [19].

Figure 10 shows the main process used to build our IPA-to-speech system. The English sentences in LJSpeech are converted to their IPA format in batches by the G2P converter. The format of the data in the LJSpeech dataset is transformed to *<IPASentence, Speech-samples>*. The Mel spectrograms are predicted and calculated by the Tacotron 2, and we employed WaveGlow as the Vocoder. The Vocoder can generate high-quality speech sounds from Mel spectrograms.

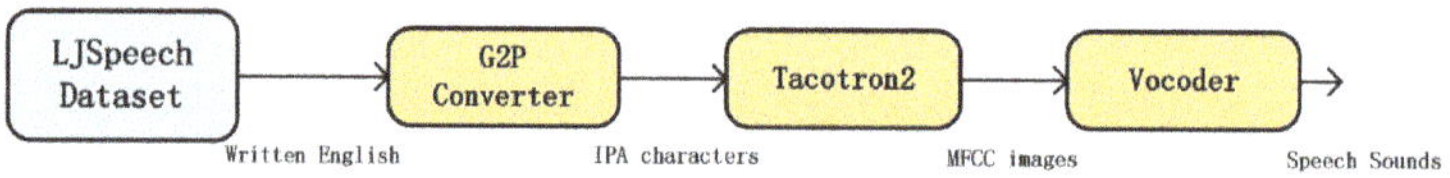

Figure 10. The main process of the IPA-to-speech system.

5.1. Data Preprocess

Speech synthesis, also known as text-to-speech (TTS), is a comprehensive technology that involves linguistics, digital signal processing, and acoustics. The main task of TTS conversion is to convert text into speech sounds. We employed the LJSpeech dataset to build our IPA-to-speech system. LJSpeech is a public domain speech dataset consisting of 13,100 speech audio files of a single speaker. Clips vary in length from 1 to 10 s and have a total length of approximately 24 h. The LJSpeech data are constructed in pairs of <English Sentences, Speech Samples>. To build our IPA-to-speech system, all the written English sentences in LJSpeech dataset were converted to their IPA format; this data preprocessing step required our English G2P converter discussed in Section 4. Table 6 gives several converted samples from the LJSpeech dataset [27].

Table 6. The original text in the LJSpeech dataset and the converted text.

Original Text in LJSpeech	Converted Text
The overwhelming majority of people in this country know how to sift the wheat from the chaff in what they hear and what they read.	ðə oʊvɜwɛlmɪŋ mədʒɔrəti ʌv pi:pəl iŋ ðɪs kʌntri noʊ haʊ tu: sɪft ðə wi:t frʌm ðə tʃaef ɪn wʌt ðeɪ hi:r ənd wʌt ðeɪ rɛd.
All the committee could do in this respect was to throw the responsibility on others.	ɔl ðə kəmɪti kʊd du: ɪn ðɪs rɪspɛkt wɑ:z tu: θroʊ ðə ri:spɑ:nsəbɪləti a:n ʌðɜz.
since these agencies are already obliged constantly to evaluate the activities of such groups	sɪns ði:z eɪdʒəsi:z ɑ:r ɔlrɛdi əblɑɪdʒd kɑ:nstəntli tu: ɪvaelju:eɪt ðə aektɪvɪti:z ʌv sʌtʃ gru:ps.

5.2. Tacotron2-Based IPA-to-Speech System

Tacotron is an end-to-end text-to-speech synthesis system that synthesizes speech from the characters introduced by the Google team. The input of the Tacotron model is characters, and the output of the model is the corresponding raw spectrograms. The limitation of the Tacotron model is the vocoder part: the sound generated by the Griffin–Lim algorithm is not of high quality. Thus, in this study, we employed the Tacotron2 model to build our IPA-to-Speech system. Compared with the original Tacotron model, Tacotron 2 uses simpler building blocks, using vanilla LSTM and convolutional layers in the encoder and decoder instead of CBHG stacks and GRU recurrent layers. It consists of two components:

- A recurrent sequence-to-sequence feature prediction network with attention, which predicts a sequence of Mel spectrogram frames from an input character sequence;
- A modified version of WaveNet, which generates time-domain waveform samples conditioned on the predicted Mel spectrogram frames.

Figure 11 illustrates the structure of the Tacotron2 model [21].

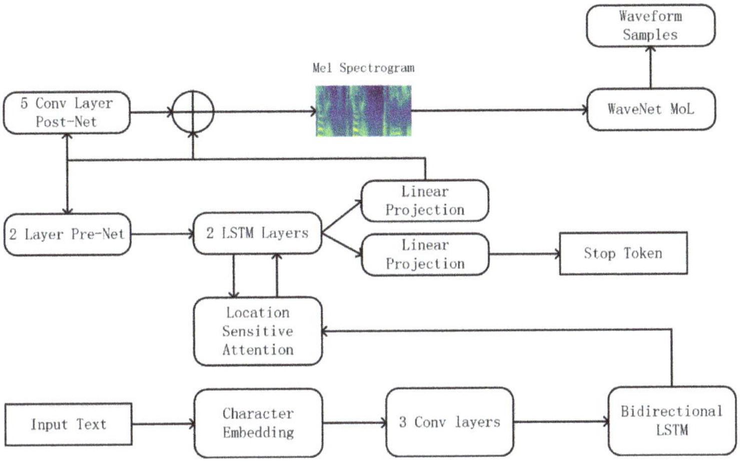

Figure 11. The structure of the Tacotron2 TTS model.

5.3. Implementation and Evaluation

A Nvidia Tesla P100 GPU was employed to train the Tacotron2 model. We selected a batch size of 48 on a single GPU with a 0.0001 learning rate. All the audio in LJSpeech was used for training, which contains about 24.6 h of audio recorded by a woman. Every text in the dataset needs to be spelled out; for example, the number "10" should be represented as "ten". This required about 20 h to finish 180,000 steps, which was about 200 epochs. Figures 12 and 13 show the training loss, validation loss, target Mel, and predicted Mel.

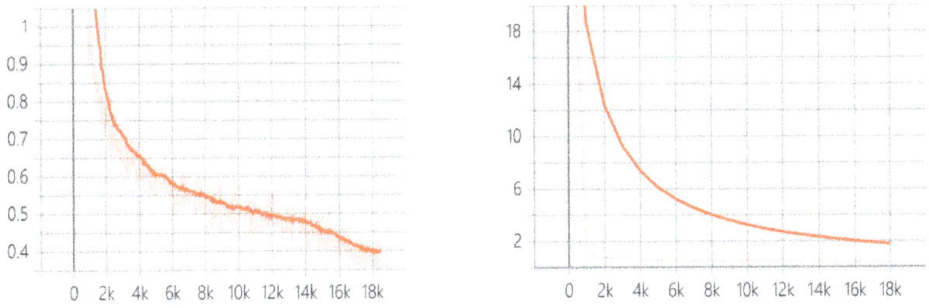

Figure 12. The training and validation losses of our Tacotron2 model.

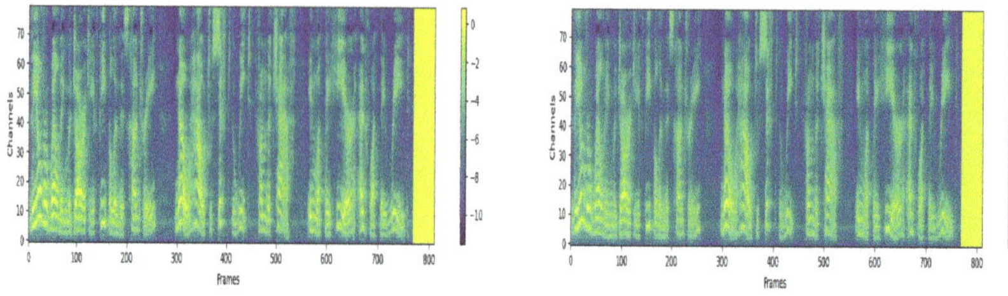

Figure 13. The target−Mel and predicted−Mel.

The mean opinion score (MOS) was utilized to evaluate the performance of our IPA-to-Speech system. The MOS is a metric that is widely used to measure the quality of speech generated by a TTS system. Generally, the MOS score is a rating from one to five, which refers to the perceived quality of audio from worst to best. After rating by human subjects, the MOS is calculated as the arithmetic mean:

$$MOS = \frac{\sum_{n=1}^{N} R_n}{N}$$

where R_n is the single rate score, and N is the total number of participants.

We employed 20 students from a linguistic department to help us evaluate our system using the MOS. The mean opinion score of our IPA-to-speech system is about 4.05 [28].

5.4. Discussion

In this part, we developed the last enhancement, which is a Tacotron2-based IPA-to-speech system. This system aims to directly generate speech from IPA characters. It has two advantages for those teaching phonetic transcription: the first is that with this speech synthesis system, teachers do not need to record their or other volunteers' speech, which helps to avoid the recorder accents problem and potential privacy problems; second, from the pedagogical aspect, the direct process of conversion from IPA characters to speech can also help students to better understand the pronunciation of single IPA characters and combined words.

6. Conclusions and Future Work

In this paper, we proposed three artificial intelligence enhancements for our linguistic E-learning system. Firt, the disordered speech classification module utilizes MFCCs to represent the features of speech, and a CNN model is used to build the classification function, which achieves the classification of disordered and nondisordered speech. Second, the grapheme-to-phoneme module uses a Transformer model, which provides high-accuracy G2P conversion. Finally, the IPA-to-speech module employs the Tacotron2 model and generates high-quality speech sounds from IPA characters.

All of these enhancements improve the functionality of the system compared with that of the traditional methods. The deep-learning-based speech classification module can better extract the features of disordered and nondisordered speech, and the CNN model provides better classification performance. Transformer-based G2P has a notably higher conversion accuracy than the statistical and other DL approaches. Moreover, IPA-to-speech provides a new idea to directly generate speech sounds from IPA characters. We found that with the development of artificial intelligence, more and more deep-learning-based technologies can be utilized in the education field, especially in linguistics, since linguistics is related to the natural language process in the field of AI. Our original E-learning system is a simple system focusing on phonetic transcription exams: the questions are speech sounds, while the answers are IPA characters. After the implementation of our three AI enhancements, our system is now a comprehensive linguistic E-learning system. It not only distinguishes disordered speech but also generates IPA characters from regular English words and creates high-quality speech sounds from IPA characters. The last two enhancements will greatly improve the experience for teachers. Also, these deep learning-based models can be pretrained and embedded into the system, so new teachers or educators do not need to complete any extra work to familiarize themselves with the system.

There are still other tasks that should be performed after this study. The first is evaluation: as we could not find a baseline against which to evaluate the performance of our IPA-to-speech system, we hired 20 volunteers to evaluate the system using the MOS. Future studies should include performing a more objective evaluation of the system. Also, the linguistic system may suffer from potential privacy problems. Since some of the speech sounds were recorded by students or teachers, we will try to replace all the speech sounds that may be related to these privacy problems with the generated speech sounds.

Author Contributions: Conceptualization, J.L.; methodology, Y.L.; software, S.L. and C.R.; validation, T.X.; formal analysis, T.X.; investigation, J.L.; writing—original draft preparation, J.L.; writing—review and editing, Z.W. and W.C.; supervision, W.C. All authors have read and agreed to the published version of the manuscript.

Funding: This study was supported by the Fundamental Research Funds for the Central Universities, grant number 2023QN1079.

Institutional Review Board Statement: Not applicable.

Informed Consent Statement: Not applicable.

Data Availability Statement: We leveraged two open datasets for evaluation, named The LJSpeech Dataset and SEED. The LJSpeech can be downloaded at https://keithito.com/LJ-Speech-Dataset/ (accessed on 1 January 2017) and the SEED is not available due to privacy.

Conflicts of Interest: The authors declare no conflict of interest.

References

1. Brown, A. International phonetic alphabet. In *The Encyclopedia of Applied Linguistics*; John Wiley & Sons: Hoboken, NJ, USA, 2012.
2. Howard, S.J.; Heselwood, B.C. Learning and teaching phonetic transcription for clinical purposes. *Clin. Linguist. Phon.* **2002**, *16*, 371–401. [CrossRef] [PubMed]
3. Seals, C.D.; Li, S.; Speights Atkins, M.; Bailey, D.; Liu, J.; Cao, Y.; Bassy, R. Applied webservices platform supported through modified edit distance algorithm: Automated phonetic transcription grading tool (APTgt). In *Learning and Collaboration Technologies. Designing, Developing and Deploying Learning Experiences: 7th International Conference, LCT 2020, Held as Part of the 22nd HCI International Conference, HCII 2020, Copenhagen, Denmark, 19–24 July 2020*; Springer: Cham, Switzerland, 2020.
4. Liu, J.; Speights, M.; Bailey, D.; Li, S.; Luan, Y.; Mishra, I.; Cao, Y.; Seals, C. Optimization to automated phonetic transcription grading tool (APTgt)—Automatic exam generator. In Proceedings of the International Conference on Human-Computer Interaction, Virtual, 24–29 July 2021; Springer: Cham, Switzerland, 2021.
5. Liu, J.; Ren, C.; Luan, Y.; Li, S.; Xie, T.; Seals, C.; Speights Atkins, M. Transformer-Based Multilingual G2P Converter for E-Learning System. In Proceedings of the International Conference on Human-Computer Interaction, Virtual, 26 June–1 July 2022; Springer: Cham, Switzerland, 2022.
6. Liu, J.; Ren, C.; Luan, Y.; Li, S.; Xie, T.; Seals, C.; Speights Atkins, M. Speech Disorders Classification by CNN in Phonetic E-Learning System. In Proceedings of the International Conference on Human-Computer Interaction, Virtual, 26 June–1 July 2022; Springer: Cham, Switzerland, 2022.
7. Schwarz, P.; Matějka, P.; Černocký, J. Towards lower error rates in phoneme recognition. In Proceedings of the International Conference on Text, Speech and Dialogue, Brno, Czech Republic, 8–11 September 2004; Springer: Berlin/Heidelberg, Germany, 2004.
8. Wu, X.; Fitzgerald, R. Reaching for the stars: DingTalk and the Multi-platform creativity of a 'one-star' campaign on Chinese social media. *Discourse Context Media* **2021**, *44*, 100540. [CrossRef]
9. Downes, S. E-learning 2.0. *ELearn* **2005**, *10*, 1. [CrossRef]
10. Madan, A.; Gupta, D. Speech feature extraction and classification: A comparative review. *Int. J. Comput. Appl.* **2014**, *90*, 20–25. [CrossRef]
11. Mohan, B.J. Speech recognition using MFCC and DTW. In Proceedings of the 2014 International Conference on Advances in Electrical Engineering (ICAEE), Vellore, India, 9–11 January 2014.
12. Lin, Y.-L.; Wei, G. Speech emotion recognition based on HMM and SVM. In Proceedings of the 2005 International Conference on Machine Learning and Cybernetics, Guangzhou, China, 18–21 August 2005; Volume 8.
13. Hunnicutt, S. *Grapheme-to-Phoneme Rules: A review*; Speech Transmission Laboratory, Royal Institute of Technology: Stockholm, Sweden, 1980; Volume QPSR 2–3. pp. 38–60.

14. Taylor, P. Hidden Markov models for grapheme to phoneme conversion. In Proceedings of the Ninth European Conference on Speech Communication and Technology, Lisbon, Portugal, 4–8 September 2005.
15. Bisani, M.; Ney, H. Joint-sequence models for grapheme-to-phoneme conversion. *Speech Commun.* **2008**, *50*, 434–451. [CrossRef]
16. Yao, K.; Zweig, G. Sequence-to-sequence neural net models for grapheme-to-phoneme conversion. *arXiv* **2015**, arXiv:1506.00196.
17. Rao, K.; Peng, F.; Sak, H.; Beaufays, F. Grapheme-to-Phoneme Conversion Using Long Short-Term Memory Recurrent Neural Networks. In Proceedings of the 2015 IEEE International Conference on Acoustics, Speech and Signal Processing (ICASSP), South Brisbane, Australia, 19–24 April 2015.
18. Yolchuyeva, S.; Németh, G.; Gyires-Tóth, B. Transformer based grapheme-to-phoneme conversion. *arXiv* **2020**, arXiv:2004.06338.
19. Tan, X.; Qin, T.; Soong, F.; Liu, T.-Y. A survey on neural speech synthesis. *arXiv* **2021**, arXiv:2106.15561 .
20. Wang, Y.; Skerry-Ryan, R.J.; Stanton, D.; Wu, Y.; Weiss, R.J.; Jaitly, N.; Yang, Z.; Xiao, Y.; Chen, Z.; Bengio, S.; et al. Tacotron: Towards end-to-end speech synthesis. *arXiv* **2017**, arXiv:1703.10135.
21. Shen, J.; Pang, R.; Weiss, R.J.; Schuster, M.; Jaitly, N.; Yang, Z.; Chen, Z.; Zhang, Y.; Wang, Y.; Skerrv-Ryan, R.; et al. Natural tts synthesis by conditioning wavenet on mel spectrogram predictions. In Proceedings of the 2018 IEEE International Conference on Acoustics, Speech and Signal Processing (ICASSP), Calgary, AB, Canada, 15–20 April 2018.
22. Arık, S.Ö.; Chrzanowski, M.; Coates, A.; Diamos, G.; Gibiansky, A.; Kang, Y.; Li, X.; Miller, J.; Ng, A.; Raiman, J.; et al. Deep voice: Real-time neural text-to-speech. In Proceedings of the International Conference on Machine Learning, PMLR, Sydney, NSW, Australia, 6–11 August 2017.
23. Ren, Y.; Ruan, Y.; Tan, X.; Qin, T.; Zhao, S.; Zhao, Z.; Liu, T.-Y. Fastspeech: Fast, robust and controllable text to speech. In Proceedings of the 33rd Conference on Neural Information Processing Systems, Vancouver, QC, Canada, 8–14 December 2019.
24. Gupta, S.; Jaafar, J.; Ahmad, W.W.; Bansal, A. Feature extraction using MFCC. *Signal Image Process. Int. J.* **2013**, *4*, 101–108. [CrossRef]
25. Speights Atkins, M.; Bailey, D.J.; Boyce, S.E. Speech exemplar and evaluation database (SEED) for clinical training in articulatory phonetics and speech science. *Clin. Linguist. Phon.* **2020**, *34*, 878–886. [CrossRef]
26. Vaswani, A.; Shazeer, N.; Parmar, N.; Uszkoreit, J.; Jones, L.; Gomez, A.N.; Kaiser, Ł.; Polosukhin, I. Attention is all you need. *Adv. Neural Inf. Process. Syst.* **2017**, *30*, 1–11.
27. The LJ Speech Dataset. Available online: https://keithito.com/LJ-Speech-Dataset/ (accessed on 1 January 2017).
28. Streijl, R.C.; Winkler, S.; Hands, D.S. Mean opinion score (MOS) revisited: Methods and applications, limitations and alternatives. *Multimed. Syst.* **2016**, *22*, 213–227. [CrossRef]

applied sciences

MDPI

Review

Emerging Drivers of Adoption of Generative AI Technology in Education: A Review

Andrina Granić

Faculty of Science, University of Split, 21000 Split, Croatia; andrina.granic@pmfst.hr

Abstract: This concept-centric review identifies and synthesizes emerging drivers of Generative AI (GenAI) adoption in education, addressing a critical gap by offering the first structured integration of empirically supported predictors. Based on 27 peer-reviewed studies featuring validated research models, the review distils 11 predictors into a Three-Tier Framework. Core predictors—Performance Expectancy and Trust—consistently influence adoption across contexts. Moderate predictors—Effort Expectancy, Facilitating Conditions, Social Influence, Perceived Behavioral Control, and Perceived Compatibility—show variable relevance depending on technological and institutional factors. Emerging predictors—Habit, AI Literacy, Anxiety, and Playfulness—capture evolving socio-technical and individual dynamics, reflecting the rapid development of GenAI technologies. While the current literature offers valuable insights, gaps remain in addressing ethical concerns, barriers to adoption, teacher professional development, student engagement, and the influence of cultural and contextual diversity. The findings emphasize the need to iteratively refine the Three-Tier Framework by incorporating these dimensions and adapting to technological advancements. By consolidating empirical evidence and distinguishing between mature and emerging predictors, this review advances theoretical understanding of technology acceptance in education. It provides a structured foundation for guiding future research, informing policy and practice, and supporting responsible, context-sensitive GenAI integration across diverse educational settings.

Keywords: generative artificial intelligence (GenAI); GenAI technology adoption; predictive factors; three-tier framework; education; review

1. Introduction

The integration of generative artificial intelligence (GenAI) into education represents a transformative shift in pedagogical practices and learning experiences. As educational institutions increasingly explore its potential to support personalization, enhance learner engagement, and optimize instructional strategies, understanding the drivers of its acceptance and adoption becomes a critical academic and practical pursuit. However, despite growing interest, the current discourse remains fragmented, lacking a systematic framework that comprehensively consolidates the factors influencing GenAI uptake in educational settings.

This review addresses this gap by synthesizing the existing literature to uncover key empirically supported predictors—such as performance expectancy, trust, habit, and AI literacy—that shape the adoption process. By leveraging a concept-centric approach [1], the study avoids simple narrative listings and instead builds its analysis around validated theoretical models and core constructs, offering a structured and theoretically grounded understanding. Importantly, the review does not limit itself to specific publication venues or temporal scopes, ensuring broad and unbiased coverage of relevant empirical and theoretical contributions.

Although several domains have begun to explore GenAI's application, including healthcare, mental health, and human activity recognition, education and training contexts remain among the most investigated fields. Studies such as systematic meta-review by Kiuchi et al. [2] and survey by Ferrara [3] exemplify this interdisciplinary reach, examining conversational agents and large language models (LLMs) in contexts ranging from elderly care and mental health to behavioral modeling using wearable data. Across these fields, ethical concerns—including privacy, transparency, psychological impacts, and interpretability—frequently surface as barriers to adoption.

Focusing on healthcare, reviews by Hirata et al. [4] and Nerella et al. [5] illustrate both the promise and limitations of LLMs in diagnostic imaging, clinical NLP, and decision support, noting the need for technological refinement and ethical safeguards. Similarly, review studies by Yim et al. [6] and Goktas and Grzybowski [7] reveal that GenAI currently plays a supportive role, enhancing diagnostic decision-making and patient engagement but rarely replacing human judgment. These insights echo broader concerns raised by Hobensack et al. [8] and Ng [9] regarding the reliability and transparency of GenAI tools.

In education, recent systematic reviews and scoping studies have begun to trace both the opportunities and risks associated with GenAI. Ogunleye et al. [10] and Yusuf et al. [11] identify key adoption trends, ethical challenges, and themes such as academic integrity, assessment redesign, and institutional readiness. Sekli et al. [12] and Drobnjak et al. [13] highlight GenAI's potential in supporting personalized feedback, adaptive content generation, and conversational tutoring while also noting constraints such as the hallucination effect and the need for human oversight.

At a pedagogical level, in their review study Sengul et al. [14] emphasize the gap between educational and industrial GenAI adoption, particularly in computing education, and call for alignment of curricular design with real-world AI practices. These concerns are further echoed in the scoping review of Ali et al. [15] who offer a categorization of user, operational, technological, and ethical challenges in ChatGPT's educational use. Meanwhile, Amarathunga's [16] systematic review contributes bibliometric insight, mapping global trends and flagging the technology's potential impact on learner creativity and academic dishonesty.

The ethical dimension remains central to the discourse. Scoping reviews by Hagendorff [17] and Yan et al. [18] explore the normative landscape of GenAI, with detailed taxonomies of risks related to fairness, privacy, hallucinations, and societal impacts. They call for open-source, human-centered approaches to ensure trustworthy and transparent applications, particularly when sensitive student data and evaluative processes are involved.

These studies together illuminate the cross-domain interest in GenAI, the educational sector's unique adoption challenges, and the urgency for a unified framework to understand the underlying psychological, technological, and contextual factors driving user acceptance.

Research Question and Review Objective

Despite a rapidly growing body of research, to the best of the author's knowledge, no comprehensive synthesis currently exists that consolidates empirically supported factors predicting the acceptance and adoption of GenAI in education. While existing reviews often highlight trends, challenges, or ethical implications, they seldom examine validated theoretical models or focus systematically on adoption predictors. To fill this gap, the present review formulates the following research question:

RQ: What are the key factors that influence the acceptance and adoption of generative AI (GenAI) technologies in educational settings?

The goal is to identify emerging drivers and to analyze how core predictors are conceptualized, operationalized, and empirically evaluated in education-related research. Particular attention is paid to theoretical underpinnings, methodological rigor, participant demographics, and contextual diversity. In doing so, the study offers practical and theoretical insights for researchers, educators, and policymakers aiming to facilitate responsible and effective GenAI integration in learning environments.

The subsequent sections detail the review methodology, present a conceptually organized synthesis of findings, and discuss implications for future educational practices and research directions.

2. Research Approach

This review aims to uncover the key factors influencing the acceptance and adoption of generative AI (GenAI) technologies in educational settings. This research focus stems from a gap-oriented analysis of the literature, which identified a lack of consolidated, empirically grounded insights into the predictors of GenAI adoption. While general technology acceptance research is well-established, studies specifically targeting GenAI remain fragmented, particularly regarding emerging socio-technical and individual-level factors.

2.1. Database and Search Strategy

The review adopts a concept-centric approach, structuring the analysis around key theoretical models and predictors relevant to educational technology adoption rather than being limited to specific journals or timeframes (cf. [1]). This strategy supports a deeper exploration of key constructs while minimizing biases associated with dominant publication venues or temporal trends.

To ensure coverage of high-quality, peer-reviewed research, the Web of Science (WoS) Core Collection database was exclusively used, with no restrictions on category or publication date. Both journal articles and conference proceedings were included. This approach enabled the identification of relevant studies that offer insights into the factors, impacts, and practical considerations shaping the integration of GenAI in education.

The search strategy was designed to capture studies involving generative AI technologies—such as large language models (LLMs), GPT, DALL·E, and other tools capable of producing human-like content—and to focus on theoretical and empirical discussions of their adoption. The search string used was as follows:

("generative AI" OR "generative artificial intelligence" OR "large language model*" OR "LLM*") AND (("theor*" OR "model") AND ("adoption" OR "acceptance")) AND ("education*" OR "learn*")

The search was applied using the "TOPIC" filter (title, abstract, author keywords, and Keywords Plus) in the WoS Core Collection. Truncation (e.g., "LLM*") was used to include multiple relevant term variants.

2.2. Inclusion and Exclusion Criteria

This review focused on empirical studies that apply theoretical models or acceptance frameworks to investigate the adoption of generative AI (GenAI) technologies in educational contexts, particularly those supporting teaching and learning. GenAI technologies, as defined here, include advanced tools such as large language models (LLMs) like GPT and image generators like DALL·E that produce human-like outputs across modalities.

To ensure methodological rigor, the following inclusion criteria were applied:

- Empirical, peer-reviewed studies published in English (journal articles and conference proceedings),
- Application of theoretical models (e.g., TAM, UTAUT) to examine GenAI adoption or acceptance in education,
- Clearly formulated hypotheses and empirically tested relationships between constructs.

These criteria ensured the selection of studies providing validated, model-based evidence on predictors such as perceived usefulness, ease of use, and behavioral intention—critical for understanding adoption dynamics in educational settings.

The exclusion criteria were as follows:

- Purely theoretical, conceptual, or review papers,
- Meta-analyses or studies without primary data analysis,
- Studies using surveys alone, even when employing validated instruments, without deeper model-driven examination of interrelationships between constructs,
- Non-peer-reviewed works, such as book chapters.

In particular, studies that lacked explicit modeling or did not operationalize constructs within a theoretical framework were excluded due to concerns about the reliability and interpretability of their findings. While survey-based studies offer preliminary insights, their omission ensures consistency in analytical depth and alignment with the review's concept-centric approach. Among the excluded works—despite their use of validated scales—are studies by Huang et al. [19], Wang et al. [20], Wu et al. [21], Herani and Angela [22], Smith et al. [23], Petrič [24], Chan and Zhou [25], Tangadulrat et al. [26], and Sallam et al. [27].

2.3. Screening and Selection Process

The literature search was conducted in mid-December 2024 using the WoS Core Collection, with no restrictions on category or publication year. This initial search yielded 161 records containing the specified terms in the title, abstract, keywords, or author keywords fields. After excluding one book chapter and non-relevant entries, titles, abstracts, and full texts were rigorously screened for relevance and alignment with the inclusion criteria.

From this process, 63 publications were retained for in-depth review. Ultimately, 27 studies met the final inclusion criteria: empirical investigations grounded in theoretical models, presenting hypotheses and model-based analysis specific to GenAI adoption in education.

By narrowing the selection to studies with clearly operationalized constructs and validated relationships, this review provides a transparent, methodologically robust foundation for identifying emerging drivers of acceptance and adoption—in line with the goals and standards of a concept-centric review.

3. Results and Discussion

This section presents the main findings of the review, focusing on empirical studies that examine the factors influencing the acceptance and adoption of Generative AI (GenAI) technologies in educational settings. To ensure clarity and coherence, the results are organized into four thematic subsections:

- Regional and Cultural Dimensions of GenAI Adoption in Education, which examines the geographic distribution of the selected studies and explores how regional and cultural contexts shape the conceptualization and adoption of GenAI technologies;

- Participant Demographics and Sampling Approaches, which provides an overview of participant types and sample sizes, highlighting the diversity of educational levels, roles, and research designs;
- Adoption Models and Theoretical Frameworks Used in GenAI Research, which analyzes the conceptual foundations and theoretical models employed across the studies; and
- What Influences GenAI Adoption in Education? A Closer Look, which explores the key predictors of adoption across individual, institutional, and socio-technical domains.

3.1. Regional and Cultural Dimensions of GenAI Adoption in Education

A review of the publication timeline shows that 26 of the 27 included studies were published in 2024, with only 1 study—Duong et al. [28]—appearing earlier. This concentration underscores both the recency of GenAI integration in education and the rapid academic response to its emerging significance. It also reflects the novelty and dynamism of the field, still in the early stages of theoretical consolidation and empirical investigation. Since the review was conducted in mid-December 2024, it is likely that additional relevant publications have since emerged, indicating a continually evolving research landscape.

Geographically, the studies reflect broad international engagement. The majority originated from China (11 studies), reflecting the country's strong emphasis on AI research and its integration in education. Other active contributors include India, Thailand, the United Arab Emirates (UAE), and Vietnam, each with two studies. Additional countries represented by single studies include Malaysia, Norway, Poland, Singapore, South Korea, and Tanzania. One study involved a cross-national comparison between Egypt and Poland, and another encompassed multiple countries. This geographical diversity underscores the global relevance of GenAI in education and signals growing international interest in its adoption across varied educational systems.

These trends are visualized in Figure 1, which maps the geographic distribution of the reviewed studies. The figure illustrates China's leading position, followed by a broader representation across Asia, Europe, and Africa, confirming the widespread attention GenAI is receiving worldwide.

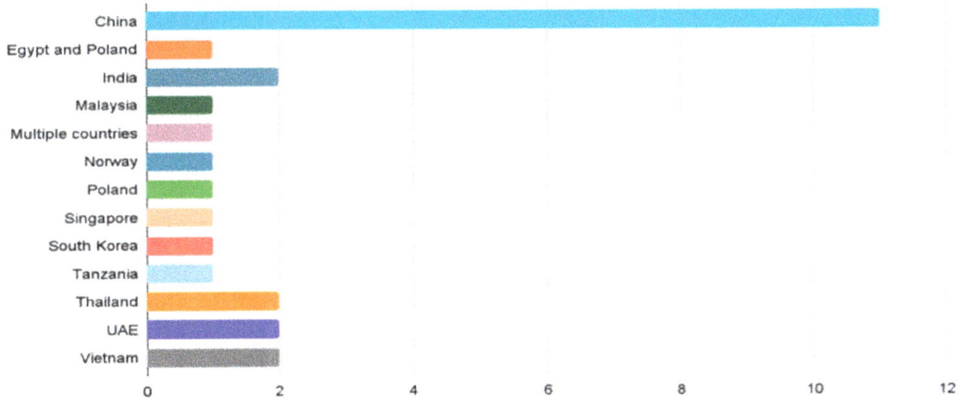

Figure 1. Geographic distribution of empirical studies on GenAI adoption in education.

Beyond geographic distribution, the reviewed studies reflect diverse educational, cultural, and socio-technical contexts that shape how GenAI is perceived and adopted in practice. A substantial proportion of studies were conducted in East and Southeast Asia, particularly China (11 studies), followed by India (2), Thailand (2), and Vietnam (2). These studies frequently emphasize performance-oriented educational goals, AI literacy, and institutional support, aligning with national policies focused on digital transformation. For example, studies from China and Malaysia (e.g., [29,30]) incorporate predictors such as trust in AI, personal innovativeness, and perceived learning performance, suggesting an adoption discourse oriented toward technological efficacy and learner outcomes.

In contrast, European studies, such as those from Norway [31] and Poland [32,33], often integrate predictors like habit, hedonic motivation, and facilitating conditions, and in some cases challenge established assumptions (e.g., non-significance of effort expectancy or social influence). These patterns may reflect greater emphasis on individual autonomy, user experience, and long-term engagement within European educational settings.

Studies from Africa (e.g., Tanzania [34]) emphasize factors such as information quality and social influence, which are particularly salient in contexts where access to reliable digital resources and peer validation play a key role in shaping technology use. These findings highlight the importance of adapting GenAI tools to local infrastructure and educational needs, though direct discussion of systemic infrastructural constraints is limited in the reviewed literature.

In the Middle East, research from the United Arab Emirates (e.g., [35]) reflects a hybrid integration of global acceptance frameworks (e.g., TAM, UTAUT) with regionally relevant constructs such as user satisfaction, perceived efficiency, and learning performance. However, another UAE-based study [36] focuses more on perceived anthropomorphism and social sustainability, demonstrating how local values and social perceptions of AI can shape the design of adoption models.

These emerging patterns suggest that while core predictors of GenAI adoption—such as performance expectancy, perceived usefulness, and social influence—remain broadly applicable, their relative influence, interpretation, and model integration differ across cultural and institutional settings. These insights affirm the need for culturally responsive research that accounts for local pedagogical norms, technological readiness, and educational policy when studying GenAI adoption.

3.2. Participant Demographics and Sampling Approaches

Another important aspect concerns the types of participants and users examined in the analyzed studies. A significant majority of the research ($n = 21$) focused on higher education students, including bachelors, masters, and doctoral students (e.g., [36–41]), with data primarily collected through web-based surveys conducted in university settings (e.g., [31]). Among these, several studies specifically targeted business students (e.g., [34,42,43]) and English as a Foreign Language (EFL) students (e.g., [29,44,45]). Additionally, individual studies included healthcare students [46] and art and design college students [47], while others did not specify the students' study programs. Only one study focused on elementary education students [48].

Beyond students, a smaller number of studies examined educators, particularly higher education teachers [32,49], pre-service teachers [50], or a mix of educators and higher education students [35,51]. Figure 2 illustrates the distribution of participant types in studies on GenAI adoption in education.

Sample sizes varied significantly, ranging from the smallest group of 16 healthcare students [46] to the largest group of 1389 higher education students in a study by Duong et al. [28]. Notably, smaller sample sizes—typically up to 400 participants—were more common, highlighting a tendency toward moderate-sized participant groups in these studies.

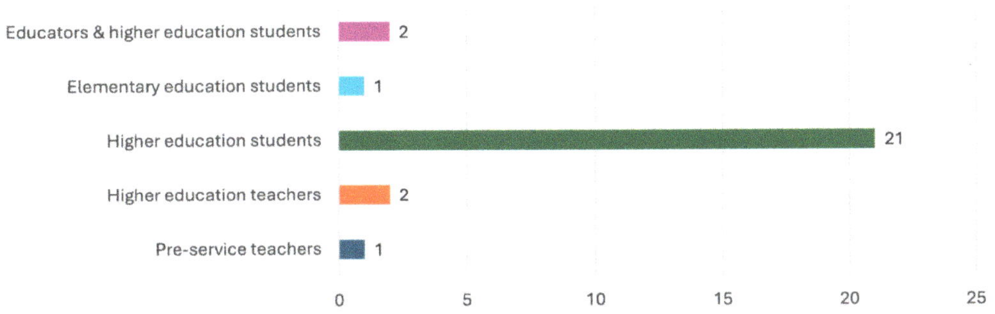

Figure 2. Participant types in studies on GenAI adoption in education.

3.3. Adoption Models and Theoretical Frameworks Used in GenAI Research

The review highlights the application of various technology acceptance and adoption models in the selected studies, reflecting a diverse approach to understanding the factors influencing the adoption of Generative AI (GenAI) in education. Among the 27 studies analyzed, the most frequently applied models were the Technology Acceptance Model (TAM) [52,53], the Unified Theory of Acceptance and Use of Technology (UTAUT) [54], and the extended UTAUT (UTAUT2) [55].

Other frameworks, such as the Innovation Diffusion Theory (IDT) [56,57] and the Expectation–Confirmation Model (ECM) [58], were also applied in individual studies. Specifically, TAM was employed in seven studies, UTAUT/UTAUT2 in eight studies, and the Theory of Planned Behavior (TPB) [59,60] in three studies. Additionally, seven studies combined multiple theoretical frameworks to enhance the explanatory power of individual models. The breakdown of model utilization is as follows (in alphabetical order):

- Expectation–Confirmation Model (ECM) [58], a model based on Oliver's [61] leading cognitive theory in the field of consumer satisfaction that explains post-adoption satisfaction and behavior through Expectation–Confirmation Theory (ECT), was applied in one study [38].
- Innovation Diffusion Theory (IDT), proposed by Rogers [56,57] as one of the most popular models for studying the adoption and use of innovations, was adopted in a single study [39].
- Technology Acceptance Model (TAM), introduced by Davis [52,53], a leading established theory and a parsimonious yet powerful model for explaining, predicting, and enhancing user acceptance in diverse technological deployments, was utilized in seven studies [28,29,37,40,44,46,62].
- Integration of TAM and Self-Determination Theory (SDT) [63], a theoretical approach widely used in studies of motivational behavior that examines relationships among the social environment, psychological needs, motivation, outcomes, and well-being, was explored in two studies [41,47].
- Combination of TAM and Theory of Planned Behavior (TPB) [59,60], a theory stating that behavior is a direct function of behavioral intention and perceived behavioral control, was investigated in one study [49].
- Integration of TAM and UTAUT, a rich theoretical approach to explaining behavioral intentions and technology use, was analyzed in one study [35].
- TPB was employed in three studies [36,51,64].
- Combination of TPB and Behavioral Reasoning Theory (BRT), a theory that explores the cognitive processes and reasoning that bridge the gap between intentions and actual behavior, was examined in one study [30].

- UTAUT was utilized in three studies [33,48,50].
- UTAUT2, an extension of the original UTAUT that emphasizes the context of consumer use, was applied in five studies [31,32,34,42,43].
- Combination of UTAUT and SDT was investigated in two studies [45,65].

It is also important to note that the studies using TAM, UTAUT, UTAUT2, and TPB did not employ the core models in their original forms but instead extended or modified them to address specific aspects of GenAI adoption. These adjustments were made to improve the models' relevance and accuracy in the context of education and technology integration:

- Extended Models: These models expand the original framework by adding new constructs or pathways while maintaining the core structure to address gaps or new challenges specific to the context.
- Modified Models: These models fundamentally alter or redefine the original framework by removing constructs, changing relationships, or integrating elements from other models to better fit the research focus.

These adjustments in the use of models reflect the dynamic nature of GenAI adoption research, where traditional technology acceptance models are often reshaped to capture the unique factors at play in the context of education. Figure 3 illustrates the utilization patterns of models and theoretical frameworks in exploring the adoption of GenAI technology in education.

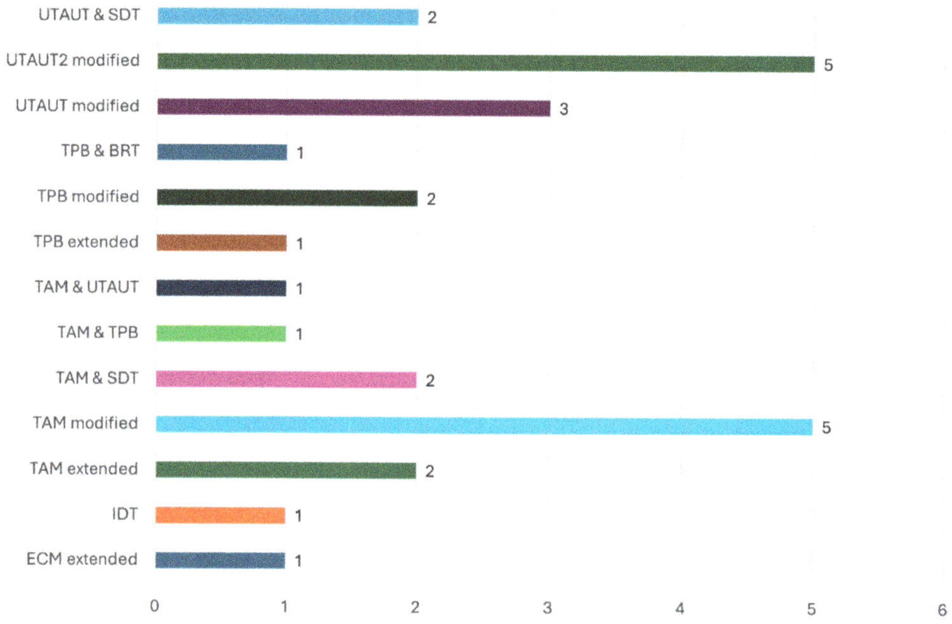

Figure 3. Overview of models and theories employed in GenAI technology adoption studies in education.

3.4. What Influences GenAI Adoption in Education? A Closer Look

The adoption of Generative AI (GenAI) technologies in education has attracted significant scholarly interest due to their potential to enhance teaching practices and learning experiences. However, the successful integration of these technologies depends on a range of interrelated factors that reflect both opportunities and challenges. Drawing from the 27 empirically grounded studies included in the review, eleven key predictors were identified, illustrating the complex interplay of individual, institutional, and socio-technical dimensions.

These predictors include individual-level factors, including individual attitudes and readiness to adopt GenAI, perceptions of its ease of use and usefulness, and emotional factors such as anxiety and playfulness. Technological and skill-based enablers—namely, AI literacy, technological infrastructure, and facilitating conditions—further influence adoption, as do institutional support mechanisms. Additionally, social and cultural influences, such as social influence, trust in AI, and perceived compatibility with existing practices, play a pivotal role in determining behavioral intention and actual use.

Together, these predictors offer a comprehensive lens through which the adoption of GenAI in education can be understood. While this subsection provides a high-level synthesis, a more detailed categorization and interpretation is presented in the following section: "The Three-Tier Framework for GenAI Adoption in Education".

By offering this closer look, the subsection emphasizes the context-sensitive and multidimensional nature of GenAI adoption. Educators and learners do not operate in a vacuum; their decisions are shaped by internal dispositions, external support, and broader socio-technical environments. Understanding these nuances is essential for fostering effective adoption strategies tailored to diverse educational contexts.

- *Performance Expectancy* (*Perceived Usefulness*): Users adopt GenAI when they perceive it as beneficial for achieving educational goals. Grassini et al. [31], Jang [42], and Zheng et al. [45] consistently highlight its role in shaping Behavioral Intention (BI). Additionally, Duong et al. [28] and Liu et al. [44] emphasize that Performance Expectancy influences not only BI but also actual use, making it essential for both initial adoption and sustained engagement.
- *Trust*: Trust plays a central role in GenAI adoption by fostering user confidence and mitigating concerns about privacy, ethics, and transparency. Shahzad et al. [40] reveal that trust plays a crucial role in GenAI awareness, acceptance, and adoption. Similarly, Ma [29] and Yap et al. [46] highlight that perceived trust directly influences BI to use GenAI, highlighting its central role in fostering adoption
- *Effort Expectancy* (*Perceived Ease of Use*): While simplicity enhances adoption in some contexts, its influence is often secondary to functionality. Du and Lv [48] and Lu et al. [49] show that Effort Expectancy indirectly boosts perceptions of usefulness, whereas Grassini et al. [31] and Wang et al. [50] suggest that ease of use has limited impact in educational settings where performance and trust are prioritized.
- *Facilitating Conditions*: External resources and institutional support are critical for bridging the gap between BI and actual use. Zheng et al. [45] and Du and Lv [48] highlight the importance of accessible infrastructure, while Grassini et al. [31] report that Facilitating Conditions may be less impactful in resource-rich environments, emphasizing their context-dependent nature.
- *Social Influence*: Social norms and external encouragement can shape adoption decisions. Changalima et al. [34], Jang [42], and Zheng et al. [45] find that Social Influence significantly impacts BI, particularly in environments where peer or societal expectations drive behavior. Wang and Reynolds [65] find SI to significantly influence both BI and usage intentions. However, studies like Sudan et al. [43] and Strzelecki et al. [32] report minimal effects, suggesting that intrinsic motivators often outweigh external pressures.
- *Perceived Behavioral Control*: Confidence in one's ability to use GenAI effectively is an important determinant of BI. Lu et al. [49], Ivanov et al. [51], and Al-Qaysi et al. [30] underscore its role in empowering users, while Al-Emran et al. [36] observe that its

significance may diminish when external enablers, such as of facilitating conditions or social pressures, are strong.

- *Perceived Compatibility*: Compatibility between GenAI and users' needs, workflows, and values significantly foster adoption. Raman et al. [39] demonstrate that seamless integration into educational practices enhances BI, while Yap et al. [46] validate its role in making GenAI intuitive and relevant for users.
- *Habit*: Habit strongly predicts both BI and continued use. Zheng et al. [45], Grassini et al. [31], and Strzelecki et al. [32] emphasize that routine engagement with GenAI fosters sustained adoption. Sudan et al. [43] further highlight the role of prior usage experiences in enhancing the chance of continued engagement and long-term use.
- *AI Literacy*: Greater AI literacy equips users with the knowledge needed to engage confidently with GenAI. Jang [42] and Wang et al. [64] find that understanding AI principles enhances user confidence and perceived benefits, leading to positive adoption intentions.
- *Anxiety*: Concerns about job displacement, ethical implications, or technological complexity can act as barriers to adoption. Shen et al. [47], Li et al. [62], and Wang et al. [50] show that anxiety negatively impacts BI by reducing perceptions of usefulness and ease of use. Addressing these fears is essential for fostering acceptance.
- *Playfulness* (*Hedonic Motivation*): Enjoyment and intrinsic satisfaction derived from GenAI usage strongly influence BI. Zheng et al. [45] and Yap et al. [46] highlight that enjoyment and playfulness foster user engagement and enhance the likelihood of continued use, while Strzelecki et al. [32] emphasize the enjoyment and pleasure of GenAI use as strong predictors of BI.

These findings collectively underscore the multifaceted nature of GenAI adoption in education, shaped by a dynamic interplay of technological, individual, and social factors. The relative weight of these predictors often varies by context, depending on institutional type, learner demographics, or resource availability. This variation points to the importance of context-aware approaches when planning and evaluating GenAI adoption initiatives.

To support both research and practice, a systematic categorization of these predictors may prove valuable. Such a framework can guide future investigations, enabling researchers to extend existing models, accommodate emerging evidence, and identify new variables. Ultimately, this structured approach will help develop a more comprehensive and adaptable understanding of GenAI adoption across diverse educational settings. To address the complexity and contextual variability of the identified predictors, a more structured representation is essential. Accordingly, the following section proposes a conceptual framework that organizes these predictors into coherent categories. This framework aims not only to clarify the relationships among key adoption factors but also to support researchers and practitioners in systematically analyzing and applying them within diverse educational environments.

4. The Three-Tier Framework for GenAI Adoption in Education

Building on the synthesized findings of this review, this section introduces the *Three-Tier Framework for GenAI Technology Adoption in Education*. Designed to offer conceptual clarity and practical utility, the framework organizes the 11 key predictors into three interconnected tiers—*Core Predictors*, *Moderate Predictors*, and *Emerging Predictors*. This tiered structure enables a nuanced understanding of the diverse and interdependent factors that influence educators' and learners' acceptance and integration of generative AI tools in educational settings.

The framework provides a holistic perspective by clustering predictors based on their conceptual centrality, empirical support, and frequency of appearance across the reviewed studies. Each tier reflects a different level of influence and maturity in the research landscape, thereby offering both theoretical grounding and practical insight into the dynamics of GenAI adoption.

1. Core Predictors (consistently significant across studies):
   These factors consistently exert a strong influence on the adoption of GenAI technologies, serving as essential drivers rooted in individual perceptions and socio-technical dynamics. They are the most widely recognized and stable determinants of how individuals perceive and engage with AI technologies:
   - Performance Expectancy (Perceived Usefulness)
   - Trust
2. Moderate Predictors (context-dependent):
   These predictors vary in their influence depending on specific educational or technological contexts, emphasizing the importance of task and environmental factors in shaping GenAI adoption. While significant, their impact is less consistent and often contingent on situational factors:
   - Effort Expectancy (Perceived Ease of Use)
   - Facilitating Conditions
   - Social Influence
   - Perceived Behavioral Control
   - Perceived Compatibility
3. Emerging Predictors (increasing relevance):
   Reflecting evolving individual and socio-technical dimensions, these factors are gaining prominence as GenAI technologies mature and expand their applications in education. Although less established than core and moderate predictors, they are increasingly supported by empirical research and are expected to grow in relevance:
   - Habit
   - AI Literacy
   - Anxiety
   - Playfulness (Hedonic Motivation)

Each predictor is analyzed in terms of its definition, underlying rationale, and empirical evidence, as synthesized from the 27 studies included in this review (see Table 1). This structured approach ensures the framework's relevance across contexts and its usefulness as a foundation for future research, policy formulation, and institutional implementation.

The *Three-Tier Framework for GenAI Adoption in Education* is both adequate and credible, grounded in established models such as TAM, UTAUT, and UTAUT2, while integrating emerging predictors specific to educational contexts. Core Predictors such as Performance Expectancy and Trust consistently demonstrate strong influence on adoption. Moderate Predictors emphasize the context-dependent nature of factors like infrastructure, social dynamics, and compatibility with existing practices. Meanwhile, Emerging Predictors—including Habit, AI Literacy, and Anxiety—are gaining relevance as GenAI tools become increasingly embedded in educational settings.

Although further exploration is needed to understand the interplay and relative influence of these predictors, the framework offers a sound foundation for investigating GenAI adoption in education. As research progresses and technologies evolve, Emerging Predictors are expected to gain greater prominence, calling for ongoing refinement of the framework to accommodate new insights and address emerging challenges.

Table 1. The Three-Tier Framework for GenAI Technology Adoption in Education.

The Three-Tier Framework for GenAI Technology Adoption in Education
Performance Expectancy/Perceived Usefulness • Definition Performance Expectancy, or Perceived Usefulness, refers to the degree to which users believe that using GenAI technologies will enhance their performance and effectiveness in educational tasks. It encompasses users' perceptions of how GenAI can improve productivity, learning efficiency, task quality, and overall educational outcomes, making it a consistently significant predictor of technology adoption. • Empirical Evidence Studies highlight the strong role of Performance Expectancy, or Perceived Usefulness, in shaping Behavioral Intention (BI), with users more likely to adopt GenAI when they perceive it as beneficial for their educational goals [31,33,42,45,50]. Numerous studies further validate the significance of Performance Expectancy in predicting BI and usage intentions, including Strzelecki et al. [32], Wang and Reynolds [65], Du and Lv [48], and Soliman et al. [41]. Additionally, Duong et al. [28] and Liu et al. [44] demonstrate its influence not only on BI but also on actual use, highlighting its pivotal role in both initial adoption and sustained engagement. These findings position Performance Expectancy as a foundational element due to its consistent significance in driving GenAI adoption in educational contexts. • Rationale Performance Expectancy, or Perceived Usefulness, is a universally recognized core predictor of GenAI adoption in education. Both students and educators prioritize technologies that deliver tangible benefits, such as enhancing academic performance, improving productivity, or increasing instructional effectiveness. Empirical evidence consistently highlights that users who perceive GenAI as capable of improving learning outcomes or task quality are more likely to develop positive attitudes and behavioral intentions toward its adoption. This factor not only drives initial acceptance but also supports sustained usage, as users are intrinsically motivated by tools that align with their goals for academic and instructional improvement.
Trust • Definition Trust refers to users' confidence in the reliability, security, accuracy, integrity, and fairness of GenAI technologies. It encompasses the belief that these technologies will perform as expected, provide unbiased results, and handle sensitive data responsibly. Trust also includes transparency regarding the technology's functioning and adherence to ethical guidelines, particularly in terms of privacy, security, and avoiding misuse. • Empirical Evidence Shahzad et al. [40] underscore the crucial role of perceived trust in GenAI awareness, acceptance, and adoption in higher education. Their study reveals that perceived trust significantly moderates the relationship between GenAI awareness and its perceived ease of use, usefulness, and intelligence. Furthermore, perceived trust directly influences Behavioral Intention (BI) to use GenAI, highlighting its central role in fostering adoption [29,46]. In addition, trust also emerges as a critical factor in other studies, albeit implicitly. Shen et al. [47] link trust to user confidence by addressing concerns related to privacy, ethics, and originality. Wang et al. [64] implicitly tie trust to transparency and accountability through their focus on AI ethics. Similarly, Li et al. [62] identify AI confidence, interactivity, and content design as factors fostering positive attitudes, further underscoring trust's implicit role in GenAI acceptance. These findings collectively position trust as a foundational element due to its consistent significance in fostering adoption across various contexts. • Rationale Trust is consistently recognized as a core predictor of GenAI adoption, particularly in educational settings where the technology is relied upon for decision-making, content generation, and sensitive processes. Users' confidence in the reliability, security, transparency, and ethical integrity of GenAI directly influences their willingness to adopt and integrate it into their workflows. In education, where AI is increasingly used for personalized learning, assessments, and content creation, it is essential for users to trust that the technology will deliver accurate, unbiased results and safeguard privacy. Establishing trust mitigates skepticism and concerns about misuse or misinformation, thereby lowering adoption barriers and supporting the long-term success and integration of GenAI tools in educational environments.

Table 1. *Cont.*

The Three-Tier Framework for GenAI Technology Adoption in Education
Effort Expectancy/Perceived Ease of Use • Definition Effort Expectancy, also referred to as Perceived Ease of Use, represents the degree to which users believe that using GenAI technologies will be easy and effortless. This factor reflects the perceived simplicity and learnability of the technology, influencing users' willingness to adopt and use it. • Empirical Evidence Studies demonstrate that when users perceive GenAI as easy to use, they are more likely to adopt it, as evidenced by its significant influence on Behavioral Intention (BI) [33,45,48,65]. Effort Expectancy also indirectly shapes perceptions of usefulness and control; for instance, Lu et al. [49] found that ease of use enhances users' views of GenAI's utility, while Saihi et al. [35] emphasized the adoption benefits of simpler interfaces. Additionally, Duong et al. [28] linked Effort Expectancy to both BI and actual use, highlighting its role in sustaining long-term engagement. Conversely, several studies suggest that Effort Expectancy has limited significance in driving BI for GenAI technologies. Grassini et al. [31], Jang [42], and Wang et al. [50] observed no significant impact of Effort Expectancy on BI in educational contexts, where users prioritize functionality and performance over ease of use. Similarly, Shen et al. [47] and Sudan et al. [43] reported weak or negligible correlations, suggesting that simplicity plays a minor role when trust or perceived benefits dominate decision-making. These findings suggest that Effort Expectancy's influence on GenAI adoption is moderate and context-dependent, particularly in settings where users possess baseline digital skills or prioritize performance over usability. • Rationale Effort Expectancy is a moderate but important predictor of GenAI adoption, particularly in contexts where ease of use reduces cognitive load and supports broader adoption. For novice users or those with lower digital literacy, intuitive designs and simplified interfaces are critical for ensuring a smooth onboarding process and effective engagement with the technology. In educational settings, ease of use enhances users' ability to achieve learning outcomes efficiently, making it a valuable attribute. However, for experienced users or those focused on performance and functionality, the importance of Effort Expectancy diminishes, as they prioritize utility over usability. Despite this variability, ensuring the technology is user-friendly remains vital for fostering adoption among diverse user groups with varying levels of proficiency.
Facilitating Conditions • Definition Facilitating Conditions refers to the availability of resources, infrastructure, and support necessary for users to effectively adopt and utilize GenAI technologies. This includes access to devices, platforms, training, and institutional support. • Empirical Evidence Zheng et al. [45] emphasize the critical role of Facilitating Conditions (FCs) in both shaping Behavioral Intention (BI) and sustaining actual use, highlighting their prominence across the adoption lifecycle. Similarly, Strzelecki and ElArabawy [33] and Du and Lv [48] find that FCs significantly influence actual usage behavior, demonstrating their impact in translating intention into action. Saihi et al. [35] and Ma [29] further validate FCs as a determinant of both initial adoption and continued use, stressing the worth of accessible resources, infrastructure, and institutional support in ensuring successful GenAI integration in educational contexts. Conversely, other studies suggest that Facilitating Conditions may have a limited role in predicting BI for GenAI technologies. Wang and Reynolds [65] report that while autonomy, competence, and relatedness strongly influence BI and usage, FCs lack significant impact on usage intentions. Grassini et al. [31] and Wang et al. [50] similarly find no statistical association between FCs and BI, emphasizing that users may prioritize intrinsic motivators like performance expectancy and habit. Strzelecki et al. [32] further note that while BI strongly predicts usage, FCs do not directly influence BI.

Table 1. *Cont.*

The Three-Tier Framework for GenAI Technology Adoption in Education
These findings indicate that the role of FCs in GenAI adoption may depend on contextual factors, such as resource availability and the prominence of other adoption drivers. While FCs may be less influential in resource-rich environments, they remain essential in under-resourced settings where external support is critical for bridging the gap between intention and use. • Rationale Facilitating Conditions serve as a moderate yet fundamental predictor of GenAI adoption by enabling users to translate intentions into actual use. In educational settings, access to resources such as platforms, training, and institutional support is crucial, especially in under-resourced environments, where these conditions can bridge gaps and remove barriers to adoption. While their impact may be less pronounced in resource-rich settings where intrinsic motivators dominate, Facilitating Conditions remain critical for ensuring equitable adoption. By providing the necessary tools and support, FCs empower users—regardless of motivation levels—to fully engage with GenAI technologies, making them a vital component of the adoption process.
Social Influence • Definition Social Influence refers to the degree to which individuals perceive that important others (such as peers, instructors, or societal norms) believe they should use GenAI technologies. It reflects the impact of social pressure or encouragement on an individual's decision to adopt and use technology. • Empirical Evidence Numerous studies highlight the significant role of Social Influence (SI) in shaping Behavioral Intention (BI) to adopt GenAI technologies. Changalima et al. [34], Jang [42], Zheng et al. [45], and Wang et al. [50] report that SI positively and significantly impacts BI, emphasizing the value of external encouragement and societal norms in adoption decisions. Strzelecki and ElArabawy [33], Saihi et al. [35], and Du and Lv [48] further corroborate these findings, illustrating that SI, often reflected through peer or societal expectations, plays a crucial role in adoption decisions. Additionally, Wang and Reynolds [65] find SI to significantly influence both BI and usage intentions, particularly in environments where social validation drives behavior. Conversely, several studies challenge the influence of Social Influence in predicting GenAI adoption. Grassini et al. [31] report no significant impact of SI on BI, with factors like performance expectancy and habit being more influential. Similarly, Sudan et al. [43] and Strzelecki et al. [32] found minimal or negligible effects of SI, highlighting that intrinsic motivators, such as perceived benefits and personal habits, often outweigh external social pressures. These findings suggest that while Social Influence can play a pivotal role in specific cultural or institutional contexts, its overall impact on GenAI adoption is inconsistent and highly context-dependent. • Rationale Social Influence emerges as a moderate predictor of GenAI adoption, reflecting its context-dependent nature. In educational settings, peers, instructors, and societal expectations can significantly shape individuals' decisions to adopt and use GenAI technologies. This influence is particularly pronounced in collectivist cultures, where group norms and social validation strongly motivate behavior. In these contexts, SI enhances both initial adoption and continued use, as individuals are driven by the expectations and encouragement of their social circles. Conversely, in individualistic cultures or performance-oriented environments, the impact of SI may diminish, as personal autonomy and intrinsic motivators, such as perceived utility and habit, take precedence. While social support or pressure can facilitate or hinder adoption based on cultural and institutional attitudes, its variability across contexts positions Social Influence as a moderate yet necessary predictor of GenAI adoption.
Perceived Behavioral Control • Definition Perceived Behavioral Control (PBC) refers to the perceived ease or difficulty of performing a behavior, which is influenced by both internal factors (such as personal control and self-efficacy) and external factors (such as resources, obstacles, and support available for technology adoption). It reflects the users' belief in their ability to control their interaction with GenAI technologies.

Table 1. *Cont.*

The Three-Tier Framework for GenAI Technology Adoption in Education
• Empirical Evidence Research highlights the dual nature of Perceived Behavioral Control (PBC) in influencing GenAI adoption. Lu et al. [49] and Ivanov et al. [51] demonstrate that PBC significantly impacts Behavioral Intention (BI), emphasizing that users' confidence in their ability to manage GenAI usage plays a critical role in adoption decisions. Similarly, Al-Qaysi et al. [30] identify PBC, alongside attitudes and subjective norms, as a strong predictor of technology use, underscoring its importance in shaping both intention and behavior. Conversely, other studies question the consistent significance of PBC in predicting adoption. Wang et al. [64] report that while attitudes strongly influenced usage intentions, PBC did not significantly impact BI. Likewise, Al-Emran et al. [36] observed that attitudes and subjective norms were more dominant predictors of GenAI adoption, with PBC playing a minimal role. These findings suggest that the influence of PBC may depend on external factors such as facilitating conditions or social pressures, which can overshadow users' perceptions of control. Overall, while PBC is relevant in certain contexts, its role in GenAI adoption is highly variable, shaped by the interplay of individual and environmental factors. • Rationale Perceived Behavioral Control (PBC) is a moderate predictor of GenAI adoption, shaped by its context-dependent nature. In settings where users feel confident in their ability to manage the adoption process—such as those with robust self-efficacy and access to necessary resources—PBC significantly influences Behavioral Intention (BI) and actual usage. It empowers users to overcome barriers and navigate challenges, particularly in environments where personal agency plays a central role. However, in contexts dominated by external factors like institutional support, social norms, or organizational mandates, the influence of PBC diminishes, as external determinants take precedence. This variability highlights the interplay between PBC and other factors, such as facilitating conditions and social influence. While not universally strong, PBC remains influential in specific contexts where individual control and resource availability are critical. This makes it a moderate but fundamental factor in the framework, addressing diverse user needs and supporting adoption in scenarios where autonomy and self-efficacy drive engagement.
Perceived Compatibility • Definition Perceived Compatibility refers to the degree to which GenAI aligns with users' needs, values, and educational practices. Basically, it reflects how well GenAI fits into users' existing systems, workflows, and expectations within an educational context. • Empirical Evidence Studies consistently show that Perceived Compatibility significantly influences the adoption of GenAI, with alignment between the technology and users' needs, values, and educational practices playing a key role. Raman et al. [39] found that GenAI is more likely to be adopted when it fits well with users' educational workflows and needs, emphasizing the technology's relevance and usability. Similarly, Yap et al. [46] highlighted that Perceived Compatibility facilitates Behavioral Intention (BI) by making GenAI more intuitive and useful, particularly when it integrates seamlessly into existing systems and practices. The absence of opposing findings suggests that, within the contexts studied, compatibility with existing practices consistently supports adoption. However, this could reflect a limited range of studies in the field, and future research may reveal more complex dynamics or variations depending on specific educational environments. Although Perceived Compatibility has not been challenged significantly in the reviewed studies, its influence remains context-dependent, and further exploration of its role across diverse educational settings is warranted. • Rationale Perceived Compatibility is a moderate predictor of GenAI adoption, reflecting its influence in aligning technology with users' existing practices, systems, and values. In educational contexts, when GenAI integrates seamlessly into established workflows and addresses specific needs, it enhances BI and usage. As demonstrated by [39,46], this alignment fosters adoption by making GenAI intuitive and contextually valuable. While the absence of opposing findings supports its consistent positive influence, the limited body of research suggests the need for further exploration across diverse settings. The context-sensitive nature of Perceived Compatibility, coupled with its critical role in reducing resistance and increasing acceptance, establishes it as a moderate yet essential factor within the framework for GenAI adoption.

Table 1. *Cont.*

The Three-Tier Framework for GenAI Technology Adoption in Education
Habit • Definition Habit refers to the extent to which users perform behaviors automatically based on prior experiences, with actions becoming ingrained through repeated use. In the context of GenAI, this involves users becoming accustomed to using the technology regularly, such that future use is influenced by prior interactions. • Empirical Evidence Several studies [31,32,45] consistently highlight that Habit significantly influences Behavioral Intention (BI) and the sustained use of GenAI, with repeated interactions and integration into users' routines being key drivers of continued adoption. The findings from Sudan et al. [43] also reinforce this by showing that habit influences not only BI but also adoption decisions, emphasizing the importance of prior usage experiences in enhancing the likelihood of continued engagement. The absence of opposing findings further strengthens the argument that habit plays a crucial role in shaping GenAI adoption, particularly regarding sustained use after the initial adoption phase. • Rationale Habit is considered an emerging predictor of GenAI adoption, gaining significance over time as users interact with the technology regularly. Unlike traditional predictors such as perceived usefulness and ease of use, which primarily influence initial adoption, habit becomes more influential as users incorporate GenAI into their daily routines. Studies show that users who develop habits around GenAI usage tend to demonstrate higher levels of sustained adoption, driven by the reinforcement of positive outcomes through consistent engagement. In educational settings, regular use of GenAI for learning, research, or problem-solving fosters automaticity, ensuring the technology's long-term integration. As such, habit is a fundamental emerging factor in the Three-Tier Framework, contributing significantly to long-term GenAI adoption in education.
AI Literacy • Definition AI Literacy refers to the knowledge and understanding of Generative AI (GenAI) concepts, applications, and limitations, as well as the ability to evaluate and effectively use GenAI tools. It encompasses both the theoretical knowledge of how these systems work and the practical skills needed to interact with them in a meaningful and informed way. • Empirical Evidence AI literacy has emerged as a meaningful predictor of GenAI adoption, with studies demonstrating its positive influence on Behavioral Intention (BI). Jang [42] found that higher AI literacy significantly shapes BI, suggesting that users with a better understanding of AI are more inclined to adopt GenAI technologies. Similarly, Wang et al. [64] identified AI literacy as a key factor in enhancing user confidence, with greater AI knowledge leading to increased trust in the technology and its perceived benefits. These findings highlight AI literacy as a central factor in fostering positive intentions and confidence toward GenAI adoption, underscoring its importance as an emerging predictor. • Rationale AI Literacy is an emerging predictor of GenAI adoption, as it provides users with both the theoretical understanding and practical skills necessary for effective interaction with the technology. Informed users are more likely to trust and engage with GenAI tools in meaningful ways. By reducing misconceptions, alleviating concerns about misuse, and fostering confidence, AI literacy increases BI and facilitates adoption. In educational contexts, where critical thinking and informed decision-making are vital, AI literacy helps users navigate the complexities of GenAI technologies, supporting their long-term integration into academic workflows. As AI tools become more pervasive, enhancing AI literacy will likely become an even more critical factor in ensuring successful and sustained adoption, positioning it as a crucial emerging predictor within the Three-Tier Framework.
Anxiety • Definition Anxiety refers to concerns or fears regarding potential negative outcomes of using Generative AI (GenAI) technologies, such as job displacement, ethical dilemmas, privacy breaches, or misuse of data. This factor encompasses users' apprehensions about the risks associated with adopting and integrating GenAI into their personal, educational, or professional environments.

Table 1. *Cont.*

The Three-Tier Framework for GenAI Technology Adoption in Education
• Empirical Evidence Shen et al. [47] found that job replacement anxiety notably reduces Behavioral Intention (BI) to adopt GenAI, as users fear that the technology may replace their current or future job roles. Wang et al. [50] reported that GenAI anxiety significantly diminishes users' intentions to adopt the technology, particularly by negatively affecting performance expectancy and effort expectancy. Similarly, Li et al. [62] demonstrated that AI anxiety leads to a lower perception of the technology's usefulness and a reduced likelihood of adoption. These findings underscore that anxiety can create substantial barriers to GenAI adoption, highlighting the imperative of addressing these concerns for successful implementation and engagement with the technology. • Rationale Anxiety is an emerging predictor of GenAI adoption, as concerns about job displacement, ethical dilemmas, privacy risks, and the misuse of data can significantly deter users from adopting the technology. These anxieties often undermine trust, reduce perceived usefulness, and heighten resistance, especially in contexts where users are unfamiliar with or skeptical of AI technologies. Fears about GenAI can create substantial barriers to adoption, hindering its integration into workflows and educational practices. As GenAI continues to evolve and become more widespread, these concerns are likely to intensify, making it an increasingly critical factor in shaping adoption decisions. Addressing these anxieties through transparent communication, clear ethical guidelines, and trust-building measures will be necessary for mitigating fears and fostering a more positive outlook toward GenAI adoption, particularly in educational settings where such concerns are often amplified.
Playfulness/Hedonic Motivation • Definition Playfulness or Hedonic Motivation refers to the degree to which users perceive using GenAI as enjoyable, fun, or intrinsically satisfying. This factor highlights the enjoyment and creative fulfilment that users derive from interacting with GenAI technologies, which can enhance the overall user experience beyond practical or utilitarian outcomes. • Empirical Evidence Zheng et al. [45] found that Hedonic Motivation significantly predicts Behavioral Intention (BI), emphasizing that the enjoyment users experience while engaging with GenAI plays a crucial role in their decision to adopt the technology. Similarly, Yap et al. [46] identified perceived playfulness as a top predictor of BI, further reinforcing the idea that the intrinsic satisfaction users derive from interacting with GenAI enhances their likelihood of continued use. Strzelecki et al. [32] also highlighted that hedonic motivation strongly influences BI. These studies consistently emphasize that the fun and creative aspects of interacting with GenAI are major factors in fostering user engagement, which is fundamental for adoption and long-term integration. • Rationale Playfulness or hedonic motivation is an emerging predictor of GenAI adoption, as the enjoyment, creativity, and intrinsic satisfaction users derive from interacting with the technology significantly enhance their BI and likelihood of continued use. Studies suggest that when GenAI is perceived as fun and engaging, users are more inclined to incorporate it into their routines, leading to deeper and more sustained engagement. In educational contexts, hedonic motivation plays a particularly important role by stimulating creativity and exploration, which are vital for dynamic and innovative learning experiences. While playfulness alone may not be sufficient for widespread adoption, it supports ongoing user engagement and fosters a more positive attitude toward GenAI. This makes hedonic motivation an essential, though emerging, factor that strengthens adoption by making interactions both enjoyable and rewarding.

Importantly, the Three-Tier Framework complements the *Triad of Predictors* proposed by Granić [66,67], originally introduced in Granić [68], which categorizes determinants of behavioral intention into three thematic dimensions:

- User Aspects: Personal traits or characteristics that influence technology acceptance.
- Task and Technological Aspects: Features of the technology and the specific tasks it supports.
- Social and Environmental Aspects: Broader contextual and social influences on adoption behavior.

Building on this foundation, Davis and Granić [69] further developed the Triad by integrating constructs from various extensions and modifications of the Technology Acceptance Model (TAM).

Together, the Three-Tier Framework and the *Triad of Predictors* provide complementary perspectives on GenAI adoption. The Triad offers a macro-level conceptual structure emphasizing the interplay between individual, technological, and social dimensions. In contrast, the Three-Tier Framework delivers a more granular categorization—Core, Moderate, and Emerging—designed to support targeted analysis and intervention.

By integrating both approaches, researchers and practitioners gain a comprehensive understanding of adoption dynamics. The Triad of Predictors offers strategic, system-level insight, while the Three-Tier Framework identifies actionable predictors for specific educational scenarios. Combined, they form a robust analytical foundation to advance research, guide implementation, and support the effective integration of GenAI technologies in education.

5. Conclusions

This review advances the understanding of Generative AI (GenAI) adoption in education by proposing a Three-Tier Framework comprising 11 key predictors, synthesized from 27 empirical studies. The framework provides a valuable conceptual foundation for interpreting the multifaceted nature of GenAI adoption. However, to ensure its continued relevance, it is necessary to acknowledge several limitations, identify opportunities for refinement, and outline promising directions for future research.

5.1. Limitations of the Review

While this review offers important insights into GenAI adoption, certain methodological constraints and literature gaps must be acknowledged.

First, several relevant dimensions remain underexplored in the reviewed studies. These include ethical concerns (e.g., data privacy, bias), the role of teacher professional development, the impact of student autonomy, and contextual factors such as institutional policies and socio-cultural dynamics. These elements deserve further attention to understand how they interact with adoption predictors.

Second, the review exclusively used the Web of Science Core Collection to ensure access to high-quality, peer-reviewed literature. While this increased methodological rigor, it may have excluded relevant studies from other databases or grey literature—particularly those representing emerging markets or non-English-speaking regions. As such, the findings may not fully capture the global diversity of GenAI adoption.

Third, although the Boolean search strategy supported precision, it may have omitted studies using alternative terminologies or conceptual framings. Moreover, the review focused primarily on studies reporting significant adoption predictors, placing less emphasis on conflicting or null findings, which may limit insights into barriers or resistance to adoption.

Future research should address these limitations by incorporating a broader range of sources, including non-English publications and studies from underrepresented regions. Greater attention to barriers, ethical considerations, and diverse contextual variables will offer a more balanced and comprehensive understanding of GenAI integration in education.

5.2. Potential Extension of the Framework

The Three-Tier Framework provides a structured and theory-driven approach to understanding GenAI adoption. However, as technologies and educational contexts evolve, the framework must remain dynamic, inclusive, and adaptable. Future research should focus on refining and updating the framework by incorporating new predictors—such as ethics and emotional factors—and by accounting for teacher development, student engagement, and contextual and cultural diversity. These extensions are especially relevant as GenAI tools increasingly support personalized learning, automate instructional tasks, and reshape classroom dynamics.

Currently, the framework is based on studies from a relatively limited number of countries and education systems. Expanding the empirical base to include diverse cultural, regional, and socio-economic contexts will enhance its generalizability and global relevance. Understanding how local values, policy environments, and educational norms affect adoption is essential to making the framework broadly applicable.

Additionally, as new GenAI applications—such as multimodal AI, adaptive systems, and emotion-aware interfaces—are introduced, the relevance of existing predictors may shift. Regular empirical validation and theoretical refinement will be necessary to ensure that the framework reflects these ongoing technological changes.

Finally, future studies should actively include research reporting null or conflicting results to provide a more balanced view of the adoption landscape. Incorporating such findings will strengthen the framework's robustness and prevent overly optimistic or one-sided interpretations.

In summary, extending the Three-Tier Framework involves both addressing current limitations and preparing it for future technological and contextual developments. By doing so, it can continue to serve as a reliable and adaptable tool for researchers, educators, and policymakers working toward responsible and effective GenAI integration across diverse educational environments.

5.3. Directions for Future Research

Building on the findings and limitations identified in this review, several key areas warrant further exploration to refine theoretical models, inform policy, and guide practical implementation. These research directions focus on underexplored predictors, cultural and contextual variation, evolving technologies, and interdisciplinary perspectives. Collectively, they aim to ensure the continued relevance and robustness of the proposed Three-Tier Framework for GenAI adoption in education.

- Refining the Three-Tier Framework and longitudinal studies: Future research should refine and expand the Three-Tier Framework by incorporating emerging predictors and examining their stability over time. As GenAI technologies mature, predictors such as trust, performance expectancy, and social influence may evolve in significance. Longitudinal studies are particularly valuable for capturing temporal shifts in adoption dynamics and understanding how sustained exposure to GenAI tools reshapes user attitudes and behaviors.
- Cultural, contextual, and demographic variations in GenAI adoption: The review highlights that GenAI adoption is shaped by regional, cultural, and institutional contexts. Studies across regions such as East Asia, Europe, and Africa suggest differences in the conceptualization of adoption drivers, reflecting variations in educational values, technological infrastructure, and policy environments. Future research should investigate how cultural dimensions—such as collectivism vs. individualism, or institutional centralization—affect perceptions and adoption. In parallel, studies should examine how contextual factors (e.g., educational level) and demographic characteristics (e.g., age, gender, socio-economic status) influence adoption behavior. Special attention should be given to teacher-specific factors, such as AI literacy, pedagogical readiness, and access to professional development, particularly in under-resourced settings.

- Ethical considerations and adoption barriers: Ethical concerns, including data privacy, algorithmic bias, and fairness, remain insufficiently addressed in the existing literature. These issues can significantly influence trust, perceived risk, and ultimately, adoption intentions. Future studies should explore how ethical considerations interact with predictors such as perceived usefulness, trust, and social influence. Additionally, research should identify barriers such as GenAI-related anxiety, fear of obsolescence, and lack of infrastructure to develop interventions that promote equitable and responsible adoption.
- Student engagement and learner-centered adoption: Student-related factors—such as engagement, autonomy, and motivation—are underrepresented in current GenAI adoption studies. As GenAI tools increasingly support personalized and student-centered learning, research should focus on how these technologies influence learner engagement, self-regulation, and educational outcomes. Understanding these dynamics will be essential for maximizing the pedagogical potential of GenAI while mitigating risks such as over-reliance or reduced critical thinking.
- Technological advancements and predictor evolution: Rapid advancements in GenAI technologies, such as multimodal interfaces and adaptive AI systems, may alter the relevance and impact of existing adoption predictors. Research should investigate how these new technological affordances reshape user experiences and expectations, and whether they give rise to new predictors not yet captured by current models. This continuous adaptation is essential to ensure that theoretical frameworks remain relevant in a fast-evolving landscape.
- Interdisciplinary collaboration and model development: The complexity of GenAI adoption necessitates interdisciplinary collaboration across fields such as education, computer science, psychology, and sociology. Future research should draw on diverse disciplinary insights to build more comprehensive models that reflect the cognitive, emotional, social, and technological dimensions of adoption. Such collaboration will help address multifaceted challenges and contribute to the development of holistic and sustainable GenAI integration strategies.

By pursuing these research directions, future studies can enhance the explanatory and predictive power of GenAI adoption models; improve the adaptability of the Three-Tier Framework; and inform inclusive, effective, and ethically grounded integration of GenAI technologies across diverse educational settings.

Funding: No funding was received to assist with the preparation of this manuscript.

Institutional Review Board Statement: Not applicable.

Informed Consent Statement: Not applicable.

Data Availability Statement: The data of this review consist of articles published in journals and conferences. Many of these are freely available online; others can be accessed for a fee or through subscription.

Conflicts of Interest: The author declares no conflicts of interest.

References

1. Webster, J.; Watson, R.T. Analyzing the Past to Prepare for the Future: Writing a Literature Review. *MIS Q.* **2002**, *26*, xiii–xxiii. Available online: https://www.jstor.org/stable/4132319 (accessed on 9 January 2025).
2. Kiuchi, K.; Otsu, K.; Hayashi, Y. Psychological insights into the research and practice of embodied conversational agents, chatbots and social assistive robots: A systematic meta-review. *Behav. Inf. Technol.* **2023**, *43*, 3696–3736. [CrossRef]
3. Ferrara, E. Large Language Models for Wearable Sensor-Based Human Activity Recognition, Health Monitoring, and Behavioral Modeling: A Survey of Early Trends, Datasets, and Challenges. *Sensors* **2024**, *24*, 5045. [CrossRef] [PubMed]

4. Hirata, K.; Matsui, Y.; Yamada, A.; Fujioka, T.; Yanagawa, M.; Nakaura, T.; Ito, R.; Ueda, D.; Fujita, S.; Tatsugami, F.; et al. Generative AI and large language models in nuclear medicine: Current status and future prospects. *Ann. Nucl. Med.* **2024**, *38*, 853–864. [CrossRef]
5. Nerella, S.; Bandyopadhyay, S.; Zhang, J.; Contreras, M.; Siegel, S.; Bumin, A.; Silva, B.; Sena, J.; Shickel, B.; Bihorac, A.; et al. Transformers and large language models in healthcare: A review. *Artif. Intell. Med.* **2024**, *154*, 102900. [CrossRef]
6. Yim, D.; Khuntia, J.; Parameswaran, V.; Meyers, A. Preliminary Evidence of the Use of Generative AI in Health Care Clinical Services: Systematic Narrative Review. *JMIR Med. Inform.* **2024**, *20*, e52073. [CrossRef] [PubMed]
7. Goktas, P.; Grzybowski, A. Assessing the Impact of ChatGPT in Dermatology: A Comprehensive Rapid Review. *J. Clin. Med.* **2024**, *13*, 5909. [CrossRef] [PubMed]
8. Hobensack, M.; von Gerich, H.; Vyas, P.; Withall, J.; Peltonen, L.M.; Block, L.J.; Davies, S.; Chan, R.; Van Bulck, L.; Cho, H.; et al. A rapid review on current and potential uses of large language models in nursing. *Int. J. Nurs. Stud.* **2024**, *154*, 104753. [CrossRef]
9. Ng, C.K.C. Generative Adversarial Network (Generative Artificial Intelligence) in Pediatric Radiology: A Systematic Review. *Children* **2023**, *10*, 1372. [CrossRef]
10. Ogunleye, B.; Zakariyyah, K.I.; Ajao, O.; Olayinka, O.; Sharma, H. A Systematic Review of Generative AI for Teaching and Learning Practice. *Educ. Sci.* **2024**, *14*, 636. [CrossRef]
11. Yusuf, A.; Pervin, N.; Román-González, M.; Noor, N.M. Generative AI in education and research: A systematic mapping review. *Rev. Educ.* **2024**, *12*, e3489. [CrossRef]
12. Sekli, G.M.; Godo, A.; Véliz, J.C. Generative AI Solutions for Faculty and Students: A Review of Literature and Roadmap for Future Research. *J. Inf. Technol. Educ. Res.* **2024**, *23*, 14. [CrossRef]
13. Drobnjak, A.; Botički, I.; Seow, P.; Kahn, K. Learning with Conversational AI and Personas: A Systematic Literature Review. In Proceedings of the the 31st International Conference on Computers in Education, Matsue, Japan, 4–8 December 2023; Asia-Pacific Society for Computers in Education: Taoyuan, Taiwan, 2023. [CrossRef]
14. Sengul, C.; Neykova, R.; Destefanis, G. Software engineering education in the era of conversational AI: Current trends and future directions. *Front. Artif. Intell.* **2024**, *7*, 1436350. [CrossRef]
15. Ali, O.; Murray, P.A.; Momin, M.; Al-Anzi, F.S. The knowledge and innovation challenges of ChatGPT: A scoping review. *Technol. Soc.* **2023**, *75*, 102402. [CrossRef]
16. Amarathunga, B. ChatGPT in education: Unveiling frontiers and future directions through systematic literature review and bibliometric analysis. *Asian Educ. Dev. Stud.* **2024**, *13*, 412–431. [CrossRef]
17. Hagendorff, T. Mapping the Ethics of Generative AI: A Comprehensive Scoping Review. *Minds Mach.* **2024**, *34*, 39. [CrossRef]
18. Yan, L.; Sha, L.; Zhao, L.; Li, Y.; Martinez-Maldonado, R.; Chen, G.; Li, X.; Jin, Y.; Gašević, D. Practical and ethical challenges of large language models in education: A systematic scoping review. *Br. J. Educ. Technol.* **2024**, *55*, 90–112. [CrossRef]
19. Huang, F.; Wang, Y.; Zhang, H. Modelling Generative AI Acceptance, Perceived Teachers' Enthusiasm and Self-Efficacy to English as a Foreign Language Learners' Well-Being in the Digital Era. *Eur. J. Educ.* **2024**, *59*, e12770. [CrossRef]
20. Wang, Q.; Gao, Y.; Wang, X. Exploring Engagement, Self-Efficacy, and Anxiety in Large Language Model EFL Learning: A Latent Profile Analysis of Chinese University Students. *Int. J. Hum.–Comput. Interact.* **2024**, *41*, 7815–7824. [CrossRef]
21. Wu, D.; Zhang, S.; Ma, Z.; Yue, X.-G.; Dong, R.K. Unlocking Potential: Key Factors Shaping Undergraduate Self-Directed Learning in AI-Enhanced Educational Environments. *Systems* **2024**, *12*, 332. [CrossRef]
22. Herani, R.; Angela, J. Navigating ChatGPT: Catalyst or challenge for Indonesian youth in digital entrepreneurship? *J. Entrep. Emerg. Econ.* **2024**, *17*, 602–628. [CrossRef]
23. Smith, C.E.; Shiekh, K.; Cooreman, H.; Rahman, S.; Zhu, Y.; Siam, M.K.; Ivanitskiy, M.; Ahmed, A.M.; Hallinan, M.; Grisak, A.; et al. Early Adoption of Generative Artificial Intelligence in Computing Education: Emergent Student Use Cases and Perspectives in 2023. In Proceedings of the ITiCSE 2024 Conference on Innovation and Technology in Computer Science Education, Milan, Italy, 8–10 July 2024; ACM: New York, NY, USA, 2024; Volume 1, pp. 3–9. [CrossRef]
24. Petrič, G. Everyone Talks Everything with ChatGPT: Students' Uses of ChatGPT and Their Impact on Learning Performance. *Int. J. Technol. Hum. Interact.* **2024**, *20*, 1–22. [CrossRef]
25. Chan, C.K.Y.; Zhou, W. An expectancy value theory (EVT) based instrument for measuring student perceptions of generative AI. *Smart Learn. Environ.* **2023**, *10*, 64. [CrossRef]
26. Tangadulrat, P.; Sono, S.; Tangtrakulwanich, B. Using ChatGPT for Clinical Practice and Medical Education: Cross-Sectional Survey of Medical Students' and Physicians' Perceptions. *JMIR Med. Educ.* **2023**, *22*, e50658. [CrossRef] [PubMed]
27. Sallam, M.; Salim, N.A.; Barakat, M.; Al-Mahzoum, K.; Al-Tammemi, A.B.; Malaeb, D.; Hallit, R.; Hallit, S. Assessing Health Students' Attitudes and Usage of ChatGPT in Jordan: Validation Study. *JMIR Med. Educ.* **2023**, *9*, e48254. [CrossRef]
28. Duong, C.D.; Vu, T.N.; Ngo, T.V.N. Applying a modified technology acceptance model to explain higher education students' usage of ChatGPT: A serial multiple mediation model with knowledge sharing as a moderator. *Int. J. Manag. Educ.* **2023**, *21*, 100883. [CrossRef]

29. Ma, M. Exploring the acceptance of generative artificial intelligence for language learning among EFL postgraduate students: An extended TAM approach. *Int. J. Appl. Linguist.* **2024**, *35*, 91–108. [CrossRef]
30. Al-Qaysi, N.; Al-Emran, M.; Al-Sharafi, M.A.; Iranmanesh, M.; Ahmad, A.; Mahmoud, M.A. Determinants of ChatGPT Use and its Impact on Learning Performance: An Integrated Model of BRT and TPB. *Int. J. Hum.–Comput. Interact.* **2024**, *41*, 5462–5474. [CrossRef]
31. Grassini, S.; Aasen, M.L.; Møgelvang, A. Understanding University Students' Acceptance of ChatGPT: Insights from the UTAUT2 Model. *Appl. Artif. Intell.* **2024**, *38*, 2371168. [CrossRef]
32. Strzelecki, A.; Cicha, K.; Rizun, M.; Rutecka, P. Acceptance and use of ChatGPT in the academic community. *Educ. Inf. Technol.* **2024**, *29*, 22943–22968. [CrossRef]
33. Strzelecki, A.; ElArabawy, S. Investigation of the moderation effect of gender and study level on the acceptance and use of generative AI by higher education students: Comparative evidence from Poland and Egypt. *Br. J. Educ. Technol.* **2024**, *55*, 1209–1230. [CrossRef]
34. Changalima, I.A.; Amani, D.; Ismail, I.J. Social influence and information quality on Generative AI use among business students. *Int. J. Manag. Educ.* **2024**, *22*, 101063. [CrossRef]
35. Saihi, A.; Ben-Daya, M.; Hariga, M. The moderating role of technology proficiency and academic discipline in AI-chatbot adoption within higher education: Insights from a PLS-SEM analysis. *Educ. Inf. Technol.* **2024**, *30*, 5843–5881. [CrossRef]
36. Al-Emran, M.; Abu-Hijleh, B.; Alsewari, A.A. Exploring the Effect of Generative AI on Social Sustainability Through Integrating AI Attributes, TPB, and T-EESST: A Deep Learning-Based Hybrid SEM-ANN Approach. *IEEE Trans. Eng. Manag.* **2024**, *71*, 14512–14524. [CrossRef]
37. Kanont, K.; Pingmuang, P.; Simasathien, T.; Wisnuwong, S.; Wiwatsiripong, B.; Poonpirome, K.; Songkram, N.; Khlaisang, J. Generative-AI, a Learning Assistant? Factors Influencing Higher-Ed Students' Technology Acceptance. *Electron. J. E-Learn.* **2024**, *22*, 6. [CrossRef]
38. Ngo, T.T.A.; Tran, T.T.; An, G.K.; Nguyen, P.T. ChatGPT for Educational Purposes: Investigating the Impact of Knowledge Management Factors on Student Satisfaction and Continuous Usage. *IEEE Trans. Learn. Technol.* **2024**, *17*, 1367–1378. Available online: https://api.semanticscholar.org/CorpusID:268865544 (accessed on 19 December 2024). [CrossRef]
39. Raman, R.; Mandal, S.; Das, P.; Kaur, T.; Sanjanasri, J.P.; Nedungadi, P. Exploring University Students' Adoption of ChatGPT Using the Diffusion of Innovation Theory and Sentiment Analysis with Gender Dimension. *Hum. Behav. Emerg. Technol.* **2024**, *2024*, 3085910. [CrossRef]
40. Shahzad, M.F.; Xu, S.; Javed, I. ChatGPT awareness, acceptance, and adoption in higher education: The role of trust as a cornerstone. *Int. J. Educ. Technol. High. Educ.* **2024**, *21*, 46. [CrossRef]
41. Soliman, M.; Ali, R.A.; Khalid, J.; Mahmud, I.; Ali, W.B. Modelling continuous intention to use generative artificial intelligence as an educational tool among university students: Findings from PLS-SEM and ANN. *J. Comput. Educ.* **2024**, 1–32. [CrossRef]
42. Jang, M. AI Literacy and Intention to Use Text-Based GenAI for Learning: The Case of Business Students in Korea. *Informatics* **2024**, *11*, 54. [CrossRef]
43. Sudan, T.; Hans, A.; Taggar, R. Transformative learning with ChatGPT: Analyzing adoption trends and implications for business management students in India. *Interact. Technol. Smart Educ.* **2024**, *21*, 735–772. [CrossRef]
44. Liu, G.L.; Darvin, R.; Ma, C. Exploring AI-mediated informal digital learning of English (AI-IDLE): A mixed-method investigation of Chinese EFL learners' AI adoption and experiences. *Comput. Assist. Lang. Learn.* **2024**, 1–29. [CrossRef]
45. Zheng, Y.; Wang, Y.; Liu, K.S.X.; Jiang, M.Y.C. Examining the moderating effect of motivation on technology acceptance of generative AI for English as a foreign language learning. *Educ. Inf. Technol.* **2024**, *29*, 23547–23575. [CrossRef]
46. Yap, K.Y.L.; Ho, J.; Toh, P.S.T. Development of a Metaverse Art Gallery of Image Chronicles (MAGIC) for Healthcare Education: A Digital Health Humanities Approach to Patients' Medication Experiences. *Information* **2024**, *15*, 431. [CrossRef]
47. Shen, X.; Mo, X.; Xia, T. Exploring the attitude and use of GenAI-image among art and design college students based on TAM and SDT. *Interact. Learn. Environ.* **2024**, *33*, 1198–1215. [CrossRef]
48. Du, L.; Lv, B. Factors influencing students' acceptance and use generative artificial intelligence in elementary education: An expansion of the UTAUT model. *Educ. Inf. Technol.* **2024**, *29*, 24715–24734. [CrossRef]
49. Lu, H.; He, L.; Yu, H.; Pan, T.; Fu, K. A Study on Teachers' Willingness to Use Generative AI Technology and Its Influencing Factors: Based on an Integrated Model. *Sustainability* **2024**, *16*, 7216. [CrossRef]
50. Wang, K.; Ruan, Q.; Zhang, X.; Fu, C.; Duan, B. Pre-Service Teachers' GenAI Anxiety, Technology Self-Efficacy, and TPACK: Their Structural Relations with Behavioral Intention to Design GenAI-Assisted Teaching. *Behav. Sci.* **2024**, *14*, 373. [CrossRef]
51. Ivanov, S.; Soliman, M.; Tuomi, A.; Alkathiri, N.A.; Al-Alawi, A.N. Drivers of generative AI adoption in higher education through the lens of the Theory of Planned Behaviour. *Technol. Soc.* **2024**, *77*, 102521. [CrossRef]
52. Davis, F.D. A Technology Acceptance Model for Empirically Testing New End-User Information Systems: Theory and Results. Ph.D. Thesis, MIT Sloan School of Management, Cambridge, MA, USA, 1986.

53. Davis, F.D. Perceived Usefulness, Perceived Ease of Use, and User Acceptance of Information Technology. *MIS Q.* **1989**, *13*, 319–340. [CrossRef]
54. Venkatesh, V.; Morris, M.G.; Davis, G.B.; Davis, F.D. User acceptance of information technology: Towards a unified view. *MIS Q.* **2003**, *27*, 425–478. Available online: https://www.jstor.org/stable/30036540 (accessed on 9 January 2025). [CrossRef]
55. Venkatesh, V.; Thong, J.Y.; Xu, X. Consumer Acceptance and Use of Information Technology: Extending the Unified Theory of Acceptance and Use of Technology. *MIS Q.* **2012**, *36*, 157–178. [CrossRef]
56. Rogers, E. *Diffusion of Innovations;* The Free Press: New York, NY, USA, 1962.
57. Rogers, E. *Diffusion of Innovations*, 4th ed.; The Free Press: New York, NY, USA, 1995.
58. Bhattacherjee, A. Understanding Information Systems Continuance: An Expectation-Confirmation Model. *MIS Q.* **2001**, *25*, 351–370. [CrossRef]
59. Ajzen, I. From Intentions to Actions: A Theory of Planned Behavior. In *Action Control. From Cognition to Behavior*; Kuhl, J., Beckmann, J., Eds.; Springer: Berlin/Heidelberg, Germany, 1985. [CrossRef]
60. Ajzen, I. The Theory of Planned Behavior. *Organ. Behav. Hum. Decis. Process.* **1991**, *50*, 179–211. [CrossRef]
61. Oliver, R.L. A Cognitive Model of the Antecedents and Consequences of Satisfaction Decisions. *J. Mark. Res.* **1980**, *17*, 460–469. [CrossRef]
62. Li, W.; Zhang, X.; Li, J.; Yang, X.; Li, D.; Liu, Y. An explanatory study of factors influencing engagement in AI education at the K-12 Level: An extension of the classic TAM model. *Sci. Rep.* **2024**, *14*, 13922. [CrossRef]
63. Deci, E.L.; Ryan, R.M. *Intrinsic Motivation and Self-Determination in Human Behavior*; Plenum: New York, NY, USA, 1985. [CrossRef]
64. Wang, C.; Wang, H.; Li, Y.; Dai, J.; Gu, X.; Yu, T. Factors Influencing University Students' Behavioral Intention to Use Generative Artificial Intelligence: Integrating the Theory of Planned Behavior and AI Literacy. *Int. J. Hum.–Comput. Interact.* **2024**, *41*, 6649–6671. [CrossRef]
65. Wang, X.; Reynolds, B.L. Beyond the Books: Exploring Factors Shaping Chinese English Learners' Engagement with Large Language Models for Vocabulary Learning. *Educ. Sci.* **2024**, *14*, 496. [CrossRef]
66. Granić, A. Technology adoption at individual level: Toward an integrated overview. *Univers. Access Inf. Soc.* **2024**, *23*, 843–858. [CrossRef]
67. Granić, A. User Acceptance of Interactive Technologies. In *Foundations and Fundamentals in Human-Computer Interaction, Volume I, Handbook of Human-Computer Interaction: Foundations and Advances, 6-Volume Set*, 1st ed.; Stephanidis, C., Salvendy, G., Eds.; CRC Press: Boca Raton, FL, USA, 2024; pp. 356–389. [CrossRef]
68. Granić, A. Educational Technology Adoption: A systematic review. *Educ. Inf. Technol.* **2022**, *27*, 9725–9744. [CrossRef]
69. Davis, F.D.; Granić, A. *The Technology Acceptance Model—30 Years of TAM*; Springer: Cham, Switzerland, 2024. [CrossRef]

applied sciences

MDPI

Systematic Review

Machine Learning and Generative AI in Learning Analytics for Higher Education: A Systematic Review of Models, Trends, and Challenges

Miguel Ángel Rodríguez-Ortiz [1,2], Pedro C. Santana-Mancilla [2] and Luis E. Anido-Rifón [1,*]

1 atlanTTic Research Center, School of Telecommunications Engineering, University of Vigo, 36310 Vigo, Spain; maro@ucol.mx
2 School of Telematics, Universidad de Colima, Colima 28040, Mexico
* Correspondence: lanido@det.uvigo.es

Featured Application: This review supports the design of hybrid learning analytics systems that combine ML and GenAI to enable early risk detection and personalized feedback in higher education.

Abstract: This systematic review examines how machine learning (ML) and generative AI (GenAI) have been integrated into learning analytics (LA) in higher education (2018–2025). Following PRISMA 2020, we screened 9590 records and included 101 English-language, peer-reviewed empirical studies that applied ML or GenAI within LA contexts. Records came from 12 databases (last search 15 March 2025), and the results were synthesized via thematic clustering. ML approaches dominate LA tasks, such as engagement prediction, dropout-risk modelling, and academic-performance forecasting, whereas GenAI—mainly transformer models like GPT-4 and BERT—is emerging in real-time feedback, adaptive learning, and sentiment analysis. Studies spanned world regions. Most ML papers ($n = 75$) examined engagement or dropout, while GenAI papers ($n = 26$) focused on adaptive feedback and sentiment analysis. No formal risk-of-bias assessment was conducted due to heterogeneity. While ML methods are well-established, GenAI applications remain experimental and face challenges related to transparency, pedagogical grounding, and implementation feasibility. This review offers a comparative synthesis of paradigms and outlines future directions for responsible, inclusive, theory-informed AI use in education.

Keywords: AI-enhanced learning; generative artificial intelligence; higher education; learning analytics; machine learning

1. Introduction

Amid the growing integration of AI in education, machine learning (ML) and generative AI (GenAI) have gained prominence for their ability to personalize learning, predict outcomes, and support institutional decision-making [1–3]. While higher education fosters critical thinking and innovation [4], it continues to face challenges such as dropout, inequality, and the demand for adaptive environments [5]. In response, learning analytics (LA), enhanced by ML, is increasingly used to promote equity and student success through data-driven practices [6,7].

Defined by Siemens [8] as the analysis of data about learners and their contexts to optimize learning, LA has embraced ML techniques—supervised, unsupervised, and semi-supervised—to forecast performance, detect disengagement, and enable timely interventions [9–12]. Recent advances in GenAI, especially large language models (LLMs), are expanding LA's capabilities through automated feedback, intelligent tutoring, and interaction analysis [13–15], although empirical applications remain limited and underexplored in terms of pedagogy and ethics. Some studies caution that GenAI's "black-box" nature and lack of pedagogical grounding may undermine trust and educational validity, contrasting with perspectives that emphasize its potential to enhance learner engagement and scalability.

Existing reviews on ML in LA [16–19] rarely address GenAI or offer an integrated synthesis of both approaches. To fill this gap, this review analyzes empirical studies from 2018 to 2025, including recent developments from major conferences.

The study addresses two research questions:

RQ1: How are ML and GenAI applied in LA within higher education?
RQ2: What benefits arise from their integration in this context?

The results indicate that, while ML remains dominant in predictive analytics, GenAI is emerging in real-time feedback and adaptive learning but still faces critical challenges around transparency and pedagogical validity. By examining methodological trends, educational applications, and broader implications, this review offers a timely perspective to guide the responsible adoption of AI in higher education.

2. Materials and Methods

2.1. Study Eligibility Criteria

This review adheres to PRISMA 2020 guidelines [20] to ensure methodological transparency and rigor. A completed PRISMA 2020 checklist is provided as Supplementary Materials. It synthesizes empirical studies on the use of ML and GenAI in learning analytics (LA) within higher education, applying stream-specific inclusion and exclusion criteria as follows:

- Inclusion Criteria (both streams);
    - Language: Articles must be published in English;
    - Accessibility: Full-text availability is required;
    - Study type: Only empirical studies with a clearly stated research question;
    - Context: The study must explicitly address a learning analytics objective in higher education.
- For LA + ML (2018–2023);
    - Must apply Machine Learning techniques within LA;
    - Must report empirical results, such as predictive performance or analytics-based outcomes.
- For LA + GenAI (2020–2025);
    - Must apply generative AI models in LA;
    - Applications must involve generation, adaptation, or feedback, aligned with LA goals.
- Exclusion Criteria.
    - Non-empirical works;
    - Preprints and non-peer-reviewed documents;
    - Studies lacking either an LA objective or the use of ML/GenAI;
    - Conference papers were not formally included in the final dataset to ensure consistent peer-review standards and methodological rigor across all selected studies. This decision aimed to maintain a homogeneous level of academic scrutiny, focusing exclusively on peer-reviewed journal articles.

This stream-specific distinction ensures a consistent comparative analysis and captures the emergence of GenAI as a distinct methodological paradigm within LA.

2.2. Data Sources

The literature search was conducted in two phases to reflect the evolution of the research questions and the emergence of generative AI (GenAI) in education.

In the first phase, covering ML-based learning analytics (LA + ML) from 2018 to 2023, 11 scholarly databases were consulted to construct a comprehensive and systematic corpus: ACM Digital Library, IEEE Xplore, Emerald, ERIC, ProQuest, Sage Journals, Web of Science, ScienceDirect, Wiley Online Library, Taylor & Francis, and Scopus.

In a subsequent phase, an updated search strategy was implemented to include studies on generative AI in learning analytics (LA + GenAI), published between 2020 and early 2025. Given the novelty and rapid development of GenAI applications, the search focused on sources with high coverage of emerging AI research: ACM Digital Library, IEEE Xplore, SpringerLink (EC-TEL), and Scopus. This approach responded to the more limited but rapidly evolving body of GenAI literature in education, which tends to be concentrated in fewer venues.

While the formal PRISMA-based corpus was limited to peer-reviewed journal articles, an exploratory review of flagship conference proceedings, including LAK, L@S, and EC-TEL, was conducted for the full period (2018–2025) to identify recent trends not yet indexed. Selected contributions, particularly from 2024 and 2025, are discussed in the Related Work section to enrich the contextual interpretation.

2.3. Search Strategy

Search queries combined three conceptual pillars—learning analytics (including social learning analytics), machine learning or GenAI, and higher education. Synonyms and Boolean operators ensured comprehensive yet focused retrieval. The search spanned 2018–2025 (2018–2023 for ML in LA). Full queries are detailed in the Supplementary Materials (Table S1) and available on Zenodo (https://doi.org/10.5281/zenodo.15233231). Of the 9590 records retrieved, 8263 were excluded after deduplication and screening (see PRISMA flowchart, Figure 1).

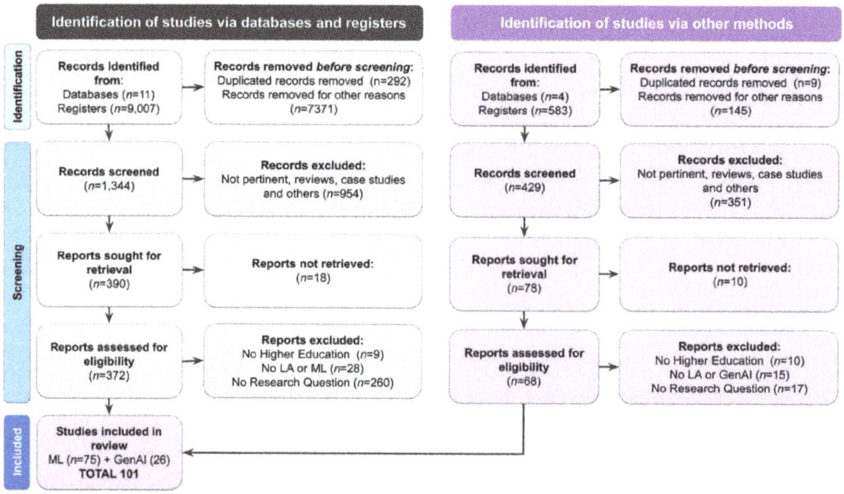

Figure 1. PRISMA flowchart for study selection (2018–2025), detailing inclusion of 101 studies from an initial pool of 9590 records.

2.4. Data Extraction and Collection Process

All references were managed using RefWorks ProQuest. After deduplication, two parallel screening processes were conducted. For LA + ML, 1344 records were screened, 954 excluded, and 372 assessed for eligibility, yielding 75 empirical studies. For LA + GenAI, 429 records were screened, 351 excluded due to irrelevance or lack of empirical content, and 68 assessed, resulting in 26 included studies. Figure 1 summarizes the inclusion and exclusion process.

Screening was conducted using a structured coding protocol aligned with the review's research questions. All screening stages (titles/abstracts and full texts) were performed by a single reviewer. Each record was evaluated against the predefined eligibility criteria, with particular emphasis on the presence of an explicit research question and the study's relevance to addressing the review objectives.

2.5. Quality Appraisal

To assess the methodological quality of the included studies, we applied the Mixed Methods Appraisal Tool (MMAT, 2018 version) [21]. Each study was classified according to MMAT study type (qualitative, quantitative, mixed methods) and assessed using its corresponding checklist. Screening criteria (S1 and S2) and methodological dimensions (e.g., sampling, measurement, data analysis) were evaluated and summarized. The complete dataset with MMAT ratings, checklist responses, study types, and reviewer notes for all 101 studies is openly available on Zenodo (https://doi.org/10.5281/zenodo.16416487). This open dataset enhances transparency and supports reproducibility of the review process.

2.6. Final Dataset

The final review includes 101 empirical studies: 75 focused on traditional ML applications in learning analytics and 26 integrating GenAI models, such as GPT, BERT, or FLAVA, into LA contexts. This dual corpus enables analysis of both the evolution of ML and the rise of GenAI within the LA research landscape.

2.7. Declaration of GenAI Use

During the study, ChatGPT-4.5 (OpenAI) was used to improve language clarity and to assist in refining the Python 3.10 code. All AI-generated output was reviewed and edited by the authors, who take full responsibility for the final content.

2.8. Data Availability

The structured dataset (metadata extracted from 101 studies, including AI models, contexts, and techniques) is publicly available via Zenodo (https://doi.org/10.5281/zenodo.16416465) to ensure transparency and reproducibility.

2.9. Ethical Considerations

No human or animal subjects were involved; ethical approval was not required.

2.10. Registration and Protocol

This review was not prospectively registered, and no separate protocol was prepared. Consequently, no protocol amendments apply.

3. Results

To structure the results, the corpus was divided into two groups—LA&ML (traditional machine learning) and LA&GenAI (generative AI)—based on each study's core methodology. Within these groups, unsupervised clustering was applied to identify thematic patterns using three categorical variables: AI models, application, and educational context, encoded via one-hot encoding. Dimensionality reduction and visualization were performed with principal component analysis (PCA), and the optimal number of clusters was determined using silhouette scores. The analysis, conducted in Python via Google Colab, informed the thematic organization of Sections 3.2 and 3.3, enabling a data-driven presentation of the findings across both paradigms.

3.1. Temporal and Geographical Distribution of Publications

This section analyzes the evolution of learning analytics (LA) research using machine learning (ML) and generative AI (GenAI) in higher education between 2018 and 2025, focusing on temporal trends and geographical spread.

The annual trends (Figure 2) show consistent growth in ML-based LA studies, with a peak in 2022. GenAI studies first appeared in 2022 and accelerated in 2024, coinciding with the widespread release and adoption of large language models (LLMs) such as GPT-4. This pattern reflects two coexisting trajectories: the consolidation of traditional ML approaches—classification, regression, and ensemble models applied to LMS data—and the experimental rise of GenAI applications for feedback generation, engagement modeling, and content personalization. While ML remains foundational, GenAI is rapidly gaining traction, though it is still in the early stages of methodological standardization and empirical validation.

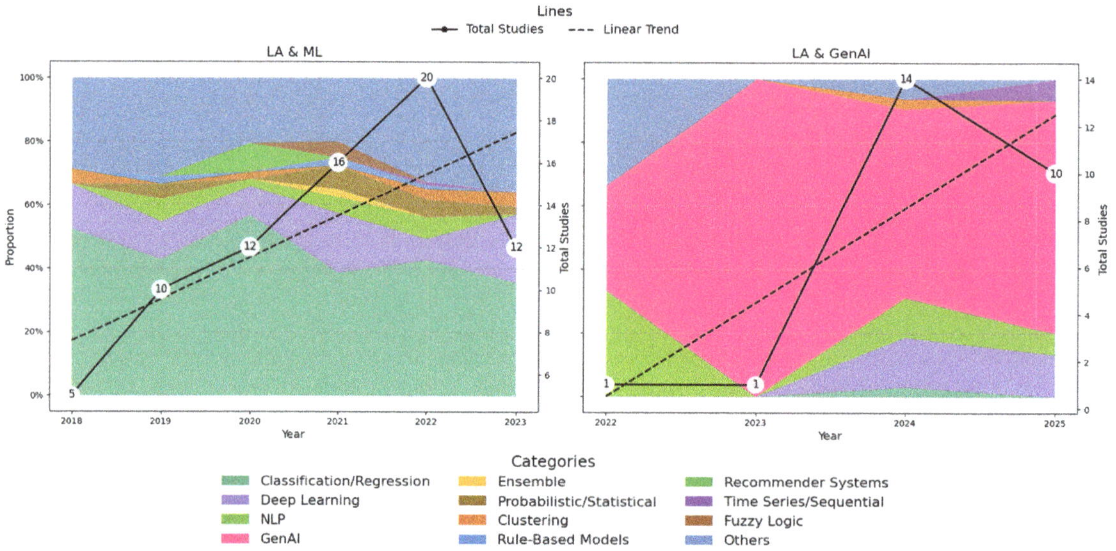

Figure 2. Temporal trends in learning analytics studies using traditional machine learning (**left**) and generative AI (**right**) in higher education between 2018 and 2025.

Geographical distribution (Figure 3) reveals notable disparities. The United States leads with 17 studies (10 ML, 7 GenAI), followed by Australia, China, and Germany, which show relatively balanced activity across both paradigms. However, GenAI research remains highly concentrated in high-income regions, with limited representation from Latin America, Sub-Saharan Africa, and parts of Asia. These gaps raise concerns about equity, contextual relevance, and disparities in access to GenAI infrastructure for research and implementation in education.

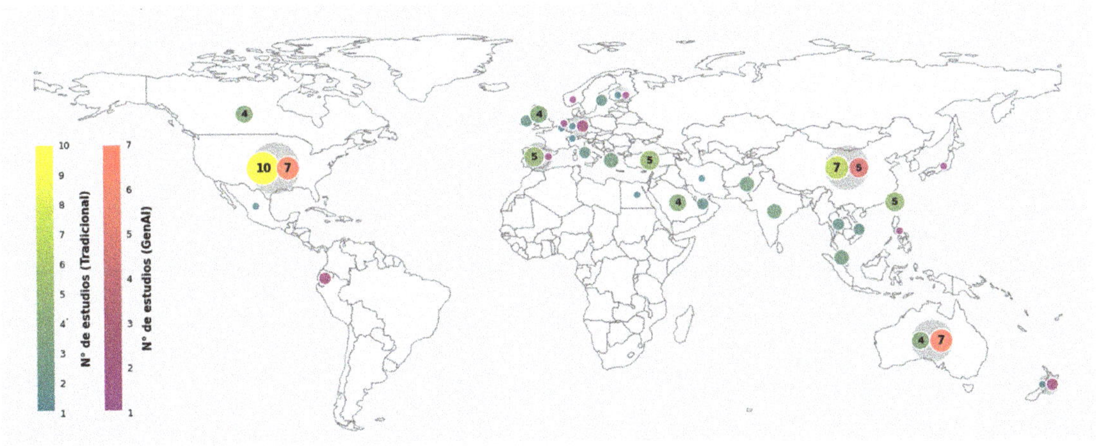

Figure 3. Geographical distribution of learning analytics studies in higher education (2018–2025) by AI model type.

The coexistence of ML and GenAI suggests a transitional phase in LA research. ML offers robustness and interpretability, while GenAI brings adaptability and personalization. Hybrid approaches could combine their strengths to enhance flexibility and inclusiveness.

Three key observations emerge: (1) GenAI adoption aligns with the release of tools like GPT-4, highlighting the role of accessibility; (2) its concentration in high-income regions may deepen epistemic inequalities; and (3) the parallel rise of both paradigms invites integrated, transparent, and context-aware frameworks.

3.2. *Learning Analytics with Traditional Models*

To analyze the application of traditional machine learning (ML) in learning analytics (LA), we clustered 75 peer-reviewed studies from 2018 to 2023. Using three categorical variables—AI models, application type, and educational context—encoded and reduced via principal component analysis (PCA), we identified thematic groupings centered on engagement prediction, dropout modeling, academic performance forecasting, and feedback systems across online, blended, and face-to-face settings.

Figure 4 presents the top 10 ML models by context. Random forest, support vector machine (SVM), and decision tree are most prevalent, particularly in online and MOOC environments, due to their robustness, interpretability, and compatibility with structured behavioral data. Logistic regression, naive Bayes, and artificial neural networks also appear frequently, indicating a reliance on established methods.

3.2.1. Engagement Prediction in Online Learning with Traditional ML Models

The analysis of student engagement in online and hybrid environments has progressed through the integration of ML within LA. This cluster, comprising 11 studies, focuses on predicting engagement, academic performance, and dropout risk using structured and unstructured data sources.

Most studies rely on classical ML algorithms, such as random forest, decision tree, SVM, and logistic regression, often enhanced with ensemble techniques like AdaBoost and Bagging. These models are widely used in MOOCs and online courses to support an early detection of at-risk behavior. However, few studies, including those addressing engagement and behavioral prediction [22–26], assess whether such predictions lead to real-time, actionable interventions.

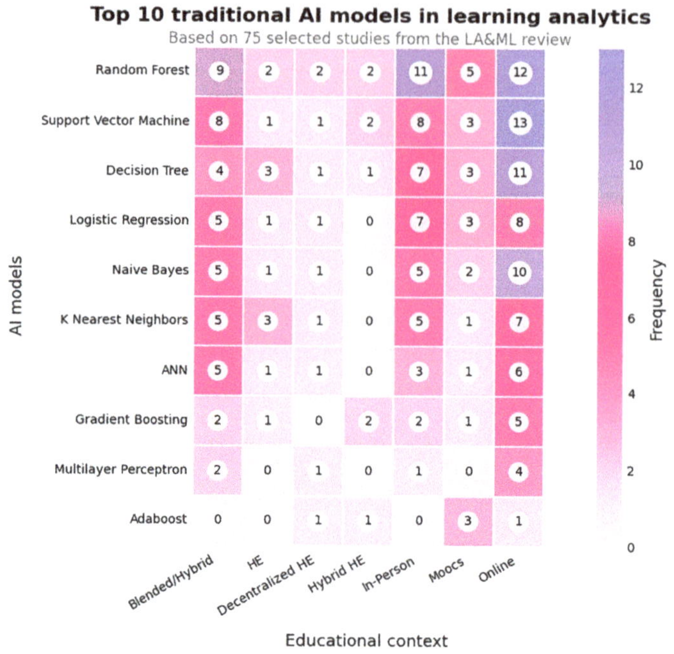

Figure 4. Top 10 AI models applied in learning analytics using traditional machine learning techniques, categorized by educational context.

The choice of algorithm often reflects the nature of the data: SVM and artificial neural networks (ANN) are used for fine-grained engagement prediction [24], while decision tree and logistic regression are suited to interpret behavioral traces such as video interaction logs [25]. Feature engineering, clustering, and regression remain central analytical techniques [27].

A notable methodological trend is the use of natural language processing (NLP) to analyze forum content and course reviews. Word embeddings—Doc2Vec, Word2Vec, FastText, and GloVe—support sentiment classification and behavioral inference [28,29]. Deep learning models, such as LSTM and GRU, enhance performance in text-based engagement tasks, with GloVe consistently yielding superior accuracy [28].

Among the most comprehensive approaches is the work of Onan [28], which combines traditional classifiers, ensemble techniques, and deep architectures like CNN, GRU, LSTM, and attention-based RNN. The model achieves 95.80% accuracy using LSTM with GloVe embeddings, demonstrating the synergy between affective analytics and attention mechanisms—an intersection between traditional ML and GenAI.

Emerging studies also explore spatiotemporal learning behaviors. Du et al. [30] show that consistent study patterns correlate with academic success. Moubayed et al. [31] apply clustering to blended learning profiles, while Lahza et al. [32] investigate strategy use in learner sourcing platforms, linking it to instructional design quality.

Key applications include:

- Dropout detection and risk assessment [22,25];
- Engagement forecasting and behavioral profiling [24,26,27];
- Sentiment analysis of learner feedback [28];
- Modeling learning routines and platform interactions [30–32].

Challenges and opportunities: Multimodal integration remains limited, with most studies focusing on a single data type. Longitudinal analyses are rare, limiting insights into lasting engagement patterns. Although XAI is occasionally applied [23,26], most models lack transparency, reducing their practical utility in educational settings.

Recommendations for future research:

- Combine behavioral, textual, and emotional data sources;
- Conduct longitudinal studies on sustained engagement;
- Enhance model explainability through tailored XAI tools;
- Link predictions to adaptive instructional strategies.

In summary, progressing toward multimodal and learner-centric LA systems is essential for fully leveraging ML in higher education contexts.

3.2.2. Dropout Prediction in Digital Education with Traditional ML Models

Student attrition remains a persistent concern in digital higher education, often preceded by disengagement—behavioral, cognitive, or emotional. Predicting these early indicators is crucial for intervention. Predictive learning analytics (PLA) and machine learning (ML) are widely used to identify at-risk students before dropout. This section synthesizes findings from 23 peer-reviewed studies applying AI models across online, blended, and MOOC environments.

Random forest and SVM are the most used models, with each in nine online learning studies. Decision tree, naive Bayes, and logistic regression remain favored for their simplicity and interpretability. Ensemble methods such as random forest and boosting reduce overfitting and handle feature interactions effectively [33–35]. In MOOCs, random forest reached F1-scores above 0.88 based on daily progress [34]. Hybrid ensembles combining classification and regression enabled early risk detection before midterms [36]. During the pandemic, extra trees and logistic regression surpassed 90% specificity in asynchronous learning [37].

Neural models like LSTM and GRU outperform traditional classifiers. LSTM exceeded CNNs and MLPs from week six in STEM courses [38]; GRU achieved 98% accuracy in remedial English [39]. IOHMM and sequential logistic regression predicted dropout by week eight, outperforming SVM and logistic models [40].

Interpretable models like decision trees and logistic regression remain valuable for deployment [41,42], often as base learners in ensembles [43]. Semi-supervised models with SHAP explanations combine diverse data types [44]. Dashboards like OU analyse supported instructor monitoring and improved student outcomes [45]. Behavioral profiling through LMS logs and motivational indicators guided personalization [46,47]. Ensemble models also supported post-pandemic readiness [48], and SHAP facilitated fairness-aware predictions [49]. Self-reports on motivation were included in some models [50]. Sentiment analysis with SVM and decision trees reached over 90% accuracy [51]. Random matrix theory and community detection uncovered engagement-success links in programming education [52]. SMOTE addressed class imbalance [39]. Hybrid ensembles also explored fuzzy logic [53]. Generative models remain rare [54], and model generalizability is still limited [55].

Key applications include:

- Early warning systems in initial weeks [35,38];
- Instructor dashboards for real-time monitoring [45];
- Adaptive feedback based on predictions [35];
- Behavioral profiling using LMS and motivation data [46,47];
- Post-pandemic readiness via ensembles [48];
- Fairness-aware interventions using SHAP [49].

Challenges and opportunities: Generative AI is underused [38,54]. LMS logs dominate, while multimodal inputs are scarce [47]. Few systems operate in real time or are deployed in institutions [35,45]. Explainability tools lack integration into actionable dashboards [49]. Generalizability is limited due to narrow datasets [55].

Recommendations for future research:

- Apply generative models for feedback and simulation;
- Incorporate multimodal sources (e.g., forums, self-reports);
- Deploy real-time tools in institutional settings;
- Embed explainability in instructor-facing systems;
- Expand validation across contexts and institutions.

3.2.3. Academic Performance Prediction in Face-to-Face Classrooms

Predicting academic performance in face-to-face higher education settings remains a core objective within learning analytics (LA), primarily aiming at the early identification of students at risk. This chapter synthesizes findings from 26 studies utilizing traditional machine learning (ML) models in classroom contexts.

The most frequently employed ML algorithms include random forest (RF), support vector machines (SVM), decision trees (DT), and logistic regression, each favored due to their effectiveness with structured academic data. Specifically, RF has reliably detected underperformance through historical grades [56]. However, its predictive accuracy is notably reduced in small or homogenous student groups [57]. Conversely, SVM demonstrated strong performance with smaller datasets [58,59], though its applicability to larger classrooms remains inadequately explored. Decision trees are valued for their interpretability in guiding interventions [60] but often face challenges with imbalanced datasets [61]. Meanwhile, artificial neural networks (ANNs) effectively model complex multimodal datasets [62,63], though their inherent opacity complicates practical deployment.

Supervised learning methodologies—primarily classification and regression—dominated the reviewed studies, largely based on academic records and limited learning management system (LMS) logs. Despite the inclusion of innovative approaches such as natural language processing (NLP) for feedback analysis [62,64] and sensor-based physical interaction data [65], face-to-face contexts continue to present significant barriers due to insufficient digital interaction data for adaptive and longitudinal modeling.

Advanced analytical techniques, including concept inventories [66], gesture analytics [67], and LMS-integrated predictive models [68], contributed positively to predictive accuracy but were often limited by small sample sizes, affecting generalizability. Additionally, early assessments [69,70], demographic indicators (e.g., socioeconomic status, language proficiency, background) [71,72], and institutional data emerged as reliable predictors. Curriculum personalization emerged as a notable application. Recommendation systems were proposed based on academic pathways [73,74], aiding program design.

Key applications include:

- Performance forecasting enabling timely academic interventions [57,62,63,65,75];
- Dropout prediction identifying risk patterns for preventive actions [56,60,61,76–79];
- Personalized feedback via NLP [64];
- Curriculum personalization through predictive analytics informing tailored educational pathways [73,74,80,81].

Challenges and opportunities: Despite significant progress, key limitations persist. Generalizability remains constrained by small or imbalanced datasets and institutional variability [57,58,67]. Face-to-face contexts lack continuous digital traces, limiting adaptive and multimodal analytics. Moreover, critical variables such as motivation, emotion, and socio-economic status are often excluded despite their predictive value [60,66,71,75,76].

Addressing these issues demands not only methodological refinement but also ethical vigilance. Incorporating sensitive data raises concerns about privacy, consent, and bias. Future research must adopt privacy-preserving practices, transparent data governance, and inclusive frameworks to ensure that predictive systems are equitable, respectful of student rights, and contextually appropriate across diverse educational environments.

Recommendations for future research:

- Enhance multimodal analytics by incorporating sensor-based data (e.g., physical interactions) alongside emotional, motivational, and socio-economic variables;
- Explore advanced methodologies such as reinforcement learning and generative AI to develop dynamic, adaptive curricular recommendations;
- Prioritize explainable AI (XAI) development to enhance interpretability and facilitate educator acceptance, rigorously validating effectiveness in authentic educational environments.

3.2.4. Feedback and Performance Modeling in Hybrid Learning with ML

Machine learning is essential in hybrid learning contexts, especially for modeling academic performance and automating feedback. Artificial neural networks (ANN), random forest (RF), and support vector machines (SVM) dominate due to their effectiveness with varied educational data. ANN frequently models complex behaviors like cognitive engagement, though it faces interpretability challenges [82–84]. SVM excels at classifying high-dimensional textual data, identifying cognitively relevant content [84,85]. RF balances predictive accuracy and interpretability, analyzing variables such as peer interactions and digital distractions [83,86].

Recent use of transformer-based models like BERT has advanced analysis of emotional and behavioral signals in collaborative settings, enabling real-time personalized feedback and co-regulated learning [87].

Multimodal datasets typically combine structured data (grades, surveys) with unstructured inputs (forum posts, interaction logs, keystroke dynamics). Natural language processing (NLP) is crucial for preprocessing text, classifying engagement, and assessing discourse relevance [83,84]. Keystroke analytics reveal writing fluency and cognitive processes, while LMS logs highlight self-regulated learning patterns [88,89]. Hierarchical regression isolates cognitive, affective, and behavioral influences on student performance [90].

Explainable AI (XAI) techniques like LIME and SHAP have begun enhancing transparency and informing pedagogical decisions, yet broader integration is underdeveloped [91,92].

Key applications include:

- Cognitive engagement via discourse analysis [84];
- Analyzing distraction and peer interactions using NLP-classified discussions [83,89];
- Real-time, emotion-sensitive feedback leveraging transformer-based models [87];
- Early identification of academic risks using RF with academic and behavioral data [93–95];
- Instructional design optimization through behavioral predictors of satisfaction and performance [96].

Challenges and opportunities: Despite progress, interpretability and adaptive responsiveness remain challenges. Consistent use and pedagogical value of XAI techniques require further integration and validation [85,89,90]. Current multimodal integration mostly focuses on textual or structured inputs, limiting personalization and scalability. Incorporating audio, video, and physiological signals within diverse, multi-institutional datasets could significantly enhance analytical robustness [81,92,93]. Moreover, personalized interventions based on predictive analytics need rigorous longitudinal evaluation to confirm their educational impact [94].

Recommendations for future research:

- Integrate richer multimodal data (audio, video, physiological signals) to enhance analytical capabilities;
- Systematically apply advanced XAI techniques for increased model interpretability;
- Develop and validate adaptive, real-time feedback systems through robust longitudinal and experimental research.

Emerging technologies such as generative AI, multimodal analytics, and advanced XAI offer new opportunities to overcome current limitations in feedback and performance modeling. Their potential lies in enhancing accuracy, personalization, and pedagogical value, but this requires rigorous validation, integration into real hybrid learning contexts, and alignment with instructional goals. Future systems must evolve toward real-time, interpretable, and adaptive support that scales to diverse learner needs.

3.3. Expanding Learning Analytics Through Generative AI

The integration of generative artificial intelligence into learning analytics represents a major methodological shift in higher education. To explore this landscape, we clustered 26 GenAI-focused studies into two groups. The first cluster (n = 14) centers on large language models (LLMs) like GPT-4, GPT-3.5, BERT, and FLAVA, applied primarily in higher education. These studies focus on engagement analytics, adaptive learning, and ethical bias detection, often using multimodal inputs and transformer-based architectures. The second cluster (n = 12) features diverse approaches combining GenAI with emotion classifiers, GANs, or custom copilots, deployed across online, hybrid, and in-person contexts.

Figure 5 presents the most used GenAI models and their distribution across learning environments. GPT-4 dominates in both frequency and application scope, followed by BERT, GPT-3.5, and hybrid models (e.g., ChatGPT with clustering). Despite advances in multimodal GenAI, implementations remain largely text-based, with emphasis on automated feedback and engagement modeling.

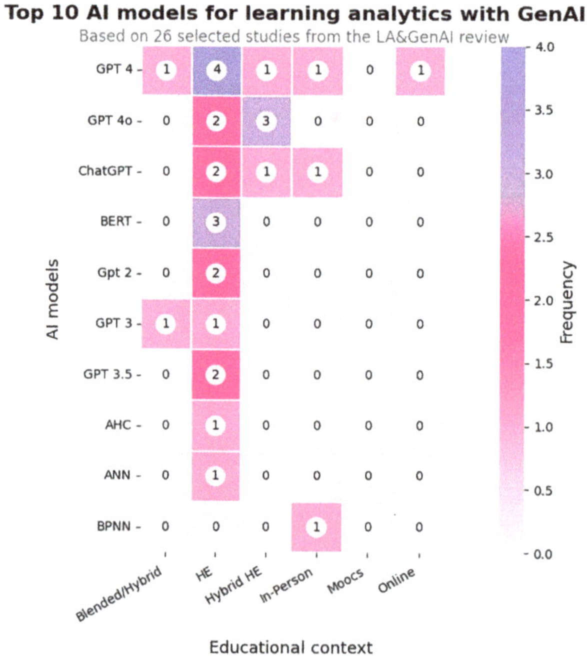

Figure 5. Top generative AI models used in learning analytics research.

These trends reflect innovation but also surface critical concerns around equity, transparency, and practical feasibility in varied educational settings.

3.3.1. Generative AI Applications for Engagement, Feedback, and Adaptation

The growing integration of generative AI (GenAI) into learning analytics (LA) underscores a methodological shift toward transformer-based models, primarily GPT-3.5, GPT-4, BERT, FLAVA, and Whisper. Frequently, these models operate within hybrid frameworks combining supervised classifiers (e.g., SVM, random forest, logistic regression) and unsupervised techniques, such as clustering and topic modeling. Among the 14 reviewed studies, GPT and BERT variants dominate, occasionally supported by recurrent neural networks (RNNs) or long short-term memory (LSTM) networks. Classical methods like decision trees and naive Bayes are occasionally used for simpler classification tasks [97]. However, the general absence of lightweight models raises scalability and equity concerns, particularly for low-resource educational contexts.

Quantitative analysis reveals GPT-4 as the most widely used model, followed by BERT and other GPT variants. Diverse methodologies, including neural networks and clustering algorithms, highlight varied applications, ranging from engagement analytics and adaptive feedback to personalized learning experiences. Large language models (LLMs) facilitate real-time reflective scaffolding [98], the detection of inclusive interactions [99], personalized feedback [100], and improved learner engagement in classroom integration contexts [101]. Some longitudinal research indicates that interaction frequency with GenAI tools enhances learner autonomy via increased social presence [102], yet rigorous empirical evidence on instructional effectiveness or scalability remains sparse.

Mixed-method approaches, integrating digital learning traces, student-generated texts, GenAI interactions, and multimodal data (audio, video, gesture), are prevalent. NLP techniques, especially sentiment analysis and topic modeling using transformers, effectively analyze textual data from learners and AI-generated content [103,104]. Clustering, combined with GPT/BERT embeddings, supports learner profiling and targeted feedback strategies [105]. Sequential models, including RNNs and LSTMs, aid the analysis of trajectories related to self-regulated learning and help-seeking behaviors [101,106].

While explainable AI (XAI) is gaining attention, its application remains limited. Fahl [107] uniquely integrates GPT-4 with semantic knowledge graphs to enhance explainability. However, broader transparency and accountability issues persist, amplifying ethical concerns such as hallucination and trustworthiness.

Key applications include:

- Formative feedback and automated assessment [100,108];
- Adaptive prompts supporting self-regulated learning [98];
- Visual analytics via GenAI-enhanced dashboards [107];
- Equity-aware collaborative learning tools [99];
- Detection of help-seeking behaviors [109];
- Adaptive content delivery in gamified or flipped learning contexts [101].

Challenges and opportunities: Although GenAI offers considerable promise, significant limitations remain. The predominant opacity of GenAI systems hinders interpretability, with rare exceptions employing explainability frameworks [107]. Additionally, the frequent misalignment with self-regulated learning (SRL) principles risks fostering metacognitive passivity, as discussed in existing critiques [110]. Real-time adaptability is another concern, as many GenAI tools provide delayed or static feedback, inadequately addressing immediate learner disengagement—though real-time adaptive scaffolding offers potential solutions [98]. The underutilization of multimodal inputs, such as audio, video, or physiological data, limits the comprehensive understanding of learner behavior and context [99,103]. Lastly, weak theoretical grounding and limited alignment with established instructional frameworks diminish practical pedagogical effectiveness, highlighting the need for evidence-centered design approaches [106].

Recommendations for future research:

- Incorporate explainable AI techniques extensively to enhance transparency and trustworthiness;
- Expand multimodal GenAI applications to capture affective, cognitive, and embodied learning experiences;
- Employ ontological models to systematically structure learning progression and knowledge monitoring [107];
- Develop reflective GenAI agents supporting deep engagement and co-regulated learning dynamics.

Addressing these recommendations, along with methodological, ethical, and institutional considerations, will significantly advance the sustainable and effective integration of GenAI within higher education.

3.3.2. Technical Approaches and Emerging Trends in GenAI for Learning Analytics

The integration of generative AI (GenAI) into learning analytics (LA) is significantly reshaping methodological frameworks in higher education. Across twelve studies [111–113], large language models (LLMs), including GPT-4, GPT-4o, DistilBERT, and Gemini 1.5 Pro, have emerged as central tools for tasks such as feedback generation, affective state detection, and instructional scaffolding. Despite broad usage, most implementations remain preliminary, lacking comprehensive validation in real-world educational environments. Furthermore, the influence of instructional context on model effectiveness and learning outcomes remains underexamined.

GPT-4 variants dominate, appearing in at least eight studies, often integrated with emotion detection tools (HSEmotion), facial recognition (MTCNN), or generative adversarial networks (GANs) for enhanced multimodal analysis [114]. DistilBERT specifically excels in detecting confusion within discourse data [112]. Applications leveraging ChatGPT and GitHub Copilot have effectively captured learner strategies in coding activities [113]. Additionally, collaborative interactions supported by GPT-4 have been linked to improved hint quality and heightened critical thinking [115], while CustomGPT offers tailored insights by analyzing interactions with digital educational resources [116].

Application domains are concentrated mainly around engagement analytics, automated feedback, and risk detection. GPT-driven dashboards and conversational agents provide adaptive support and multimodal instructional scaffolding in real-time interactions [117,118]. Nevertheless, the recurrent usage of basic LLM pipelines—primarily prompt engineering combined with simple classifiers—highlights limitations in methodological innovation and adaptation to diverse instructional contexts.

The optimization of LLM outputs typically involves fine-tuning and strategic prompt design. For instance, GPT-4o has been customized to interpret natural language queries and generate educationally relevant SQL queries and visualizations [118]. Similarly, DistilBERT has successfully facilitated efficient confusion classification in MOOCs [112], while GAN-based techniques have enriched otherwise sparse datasets [114].

Some studies prioritize explainability. For instance, NLP and ChatGPT have been utilized to uncover patterns in collaboration and bias in peer-generated feedback [119]. Furthermore, GitHub Copilot interaction logs have been employed to track cognitive transitions during learning activities [113]. However, instances of feedback inaccuracies or reinforcement of misconceptions indicate the crucial role of human oversight [120].

Hybrid methodologies that combine clustering, regression, and sequential analysis with NLP techniques frequently analyze multimodal data, including LMS logs, reading behaviors, keystrokes, and emotional inputs [111,121]. Although explainable AI methods such as integrated gradients and anchors have been incorporated, their practical impact on instructional decision-making remains inadequately assessed [92,112].

Key applications include:

- Automated formative feedback in STEM disciplines through GPT4 and ChatGPT [113,115,122];
- Real-time confusion detection in online learning environments using DistilBERT [112];
- Multilingual engagement analytics leveraging bilingual prompts and chat logs [120,121];
- Bias detection and fairness evaluation in peer feedback through explainable GenAI models [119].

Challenges and opportunities: Despite promising advances, significant challenges persist. Many GenAI implementations focus predominantly on retrospective evaluation rather than proactive, real-time instructional utility. Although explainability methods are increasingly employed, their pedagogical grounding and effectiveness in supporting actual learning processes remain unclear. Ethical concerns related to inaccurate or hallucinated LLM outputs are acknowledged but insufficiently addressed through rigorous validation or expert verification [120]. Additionally, evidence supporting sustained cognitive or motivational impacts from GenAI-facilitated interactions is limited, emphasizing a critical need for longitudinal studies. Furthermore, engagement analytics tools frequently demonstrate limited transferability across educational contexts and disciplines, indicating a pressing need for scalable and adaptable solutions.

Recommendations for Future Research:

- Develop multimodal, theory-driven GenAI models integrating gaze, speech, emotions, and learner behaviors aligned explicitly with frameworks like self-regulated learning (SRL) and feedback literacy;
- Promote participatory design processes involving educators and students to define meaningful explanations, safety measures, and effective revision practices;
- Expand the evaluation criteria beyond accuracy to emphasize educational utility, fairness, epistemic validity, and ethical rigor;
- Foster scalability and transferability through adaptive models capable of continuous learning and refinement across diverse instructional contexts.

In summary, addressing foundational ethical and methodological barriers is essential to unlock the full pedagogical potential of GenAI in learning analytics.

4. Contextual Analysis of ML and GenAI in LA

The integration of artificial intelligence into learning analytics has advanced rapidly, driven by growing interest in traditional machine learning (ML) and the emerging potential of generative AI (GenAI). While ML has supported prediction and classification in education, GenAI enables real-time interaction, feedback, and adaptive content. Yet, the literature highlights persistent gaps in theoretical grounding, teacher involvement, and empirical validation. This section contrasts contributions from both domains to uncover the key trends, challenges, and future directions.

4.1. Generative AI in Learning Analytics

Recent studies emphasize GenAI's potential to transform LA from passive prediction to real-time personalization. Borah et al. [13] proposed a framework that adapts learning paths and feedback based on cognitive and emotional profiles, integrating multimodal data, contextual rules, and conversational interfaces for co-evolving learner–instructor interaction.

Complementary analyses from the LAK and EC-TEL proceedings reinforce this trend. Huang et al. [123] found RoBERTa the most accurate and explainable model for peer feedback classification, while Pishtari et al. [124] reported significant gains in instructional design quality through GenAI-generated feedback.

Yan et al. [15] mapped GenAI's integration across the LA cycle—from data augmentation to intervention—highlighting synthetic data and agent-based tools as enablers of engagement. Khosravi et al. [14] addressed ethical and pedagogical considerations, advocating codesign with educators, transparency, and models that safeguard student agency.

Qu and Yang [125] surveyed LLM applications such as ChatGPT in language and medical education, noting limited empirical validation and weak integration into instructional practice.

Despite these developments, most GenAI applications prioritize technical innovation over pedagogical grounding. Few incorporate learning theories or evaluate long-term outcomes, underscoring the need for more critically aligned, education-centered approaches.

4.2. Traditional Machine Learning in Learning Analytics

Traditional machine learning techniques have played a pivotal role in the development of learning analytics, particularly in areas such as prediction, classification, clustering, and recommendation. PeñaAyala's taxonomy [126] is frequently cited for its comprehensive organization of these functions, focusing on learner performance, engagement, and dropout analysis.

Zawacki-Richter et al. [127] found that most ML-based LA systems focus on performance prediction and early warning but often lack pedagogical alignment, limiting their educational relevance. Similarly, Renz and Hilbig [128] showed that commercial EdTech tools prioritize algorithmic efficiency over interpretability, reducing their utility for instructors.

Regionally, Salas-Pilco and Yang [129] documented promising ML applications in Latin America, such as identifying at-risk students, but the highlighted challenges included limited infrastructure, institutional support, and the need for culturally responsive models. Glandorf et al. [130] revealed that dropout prediction varies by demographic group, while Poellhuber et al. [131] improved predictive efficiency by clustering course structures in Moodle before modeling.

Buitrago-Ropero et al. [132] reframed learner data as sociopedagogical artifacts—"data, action, and service"—arguing for a shift from purely computational to sociotechnical perspectives. Baek and Doleck [133] compared educational data mining and LA, noting shared methods but differing goals: EDM focuses on algorithmic development, while LA emphasizes educational impact—though both often lack theoretical and teacher integration.

Lastly, Ley et al. [134] and Aguilar-Esteva et al. [135] called for human-centered and equity-oriented LA, promoting models that are transparent to educators and responsive to sociocultural and sustainability concerns.

4.3. Critical Synthesis and Research Gaps

Despite technological advances in both traditional ML- and GenAI-based learning analytics (LA), several persistent challenges limit their educational impact. A major concern is the lack of pedagogical integration: many systems operate independently of learning theories or instructional strategies, reducing their capacity to support meaningful learning [131]. Additionally, both ML and GenAI tools often function as opaque black boxes, hindering interpretability and diminishing educator trust in system outputs [126].

Educator involvement also remains limited. Teachers are frequently positioned as passive end users rather than active codesigners, which compromises contextual relevance and adoption [132]. Moreover, LA systems show insufficient adaptation to local constraints, including infrastructure, culture, and language, particularly in underrepresented or resource-constrained educational settings [127]. These systemic limitations underscore the need for more inclusive and transparent LA design approaches.

To address these gaps, future research should prioritize the development of hybrid AI systems that combine the predictive strengths of ML with the interactive potential of GenAI. Such systems could better balance accuracy with personalization. Additionally, codesign practices involving both educators and learners are critical to ensure that LA tools align with real-world instructional needs. Enhancing model explainability is also essential, particularly for translating algorithmic complexity into actionable insights for educators.

Theory-driven design remains an underdeveloped area. Embedding constructs such as self-regulated learning, motivation, and feedback theory into system logic could enhance pedagogical alignment. Finally, advancing culturally responsive LA is imperative, particularly in settings with limited resources and diverse learner populations. Integrating these considerations will be key to developing LA systems that are not only innovative but also equitable, interpretable, and educationally meaningful.

These insights inform the discussion in Section 5, where we outline the broader implications for research and practice in learning analytics.

5. Discussion and Implications

This section critically synthesizes the findings of the review, addressing the two research questions and outlining their implications for educational practice, research, and institutional policy. The discussion is organized into four parts: current implementation (RQ1), potential benefits (RQ2), cross-cutting challenges, and future recommendations.

5.1. Current Implementation of ML and GenAI in Higher Education (RQ1)

The integration of machine learning and generative AI into learning analytics follows two methodological paths: the consolidation of traditional ML models and the exploratory adoption of GenAI systems.

ML-based LA systems are well-established for predicting academic performance (e.g., random forest, SVM, logistic regression), detecting dropout (e.g., LSTM, GRU, decision trees), modeling engagement (clustering, sentiment analysis), and optimizing feedback using k-means or regression. These tools operate mainly on structured LMS data—clickstreams, assessments, participation logs—and are valued for accuracy and simplicity, though they often lack alignment with classroom practice.

GenAI-based LA systems are emerging in personalized feedback (LLMs like GPT-4, BERT), emotional and cognitive engagement modeling (sentiment/discourse analysis), SRL support (tutoring agents), and real-time content scaffolding. These implementations are typically experimental, text-focused, and integrated through prompt engineering or fine-tuning. While studies—such as RoBERTa for peer feedback [123] and GenAI-enhanced instructional design [124]—show promise, most lack empirical validation, scalability, or connection to learning theory. Additionally, multimodal inputs (e.g., gaze, audio, gesture) remain underutilized.

ML-based LA systems are mature, data-driven, and prediction-focused; GenAI systems are flexible, interaction-oriented, and still experimental. Despite their complementarity, both paradigms face common limitations: fragmented implementation, weak pedagogical integration, and minimal educator involvement.

5.2. Potential Benefits of Integrating ML and GenAI into Learning Analytics (RQ2)

The integration of machine learning and generative AI into learning analytics in higher education presents benefits across pedagogical, institutional, and research domains.

Pedagogically, ML enables early risk detection, allowing timely interventions, while adaptive systems personalize feedback, pacing, and resources. GenAI tools enhance formative feedback by generating scalable, context-aware responses to student submissions. Additionally, conversational agents support metacognition and self-regulated learning (SRL) through the scaffolding of reflection, planning, and help-seeking behaviors.

Institutionally, predictive dashboards aid retention monitoring and academic advising, while LA-informed instructional design improves curriculum responsiveness. Scalable intervention systems, such as real-time alerts and adaptive prompts, enhance support in large or asynchronous courses. Cluster-based modeling of LMS data has also improved generalizability in early-warning systems, particularly when courses are grouped by structural similarity [131].

For research and innovation, multimodal learner modeling—integrating behavioral, affective, textual, and biometric data—offers a richer view of the learning process. Advances in explainable AI (XAI), including SHAP, LIME, and knowledge graphs, contribute to transparency and increase educator trust. Furthermore, hybrid LA systems that combine the predictive power of ML with the generative capabilities of GenAI enable more personalized and dynamic learning environments.

5.3. *Cross-Cutting Challenges and Gaps*

Despite advances in ML and GenAI for learning analytics (LA), several structural limitations persist that hinder their educational effectiveness.

A key cross-cutting gap lies in the limited grounding of LA systems—particularly those powered by GenAI—in established learning theories. For instance, several tools in the corpus provided personalized feedback using large language models, such as the explainable feedback dashboard by Afzaal et al. [91] and the fine-tuned GPT-4 system for hint revision explored by Singh et al. [115]. While both systems offer scalability and automation, they rarely aim to develop students' feedback literacy—a construct encompassing the capacity to interpret, act on, and benefit from feedback [136,137]. Moreover, the feedback remains largely unidirectional, with little scaffolding for student agency or dialogic interaction [138].

Similarly, only a subset of studies explicitly aligned with the dimensions of self-regulated learning (SRL), such as metacognitive monitoring, goal-setting, and reflection [139,140]. For example, while Dai et al. [100] focused on evaluating the quality of GPT-generated feedback for open-ended writing tasks, they did not examine how such feedback supports SRL processes or learner engagement. In contrast, Li et al. [98] integrated GenAI to deliver adaptive scaffolds based on real-time analytics of SRL processes, showing improved metacognitive strategies—yet still reported variability in learner compliance and limited generalizability.

These gaps suggest that, while ML and GenAI can technically model learner behavior, their educational impact remains limited without deliberate integration into pedagogical frameworks, such as SRL, feedback literacy, and formative assessment [141]. Additionally, many ML and GenAI models remain opaque, functioning as black boxes that impede interpretability and educator trust. While explainable AI (XAI) techniques, such as RoBERTa with LIME for peer feedback classification [123], have shown promise, they are still underused.

Educators are rarely involved as co-designers, resulting in systems poorly aligned with classroom realities. GenAI implementations are often tested in artificial or highly controlled environments. Even when benefits like improved instructional design are observed [124], few studies assess long-term impact in authentic educational settings.

Geographic disparities also persist. Research and deployment efforts are concentrated in high-income regions, limiting global representativeness and inclusivity. Latin America and other underrepresented regions remain largely absent from the discourse [129].

Finally, multimodal and real-time data, such as emotional, behavioral, or biometric streams, are rarely integrated into adaptive feedback loops, which constrains the responsiveness and depth of learner modeling.

Table 1 summarizes these cross-cutting challenges and outlines the recommended research directions to inform the development of pedagogically grounded, explainable, and context-sensitive LA systems.

Table 1. Challenges and future directions in ML and GenAI for learning Analytics.

Challenge	Description	Future Direction
Model opacity	Deep models lack transparency, limiting educational use.	Advance Explainable AI (XAI) in LA.
Post hoc modeling	Retrospective models restrict real-time interventions.	Develop online/incremental modeling and real-time feedback systems.
Modality-agnostic design	Methods are reused without context adaptation across modalities.	Create modality-aware models (e.g., sensor, LMS, multimodal).
Geographic bias	Research is concentrated in high-income regions.	Broaden global datasets and include non-Western case studies.
Limited institutional integration	Real-world deployment and constraints are rarely addressed.	Study adoption, co-design with educators, and report implementations.
Narrow evaluation metrics	Accuracy often outweighs pedagogical value in evaluations.	Integrate educational impact and instructional alignment.
Ethical blind spots	Fairness, consent, and bias remain underexplored.	Apply ethical audits and promote fairness-aware ML and data literacy.
Underdeveloped GenAI use	GenAI lacks pedagogical grounding and deep integration into LA.	Explore adaptive, generative, and co-creative GenAI applications.

5.4. Implications and Recommendations

5.4.1. Implications of Generative AI for Engagement, Feedback, and Adaptation

Instructors should be actively involved in the co-design of LA systems to ensure alignment with pedagogical goals. Hybrid models that combine ML's predictive strength with GenAI's adaptive feedback capabilities offer promise for more responsive learning environments. Embedding explainability into system interfaces can enhance transparency, trust, and interpretability for both educators and learners.

5.4.2. Research

Future studies should move beyond predictive accuracy and assess actual learning outcomes through longitudinal and mixed methods designs. Aligning model architectures with learning theories, such as SRL, motivation, and engagement, can improve pedagogical relevance. Additionally, the instructional impact of GenAI tools on reflection, autonomy, and knowledge construction warrants deeper exploration.

5.4.3. Institutions and Policy

Institutions must establish ethical frameworks and data governance policies to guide the responsible use of LA and GenAI. Bridging digital gaps in under-resourced contexts is essential to prevent widening educational inequities. Professional development in AI and data literacy should be prioritized for educators, advisors, and policymakers to support informed implementation.

5.5. Looking Forward: Toward Human-Centered, Hybrid Learning Analytics

This review identifies a methodological shift in learning analytics (LA), from retrospective analysis to adaptive, generative, and participatory systems. While ML contributes structure and predictive rigor, GenAI adds adaptability and interactive potential. Their convergence—when guided by educational theory and ethical principles—can foster LA tools that are transparent, equitable, and pedagogically aligned.

Achieving this vision requires moving beyond technical prototypes toward empirical validation, contextual deployment, and active stakeholder collaboration. Future systems must be co-designed, responsive to real educational settings, and focused on empowering both educators and learners.

Ultimately, the future of LA lies not only in algorithmic advancement but in designing systems that serve learning with purpose, transparency, and equity at their core.

6. Study Limitations

Achieving the transformative potential of ML and GenAI in education requires moving beyond technical prototypes toward empirical validation, contextual deployment, and active stakeholder collaboration. Future systems must be co-designed, responsive to authentic educational settings, and focused on empowering both educators and learners.

First, all screening stages (titles/abstracts and full texts) were performed by a single reviewer. While eligibility criteria were clearly defined, this introduces potential selection bias.

Second, although a structured quality appraisal was conducted using the Mixed Methods Appraisal Tool (MMAT), the high methodological heterogeneity across studies limited the possibility of assigning comparable quality scores or excluding low-quality papers. Instead, the appraisal was used descriptively to inform interpretations and increase transparency.

Third, restricting the search to English-language publications may have excluded relevant studies from non-English-speaking regions, potentially limiting the cultural and geographical representativeness of the synthesis. Expanding future reviews to include multilingual sources would enhance global inclusivity.

Finally, although the main corpus focused on peer-reviewed journal articles indexed in major databases, an exploratory scan of LAK, L@S, and EC-TEL proceedings (2018–2025) was conducted. While these records were not formally included in the PRISMA dataset, selected studies were referenced to triangulate findings and highlight emerging developments.

Acknowledging these limitations is key to contextualizing the scope of this review and guiding future research toward greater methodological rigor, inclusiveness, and practical relevance.

7. Conclusions

This review offers a critical synthesis of how machine learning (ML) and generative AI (GenAI) are shaping learning analytics (LA) in higher education. Based on 101 empirical studies (2018–2025), along with insights from recent conferences, it highlights both methodological advances and persistent challenges across academic, technological, and pedagogical dimensions.

Traditional ML models remain central to LA, particularly for performance prediction, dropout detection, and engagement analysis. While robust and interpretable, these models often rely on retrospective, mono-modal data and show limited integration with pedagogical practices. In contrast, GenAI systems—particularly those using large language models (LLMs) like GPT-4—offer promising innovations in personalized feedback and affective modeling, yet remain largely experimental, with minimal real-world validation or theoretical grounding.

Recent efforts to enhance explainability, leverage clustering, and support instructional design mark progress, but critical gaps persist. These include limited use of multimodal and real-time data, weak ethical oversight, minimal educator involvement, and a lack of evidence for sustained impact on learning.

Rather than a paradigm shift, the current trends reveal a transitional phase in which ML and GenAI approaches coexist. This duality calls for hybrid, human-centered systems that combine predictive accuracy with adaptive feedback, grounded in learning theory and responsive to educational context.

Future progress requires a shift from technical performance to pedagogical impact—from opaque experimentation to scalable, explainable, and inclusive implementations. Advancing LA in line with Sustainable Development Goals 4.3 and 4.4 depends not solely on technological innovation, but on institutional support, educator engagement, and critical reflection.

Supplementary Materials: The following supporting information can be downloaded at https://www.mdpi.com/article/10.3390/app15158679/s1, https://doi.org/10.5281/zenodo.15233231: Table S1: Summary of Queries Used in Databases; https://doi.org/10.5281/zenodo.16422602: PRISMA Checklist. Available and https://doi.org/10.5281/zenodo.16416487: MMAT Quality Appraisal at Zenodo.

Author Contributions: Conceptualization, L.E.A.-R. and P.C.S.-M.; methodology, P.C.S.-M.; validation, P.C.S.-M.; formal analysis, M.Á.R.-O.; investigation, M.Á.R.-O., L.E.A.-R. and P.C.S.-M.; data curation, M.Á.R.-O.; writing—original draft preparation, M.Á.R.-O.; writing—review and editing, L.E.A.-R. and P.C.S.-M.; supervision, L.E.A.-R.; project administration, L.E.A.-R. All authors have read and agreed to the published version of the manuscript.

Funding: This publication has been partially funded by the project R+D+I PID2023-147396OB-I00, funded by MCIN/AEI/10.13039/501100011033 and by ERDF, EU.

Institutional Review Board Statement: Not applicable.

Informed Consent Statement: Not applicable.

Data Availability Statement: The dataset supporting this study is openly available at Zenodo: https://doi.org/10.5281/zenodo.16416465.

Acknowledgments: During the preparation of this manuscript, the authors used ChatGPT-4.5 (OpenAI) to improve language clarity and ChatGPT o4-mini-high to assist with Python code for data analysis. The authors have reviewed and edited the output and take full responsibility for the content of this publication. This research was partially supported by institutional resources.

Conflicts of Interest: The authors declare no conflicts of interest.

Abbreviations

The following abbreviations are used in this manuscript:

AHC	Agglomerative Hierarchical Clustering
ANN	Artificial Neural Network
BERT	Bidirectional Encoder Representations from Transformers
DT	Decision Tree
GenAI	Generative Artificial Intelligence
GPT	Generative Pre-trained Transformer
GRU	Gated Recurrent Unit
LA	Learning Analytics
LIME	Local Interpretable Model-Agnostic Explanations
LLM	Large Language Model
LMS	Learning Management System
LSTM	Long Short-Term Memory
ML	Machine Learning
MOOC	Massive Open Online Course
NLP	Natural Language Processing

PLA	Predictive Learning Analytics
RF	Random Forest
SHAP	SHapley Additive exPlanations
SRL	Self-Regulated Learning
SVM	Support Vector Machine
XAI	Explainable Artificial Intelligence

References

1. Adoption of Data Analytics in Higher Education Learning and Teaching. In *Advances in Analytics for Learning and Teaching*; Ifenthaler, D., Gibson, D., Eds.; Springer International Publishing: Cham, Switzerland, 2020; ISBN 978-3-030-47391-4.
2. Sghir, N.; Adadi, A.; Lahmer, M. Recent Advances in Predictive Learning Analytics: A Decade Systematic Review (2012–2022). *Educ. Inf. Technol.* **2023**, *28*, 8299–8333. [CrossRef]
3. Umer, R.; Susnjak, T.; Mathrani, A.; Suriadi, L. Current Stance on Predictive Analytics in Higher Education: Opportunities, Challenges and Future Directions. Interact. *Learn. Environ.* **2021**, *31*, 3503–3528. [CrossRef]
4. United Nations Educational, Scientific and Cultural Organization (UNESCO). *Educación 2030: Declaración de Incheon y Marco de Acción para la Realización del Objetivo de Desarrollo Sostenible 4*; United Nations Educational, Scientific and Cultural Organization (UNESCO): Paris, France, 2016.
5. Organisation for Economic Co-operation and Development (OECD). Education at a Glance 2022: OECD Indicators. Paris, France, 2022. Available online: https://www.oecd.org/en/publications/education-at-a-glance-2022_3197152b-en.html (accessed on 3 August 2025). ISBN 978-92-64-58258-3.
6. Castro, R. Blended Learning in Higher Education: Trends and Capabilities. *Educ. Inf. Technol.* **2019**, *24*, 2523–2546. [CrossRef]
7. Di Pietro, G. The Determinants of University Dropout in Italy: A Bivariate Probability Model with Sample Selection. *Appl. Econ. Lett.* **2004**, *11*, 187–191. [CrossRef]
8. Siemens, G. Learning Analytics: The Emergence of a Discipline. *Am. Behav. Sci.* **2013**, *57*, 1380–1400. [CrossRef]
9. El Naqa, I.; Murphy, M.J. What Is Machine Learning? In *Machine Learning in Radiation Oncology*; El Naqa, I., Li, R., Murphy, M.J., Eds.; Springer International Publishing: Cham, Switzerland, 2015; pp. 3–11. ISBN 978-3-319-18304-6.
10. Alalawi, K.; Athauda, R.; Chiong, R. Contextualizing the Current State of Research on the Use of Machine Learning for Student Performance Prediction: A Systematic Literature Review. *Eng. Rep.* **2023**, *5*, e12699. [CrossRef]
11. Albreiki, B.; Zaki, N.; Alashwal, H. A Systematic Literature Review of Student' Performance Prediction Using Machine Learning Techniques. *Educ. Sci.* **2021**, *11*, 552. [CrossRef]
12. Namoun, A.; Alshanqiti, A. Predicting Student Performance Using Data Mining and Learning Analytics Techniques: A Systematic Literature Review. *Appl. Sci.* **2021**, *11*, 237. [CrossRef]
13. Borah, A.R.; Nischith, T.N.; Gupta, S. Improved Learning Based on GenAI. In Proceedings of the 2024 2nd International Conference on Intelligent Data Communication Technologies and Internet of Things (IDCIoT), Bengaluru, India, 4–6 January 2024; pp. 1527–1532.
14. Khosravi, H.; Viberg, O.; Kovanovic, V.; Ferguson, R. Generative AI and Learning Analytics. *J. Learn. Anal.* **2023**, *10*, 1–6. [CrossRef]
15. Yan, L.; Martinez-Maldonado, R.; Gasevic, D. Generative Artificial Intelligence in Learning Analytics: Contextualising Opportunities and Challenges through the Learning Analytics Cycle. In Proceedings of the 14th Learning Analytics and Knowledge Conference, Kyoto, Japan, 18–22 March 2024; pp. 101–111.
16. Moonsamy, D.; Naicker, N.; Adeliyi, T.T.; Ogunsakin, R.E. A Meta-Analysis of Educational Data Mining for Predicting Students Performance in Programming. Int. J. Adv. Comput. *Sci. Appl.* **2021**, *12*, 97–104. [CrossRef]
17. Moreno-Marcos, P.M.; Alario-Hoyos, C.; Munoz-Merino, P.J.; Kloos, C.D. Prediction in MOOCs: A Review and Future Research Directions. *IEEE Trans. Learn. Technol.* **2019**, *12*, 384–401. [CrossRef]
18. Topali, P.; Chounta, I.; Martínez-Monés, A.; Dimitriadis, Y. Delving into Instructor-led Feedback Interventions Informed by Learning Analytics in Massive Open Online Courses. *J. Comput. Assist. Learn.* **2023**, *39*, 1039–1060. [CrossRef]
19. Zhu, M.; Sari, A.R.; Lee, M.M. Trends and Issues in MOOC Learning Analytics Empirical Research: A Systematic Literature Review (2011–2021). *Educ. Inf. Technol.* **2022**, *27*, 10135–10160. [CrossRef]
20. Page, M.J.; McKenzie, J.E.; Bossuyt, P.M.; Boutron, I.; Hoffmann, T.C.; Mulrow, C.D.; Shamseer, L.; Tetzlaff, J.M.; Akl, E.A.; Brennan, S.E.; et al. The PRISMA 2020 Statement: An Updated Guideline for Reporting Systematic Reviews. *BMJ* **2021**, *372*, n71. [CrossRef]
21. Hong, Q.N.; Fàbregues, S.; Bartlett, G.; Boardman, F.; Cargo, M.; Dagenais, P.; Gagnon, M.-P.; Griffiths, F.; Nicolau, B.; O'Cathain, A.; et al. The Mixed Methods Appraisal Tool (MMAT) Version 2018 for Information Professionals and Researchers. *Educ. Inf.* **2018**, *34*, 285–291. [CrossRef]

22. Abou Gamie, E.; Abou El-Seoud, S.; Salama, M.A. Comparative Analysis for Boosting Classifiers in the Context of Higher Education. *Int. J. Emerg. Technol. Learn. IJET* **2020**, *15*, 16. [CrossRef]
23. Arslan, O.; Xing, W.; Inan, F.A.; Du, H. Understanding Topic Duration in Twitter Learning Communities Using Data Mining. *J. Comput. Assist. Learn.* **2022**, *38*, 513–525. [CrossRef]
24. Ayouni, S.; Hajjej, F.; Maddeh, M.; Al-Otaibi, S. A New ML-Based Approach to Enhance Student Engagement in Online Environment. *PLoS ONE* **2021**, *16*, e0258788. [CrossRef]
25. Lemay, D.J.; Doleck, T. Predicting Completion of Massive Open Online Course (MOOC) Assignments from Video Viewing Behavior. *Interact. Learn. Environ.* **2022**, *30*, 1782–1793. [CrossRef]
26. Moreno-Marcos, P.M.; Muñoz-Merino, P.J.; Alario-Hoyos, C.; Estévez-Ayres, I.; Delgado Kloos, C. Analysing the Predictive Power for Anticipating Assignment Grades in a Massive Open Online Course. *Behav. Inf. Technol.* **2018**, *37*, 1021–1036. [CrossRef]
27. Cristea, A.I.; Alamri, A.; Alshehri, M.; Pereira, F.D.; Toda, A.M.; de Oliveira, E.H.T.; Stewart, C. The Engage Taxonomy: SDT-Based Measurable Engagement Indicators for MOOCs and Their Evaluation. *User Model. User Adapt. Interact.* **2023**, *34*, 323–374. [CrossRef]
28. Onan, A. Sentiment Analysis on Massive Open Online Course Evaluations: A Text Mining and Deep Learning Approach. *Comput. Appl. Eng. Educ.* **2021**, *29*, 572–589. [CrossRef]
29. Onan, A.; Toçoğlu, M.A. Weighted Word Embeddings and Clustering-based Identification of Question Topics in MOOC Discussion Forum Posts. *Comput. Appl. Eng. Educ.* **2021**, *29*, 675–689. [CrossRef]
30. Du, X.; Zhang, M.; Shelton, B.E.; Hung, J.-L. Learning Anytime, Anywhere: A Spatio-Temporal Analysis for Online Learning. *Interact. Learn. Environ.* **2022**, *30*, 34–48. [CrossRef]
31. Moubayed, A.; Injadat, M.; Shami, A.; Lutfiyya, H. Student Engagement Level in E-Learning Environment: Clustering Using K-Means 2020. *Am. J. Distance Educ.* **2020**, *34*, 137–156. [CrossRef]
32. Lahza, H.; Khosravi, H.; Demartini, G. Analytics of Learning Tactics and Strategies in an Online Learnersourcing Environment. *J. Comput. Assist. Learn.* **2023**, *39*, 94–112. [CrossRef]
33. Akçapınar, G.; Altun, A.; Aşkar, P. Using Learning Analytics to Develop Early-Warning System for at-Risk Students. *Int. J. Educ. Technol. High. Educ.* **2019**, *16*, 40. [CrossRef]
34. Dass, S.; Gary, K.; Cunningham, J. Predicting Student Dropout in Self-Paced MOOC Course Using Random Forest Model. *Information* **2021**, *12*, 476. [CrossRef]
35. Gupta, A.; Garg, D.; Kumar, P. An Ensembling Model for Early Identification of At-risk Students in Higher Education. *Comput. Appl. Eng. Educ.* **2022**, *30*, 589–608. [CrossRef]
36. Kostopoulos, G.; Kotsiantis, S.; Pierrakeas, C.; Koutsonikos, G.; Gravvanis, G.A. Forecasting Students' Success in an Open University. *Int. J. Learn. Technol.* **2018**, *13*, 26. [CrossRef]
37. Karalar, H.; Kapucu, C.; Gürüler, H. Predicting Students at Risk of Academic Failure Using Ensemble Model during Pandemic in a Distance Learning System. *Int. J. Educ. Technol. High. Educ.* **2021**, *18*, 63. [CrossRef]
38. Yu, C.-C.; Wu, Y. (Leon) Early Warning System for Online STEM Learning—A Slimmer Approach Using Recurrent Neural Networks. *Sustainability* **2021**, *13*, 12461. [CrossRef]
39. Al-Sulami, A.; Al-Masre, M.; Al-Malki, N. Predicting At-Risk Students' Performance Based on LMS Activity Using Deep Learning. *Int. J. Adv. Comput. Sci. Appl.* **2023**, *14*, 1210–1220. [CrossRef]
40. Mubarak, A.A.; Cao, H.; Zhang, W. Prediction of Students' Early Dropout Based on Their Interaction Logs in Online Learning Environment. *Interact. Learn. Environ.* **2022**, *30*, 1414–1433. [CrossRef]
41. Bayazit, A.; Apaydin, N.; Gonullu, I. Predicting At-Risk Students in an Online Flipped Anatomy Course Using Learning Analytics. *Educ. Sci.* **2022**, *12*, 581. [CrossRef]
42. Esteban, A.; Romero, C.; Zafra, A. Assignments as Influential Factor to Improve the Prediction of Student Performance in Online Courses. *Appl. Sci.* **2021**, *11*, 10145. [CrossRef]
43. Talamás-Carvajal, J.A.; Ceballos, H.G. A Stacking Ensemble Machine Learning Method for Early Identification of Students at Risk of Dropout. *Educ. Inf. Technol.* **2023**, *28*, 12169–12189. [CrossRef]
44. Karlos, S.; Kostopoulos, G.; Kotsiantis, S. Predicting and Interpreting Students' Grades in Distance Higher Education through a Semi-Regression Method. *Appl. Sci.* **2020**, *10*, 8413. [CrossRef]
45. Herodotou, C.; Hlosta, M.; Boroowa, A.; Rienties, B.; Zdrahal, Z.; Mangafa, C. Empowering Online Teachers through Predictive Learning Analytics. *Br. J. Educ. Technol.* **2019**, *50*, 3064–3079. [CrossRef]
46. Sousa-Vieira, M.E.; López-Ardao, J.C.; Fernández-Veiga, M.; Rodríguez-Rubio, R.F. Study of the Impact of Social Learning and Gamification Methodologies on Learning Results in Higher Education. *Comput. Appl. Eng. Educ.* **2023**, *31*, 131–153. [CrossRef]
47. Zamecnik, A.; Kovanović, V.; Joksimović, S.; Liu, L. Exploring Non-Traditional Learner Motivations and Characteristics in Online Learning: A Learner Profile Study. *Comput. Educ. Artif. Intell.* **2022**, *3*, 100051. [CrossRef]
48. Asad, R.; Altaf, S.; Ahmad, S.; Mahmoud, H.; Huda, S.; Iqbal, S. Machine Learning-Based Hybrid Ensemble Model Achieving Precision Education for Online Education Amid the Lockdown Period of COVID-19 Pandemic in Pakistan. *Sustainability* **2023**, *15*, 5431. [CrossRef]

49. Deho, O.B.; Joksimovic, S.; Li, J.; Zhan, C.; Liu, J.; Liu, L. Should Learning Analytics Models Include Sensitive Attributes? Explaining the Why. *IEEE Trans. Learn. Technol.* **2023**, *16*, 560–572. [CrossRef]
50. Imhof, C.; Comsa, I.-S.; Hlosta, M.; Parsaeifard, B.; Moser, I.; Bergamin, P. Prediction of Dilatory Behavior in eLearning: A Comparison of Multiple Machine Learning Models. *IEEE Trans. Learn. Technol.* **2022**, *16*, 648–663. [CrossRef]
51. Dake, D.K.; Gyimah, E. Using Sentiment Analysis to Evaluate Qualitative Students' Responses. *Educ. Inf. Technol.* **2023**, *28*, 4629–4647. [CrossRef]
52. Mai, T.T.; Bezbradica, M.; Crane, M. Learning Behaviours Data in Programming Education: Community Analysis and Outcome Prediction with Cleaned Data. *Future Gener. Comput. Syst.* **2022**, *127*, 42–55. [CrossRef]
53. Tsiakmaki, M.; Kostopoulos, G.; Kotsiantis, S.; Ragos, O. Fuzzy-Based Active Learning for Predicting Student Academic Performance Using autoML: A Step-Wise Approach. *J. Comput. High. Educ.* **2021**, *33*, 635–667. [CrossRef]
54. Flor, M.; Andrews-Todd, J. Towards Automatic Annotation of Collaborative Problem-solving Skills in Technology-enhanced Environments. *J. Comput. Assist. Learn.* **2022**, *38*, 1434–1447. [CrossRef]
55. Cui, Y.; Chen, F.; Shiri, A. Scale up Predictive Models for Early Detection of At-Risk Students: A Feasibility Study. *Inf. Learn. Sci.* **2020**, *121*, 97–116. [CrossRef]
56. Polyzou, A.; Karypis, G. Feature Extraction for Next-Term Prediction of Poor Student Performance. *IEEE Trans. Learn. Technol.* **2019**, *12*, 237–248. [CrossRef]
57. Wakelam, E.; Jefferies, A.; Davey, N.; Sun, Y. The Potential for Student Performance Prediction in Small Cohorts with Minimal Available Attributes. *Br. J. Educ. Technol.* **2020**, *51*, 347–370. [CrossRef]
58. Abu Zohair, L.M. Prediction of Student's Performance by Modelling Small Dataset Size. *Int. J. Educ. Technol. High. Educ.* **2019**, *16*, 27. [CrossRef]
59. Damuluri, S.; Islam, K.; Ahmadi, P.; Qureshi, N.S. Analyzing Navigational Data and Predicting Student Grades Using Support Vector Machine. *Emerg. Sci. J.* **2020**, *4*, 243–252. [CrossRef]
60. Nuankaew, P. Dropout Situation of Business Computer Students, University of Phayao. *Int. J. Emerg. Technol. Learn. IJET* **2019**, *14*, 115. [CrossRef]
61. Azcona, D.; Hsiao, I.-H.; Smeaton, A.F. Detecting Students-at-Risk in Computer Programming Classes with Learning Analytics from Students' Digital Footprints. *User Model. User Adapt. Interact.* **2019**, *29*, 759–788. [CrossRef]
62. Wu, J.-Y.; Hsiao, Y.-C.; Nian, M.-W. Using Supervised Machine Learning on Large-Scale Online Forums to Classify Course-Related Facebook Messages in Predicting Learning Achievement within the Personal Learning Environment. *Interact. Learn. Environ.* **2018**, *28*, 65–80. [CrossRef]
63. Crivei, L.M.; Ionescu, V.-S.; Czibula, G. An Analysis of Supervised Learning Methods for Predicting Students' Performance in Academic Environments 2019. *ICIC Express Lett* **2019**, *13*, 181–189.
64. Raković, M.; Winne, P.H.; Marzouk, Z.; Chang, D. Automatic Identification of Knowledge-transforming Content in Argument Essays Developed from Multiple Sources. *J. Comput. Assist. Learn.* **2021**, *37*, 903–924. [CrossRef]
65. Spikol, D.; Ruffaldi, E.; Dabisias, G.; Cukurova, M. Supervised Machine Learning in Multimodal Learning Analytics for Estimating Success in Project-based Learning. *J. Comput. Assist. Learn.* **2018**, *34*, 366–377. [CrossRef]
66. Bertolini, R.; Finch, S.J.; Nehm, R.H. Testing the Impact of Novel Assessment Sources and Machine Learning Methods on Predictive Outcome Modeling in Undergraduate Biology. *J. Sci. Educ. Technol.* **2021**, *30*, 193–209. [CrossRef]
67. Ding, X.; Larson, E.C.; Doyle, A.; Donahoo, K.; Rajgopal, R.; Bing, E. EduAware: Using Tablet-Based Navigation Gestures to Predict Learning Module Performance. *Interact. Learn. Environ.* **2021**, *29*, 720–732. [CrossRef]
68. Okike, E.U.; Mogorosi, M. Educational Data Mining for Monitoring and Improving Academic Performance at University Levels. *Int. J. Adv. Comput. Sci. Appl.* **2020**, *11*, 570–581. [CrossRef]
69. Everaert, P.; Opdecam, E.; van der Heijden, H. Predicting First-Year University Progression Using Early Warning Signals from Accounting Education: A Machine Learning Approach. *Account. Educ.* **2022**, *33*, 1–26. [CrossRef]
70. Nkhoma, C.; Dang-Pham, D.; Hoang, A.-P.; Nkhoma, M.; Le-Hoai, T.; Thomas, S. Learning Analytics Techniques and Visualisation with Textual Data for Determining Causes of Academic Failure. *Behav. Inf. Technol.* **2020**, *39*, 808–823. [CrossRef]
71. Li, F.; Lu, Y.; Ma, Q.; Gao, J.; Wang, Z.; Bai, L. SPOC Online Video Learning Clustering Analysis: Identifying Learners' Group Behavior Characteristics. Comput. *Appl. Eng. Educ.* **2023**, *31*, 1059–1077. [CrossRef]
72. Verdú, M.J.; Regueras, L.M.; de Castro, J.P.; Verdú, E. Clustering of LMS Use Strategies with Autoencoders. *Appl. Sci.* **2023**, *13*, 7334. [CrossRef]
73. Alsayed, A.O.; Rahim, M.S.M.; AlBidewi, I.; Hussain, M.; Jabeen, S.H.; Alromema, N.; Hussain, S.; Jibril, M.L. Selection of the Right Undergraduate Major by Students Using Supervised Learning Techniques. *Appl. Sci.* **2021**, *11*, 10639. [CrossRef]
74. Mai, T.L.; Chung, M.T.; Le, V.T.; Thoai, N. From Transcripts to Insights for Recommending the Curriculum to University Students. *SN Comput. Sci.* **2020**, *1*, 323. [CrossRef]
75. Zabriskie, C.; Yang, J.; DeVore, S.; Stewart, J. Using Machine Learning to Predict Physics Course Outcomes. *Phys. Rev. Phys. Educ. Res.* **2019**, *15*, 020120. [CrossRef]

76. Fateh Allah, A.G. Using Machine Learning To Support Students' Academic Decisions. *J. Theor. Appl. Inf. Technol.* **2020**, *98*, 3778–3796.
77. Hassan, H.; Ahmad, N.B.; Anuar, S. Improved Students' Performance Prediction for Multi-Class Imbalanced Problems Using Hybrid and Ensemble Approach in Educational Data Mining. *J. Phys. Conf. Ser.* **2020**, *1529*, 052041. [CrossRef]
78. Hussain, S.; Ayoub, M.; Jilani, G.; Yu, Y.; Khan, A.; Wahid, J.A.; Butt, M.F.A.; Yang, G.; Moller, D.P.F.; Weiyan, H. Aspect2Labels: A Novelistic Decision Support System for Higher Educational Institutions by Using Multi-Layer Topic Modelling Approach. *Expert Syst. Appl.* **2022**, *209*, 118119. [CrossRef]
79. Nasa-Ngium, P.; Nuankaew, W.S.; Nuankaew, P. Analyzing and Tracking Student Educational Program Interests on Social Media with Chatbots Platform and Text Analytics. *Int. J. Interact. Mob. Technol. IJIM* **2023**, *17*, 4–21. [CrossRef]
80. Liu, Z.; Yang, C.; Rüdian, S.; Liu, S.; Zhao, L.; Wang, T. Temporal Emotion-Aspect Modeling for Discovering What Students Are Concerned about in Online Course Forums. *Interact. Learn. Environ.* **2019**, *27*, 598–627. [CrossRef]
81. Maldonado, E.; Seehusen, V. Data Mining Student Choices: A New Approach to Business Curriculum Planning. *J. Educ. Bus.* **2018**, *93*, 196–203. [CrossRef]
82. Guerrero-Higueras, Á.M.; DeCastro-García, N.; Rodriguez-Lera, F.J.; Matellán, V.; Conde, M.Á. Predicting Academic Success through Students' Interaction with Version Control Systems. *Open Comput. Sci.* **2019**, *9*, 243–251. [CrossRef]
83. Liao, C.-H.; Wu, J.-Y. Deploying Multimodal Learning Analytics Models to Explore the Impact of Digital Distraction and Peer Learning on Student Performance. *Comput. Educ.* **2022**, *190*, 104599. [CrossRef]
84. Wu, J.-Y. Learning Analytics on Structured and Unstructured Heterogeneous Data Sources: Perspectives from Procrastination, Help-Seeking, and Machine-Learning Defined Cognitive Engagement. *Comput. Educ.* **2021**, *163*, 104066. [CrossRef]
85. Cagliero, L.; Canale, L.; Farinetti, L.; Baralis, E.; Venuto, E. Predicting Student Academic Performance by Means of Associative Classification. *Appl. Sci.* **2021**, *11*, 1420. [CrossRef]
86. Sassirekha, M.S.; Vijayalakshmi, S. Predicting the Academic Progression in Student's Standpoint Using Machine Learning. *Automatika* **2022**, *63*, 605–617. [CrossRef]
87. Zheng, L.; Zhong, L.; Niu, J. Effects of Personalised Feedback Approach on Knowledge Building, Emotions, Co-Regulated Behavioural Patterns and Cognitive Load in Online Collaborative Learning. *Assess. Eval. High. Educ.* **2022**, *47*, 109–125. [CrossRef]
88. Talebinamvar, M.; Zarrabi, F. Clustering Students' Writing Behaviors Using Keystroke Logging: A Learning Analytic Approach in EFL Writing. *Lang. Test. Asia* **2022**, *12*, 6. [CrossRef]
89. Araka, E.; Oboko, R.; Maina, E.; Gitonga, R. Using Educational Data Mining Techniques to Identify Profiles in Self-Regulated Learning: An Empirical Evaluation. *Int. Rev. Res. Open Distrib. Learn.* **2022**, *23*, 131–162. [CrossRef]
90. Choi, Y.; Kim, J. Learning Analytics for Diagnosing Cognitive Load in E-Learning Using Bayesian Network Analysis. *Sustainability* **2021**, *13*, 10149. [CrossRef]
91. Afzaal, M.; Nouri, J.; Zia, A.; Papapetrou, P.; Fors, U.; Wu, Y.; Li, X.; Weegar, R. Explainable AI for Data-Driven Feedback and Intelligent Action Recommendations to Support Students Self-Regulation. *Front. Artif. Intell.* **2021**, *4*, 723447. [CrossRef]
92. Ramaswami, G.; Susnjak, T.; Mathrani, A. Supporting Students' Academic Performance Using Explainable Machine Learning with Automated Prescriptive Analytics. *Big Data Cogn. Comput.* **2022**, *6*, 105. [CrossRef]
93. Dirin, A.; Saballe, C.A. Machine Learning Models to Predict Students' Study Path Selection. *Int. J. Interact. Mob. Technol. IJIM* **2022**, *16*, 158–183. [CrossRef]
94. Hung, H.-C.; Liu, I.-F.; Liang, C.-T.; Su, Y.-S. Applying Educational Data Mining to Explore Students' Learning Patterns in the Flipped Learning Approach for Coding Education. *Symmetry* **2020**, *12*, 213. [CrossRef]
95. Qushem, U.B.; Oyelere, S.S.; Akçapınar, G.; Kaliisa, R.; Laakso, M.-J. Unleashing the Power of Predictive Analytics to Identify At-Risk Students in Computer Science. *Technol. Knowl. Learn.* **2023**, *29*, 1385–1400. [CrossRef]
96. Almasri, A.R.; Yahaya, N.A.; Abu-Naser, S.S. Instructor Performance Modeling For Predicting Student Satisfaction Using Machine Learning—Preliminary Results. *J. Theor. Appl. Inf. Technol.* **2022**, *100*, 5481–5496.
97. Samadi, M.A.; Jaquay, S.; Lin, Y.; Tajik, E.; Park, S.; Nixon, N. Minds and Machines Unite: Deciphering Social and Cognitive Dynamics in Collaborative Problem Solving with AI. In Proceedings of the 14th Learning Analytics and Knowledge Conference, Kyoto, Japan, 18–22 March 2024; pp. 885–891.
98. Li, T.; Nath, D.; Cheng, Y.; Fan, Y.; Li, X.; Raković, M.; Khosravi, H.; Swiecki, Z.; Tsai, Y.-S.; Gašević, D. Turning Real-Time Analytics into Adaptive Scaffolds for Self-Regulated Learning Using Generative Artificial Intelligence. In Proceedings of the 15th International Learning Analytics and Knowledge Conference, Dublin, Ireland, 3–7 March 2025; pp. 667–679.
99. Lewis, A. Multimodal Large Language Models for Inclusive Collaboration Learning Tasks. In Proceedings of the 2022 Conference of the North American Chapter of the Association for Computational Linguistics: Human Language Technologies: Student Research Workshop, Seattle, WA, USA, 10–15 July 2022; pp. 202–210.
100. Dai, W.; Tsai, Y.-S.; Lin, J.; Aldino, A.; Jin, H.; Li, T.; Gašević, D.; Chen, G. Assessing the Proficiency of Large Language Models in Automatic Feedback Generation: An Evaluation Study. *Comput. Educ. Artif. Intell.* **2024**, *7*, 100299. [CrossRef]
101. Civit, M.; Escalona, M.J.; Cuadrado, F.; Reyes-de-Cozar, S. Class Integration of ChatGPT and Learning Analytics for Higher Education. *Expert Syst.* **2024**, *41*, e13703. [CrossRef]

102. Xie, Z.; Wu, X.; Xie, Y. Can Interaction with Generative Artificial Intelligence Enhance Learning Autonomy? A Longitudinal Study from Comparative Perspectives of Virtual Companionship and Knowledge Acquisition Preferences. *J. Comput. Assist. Learn.* **2024**, *40*, 2369–2384. [CrossRef]
103. Milesi, M.E.; Alfredo, R.; Echeverria, V.; Yan, L.; Zhao, L.; Tsai, Y.-S.; Martinez-Maldonado, R. "It's Really Enjoyable to See Me Solve the Problem like a Hero": GenAI-Enhanced Data Comics as a Learning Analytics Tool. In Proceedings of the Extended Abstracts of the CHI Conference on Human Factors in Computing Systems, Honolulu, HI, USA, 11–16 May 2024; pp. 1–7.
104. Ilagan, J.B.R.; Ilagan, J.R.S.; Rodrigo, M.M.T. Ethical Education Data Mining Framework for Analyzing and Evaluating Large Language Model-Based Conversational Intelligent Tutoring Systems for Management and Entrepreneurship Courses. In Proceedings of Ninth International Congress on Information and Communication Technology, London, UK, 19–22 February 2024; Yang, X.-S., Sherratt, S., Dey, N., Joshi, A., Eds.; Lecture Notes in Networks and Systems. Springer Nature: Singapore, 2024; Volume 1011, pp. 61–71, ISBN 978-981-9745-80-7.
105. Nguyen, A.; Ilesanmi, F.; Dang, B.; Vuorenmaa, E.; Järvelä, S. Hybrid Intelligence in Academic Writing: Examining Self-Regulated Learning Patterns in an AI-Assisted Writing Task. In *Frontiers in Artificial Intelligence and Applications*; Lorig, F., Tucker, J., Dahlgren Lindström, A., Dignum, F., Murukannaiah, P., Theodorou, A., Yolum, P., Eds.; IOS Press: Amsterdam, The Netherlands, 2024; ISBN 978-1-64368-522-9.
106. Cheng, Y.; Lyons, K.; Chen, G.; Gašević, D.; Swiecki, Z. Evidence-Centered Assessment for Writing with Generative AI. In Proceedings of the 14th Learning Analytics and Knowledge Conference, Kyoto, Japan, 18–22 March 2024; pp. 178–188.
107. Fahl, W. GraphWiseLearn: Personalized Learning Through Semantified TEL, Leveraging QA-Enhanced LLM-Generated Content. In *The Semantic Web: ESWC 2024 Satellite Events*; Meroño Peñuela, A., Corcho, O., Groth, P., Simperl, E., Tamma, V., Nuzzolese, A.G., Poveda-Villalón, M., Sabou, M., Presutti, V., Celino, I., Eds.; Lecture Notes in Computer Science; Springer Nature: Cham, Switzerland, 2025; Volume 15345, pp. 74–83. ISBN 978-3-031-78954-0.
108. Pozdniakov, S.; Brazil, J.; Abdi, S.; Bakharia, A.; Sadiq, S.; Gašević, D.; Denny, P.; Khosravi, H. Large Language Models Meet User Interfaces: The Case of Provisioning Feedback. *Comput. Educ. Artif. Intell.* **2024**, *7*, 100289. [CrossRef]
109. Chen, A.; Xiang, M.; Zhou, J.; Jia, J.; Shang, J.; Li, X.; Gašević, D.; Fan, Y. Unpacking Help-Seeking Process through Multimodal Learning Analytics: A Comparative Study of ChatGPT vs Human Expert. *Comput. Educ.* **2025**, *226*, 105198. [CrossRef]
110. Fan, Y.; Tang, L.; Le, H.; Shen, K.; Tan, S.; Zhao, Y.; Shen, Y.; Li, X.; Gašević, D. Beware of Metacognitive Laziness: Effects of Generative Artificial Intelligence on Learning Motivation, Processes, and Performance. *Br. J. Educ. Technol.* **2025**, *56*, 489–530. [CrossRef]
111. Cohn, C.; Snyder, C.; Fonteles, J.H.; TS, A.; Montenegro, J.; Biswas, G. A Multimodal Approach to Support Teacher, Researcher and AI Collaboration in STEM+C Learning Environments. *Br. J. Educ. Technol.* **2025**, *56*, 595–620. [CrossRef]
112. Hu, Y.; Giacaman, N.; Donald, C. Enhancing Trust in Generative AI: Investigating Explainability of LLMs to Analyse Confusion in MOOC Discussions. In Proceedings of the Joint Proceedings of LAK 2024 Workshops, Kyoto, Japan, 18–22 March 2024; Volume 3667, pp. 195–204.
113. Valle Torre, M. Learning Sequence Analytics for Support in Learning Tasks. *CEUR Workshop Proc. CEUR-WSorg* **2023**, *3539*, 57–62.
114. Zhang, L.; Lin, J.; Sabatini, J.; Borchers, C.; Weitekamp, D.; Cao, M.; Hollander, J.; Hu, X.; Graesser, A.C. Data Augmentation for Sparse Multidimensional Learning Performance Data Using Generative AI. *IEEE Trans. Learn. Technol.* **2025**, *18*, 145–164. [CrossRef]
115. Singh, A.; Brooks, C.; Wang, X.; Li, W.; Kim, J.; Wilson, D. Bridging Learnersourcing and AI: Exploring the Dynamics of Student-AI Collaborative Feedback Generation. In Proceedings of the 14th Learning Analytics and Knowledge Conference, Kyoto, Japan, 18–22 March 2024; pp. 742–748.
116. Opanasenko, Y.; Bardone, E.; Pedaste, M.; Siiman, L.A. Sequence Analysis-Enhanced AI: Transforming Interactive E-Book Data into Educational Insights for Teachers. *Educ. Sci.* **2024**, *15*, 28. [CrossRef]
117. Jin, Y.; Yang, K.; Yan, L.; Echeverria, V.; Zhao, L.; Alfredo, R.; Milesi, M.; Fan, J.X.; Li, X.; Gasevic, D.; et al. Chatting with a Learning Analytics Dashboard: The Role of Generative AI Literacy on Learner Interaction with Conventional and Scaffolding Chatbots. In Proceedings of the 15th International Learning Analytics and Knowledge Conference, Dublin, Ireland, 3–7 March 2025; pp. 579–590.
118. Wang, Z.; Lin, W.; Hu, X. Self-Service Teacher-Facing Learning Analytics Dashboard with Large Language Models. In Proceedings of the 15th International Learning Analytics and Knowledge Conference, Dublin, Ireland, 3–7 March 2025; pp. 824–830.
119. Volkmann, N. EduBot Unleashed—Elevating Digital Competence in Online Collaborative Learning. In Proceedings of the 2024 21st International Conference on Information Technology Based Higher Education and Training (ITHET), Paris, France, 6–8 November 2024; pp. 1–9.
120. Misiejuk, K.; Kaliisa, R.; Scianna, J. Augmenting Assessment with AI Coding of Online Student Discourse: A Question of Reliability. *Comput. Educ. Artif. Intell.* **2024**, *6*, 100216. [CrossRef]
121. Woollaston, S.; Flanagan, B.; Ocheja, P.; Toyokawa, Y.; Ogata, H. ARCHIE: Exploring Language Learner Behaviors in LLM Chatbot-Supported Active Reading Log Data with Epistemic Network Analysis. In Proceedings of the 15th International Learning Analytics and Knowledge Conference, Dublin, Ireland, 3–7 March 2025; pp. 642–654.

122. Ouyang, F.; Guo, M.; Zhang, N.; Bai, X.; Jiao, P. Comparing the Effects of Instructor Manual Feedback and ChatGPT Intelligent Feedback on Collaborative Programming in China's Higher Education. *IEEE Trans. Learn. Technol.* **2024**, *17*, 2173–2185. [CrossRef]
123. Huang, K.; Ferreira Mello, R.; Pereira Junior, C.; Rodrigues, L.; Baars, M.; Viberg, O. That's What RoBERTa Said: Explainable Classification of Peer Feedback. In Proceedings of the 15th International Learning Analytics and Knowledge Conference, Dublin, Ireland, 3–7 March 2025; pp. 880–886.
124. Pishtari, G.; Sarmiento-Márquez, E.M.; Rodríguez-Triana, M.J.; Wagner, M.; Ley, T. Evaluating the Impact and Usability of an AI-Driven Feedback System for Learning Design. In *Responsive and Sustainable Educational Futures*; Viberg, O., Jivet, I., Muñoz-Merino, P.J., Perifanou, M., Papathoma, T., Eds.; Lecture Notes in Computer Science; Springer Nature: Cham, Switzerland, 2023; Volume 14200, pp. 324–338. ISBN 978-3-031-42681-0.
125. Qu, T.; Yang, Z. Overview of Artificial Intelligence Applications in Educational Research. In Proceedings of the 2024 International Symposium on Artificial Intelligence for Education, Xian, China, 6–8 September 2024; pp. 101–108.
126. Peña-Ayala, A. Learning Analytics: A Glance of Evolution, Status, and Trends According to a Proposed Taxonomy. *WIREs Data Min. Knowl. Discov.* **2018**, *8*, e1243. [CrossRef]
127. Zawacki-Richter, O.; Marín, V.I.; Bond, M.; Gouverneur, F. Systematic Review of Research on Artificial Intelligence Applications in Higher Education—Where Are the Educators? *Int. J. Educ. Technol. High. Educ.* **2019**, *16*, 39. [CrossRef]
128. Renz, A.; Hilbig, R. Prerequisites for Artificial Intelligence in Further Education: Identification of Drivers, Barriers, and Business Models of Educational Technology Companies. *Int. J. Educ. Technol. High. Educ.* **2020**, *17*, 14. [CrossRef]
129. Salas-Pilco, S.Z.; Yang, Y. Artificial Intelligence Applications in Latin American Higher Education: A Systematic Review. *Int. J. Educ. Technol. High. Educ.* **2022**, *19*, 21. [CrossRef]
130. Glandorf, D.; Lee, H.R.; Orona, G.A.; Pumptow, M.; Yu, R.; Fischer, C. Temporal and Between-Group Variability in College Dropout Prediction. In Proceedings of the 14th Learning Analytics and Knowledge Conference, Kyoto, Japan, 18–22 March 2024; pp. 486–497.
131. Poellhuber, L.-V.; Poellhuber, B.; Desmarais, M.; Leger, C.; Roy, N.; Manh-Chien Vu, M. Cluster-Based Performance of Student Dropout Prediction as a Solution for Large Scale Models in a Moodle LMS. In Proceedings of the LAK23: 13th International Learning Analytics and Knowledge Conference, Arlington, TX, USA, 13–17 March 2023; pp. 592–598.
132. Buitrago-Ropero, M.E.; Ramírez-Montoya, M.S.; Laverde, A.C. Digital Footprints (2005–2019): A Systematic Mapping of Studies in Education. *Interact. Learn. Environ.* **2023**, *31*, 876–889. [CrossRef]
133. Baek, C.; Doleck, T. Educational Data Mining versus Learning Analytics: A Review of Publications from 2015 to 2019. *Interact. Learn. Environ.* **2021**, *31*, 3828–3850. [CrossRef]
134. Ley, T.; Tammets, K.; Pishtari, G.; Chejara, P.; Kasepalu, R.; Khalil, M.; Saar, M.; Tuvi, I.; Väljataga, T.; Wasson, B. Towards a Partnership of Teachers and Intelligent Learning Technology: A Systematic Literature Review of Model-based Learning Analytics. *J. Comput. Assist. Learn.* **2023**, *39*, 1397–1417. [CrossRef]
135. Aguilar-Esteva, V.; Acosta-Banda, A.; Carreño Aguilera, R.; Patiño Ortiz, M. Sustainable Social Development through the Use of Artificial Intelligence and Data Science in Education during the COVID Emergency: A Systematic Review Using PRISMA. *Sustainability* **2023**, *15*, 6498. [CrossRef]
136. Carless, D.; Boud, D. The Development of Student Feedback Literacy: Enabling Uptake of Feedback. *Assess. Eval. High. Educ.* **2018**, *43*, 1315–1325. [CrossRef]
137. Molloy, E.; Boud, D.; Henderson, M. Developing a Learning-Centred Framework for Feedback Literacy. *Assess. Eval. High. Educ.* **2020**, *45*, 527–540. [CrossRef]
138. Winstone, N.E.; Nash, R.A.; Parker, M.; Rowntree, J. Supporting Learners' Agentic Engagement With Feedback: A Systematic Review and a Taxonomy of Recipience Processes. *Educ. Psychol.* **2017**, *52*, 17–37. [CrossRef]
139. Zimmerman, B.J. Becoming a Self-Regulated Learner: An Overview. *Theory Pract.* **2002**, *41*, 64–70. [CrossRef]
140. Panadero, E. A Review of Self-Regulated Learning: Six Models and Four Directions for Research. *Front. Psychol.* **2017**, *8*, 422. [CrossRef]
141. Black, P.; Wiliam, D. Developing the Theory of Formative Assessment. *Educ. Assess. Eval. Account.* **2009**, *21*, 5–31. [CrossRef]

MDPI AG
Grosspeteranlage 5
4052 Basel
Switzerland
Tel.: +41 61 683 77 34

Applied Sciences Editorial Office
E-mail: applsci@mdpi.com
www.mdpi.com/journal/applsci

www.ingramcontent.com/pod-product-compliance
Lightning Source LLC
LaVergne TN
LVHW070540071126
851858LV00011BA/285
9783725850372